Metal Oxides for Next-Generation Optoelectronic, Photonic, and Photovoltaic Applications

The Metal Oxides Book Series Edited by Ghenadii Korotcenkov

Transition Metal Oxide Thin Film-Based Chromogenics and Devices, Pandurang Ashrit, 978-0-08-101899-6
Metal Oxides in Supercapacitors, Deepak Dubal, Pedro Gomez-Romero, 978-0-12-811169-7
Metal Oxide-Based Thin Film Structures, Nini Pryds, Vincenzo Esposito, 978-0-12-811166-6
The Future of Semiconductor Oxides in Next-Generation Solar Cells, Monica Lira-Cantu, 978-0-12-811165-9
Iron Oxide Nanoparticles for Biomedical Applications, Morteza Mahmoudi, Sophie Laurent, 978-0-08-101925-2
Magnetic, Ferroelectric, and Multiferroic Metal Oxides, Biljana D. Stojanovic, 978-0-12-811180-2
Metal Oxides in Heterogeneous Catalysis, Jacques C. Vedrine, 978-0-12-811631-9
Metal Oxide-Based Photocatalysis, Adriana Zaleska-Medynska, 978-0-12-811634-0
Metal Oxides in Energy Technologies, Yuping Wu, 978-0-12-811167-3
Gallium Oxide, Stephen Pearton, Fan Ren, Michael Mastro, 978-0-12-814521-0
Gas Sensors Based on Conducting Metal Oxides, Nicolae Barsan, Klaus Schierbaum, 978-0-12-811224-3
Metal Oxide Nanostructures, Daniela Nunes, Ana Pimentel, Lidia Santos, Pedro Barquinha, Luis Pereira, Elvira Fortunato, Rodrigo Martins, 978-0-12-811512-1
Cerium Oxide (CeO2): Synthesis, Properties and Applications, Salvatore Scire, Leonardo Palmisano, 978-0-12-815661-2
Tin Oxide Materials, Marcelo Ornaghi Orlandi, 978-0-12-815924-8
Colloidal Metal Oxide Nanoparticles, Sabu Thomas, Anu Tresa Sunny, Prajitha Velayudhan, 978-0-12-813357-6
Metal Oxide Glass Nanocomposites, Sanjib Bhattacharya, 978-0-12-817458-6
Metal Oxide Powder Technologies, Yarub Al-Douri, 978-0-12-817505-7
Solution Processed Metal Oxide Thin Films for Electronic Applications, Zheng Cui, 978-0-12-814930-0
Titanium Dioxide (TiO2) and Its Applications, Francesco Parrino, Leonardo Palmisano, 978-0-12-819960-2
Metal Oxide-Based Nanostructured Electrocatalysts for Fuel Cells, Electrolyzers, and Metal-Air Batteries, Teko Napporn, Yaovi Holade, 978-0-12-818496-7
Metal Oxides in Nanocomposite-Based Electrochemical Sensors for Toxic Chemicals, Alagarsamy Pandikumar, Perumal Rameshkumar, 978-0-12-820727-7
Nanostructured Zinc Oxide, Kamlendra Awasthi, 978-0-12-818900-9
Metal Oxide-Based Nanofibers and Their Applications, Vincenzo Esposito, Debora Marani, 978-0-12-820629-4
Metal Oxides for Biomedical and Biosensor Applications, Kunal Mondal, 978-0-12-823033-6
Metal Oxides for Non-volatile Memory, Panagiotis Dimitrakis, Ilia Valov, Stefan Tappertzhofen, 978-0-12-814629-3
Renewable Polymers and Polymer-Metal Oxide Composites, Sajjad Haider, Adnan Haider, 978-0-323-85155-8
Metal Oxide-Carbon Hybrid Materials, Muhammad Akram Chaudhry, Rafaqat Hussain, Faheem Butt, 978-0-12-822694-0
Metal Oxides and Related Solids for Electrocatalytic Water Splitting, Junlei Qi, 978-0-323-85735-2
Graphene Oxide-Metal Oxide and other Graphene Oxide-Based Composites in Photocatalysis and Electrocatalysis, Jiaguo Yu, Liuyang Zhang, Panyong Kuang, 978-0-12-824526-2
Metal Oxides for Optoelectronics and Optics-Based Medical Applications, Suresh Sagadevan, Jiban Podder, Faruq Mohammad, 978-0-323-85824-3
Advances in Metal Oxides and Their Composites for Emerging Applications, Sagar D. Delekar, 978-0-323-85705-5
Metallic Glasses and Their Oxidation, Xinyun Wang, Mao Zhang, 978-0-323-90997-6
Metal Oxide-Based Heterostructures, Naveen Kumar, Bernabé Marí Soucase, 978-0-323-85241-8
Metal Oxide Defects, Vijay Kumar, Sudipta Som, Vishal Sharma, Hendrik Swart, 978-0-323-85588-4

Metal Oxides Series

Metal Oxides for Next-Generation Optoelectronic, Photonic, and Photovoltaic Applications

Series Editor

Ghenadii Korotcenkov

Edited by

Vijay Kumar

Department of Physics, National Institute of Technology Srinagar (J&K), Srinagar, India

Vishal Sharma

Institute of Forensic Science and Criminology, Panjab University, Chandigarh, India

Hendrik C. Swart

Department of Physics, University of Free State, Bloemfontein, South Africa

Subrata Das

CSIR-National Institute for Interdisciplinary Science and Technology, Thiruvananthapuram, India

ELSEVIER

Elsevier
Radarweg 29, PO Box 211, 1000 AE Amsterdam, Netherlands
The Boulevard, Langford Lane, Kidlington, Oxford OX5 1GB, United Kingdom
50 Hampshire Street, 5th Floor, Cambridge, MA 02139, United States

ISBN: 978-0-323-99143-8

For information on all Elsevier publications
visit our website at https://www.elsevier.com/books-and-journals

Publisher: Matthew Deans
Acquisitions Editor: Stephen Jones
Editorial Project Manager: Fernanda A. Oliveira
Production Project Manager: Prasanna Kalyanaraman
Cover Designer: Christian J. Bilbow

Typeset by STRAIVE, India

Contents

Contributors

T.K. Abhilash Materials Science and Technology Division, CSIR-National Institute for Interdisciplinary Science and Technology (NIIST), Thiruvananthapuram; Academy of Scientific and Innovative Research (AcSIR), Ghaziabad, India

Shahzad Ahmad Department of Chemistry, Zakir Husain Delhi College, University of Delhi, Delhi, India

Towseef Ahmad Department of Physics, National Institute of Technology Srinagar, Srinagar, Jammu and Kashmir, India

Lorena Alvarez-Contreras Centro de Investigación en Materiales Avanzados, Chihuahua, Chihuahua, Mexico

S. Anas Materials Science and Technology Division, CSIR-National Institute for Interdisciplinary Science and Technology (NIIST), Thiruvananthapuram, India

Mohd Zubair Ansari Department of Physics, National Institute of Technology Srinagar, Srinagar, Jammu and Kashmir, India

Noé Arjona Centro de Investigación y Desarrollo Tecnológico en Electroquímica, Pedro Escobedo, Querétaro, Mexico

Adersh Asok Photosciences and Photonics, Chemical Sciences and Technology Division, CSIR-National Institute for Interdisciplinary Science and Technology (NIIST), Thiruvananthapuram, Kerala; Academy of Scientific and Innovative Research (AcSIR), Ghaziabad, Uttar Pradesh, India

Irfan Ayoub Department of Physics, National Institute of Technology Srinagar, Hazratbal, Srinagar, Jammu and Kashmir, India

P. Abdul Azeem Department of Physics, National Institute of Technology, Warangal, TS, India

Achu Chandran Materials Science and Technology Division, CSIR-National Institute for Interdisciplinary Science and Technology (NIIST), Thiruvananthapuram; Academy of Scientific and Innovative Research (AcSIR), Ghaziabad, India

Vishnu Chauhan Materials Science Group, Inter-University Accelerator Centre, New Delhi, India; Materials Research Department, GSI Helmholtz Centre for Heavy Ion Research, Darmstadt, Germany

E. Coetsee Department of Physics, University of the Free State, Bloemfontein, Free State, South Africa

Ranjit De Department of Materials Science and Engineering, Pohang University of Science and Technology, Pohang, South Korea

Jesús Adrián Díaz-Real Centro de Investigación y Desarrollo Tecnológico en Electroquímica, Pedro Escobedo, Querétaro, Mexico

J. Divya Department of Physics, University of the Free State, Bloemfontein, Free State, South Africa

Jean Maria Fernandes Department of Physics, National Institute of Technology Warangal, Warangal, Telangana, India

Catalina González-Nava Universidad Politécnica de Guanajuato, Cortázar, Guanajuato, Mexico

Minerva Guerra-Balcázar Facultad de Ingeniería, División de Investigación y Posgrado, Universidad Autónoma de Querétaro, Santiago de Querétaro, Querétaro, Mexico

Deepika Gupta University School of Basic and Applied Sciences, Guru Gobind Singh Indraprastha University, New Delhi, India

Swadesh Kumar Gupta Department of Physics, DBS PG College, Kanpur, Uttar Pradesh, India

D. Paul Joseph Department of Physics, National Institute of Technology Warangal, Warangal, Telangana, India

Harjot Kaur Department of Chemistry, M.M. Engineering College, Maharishi Markandeshwar (Deemed to be University), Mullana-Ambala, Haryana, India

M. Kovendhan Department of Physics and Nanotechnology, SRM Institute of Science and Technology, Kattankulathur, Tamilnadu, India

Ajit R. Kulkarni Centre for Research in Nano Technology and Science; Metallurgical Engineering and Materials Science, Indian Institute of Technology Bombay, Mumbai, Maharashtra, India

Pragati Kumar Nano Materials and Device Lab, Department of Nanoscience and Materials, Central University of Jammu, Samba, Jammu and Kashmir, India

Rajesh Kumar University School of Basic and Applied Sciences, Guru Gobind Singh Indraprastha University, New Delhi, India

Vijay Kumar Department of Physics, University of the Free State, Bloemfontein, Free State, South Africa; Department of Physics, National Institute of Technology Srinagar, Hazratbal, Srinagar, Jammu and Kashmir, India

Anita Kumari Department of Physics, Indian Institute of Technology (Indian School of Mines), Dhanbad, India

Sangeetha M.S. Photosciences and Photonics, Chemical Sciences and Technology Division, CSIR-National Institute for Interdisciplinary Science and Technology (NIIST), Thiruvananthapuram, Kerala; Academy of Scientific and Innovative Research (AcSIR), Ghaziabad, Uttar Pradesh, India

Manoj Kumar Mahata Third Institute of Physics, Georg-August-Universität Göttingen, Göttingen, Germany

Sanjeev Kumar Mishra Department of Chemistry, Zakir Husain Delhi College, University of Delhi, Delhi, India

Umer Mushtaq Department of Physics, National Institute of Technology Srinagar, Hazratbal, Srinagar, Jammu and Kashmir, India

Sibel Oguzlar Center for Fabrication and Application of Electronic Materials, Dokuz Eylul University, Izmir, Turkey

Animesh M. Ramachandran Photosciences and Photonics, Chemical Sciences and Technology Division, CSIR-National Institute for Interdisciplinary Science and Technology (NIIST), Thiruvananthapuram, Kerala; Academy of Scientific and Innovative Research (AcSIR), Ghaziabad, Uttar Pradesh, India

Sayoni Sarkar Centre for Research in Nano Technology and Science, Indian Institute of Technology Bombay, Mumbai, Maharashtra, India

Nupur Saxena Organisation for Science Innovations and Research, Bah, Uttar Pradesh, India

Asmita Shah Unité de Dynamique et Structure des Matériaux Moléculaires (UDSMM), Université du Littoral Côte d'Opale, Calais, France

Savita Sharma Department of Physics, Kalindi College, University of Delhi, New Delhi, Delhi, India

Vishal Sharma Institute of Forensic Science & Criminology, Panjab University, Chandigarh, India

N.J. Shivaramu Department of Physics, University of the Free State, Bloemfontein, Free State, South Africa

Dharmendra Pratap Singh Unité de Dynamique et Structure des Matériaux Moléculaires (UDSMM), Université du Littoral Côte d'Opale, Calais; Department of Industrial Engineering, EIL Côte d'Opale, Longuenesse, France

Manjit Singh Photosciences and Photonics, Chemical Sciences and Technology Division, CSIR-National Institute for Interdisciplinary Science and Technology (NIIST), Thiruvananthapuram, Kerala; Academy of Scientific and Innovative Research (AcSIR), Ghaziabad, Uttar Pradesh, India

Narinder Singh Department of Physics, Sardar Patel University, Mandi, Himachal Pradesh, India

Shriya Sinha Department of Physics, Indian Institute of Technology (Indian School of Mines), Dhanbad, Jharkhand; Department of Physics, Shahid Chandrashekhar Azad Govt. P. G. College, Jhabua, Madhya Pradesh, India

Samarjeet Singh Siwal Department of Chemistry, M.M. Engineering College, Maharishi Markandeshwar (Deemed to be University), Mullana-Ambala, Haryana, India

Hendrik C. Swart Department of Physics, University of the Free State, Bloemfontein, Free State, South Africa

Manish Taunk Department of Physics, Maharaja Agrasen University, Baddi, Himachal Pradesh, India

Adhithya S. Thampi Photosciences and Photonics, Chemical Sciences and Technology Division, CSIR-National Institute for Interdisciplinary Science and Technology (NIIST), Thiruvananthapuram, Kerala; Academy of Scientific and Innovative Research (AcSIR), Ghaziabad, Uttar Pradesh, India

Jyoti Tyagi Department of Chemistry, Zakir Husain Delhi College, University of Delhi, Delhi, India

Sonica Upadhyay Department of Computer Science & Engineering, Maharaja Surajmal Institute of Technology, New Delhi, India

Harris Varghese Materials Science and Technology Division, CSIR-National Institute for Interdisciplinary Science and Technology (NIIST), Thiruvananthapuram; Academy of Scientific and Innovative Research (AcSIR), Ghaziabad, India

Garima Vashisht Department of Physics and Astrophysics, University of Delhi, Delhi, India

Pooja Yadav Department of Physics, National Institute of Technology, Warangal, TS, India

M.Y.A. Yagoub Department of Physics, University of the Free State, Bloemfontein, Free State, South Africa

Merve Zeyrek Ongun Chemistry Technology Program, Izmir Vocational High School, Dokuz Eylul University, Izmir, Turkey

Series editor biography

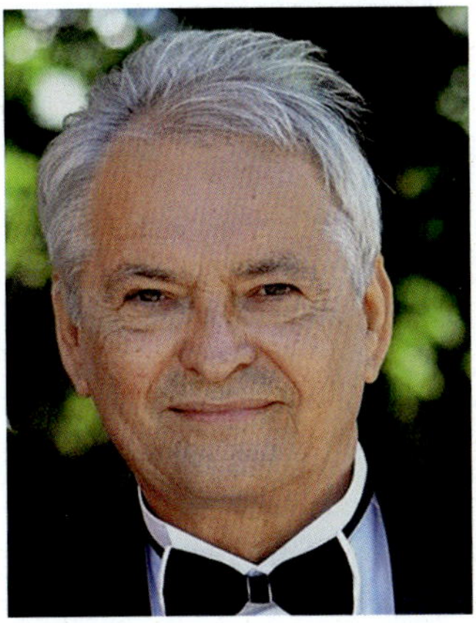

Ghenadii Korotcenkov received his PhD in the physics and technology of semiconductor materials and devices in 1976 and his Doctor of Science degree (Dr. habil.) in the physics of semiconductors and dielectrics in 1990. He has over 50 years of experience as a teacher and scientific researcher. He was a leader of the gas sensor group and the manager of various national and international scientific and engineering projects carried out in the Laboratory of Micro- and Optoelectronics, Technical University of Moldova, Chisinau, Moldova. International foundations and programs such as the CRDF, the MRDA, the ICTP, the INTAS, the INCO-COPERNICUS, the COST, and NATO have supported his research. From 2007 to 2008, he carried out his research as an invited scientist at the Korea Institute of Energy Research (Daejeon). Then, from 2008 to 2018, he was a research professor in the School of Materials Science and Engineering at Gwangju Institute of Science and Technology (GIST) in Korea. Currently, Dr. Korotcenkov is a chief scientific researcher at Moldova State University, Chisinau, Moldova.

Scientists from the former Soviet Union know the results of Dr. Korotcenkov's research in the study of Schottky barriers, MOS structures, native oxides, and photoreceivers based on III–Vs compounds such as InP, GaP, AlGaAs, and InGaAs. His research interests since 1995 include materials science, focusing on metal oxide film deposition and characterization (In_2O_3, SnO_2, ZnO, TiO_2), surface science, thermoelectric conversion, and design of physical and chemical sensors, including thin-film gas sensors.

Dr. Korotcenkov is the author or editor of 45 books and special issues, including the 11-volume *Chemical Sensors* series published by Momentum Press, the 2-volume *Handbook of Gas Sensor Materials* published by Springer, the 15-volume *Chemical Sensors* series published by Harbin Institute of Technology Press, the 3-volume *Porous Silicon: From Formation to Application* published by CRC Press, the 3-volume *Handbook of Humidity Measurements* published by CRC Press, the 3-volume *Handbook of II-VI Semiconductor-based Sensors and Radiation Detectors* published by Springer, and six proceedings of international conferences published by Trans Tech Publications, Elsevier, and EDP Sciences. Currently he is series editor of

Metal Oxides published by Elsevier. Since 2017, more than 35 volumes have been published within this series.

Dr. Korotcenkov is the author and coauthor of more than 650 scientific publications, including 31 review papers, 38 book chapters, more than 200 peer-reviewed articles published in scientific journals (h-factor=43 (Web of Science), h=44 (Scopus), and h=59 (Google Scholar), as of 2022). He is the holder of 17 patents. He has presented more than 250 reports at national and international conferences, including 17 invited talks. Dr. Korotcenkov, as a cochair or member of program, scientific, and steering committees, has participated in the organization of more than 40 international scientific conferences. He is a member of the editorial boards of five scientific international journals. His name and activities have been listed by many biographical publications, including *Who's Who*, and he has also been listed in the Stanford University World's Top 2% Scientists in Applied Physics/Analytical Chemistry in the Physics and Astronomy Cluster. His research activities have been honored by the National Prize of the Republic of Moldova (2022), the Honorary Diploma of the Government of the Republic of Moldova (2020), an Award of the Academy of Sciences of Moldova (2019), and an Award of the Supreme Council of Science and Advanced Technology of the Republic of Moldova (2003); the Prize of the Presidents of the Ukrainian, Belarus, and Moldovan Academies of Sciences (2004); Senior Research Excellence Award of the Technical University of Moldova (2001; 2003; 2005); and the National Youth Prize of the Republic of Moldova in the field of science and technology (1980), among others. Some of his research results and published books have won awards at international exhibitions. Dr. Korotcenkov also received a fellowship from the International Research Exchange Board (IREX, United States, 1998), Brain Korea 21 Program (2008–2012), and BrainPool Program (Korea, 2007–2008 and 2015–2017).

https://www.scopus.com/authid/detail.uri?authorId=6701490962
https://publons.com/researcher/1490013/ghenadii-korotcenkov/
https://scholar.google.com/citations?user=XR3RNhAAAAAJ&hl
https://www.researchgate.net/profile/G_Korotcenkov

Preface to the series

The field of synthesis, study, and application of metal oxides is one of the most rapidly progressing areas of science and technology. Metal oxides are one of the most ubiquitous compound groups on earth, which has a large variety of chemical compositions, atomic structures, and crystalline shapes. In addition, metal oxides are known to possess unique functionalities that are absent or inferior in other solid materials. In particular, metal oxides represent an assorted and appealing class of materials, properties of which exhibit a full spectrum of electronic properties—from insulating to semi-conducting, metallic, and superconducting. Moreover, almost all the known effects including superconductivity, thermoelectric effects, photoelectrical effects, luminescence, and magnetism can be observed in metal oxides. Therefore, metal oxides have emerged as an important class of multifunctional materials with a rich collection of properties, which have great potential for numerous device applications. Specific properties of the metal oxides, such as the wide variety of materials with different electrophysical, optical, and chemical characteristics, their high thermal and temporal stability, and their ability to function in harsh environments, make metal oxides very suitable materials for designing transparent electrodes, high-mobility transistors, gas sensors, actuators, acoustical transducers, photovoltaic and photonic devices, photo-and heterogeneous catalysts, solid-state coolers, high-frequency and micromechanical devices, energy harvesting and storage devices, nonvolatile memories, and many others in the electronics, energy, and health sectors. In these devices, metal oxides can be successfully used as sensing or active layers, substrates, electrodes, promoters, structure modifiers, membranes, and fibers, that is, can be used as active and passive components.

Metal oxides also have low fabrication costs and are robust in practical applications. Furthermore, they can be prepared in various forms such as ceramics, thick films, and thin films. Deposition techniques compatible with standard microelectronic technology can be used for thin film deposition. The microelectronic approach promotes low costs for mass production, offers the possibility to manufacture devices on a chip, and guarantees good reproducibility, all of which are critical for large-scale production. Various metal oxides nanostructures, including nanowires, nanotubes, nanofibers, core-shell structures, and hollow nanostructures, can also be synthesized. The field of metal oxide nanostructured morphologies (e.g., nanowires, nanorods, and nanotubes) has become one of the most active research areas within the nanoscience community.

The ability to create a variety of metal oxide-based composites and the ability to synthesize various multicomponent compounds significantly expand the range of properties that metal oxide-based materials can have, making metal oxides a truly versatile multifunctional material for widespread use. As it is known, small changes in their chemical composition and atomic structure can be accompanied by spectacular variations in the properties and behavior of metal oxides. Even now, advances in synthesizing and characterizing techniques are revealing numerous new functions of metal oxides.

Taking into account the importance of metal oxides for progress in microelectronics, optoelectronics, photonics, energy conversion, sensor, and catalysis, a large number of various books devoted to this class of materials have been published. However, one should note that some books from this list are too general, some books are collections of various original works without any generalizations, and others were published many years ago. But, during the past decade, great progress has been made in the synthesis as well as in the structural, physical, and chemical characterization and application of metal oxides in various devices, and a large number of papers have been published on metal-oxides. In addition, till now, many important topics related to metal oxides study and application have not been discussed. To remedy the situation in this area, we decided to generalize and systematize the results of research in this direction and to publish a series of books devoted to metal oxides.

One should note that the proposed book series "Metal Oxides" is the first one, devoted to the consideration of metal oxides only. We believe that combining books on metal oxides in a series could help readers in searching the required information on the subject. In particular, we plan that the books from our series, which have a clear specialization by their content, will provide interdisciplinary discussion for various oxide materials with a wide range of topics, from material synthesis and deposition to characterizations, processing and then to device fabrications and applications. This book series is prepared by a team of highly qualified experts, which guarantees it a high quality.

I hope that our books will be useful and comfortable to use. I would also like to hope that readers will consider this "Metal Oxides" book series like an encyclopedia of metal oxides that enables them to understand the present status of metal oxides, to estimate the role of multifunctional metal oxides in the design of advanced devices, and then based on observed knowledge to formulate new goals for the further research.

The intended audience of this book series is scientists and researchers, working or planning to work in the field of materials related to metal oxides, that is, scientists and researchers whose activities are related to electronics, optoelectronics, energy, catalysis, sensors, electrical engineering, ceramics, biomedical designs, etc. I believe that this "Metal Oxides" book series will also be interesting for practicing engineers or project managers in industries and national laboratories, which would like to design metal oxide-based devices, but do not know how to do it, and how to select optimal metal oxide for specific applications. With many references to the vast resource of recently published literature on the subject, this book series will be serving as a significant and insightful source of valuable information, providing scientists and engineers with new insights for understanding and improving existing metal oxide-based devices and for designing new metal oxide-based materials with new and unexpected properties.

I believe that this "Metal Oxides" book series would be very helpful for university students, post-docs, and professors. The structure of these books offers a basis for courses in the field of material sciences, chemical engineering, electronics, electrical engineering, optoelectronics, energy technologies, environmental control, and many others. Graduate students could also find the book series to be very useful in their research and understanding features of metal oxides synthesis, study and application of this multifunctional material in various devices. We are sure that all of them will find the information useful for their activity.

Finally, I thank all contributing authors and book editors who have been involved in the creation of these books. I am thankful that they agreed to participate in this project and for their efforts in the preparation of these books. Without their participation, this project would have not been possible. I also express my gratitude to Elsevier for giving us the opportunity to publish this series. I especially thank all teams of the editorial office at Elsevier for their patience during the development of this project and for encouraging us during the various stages of preparation.

Ghenadii Korotcenkov

Section A

Metal oxide-based transparent electronics

Optical transparency combined with electrical conductivity: Challenges and prospects

1

Towseef Ahmad and Mohd Zubair Ansari
Department of Physics, National Institute of Technology Srinagar, Srinagar, Jammu and Kashmir, India

Chapter outline

1 Introduction

From a technological and scientific perspective, metal oxides are the most important class of materials. These have applications in a variety of technological disciplines. Oxide semiconductors are quickly gaining attraction as the newest materials to threaten silicon's dominance [1]. Metal oxide thin films have been recognized for many years for the industrial interest in their unique features. Preparation techniques and operation conditions have a strong influence on the physicochemical properties of metal oxide thin films. It is feasible to produce thin films with either an "amorphous" or "crystalline" structure. After that, the film's structural-electrical and optical

Metal Oxides for Next-Generation Optoelectronic, Photonic, and Photovoltaic Applications
https://doi.org/10.1016/B978-0-323-99143-8.00002-X

properties can be altered by altering the environment and deposition method. Controlling film characteristics is thus a critical aspect of metal oxide film creation for a variety of applications, including microelectronic circuit construction, fuel cells, catalysts, piezoelectric cells, and sensors [2]. Prior to the development of interest in thin films, bulk metal oxides were the focus of research. Metal oxides are ZnO (zinc oxide), TiO_2 (titanium oxide), WO_3 (tungsten oxide), SnO_2 (tin oxide), Cu_2O (cuprous oxide), and CdO (cadmium oxide). Recently, several metallic oxide thin films have emerged, such as V_2O_5 (vanadium oxide), CuO (cupric oxide), NiO (nickel oxide), and MoO_2 (molybdenum oxide), as displayed in Fig. 1. Further, Fig. 1 also depicts the different synthesis methods used to prepare these metal oxide nanoparticles, as well as the different characterization methods used to analyze these metal oxides.

Metal oxide nanoparticles have shown promising results in terms of "physical and chemical" properties due to their high density and small size; consequently, it is essential to comprehend their many features in terms of production, properties, and applications [3]. Nanomaterials are a major milestone not only in the field of miniaturization, but also in the atomic and quantum realms. When a material is reduced to the nanodomain, its physical, biological, and chemical properties are largely determined by quantum physics, as opposed to classical physics, which governs the bulk phase [4]. Such distinctions come primarily as a result of greater surface area, chemical reactivity, and mechanical strength [1,5]. Metallic oxide semiconductors have long been regarded as the most promising nanomaterials, thanks to their wide range of applications in "electronics," "optoelectronics," "biosensors," "piezoelectricity," and "catalysis" [6].

Nanotechnology has enormous potential for improving the efficiency of "water purification" and "decontamination." Nanomaterials are effective at removing organic-inorganic pollutants, as well as heavy metals, from wastewater while also

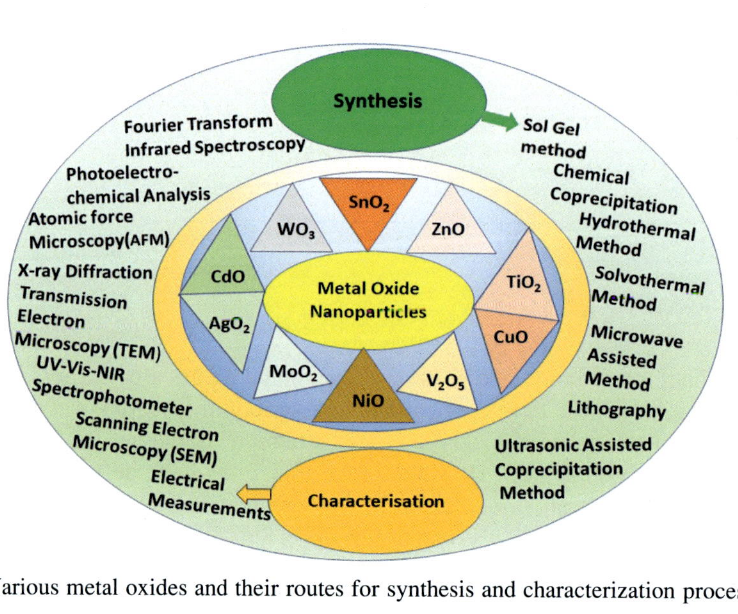

Fig. 1 Various metal oxides and their routes for synthesis and characterization processes.

killing microorganisms. All living organisms require clean water to survive. However, global contamination of existing water resources has increased as a result of swift industrialization and huge population growth. Agriculture has seen a significant increase in the demand for and consumption of clean water. Fresh and clean water containing a wide variety of impurities is widely used in many different fields of consumption, including industry, households, and agriculture. is approximately 70%, 22%, and 8%, respectively [7–12].

Researchers are interested in the symbiotic features of metal oxide nanoparticle-polymer nanocomposites. The hybrid nanocomposites have excellent elastomeric stiffness, wear resistance, and improved thermal and conductivity characteristics. The interfacial interactions between the matrices and nanomaterials, as well as the volume percentage of the components, are critical parameters for producing the effective properties of nanocomposites. Effective dispersion of nanomaterials in polymer matrixes is always a difficult issue due to their aggregation nature [13].

The foundations of nanotechnology and nanoscience are nanomaterials. A nanomaterial's dimensionality determines the range of possible applications for that material, sorted into categories of zero, one, two, and three dimensions. The 1D nanomaterials in these examples are influential substances because of their appealing physical characteristics. For the most part, there are three distinct shapes that may be found in one-dimensional nanomaterials; these are nanorods, nanowires, and nanotubes [14]. The different applications of metal oxide nanoparticles are displayed in Fig. 2 in electronics, texture, security, and defense, etc., and also have environmental applications in wastewater treatment, healthcare, food and agriculture, pollution monitoring, energy, environmental catalysis, and heavy metal remediation.

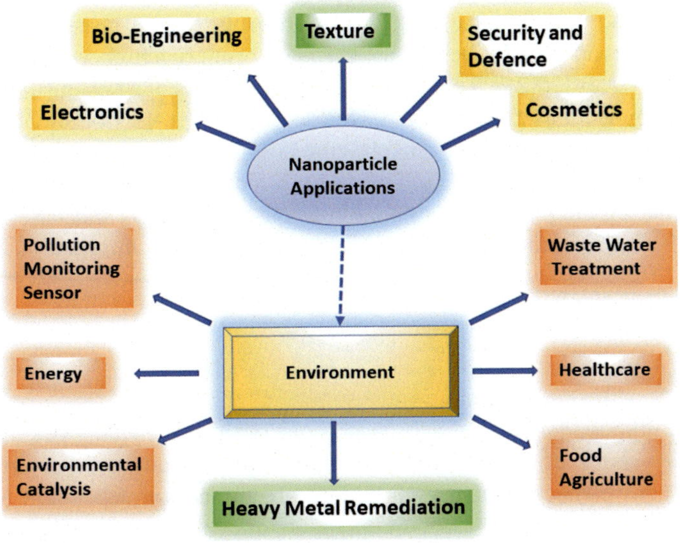

Fig. 2 Different metal oxide nanoparticles and their applications.

2 Optical properties of metal oxides

When it comes to the advancement of science and technology, metal oxides (MO) are an indispensable class of materials. Some metal oxides are classified as semiconductors, while others are classified as insulators based on the size of their bandgaps. Insulators are non- or poor thermal and electrical conductors, and they typically have a colorless or white appearance. The presence of impurities or dopants, however, can cause them to take on a hue. The presence of metal ions like Cr and Ti in insulators like Al_2O_3 and SiO_2 is responsible for the vivid hues of many gemstones, including ruby and sapphire. The bandgap of MO can be quite different from one metal to another. The small bandgap responsible for Fe_2O_3's vibrant color is also responsible for the absence of color in ZnO, which has a much larger bandgap. However, unlike II-VI or III-V semiconductors, which typically have a large exciton Bohr radius and thus a significant quantum confinement effect for nanoparticles with a radius of a few nanometers, most metal oxides have a small exciton Bohr radius, such as 3 for TiO_2 [15,16]. Direct bandgap transitions, like those in semiconductors, are often robust and have a prominent excitonic peak, while indirect bandgap transitions are characterized by weaker and more featureless absorption. This is due to the large bandgap, which causes even the lowest-energy electronic transitions to occur in the near UV region. The samples typically have no discernible color or seem white because of low or nonexistent visual absorption [17–19]. Nanoparticles of common metal oxides exhibit relatively low visual absorbance in their pure forms. These include TiO_2, SnO_2, WO_3, and ZnO, and impurities or dopants may cause them to change their light-absorbing behavior. Metal oxide nanoparticles such as SnO_2 and zinc oxide (ZnO) stand out due to their stability, accessibility, and potential for usage in areas such as solar energy conversion. Just like other nonmetal oxide semiconductors, MO nanomaterials typically exhibit photoemission or photoluminescence (PL) when excited above their bandgap. This is probably due to charge carriers being trapped and the band edge PL being quenched as a result of a higher density of defects or surface trap states [20,21]. To modify the characteristics of a host insulator or semiconductor, doping is a useful technique. The idea is to insert electronic states inside the bandgap to cause new transitions and alter the host material's properties and capabilities. The electronic industry relies heavily on p- and n-doped Si, both of which are made possible by the practice of doping bulk materials. SnO_2 and other MOs have been doped with elements including aluminum, sulfur, and nitrogen (C, S, and N) [22,23].

2.1 SnO_2

Metal oxide semiconductor nanoparticles, as one of the most vital types of materials, appear in a variety of fields of science and industry due to their structure and properties, both physical and chemical, that are dimension-dependent [24]. SnO_2 nanoparticles have received a lot of attention due to their remarkable chemical resistance and mechanical strength (inert to moderate acids and films coatings are stress resistant), little visible absorption due to a large bandgap (transparent materials), and high mechanical strength [25,26]. Free carriers in nonstoichiometric oxides with oxygen vacancy holes (n-type) cause a considerable level of electrical conductivity; pure SnO_2 has a bandgap of 3.64 eV; hence, it is a semiconductor. The ability to tailor its n-type carrier capacity by introducing impurities or dopants makes it a critical semiconductor material for use in a wide range of fields, including optics, electronics (in Lithium-batteries, supercapacitors, etc.), gas-sensing devices, conducting solar-window materials in solar cells, and catalysis [27–30]. The composition and microstructure of SnO_2 nanomaterials (porosity, presence of defects, and particle size) are mostly determined by the technique of synthesis [31]. The "chemical vapor deposition," "hydrothermal," or "solvo-thermal approaches," "sol-gel technology," "magnetron sputtering," and "spray pyrolysis" are some of the well-established synthetic methods for producing SnO_2 nanoparticles or thin films (aggregates of nanoparticles on a substrate) [32–35]. Nonstoichiometric polycrystalline SnO_2 nanoparticle thin films are used in heat-reflecting mirrors because of their transparency in the visible range and high reflectivity in the near-infrared (IR). They have a high IR absorption coefficient because free carriers are present in the crystal structure. The IR spectra of SnO_2 or doped SnO_2 nanoparticles can be explained using the Drude Theory (a drop in the free carrier density causes a decrease in electrical conductivity, which causes a reduction in IR (infrared) absorption and vice versa) [36–38].

Sun et al. [39] report UV-Vis-NIR transmittance although within a narrow range visible transmittance remains consistently above 80% (Fig. 3C). Film near-infrared transmittance continues to drop from an undoped state of 92.55% to 60.48% after 11 mol% Sb doping. At 11 mol% Sb doping, the transmittance is maximized and rises to 68.95%. This suggests that a 32.07% decrease in NIR transmittance occurs with a Sb doping level of 11%. The conclusion that the film's optical properties change in the NIR (near-infrared region) is supported by the fact that the film's electrical properties also change with time. Increases in carrier concentration and conductivity result in decreased infrared light transmission. The cut-off wavelength λ_p which is positioned at visible near-infrared light, determines the highest wavelength limit of the transparent conductive film.

$$\lambda_p = 2\pi c_0 \left(\frac{Ne^2}{\varepsilon_0 \varepsilon_l m^*} - \gamma^2 \right)^{-\frac{1}{2}} \tag{1}$$

$$\gamma = \frac{e}{m^* \mu} \tag{2}$$

Where N is the number of free electrons, ε_0 is the vacuum dielectric constant, ε_l is the high frequency dielectric constant, m^* is the effective mass of free electrons in the conduction band, e is the electron charge, and μ is the electron mobility.

It can be seen from the expression (1) that when the carrier concentration in the ATO film increases, the plasma oscillation wavelength λ_p shifts from the long wave to the short wave region when the Sb doping quantity is kept to a reasonable level. As for the wavelengths of light, the film is highly reflective to infrared light longer than λ_p

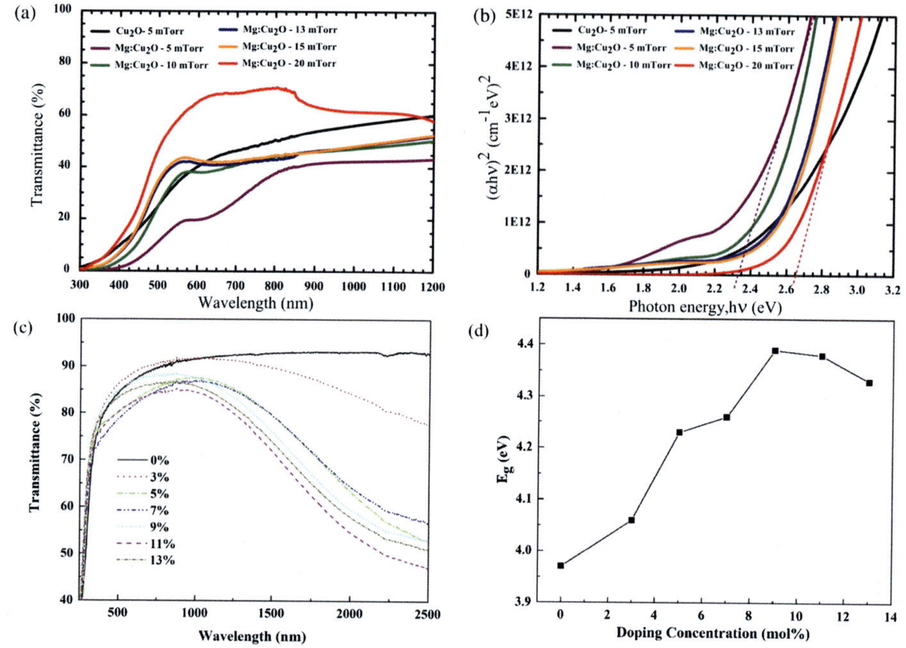

Fig. 3 (A) and (B) Variation in the transmittance spectra undoped and Mg-doped Cu_2O along with a Tauc plot of undoped and Mg-doped at different Torr. (C) Changes in the transmission spectra of thin films due to Sb doping; and (D) the relationship between doping level and optical bandgap.
Data from N. Sliti, E. Fourneau, T. Ratz, S. Touihri, N.D. Nguyen, Mg-doped Cu_2O thin films with enhanced functional properties grown by magnetron sputtering under optimized pressure conditions, Ceram. Int. 48 (2022) 23748–23754 and M. Sun, J. Liu, B. Dong, Effects of Sb doping on the structure and properties of SnO_2 films, Curr. Appl. Phys. 20 (2020) 462–469.

and very transparent to visible light shorter than λ_p. The optical bandgap measurements of undoped SnO_2 film and ATO films doped with Sb concentrations ranging from 3.0 to 13.0 mol% are displayed in Fig. 3D. The optical bandgap of pure SnO_2 film is the narrowest, coming in at 3.97 eV. When the amount of Sb doped into the film is raised to 9 mol%, the optical bandgap reaches a maximum of 4.39 eV. With increasing Sb doping quantity, the optical bandgap narrows ever so slightly. The optical bandgap of ATO film is reduced to 4.33 eV when the Sb doping level reaches 13 mol%. The combination of great conductivity and transparency in the visible range of the electromagnetic radiation spectrum has sparked interest in SnO_2 coatings. A small concentration of midgap states, which are normally responsible for the absorption of photons with energies below the bandgap value, can explain the high transparency. SnO_2 thin films generally have n (refractive index) values between 1.9 and 2 [40–42]. Thomas et al. [43] report the transmittance, reflectance, and absorption optical characteristics of pure and FTO:Er films. The maximum transmittance of the film is seen to be around 79% for Er (1.5 wt%), and it rises steadily from undoped film to doped film. Jo et al. [44] demonstrate how the optical transmittances of FTO films change as a function of wavelength due to the Fe co-doping effect. The T (average optical transmittance) between 400 and 700 nm is shown to fall from 85.1% for undoped FTO to 82.5% for 3Fe-FTO. Because of the reduced bandgap of FTO films, we can pinpoint this average T fluctuation. As a result, 2Fe-FTO has the highest figure of merit (T^{10}/R_{sh}) among the other films, owing primarily to the lower R_{sh} caused by improvements in N and demonstrating the implementation of excellent transparent conduction in a TCO film for EC devices. Go et al. [45] prepared FTO codoped with graphene and observed that the graphene/FTO coated sheet resistance was 52.4 Ω/cm^2 compared to plain FTO's 118.4 Ω/cm^2. Graphene/FTO coating (85%) was also more transparent than ordinary FTO (73.5%). Ramarajan et al. [46] prepared Sb doped SnO_2 at different substrate temperatures. Due to its improved surface and crystalline characteristics, the ATO450 film had the highest transmittance (Tr = 89%) at 550 nm and decreased for films produced at temperatures above 450°C, indicating increased surface roughness and light scattering by grain boundaries. ATO450 film's NIR transmittance exceeds 90%. Li et al. [47] prepared reducing graphene oxide aerogels (rGOA), which were used to create antimony tin oxide (ATO) nanoparticles. Joshi et al. [48] reported Sb doped SnO_2 samples and reported 200–800 nm transmittance spectra Sb doping reduces SnO_2 film transparency. This may be caused by light scattering on the film's surface. The maximum visible transmittance was 83% for an undoped SnO_2 thin film. Islam et al. [49] prepared Ba-doped SnO_2 with different annealing temperatures; the 873 K-annealed film transmits 90% of visible light. Agglomeration of crystallite grains may reduce surface scattering, increasing transmittance. Kim et al. [50] prepared Sb-doped SnO_2 coatings with visible band transmittances ranging from 80% to 90% and sheet resistances of 10–20 Ω/sq.

2.2 CuO

The two possible oxidation states of copper are +2 and +1. However, under certain circumstances, some compounds are reported to have an oxidation state of +3. Furthermore, trivalent copper has a short lifespan. Consequently, copper is found in two stable forms: cuprous oxide (Cu_2O) and cupric oxide (CuO). Cupric oxide (CuO), on the other hand, is a gas-sensitive material with a variety of intriguing properties, according to reports [51–53]. Cupric oxide has gotten a lot of attention because it is the simplest copper compound and has a lot of interesting physical properties like "electron correlation," "spin dynamics," and "superconductivity." This resulted in a surge in CuO-based device studies conducted in the second half of the 20th century, including theoretical studies, fabrication, characterization, and applications [54,55].

When compared to bulk materials, metal oxide nanoparticles exhibit unique "optical and electrical" properties. Researchers have long been interested in metal oxide-based nanocomposites because they offer excellent "electrical and optical" properties. The optical characteristics of thin films used in optoelectronic devices are critical. CuO's optical characteristics make it useful as an absorber layer in "photovoltaic devices." This function necessitates good visible solar spectrum absorption

[53,56,57]. The preparation circumstances impact the film's surface morphology and optical qualities, just as the deposition procedure and experimental settings impact the film's structural properties. Films fabricated through the sputtering method typically exhibit interference fringe in the transmission spectra, but fringe is rarely seen in the transmission spectra of CuO films. Fringe is a characteristic of the deposited film's surface smoothness. According to the transmission spectrum, we can see that there is no fringe for the films made by the spraying method. CuO thin films with varying precursor concentrations and Mn doping concentrations are deposited using the spray and SILAR techniques [58–60]. Saravanan et al. [61] discovered a correlation between annealing temperature and transmittance of undoped CuO thin films made by spray pyrolysis. Gulen et al. [62] discovered that increasing the Mn doping concentrations in CuO films made by the SILAR technique improved the film transmittance. They discovered that as Mn doping concentrations rise from 1% to 5%, transmittance rises rapidly from 10% to 80%. Chafi et al. [63] observed a decrease in transmittance in CuO thin films produced by spray pyrolysis with varying precursor concentrations; the decline in transmittance of the deposited film is most likely due to an increase in the molar concentration of the solution. Furthermore, the optical bandgap (Eg) is usually calculated using optical transmittance, CuO has a direct bandgap of 1.2 eV in bulk, which varies in range from 1 to 8 eV depending on the deposition procedures and circumstances. Diachenko et al. [64] use the spray pyrolysis process to make copper oxide thin films. The absorption coefficient of the copper oxide films developed was found to be high, indicating that they might be used as absorbing layers in solar cells. The optical bandgap of the material was measured and found to be between 1.45 and 1.60 eV. In Fig. 3A, Sliti et al. [65] show the formation of a thin film after Mg doping in Cu_2O as a function of plasma pressure during deposition. A typical transmission value was derived from a number of observations made between 500 nm and 800 nm. In the case of the 5-mTorr samples, the film transmittance value drops after being doped with the Cu_2O compound. On average, the pure Cu_2O thin film has a transmittance of about 46%. Once Mg is introduced, the average transmission of Mg:Cu_2O thin films increases from 30% to 63% over the visible spectrum as the plasma pressure is changed from 5 to 20 mTorr. These findings are consistent with those of prior studies on RF-produced, undoped Cu_2O. Bandgap values were found to be lower for the Mg-doped Cu_2O thin film compared to the undoped Cu_2O thin film at a low Ar plasma pressure (Fig. 3B). To gain perspective on this finding, the lower transmittance values observed at lower plasma pressures. However, the bandgap increases from 2.32 eV to 2.65 eV when the plasma pressure is increased. Because of the high carrier concentration that results from the degradation of the semiconductor, the Moss-Burstein effect (also known as the Burstein-Moss shift or BM shift) may be responsible for the widening of the bandgap in the thin films. As a result, the value of the perceived bandgap increases. Samuel et al. [66] synthesized Cu_2O films at a substrate temperature of 100°C using spray pyrolysis to create CNT(carbon nanotubes)-impregnated Cu_2O thin films on a glass substrate. In the visible (380–780 nm) zone, the average transmittance is greater than 65%. The transmittance reached a maximum of 70%, whereas the absorption increased in the wavelength range 300–600 nm. The bandgap of CNT impregnated Cu_2O thin films ranged from

1.80 to 2.38 eV for samples annealed at 200°C and 1.68 to 2.17 eV for samples annealed at 230°C.

2.3 ZnO

It has been established that semiconducting metal oxides are the most significant nanostructured materials for use in high-performance electronics, energy conversion/storage, and environmental cleanup. ZnO has garnered a considerable amount of attention among the plethora of metal oxide materials [67]. This substance has some aspects that make it an appropriate contender for use in optoelectronics, laser technology, and electronics, including outstanding thermomechanical stability at room temperature and a binding energy of 60 meV [68,69]. Due to its piezo- and pyroelectric properties, ZnO can also be used as a photocatalyst, converter, sensor, and energy generator in the manufacture of hydrogen. Moreover, it is utilized in pro-ecological systems and biomedicine because of its biocompatibility, biodegradability, and low toxicity characteristics [6,70]. Astonishing characteristics of ZnO nanostructures are their wide bandgap, high excitation binding energy, nontoxicity, biocompatibility, chemical and photochemical stability, and strong electron communication characteristics, making it a desirable substance for producing effective sensors and biosensors [71–73].

Due to ZnO's unique characteristics, it is given much thought, particularly in structures with hexagonal wurtzite. Cubic rock salt and blende are two more structural forms of ZnO. It displays the thermodynamically stable wurtzite structure at reasonable temperature and pressure settings. Under high pressure conditions, ZnO forms a cubic structure with an indirect bandgap semiconductor (Eg = 2.7 eV) [74,75]. The numerous structures of zinc oxide at nanometer scales are employed in various nanotechnology fields. The structure of ZnO is found to be one-dimensional (1D), two-dimensional (2D), and three-dimensional (3D). The majority of these are 1D structures, including wire and combs, needles, springs, and rings, as well as nanorods, tubes, belts, helixes, and ribbons. Another form of ZnO developed in 2D structures is nanoplate, nanosheet, and nanopellets. On the other side, ZnO can also be found in 3D formations, including snowflakes, coniferous urchin-like structures, flowers, and dandelion-like structures. ZnO, of all known materials, provides one of the broadest ranges of diverse nanoparticle structures [1,76]. Since ZnO nanowires, nanobelts, nanoflowers, nanorods, and nanotubes exhibit excellent electron transport routes along the length direction and a sufficient surface-to-volume ratio, they are particularly well suited for electrochemical sensing. ZnO-based matrices might be thought of as a good platform for inexpensive biosensors because fabrication with low-cost techniques is remarkably simple and produces a large number of nanostructures. In order to advance in this field, we must further characterize these nanostructured materials by combining electrochemical sensors and biosensors to achieve the goal of developing an immediate, sensitive and affordable point-of-care diagnostic tool [53,72,77].

Semiconductor materials have offered a wealth of knowledge on a variety of topics based on their physical behavior, such as their electrical and vibrational states, the presence and nature of faults and impurities. Semiconductors are related to both their external and intrinsic properties [78]. The intrinsic properties of a semiconductor are based on interactions between holes in the VB (valence band) and electrons in the CB (conduction band), as well as excitonic causes caused by the Coulomb interaction [79]. Extrinsic optical properties are reliant on dopants or defects introduced into the semiconductor, which provide distinct electronic-states between VB and CB. A ZnO semiconductor optical transition has been investigated using a variety of methods, including photo-luminescence (PL), transmittance and optical absorption, reflection, and cathode luminescence [80,81]. Different nanostructures of ZnO are illustrated by their respective PL spectra to emit UV light, one or two bands of visible light, interstitial emissions, antisite emissions, vacant positions, and complicated defects [82,83]. With a larger exciton energy of 60 meV at ambient temperature, ZnO displays a bandgap of 3.37 eV. It is capable of effective exciton emission at RT (room temperature) below low-excitation energy since its exciton energy is higher than that of GaN (25 meV) [84,85]. ZnO is thus among the most promising photonic materials in the blue-UV spectrum. When studying the optical characteristics of ZnO nanorods, PL spectroscopy has been used, providing information on the bandgap, defects, and crystallographic features. ZnO nanorods with reduced impurity concentrations have demonstrated deep level emissions (DLE), and a single emission has been observed during UV emission (from 3.23 to 3.30 eV) [86,87]. The imperfections in the nanostructured materials, such as zinc vacancies (VZn), oxygen vacancies (VO), zinc interstitials (Zni), oxygen interstitials (Oi), and extrinsic impurities, are responsible for NBE (near band edge) emission and DLE band emissions in the UV emission and visible ranges, respectively [88,89]. ZnO's optical quality can be determined by dividing the NBE emission intensity by the DLE emission intensity (INBE/IDLE). The deep level defect is detected by a large INBE/IDLE value and a lower concentration [75,90,91].

Caudra et al. [92] investigate the optical transmittance spectra of ZnO and doped-ZnO thin films deposited on soda lime substrates. In the visible region, good transmittance was observed, with average transmittances of 80% and 78% for Na and K-dopants, respectively and transmittance found to be related to the thickness of the film; a reduction in transmittance was observed with a rise in film thickness. Nasir et al. [93] prepared ZnO thin films via the sol-gel technique with different annealing temperatures of 400°C, 450°C, 500°C, and 550°C. All samples exhibit good transmittance of visible and infrared transmission greater than 95%; however, a little lower optical transmittance less than 95% was observed for the sample annealed at 400°C. Yesilkaya et al. [94] investigate the effects of Na doping and heat treatment on ZnO thin films. The bandgap of undoped material was 3.36 eV and was reduced to 3.35 eV by thermal annealing, while the bandgap of unannealed material was reduced from 3.32 eV to 3.31 eV. Hammad et al. [95] prepared copper doped zinc oxide via the sputtering process. The increase in Cu doping causes decrease in visible transmittance and causes a shift of the absorption edge toward a higher wavelength region, along with a decrease in the bandgap from 4.105 eV to 3.444 eV. Asaithambi et al. [96] prepared transition metals (Cu, Fe, and Zn) doped SnO_2 using the co-precipitation method. The variation in absorption edge and peak shift toward higher wavelength was observed. The bandgaps obtained using the Tauc plot are 3.38 eV, 2.64 eV,

2.79 eV, and 2.89 eV for SnO_2, Fe-SnO_2, Cu-SnO_2, and Zn-SnO_2, respectively. Ahmad et al. [97] prepared boron-doped ZnO and observed good retention of visible transmittance and an absorption region shift toward a lower energy region due to increased B content in ZnO, resulting in a significant decrease in bandgap energy.

Goktas et al. [98] discovered that the transmittance of ZnO varies with doping and temperature. Other than the 30% Mg-doped ZnO thin film, all of the others had transmittances in the visible area of 85%–98% as shown in Fig. 4A. Transmission of ZnO: Mg20% at 571 nm is 98.2%, the greatest of any ZMO thin film. Due to the film's great regularity and well-oriented nanograins, less light is reflected or scattered from its surfaces. In the case of optoelectronic applications, such as window layers in solar cell devices, this high optical transparency is essential. The transmission spectra of the 5% Mg-doped-ZnO thin film are shown in Fig. 4B. When the T is raised from 773 to 973 K, the films become less transparent, caused by the higher dispersion of incident light due to the rougher surface. Compared to ZMO thin films, 30% Mg-doped-ZnO thin films had a lower orientation degree and smaller nanograin size. The absorption spectrum (Fig. 4C) confirms the obvious rise in light scattering and subsequent decrease in transmittance. The enhancement of the photoluminescence property of ZnO was done by Li doping, as reported by Chen and Ding [99]. The decrease in visible transmittance from 70% (undoped ZnO) to 10% (16% Li doped) with an increase in Li doping and further annealing temperature caused an increase in transmittance to 15%. The sharp absorption edge for all films exists between 380 nm and 400 nm. Doping reduces the bandgap from 3.36 eV to 3.14 eV; however, with an annealing temperature in the 400–500°C range, the bandgap rises from 3.14 eV to 3.1 eV.

3 Electrical properties of metal oxides

The materials used to make transparent conductors have some degree of transparency but are not completely transparent to light. Transparency and conductivity seem to be at odds with one another from the perspective of the band structure; hence, a material with both characteristics seems unlikely. Metallic conductivity emerges when the Fermi level lies within a band with a substantial density of states to offer high carrier concentration, in contrast to a fully filled valence band and an empty conduction band. Effective transparent conductors strike a balance between transmitting enough light in the visible spectrum and having enough conductivity to be practical. Several widely-used oxides, including In_2O_3, SnO_2, ZnO, and CdO, are capable of achieving this combination. This class of materials is an insulator with an optical bandgap of roughly 3 eV in its undoped stoichiometric condition. In order to evolve into a transparent conductor, oxide (TCO), the Fermi level can only be displaced upward in these TCO hosts by degenerately doping them within the range of acceptable conductance. Conventional n-type TCO hosts are characterized by a widely scattered free electron, just like a conductor's band. Then, degenerate doping results in (i) the small effective mass of electrons, which contributes to their great mobility, and (ii) the low density of states in the conduction band, which results in minimal optical absorption. The Burstein-Moss (BM) shift, where the Fermi energy is shifted to be higher than the conduction band

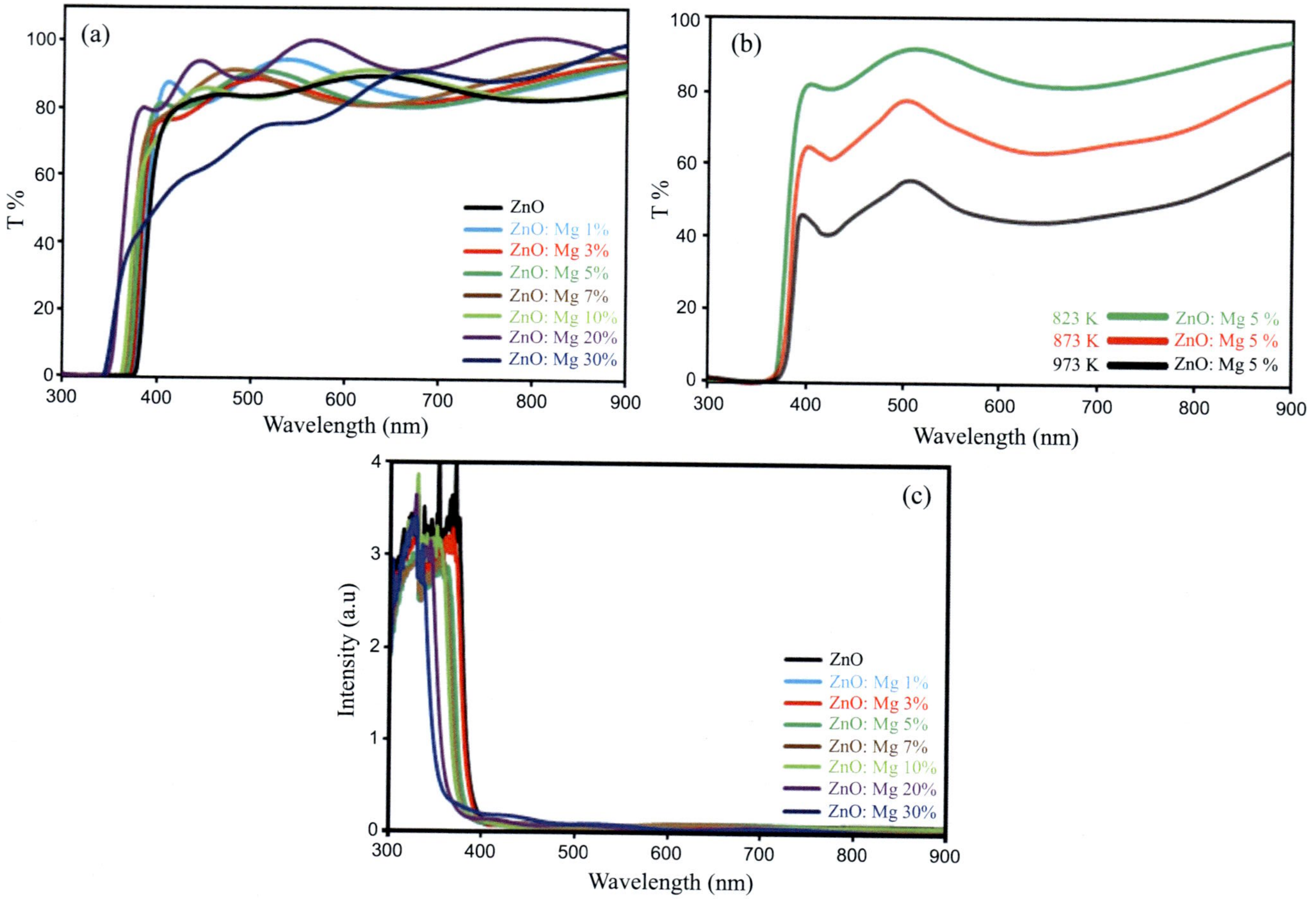

Fig. 4 At room temperature, a UV/VIS spectrophotometer was used to analyze the optical transmittance and absorbance of ZnO and ZMO thin films. (A) The transmittance of ZnO:Mg with different Mg doping concentrations. (B) ZnO:Mg5% transmittance at various temperatures; and (C) absorption spectra for ZMO thin films.

Data from A. Goktas, A. Tumbul, Z. Aba, M. Durgun, Mg doping levels and annealing temperature induced structural, optical and electrical properties of highly c-axis oriented ZnO:Mg thin films and Al/ZnO:Mg/p-Si/Al heterojunction diode, Thin Solid Films 680 (2019) 20–30.

minimum, is a consequence of the high energy dispersion of the conduction band. Because of the intricate relationship between electronic and optical features, optimizing a TCO's performance is difficult [100,101]. High levels of carrier concentration sought for high conductivity may lead to a rise in the optical absorption (i) at relatively short wavelengths due to band-to-band transitions in the partially full conduction band, and (ii) at long wavelengths, as a result of intra-band transitions in this material band. As a corollary, plasma oscillations may alter optical characteristics by reflecting electromagnetic waves with a lower frequency than the plasmon. As an added bonus, ionized contamination of electron donors by impurities (native point defects or substitutional dopants) impacts charge transfer negatively, while structural relaxation of these impurities may cause a change in the host material's electrical and optical characteristics. The Fermi level is undergoing a nonrigid-band change. At a critical temperature, semiconductor materials experience a transition from semiconductor to metal, where the conductivity abruptly increases due to the formation of unoccupied d or f electron energy bands as a result of overlapping electron orbitals [102,103].

3.1 SnO_2

When heated in a reducing atmosphere, bulk tin oxide easily transforms into an n-type semiconductor. The existence of oxygen vacancies in SnO_2 crystals causes an electron donor state to rise immediately below the CB (conduction band). When a semiconducting system is exposed to an external electric field for the purpose of measuring some electrical parameter, as a result of scattering, properties like conductance and charge carrier density will revert to equilibrium. This scattering is caused by free carriers interacting with some scattering center (for example, ionized impurities or dopants, thermal vibrations of lattice points, structural flaws, and so on). This scattering reduces the movement of free charge carriers. By estimating mobility, Thangaraju stated that these values match observed mobility using the mean free path and carrier or dopant concentration values. The film's resistivity is determined by the carrier concentration [104,105]. Pure and Zn-doped SnO_2 have almost linear I-V curves, showing that the material is ohmic. Ye Tao et al. [106] prepared tin oxide films to observe the effect of substrate temperature and O_2 flow rate. A range of optimum O_2 flow rates for producing films with improved transparent conductive characteristics exists at three substrate temperatures. The films show the lowest resistivity $3.65 \times 10^{-3}\,\Omega\,cm$ at RT for the substrate, which corresponds to a high carrier concentration $1.41 \times 10^{20}/cm^3$ and mobility $12.11\ cm^2 V^{-1} s^{-1}$, along with a typical transmittance in the visible light spectrum of 81.50%.

Sun et al. [39] reported (Fig. 5A) that sheet resistance and resistivity are both maximum for the pure SnO_2 film at $4.98 \times 10^5/cm^2$. So, the pure SnO_2 coating is the least conductive option. As Sb doping rises from 0 to 5 mol%, both sheet resistance and resistivity drop dramatically. Resistance and sheet resistance both gradually decrease as Sb doping is increased from 5% to 11%. An increase in sheet resistance and resistivity is observed for Sb doping levels above 11 mol%. Fig. 5B demonstrates that the maximum conductivity of the film is achieved with 11 mol% Sb doping. Increasing Sb doping results in a rise in free carriers, which in turn lowers the film's sheet resistance

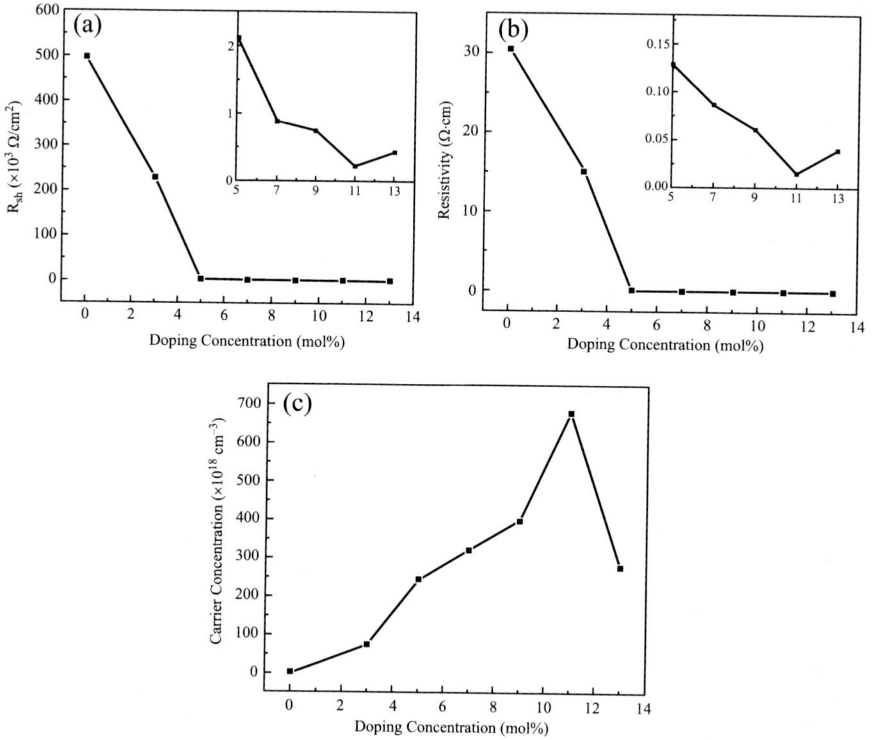

Fig. 5 Impact of Sb doping on the interplay between sheet resistance, resistivity, carrier concentration, and resistivity. (A) The correlation between sheet resistance and Sb doping with an expanded section in the top right corner; (B) the correlation between resistivity and Sb doping, with an expanded section in the top right corner; and (C) the correlation between carrier concentration and Sb doping.
Data from M. Sun, J. Liu, B. Dong, Effects of Sb doping on the structure and properties of SnO_2 films, Curr. Appl. Phys. 20 (2020) 462–469.

and boosts its conductivity as shown in Fig. 5C. The substitution of Sb^{5+} ions for Sn^{4+} ions in the crystal lattice creates free carriers because of the release of surplus electrons. When the Sb content of the film rises over the crucial value of 11 mol%, the carrier concentration drops. This is because, above this threshold, the carrier concentration is dominated by trap states caused by an excessive number of lattice defects. Additionally, the carrier mobility in the film and hence the film's conductivity will be affected by the scattering effect of a high concentration of dopant ions on carriers. Peddavarapu Sivakumar et al. [107] prepared Ga doped SnO_2 films to observe the influence of Ga doping on different properties of SnO_2. As can be seen, all of the deposited films have resistivity values in the $10^{-3}\,\Omega$ cm range. The measured values for pure SnO_2 and 1, 2, 3, 5 at.% Ga-doped-SnO_2 were $2.63 \times 10^{-3}\,\Omega$cm, $2.8 \times 10^{-3}\,\Omega$cm, $2.95 \times 10^{-3}\,\Omega$cm, and $3.2 \times 10^{-3}\,\Omega$cm, respectively. Yinzhu Jiang et al. [108] prepared Zn doped tin oxide films using the chemical-vapor-deposition method (CVD). According to the electrical-resistivity trend as a function of zinc-doping, the charge-carrier concentration displays a maximum value of 0.12 which is $2.9 \times 10^{20} cm^{-3}$. However, the Hallmobility exhibits a steady rise with zinc doping up to $k = 0.12$, while minimal change is observed with higher zinc doping ($k > 0.12$). Differences in electrical characteristics as a function of the preparation process and other parameters are shown in Table 1.

3.2 CuO

Copper oxide is thought to be an inherently p-type semiconductor because copper vacancies serve as acceptors for the entire conduction process. The dominant inherent defect, such as oxygen or copper vacancy, usually reveals the electrical characteristics of pure CuO. Because of its volatility, a nonstoichiometric cupric oxide is known to have dominating copper vacancy defects. Indeed, copper vacancies cause shallow acceptor levels to rise above the VB (valence band), resulting in p-type behaviors [115,116]. Several experiments were conducted with the purpose of determining the impact of deposition procedures and circumstances on CuO thin film electrical properties. CuO thin films have been found to have a dark electrical resistivity anywhere from 10 to 10^8 (Ωcm), depending on their fabrication process. The concentration of free carriers and the variance in hole-mobility vary from 10^{10} to $10^{16}/cm^3$ and 10 to 100 ($cm^2 V^{-1} s^{-1}$), respectively [117–119]. Thin films' electrical characteristics are affected by structural changes. Phase shifts, doping, crystallite size growth (D), grain boundaries, and stoichiometry departure are all examples of these alterations. The resistivity of the films is affected by these modifications to the charge carrier population, carrier mobility, and transport mechanism. In addition, the resistivity of a material decreases as the film grain size decreases. In comparison to other approaches such as "thermal oxidation" and "reactive RF sputtering," Dhanasekaran and Mahalingam [120] found that CuO thin-films generated using the "sol-gel method" have a lower resistivity of $10^2 \Omega$ cm.

Fig. 6C depicts how the carrier concentration and mobility change as a function of the deposition plasma pressure. When the pressure of the sputtering plasma is raised, the concentration of hole carriers rises from $5.37 \times 10^{15} cm^{-3}$ to $7.5 \times 10^{17} cm^{-3}$, but the mobility of the charge carriers drops from $8.31 cm^2 V^{-1} s^{-1}$ to $0.11 cm^2 V^{-1} s^{-1}$. We show how the electrical resistivity has changed over time, and how it has been related to the pressure in the plasma [65]. The insertion of Mg dopants into the Cu_2O thin films causes a drop in the film's resistivity, from 228Ω cm to 140Ω cm. At 13 mTorr film deposition yielded the lowest resistivity value of 10.2Ω cm, where the carrier concentration is $1.72 \times 10^{17} cm^{-3}$ and the mobility is $3.13 cm^2 V^{-1} s^{-1}$. The optical gap and carrier mobility of the materials produced in this work are the highest reported for thin films generated using AA-CVD, and the concentrations of carriers are roughly $10^{17} cm^{-3}$ (Fig. 6B). Wen-Jen Lee prepared copper oxide films using the SILAR method. The results illustrate that the crystal structure, bandgap, resistivity, carrier concentration, and mobility of the films are all affected by the post

Table 1 Different synthesis methods of undoped and doped SnO_2 and their electrical measurements.

S. no.	Deposition tech.	Parameters	ρ (Ωcm)	n (cm^{-3})	μ ($cm^2V^{-1}s^{-1}$)	Bandgap (eV)	Ref.
01	Spray pyrolysis method	Deposited on glass substrate at 400°C	TO—4.11×10^{-2} ATO—6.28×10^{-4} BATO2—4.41×10^{-4} BATO4—8.72×10^{-4} BATO6—2.93×10^{-3}	1.04×10^{19} 31.2×10^{20} 47.3×10^{20} 1.42×10^{20} 1.06×10^{19}	16.30 21.47 22.20 19.47 17.58	3.52 3.69 3.71 3.58 3.57	[46]
02	Chemical coprecipitation method	Sintered samples at 400°C, 600°C, 800°C, 1000°C	5.42×10^2 5.42×10^2 5.42×10^2 5.42×10^2	–	–	–	[109]
03	Thermal evaporation method	ATO (2.6×10^{-6}m bar) 35 nm as deposited 35 nm annealed 45 nm as deposited 45 nm annealed	 2.101×10^{-6} 1.525×10^{-2} 1.935×10^{-6} 1.300×10^{-3}	–	–	 2.86 – 2.59 3.08	[110]
04	Streaming Process for Electrodeless Electrochemical Deposition (SPEED)	Substrate temperature was varied between 440°C and 500°C.	6×10^{-4}	1×10^{21}	8.3	4.0	[111]

05	Spray pyrolysis method	Variation of antimony doping.	11.1×10^{-3} 4.43×10^{-3} 2.17×10^{-3} 3.92×10^{-3}	1.8×10^{19} 4.65×10^{19} 11.8×10^{19} 3.07×10^{19}	–	3.65 3.82 3.92 3.86	[112]
06	Spray pyrolysis method	Variation of substrate temperature (450, 475, 500, 525, 550°C) resp.	12.9×10^{-4} 10.6×10^{-4} 6.9×10^{-4} 8.1×10^{-4} 9.7×10^{-4}	2.46×10^{20} 2.81×10^{20} 3.73×10^{20} 3.36×10^{20} 3.26×10^{20}	–	–	[113]
07	Chemical spray pyrolysis method	Substrate temperature 400°C.	Undoped SnO_2 6.78×10^{-2} ATO 6.715×10^{-4}	Undoped SnO_2 1.70×10^{18} ATO 7.899×10^{20}	ATO 28.45	3.82	[114]

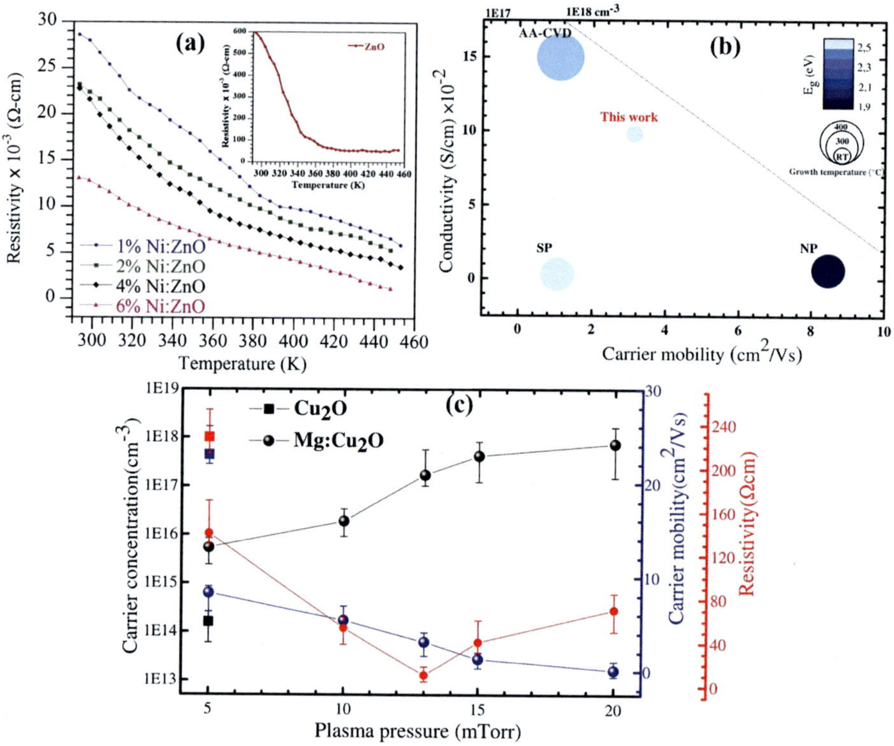

Fig. 6 (A) Resistance changes for Ni-doped ZnO films as a function of temperature. The ZnO film's variation is depicted in the inset. (B) When comparing the hole mobility and electrical conductivity of $Mg{:}Cu_2O$ thin films prepared using spray pyrolysis (SP), nebulizer pyrolysis (NP), and aerosol-assisted metal organic chemical vapor (AA-CVD), data points are larger for higher growth temperatures. The best achievable outcome of this effort is demonstrated. Optical bandgap values are represented by the *disc colors*, while *dashed lines* denote regions of uniform carrier concentration. (C) The dependence of the resistivity, mobility, and concentration of charge carriers on the deposition plasma pressure in Cu_2O and $Mg{:}Cu_2O$ thin films. The error bars show the range of values that were recorded after multiple measurements were taken. Data from M.Y. Ali, M.K.R. Khan, A.M.M.T. Karim, M.M. Rahman, M. Kamruzzaman, Effect of Ni doping on structure, morphology and opto-transport properties of spray pyrolised ZnO nano-fiber, Heliyon 6 (2020) e03588 and Reused with permission from N. Sliti, E. Fourneau, T. Ratz, S. Touihri, N.D. Nguyen, Mg-doped Cu2O thin films with enhanced functional properties grown by magnetron sputtering under optimized pressure conditions, Ceram. Int. 48 (2022) 23748–23754.

annealing temperature and time. Furthermore, due to the creation of a Cu_2O-CuO heterojunction, the film with a Cu_2O-CuO mixed crystal exhibits a high carrier mobility of $93.7\,cm^2\,V^{-1}s^{-1}$ and a low carrier concentration of ($1.8 \times 10^{12}\,cm^3$). In a commercial dry air atmosphere at temperatures of 100, 150, 200, and 250°C, the I-V behavior of copper oxide nanoparticles was evaluated from −40 mV to 40 mV with a step of 50 mV and a high sensitivity of copper oxide with temperature, along with I-V characteristics observed with oxygen concentration variations of 5%, 10%, 15%, and 20%. At the same voltage, an increase in current was observed with an increase in oxygen concentration. In copper oxide, a p-type semiconductor, the dominant carrier is "hole," and the minority carrier is "electron." The adsorption of oxygen reduces the electron concentration in copper oxide while increasing the hole concentration, lowering resistance [121]. Differences in electrical characteristics as a function of the preparation process and other parameters are shown in Table 2.

3.3 ZnO

To increase ZnO nanoparticles' potential for use in nanoelectronics, significant changes must be made to their electrical characteristics. The development of nanoelectronics necessitates a fundamental investigation into the electrical characteristics of ZnO nanostructures. ZnO nanostructures are well suited for use as highly sensitive biosensors and chemical sensors due to their high S/V ratio [126,127]. In order to fabricate devices with high reliability, it is essential to have an understanding of surface features like geometrical properties, surface passivation, surface roughness, and the transport behavior of ZnO nanostructures. The electron transport properties of microscale devices are critical in the new research [77,128]. Using a 3 V bias voltage, Harnock et al. [14] demonstrated that the current-voltage (I-V) curve of ZnO nanorods is nonlinear and asymmetric, with an asymmetry factor of greater than 25. Lee et al. [129] determined that the average resistivity of a ZnO nanowire was roughly an order of magnitude higher than the resistivity of exposed single ZnO-nanowires in anodic aluminum oxide (AAO) templates. In order to probe the electrical properties of AZO thin films, it has been observed that AZO and ZnO were n-type materials. The carrier concentration and mobility of pure ZnO are$1.0 \times 10^{19}/cm^3$ and $17.7\, cm^2\, V^{-1}\, s^{-1}$, respectively. As the Al concentration increases, so does the carrier concentration, which reaches saturation at $1.7 \times 10^{20}\, cm^{-3}$. Caglar et al. [130] report dependency of electrical conductivity on temperature, observe that more charge carriers are able to overcome the activation energy barrier and contribute to the electrical conductivity, leading to an increase in conductivity. The magnitudes of activation energy for different regions were found to be 25 meV and 1.32 eV, which correspond to shallow donor level and deep donor level, respectively. The conductivity of prepared nanophase zinc oxide was found to be 7.261×10^{-7} S/cm. In addition, Lee et al. [129] reported that at temperatures between 450°C and 600°C, the conductivity of nanophase ZnO with a particle size of 60 nm is between 2×10^{-6} and 2×10^{-4} S/cm. ZnO nanostructured thin film's elevated conductivity can be traced to the capture of electrons at its grain boundaries. Nasir et al. [93] investigated the effect of annealing temperature on ZnO resistivity in the range of 400–550°C. As annealing temperature increased, resistivity dropped and was found to be lowest at 500°C with a value of $5.36 \times 10^4\, \Omega$/cm. Ali et al. [131] used Van-der-Pauw's approach to measure the resistivity of ZnO and Ni:ZnO thin films from room temperature to 455°C.

Table 2 Different synthesis routes of undoped and doped CuO and their electrical measurements.

S. no.	Deposition tech.	Parameters	ρ (Ωcm)	n (cm^{-3})	μ ($cm^2 V^{-1} s^{-1}$)	Bandgap (eV)	Ref.
01	Chemical solution deposition method	Annealing temperatures 350°C, 450°C, and 550°C	–	–	–	2.05–1.98	[122]
02	Magnetron sputtered	Annealing temperature 700°C		1.5×10^{15}	51	2.4	[123]
03	RF magnetron sputter	As deposited	0.52	6.25×10^{20}	0.02	2.0	[124]
		Annealed at 500°C	446	2.95×10^{14}	47.5	2.6	
04	Magnetron sputtering	As deposited	490	–	–	2.38	[125]
		Annealed at					
		180°C	61.8	–	–	2.46	
		260°C	12.5	7.12×10^{17}	0.69	2.49	
		280°C	7.32	3.19×10^{17}	2.67	2.51	
		300°C	7.78	2.54×10^{17}	3.16	–	
		320°C	10.4	8.30×10^{16}	7.6	–	
		Vacuum—200°C	549	–	–	2.38	
05	Pulsating spray pyrolysis	Substrate temperature					[64]
		600 K	12	–	–	1.45	
		725 K	5.9			1.60	
06	Successive ionic layer adsorption and reaction (SILAR) method	As deposited	1.22×10^3	2.9×10^{15}		1.90	[121]
		Annealed at					
		200°C—1 h	1.27×10^3	1.0×10^{14}		1.87	
		300°C—1 h	3.72×10^4	1.8×10^{12}		1.40	
		300°C—2 h	7.94×10^2	8.0×10^{14}		1.36	
		300°C—4 h	1.02×10^2	7.4×10^{16}		1.34	
07	Magnetron sputtering method	As deposited	0.26	7.4×10^{19}	0.12	2.0	[59]
		Vacuum annealed at 700 K	149.00	1.5×10^{15}	51.00	2.5	

Each measurement was taken as the temperature gradually increased to maintain a constant temperature throughout the film. The relationship between resistivity and temperature is depicted in Fig. 6A. The resistivity of all samples is found to drop monotonically with increasing temperature, a property characteristic of semiconducting materials having a negative temperature coefficient of resistance (TCR). Additionally, the inset of Fig. 6A shows that the resistivity of ZnO thin film reduces in two stages with temperature: first, it decreases abruptly in the low temperature zone (300–340 K), and then it declines slowly with temperature. This is because at low temperatures, a greater amount of activation energy for the donor transition is needed than at high temperatures. The resistivity goes down predictably as the Ni doping concentration goes up. It is important to note that the RT resistivity of Ni-doped samples is much lower than that of undoped ZnO. Doping causes a decrease in resistivity because part of the Zn atoms, which have full 3*d* shells (offering only semi core valance band states), are swapped with Ni, which has a 3*d* shell that is not full ($3d^8$) and has states near the bottom of the conduction band. Incorporating extra conduction band states into a material reduces its resistivity by making room for more conduction electrons. The resistivity goes down predictably as the Ni doping concentration goes up. It is important to note that the RT resistivity of Ni-doped samples is much lower than that of undoped ZnO. Table 3 shows the differences in electrical characteristics as a function of the preparation process and other parameters.

Table 3 Various routes for synthesis of ZnO and doped ZnO and their electrical measurements.

Sample name	Substrate resistivity (Ω cm)/ sheet resistance	Carrier type	Carrier concentration (cm^{-3})	Resistivity (Ω cm)	Mobility ($cm^2 V^{-1} s^{-1}$)	Ref.
101	10	*n*	7.7×10^{16}	0.53	153	[132]
102	3×10^{-3}	*n*	7.2×10^{15}	2.86	304	
201	2×10^{-3}	*p*	8.9×10^{15}	13.7	51	
202	1×10^{-4}	*p*	1.2×10^{16}	14.6	35	
301	10	*p*	2.6×10^{16}	12.8	18	
302	7.5	*p*	3.7×10^{16}	12.1	14	
401	1×10^{-3}	*p*	1.1×10^{18}	0.26	24	
402	3×10^{-3}	*p*	5.4×10^{17}	0.33	35	
Pure ZnO	–	–	–	1.39×10^{-2}	–	[133]
ZnO 5% Cr	–	–	–	8.16×10^{-3}	–	
ZnO 10% Cr	–	–	–	1.43×10^{-2}	–	
ZnO 15% Cr	–	–	–	9.53×10^{-2}	–	

Continued

Table 3 Continued

Sample name	Substrate resistivity (Ωcm)/ sheet resistance	Carrier type	Carrier concentration (cm^{-3})	Resistivity (Ωcm)	Mobility ($cm^2V^{-1}s^{-1}$)	Ref.
ZnO+ Li 0%	8.1	*n*	4.28×10^{13}	102	–	[134]
ZnO+ Li 5%	4	*p*	6×10^{13}	64	–	
ZnO+ Li 10%	3.3	*p*	34.2×10^{13}	49	–	
ZnO+ Li 15%	6	*p*	6×10^{13}	54	–	
ZnO+ Li 20%	3.8	*p*	13.2×10^{13}	30	–	
ZnO+ Li 40%	3.9	*p*	1.3×10^{13}	44	–	
ZnO+ Li 50%	15	*p*	0.5×10^{13}	22	–	
ZnO+ Li 60%	33	*p*	3×10^{13}	310	–	
ZnO+ Li 70%	1.4	*n*	29×10^{13}	23	–	
ZnO		*n*	4.50×10^{18}	2164	25.3	[135]
ZnO+ Cu3%		*p*	7.58×10^{14}	304	4.5	
ZnO+ Cu6%		*p*	1.34×10^{13}	251	1.2	

Malek et al. [136] prepared ZnO thin films and investigated the effect of withdrawal speed on electrical properties. The electrical measurement showed that increasing the rate of withdrawal causes a decrease in current. A high conductivity of 5.87×10^{-4} S/cm and a low resistivity of about 1.7×10^{3} Ω/cm were measured for a withdrawal speed of 1 mm/s. Barir et al. [137] measured the resistivity of ZnO samples with different precursor concentrations from 0.1 to 0.4 M. The decrease of resistivity was observed to decrease from 49.80 to 2.09 Ωcm for 0.3 M, then increase for 3.41 Ωcm with increase of precursor concentration to 0.4 M. Cuadra et al. [92] prepared Na and K doped ZnO by using spray pyrolysis method to enhance electrical properties using these alkali metals. The decrease of resistivity was observed from 1.03×10^{-1} Ωcm to 5.64×10^{-2} Ωcm for K-doped ZnO and 3.18×10^{-2} Ωcm for Na-doped ZnO. The carrier concentration was found to increase 5.17×10^{17} (undoped ZnO) to 1×10^{18} (of doped ZnO).

4 Application

4.1 Sensors

Researchers in the field of gas sensors are increasingly placing an emphasis on creating high-performance, long-lasting sensors capable of precise monitoring at or near RT. For decades now, metal oxide semiconductors (MOS) have been the go-to sensing material due to their rapid reaction, low cost, user friendliness, versatility in target gases, and extended lifetimes. They do, however, have drawbacks, such as increased energy requirements and decreased selectivity. In order to detect gases, scientists have turned to nano-metal oxides and other materials with high surface areas, specific surface energies, and surface reactivity [51,138]. For optimal performance, MONP-based gas sensors typically include a heating layer or wire, electrodes that can detect resistance, and a sensing film that undergoes a change in resistance when exposed to gas. When exposed to the gas of interest, conductometric gas sensors undergo a change in electrical conductivity. However, the receptor function is influenced by gas-solid interactions and can be altered by combining different oxides or adding noble metals, whereas the transducer function is determined by the microstructure of the oxide. Adsorption of oxygen species, which increases the number of reactive surface sites, is fundamental to the gas sensing process [139]. Conductivity drops when molecules like O_2 or NO_2 are adsorbed at vacancy sites in the oxide, but rises when CO or H_2 react with deposited O_2, releasing electrons and boosting conductance in the oxygen-containing atmosphere. The conductance is continuously changing due to the adsorption reactions and the sudden temperature change of the sensor. [140,141].

The conductivity of a MO_x-based sensor depends on many things, such as the size of the particles, how their edges connect (barriers, cross sections of the channels, etc.), how the nanoparticles are arranged in space (hierarchical, branching dendritic), the p-n junction between the nanocomposites (heterostructures), etc. [142].

For better gas sensing performance, it is necessary to expand the sensing film's surface area, leading to the development of porous nanostructures such as nanotubes, nanospheres, nanowires, nanosheets, etc. [143]. Doping can decrease electrical resistance and boost catalytic characteristics by producing smaller-sized doped metal oxides, catalytic activity, p-n heterojunctions, or increasing the density of functional groups [144]. Adding a second component (noble or transitional metal, nano-oxide, or contaminant) causes structural abnormalities in the metal oxide, reducing the energy gap. Gas sensing is used in industry, healthcare, biomedicine, and environmental monitoring [145]. Strong gas sensor performance includes high sensitivity, a fast response time, low energy consumption, signal stability, long-term monitoring, and good reproducibility [145]. The examples below are of gas sensors based on semiconductor nano oxides for various gases.

4.1.1 Carbon dioxide gas sensors

CO_2 sensors are vital for air quality monitoring, hospital interiors, and food packing. Performance, linearity range, detection limit, and pricing for these sensors vary greatly per field. CO_2 sensors are used indoors, in industry, and for environmental monitoring. Chemically stable metals and binary metal oxides are insensitive to inert-gases like CO_2. Doping most semiconductor nanostructures for CO_2 sensing was ineffective. Researchers created nanocomposite MO_x to solve this challenge. CuO-$Cu_xFe_{3-x}O$ nanocomposites (with $0 \leq x \leq 1$) was synthesized by radio frequency (RF) sputtering from $CuFeO_2$ and employed as a novel CO_2 detecting active layer [146]. The response of the sensor in a carbon dioxide atmosphere was tested at various temperatures (130–475°C) and frequencies (0.5–250 kHz). The ZnO sensor doped with 50 at.% La performed better than undoped ZnO when heated to 400°C. Large lattice distortion and the synergistic effect of Zn and La-active materials are responsible for the enhanced performance, which in turn is due to increased CO_2 adsorption and surface reactivity [147]. The sensor's reaction in a carbon dioxide atmosphere was measured at variable concentration up to 5000 ppm and frequencies ranging from 0.5 kHz to 250 kHz across a temperature range of 130–475°C. For a CO_2 level of 5000 ppm, the results show a strong reaction of 50% ($R_{air}/R_{CO_2} = 1.9$) at temperature and frequency (250°C and 700 Hz) [148].

4.1.2 Carbon monoxide gas sensors

Toxic CO (carbon monoxide) gas has no discernible odor, color, or taste and is consequently difficult to detect. Usual animal metabolism produces it in extremely small quantities, and CO is assumed to have certain biological effects when encountered at concentrations of 25 ppm. Partial oxidation of carbon-containing molecules, such as in an interior stove or an internal combustion (IC) engine operating in a low-oxygen environment, yields carbon monoxide. Toxic carbon monoxide can be detected with a CO gas sensor, a crucial main instrument. Dopants made from noble metals such as, Au, Ag, and Pd are added to metal oxide gas sensors to improve their sensitivity, stability, and response time while also lowering their operating temperature. Thick-film technologies have enabled the realization of materials like SnO_2 doped with Pd, which can be used to detect CO. The use of SnO_2 sensors for CO sensing in humid environments appears to be excellent and will help in deducing the fundamentals of metal oxide-based gas sensors. Across a wide concentration range (from 0.25 to 1000 ppm), the CuO-graphene nanocomposite demonstrated not only great detection capability but also good stability and selectivity at room temperature [149]. Silver-zinc-oxide-metal sulfide, the performance of ternary nanocomposite CO sensors was evaluated in comparison to those of ZnO, ZnO-MoS_2, and Pt-ZnO/MoS_2 using a layer-by-layer self-assembly method. Among the four tested sensors for CO gas at RT, the Ag-ZnO-MoS_2 nanocomposite film showed the best response due to the catalytic activity of Ag and the synergistic effect of ZnO and MoS_2. This nanocomposite sensor had demonstrated remarkable responsiveness, rapid response/recovery, high selectivity, and great repeatability [150].

4.1.3 Oxygen gas sensors

Since oxygen is essential to life, it is measured and monitored in a wide variety of industrial settings, including factories, laboratories, foundries, monitoring and controlling large-scale combustion furnaces, and hospitals. The automotive and healthcare sectors have benefited most from the recent advancements in O_2 sensor technology. At the moment, there are three main types of O_2 sensors available for regulating the air-fuel ratio in engines: semiconductor oxides (TiO_2 sensors), concentration cells (Zr-based sensors), and electrochemically pumped O_2-sensors (based on limiting-current). Semiconductor-oxide sensors (TiO_2 sensors) and solid-state sensors (Zr sensors) are preferred for controlling and measuring O_2 because of their compact size, low cost, exceptional dependability, and great stability [151].

4.1.4 Nitric oxide gas sensor

NO (nitrogen monoxide) is a highly reactive gas that is involved in a wide variety of different chemistry and biology activities. It has been the molecule of focus for decades, and for good reason, it is a crucial messenger molecule. In a word, NO has advantages and disadvantages. Because of its extreme reactivity, it plays a pivotal role as a signal molecule in the chemical industry and causes severe toxicity in mammals. The highest reaction to nitrous oxide makes WO_3 the most popular oxide material for use in sensing. Several nanocomposites with promising performance have been created with the aim of enhancing the capabilities of WO_3-based sensors. For instance, NO gas at RT (300 K) was monitored using screen-printed gas sensor chips that contained In_2O_3-WO_3 nanoparticles tuned at a weight-to-weight ratio of 4:1 (preparation by sol-gel and calcination). Two comb-like interdigitated gold electrodes with fully platinized edges were covered with a thin sheet of nano-In_2O_3-WO_3 composite. Results indicated that this NO gas sensor had very good linearity even in the 100–1000 ppb concentration range, with a reaction time of less than 30 s and a recovery time of less than 40 s. At RT, the nanocomposite In_2O_3-WO_3 sensor detected NO concentrations in the sub-ppb to ppb range [152].

4.1.5 Ammonia gas sensors

Ammonia (NH_3) is a colorless gas with a strong, unpleasant odor; it is the simplest pnictogen hydride. N_2O is produced in large quantities by aquatic species and serves as a precursor to both food and chemical fertilizers, making it an important component in meeting the dietary needs of organisms at the earth's surface. As such, several MONPs, including ZnO, SnO_2, and In_2O_3, have been tested as potential NH_3 gas sensors. Among these, SnO_2 has emerged as a candidate for use in ammonia sensors. As an illustration, ammonia detection makes use of a bunch matrix of SnO_2 NPs made up of thousands of extremely small-sized 3.0 nm (SnO_2) NPs and randomly distributed wormhole like pores linking the oxide-NPs. As was discussed before, a noble metal catalyst may enhance the sensing characteristics. Therefore, metal (Pt) and metal oxide (PtO_2) were synthesized to create Pt-activated SnO_2 NP clusters. Pt-activated clusters showed increased gas responsiveness from 6.50 to 203.50

toward 500 ppm of "NH_3" in contrast to the as-synthesized pure SnO_2 nanoparticles via a solvothermal technique. Researchers found that Pt-doped SnO_2 has promising properties for NH_3 sensing, including fast response recovery times, selectivity, linear dependency-rates, repeatability, high sensitivity, and long-term stability [153,154].

4.1.6 Ozone gas sensors

The oxidizing agent ozone (O_3) has innumerable practical applications. Ozone concentration readings are typically needed for medicinal purposes, pharmaceutical, biotechnological, and chemical operations, packaging, research facilities, water decontamination, and other similar uses. However, ozone's presence in the atmosphere is extremely harmful to human health because of its potent oxidizing reagent properties. It could irritate the eyes, give you a headache, make your chest hurt, and even damage your lungs. Spirometry readings dropped by 20% after being exposed to 0.1 ppm O_3 for 2 h continuously [155]. Therefore, there is a pressing need for a straightforward yet extremely sensitive method of measuring atmospheric ozone levels, given the widespread application of O_3. Electrochemical, optical, and resistance techniques, as well as other technologies like impedance spectroscopy, and photo-reduction under UV light, have all been presented as methods for measuring O_3 concentrations. Ozone affects the conductivity of n-type oxide by filling vacancies and decreasing MO_x. Composite-based semiconductor gas sensors with improved sensitivity and response time at room temperature are being investigated as a means of synthesizing sensors at a lower cost than those doped with noble metals. Multiple TiO_2-SnO_2 sensing materials with varying ratios were tested for their ability to detect ozone in the air. The 1:4 TiO_2-SnO_2 blend produced the highest sensor-response units (327) when operating at RT. The ozone response was increased to 6.6×10^6 when Au (0.5 wt%) was added to TiO_2-SnO_2 (1:4) to sense O_3 by photoreduction and enhance the signal. In addition, the response signal and recovery durations were decreased from 35 to 5 s [156–158].

4.2 Batteries

Recent years have seen a significant increase in interest in rechargeable batteries as a solution to growing environmental problems. Lithium-ion batteries are extremely important to our daily lives because of their vast applications in the sectors of portable electronics, electric automobiles, and hybrid vehicles. Scientists have placed a heavy emphasis on developing innovative electrode materials employing nano MO_x, a standard tool used for decades, in order to improve the performance of LIBs. Electrode materials were widely researched, and many TMOs were explored in LIBs. TMOs function according to two primary principles in terms of reaction mechanisms: (a) intercalation/deintercalation and (b) conversion reaction [159].

4.3 Solar cell

One of a solar cell's primary roles is to photogenerate charge carriers in a material that can absorb light, with the other being to segregate those carriers into conductive materials that can then convey the electricity. The majority of the carriers are created in the surface layer. The conductivity of a semiconductor is greatly improved when a single photon from a light source strikes the material, since electron-hole pairs are created. They may perhaps make it to the p-n junction, where the electric field creates a gap between the charges if recombination is avoided. Semiconductors made from metal oxides have the advantages of being inexpensive, stable, and kind to the environment. They have been used in photovoltaics for the past decade. Dye solar cells (DSCs) utilize photovoltaic (PV) cells as photoelectrodes, and metal oxide p-n junctions have been developed [160]. Metal oxides are a great option for next-generation photovoltaics because of their excellent adaptability and the ease with which they can be produced using simple, inexpensive, and scalable processes. Several metal oxides may be more useful as photo-harvesters than others because of their wide bandgap energy range and their strong tunability. While the study of silicon (first generation solar cells) and other III-V semiconductors has received more attention, only the most promising metal oxides have received serious consideration as potential photon absorbers. In photovoltaics, these oxides serve as light absorbers, transparent electrodes, transport layers, and unique resources with their own set of features and abilities.

4.3.1 CuO solar cells

Copper oxides (CuO) are by far the most common MO_x chemical currently in use. By fine-tuning the three stable binary oxide phases of copper-oxides, cuprous oxide (Cu_2O), CuO tenorite or cupric oxide (CuO), and cuprous oxide (Cu_4O_3), the full bandgap range of 1.4 to 2.2 eV may be covered by CuOx (paramelaconite) [161]. Among the several p-type semiconductors available, cuprous oxide (Cu_2O) stands out as a promising candidate for use in photovoltaics due to its inexpensive cost, high absorption coefficient, good mobility, wide availability, and lack of toxicity. The usage of Cu_2O as an electronic material began in the 1920s and reached its zenith in the early 1970s [162].

4.3.2 Binary heterojunction solar cells

Excitation and current are generated locally due to the electric field's enhancement at a heterojunction (p-n junction) between two different types of semiconducting materials, one of which has a larger affinity for electrons and the other for holes. Because n-type Cu_2O is difficult to dope, an oxide window for Cu_2O absorbers was created using a suitable bilayer ZnO oxide, either planar or nanostructured [163,164]. The band offset of Cu_2O is different from that of ZnO and Ga_2O_3, which shows how specialized materials can be used. ZnO is unique among n-type windows due to its low temperature synthesis, naturally low-cost approach, strong electron-mobility ($120\,cm^2V^{-1}s^{-1}$), and large direct-bandgap (3.37 eV). Still, the power conversion efficiency of solar cells made with electrodeposited Cu_2O layers is only 1.43%, well below that of Cu_2O/ZnO cells. Hole diffusion against the barrier formed by the band bending to recombine with electrons trapped in interfacial states was shown to be the major current flow mechanism across the heterojunction in a ZnO (transparent window layer)/Cu_2O solar cell. Ga_2O_3/Cu_2O performed better than ZnO by 3.97% and 5.38%, respectively, depending on the deposition technique used [165].

4.3.3 Thin film solar cells

To create a thin film solar cell, also known as a second generation solar cell, photovoltaic material is deposited as many thin coatings or a monolayer on various substrates such as glass, metal, or plastic. Ferroelectric solar cells can control their photovoltaic characteristics with an applied electric field, and their photovoltaic voltage is several orders of magnitude larger than the equivalent bandgap of the material, making them superior to traditional semiconductor solar cells [166]. The most widely used ferroelectric substance is barium titanate (also known as $BaTiO_3$ or BTO). However, until recently, the PZT (Pb (Zr, Ti) O_3) family of oxygen octahedra ferroelectric thin films had received a great deal of attention. Under the influence of visible and near ultraviolet (NUV) light, various ferroelectric materials, including $Pb(Zr_{1-x}Ti_x)O_3$ and $LiNbO_3$, show photoelectric and photovoltaic effects; however, the magnitude of photocurrent and voltage obtained for the device application is far below the photoelectronics requirements [167].

4.4 Antennas

The antenna transforms the EM radiation from space into a DC current. The primary function of an antenna is to transmit electrical signals (radio waves being a type of electrical current), So, it is a given that antennas need to be conductive and should not oxidize. Pure metals and other traditionally hard materials are excellent conductors, but many metals are vulnerable to oxidation in a number of ways. When exposed to air, pure metals like silver and copper oxidize into oxides, while gold hardly changes at all; yet, gold's higher price, is also an important consideration. Uncoated iron rusts quickly. Incredibly, aluminum oxide, the thin gray film that forms over aluminum, has almost no effect on the metal's conductivity. There are conductive materials that can attain desirable electrical antenna characteristics, but they are typically opaque and difficult to incorporate onto flexible carriers. The most reliable antennas may be made from metal oxides; hence, they are the greatest solution. They look like a potential replacement material for antennas. At 66.85 °C, vanadium dioxide (VO_2) undergoes a sudden transformation from an insulator to a conductor, a phenomenon known as the reversible semiconductor-metal phase transition [168]. During this phase transition, the VO_2 crystal structure changes to a tetragonal metallic one from a monoclinic insulating phase. Due to this, scientists have recorded the IMT (insulator-metal transition) in a vanadium dioxide film subjected to a well-calibrated electric field by depositing a sequence of gold resonators on the film. Moreover, VO_2-containing composites have shown the potential for remarkably tunable electrical and optical properties. When mixed, Au and VO_2 form a metamaterial, an artificial material that can alter the behavior of electromagnetic waves [169].

4.5 Optoelectronic and electronics

When it comes to the interest of materials scientists, semiconducting transparent oxides stand out among the many types of oxides available. Thin coatings are also called transparent conducting oxides (TCOs). TCOs are unique in that they combine transparency with conductivity, making them both useful and interesting materials. They have specific features that make them ideal for long-term use in the optoelectronics industry, such as flat panel displays, light emitting diodes (LED), touch screens, low emissivity windows, etc. [170]. Most commercially available TCOs are n-type, such as FTO (F-doped SnO_2) or Sn-doped In_2O_3, and it is currently very difficult to develop functional p-TCOs. The rising price of indium has prompted a substantial study aimed at discovering entirely transparent, hole-conducting alternatives that might be used in place of indium-rich TCOs. The majority of the current p-TCOs technology study is geared toward developing high-quality films for a device through significantly enhanced electrical features [171]. Low pore-mobility and electric conductivity are seen due to the presence of pores that are confined around oxygen atoms and need considerable energy to migrate inside the lattice. Cu-containing p-type TCOs, such as M_2 Cu_2O_2 ("M_2—bivalent ions like Mg, Ca, Sr, Ba") and CuM_3O_2 ("M_3—trivalent ions like Al, Ga, In") have recently been proposed [172]. By exchanging trivalent cations for divalent ones, a vacant state in the valence-band was formed, which served as a hole and boosted conductivity. For instance, a large number of hole-carriers accumulated in the upper valence band of $La_{1-x}Sr_xVO_3$ compounds when Sr^{2+} replaced La^{3+}. $La_{2/3}Sr_{1/3}VO_3$ p-type TCOs films had low visible-range absorption, high visible-range transmittance (between 53.9% and 70.1%), low carrier mobility, and high conductivity (between 742.3 S/cm and 872.3 S/cm) at room temperature. So, thin films of codoped p-TCO could offer enhanced optical and electrical capabilities in addition to high conductivity without compromising any of the material's transparency in the visible spectrum [173].

5 Future challenges and aspects

Metal oxide nanoparticles have quickly gained popularity in practical gas sensors, health foods and agriculture, textiles, cosmetics, and medicine. Electronics and optoelectronics dramatically improve the results, especially in terms of sensitivity and detection limits, to the point where single molecules can be detected. Optimizing the production of nanoparticles, which means shortening the time it takes to get the right size and shape to make the nanoparticles more stable, and customizing the properties of individual nanoparticles to meet the needs of a wide range of applications, will be future challenges. Better microstructural and sizing control of NPs, as well as more precise control over their composition, can significantly boost the performance of already available metal oxides. Despite recent advances, the issue of biocompatibility, toxicity, and long-term stability of metal oxide nanoparticles remains. Solutions for replacing single-band antennas with multiband antennas that offer improved radiation, a higher data rate, and a smaller form factor are required by today's wireless communication systems. Future products with high optical transmittance in the IR (infrared) region and operation in the 4G, 5G frequency ranges hold great promise. The efficiency of quantum devices is constrained by the bandgap energy of the nano-oxide layer. Energy may be lost during light harvesting due to the low energy of the passing photons or as heat from high-frequency photons. Nanostructures need to be able to upconvert in a wide spectral range when exposed to sunlight. Bandgap tenability and the creation of numerous excitations are crucial factors in photon adsorption and photocurrent generation. Despite the recent quick progress, several obstacles regarding the synthesis of 2D MO_x are presented. In the first place, current top-down and bottom-up methods have a poor degree of controllability for the thickness and lateral dimensions of 2D MO_x. However, effective solutions must be found for the problems that arise during practical implementation.

6 Conclusion

Metal oxides' optical and electrical properties, as well as their varied practical uses, have been explored in this chapter and have been found to be useful in a variety of applications. To improve the optoelectrical properties of metal oxides, a wide range of preparation techniques and parameter tweaks have been tried. Enhanced optical and electrical properties have led to numerous useful applications for the metal oxides CuO, SnO_2, and ZnO covered in this chapter. Over the past few decades, there have been significant advancements in the fields of solar cells, sensors, batteries, and optoelectronics. It is clear from this chapter that there is still room for improvement in the optical and electrical properties of metal oxides, despite the fact that these properties have been enhanced by the use of various techniques, such as doping, temperature change, and other variations. Due to the high potential for structural and functional diversity in metal oxides, they remain an attractive subject of study for applications ranging from environmental remediation and energy crisis resolution to decarbonization. The search for low-cost, environmentally benign organic ligands, metal oxides, solvents, dopants, and other compounds continues. Optoelectronics, photovoltaics, and energy storage are just a few examples of promising new technologies that could benefit from inexpensive, transparent metals with high conductivities. Metal oxides are a promising material for these purposes due to their inexpensive cost, nontoxic nature, high specific capacity, high transparency in the visual region, and high electron mobility. Collectively, this field of study indicates optimism for a future in which smart materials are prioritized to maintain the sustainability of the world's energy and ecological systems. With their superior semiconducting properties, high surface-to-volume ratio, synergistic effects with other materials, and accomplishments of flexibility and stretchability, MONs are still attractive and promising materials for future applications despite the aforementioned challenges. Furthermore, the performance of the functional device may be enhanced by combining different nanomaterials. As was noted above, MONs would be a more appealing candidate material for future electronics if their constraints were overcome through further development.

References

[1] R.S. Devan, R.A. Patil, J.H. Lin, Y.R. Ma, One-dimensional metal-oxide nanostructures: recent developments in synthesis, characterization, and applications, Adv. Funct. Mater. 22 (2012) 3326–3370, https://doi.org/10.1002/adfm.201201008.

[2] F. Emadi, A. Gholami, A. Amini, G. Younes, Graphene: recent advances in engineering, medical and biological sciences, and future prospective, Trends Pharmacol. Sci. 4 (2020) 131–138.

[3] A.G. Alsultan, N.A. Mijan, Y.H. Taufiq-Yap, Nanomaterials: an overview of nanorods synthesis and optimization, in: Nanorods and Nanocomposites, 2019, https://doi.org/10.5772/intechopen.84550.

[4] P. Kumar, P. Kumar, A. Deep, L.M. Bharadwaj, Synthesis and conjugation of ZnO nanoparticles with bovine serum albumin for biological applications, Appl. Nanosci. 3 (2013) 141–144, https://doi.org/10.1007/s13204-012-0101-0.

[5] P.K. Singh, P. Kumar, M. Hussain, A.K. Das, G.C. Nayak, Synthesis and characterization of CuO nanoparticles using strong base electrolyte through electrochemical discharge process, Bull. Mater. Sci. 39 (2016) 469–478, https://doi.org/10.1007/s12034-016-1159-1.

[6] T.C. Bharat, M.S. Shubham, H.S. Gupta, P.K. Singh, A.K. Das, Synthesis of doped zinc oxide nanoparticles: a review, Mater. Today Proc. 11 (2019) 767–775, https://doi.org/10.1016/j.matpr.2019.03.041.

[7] O. Autin, J. Hart, P. Jarvis, J. MacAdam, S.A. Parsons, B. Jefferson, The impact of background organic matter and alkalinity on the degradation of the pesticide metaldehyde by two advanced oxidation processes: UV/H2O2 and UV/TiO2, Water Res. 47 (2013) 2041–2049, https://doi.org/10.1016/j.watres.2013.01.022.

[8] J. Fenoll, E. Ruiz, P. Hellín, P. Flores, S. Navarro, Heterogeneous photocatalytic oxidation of cyprodinil and fludioxonil in leaching water under solar irradiation, Chemosphere 85 (2011) 1262–1268, https://doi.org/10.1016/j.chemosphere.2011.07.022.

[9] H.X. Guo, K.L. Lin, Z.S. Zheng, F.B. Xiao, S.X. Li, Sulfanilic acid-modified P25 TiO2 nanoparticles with improved photocatalytic degradation on Congo red under visible light, Dyes Pigments 92 (2012) 1278–1284, https://doi.org/10.1016/j.dyepig.2011.09.004.

[10] M.H. Plumlee, J. Larabee, M. Reinhard, Perfluorochemicals in water reuse, Chemosphere 72 (2008) 1541–1547, https://doi.org/10.1016/j.chemosphere.2008.04.057.

[11] R. Qiu, D. Zhang, Y. Mo, L. Song, E. Brewer, X. Huang, et al., Photocatalytic activity of polymer-modified ZnO under visible light irradiation, J. Hazard. Mater. 156 (2008) 80–85, https://doi.org/10.1016/j.jhazmat.2007.11.114.

[12] X. Wang, S. Zhang, B. Peng, H. Wang, H. Yu, F. Peng, Enhancing the photocatalytic efficiency of TiO2 nanotube arrays for H2 production by using non-noble metal cobalt as co-catalyst, Mater. Lett. 165 (2016) 37–40, https://doi.org/10.1016/j.matlet.2015.11.103.

[13] Z. Zhang, J. Liu, J. Gu, L. Su, L. Cheng, An overview of metal oxide materials as electrocatalysts and supports for polymer electrolyte fuel cells, Energy Environ. Sci. 7 (2014) 2535–2558, https://doi.org/10.1039/c3ee43886d.

[14] P.K. Aspoukeh, A.A. Barzinjy, S.M. Hamad, Synthesis, properties and uses of ZnO nanorods: a mini review, Int. Nano Lett. 12 (2022) 153–168, https://doi.org/10.1007/s40089-021-00349-7.
[15] T. Hyeon, S.S. Lee, J. Park, Y. Chung, H.B. Na, Synthesis of highly crystalline and monodisperse maghemite nanocrystallites without a size-selection process, J. Am. Chem. Soc. 123 (2001) 12798–12801, https://doi.org/10.1021/ja016812s.
[16] Y.S. Kang, S. Risbud, J.F. Rabolt, P. Stroeve, Synthesis and characterization of nanometer-size Fe3O4 and γ-Fe2O3 particles, Chem. Mater. 8 (1996) 2209–2211, https://doi.org/10.1021/cm960157j.
[17] Z. Li, X. Lai, H. Wang, D. Mao, C. Xing, D. Wang, Direct hydrothermal synthesis of single-crystalline hematite nanorods assisted by 1,2-propanediamine, Nanotechnology 20 (2009), https://doi.org/10.1088/0957-4484/20/24/245603.
[18] L. Xue, L. Xaing, L. Peng-Ting, C. Xing-Wang, L. Ying, C. Chuan-Bao, Mg doping reduced full width at half maximum of the near-band-edge emission in Mg doped ZnO films, Chin. Phys. B 19 (2010), https://doi.org/10.1088/1674-1056/19/2/027202.
[19] B.P. Zhang, N.T. Binh, Y. Segawa, Y. Kashiwaba, K. Haga, Photoluminescence study of Zno nanorods epitaxially grown on sapphire (11$\bar{2}$0) substrates, Appl. Phys. Lett. 84 (2004) 586–588, https://doi.org/10.1063/1.1642755.
[20] L. Liu, H. Zhao, J.M. Andino, Y. Li, Photocatalytic CO2 reduction with H2O on TiO2 nanocrystals: comparison of anatase, rutile, and brookite polymorphs and exploration of surface chemistry, ACS Catal. 2 (2012) 1817–1828, https://doi.org/10.1021/cs300273q.
[21] P.V. Radovanovic, N.S. Norberg, K.E. McNally, D.R. Gamelin, Colloidal transition-metal-doped ZnO quantum dots, J. Am. Chem. Soc. 124 (2002) 15192–15193, https://doi.org/10.1021/ja028416v.
[22] V. Balek, J. Ŝubrt, I.M. Bountseva, H. Irie, K. Hashimoto, Emanation thermal analysis study of N-doped titania photoactive powders, J. Therm. Anal. Calorim. 92 (2008) 161–167, https://doi.org/10.1007/s10973-007-8755-7.
[23] W. Chen, J.Z. Zhang, A.G. Joly, Optical properties and potential applications of doped semiconductor nanoparticles, J. Nanosci. Nanotechnol. 4 (2004) 919–947, https://doi.org/10.1166/jnn.2004.142.
[24] P. Khandel, S.K. Shahi, Microbes mediated synthesis of metal nanoparticles: current status and future prospects, Int. J. Nanomater. Biostructures 6 (2016) 1–24.
[25] M.H. Ahn, E.S. Cho, S.J. Kwon, Effect of the duty ratio on the indium tin oxide (ITO) film deposited by in-line pulsed DC magnetron sputtering method for resistive touch panel, Appl. Surf. Sci. 258 (2011) 1242–1248, https://doi.org/10.1016/j.apsusc.2011.09.081.
[26] M. Herrera, A. Cremades, D. Maestre, J. Piqueras, Growth and characterization of Mn-doped In2O3 nanowires and terraced microstructures, Acta Mater. 75 (2014) 51–59, https://doi.org/10.1016/j.actamat.2014.04.069.
[27] A. Ganguly, O. Anjaneyulu, K. Ojha, A.K. Ganguli, Oxide-based nanostructures for photocatalytic and electrocatalytic applications, CrystEngComm 17 (2015) 8978–9001, https://doi.org/10.1039/c5ce01343g.
[28] F. Mei, R. Li, T. Yuan, Transparent and conductive applications of tin oxide, Tin Oxide Mater. (2020) 579–597, https://doi.org/10.1016/b978-0-12-815924-8.00020-7. Elsevier.
[29] E. Muhire, J. Yang, X. Huo, M. Gao, Dependence of electrical and optical properties of sol-gel-derived SnO2 thin films on Sb-substitution, Mater. Sci. 25 (2019) 21–27.
[30] M. Wang, J. Bu, Y. Xu, Y. Liu, Y. Yang, Sb-doped SnO2 (ATO) hollow submicron spheres for solar heat insulation coating, Ceram. Int. 47 (2021) 547–555, https://doi.org/10.1016/j.ceramint.2020.08.162.

[31] W. Mao, B. Xiong, Y. Liu, C. He, W. Mao, B. Xiong, et al., Correlation between defects and conductivity of Sb-doped tin oxide thin films Correlation between defects and conductivity of Sb-doped tin oxide thin films, Appl. Phys. Lett. (2013) 031915. 1–5 https://doi.org/10.1063/1.4816084.

[32] A. Abdel-Galil, M.R. Balboul, A. Atta, I.S. Yahia, A. Sharaf, Preparation, structural and optical characterization of nanocrystalline CdS thin film, Phys. B Phys. Condens. Matter 447 (2014) 35–41, https://doi.org/10.1016/j.physb.2014.04.064.

[33] A.R. Babar, K.Y. Rajpure, Effect of intermittent time on structural, optoelectronic, luminescence properties of sprayed antimony doped tin oxide thin films, J. Anal. Appl. Pyrolysis 112 (2015) 214–220, https://doi.org/10.1016/j.jaap.2015.01.024.

[34] S. Gürakar, T. Serin, N. Serin, Studies on optical properties of antimony doped SnO2 films, Appl. Surf. Sci. 352 (2015) 16–22, https://doi.org/10.1016/j.apsusc.2015.03.057.

[35] S. Yu, L. Zhao, R. Liu, M. Wu, Y. Sun, L. Li, Electrical properties of bulk and interface layers in Sb doped SnO2 thin films, Ceram. Int. 45 (2019) 2201–2206, https://doi.org/10.1016/j.ceramint.2018.10.131.

[36] T. Abendroth, B. Schumm, S.A. Alajlan, A.M. Almogbel, G. Mäder, P. Härtel, et al., Optical and thermal properties of transparent infrared blocking antimony doped tin oxide thin films, Thin Solid Films 624 (2017) 152–159, https://doi.org/10.1016/j.tsf.2017.01.028.

[37] H. Liu, Q. Li, L. Wang, Y. Mao, C. Wu, Effect of SnO2 and Sb doped SnO2 on the structure and electrical conductivity of epichlorohydrin rubber morphological analysis of particles, Polym. Compos. (2015), https://doi.org/10.1002/pc.

[38] B. Shen, Y. Wang, L. Lu, H. Yang, Synthesis and characterization of Sb-doped SnO2 with high near-infrared shielding property for energy-efficient windows by a facile dual-titration co-precipitation method, Ceram. Int. (2020), https://doi.org/10.1016/j.ceramint.2020.04.157. 0–1.

[39] M. Sun, J. Liu, B. Dong, Effects of Sb doping on the structure and properties of SnO2 films, Curr. Appl. Phys. 20 (2020) 462–469, https://doi.org/10.1016/j.cap.2020.01.009.

[40] S.C. Dixon, D.O. Scanlon, J. Carmalt, I.P. Parkin, Oxides: an overview, J. Mater. Chem. C (2016), https://doi.org/10.1039/C6TC01881E.

[41] F. Hossain, A.H. Shah, A. Islam, Structural, morphological and opto-electrical properties of transparent conducting SnO2 thin films: influence of Sb doping, Mater. Res. Innov. (2020) 1–10, https://doi.org/10.1080/14328917.2020.1801272.

[42] H. Sun, B. Liu, X. Liu, Z. Yin, I. Introduction, Dispersion of antimony doped tin oxide nanopowders for preparing transparent thermal insulation water-based coatings, J. Mater. Res. (2017), https://doi.org/10.1557/jmr.2017.211.

[43] R. Thomas, T. Mathavan, V. Ganesh, I.S. Yahia, H.Y. Zahran, S. AlFiafy, et al., Investigation of erbium co-doping on fluorine doped tin oxide via nebulizer spray pyrolysis for optoelectronic applications, Opt. Quant. Electron. 52 (2020) 1–15, https://doi.org/10.1007/s11082-020-02376-8.

[44] M.H. Jo, B.R. Koo, H.J. Ahn, Fe co-doping effect on fluorine-doped tin oxide transparent conducting films accelerating electrochromic switching performance, Ceram. Int. 46 (2020) 10578–10584, https://doi.org/10.1016/j.ceramint.2020.01.061.

[45] L. Go, L. Macaraig, E. Enriquez, Few-layer-graphene as intercalating agent for spray-pyrolysed fluorine-doped tin oxide transparent conducting electrode, Bull. Mater. Sci. 43 (2020), https://doi.org/10.1007/s12034-019-2023-x.

[46] R. Ramarajan, M. Kovendhan, K. Thangaraju, D.P. Joseph, R.R. Babu, V. Elumalai, Enhanced optical transparency and electrical conductivity of Ba and Sb co-doped SnO2 thin films, J. Alloys Compd. 823 (2020), https://doi.org/10.1016/j.jallcom.2020.153709.

[47] H. Li, M. Sztukowska, H. Liu, H. Dong, D. Małaszkiewicz, H. Zhang, et al., Infrared barriering behavior of reduced graphene oxide aerogel/antimony tin oxide-polyaniline hybrids, Ceram. Int. 46 (2020) 10971–10978, https://doi.org/10.1016/j.ceramint.2020.01.112.
[48] G. Joshi, J.K. Rajput, L.P. Purohit, Improved stability of gas sensor by inclusion of Sb in nanostructured SnO2 thin films grown on sodalime, J. Alloys Compd. 830 (2020) 2–7, https://doi.org/10.1016/j.jallcom.2020.154659.
[49] M.A. Islam, J.R. Mou, M.F. Hossain, M.S. Hossain, Highly transparent conducting and enhanced near-band edge emission of SnO2:Ba thin films and its structural, linear and nonlinear optical properties, Opt. Mater. (Amst) 106 (2020) 109996, https://doi.org/10.1016/j.optmat.2020.109996.
[50] J. Kim, B.J. Murdoch, J.G. Partridge, K. Xing, D. Qi, J. Lipton-Duffin, et al., Ultrasonic spray pyrolysis of antimony-doped tin oxide transparent conductive coatings, Adv. Mater. Interfaces (2020) 2000655. 1–11 *https://doi.org/10.1002/admi.202000655*.
[51] X. Li, W. Wei, S. Wang, L. Kuai, B. Geng, Single-crystalline α-Fe2O3 oblique nanoparallelepipeds: high-yield synthesis, growth mechanism and structure enhanced gas-sensing properties, Nanoscale 3 (2011) 718–724, https://doi.org/10.1039/c0nr00617c.
[52] A. Rydosz, The use of copper oxide thin films in gas-sensing applications, Coatings 8 (2018), https://doi.org/10.3390/coatings8120425.
[53] S. Steinhauer, Gas sensors based on copper oxide nanomaterials: a review, Chemosensors 9 (2021) 1–20, https://doi.org/10.3390/chemosensors9030051.
[54] Y. Alajlani, F. Placido, H.O. Chu, R. De Bold, L. Fleming, D. Gibson, Characterisation of Cu2O/CuO thin films produced by plasma-assisted DC sputtering for solar cell application, Thin Solid Films 642 (2017) 45–50, https://doi.org/10.1016/j.tsf.2017.09.023.
[55] T. Oku, T. Yamada, K. Fujimoto, T. Akiyama, Microstructures and photovoltaic properties of Zn(Al)O/Cu2O-based solar cells prepared by spin-coating and electrodeposition, Coatings 4 (2014) 203–213, https://doi.org/10.3390/coatings4020203.
[56] Z.N. Kayani, M. Umer, S. Riaz, S. Naseem, Characterization of copper oxide nanoparticles fabricated by the sol–gel method, J. Electron. Mater. 44 (2015) 3704–3709, https://doi.org/10.1007/s11664-015-3867-5.
[57] J. Singh, M. Rawat, A brief review on synthesis and characterization of copper oxide nanoparticles and its applications, J. Bioelectron. Nanotechnol. 1 (2016), https://doi.org/10.13188/2475-224x.1000003.
[58] A.R. Ansari, A.H. Hammad, M.S. Abdel-Wahab, M. Shariq, M. Imran, Structural, optical and photoluminescence investigations of nanocrystalline CuO thin films at different microwave powers, Opt. Quant. Electron. 52 (2020) 1–16, https://doi.org/10.1007/s11082-020-02535-x.
[59] D.S. Murali, S. Kumar, R.J. Choudhary, A.D. Wadikar, M.K. Jain, A. Subrahmanyam, Synthesis of Cu2O from CuO thin films: optical and electrical properties, AIP Adv. 5 (2015), https://doi.org/10.1063/1.4919323.
[60] P.S. Vindhya, T. Jeyasingh, V.T. Kavitha, Dielectric properties of copper oxide nanoparticles using *AnnonaMuricata* leaf, AIP Conf. Proc. 2162 (2019), https://doi.org/10.1063/1.5130231.
[61] V. Saravanan, P. Shankar, G.K. Mani, J.B.B. Rayappan, Growth and characterization of spray pyrolysis deposited copper oxide thin films: influence of substrate and annealing temperatures, J. Anal. Appl. Pyrolysis 111 (2015) 272–277, https://doi.org/10.1016/j.jaap.2014.08.008.

[62] Y. Gülen, F. Bayansal, B. Şahin, H.A. Çetinkara, H.S. Güder, Fabrication and characterization of Mn-doped CuO thin films by the SILAR method, Ceram. Int. 39 (2013) 6475–6480, https://doi.org/10.1016/j.ceramint.2013.01.077.
[63] F.Z. Chafi, L. Bahmad, N. Hassanain, B. Fares, L. Laanab, A. Mzerd, Characterization Techniques of Fe-Doped CuO Thin Films Deposited by the Spray Pyrolysis Method, ArXiv, 2018, pp. 1–13.
[64] O. Diachenko, J. Kováč, O. Dobrozhan, P. Novák, J. Kováč, J. Skriniarova, et al., Structural and optical properties of CuO thin films synthesized using spray pyrolysis method, Coatings 11 (2021), https://doi.org/10.3390/coatings11111392.
[65] N. Sliti, E. Fourneau, T. Ratz, S. Touihri, N.D. Nguyen, Mg-doped Cu2O thin films with enhanced functional properties grown by magnetron sputtering under optimized pressure conditions, Ceram. Int. 48 (2022) 23748–23754, https://doi.org/10.1016/j.ceramint.2022.05.028.
[66] O.S. Samuel, A. Olanrewaju, A.K. David, Optical properties of Cu2O thin films impregnated with carbon nanotube (CNT), Pertanika J. Sci. Technol. 30 (2022) 343–350, https://doi.org/10.47836/pjst.30.1.19.
[67] I. Ayoub, V. Kumar, R. Abolhassani, S. Rishabh, V. Sharma, S. Rakesh, et al., Advances in ZnO: manipulation of defects for enhancing their technological potentials, Nanotechnol. Rev. 11 (2022) 575–619, https://doi.org/10.1515/ntrev-2022-0035.
[68] H. Beitollahi, S. Tajik, F. Garkani Nejad, M. Safaei, Recent advances in ZnO nanostructure-based electrochemical sensors and biosensors, J. Mater. Chem. B 8 (2020) 5826–5844, https://doi.org/10.1039/d0tb00569j.
[69] M.A. Borysiewicz, ZnO as a functional material, a review, Crystals 9 (2019), https://doi.org/10.3390/cryst9100505.
[70] F. Flory, Optical properties of nanostructured materials: a review, J. Nanophotonics 5 (2011) 052502, https://doi.org/10.1117/1.3609266.
[71] B. Chander Joshi, A.K. Chaudhri, Sol-gel-derived Cu-doped ZnO thin films for optoelectronic applications, ACS Omega 7 (2022) 21877–21881, https://doi.org/10.1021/acsomega.2c02040.
[72] Y. Kang, F. Yu, L. Zhang, W. Wang, L. Chen, Y. Li, Review of ZnO-based nanomaterials in gas sensors, Solid State Ionics 360 (2021) 115544, https://doi.org/10.1016/j.ssi.2020.115544.
[73] Ö. Öztürk, E. Aşikuzun, Z.B. Hacioğlu, S. Safran, Characteristics of ZnO:Er nano thin films produced different thickness using different solvent by sol-gel method, J. Polytech. 0900 (2020) 37–45, https://doi.org/10.2339/politeknik.676184.
[74] P.J.P. Espitia, N.d.F.F. Soares, J.S.d.R. Coimbra, N.J. de Andrade, R.S. Cruz, E.A.A. Medeiros, Zinc oxide nanoparticles: synthesis, antimicrobial activity and food packaging applications, Food Bioprocess Technol. 5 (2012) 1447–1464, https://doi.org/10.1007/s11947-012-0797-6.
[75] J. Wojnarowicz, T. Chudoba, W. Lojkowski, A review of microwave synthesis of zinc oxide nanomaterials: reactants, process parameters and morphoslogies, Nanomaterials 10 (2020), https://doi.org/10.3390/nano10061086.
[76] H. Xie, Z. Li, L. Cheng, A.A. Haidry, J. Tao, Y. Xu, et al., Recent advances in the fabrication of 2D metal oxides, IScience 25 (2022) 1–30, https://doi.org/10.1016/j.isci.2021.103598.
[77] R. Medhi, M.D. Marquez, T.R. Lee, Visible-light-active doped metal oxide nanoparticles: review of their synthesis, properties, and applications, ACS Appl. Nano Mater. 3 (2020) 6156–6185, https://doi.org/10.1021/acsanm.0c01035.

[78] H.J. Fan, R. Scholz, F.M. Kolb, H.J. Fan, R. Scholz, F.M. Kolb, et al., Two-dimensional dendritic nanowires from oxidation of microcrystals of Zn microcrystals, Appl. Phys. Lett. 4142 (2011) 1–4, https://doi.org/10.1063/1.1811774.
[79] C.Y. Chen, C.A. Lin, M.J. Chen, G.R. Lin, J.H. He, ZnO/Al2O3 core – shell nanorod arrays: growth, structural characterization, and luminescent properties, Nanotechnology (2009), https://doi.org/10.1088/0957-4484/20/18/185605.
[80] L. Al-Farsi, T.M. Souier, M. Al-Hinai, M.T.Z. Myint, H.H. Kyaw, pH controlled nanostructure and optical properties of ZnO and Al-doped ZnO nanorod arrays grown by microwave-assisted hydrothermal method, Nanomaterials 12 (2022) 3735.
[81] P.N. Mishra, P.K. Mishra, D. Pathak, The influence of Al doping on the optical characteristics of ZnO nanopowders obtained by the low-cost sol-gel method, Chemistry (2022) 1136–1146.
[82] H.H. Kim, J. Jeong, Y. Yi, Realization of excitation wavelength independent blue emission of ZnO quantum dots with intrinsic defects, ACS Photonics (2020), https://doi.org/10.1021/acsphotonics.9b01587.
[83] D.F. Wang, T.J. Zhang, Study on the defects of ZnO nanowire, Solid State Commun. 149 (2009) 1947–1949, https://doi.org/10.1016/j.ssc.2009.07.038.
[84] F. Fedichkin, B. Jouault, C. Brimont, T. Bretagnon, Room-temperature transport of indirect excitons in (Al,Ga)N/GaN quantum wells, Phys. Rev. Appl. (2016), https://doi.org/10.1103/PhysRevApplied.6.014011.
[85] C. Klingshirn, R. Hauschild, J. Fallert, H. Kalt, Room-temperature stimulated emission of ZnO: alternatives to excitonic lasing, Phys. Rev. B (2007) 1–9, https://doi.org/10.1103/PhysRevB.75.115203.
[86] X. Chen, X. Qingshuang, L. Jitao, Significantly improved photoluminescence properties of ZnO thin films by lithium doping, Ceram. Int. 46 (2020) 2309–2316, https://doi.org/10.1016/j.ceramint.2019.09.220.
[87] M.T. Qureshi, Structural and optical properties of pure and copper doped zinc, Results Phys. (2018), https://doi.org/10.1016/j.rinp.2018.04.010.
[88] N.H. Alvi, K. Hasan, O. Nur, M. Willander, The origin of the red emission in n-ZnO nanotubes/p-GaN white light emitting diodes, Nanoscale Res. Lett. (2011) 1–7.
[89] A. You, M.A.Y. Be, I. In, Stimulated emission and optical gain in ZnO epilayers grown by plasma-assisted molecular-beam epitaxy with buffers, Appl. Phys. Lett. 1469 (2005) 10–13, https://doi.org/10.1063/1.1355665.
[90] S. Morkötter, Synthesis and optical properties of single crystal ZnO nanorods, Nanotechnology (2004), https://doi.org/10.1088/0957-4484/15/6/012.
[91] M. Willander, O. Nur, Q.X. Zhao, L.L. Yang, M. Lorenz, Zinc oxide nanorod based photonic devices: recent progress in growth, light emitting diodes and lasers, Nanotechnology (2009), https://doi.org/10.1088/0957-4484/20/33/332001.
[92] J.G. Cuadra, S. Porcar, D. Fraga, T. Stoyanova-Lyubenova, J.B. Carda, Enhanced electrical properties of alkali-doped ZnO thin films with chemical process, Solar 1 (2021) 30–40, https://doi.org/10.3390/solar1010004.
[93] M.F. Nasir, M.N. Zainol, M. Hannas, M.H. Mamat, S.A. Rahman, M. Rusop, Electrical properties of undoped zinc oxide nanostructures at different annealing temperature, AIP Conf. Proc. 1733 (2016), https://doi.org/10.1063/1.4948886.
[94] S. Serkis Yesilkaya, U. Ulutas, Influence of indium on the parameters of ZnS thin films and ZnS/CuInS2 cells, Mater. Lett. 315 (2022) 131959, https://doi.org/10.1016/j.matlet.2022.131959.
[95] A.B.A. Hammad, A.M. El Nahrawy, D.M. Atia, H.T. El-Madany, A.M. Mansour, Effect of Cu co-doping on the microstructure and optical properties of alumino-zinc thin films

for optoelectronic applications, Int. J. Mater. Eng. Innov. 12 (2021) 18–36, https://doi.org/10.1504/IJMATEI.2021.113214.

[96] S. Asaithambi, P. Sakthivel, M. Karuppaiah, G.U. Sankar, K. Balamurugan, R. Yuvakkumar, et al., Investigation of electrochemical properties of various transition metals doped SnO2 spherical nanostructures for supercapacitor applications, J. Energy Storage 31 (2020) 101530, https://doi.org/10.1016/j.est.2020.101530.

[97] A.A. Ahmad, A.M. Alsaad, Q.M. Al-Bataineh, M.A. Al-Naafa, Optical and structural investigations of dip-synthesized boron-doped ZnO-seeded platforms for ZnO nanostructures, Appl. Phys. A Mater. Sci. Process. 124 (2018), https://doi.org/10.1007/s00339-018-1875-z.

[98] A. Goktas, A. Tumbul, Z. Aba, M. Durgun, Mg doping levels and annealing temperature induced structural, optical and electrical properties of highly c-axis oriented ZnO:Mg thin films and Al/ZnO:Mg/p-Si/Al heterojunction diode, Thin Solid Films 680 (2019) 20–30, https://doi.org/10.1016/j.tsf.2019.04.024.

[99] H.X. Chen, J.J. Ding, Enhanced photoluminescence properties of Al doped ZnO films, IOP Conf. Ser. Mater. Sci. Eng. 292 (2018), https://doi.org/10.1088/1757-899X/292/1/012103.

[100] J. Hao, N.X. Sun, J. Qiu, D. Wang, Structural, electronic, and optical properties of functional metal oxides, Adv. Condens. Matter Phys. 2014 (2014) 2–4, https://doi.org/10.1155/2014/134951.

[101] D.L. Kim, H.J. Kim, Review on optical and electrical properties of oxide semiconductors, Oxide-Based Mater Devices 7603 (2010) 760313, https://doi.org/10.1117/12.846661.

[102] E. Fazio, S. Spadaro, C. Corsaro, N. Giulia, S.G. Leonardi, F. Neri, et al., Metal-oxide based nanomaterials: synthesis, characterization and their applications in electrical and electrochemical sensors, Sensors 21 (2021), https://doi.org/10.3390/s21072494.

[103] H.V. Kiranakumar, R. Thejas, C.S. Naveen, M.I. Khan, G.D. Prasanna, S. Reddy, et al., A review on electrical and gas-sensing properties of reduced graphene oxide-metal oxide nanocomposites, Biomass Convers. Biorefinery (2022), https://doi.org/10.1007/s13399-022-03258-7.

[104] E. Elangovan, K. Ramamurthi, Studies on micro-structural and electrical properties of spray-deposited fluorine-doped tin oxide thin films from low-cost precursor, Thin Solid Films (2005), https://doi.org/10.1016/j.tsf.2004.09.022.

[105] B. Thangaraju, Structural and electrical studies on highly conducting spray deposited fluorine and antimony doped SnO2 thin films from SnCl2 precursor, Thin Solid Films 402 (2002) 71–78.

[106] Y. Tao, B. Zhu, Y. Yang, J. Wu, X. Shi, The structural, electrical, and optical properties of SnO2 films prepared by reactive magnetron sputtering: influence of substrate temperature and O2 flow rate, Mater. Chem. Phys. 250 (2020) 123129, https://doi.org/10.1016/j.matchemphys.2020.123129.

[107] P. Sivakumar, H.S. Akkera, T. Ranjeth Kumar Reddy, G. Srinivas Reddy, N. Kambhala, N.N.K. Reddy, Influence of Ga doping on structural, optical and electrical properties of transparent conducting SnO2 thin films, Optik (Stuttg) 226 (2021) 165859, https://doi.org/10.1016/j.ijleo.2020.165859.

[108] Y. Jiang, W. Sun, B. Xu, N. Bahlawane, Unusual enhancement in electrical conductivity of tin oxide thin films with zinc doping, Phys. Chem. Chem. Phys. (2011) 5760–5763, https://doi.org/10.1039/c0cp00816h.

[109] X. Zhong, B. Yang, X. Zhang, J. Jia, G. Yi, Effect of calcining temperature and time on the characteristics of Sb-doped SnO2 nanoparticles synthesized by the sol–gel method, Particuology 10 (2012) 365–370, https://doi.org/10.1016/j.partic.2011.09.005.

[110] M. Khetri, R. Ramarajan, N.P. Reddy, M. Kovendhan, K.V.A. Srivatsav, B.S. Rao, et al., Stabilization of 5 wt % "Sb" doped SnO2 thin film by post oxidation of thermally evaporated metallic layer, AIP Conf. Proc. 2115 (2019) 1–5, https://doi.org/10.1063/1.5113117.
[111] F. Khalilzadeh-Rezaie, I.O. Oladeji, G.T. Yusuf, J. Nath, N. Nader, S. Vangala, J.W. Cleary, W.V. Schoenfeld, R.E. Peale, Optical and electrical properties of tin oxide-based thin films prepared by streaming process for electrodeless electrochemical deposition, MRS Online Proc. Library 1805 (2015) 423, https://doi.org/10.1557/opl.2015.571.
[112] A. Rahal, A. Benhaoua, C. Bouzidi, B. Benhaoua, B. Gasmi, Effect of antimony doping on the structural, optical and electrical properties of SnO2 thin films prepared by spray ultrasonic, Superlattice. Microst. 76 (2014) 105–114, https://doi.org/10.1016/j.spmi.2014.09.024.
[113] A.R. Babar, S.S. Shinde, A.V. Moholkar, C.H. Bhosale, K.Y. Rajpure, Structural and optoelectronic properties of sprayed Sb:SnO2 thin films: effects of substrate temperature and nozzle-to-substrate distance, J. Semicond. 32 (2011) 1–9, https://doi.org/10.1088/1674-4926/32/10/102001.
[114] R. Ramarajan, M. Kovendhan, K. Thangaraju, D.P. Joseph, R.R. Babu, Facile deposition and characterization of large area highly conducting and transparent Sb-doped SnO2 thin film, Appl. Surf. Sci. 487 (2019) 1385–1393.
[115] M. Chen, H. Zhu, X. Li, J. Yu, H. Cai, X. Quan, et al., The influence of atmosphere on electrical property of copper oxide nanoparticles, J. Nanomat. 2014 (2014) 1–7.
[116] N.R. Dhineshbabu, V. Rajendran, N. Nithyavathy, R. Vetumperumal, Study of structural and optical properties of cupric oxide nanoparticles, Appl. Nanosci. 6 (2016) 933–939, https://doi.org/10.1007/s13204-015-0499-2.
[117] S.M.A. Al-Dujayli, N.A. Ali, The effects of CuO doping on structural, electrical and optical properties of CdO thin films deposited by pulsed laser deposition technique, J. Ovonic Res. 18 (2022) 579–590, https://doi.org/10.15251/JOR.2022.184.579.
[118] J. Resende, V.S. Nguyen, C. Fleischmann, L. Bottiglieri, S. Brochen, W. Vandervorst, et al., Grain-boundary segregation of magnesium in doped cuprous oxide and impact on electrical transport properties, Sci. Rep. 11 (2021) 1–10, https://doi.org/10.1038/s41598-021-86969-7.
[119] G. Wisz, P. Sawicka-Chudy, M. Sibiński, D. Płoch, M. Bester, M. Cholewa, et al., TiO2/CuO/Cu2O photovoltaic nanostructures prepared by DC reactive magnetron sputtering, Nanomaterials 12 (2022) 1–11, https://doi.org/10.3390/nano12081328.
[120] V. Dhanasekaran, T. Mahalingam, Electrochemical and physical properties of electroplated CuO thin films, J. Nanosci. Nanotechnol. 13 (2013) 250–259, https://doi.org/10.1166/jnn.2013.6709.
[121] W.J. Lee, X.J. Wang, Structural, optical, and electrical properties of copper oxide films grown by the silar method with post-annealing, Coatings 11 (2021), https://doi.org/10.3390/coatings11070864.
[122] M. Dahrul, H. Alatas, Irzaman., Preparation and optical properties study of CuO thin film as applied solar cell on LAPAN-IPB satellite, Procedia Environ. Sci. 33 (2016) 661–667, https://doi.org/10.1016/j.proenv.2016.03.121.
[123] D.S. Murali, S. Kumar, R.J. Choudhary, A.D. Wadikar, M.K. Jain, A. Subrahmanyam, Synthesis of Cu2O from CuO thin films: optical and electrical properties, AIP Adv. 5 (2015) 1–6, https://doi.org/10.1063/1.4919323.
[124] J. Sohn, S.H. Song, D.W. Nam, I.T. Cho, E.S. Cho, J.H. Lee, et al., Effects of vacuum annealing on the optical and electrical properties of p-type copper-oxide thin-film transistors, Semicond. Sci. Technol. 28 (2013), https://doi.org/10.1088/0268-1242/28/1/015005.

[125] Y. Wang, P. Miska, D. Pilloud, D. Horwat, F. Mücklich, J.F. Pierson, Transmittance enhancement and optical band gap widening of Cu2O thin films after air annealing, J. Appl. Phys. 115 (2014) 2–7, https://doi.org/10.1063/1.4865957.
[126] J. Shi, J. Zhang, L. Yang, M. Qu, D.C. Qi, K.H.L. Zhang, Wide bandgap oxide semiconductors: from materials physics to optoelectronic devices, Adv. Mater. 33 (2021), https://doi.org/10.1002/adma.202006230.
[127] A. Slassi, M. Hammi, Z. Oumekloul, A. Nid-bahami, M. Arejdal, Y. Ziat, et al., Effect of halogens doping on transparent conducting properties of SnO2 rutile: an ab initio investigation, Opt. Quant. Electron. 50 (2018) 1–13, https://doi.org/10.1007/s11082-017-1262-6.
[128] S.S. Soumya, R. Vinodkumar, N.V. Unnikrishnan, Conductivity type inversion and optical properties of aluminium doped SnO2 thin films prepared by sol-gel spin coating technique, J. Sol-Gel Sci. Technol. 99 (2021) 636–649, https://doi.org/10.1007/s10971-021-05599-7.
[129] S.U. Lee, B. Hong, W.S. Choi, Structural, electrical, and optical properties of antimony-doped tin oxide films prepared at room temperature by radio frequency magnetron sputtering for transparent electrodes, J. Vac. Sci. Technol. A Vac. Surfaces Film 27 (2009) 996–1000, https://doi.org/10.1116/1.3139891.
[130] M. Caglar, S. Ilican, Y. Caglar, F. Yakuphanoglu, Electrical conductivity and optical properties of ZnO nanostructured thin film, Appl. Surf. Sci. 255 (2009) 4491–4496, https://doi.org/10.1016/j.apsusc.2008.11.055.
[131] M.Y. Ali, M.K.R. Khan, A.M.M.T. Karim, M.M. Rahman, M. Kamruzzaman, Effect of Ni doping on structure, morphology and opto-transport properties of spray pyrolised ZnO nano-fiber, Heliyon 6 (2020) e03588, https://doi.org/10.1016/j.heliyon.2020.e03588.
[132] S.U. Yuldashev, G.N. Panin, T.W. Kang, R.A. Nusretov, I.V. Khvan, S.U. Yuldashev, et al., Electrical and optical properties of ZnO thin films grown on Si substrates electrical and optical properties of ZnO thin films grown on Si substrates, J. Appl. Phys. (2006) 013704, https://doi.org/10.1063/1.2209773.
[133] R.C. Prajapati, Structural, magnetic, electrical and optical studies of Cr doped nanostructured ZnO thin films for spintronics application, Mater. Res. Express 6 (2019) 106412.
[134] M. Ardyanian, N. Sedigh, Heavy lithium-doped ZnO thin films prepared by spray pyrolysis method, Bull. Mater. Sci. 37 (2014) 1309–1314.
[135] B.C. Joshi, A.K. Chaudhri, Sol−gel-derived Cu-doped ZnO thin films for optoelectronic applications, ACS Omega (2022), https://doi.org/10.1021/acsomega.2c02040.
[136] M.F. Malek, N. Zakaria, M.Z. Sahdan, M.H. Mamat, Z. Khusaimi, M. Rusop, Electrical properties of ZnO thin films prepared by sol-gel technique, in: 2010 International Conference on Electronic Devices, Systems and Applications (ICEDSA 2010)—Proceedings, 2, 2010, p. 384, https://doi.org/10.1109/ICEDSA.2010.5503037.
[137] R. Barir, B. Benhaoua, S. Benhamida, A. Rahal, T. Sahraoui, R. Gheriani, Effect of precursor concentration on structural optical and electrical properties of NiO thin films prepared by spray pyrolysis, J. Nanomater. 2017 (2017) 433–439, https://doi.org/10.1155/2017/5204639.
[138] J.S. Wilkes, M.J. Zaworotko, Air and water stable 1-ethyl-3-methylimidazolium based ionic liquids, J. Chem. Soc. Chem. Commun. (1992) 965–967, https://doi.org/10.1039/C39920000965.
[139] A.I. Ayesh, A.F.S. Abu-Hani, S.T. Mahmoud, Y. Haik, Selective H2S sensor based on CuO nanoparticles embedded in organic membranes, Sensors Actuators B Chem. 231 (2016) 593–600, https://doi.org/10.1016/j.snb.2016.03.078.
[140] A.B. Gadkari, T.J. Shinde, P.N. Vasambekar, Ferrite gas sensors, IEEE Sensors J. 11 (2011) 849–861, https://doi.org/10.1109/JSEN.2010.2068285.

[141] M.A. Haija, A.I. Ayesh, S. Ahmed, M.S. Katsiotis, Selective hydrogen gas sensor using CuFe2O4 nanoparticle based thin film, Appl. Surf. Sci. 369 (2016) 443–447, https://doi.org/10.1016/j.apsusc.2016.02.103.

[142] G. Korotcenkov, B.K. Cho, Metal oxide composites in conductometric gas sensors: achievements and challenges, Sensors Actuators B Chem. 244 (2017) 182–210, https://doi.org/10.1016/j.snb.2016.12.117.

[143] Y.F. Sun, S.B. Liu, F.L. Meng, J.Y. Liu, Z. Jin, L.T. Kong, et al., Metal oxide nanostructures and their gas sensing properties: a review, Sensors 12 (2012) 2610–2631, https://doi.org/10.3390/s120302610.

[144] A. Gaskov, M. Rumyantseva, Metal oxide nanocomposites: synthesis and characterization in relation with gas sensing phenomena, in: M.I. Baraton (Ed.), Sensors for Environment, Health and Security, Springer, Dordrecht, 2009, pp. 3–30.

[145] C.J. Weschler, Ozone in indoor environments: concentration and chemistry, Indoor Air 10 (2000) 269–288, https://doi.org/10.1034/j.1600-0668.2000.010004269.x.

[146] A. Chapelle, M.H. Yaacob, I. Pasquet, L. Presmanes, A. Barnabé, P. Tailhades, et al., Structural and gas-sensing properties of CuO-CuxFe3-xO4 nanostructured thin films, Sensors Actuators B Chem. 153 (2011) 117–124, https://doi.org/10.1016/j.snb.2010.10.018.

[147] Y.J. Jeong, C. Balamurugan, D.W. Lee, Enhanced CO2 gas-sensing performance of ZnO nanopowder by la loaded during simple hydrothermal method, Sensors Actuators B Chem. 229 (2016) 288–296, https://doi.org/10.1016/j.snb.2015.11.093.

[148] A. Chapelle, F. Oudrhiri-Hassani, L. Presmanes, A. Barnabé, P. Tailhades, CO2 sensing properties of semiconducting copper oxide and spinel ferrite nanocomposite thin film, Appl. Surf. Sci. 256 (2010) 4715–4719, https://doi.org/10.1016/j.apsusc.2010.02.079.

[149] S. Bandi, V. Hastak, D.R. Peshwe, A.K. Srivastav, In-situ TiO2-rGO nanocomposites for CO gas sensing, Bull. Mater. Sci. 41 (2018) 1–5, https://doi.org/10.1007/s12034-018-1632-0.

[150] D. Zhang, Y. Sun, C. Jiang, Y. Yao, W. Dongyue, Y. Zhang, Room-temperature highly sensitive CO gas sensor based on Ag-loaded zinc oxide/molybdenum disulfide ternary nanocomposite and its sensing properties, Sensors Actuators B Chem. 253 (2017) 1120–1128, https://doi.org/10.1016/j.snb.2017.07.173.

[151] E.M. Logothetis, K. Park, A.H. Meitzler, K.R. Laud, Oxygen sensors using CoO ceramics, Appl. Phys. Lett. 26 (1975) 209–211, https://doi.org/10.1063/1.88118.

[152] B.Y. Chang, C.Y. Wang, H.F. Lai, R.J. Wu, M. Chavali, Evaluation of Pt/In2O3-WO3 nano powder ultra-trace level NO gas sensor, J. Taiwan Inst. Chem. Eng. 45 (2014) 1056–1064, https://doi.org/10.1016/j.jtice.2013.09.002.

[153] X. Liu, N. Chen, B. Han, X. Xiao, G. Chen, I. Djerdj, et al., Nanoparticle cluster gas sensor: Pt activated SnO2 nanoparticles for NH3 detection with ultrahigh sensitivity, Nanoscale 7 (2015) 14872–14880, https://doi.org/10.1039/c5nr03585f.

[154] D. Zhang, C. Jiang, P. Li, Y. Sun, Layer-by-layer self-assembly of Co3O4 nanorod-decorated MoS2 nanosheet-based nanocomposite toward high-performance ammonia detection, ACS Appl. Mater. Interfaces 9 (2017) 6462–6471, https://doi.org/10.1021/acsami.6b15669.

[155] G. Korotcenkov, B.K. Cho, Ozone measuring: what can limit application of SnO2-based conductometric gas sensors? Sensors Actuators B Chem. 161 (2012) 28–44, https://doi.org/10.1016/j.snb.2011.12.003.

[156] R. Knake, P.C. Hauser, Sensitive electrochemical detection of ozone, Anal. Chim. Acta 459 (2002) 199–207, https://doi.org/10.1016/S0003-2670(02)00121-6.

[157] A. Labidi, C. Jacolin, M. Bendahan, A. Abdelghani, J. Guérin, K. Aguir, et al., Impedance spectroscopy on WO3 gas sensor, Sensors Actuators B Chem. 106 (2005) 713–718, https://doi.org/10.1016/j.snb.2004.09.022.

[158] M. Mori, Y. Itagaki, Y. Sadaoka, Effect of VOC on ozone detection using semiconducting sensor with SmFe1-xCoxO3 perovskite-type oxides, Sensors Actuators B Chem. 163 (2012) 44–50, https://doi.org/10.1016/j.snb.2011.12.047.

[159] X. Liu, C. Chen, Y. Zhao, B. Jia, A review on the synthesis of manganese oxide nanomaterials and their applications on lithium-ion batteries, J. Nanomater. 2013 (2013), https://doi.org/10.1155/2013/736375.

[160] R. Jose, V. Thavasi, S. Ramakrishna, Metal oxides for dye-sensitized solar cells, J. Am. Ceram. Soc. 92 (2009) 289–301, https://doi.org/10.1111/j.1551-2916.2008.02870.x.

[161] A.Y. Anderson, Y. Bouhadana, H.N. Barad, B. Kupfer, E. Rosh-Hodesh, H. Aviv, et al., Quantum efficiency and bandgap analysis for combinatorial photovoltaics: sorting activity of Cu-O compounds in all-oxide device libraries, ACS Comb. Sci. 16 (2014) 53–65, https://doi.org/10.1021/co3001583.

[162] I. Sullivan, B. Zoellner, P.A. Maggard, Copper(I)-based p-type oxides for photoelectrochemical and photovoltaic solar energy conversion, Chem. Mater. 28 (2016) 5999–6016, https://doi.org/10.1021/acs.chemmater.6b00926.

[163] X. Chen, P. Lin, X. Yan, Z. Bai, H. Yuan, Y. Shen, et al., Three-dimensional ordered ZnO/Cu2O nanoheterojunctions for efficient metal-oxide solar cells, ACS Appl. Mater. Interfaces 7 (2015) 3216–3223, https://doi.org/10.1021/am507836v.

[164] T. Minami, T. Miyata, Y. Nishi, Relationship between the electrical properties of the n-oxide and p-Cu2O layers and the photovoltaic properties of Cu2O-based heterojunction solar cells, Sol. Energy Mater. Sol. Cells 147 (2016) 85–93, https://doi.org/10.1016/j.solmat.2015.11.033.

[165] H. Siddiqui, M.R. Parra, P. Pandey, N. Singh, M.S. Qureshi, F.Z. Haque, A review: synthesis, characterization and cell performance of Cu_2O based material for solar cells, Orient. J. Chem. 28 (2012) 1533–1545.

[166] X. Yang, X. Su, M. Shen, F. Zheng, Y. Xin, L. Zhang, et al., Enhancement of photocurrent in ferroelectric films via the incorporation of narrow bandgap nanoparticles, Adv. Mater. 24 (2012) 1202–1208, https://doi.org/10.1002/adma.201104078.

[167] N. Suzuki, M. Osada, M. Billah, Y. Bando, Y. Yamauchi, S.A. Hossain, Chemical synthesis of porous barium titanate thin film and thermal stabilization of ferroelectric phase by porosity-induced strain, J. Vis. Exp. 2018 (2018) 1–7, https://doi.org/10.3791/57441.

[168] J. Cao, E. Ertekin, V. Srinivasan, W. Fan, S. Huang, H. Zheng, et al., Strain engineering and one-dimensional organization of metal-insulator domains in single-crystal vanadium dioxide beams, Nat. Nanotechnol. 4 (2009) 732–737, https://doi.org/10.1038/nnano.2009.266.

[169] M. Liu, H.Y. Hwang, H. Tao, A.C. Strikwerda, K. Fan, G.R. Keiser, et al., Terahertz-field-induced insulator-to-metal transition in vanadium dioxide metamaterial, Nature 487 (2012) 345–348, https://doi.org/10.1038/nature11231.

[170] A.N. Banerjee, R. Maity, S. Kundoo, K.K. Chattopadhyay, Poole-Frenkel effect in nanocrystalline SnO2:F thin films prepared by a sol-gel dip-coating technique, Phys. Status Solidi Appl. Res. 201 (2004) 983–989, https://doi.org/10.1002/pssa.200306766.

[171] K.H.L. Zhang, K. Xi, M.G. Blamire, R.G. Egdell, P-type transparent conducting oxides, J. Phys. Condens. Matter 28 (2016), https://doi.org/10.1088/0953-8984/28/38/383002.

[172] S.H. Wei, X. Nie, S.B. Zhang, Electronic structure and doping of p-type transparent conducting oxides, Conf. Rec. IEEE Photovoltaic Spec. Conf. (2002) 496–499, https://doi.org/10.1109/pvsc.2002.1190610.

[173] L. Hu, R. Wei, J. Yan, D. Wang, X. Tang, X. Luo, et al., La2/3Sr1/3VO3 thin films: a new p-type transparent conducting oxide with very high figure of merit, Adv. Electron. Mater. 4 (2018) 2–7, https://doi.org/10.1002/aelm.201700476.

Transparent ceramics: The material of next generation

2

Jyoti Tyagi, Sanjeev Kumar Mishra, and Shahzad Ahmad
Department of Chemistry, Zakir Husain Delhi College, University of Delhi, Delhi, India

Chapter outline

1 Introduction

The word transparent material makes an image of glass, polymer, or alkali hydride in one's visualization. Even single crystals that belong to the class of inorganic materials also come into the domain of transparent materials. Besides being optically transparent, the properties such as chemical and mechanical stability make transparent ceramics a better choice for applications compared to their traditional counterparts. But the fabrication process of these materials, known as "growth," is slow and challenging, particularly when the large size of the crystals is desirable. These large-size single crystals do not have any substantial practical utility because of their unpreferable shapes resulting from slow thermodynamic growth processes [1]. Before 1966, when the first transparent ceramic Yttralox (a solid solution of approximately 90% yttrium oxide or yttria, Y_2O_3, and 10% thorium oxide, ThO_2) was reported, no one even realized that ceramics could travel on the path of transparency [2]. After this path-changing material was developed at "General Electric," a new era of transparent ceramics emerged, which possessed outstanding properties, such as high mechanical strength, chemically inert nature, and malleability. Furthermore, transparent ceramics are cost-effective and scalable at the production level, leading to the replacement of single crystals in various applications such as solid-state lasers, optics, electronics, lighting, scintillating, and biomedical. Conventional ceramics constitute grains, grain boundaries, pores, and impurities that lead to light scattering. The double refraction, secondary phase, and surface irregularity further contribute to light scattering. The

Metal Oxides for Next-Generation Optoelectronic, Photonic, and Photovoltaic Applications
https://doi.org/10.1016/B978-0-323-99143-8.00003-1

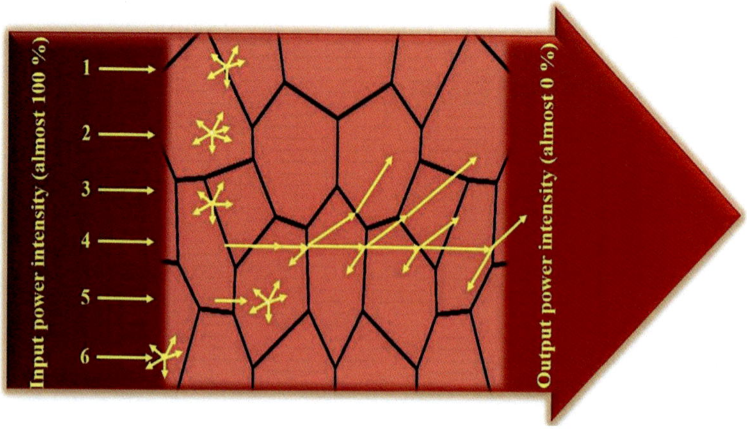

Fig. 1 Potential scatterers in ceramics: (1) grain boundary, (2) residual pores, (3) secondary phase, (4) double refraction, (5) inclusions, and (6) surface roughness.
Reprinted with permission from reference H. Shahbazi, M. Tataei, M.H. Enayati, A. Shafeiey, M.A. Malekabadi, Structure-transmittance relationship in transparent ceramics, J. Alloys Compd. 785 (2019) 260–285. https://doi.org/10.1016/j.jallcom.2019.01.124. Copyright © 2019, Elsevier.

result of these various possibilities of light scattering is opacity, as illustrated in Fig. 1 [3]. At the microstructure level of ceramics, the various scatterers behave as optical heterogeneities that lead to light scattering.

We can optimize light scattering in ceramics to attain transparency comparable to single crystals. The porosity of the ceramics, the impurities present in the ceramic samples, and the grain growth of the ceramics can be controlled by using high-purity raw materials, limited additives, and a correctly controlled sintering process. These aforementioned steps ensure the densification of ceramics with suitable grain size [4]. In one of the earliest studies, the combination of sintering medium and aids was suggested to eliminate pores and densify the crystalline solids [5]. It was also stated that a nondiffusing gas restricts the densification process as gas molecules remain in the pores. In particular, the combination of magnesium oxide (sintering aid) and hydrogen (sintering medium) was identified as an appropriate method to remove gas from pores via a high sintering rate and restrict the discontinuous growth of grains to attain transparency in Al_2O_3 [6]. However, with the advancement in sintering technologies, fully dense and nearly transparent ceramics can be fabricated with minimum or no sintering additive and very fine-grained precursor powder [7,8]. Various advanced sintering methods that are employed for transparent ceramic fabrication are vacuum sintering, microwave sintering, hot pressing (HP), hot isostatic pressing (HIP), and spark plasma sintering (SPS) [9].

Currently, oxide-based transparent ceramics such as alumina, zirconia, magnesia, spinel, yttrium aluminum garnet (YAG), and yttria are investigated across the globe and used for their outstanding properties [10–15]. Nonoxide-based ceramics, such as fluorides, aluminum oxynitride (AlON), SiAlON, and silicon nitride (Si_3N_4), are also attracting researchers to explore their applications in various fields [16–19].

We briefly present the various factors that decide the transparency of ceramics in this chapter. Afterward, the classification of transparent ceramics, viz., oxides and nonoxides, will be discussed. Then, the application of transparent ceramics in solid-state lasers, lighting, scintillating, electro-optic ceramics and devices, optical lenses, windows and domes, and biomedical fields are also briefly explored.

2 What makes the ceramics transparent?

Historically, the ceramics that are polycrystalline inorganic materials are opaque and mechanically weak. When electromagnetic radiation is incident on a solid surface, various processes disturb the radiation path and obstruct the radiation to pass through, making the surface translucent or opaque. The processes that obstruct the radiation can be refraction, reflection, birefringence, scattering, and adsorption [20]. But, with advances in material chemistry and synthesis processes, one can transform conventional ceramics into transparent ceramics. The processing of ceramics comprises three key steps, viz., raw material powder preparation, green body building via compression of powders, and sintering [21]. Other than these three steps, the posttreatment process, such as surface finish/polishing, is also followed to reduce the roughness that is usually present at the ceramic surface after the sintering process. If the group roughness is more at the surface, then it will contribute to scattering and affect the transparency. Other than polishing, the annealing and machining processes are also done to achieve the final product [9].

The powder preparation can be carried out by the solid-state reaction (SSR) method, popular at the industrial level, or via chemical routes. The SSR method is further classified into various methodologies, such as decomposition, chemical reaction, mechanochemical synthesis, and mechanochemical synthesis and activation. Chemical routes include coprecipitation, liquid evaporation methods, and gel methods, among which the sol-gel process is widely adopted for nonaqueous liquid reactions. The SSR process is cost-effective compared to other processes because of its simplicity and scalability. But while adopting the SSR method, sample homogeneity is a matter of concern as solid raw materials are used. On the other hand, chemical routes provide homogeneity as well as better sintering. However, these are not suitable for low-level production, such as laboratory studies, due to their complexity and costly chemicals. For the formation of a green body, there are dry methods (pressing) and wet methods. Under wet methods, various processes such as slip casting, gel casting, tape casting, injection molding, and electrophoretic can be employed [3]. After compaction of raw powders, the sintering process is carried out via various available advanced techniques, such as vacuum sintering, microwave sintering, HP, HIP, and SPS [9]. These various stages are crucial for the fabrication of transparent ceramics possessing high optical properties.

The essential prerequisites for a ceramic to be transparent are no free electrons, minimum or no defects, minimal porosity, and controlled grain size. Besides these preliminary requirements, some other factors further determine the optical transparency of ceramics, such as the quality of raw material powders, sintering technology, and posttreatment processes. The detailed factors for transparency in ceramics are illustrated in Fig. 2.

The transparency of a ceramic depends on the type of crystal structure of the material. The material should possess highly symmetric crystal structures, such as cubic, tetragonal, and hexagonal. These structure types have negligible birefringent effects, which ensure high transparency. Most of the transparent ceramics belong to the cubic crystal system, but ceramics with tetragonal and hexagonal structures are also reported [22]. Apart from the crystal structure, defects also affect the transparency, as they are a source of light scattering. Point defects, such as residual oxygen vacancy, line defects, dislocations, and grain boundary, should be minimal in ceramics to have transparency comparable to glass or single crystals [23]. Furthermore, the random orientation of grains in ceramics is not favorable because it reduces transparency. Crystallographic orientation is an important method for the enhancement of transparency, especially when the ceramics are birefringent (alumina, Al_2O_3) as it reduces the birefringence at the boundaries of grains. In particular, Al_2O_3 with high transparency can be fabricated when the optical axis (*c*-axis) is oriented in a strong magnetic field along with fulfilling other fabrication requirements. The fabricated textured Al_2O_3 (0.80 mm thickness) exhibited a real in-line transmittance of 70% at $\lambda = 640$ nm, higher than that of randomly aligned Al_2O_3. The c-axis alignment decreased the actual difference of the refractive index and remarkably restricted the birefringent effects [24]. Besides a highly symmetric crystal structure free from defects, high-purity raw material powder (>99.9%) is also required to fabricate transparent ceramics, because impurities create

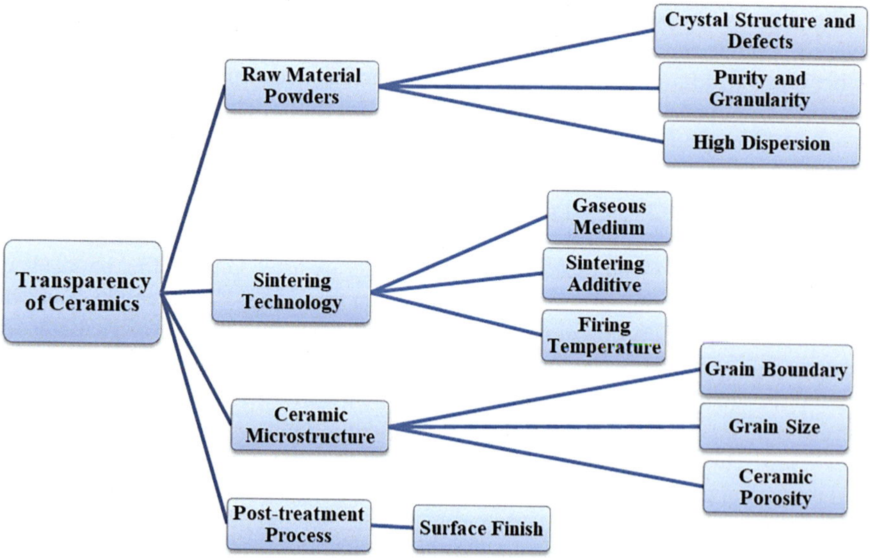

Fig. 2 Various key factors for transparency in ceramics.
Based on reference J. Green, What Does It Take To Make Opaque Ceramics Transparent?, 2021. www.preciseceramic.com (Accessed 24 March 2022).

dissimilar phases that act as light scattering sources and thus reduce the ceramic transparency. The particle size of precursor powder is also important for transparency. Small-size particles are favorable for porosity removal and a better sintering process. However, if the particle size is too small or fine, then the possibility of low molding density, agglomeration, and adsorption of some molecules from the environment increases. Thus, the size of the particles should be optimum and fall into a suitable range. Furthermore, highly dispersed powder (particles remain independent) is favorable, as it ensures better sintering and no agglomeration takes place that can lead to the formation of larger secondary particles.

The ceramic microstructure also disturbs the transparency percentage due to the scattering of light by different scatterers such as residual pores, grain boundary, and impurities that lead to the formation of the secondary phase [4]. Out of the various light scattering sources that exist in ceramics, the contribution of residual pores in decreasing the intensity of light is maximum. Owing to dissimilar optical properties of grains and air trapped in pores, they affect transparency to a large extent. Furthermore, pores can be intergrain or intragrain, while the latter is more difficult to remove compared to the former. The formation of intragrain pores is associated with raw material powder quality, particularly when hard agglomerates are used [1]. After pores, the next light scattering source, grain boundary, also affects transparency because of the unique properties of grain and grain boundary. In particular, the refractive indexes of grain and grain boundary are different, and a scattering of light occurs. So, to achieve transparency similar to single crystals, the grain boundary quality must be controlled. Besides the grain boundary, if impurities are also present at these boundaries, which is very common, light scattering further increases. One solution to control impurity is using a high-purity starting material, and thus, scattering percentage can be controlled. Moreover, during the fabrication of ceramics, sintering aids are also used for densification. However, an increase in the density and concentration of impurities must be tuned, so that there is no compromise in the transparency of the ceramics.

Apart from the aforementioned aspects, the role of sintering aids and processes becomes crucial in eliminating porosity, improving optical quality, and imparting uniform microstructure so that a transparent ceramic can be produced [4]. In the sintering step, choosing the correct gaseous medium, firing temperature, type, and the amount of sintering additive are critical to fabricate ceramics with the desired properties. Nearly 100% transparency in ceramics can be accomplished by using advanced sintering techniques at the third step of processing [9]. Before employing the final sintering process, cold isostatic pressing (CIP) or hydrostatic pressing is usually carried out to attain high density in the final sintering method. In CIP, raw powder materials are compressed into a homogeneous solid mass, which can increase the density (nearly 5%–10%) of the green bodies [25]. In vacuum and microwave sintering, the grain size lies in the range of a few hundred micrometers, as these two techniques are pressure-less sintering. While when HP, HIP, and SPS are used, small grain size (submicrometer) ceramics are obtained, desirable for high mechanical properties [26]. For further detail on sintering techniques, a review by Kong and his co-workers can be referred [9]. Besides this technique, the sintering medium also plays an important role in fabricating transparent ceramics. A transparent ceramic with no residual pores is difficult to produce in the air or inert gas atmosphere, as these gases can easily be trapped in pores. A diffusing gas atmosphere is ideal for high densification of ceramics, as gas trapping in the pores will be minimum. So, to achieve minimum porosity, sintering is done in vacuum, hydrogen, or another conducive atmosphere. Along with the sintering medium, sintering aids are equally significant as they reduce the sintering temperature and ensure high densification via pore elimination [4]. The sintering aids eliminate pores via a liquid phase presence or by inserting dislocations and promoting diffusion rates during heating processes. The grain boundaries reduce during sintering due to an energetically favorable process, and thus, the grain growth takes place. But exaggerated grain growth is also not favorable for transparency. If the mobility of the grain boundary is high, then intragrain pore formation takes place that is nearly impossible to remove. The diffusion of gas from intergrain pores (present at the grain boundary) is faster than that from intragrain pores. Furthermore, there should be no secondary phase left after sintering at the grain boundary as it leads to scattering of light, and thus, in-line transmission losses take place. Therefore, the sintering additives should not form a new solid phase (secondary phase); rather, they should leave the ceramic or fully dissolve to make a solid solution with the ceramic or main phase. Thus, it becomes necessary to identify the solubility limit of the sintering aids in the ceramic [27]. Mostly, the sintering aid quantity in the transparent ceramics is less than that in ceramics where mechanical and electrical properties need to be tuned for some specific applications.

The most frequently employed sintering aids to fabricate transparent ceramics are oxides, followed by alkali halides. In the category of oxides, SiO_2 is largely employed in the production of YAG using vacuum sintering. SiO_2 is added in three different forms, viz., tetraethyl orthosilicate (TEOS, the organic precursor), silica powder, and colloidal silica. MgO and CaO are also popular for fabricating transparent ceramics using vacuum sintering. MgO is known for restricting exaggerated grain growth in ceramics [28,29]. Mixed sintering aids such as MgO-SiO_2 and CaO-MgO have shown promising results in obtaining a uniform microstructure of ceramics along with densification of optimum grain size. A few reports on other oxide-based additives, such as La_2O_3, Sc_2O_3, B_2O_3, and ZrO_2, are also reported. Alkali halides such as LiF, LiCl, NaF, NaCl, KF, and KCl were also investigated as sintering additives. Among halides, results from LiF were most promising, especially in the pressure-assisted sintering techniques (HP and SPS). Besides densification, LiF usage also prevents carbon contamination, as HP and SPS sintering techniques use graphite equipment, such as molds and pistons. Thus, the optical properties of the ceramics remain intact when LiF is used as a sintering additive. For further specifics about sintering aids, a review by Hostaša et al. [4] can be considered. All inclusive, the sintering process, including techniques, firing temperature, sintering medium, and sintering aids, needs to be optimized to achieve suitable grain size, no impurity, and controlled grain boundary.

3 Classification of transparent ceramics

Broadly, transparent ceramics are based on oxides and nonoxides. Out of these two types, primarily oxide-based transparent ceramics are fabricated in most of the studies because of their mechanical strength and chemical stability, which are also easy to synthesize compared to nonoxide ceramics (AlON and Si_3N_4). The metal-oxide

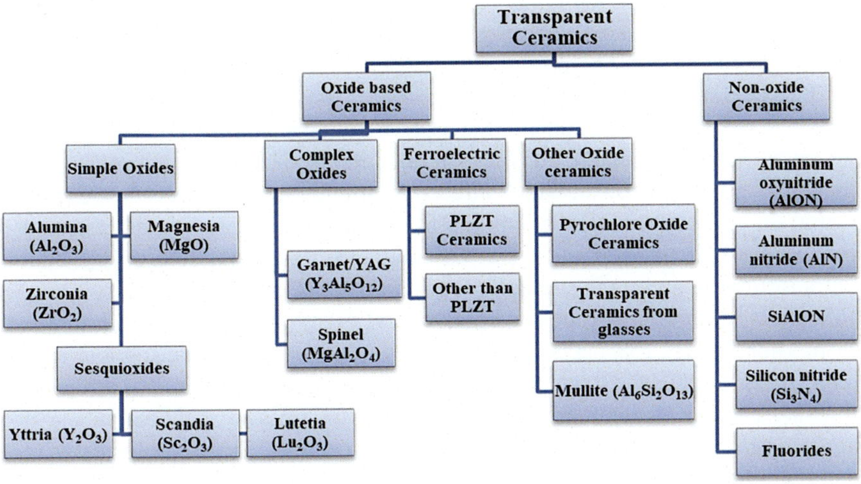

Fig. 3 Oxide and nonoxide-based transparent ceramics classification with representative oxide examples.
Based on reference Z. Xiao, S. Yu, Y. Li, S. Ruan, L.B. Kong, Q. Huang, et al., Materials development and potential applications of transparent ceramics: a review, Mater. Sci. Eng.: R: Rep. 139 (2020) 100518. https://doi.org/10.1016/j.mser.2019.100518.

ceramics can be further categorized as simple oxides, sesquioxides, and complex oxides. Transparent ferroelectric ceramics, mullite, pyrochlore oxide ceramics, and transparent ceramics from glasses are also falling into the metal-oxide-based ceramics. Nonoxide ceramics such as AlON, aluminum nitride (AlN), SiAlON, Si_3N_4, and fluorides are also fabricated owing to the advances in sintering technology. The classification of transparent ceramics based on fabrication carried out to date is represented in Fig. 3 along with some demonstrative oxide examples.

3.1 Metal-oxide ceramics

3.1.1 Alumina (Al_2O_3)

Al_2O_3, known as corundum, possesses a hexagonal crystal structure with $a = 4.754$ Å and $c = 12.99$ Å as cell parameters. In the crystal lattice, oxide (O^{2-}) ions form the close-packed hexagonal arrangement and Al^{3+} fills two-thirds of the octahedral interstices [30]. Since Al_2O_3 is optically anisotropic and possesses a noncubic crystallographic structure, there is a problem with full transparency in Al_2O_3-based ceramics because of birefringence effects. However, Al_2O_3-based transparent ceramics, owing to their high hardness, strength, and exceptional corrosion resistance, have significant uses in armors, electromagnetic windows and domes, and the casing of high-pressure metal halide lamps. Depending on the application, the mechanical properties of the Al_2O_3 can be tuned by controlling grain size. If the grain size of the Al_2O_3 is optimized properly and lies in the submicron range, then it will be the hardest material, even including sapphire. Therefore, Al_2O_3-based transparent ceramics can also be used as a promising substitute for sapphire [31]. Initially, translucent Al_2O_3 ceramics were produced owing to the available sintering methods, and thus, these ceramics found their application only as lamp envelopes. Conventional transparent Al_2O_3 ceramic fabrication using early sintering methods required very high temperatures, generally $\geq$1700°C. Production at such a high temperature leads to exaggerated grain growth, which affects the strength and hardness of the resultant ceramic. In particular, bigger grains (up to 410 μm) led to more light scattering via birefringence effects, and thus, in-line transmittance was nearly 10% only [32]. Such low transmittance and weak strength (due to high temperature) obstruct their application in areas where transparency and mechanical strength are needed. To improve the transmission to only 15%–20%, it is necessary to increase the sintering temperature, which will further reduce the mechanical resistance of the ceramics. So increasing the temperature is not the solution. However, with progress in technology and other sintering aids, better quality transparent Al_2O_3 ceramics were obtained by many researchers with higher in-line transmittance (55%–65%) as well [33]. Sintering additives such as MgO, Y_2O_3, and La_2O_3 are also employed during the sintering process to reduce the porosity in Al_2O_3 ceramics [34]. When advanced sintering techniques (HP, HIP, and SPS) were employed along with sintering aids, the sintering temperature was reduced to 1200°C while maintaining the grain size to less than 1 μm ($\sim$0.32 μm) and high density (nearly 99.96%) in ceramics. Furthermore, the in-line transmittance and microhardness values were up to approximately 80% and 22 GPa, respectively [12]. Thus, transparent Al_2O_3 ceramics with minimum residual porosity and reduced grain size are being realized by employing the latest sintering technologies as reviewed by Kong and his co-workers [1].

3.1.2 Magnesia (MgO)

MgO possesses a cubic crystal structure with $a = 4.217$ Å as a cell parameter. It is an isotropic substance that makes it free from birefringence effects, and thus, one criterion for transparency is already fulfilled. MgO-based transparent ceramics can be used as sensor protectors and might replace sapphire infrared windows owing to their outstanding thermal and mechanical properties. Because it has a high melting point (2852°C) compared to Al_2O_3 (2054°C), its sintering temperature for complete densification is high, making the production of transparent MgO ceramic a challenge. The techniques, similar to Al_2O_3 are also employed for its fabrication. With MgO, high densification was attained early in 1974 by a multistep method using vacuum sintering, taking MgO or $Mg(OH)_2$ as a high-purity precursor powder, and LiF as the sintering additive [35]. It was observed that various factors affected the transparency of the final product, such as the particle size of the raw material, LiF concentration, and calcination temperature. In particular, only 0.02 wt% of F- was enough for MgO densification. However, when the particle size of the precursor powder was >0.15 μm, F- ions were incapable of MgO densification. LiF is a well-accepted sintering additive for MgO ceramic fabrication as it decreases the sintering temperature. Another example of MgO ceramic production, though translucent, using LiF as the sintering additive via HP was shown by Fang and his co-workers [11]. For the synthesis process, commercial nanopowder of MgO with high purity, high surface area, high dispersion, and small grain size was considered for the green body building. LiF was added to MgO nanopowder after dispersing in 2-propanol to attain 2–4 wt%. Afterward, using the HP technique, thin circular discs of nanopowder were sintered in Ar or vacuum at 1100°C for up to 1 h at either 24 or 45 MPa. Highly compact and optically translucent MgO ceramic circular discs were obtained after annealing in the air. It was supposed that LiF presence promotes MgO densification via forming a liquid phase that functions as a lubricant for MgO particle rearrangement during an early stage of sintering and later acts as a medium for easy material transport to assist the sintering process [36]. Sintering aids have some negative effects on the optical properties of ceramics, so for some specific applications, there is no possibility of such aids. With advanced sintering techniques such as HIP and SPS, transparent MgO ceramics have been obtained, adding no sintering additive [10,37]. In another study, the fabrication of translucent MgO ceramics was realized using a pressure-less sintering method with no aid [38]. The wet precipitation method prepared the precursor powder of nanocrystalline MgO. Using these nanocrystalline powders, the MgO ceramics were prepared, but the transparency of the final product required more improvement. Although MgO belongs to the cubic crystal system, completely transparent MgO ceramics are not easily fabricated as at higher temperatures, the vapor pressure of MgO is high. So, presently MgO ceramics' optical properties are not comparable to their single-crystal equivalents.

3.1.3 Zirconia (ZrO_2)

ZrO_2 exists in monoclinic, tetragonal (>1100°C), and cubic (>1900°C) crystal structures. Among these forms, tetragonal and cubic forms exhibit optical properties because of the highly symmetric structure and thus, low birefringence effects. ZrO_2-based ceramics have exceptional properties such as low thermal conductivity, high toughness and strength, high oxygen diffusivity, chemical stability, and high refractive index (nearly 2.2 of the cubic form). It does not react with melts and some solutions of acids and bases, even at 2000°C. Furthermore, the possibility of carbon contamination in the final product is very low, up to 1900°C. Besides this, an oxidizing medium does not affect its composition, even at the highest temperatures. This wide range of properties exemplifies its extensive study of applications in various fields. When Y_2O_3 is introduced as a dopant to ZrO_2, its cubic and tetragonal crystal structure get stabilized via oxygen vacancy generation. Based on this fact, many researchers have widely studied yttria-stabilized zirconia (YSZ) compared to other dopants such as MgO and CaO [1]. Transparent cubic ZrO_2 ceramics with 8 mol% yttria (c-YSZ) have attracted many research groups owing to their unique fusion of high mechanical strength and fine optical properties [39–41]. The fabrication of c-YSZ ceramics has been realized using advanced sintering techniques, such as HIP and SPS. In particular, Yamashita and his co-workers have fabricated 8 mol%—YSZ via HIP that exhibited light transmission nearly equivalent to single crystals in the visible region. In this study, the microstructure of presintered and HIP substances was correlated with the transparency of the resultant ceramic. When raw materials possessed small grains and fine intergranular residual pores (in the submicrometer range), a highly transparent ceramic is obtained. They also showed that, for high transparency, a porosity of less than 100 ppm is necessary [40]. Besides HIP, SPS has also emerged as an efficient process to obtain highly dense and transparent ZrO_2 ceramics. The resultant ceramic samples have exhibited high in-line transmittance and better preserving nano-grained structures as compared to the current SPSed analogs [41,42].

The two forms of ZrO_2-based ceramics, viz., cubic and tetragonal, exhibited contrasting results when they were compared in terms of their toughness and transparency. The tetragonal ZrO_2 showed high mechanical strength (>900 MPa) compared to cubic form (<600 MPa). Furthermore, tetragonal ZrO_2 ceramics obtained from nanocrystalline structures exhibited higher fracture toughness (8 MPa m$^{1/2}$) compared to cubic ZrO_2 ceramics (2.8 MPa m$^{1/2}$). On the other hand, cubic ZrO_2 ceramics are more transparent compared to their tetragonal counterparts (due to birefringence effects) that were even nearly free from pores [43,44]. Therefore, depending on the application, tetragonal (mechanical strength desirable) and cubic (optical transparency required) ZrO_2 ceramics can be fabricated.

3.1.4 Sesquioxides

Under the category of sesquioxides, Y_2O_3, scandia (Sc_2O_3), and lutetia (Lu_2O_3) have attracted many researchers to develop transparent ceramics based on these materials owing to their excellent optical properties. All these materials possess cubic crystal structures so birefringence effects are not present. Therefore, transparent ceramics based on these substances are employed as the functional medium of solid-state lasers. Furthermore, these materials possess high refractivity, good mechanical strength, high melting temperatures, high chemical inertness, high thermal conductivity, a low thermal expansion coefficient, and high light transmission over a wide range of electromagnetic radiations [1]. Traditionally, vacuum or H_2 sintering of green bodies attained via isostatic pressing has been used to produce transparent ceramics from these sesquioxides. But to fabricate a highly dense final product, there is a need for sintering aids due to the high melting temperatures of these sesquioxides. For example, different types and combinations of sintering additives, such as ThO_2, ZrO_2, La_2O_3, HfO_2, and LiF, have been employed to obtain the desired product of transparent Y_2O_3 at reduced temperatures [1]. The effect of lower sintering temperature and combined aids is fine grain size and pore elimination in the Y_2O_3 ceramics. Furthermore, sintering aid effects can be fully utilized if incorporated with other methods like sintering in an oxygen atmosphere [45]. These additives get fully dissolved with the main phase of ceramics and thus form a solid solution with no secondary phase formation. Out of various sintering aids, ZrO_2 has been widely used as a dopant to obtain transparent Y_2O_3 ceramics with a high in-line transmittance of greater than 80% [46]. For the fabrication of transparent ceramics from sesquioxides, HP, HIP, and SPS are effective sintering processes. Recently, even hybrid sintering methods such as HP followed by HIP are being employed to obtain transparent Y_2O_3 ceramics [15] as shown in Fig. 4. Out of the three aforesaid sesquioxides, applications of Y_2O_3 are very

Fig. 4 Images of 2-mm-thick Y_2O_3 ceramic samples after HIP and HP sintering at different temperatures (from left to right): 1300°C, 1350°C, 1400°C, 1450°C, 1500°C, and 1550°C. Reprinted with permission from L.L. Zhu, Y.J. Park, L. Gan, H.N. Kim, J.W. Ko, H.D. Kim, Fabrication and characterization of highly transparent Y2O3 ceramics by hybrid sintering: a combination of hot pressing and a subsequent HIP treatment, J. Eur. Ceram. Soc. 39 (2018) 3255–3260. https://doi.org/10.1016/j.jeurceramsoc.2018.03.022. Copyright © 2018, Elsevier.

promising, and therefore, it has been more widely studied compared to the other two sesquioxides. Besides being applicable in solid-state lasers and scintillators, other significant applications of Y_2O_3-based ceramics are in IR domes, NIR upconverters, refractory substances, semiconductor parts, and gas nozzles [1].

For the fabrication of Sc_2O_3-based transparent ceramics, HIP is also used along with the most popular process of vacuum sintering. Before sintering, the production of Sc_2O_3 nanopowders by a precipitation method is also carried out due to the high sinterability of nanopowders that led to the production of high-quality ceramics [47]. Recently, Sc_2O_3 nanopowders doped with ytterbium (Yb) were also fabricated via a coprecipitation process that results in transparent Yb: Sc_2O_3 ceramics after vacuum sintering with nearly 74% transmittance at 1100nm [48].

Transparent ceramics based on Lu_2O_3 are again widely obtained from vacuum sintering. But two-step sintering processes, i.e., vacuum sintering followed by HIP and SPS, are also employed to fabricate Lu_2O_3 ceramics that possess fine grains [49]. Lu_2O_3 ceramics can be easily doped with Yb^{3+} with no compromise in their properties due to similar ionic radii and atomic masses of Lu^{3+} and Yb^{3+}. Transparent Lu_2O_3-based ceramics are doped with Nd and Er, owing to their applications in high-power solid-state lasers and scintillators, respectively [1]. For example, transparent Nd:Lu_2O_3 ceramics were fabricated via the SPS sintering method after doping with 0.2 at% LiF that can be used effectively for laser operation [50].

Apart from the various aforementioned properties of the sesquioxide-based transparent ceramics, the cost is a critical factor for their large-scale application. At present, high cost limits their application at the industrial level, and similar materials are also available at lower cost [1].

3.1.5 *Yttrium-aluminum garnet ($Y_3Al_5O_{12}$)*

YAG with the chemical formula $Y_3Al_5O_{12}$ has a cubic crystal structure. Its centrosymmetric structure makes it isotropic and thus free from birefringence effects. It finds wide application in solid-state lasers owing to exceptional uniform optical properties. However, YAG-based ceramics also have significant applications as host materials for fluorescent and structural materials at high temperatures due to high chemical inertness and high thermal stability [13]. Traditionally, YAG ceramics are fabricated by vacuum sintering the combination of oxides (Al_2O_3 and Y_2O_3) along with sintering aids at high temperatures. The final product is usually obtained via multistep processes such as initial cold pressing, then CIP in some cases, and after that sintering at high temperature by following any advanced sintering technique (HP, HIP, and SPS). For applications in solid-state lasers, YAG-based transparent ceramics are doped with various rare earth elements such as neodymium (Nd), ytterbium (Yb), erbium (Er), and cerium (Ce). For example, YAG ceramic doped with Nd (Nd:YAG) was first fabricated in 1995 by taking a mixture of Al_2O_3, Y_2O_3, and Nd_2O_3 by the SSR method. Initially, the three oxides were prepared by methods such as thermal pyrolysis and alkoxide hydrolysis. Subsequently, these oxide powders were compressed and vacuum sintered between 1600°C and 1850°C for 5 h to get the final product. The optical transmittance of the obtained YAG ceramics with 1 mm thickness was comparable to that of YAG single crystals in the UV and IR regions [51]. SSR is the most common method for the production of transparent YAG ceramics using commercial oxides as precursor materials or some other compounds of Y and Al [29]. For the YAG ceramic production, various sintering aids such as SiO_2, MgO, CaO, TEOS, and LiF were employed to attain high densification and good optical transparency along with reduced grain size. Among the different employed aids, SiO_2 and MgO are the most effective sintering additives for the fabrication of transparent YAG ceramics, especially Nd:YAG [52,53]. Besides vacuum sintering, SPS is also employed for YAG ceramic fabrication though it is not a widely used sintering technique for YAG as compared to other ceramics [54]. The transparent YAG ceramics possess high transmittance values, greater than 80% in most of the studies reviewed by Kong and his co-workers [1]. In particular, highly transparent YAG ceramics were fabricated using only 0.03 wt% of MgO as dopant via vacuum sintering at high temperatures (1820°C) that exhibited 84.4% transmittance at 1064 nm. Furthermore, the effect of the MgO concentration on the optical performance and microstructure was systematically studied. It was found that MgO favors densification when the sintering temperature is $<$1660°C, but when the temperature range was 1540–1660°C, the grain growth suddenly increases. However, the excessive increase in the MgO concentration resulted in the deterioration of optical performance and the development of the Mg-rich secondary phase along with intragranular pores formation [29].

3.1.6 Spinel ($MgAl_2O_4$)

Magnesium aluminate ($MgAl_2O_4$), also known as spinel, belongs to the cubic crystal system and is thus optically isotropic. There are no polymorphic changes, and thus, no phase transformations take place during sintering. It possesses a unique blend of exceptionally high optical and mechanical properties compared to MgO and Al_2O_3. It exhibits transparency in UV and mid-IR electromagnetic radiation between 0.2

and 5.5 μm. Furthermore, spinel-based ceramics possess high hardness values, high chemical inertness, high thermal stability, and high resistance to erosion. Because of these excellent properties, $MgAl_2O_4$-based transparent ceramics find significant applications in defense as transparent armors, protective windows for vehicles, missile domes, submarine IR sensors, and optical components as optical lenses [1,55]. Transparent spinel-based ceramic fabrication with low transmittance was realized in the early 1960s, leading to the wide attention of different research groups to investigate these ceramics. In 1974, translucent spinel ceramics were obtained via vacuum sintering using coprecipitated ultrafine spinel powder along with 0.25 wt% CaO as a sintering additive. The in-line transmittance of the obtained ceramic was only >10% between 0.3 and 6.5 μm [56]. It is difficult to get transparent spinel ceramics via traditional sintering methods due to grain growth at elevated temperatures; therefore, advanced sintering technologies (HP, HIP, and SPS) are usually employed to fabricate these transparent ceramics. Besides advanced technologies, high-quality raw material powders and an optimum amount of sintering additives such as LiF, Cao, and B_2O_3 are also needed to attain highly transparent $MgAl_2O_4$-based ceramics [56–58]. At present, LiF is the most effective sintering additive used for producing highly transparent spinel ceramics. In particular, transparent $MgAl_2O_4$ ceramics were fabricated via HP using ultrapure LiF as a sintering aid. The final product was obtained by taking Al_2O_3-MgO mixture powders in stoichiometry rather than directly considering spinel powders, which also result in low-cost fabrication. Transmittance up to 70% has been attained in the visible range with a maximum value of 78%. Moreover, a thermodynamic investigation was also carried out to understand the factors, such as pressure, that affect the transparency of the final product [57]. Another study has also used LiF (3 wt%) as a sintering aid to fabricate highly transparent spinel ceramics via the HP method. The $MgAl_2O_4$ nanopowders were synthesized using a sol-gel process in which they were hot pressed at 1600°C and 50 MPa to obtain the final product. The resultant transparent ceramics exhibited high in-line transmittance of 83.7% at 1100 nm, as shown in Fig. 5 [59].

Recently, the use of the SPS method showed the fabrication of highly dense and transparent $MgAl_2O_4$ ceramics at low sintering temperatures. For example, transparent $MgAl_2O_4$ ceramics with a high transparency of 86.7% at 1100 nm were obtained by combining the colloidal gel casting process with SPS. The highly dense ceramics of 99.97% density were fabricated using isobutylene-maleic anhydride (ISOBAM) as a sintering aid that acted as a dispersant as well as a gelation agent [14]. The SPS and other pressure sintering methods are accompanied by carbon contamination of the fabricated transparent ceramic samples, resulting in discoloration of the ceramics. Because of the fact that the green bodies are in direct contact with molds during the sintering process, which are usually made up of graphite, a recent study that fabricated transparent $MgAl_2O_4$ ceramics without any sintering additive via a two-step SPS at 1250°C shed more light on the carbon contamination of ceramics. In particular, the critical temperature effect on optical transparency and carbon contamination was studied. In the study, high carbon contamination was observed at higher critical temperatures and fast heating rates [60].

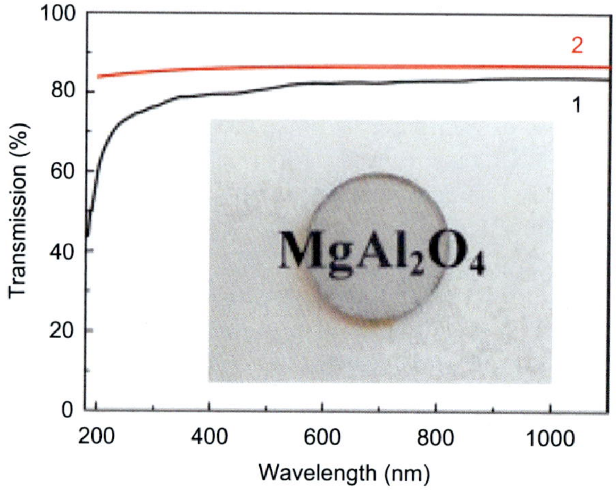

Fig. 5 Transmission spectrum of $MgAl_2O_4$ ceramics (1 mm thick) doped with 3 wt% LiF (1) and "defect-free" $MgAl_2O_4$ crystal (2, calculated). Inset shows an image of the final ceramic product. Reprinted with permission from S.S. Balabanov, R.P. Yavetskiy, A.V. Belyaev, E.M. Gavrishchuk, V.V. Drobotenko, I.I. Evdokimov, et al., Fabrication of transparent MgA12O4 ceramics by hot-pressing of sol-gel-derived nanopowders, Ceram. Int. 41 (2015) 13366–13371. https://doi.org/10.1016/j.ceramint.2015.07.123. Copyright © 2015, Elsevier.

3.1.7 Transparent ferroelectric ceramics

Transparent ferroelectric ceramics possess an ABO_3-type structure and are perovskite-type substances. These ceramics are dielectric materials along with transparency, which is a unique combination compared to dielectric polymers [61]. The significant applications of these transparent ceramics are they are used as digital light modulators, dielectric materials in transparent electronic appliances, transparent pulsed capacitors, pyroelectric security sensors, large-aperture light shutters, and medical diagnostic transducers. Moreover, because of the high thermal stability and good mechanical strength, these ceramics can be used for electricity production in very high fields [62–64]. Out of the various investigated ferroelectric ceramics such as $Pb_{1-x}La_x(Zr_yTi_{1-y})_{1-x/4}O_3$ (PLZT), $Pb(Zn_{1/3}Nb_{2/3})O_3$-$PbTiO_3$ (PZN-PT), PZN-PLZT, and $Pb(Mg_{1/3}Nb_{2/3})O_3$-$PbTiO_3$ (PMN-PT), PLZT ceramics, i.e., lanthanum-doped lead zirconate titanate, possess higher electro-optic coefficients and thus exhibit better electro-optical properties compared to single crystals of $LiNbO_3$. Furthermore, their optical transmittance is also higher when compared with other reported ferroelectric substances [1].

The transmittance of transparent PLZT ceramics can be tuned better by using high-quality raw material powders [65]. Therefore, to produce fine-grained powders with high dispersion, several wet chemical routes such as sol–gel process, hydrothermal synthesis, and co-precipitation are employed. The SSR method, like mechanochemical synthesis and activation, is also used to prepare fine PLZT powders, which result in more transparent PLZT ceramics [62,66]. Even the combination of wet and dry methods to fabricate the fine PLZT powders was also employed to obtain transparent PLZT ceramics. In particular, the final ceramic obtained via the SPS technique exhibited 31% transmittance at 700 nm [67]. For fabricating PLZT ceramics, PbO is one of the precursor raw materials that possesses low volatility. Therefore, a controlled sintering atmosphere also plays a crucial role in fabricating transparent ferroelectric ceramics via advanced sintering methods in less sintering time and also at lower temperatures. Moreover, if the sintering atmosphere is controlled properly, then high-quality transparent PLZT ceramics can also be fabricated via conventional sintering techniques. For example, using a controlled atmosphere at three stages, the obtained PLZT9/65/35 ceramics of 0.7 mm thickness possessed 51% optical transmittance at 550 nm. To obtain the final product, initially, sintering was done in an oxygen atmosphere followed by a CO_2 atmosphere to eliminate pores, and finally, to remove oxygen vacancies, an oxygen atmosphere was again employed. Moreover, in the absence of a CO_2 atmosphere, the ceramic exhibited only 34% optical transmittance [68]. Among the various advanced sintering techniques, SPS and microwave sintering are effective methods for producing transparent PLZT ceramics with improved properties. Furthermore, the sintering temperature is also low in these two processes, which favor a lower or minimum loss of PbO during sintering. The microwave sintering used to fabricate PLZT 9/65/35 transparent ceramics decreased the sintering temperature (1200–1000°C) as well as the sintering time (180–20 min) compared to conventional methods. The precursor materials were prepared by partial coprecipitation and SSR, a hydrothermal process; the former results in ultrafine, uniform, and chemically active PLZT powders. The resultant ceramics exhibited better properties, such as 53.8% transmittance at 850 nm, a dielectric constant of 3895, and a maximum relative density of 96.5%. The final PLZT ceramics can be used in optical communication switches owing to the characteristic hysteresis loop of the quadratic type [66]. In another study, the transparent ferroelectric pristine PLZT and Ni-doped PLZT (PLZTN) ceramics, $(Pb_{1-x}La_x)(Zr_{0.65(1-y)}Ti_{0.35(1-y)}Ni_y)O_3$ ($x = 0.08$, 0.09; $y = 0$, 0.025), were fabricated by HP sintering. Both the ceramics exhibited their application as self-powered photodetectors in the visible (PLZT at 405 nm) and NIR (PLZTN) regions based on the photovoltaic effect. The devices comprise PLZT and PLZTN, which exhibited high sensitivity and quick response in visible and NIR regions, respectively. Therefore, these transparent ceramics possess great potential to develop future transparent photodetectors based on the photovoltaic effect. Moreover, photodetectors in NIR regions, i.e., PLZTN, can find their significant applications in optical fiber communication, unmanned driving, and biological pattern recognition [69].

Other than PLZT, PMN-PT-based transparent ceramics are also fabricated widely due to their transparency in the whole visible and mid-IR spectrum. The electro-optic coefficient of these ceramics is also higher than PLZT, nearly by a factor of 20. Doping of PMN-PT ceramics by La results in an increment in transparency in many studies. For example, La-doped 0.75PMN-0.25PT transparent ceramics were fabricated via three different methods, viz., uniaxial HP, O_2 atmosphere sintering, and two-stage sintering. Out of the three methods, two-stage sintering in which the green bodies were initially sintered in the O_2 atmosphere between 1220°C and 1260°C for 3–5 h and later on sintered between 1200°C and 1240°C for the same time under uniaxial pressure of 60 MPa gave the best results of transparency and density. The resultant ceramics showed maximum optical transmittance (70%) at 900 nm, which is nearly the same as the theoretical transmittance of 71% [70]. Another similar study reported La-doped 0.88PMN-0.12PT transparent ferroelectric ceramics using SSR and two-stage sintering. The La doping was done at four different concentrations 0.5, 1.0, 1.5, and 2.0 mol% to study the effects on the microstructure and other properties. 1.0 mol% La doping exhibited maximum transparency of nearly 70% in the NIR region, which makes it suitable for electro-optical applications [71].

The aforesaid, transparent ferroelectric ceramics comprise nearly 60 wt% of Pb that negatively affects the environment as well as human health. Therefore, there is a requirement for transparent Pb-free ferroelectric ceramics, which can be employed in electronic devices, such as transparent pulsed capacitors. In this direction, Pb-free relaxor ferroelectric ceramics with high transparency based on $(K_{0.5}Na_{0.5})NbO_3$ were fabricated via a pressure-less solid-state sintering process. The resultant ceramics, 0.8 $(K_{0.5}Na_{0.5})NbO_3$-$0.2Sr(Sc_{0.5}Nb_{0.5})O_3$, possessed nearly 60% transmittance at 700 nm as well as 2.48 J cm^{-3} of energy storage density, which is a high value for Pb-free ceramics. Furthermore, these Pb-free ceramics open new avenues of applications, beyond the typical piezoelectric applications [63]. In another study, Pb-free ferroelectric ceramics doped with Er^{3+}, $(K_{0.5}Na_{0.5})_{1-x}Li_xNb_{1-x}Bi_xO_3$ $(0.05 \leq x \leq 0.08)$, were fabricated via pressure-less sintering. The resultant ceramics possessed a fine-grained dense structure due to a phase similar to cubic. Consequently, ceramics exhibited high transparency of 50% in NIR and 75% in MIR regions. Additionally, ceramics also possessed strong electro-optic effects that were supported by a large linear electro-optic coefficient of 128–184 pm/V. The ceramics also displayed electro-optic properties along with good photoluminescence. Furthermore, the ceramics displayed green and red up-conversion photoluminescence emissions because of the doping with Er^{3+}. The obtained ceramics can be potentially used in visible displays, optical attenuators, and MIR solid lasers owing to a large dielectric coefficient of 1400, a high piezoelectric coefficient of 70–90 pC/N, and low dielectric loss of 0.03 [72].

3.1.8 Other oxide ceramics

Ceramics that possess a cubic fluorite-type structure with a general formula $A_2B_2O_7$ are also known as pyrochlore oxide ceramics. In this category of ceramics, complex oxides with pyrochlore structures, such as $Lu_2Ti_2O_7$, $Lu_2Zr_2O_7$, $Y_2Hf_2O_7$, $Y_2Ti_2O_7$, $LaGdZr_2O_7$, $LaYZr_2O7$, Lu_3NbO_7, and Ga_3TaO_7, have been fabricated by vacuum sintering and SPS methods [1]. It is important to mention that these pyrochlore oxide transparent ceramics cannot be fabricated via conventional sintering methods. For example, transparent La gadolinium hafnate $(La_{0.8}Gd_{1.2}Hf_2O_7)$ ceramics doped with different Eu^{3+} amounts (1, 3, 5, and 10 at%) were prepared via vacuum sintering. It was observed that the annealing process improved the transparency and luminescence intensity of the fabricated ceramics compared to unannealed samples, as shown in Fig. 6. The study revealed that, with an increase in Eu^{3+} doping, the optical transmittance and luminescence intensity of the resultant

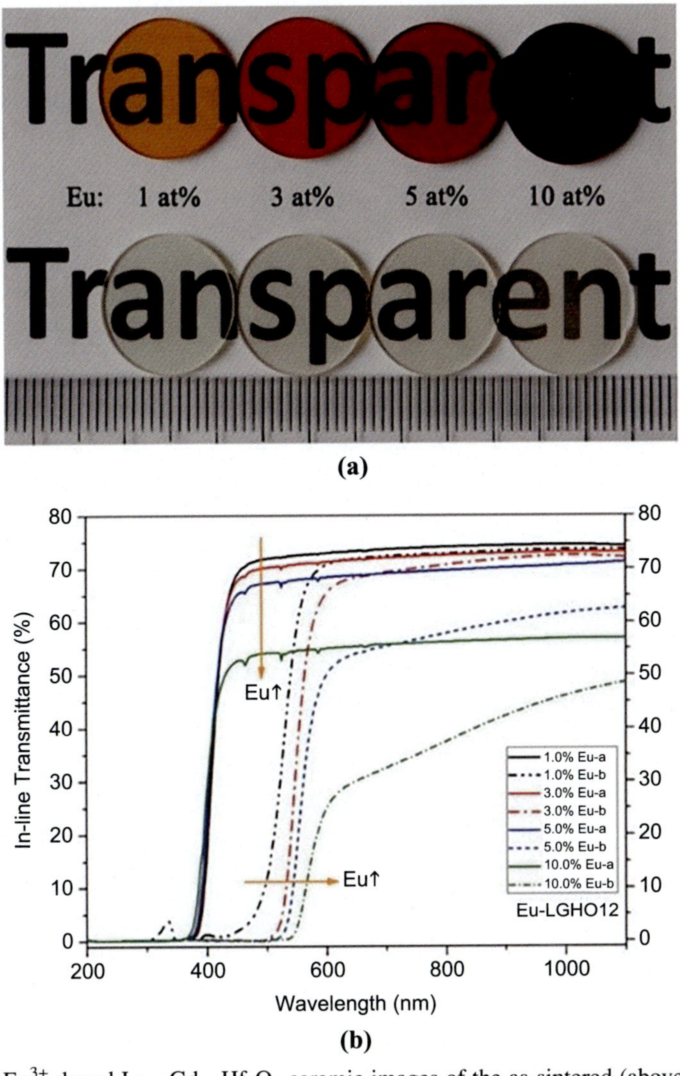

Fig. 6 (A) Eu^{3+}-doped $La_{0.8}Gd_{1.2}Hf_2O_7$ ceramic images of the as-sintered (above) and thermally annealed (below) samples and (B) in-line transmittance curves of Eu^{3+}-doped $La_{0.8}Gd_{1.2}Hf_2O_7$ ceramics (1 mm thick), as sintered (dash lines) and annealed (solid lines). Reprinted with permission from reference Z.J. Wang, G.H. Zhou, J. Zhang, X.P. Qin, S.W. Wang, Luminescence properties of Eu3+-doped Lanthanum gadolinium hafnates transparent ceramics, Opt. Mater. 71 (2017b) 5–8. https://doi.org/10.1016/j.optmat.2016.05.046. Copyright © 2016, Elsevier.

ceramics decrease owing to oxygen vacancies and some other defects caused by vacuum sintering [73].

Transparent spinel ceramics without Mg, such as zinc spinel ($ZnAl_2O_4$), are also reported in this class. The transparent $ZnAl_2O_4$ ceramics prepared via SPS at

1260°C exhibited a maximum of 64% in-line transmittance at 550 nm. The resultant ceramics possessed good microwave dielectric properties with a dielectric constant of 8.71, thus making them a suitable material for optical-electronic integration [74].

Recently, some studies have shown an alternative approach for the fabrication of transparent ceramics from glasses via a simple thermal annealing process at moderate temperatures in air. The various examples of transparent ceramics in this category that have been successfully fabricated from glass with the same composition by full crystallization are $BaAl_4O_7$, $Sr_3Al_2O_6$, $Sr_{1+x/2}Al_{2+x}Si_{2-x}O_8$ ($0 \leq x \leq 0.4$), $Sr_{1-x/2}Al_{2-x}Si_xO_4$ ($x=0.2$, 0.4, and 0.5), and $Ln_{1+x}Sr_{1-x}Ga_3O_{7+\delta}$ (Ln = Eu, Gd, or Tb) [1]. This new approach of synthesis is easy, economical, and can enable large-scale production of ceramics. Transparent $BaAl_4O_7$ ceramic is one of the earliest examples prepared from a glass of identical composition. Another addition to this type is transparent $Sr_3Al_2O_6$ ceramics, obtained from a glass of the same composition via thermal annealing at 840°C for 5 h. The resultant ceramics exhibited high density and transparency in visible and IR regions up to 6 μm with grain size at a microlevel. The ceramics were also doped with Eu^{3+}, Er^{3+}, Ho^{3+}, and Ce^{3+} to investigate their applications in photonics. The $Sr_3Al_2O_6:Eu^{3+}$ and $Sr_3Al_2O_6:Er^{3+}$ ceramics displayed luminescence with strong emissions in the visible and IR ranges, respectively. Furthermore, $Sr_3Al_2O_6:Ce^{3+}$ ceramics showed scintillation behavior under X-ray excitation. The study was also extended to obtain $Sr_3Ga_2O_6$ analogs but its hygroscopicity limited the transparency of the ceramics under normal conditions [75]. Another recent study has revealed the fabrication of transparent $Sr_{1-x/2}Al_{2-x}Si_xO_4$ ($x=0.2$, 0.4, and 0.5) ceramics doped with Eu using full and congruent crystallization of glasses possessing the same composition. The glasses were synthesized via aerodynamic levitation and laser-heating techniques. Highly transparent ceramics (visible region) were obtained from the glass samples with different values of x, i.e., 0.2, 0.4, and 0.5 via thermal annealing in the air for 6–7 h at 930°C, 985°C, and 1010°C, respectively [76]. A new class of completely dense and fairly transparent nonstoichiometric melilite ceramics, $Ln_{1+x}Sr_{1-x}Ga_3O_{7+\delta}$ (Ln = Eu, Gd, or Tb), which can conduct oxide ions, were produced from glass by full crystallization. Highly pure $SrCO_3$, Ln_2O_3, and Ga_2O_3 powders were transformed into glasses by the combination of an aerodynamic levitator and the CO_2 laser heating method. Subsequently, the obtained glass beads were fully crystallized via thermal annealing in air at 720°C for nearly 2 h. The resultant ceramics exhibited bulk oxide ion conductivity of more than 0.02 S cm^{-1} at 500°C. So, this study can open directions for the synthesis of new transparent and completely dense solid-state electrolytes [77].

Mullite, i.e., aluminum silicate ($3Al_2O_3{\cdot}2SiO_2/Al_6Si_2O_{13}$), is another popular transparent ceramic owing to its exceptional properties such as high-temperature strength, high resistance to thermal shock, low thermal expansion coefficient, a low dielectric constant of 6.7, and low thermal conductivity. It finds its application as substrates in electronic packaging, window materials at high temperatures, and also as the substrates of Si thin-film solar cells [78]. Almost 50 years ago, translucent as well as fully dense mullite was fabricated via HP at 1500–1650°C under 30–50 MPa [79]. However, with time and progress in sintering technologies, transparent mullite was produced using practically all the advanced sintering methods. For example, IR transparent mullite ceramic with an in-line transmittance of 83%–88% at 2.5–4 μm with a grain size of 200 nm was fabricated via SPS at 1350°C. The nanopowders of mullite were prepared by the combination of sol-gel and pulse current heating method. The final ceramic product exhibited high values of hardness (17.82 GPa) and toughness (3.6 MPa m$^{1/2}$) due to its high density and fine grain size [80].

3.2 Nonoxide ceramics

Transparent ceramics based on nonoxides have been also fabricated using various advanced sintering technologies. AlON, MgAlON, LiAlON, AlN, SiAlON, Si_3N_4, and fluorides are the best-known examples in the class of transparent nonoxide ceramics. But the examples in this category are limited owing to the restricted options available for starting materials that possess desirable properties required for the production of transparent ceramics, as mentioned in Section 2.

3.2.1 Aluminum oxynitride (AlON) and aluminum nitride (AlN)

Aluminum oxynitride (AlON) is represented as $Al_{(64+x)/3}O_{32-x}N_x$ ($2 \leq x \leq 5$); it possesses a defective cubic spinel crystal structure. When a small amount of nitrogen is added to Al_2O_3, its rhombohedral structure changes to a cubic spinel-structured AlON. AlON exhibits high optical transmittance, ≥80% in the near-UV to NIR region, and high mechanical strength. Owing to the fusion of these two properties, it finds its application in defense, such as transparent armors, missile domes, and IR windows, and in other commercial applications, such as semiconductor processing and laser windows. Initially, AlON ceramics were fabricated using reaction sintering of precursor materials, i.e., Al_2O_3 and AlN. Later on, the fabrication of transparent AlON ceramics has also been realized using all the aforesaid advanced sintering technologies [17]. Even a combination of sintering processes was also employed in many studies to obtain a highly transparent desired product. To synthesize AlON ceramics, first AlON powder needs to be prepared for which different methods such as the chemical reaction between Al_2O_3 and AlN, SSR, Al_2O_3 carbothermal reduction, high-temperature fast self-propagating combustion, and plasma arc synthesis are being employed in various studies. Among all, SSR and carbothermal reduction are most widely applicable due to their simple processing and cost-effectiveness. However, it is still difficult to obtain pure AlON powders via presently available methods. After the formation of AlON powders, sintering is done in an N_2 atmosphere at relatively high temperatures (>1850°C) for a long duration, normally between 20 and 100 h [1]. Sintering aids such as MgO and Y_2O_3 are also used to produce transparent AlON ceramics. Sometimes, hybrid methods such as SSR and carbothermal reduction are also used to obtain high-quality precursor powders of AlON that led to the production of highly transparent AlON ceramics. For example, AlON powders were prepared using precursor materials, commercial γ-Al_2O_3 and carbon black, via carbothermal reduction and nitridation method. Subsequently, transparent AlON ceramics were obtained using pressure-less sintering and sintering aids (0.25 wt% MgO and 0.04 wt% Y_2O_3) under an N_2 atmosphere. In the study, ball milling time effects on morphology and particle size of AlON powder and the microstructure and optical transmittance of ceramic were also investigated. Single-phase AlON powder was prepared via calcination of the γ-Al_2O_3/C mixture at 1550°C for 1 h and subsequent thermal treatment at 1750°C for 2 h. The resultant powder was ball milled for either 12 or 24 h and then sintered at 1850°C for 6 h to obtain two final samples of AlON ceramics. The sample that was ball milled for 24 h exhibited better in-line transmittance of >80% from visible to IR region [81]. In another study, the fabrication of highly transparent MgAlON with transmittance of 86.59% at 3700 nm and 82.37% at 600 nm, by following a similar synthesis strategy and sintering process, was realized [82]. Using different synthesis processes, viz., reaction sintering of AlN, Al_2O_3, and $LiAl_5O_8$ powders at 1750°C for 20 h and post-HIP at 1850°C for 3 h under180 MPa, a highly transparent LiAlON ceramic of size Φ57 mm × 6 mm was fabricated as shown in Fig. 7. The resultant AlON ceramics displayed high optical in-line transmittance with a maximum value of nearly 85.5% from the visible to MIR region. Furthermore, the final AlON ceramic possessed a flexural strength of 303 MPa and a Vickers hardness of 15.0 GPa [83].

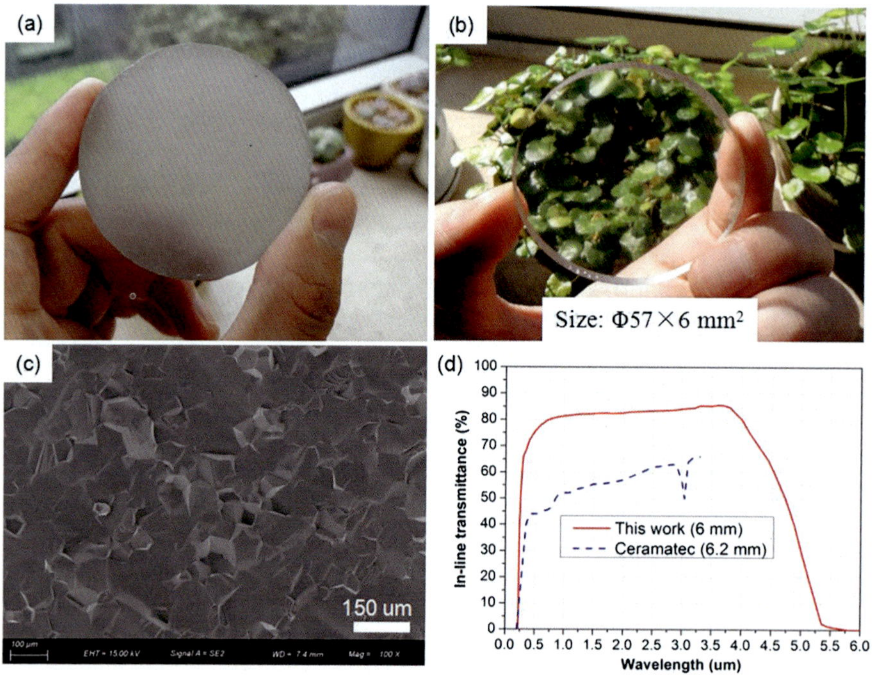

Fig. 7 LiAlON ceramics of (A) reaction sintering at 1750°C for 20 h and (B) post-HIP at 1850°C for 3 h; (C) microstructure and (D) transmittance of the ceramic sample after HIP. Reprinted with permission from reference R.S. Zhang, Y.Z. Wang, M. Tian, H. Wang, Highly transparent LiAlON ceramic prepared by reaction sintering and post hot isostatic pressing, J. Eur. Ceram. Soc. 38 (2018) 5252–5256. https://doi.org/10.1016/j.jeurceramsoc.2018.07.053. Copyright © 2018, Elsevier.

Another nonoxide ceramic, AlN, possesses high-temperature stability in inert atmospheres and high thermal conductivity. Owing to its properties, it can be used in electronics, optics, lighting, etc. But presently, highly transparent AlN ceramic fabrication is a challenge. Only AlN ceramics with low transmittance have been reported in the literature. However, a recent study reported the fabrication of fairly transparent AlN ceramics by HP sintering from AlN nanopowders, prepared via a hydrothermal process [84]. The effects of sintering additives were also investigated on the microstructure and composition of ceramics, and it was found that fluoride additives (CaF_2 and MgF_2) are better than oxide additives (Y_2O_3). In the future, this study can be used to develop transparent AlN ceramics using suitable sintering aids.

3.2.2 SiAlON and silicon nitride (Si_3N_4)

SiAlON ceramics comprise Si, Al, O, and N elements. These ceramics are solid solutions of Si_3N_4. But structurally, some of the Si—N bonds are replaced by Al—N and Al—O bonds. These ceramics are a distinct class of refractory materials at high temperatures. Furthermore, these ceramics possess high strength at ambient as well as high temperature, high resistance to thermal shock, high wear resistance, high chemical stability, and low thermal expansion. Due to remarkable properties and advanced sintering techniques, SiAlON ceramics have materialized as a novel class of engineered ceramics that can be highly transparent in the near future [1]. These ceramics can also act as phosphors when doped with heavy metal ions. For example, transparent yellow SiAlON ceramics doped with Eu_2O_3 were fabricated via HP sintering that exhibited yellow luminescence at 570nm on excitation in UV to a blue light region [16]. When a small amount of Y_2O_3 was also added to the fabricated thin yellow phosphor along with Eu_2O_3, the luminescence intensity increased significantly. The developed SiAlON phosphor also displayed good mechanical properties and can be apt for remote phosphor.

Transparent ceramics based on Si_3N_4 will be applicable in making harder as well as tougher optical windows that are even stable under severe conditions. This is the third-hardest material after diamond and cubic boron nitride. Furthermore, the metastability of cubic Si_3N_4 in the air at high temperatures (up to 1400°C) is much higher compared to diamond and cubic boron nitride. Therefore, the fabrication of transparent Si_3N_4 ceramics is a hot research topic, and success was achieved in a recent study. For example, a highly transparent cubic Si_3N_4 ceramic (the hardest spinel ceramic) was fabricated via sintering under very high pressures (>13GPa). The resultant ceramic was optically transparent between 400 and 800nm, which is below its band-gap energy, making it applicable for protective windows of small devices [18].

3.2.3 Fluorides

Fluorides have a characteristic fluorite structure possessing a cubic crystal structure with the general formula, MF_2. Transparent ceramics based on fluoride have many advantages over oxides, such as a low sintering temperature, a wide range of optical transmittance up to λ of 700 nm, low phonon frequency cut-offs, and low linear/nonlinear refractive indexes. Out of the various transparent fluoride ceramics, CaF_2 is the most widely studied and reported in the open literature [22]. The first example of transparent fluoride ceramic, Dy^{2+}:CaF_2, was reported in 1964 via the HP method, applicable to solid lasers [85]. The CaF_2 ceramics have been doped with different lanthanides (e.g., Eu, Yb, Er, Nd, Y, and Ho) to investigate their optical properties and applications, especially in a solid-state laser. For the fabrication of transparent CaF_2 ceramics doped with lanthanide ions, HP is the most widely employed sintering method. After HP, SPS is also popular for transparent CaF_2 ceramic synthesis due to the low sintering temperature. However, samples obtained from SPS exhibited slightly less optical transmittance compared to HP because of the contamination in the former method [1]. The contamination of the ceramic sample can be prevented by separating the graphite die and green body sample while sintering. For example, Mo foil utilization to separate the ceramic sample from the graphite die successfully decreased the contamination and improved the transparency as well as the average grain size of the fabricated transparent CaF_2 ceramics via SPS [19]. In particular, the optical in-line transmittance of the resultant ceramic after using Mo foil increased from 8% to 54% at 300 nm and 63% to 86% at 1100 nm. Furthermore, by using Mo foil during sintering, the average grain size of the ceramic also increased to 260 μm from 16 μm. Besides CaF_2, other alkaline earth metal-based transparent fluoride ceramics, such as MgF_2, SrF_2, and BaF_2, are also fabricated in many studies. These ceramics are also doped with lanthanide ions (e.g., Ce^{3+}, Nd^{3+}, Yb^{3+}, Pr^{3+}, and Gd^{3+}) to investigate their various applications in scintillation, luminescence, optical windows, etc. [22]. Recently, high Ce^{3+}-doped BaF_2 transparent ceramics were fabricated via the high-vacuum HP method [86]. The BaF_2 ceramics were doped with three different concentrations, 1, 10, and 50 at% Ce^{3+}, and corresponding transmittances were 77.8%, 73.2%, and 79.5% at 550 nm, respectively, as shown in Fig. 8. The sample doped with 50 at% of Ce^{3+} had a composition of $BaCeF_5$, the first time mentioned in the literature. All the fabricated ceramic samples were highly dense, and the average grain size was less than 50 μm.

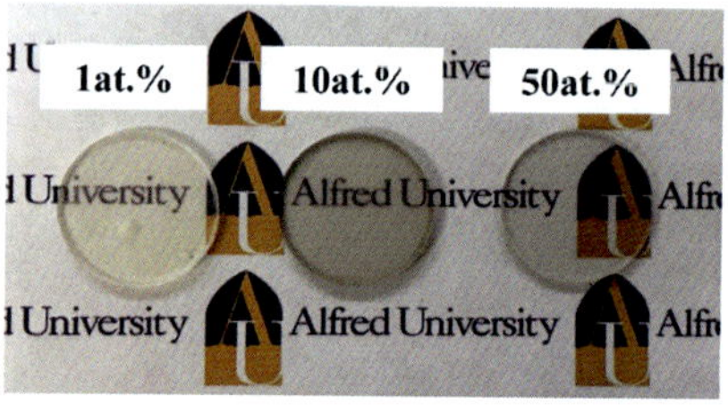

Fig. 8 Photographs of the transparent Ce: BaF_2 ceramics with different concentrations of Ce^{3+}. Reprinted with permission from reference X.Q. Chen, Y.Q. Wu, High concentration Ce3+ doped BaF2 transparent ceramics, J. Alloys Compd. 817 (2020) 153075. https://doi.org/10.1016/j.jallcom.2019.153075. Copyright © 2019, Elsevier.

4 Applications of transparent ceramics

Transparent ceramics are used for various applications (Fig. 9) due to several advantages over their transparent counterparts, especially single crystals. One of the earliest applications of these ceramics was in solid-state lasers due to the feasibility of uniform and heavy distribution of laser-active functional substances (dopants) within the microstructure. These ceramics also offer a gain medium with a high-quality beam for fiber lasers, and composite laser media can also be made using complex structures. The transparent ceramics allow easy and low-cost commercial mass production of high-power lasers compared to single crystals, which further makes them significant in this arena. The practical ceramic laser was first developed in 1995 using Nd:YAG transparent ceramics [9]. An oscillation threshold of 309 mW and a slope efficiency of 28% were obtained, equivalent to 0.9 at% Nd:YAG single crystal synthesized via Czochralski process [13]. This study presented the potential application of ceramics in solid-state lasers, and significant research has been started for the development of high-power and highly focused lasers. Single-element doped lasers were developed using various transparent ceramics, especially YAG and sesquioxides. Successful doping was done with Nd, Ho, Er, Tm, and Yb, which resulted in the application of transparent ceramics in solid-state lasers. With the advancement in synthesis technologies, composite ceramic lasers were also developed that further improved the laser performance with respect to efficiency, beam quality, and output energy [1]. The composite ceramic lasers fall into five classes: layered structures, waveguide types, cylindrical structures, gradient types, and core-shell fibers [87].

Transparent ceramics have displayed potential applications in lighting as envelopes for discharged lamps owing to their unique combination of exceptionally high thermal and mechanical properties. These ceramics can withstand high working

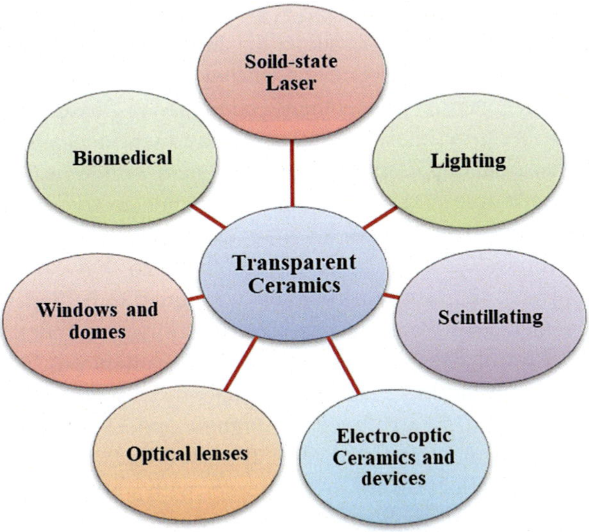

Fig. 9 Potential applications of transparent ceramics in various domains.

temperatures and thus replace the conventional quartz material used in lighting. Primarily, translucent polycrystalline Al_2O_3 ceramics, coarse grain and doped with MgO, were employed in discharge lamps by the lighting industry in cylindrical and spherical form tubes. Some other transparent ceramics such as AlON, YAG, and rare earth oxide (Y_2O_3 and Dy_2O_3) are also employed for lighting applications. Moreover, transparent ceramic-based solid-state phosphors have been synthesized and presented as potential applications in lighting as briefly reviewed by Kong and co-workers [1].

Transparent ceramics are used as scintillators that are luminescent materials capable of detecting radiations such as X-rays and gamma rays. Scintillators act as prime radiation detectors because of the emission of visible light on excitation by high-energy photons. Transparent ceramics are better scintillators than single crystals due to the presence of a higher and uniform concentration of the activators within the microstructure. Transparent ceramic scintillators have applications in spectrometers, research in high-energy physics and astrophysics, medical imaging equipment, oil-well logging, land safety monitoring, etc. Ceramic scintillators are very sensitive as their detection accuracy is $<10^{-3}$. Depending on the application field, different transparent ceramics are chosen. For example, Lu_2O_3, Eu:$(Y,Gd)_2O_3$, and $Lu_3Ga_{5-x}In_xO_{12}$ are employed as X-ray sensors in medical facilities [1].

Transparent electro-optic ceramics such as ferroelectric PLZT ceramics are used for designing various devices, which include polarization controllers, tunable optical filters, variable optical attenuator, sinusoidal filters, and Q-switches. When ferroelectric PLZT ceramics are polarized via an external electric field, they possess intrinsic and extrinsic optical uniaxial anisotropy at the microscopic and macroscopic scale, respectively. Due to this property, these are the characteristic electro-optic materials. In particular, Dy-codoped PLZT transparent ceramics exhibited electro-optic properties. The resultant ceramics (PLDZT) displayed excellent performance in variable optical attenuators and can be a competitive option for modern optical communication [64].

Currently, compact and small-size lenses are in high demand for digital cameras as well as mobile phones. Transparent ceramics that belong to $A_2B_2O_7$ and A_3BO_7 classes can be used as optical lenses in these devices as they possess large Abbe numbers and a high refractive index, generally greater than 2, compared to glasses that is between 1.50 and 1.85 [88]. Furthermore, these ceramics display high mechanical strength compared to their transparent counterparts, especially glass, which further enhances their application as scratch-free and tough material candidates for smartphone screens. In a recent study, optical properties for a series of transparent Y_3NbO_7 ceramics, $Y_{1-x}Nb_xO_{1.5+x}$ with $x=0.20$, 0.22, 0.24, 0.25, and 0.26, have been reported [88]. The refractive index of the resultant ceramics was found to be dependent on the Nb content (x); as x increased from 0.20 to 0.26, the refractive index increased from 2.04 to 2.10 at 587.6nm. The Abbe number for the ceramics was >40 and exhibited a decreasing trend with the increment in Nb content, x. Specifically, a 1-mm-thick ceramic sample with $x=0.24$ exhibited the highest optical in-line transmittance of 62% at 587.6nm and 76% at 2000nm. Thus, the $Y_{1-x}Nb_xO_{1.5+x}$ transparent ceramic series can be promising materials for optical lenses with a high refractive index.

Transparent ceramics find applications in making windows and domes for transparent armors owing to their high mechanical strength and toughness. Out of the various types of transparent ceramics that have been discussed in Section 3, AlON and $MgAl_2O_4$ are the most potential candidates for such applications. Transparent armors are frequently employed for personnel protection as face shields, screens, military vehicle windows, armored car windows, spacecraft windows, and lookdown windows in aircraft [89]. Furthermore, large windows and domes are required for multimode weapon systems that can withstand severe environments. Currently, the transparent glass windows are very thick, usually greater than 5 in., to resist the impact of multihit ballistics. Therefore, these glass windows cannot meet the weight limits without any compromise in ballistic resistance. On the other hand, transparent ceramic windows can overcome the weight limit issue along with equivalent or better resistance to multihit ballistics. Presently, transparent windows based on single crystals (sapphire) are used most widely, compared to transparent ceramic windows. But with more advances in fabrication processes and cost-effectiveness, transparent ceramics can potentially replace single crystals, such as sapphire. In a systematic study, transparent $MgAl_2O_4$ and AlON ceramics displayed higher slow crack resistance compared to a traditional material, fused silica [90]. The results obtained from the study proposed that these two ceramics can be used in spacecraft windows of the next generation that are lightweight, a crucial parameter for spacecraft.

Besides numerous applications in different areas, transparent ceramics, especially $MgAl_2O_4$ and AlON, exhibited their use in the biomedical field as well. Owing to the aforesaid exceptional properties of the $MgAl_2O_4$ and AlON transparent ceramics, these two are potentially viable options for such investigations. In a study, transparent $MgAl_2O_4$ and AlON ceramics were investigated for implant applications with prospective wear-resistant load bearing [91]. The in vitro cytotoxicity, tribological behavior, and cell-material interactions of these ceramics were studied. The high hardness of these ceramics, between 1334 and 1543 HV, exhibited an in vitro wear rate in the range of $5.3\text{–}10 \times 10^{-6}\ mm^3\ Nm^{-1}$ against an Al_2O_3 ball at a load of 20 N. Furthermore, the tested ceramics displayed no toxicity and better in vitro material-cell interactions, which makes them more biocompatible compared to commercially pure bio-inert Ti as a control. In another study, the feasibility of transparent $MgAl_2O_4$ ceramics as esthetic orthodontic brackets was assessed [92]. The resultant ceramic samples were tested for various physical and chemical properties such as density, hardness, optical transmission, flexural strength, fracture toughness, chemical composition, and phases. Furthermore, using the in vitro cell line tests, the biocompatibility of the ceramics was evaluated for cytotoxicity, genotoxicity, and apoptosis. The obtained results were in favor of fabricating orthodontic brackets using transparent $MgAl_2O_4$ ceramics due to essential mechanical, optical, physicochemical, and exceptional biocompatibility.

5 Conclusion

Transparent ceramics are the next-generation materials that find their applications in diverse arenas. The prerequisite for a ceramic to be transparent is a highly symmetrical crystal structure, namely cubic, tetragonal, and hexagonal. Furthermore, high-quality raw materials, powder, fine microstructure, an advanced sintering method, and a post-treatment process ensure the high transparency of the ceramics. Presently, transparent ceramics are fabricated using various advanced sintering techniques, such as vacuum sintering, microwave sintering, HP, HIP, and SPS. Vacuum sintering is the most widely used method for transparent ceramic fabrication. SPS and microwave sintering are relatively new techniques and require low sintering temperature as well as short sintering time. However, SPS has contamination issues during sintering, similar to HP. To achieve the maximum possible density and porosity removal, the CIP method is also adopted for green bodies in most of the studies. A variety of transparent ceramics based on oxides and nonoxides are synthesized worldwide by several research groups. The sintering aids are also employed to further enhance the optical transmittance and mechanical strength of the resultant ceramics. Sometimes, the combination of sintering aids is also used for the desired product. Depending on the application, especially for solid-state lasers, scintillation, and electro-optic devices, the transparent ceramics are doped with different lanthanide ions. Transparent ceramics are also used in lighting, optical lenses and windows, and domes of transparent armor. Solid-state phosphors based on oxide ceramics are fabricated worldwide. The biomedical application of transparent ceramics is in its infancy, but with more research in the future, these materials can become viable choices due to their outstanding properties. These materials can potentially replace glass as smartphone screens owing to their exceptional mechanical strength and toughness. Transparent ceramics are cost-effective and can be produced in large volumes in the desired shape compared to their counterparts, single crystals. However, more research is needed in this area so that fabrication costs can be further reduced.

References

[1] Z. Xiao, S. Yu, Y. Li, S. Ruan, L.B. Kong, Q. Huang, et al., Materials development and potential applications of transparent ceramics: a review, Mater. Sci. Eng.: R: Rep. 139 (2020) 100518, https://doi.org/10.1016/j.mser.2019.100518.

[2] GE, GE has transparent ceramic, Chem. Eng. News 44 (43) (1966) 38, https://doi.org/10.1021/cen-v044n043.p038a.

[3] H. Shahbazi, M. Tataei, M.H. Enayati, A. Shafeiey, M.A. Malekabadi, Structure-transmittance relationship in transparent ceramics, J. Alloys Compd. 785 (2019) 260–285, https://doi.org/10.1016/j.jallcom.2019.01.124.

[4] J. Hostaša, F. Picelli, S. Hříbalová, V. Nečina, Sintering aids, their role and behaviour in the production of transparent ceramics, Open Ceram. 7 (2021) 100137, https://doi.org/10.1016/j.oceram.2021.100137.

[5] R.L. Coble, Sintering crystalline solids. I. Intermediate and final state diffusion models, J. Appl. Phys. 32 (1961) 787–792, https://doi.org/10.1063/1.1736107.

[6] R.L. Coble, Sintering crystalline solids. II. Experimental test of diffusion models in powder compacts, J. Appl. Phys. 32 (1961) 793–799, https://doi.org/10.1063/1.1736108.

[7] A. Krell, J. Klimke, T. Hutzler, Advanced spinel and sub-μm Al_2O_3 for transparent armour applications, J. Eur. Ceram. Soc. 29 (2009) 275–281, https://doi.org/10.1016/j.jeurceramsoc.2008.03.024.

[8] M. Stuer, Z. Zhao, U. Aschauer, P. Bowen, Transparent polycrystalline alumina using spark plasma sintering: effect of Mg, Y and La doping, J. Eur. Ceram. Soc. 30 (2010) 1335–1343, https://doi.org/10.1016/j.jeurceramsoc.2009.12.001.

[9] S.F. Wang, J. Zhang, D.W. Luo, F. Gu, D.Y. Tang, Z.L. Dong, et al., Transparent ceramics: processing, materials and applications, Prog. Solid State Chem. 41 (1–2) (2013) 20–54, https://doi.org/10.1016/j.progsolidstchem.2012.12.002.

[10] R. Chaim, Z.J. Shen, M. Nygren, Transparent nanocrystalline MgO by rapid and low temperature spark plasma sintering, J. Mater. Res. 19 (2004) 2527–2531, https://doi.org/10.1557/JMR.2004.0334.

[11] Y. Fang, D. Agrawal, G. Skandan, M. Jain, Fabrication of translucent MgO ceramics using nanopowders, Mater. Lett. 58 (2004) 551–554, https://doi.org/10.1016/S0167-577X(03)00560-3.

[12] S. Ghanizadeh, S. Grasso, P. Ramanujam, B. Vaidhyanathan, J. Binner, P. Brown, et al., Improved transparency and hardness in α-alumina ceramics fabricated by high-pressure SPS of nanopowders, Ceram. Int. 43 (2017) 275–281, https://doi.org/10.1016/j.ceramint.2016.09.150.

[13] A. Ikesue, T. Kinoshita, K. Kamata, K. Yoshida, Fabrication and optical properties of high-performance polycrystalline Nd:YAG ceramics for solid-state lasers, J. Am. Ceram. Soc. 78 (1995) 1033–1040, https://doi.org/10.1111/j.1151-2916.1995.tb08433.x.

[14] H. Shahbazi, H. Shokrollahi, M. Tataei, Gel-casting of transparent magnesium aluminate spinel ceramics fabricated by spark plasma sintering (SPS), Ceram. Int. 44 (2018) 4955–4960, https://doi.org/10.1016/j.ceramint.2017.12.088.

[15] L.L. Zhu, Y.J. Park, L. Gan, H.N. Kim, J.W. Ko, H.D. Kim, Fabrication and characterization of highly transparent Y_2O_3 ceramics by hybrid sintering: a combination of hot pressing and a subsequent HIP treatment, J. Eur. Ceram. Soc. 39 (2018) 3255–3260, https://doi.org/10.1016/j.jeurceramsoc.2018.03.022.

[16] B. Joshi, Y.K. Kshetri, G. Gyawali, S.W. Lee, Transparent Mg–α/β-Sialon:Eu2+ ceramics as a yellow phosphor for pc-WLED, J. Alloys Compd. 631 (2015) 38–45, https://doi.org/10.1016/j.jallcom.2015.01.081.

[17] J.W. McCauley, P. Patel, M.W. Chen, G. Gilde, E. Strassburger, B. Paliwal, et al., AlON: a brief history of its emergence and evolution, J. Eur. Ceram. Soc. 29 (2009) 223–236, https://doi.org/10.1016/j.jeurceramsoc.2008.03.046.

[18] N. Nishiyama, R. Ishikawa, H. Ohfuji, H. Marquardt, A. Kurnosov, T. Taniguchi, et al., Transparent polycrystalline cubic silicon nitride, Sci. Rep. 7 (2017) 44755, https://doi.org/10.1038/srep44755.

[19] P. Wang, M.J. Yang, S. Zhang, R. Tu, T. Goto, L.M. Zhang, Suppression of carbon contamination in SPSed CaF_2 transparent ceramics by Mo foil, J. Eur. Ceram. Soc. 37 (2017) 4103–4107, https://doi.org/10.1016/j.jeurceramsoc.2017.04.070.

[20] A. Goldstein, A. Krell, Z. Burshtein, Transparent Ceramics: Materials, Engineering, and Applications, John Wiley and Sons, 2020, pp. 11–50, ISBN: 978-1-119-42955-5.

[21] M.N. Rahaman, Ceramic Processing and Sintering, second ed., CRC Press, New York, 2003.

[22] Y. He, K. Liu, B. Xiang, C. Zhou, L. Zhang, G. Liu, et al., An overview on transparent ceramics with pyrochlore and fluorite structures, J. Adv. Dielectr. 10 (3) (2020) 2030001, https://doi.org/10.1142/S2010135X20300017.

[23] A. Ikesue, Y. Aung, Ceramic laser materials, Nature Photon 2 (2008) 721–727, https://doi.org/10.1038/nphoton.2008.243.

[24] T. Ashikaga, B.N. Kim, H. Kiyono, T.S. Suzuki, Effect of crystallographic orientation on transparency of alumina prepared using magnetic alignment and SPS, J. Eur. Ceram. Soc. 38 (2018) 2735–2741, https://doi.org/10.1016/j.jeurceramsoc.2018.02.006.

[25] A. Ikesue, K. Kamata, K. Yoshida, Synthesis of transparent Nd-doped HfO_2-Y_2O_3 ceramics using HIP, J. Am. Ceram. Soc. 79 (1996) 359–364, https://doi.org/10.1111/j.1151-2916.1996.tb08129.x.

[26] K. Serivalsatit, J. Ballato, Submicrometer grain-sized transparent erbium-doped Scandia ceramics, J. Am. Ceram. Soc. 93 (2010) 3657–3662, https://doi.org/10.1111/j.1551-2916.2010.03954.x.

[27] S. Zamir, Solubility limit of Si in YAG at 1700 C in vacuum, J. Eur. Ceram. Soc. 37 (2017) 243–248, https://doi.org/10.1016/j.jeurceramsoc.2016.08.010.

[28] K.A. Berry, M.P. Harmer, Effect of MgO solute on microstructure development in Al_2O_3, J. Am. Ceram. Soc. 69 (1986) 143–149, https://doi.org/10.1111/j.1151- 2916.1986.tb04719.x.

[29] T.Y. Zhou, L. Zhang, S. Wei, L.X. Wang, H. Yang, Z.X. Fu, et al., MgO assisted densification of highly transparent YAG ceramics and their microstructural evolution, J. Eur. Ceram. Soc. 38 (2018) 687–693, https://doi.org/10.1016/j.jeurceramsoc.2017.09.017.

[30] L. Pauling, S.B. Hendricks, The crystal structures of hematite and corundum, J. Am. Chem. Soc. 47 (1925) 781–790.

[31] L.B. Kong, Y. Huang, W. Que, T. Zhang, S. Li, J. Zhang, et al., Transparent ceramic materials, in: Transparent Ceramics. Topics in Mining, Metallurgy and Materials Engineering, Springer, Cham, 2015, pp. 29–91, https://doi.org/10.1007/978-3-319-18956-7_2.

[32] H. Mizuta, K. Oda, Y. Shibasaki, M. Maeda, M. Machida, K. Ohshima, Preparation of high strength and translucent alumina by hot isostatic pressing, J. Am. Ceram. Soc. 75 (1992) 469–473, https://doi.org/10.1111/j.1151-2916.1992.tb08203.x.

[33] A. Krell, P. Blank, H.W. Ma, T. Hutzler, M.P.B. van Bruggen, R. Apetz, Transparent sintered corundum with high hardness and strength, J. Am. Ceram. Soc. 86 (2003) 12–18, https://doi.org/10.1111/j.1151-2916.2003.tb03270.x.

[34] A.A. Kachaev, D.V. Grashchenkov, Y.E. Lebedeva, S.S. Solntsev, O.L. Khasanov, Optically transparent ceramic (review), Glas. Ceram. 73 (2016) 117–123, https://doi.org/10.1007/s10717-016-9838-3.

[35] T. Ikegami, S.I. Matsuda, H. Suzuki, Effect of halide dopants on fabrication of transparent polycrystalline MgO, J. Am. Ceram. Soc. 57 (1974) 507, https://doi.org/10.1111/j.1151-2916.1974.tb11411.x.

[36] P.E. Hart, J.A. Pask, Effect of LiF on creep of MgO, J. Am. Ceram. Soc. 54 (1971) 315–316.

[37] K. Itatani, T. Tsujimoto, A. Kishimoto, Thermal and optical properties of transparent magnesium oxide ceramics fabricated by post hot-isostatic pressing, J. Eur. Ceram. Soc. 26 (2006) 639–645, https://doi.org/10.1016/j.jeurceramsoc.2005.06.011.

[38] D.Y. Chen, E.H. Jordan, M. Gell, Pressureless sintering of translucent MgO ceramics, Scr. Mater. 59 (2008) 757–759, https://doi.org/10.1016/j.scriptamat.2008.06.007.

[39] J.E. Alaniz, F.G. Perez-Gutierrez, G. Aguilar, J.E. Garay, Optical properties of transparent nanocrystalline yttria stabilized zirconia, Opt. Mater. 32 (1) (2009) 62–68, https://doi.org/10.1016/j.optmat.2009.06.004.

[40] K. Tsukuma, I. Yamashita, T. Kusunose, Transparent 8 mol% Y_2O_3-ZrO_2 (8Y) ceramics, J. Am. Ceram. Soc. 91 (2008) 813–818, https://doi.org/10.1111/j.1551-2916.2007.02202.x.

[41] H.B. Zhang, B.N. Kim, K. Morita, H. Yoshida, J.H. Lim, K. Hiraga, Optimization of high-pressure sintering of transparent zirconia with nano-sized grains, J. Alloys Compd. 508 (2010) 196–199, https://doi.org/10.1016/j.jallcom.2010.08.045.

[42] H. Zhang, B.-N. Kim, K. Morita, H.Y.K. Hiraga, Y. Sakka, Effect of sintering temperature on optical properties and microstructure of translucent zirconia prepared by high-pressure spark plasma sintering, Sci. Technol. Adv. Mater. 12 (2011) 055003, https://doi.org/10.1088/1468-6996/12/5/055003.

[43] J. Klimke, M. Trunec, A. Krell, Transparent tetragonal yttria-stabilized zirconia ceramics: influence of scattering caused by birefringence, J. Am. Ceram. Soc. 94 (2011) 1850–1858,- https://doi.org/10.1111/j.1551-2916.2010.04322.x.
[44] M. Trunec, Z. Chlup, Higher fracture toughness of tetragonal zirconia ceramics through nanocrystalline structure, Scripta Mater. 61 (2009) 56–59, https://doi.org/10.1016/j.scriptamat.2009.03.019.
[45] Y.H. Huang, D.L. Jiang, J.X. Zhang, Q.L. Lin, Z.G. Huang, Sintering of transparent yttria ceramics in oxygen atmosphere, J. Am. Ceram. Soc. 93 (2010) 2964–2967, https://doi.org/10.1111/j.1551-2916.2010.03940.x.
[46] L.L. Zhu, Y.J. Park, L. Gan, S.I. Go, H.N. Kim, J.M. Kim, et al., Fabrication and characterization of highly transparent Er:Y_2O_3 ceramics with ZrO_2 and La_2O_3 additives, Ceram. Int. 43 (2017) 13127–13132, https://doi.org/10.1016/j.ceramint.2017.07.004.
[47] Y. Wang, B. Lu, X. Sun, T. Sun, H. Xu, Synthesis of nanocrystalline Sc_2O_3 powder and fabrication of transparent Sc_2O_3 ceramics, Adv. Appl. Ceram. 110 (2011) 95–98, https://doi.org/10.1179/1743676110Y.0000000006.
[48] Q. Liu, Z.F. Dai, D. Hreniak, S.S. Li, W.B. Liu, W. Wang, et al., Fabrication of Yb:Sc_2O_3 laser ceramics by vacuum sintering co-precipitated nano-powders, Opt. Mater. 72 (2017) 482–490, https://doi.org/10.1016/j.optmat.2017.05.057.
[49] L.Q. An, A. Ito, T. Goto, Fabrication of transparent lutetium oxide by spark plasma sintering, J. Am. Ceram. Soc. 94 (2011) 695–698, https://doi.org/10.1111/j.1551-2916.2010.04145.x.
[50] C.W. Xu, C.D. Yang, H. Zhang, Y.M. Duan, H.Y. Zhu, D.Y. Tang, et al., Efficient laser operation based on transparent Nd:Lu_2O_3 ceramic fabricated by spark plasma sintering, Opt. Express 24 (2016) 20571–20579, https://doi.org/10.1364/OE.24.020571.
[51] A. Ikesue, I. Furusata, K. Kamata, Fabrication of polycrystalline, transparent YAG ceramics by a solid-state reaction method, J. Am. Ceram. Soc. 78 (1) (1995) 225–228, https://doi.org/10.1111/j.1151-2916.1995.tb08389.x.
[52] R. Boulesteix, L. Bonnet, A. Maitre, L. Chretien, C. Salle, Silica reactivity during reaction-sintering of Nd:YAG transparent ceramics, J. Am. Ceram. Soc. 100 (2017) 945–953, https://doi.org/10.1111/jace.14680.
[53] W.B. Liu, W.X. Zhang, J. Li, H.M. Kou, D. Zhang, Y.B. Pan, Synthesis of Nd:YAG powders leading to transparent ceramics: the effect of MgO dopant, J. Eur. Ceram. Soc. 31 (2011) 653–657, https://doi.org/10.1016/j.jeurceramsoc.2010.10.016.
[54] X.R. Zhang, G.F. Fan, W.Z. Lu, Y.H. Chen, X.F. Ruan, Effect of the spark plasma sintering parameters, LiF additive, and Nd dopant on the microwave dielectric and optical properties of transparent YAG ceramics, J. Eur. Ceram. Soc. 36 (2016) 2767–2772, https://doi.org/10.1016/j.jeurceramsoc.2016.04.029.
[55] A. Goldstein, Correlation between MgAl2O4-spinel structure, processing factors and functional properties of transparent parts (progress review), J. Eur. Ceram. Soc. 32 (2012) 2869–2886, https://doi.org/10.1016/j.jeurceramsoc.2012.02.051.
[56] R.J. Bratton, Translucent sintered $MgAl_2O_4$, J. Am. Ceram. Soc. 57 (1974) 283–286, https://doi.org/10.1111/j.1151-2916.1974.tb10901.x.
[57] L. Esposito, A. Piancastelli, P. Miceli, S. Martelli, A thermodynamic approach to obtaining transparent spinel ($MgAl_2O_4$) by hot pressing, J. Eur. Ceram. Soc. 35 (2015) 651–661, https://doi.org/10.1016/j.jeurceramsoc.2014.09.005.
[58] K. Tsukuma, Transparent $MgAl_2O_4$ spinel ceramics produced by HIP post-sintering, J. Ceram. Soc. Jpn. 114 (2006) 802–806, https://doi.org/10.2109/jcersj.114.802.
[59] S.S. Balabanov, R.P. Yavetskiy, A.V. Belyaev, E.M. Gavrishchuk, V.V. Drobotenko, I.I. Evdokimov, et al., Fabrication of transparent $MgAl_2O_4$ ceramics by hot-pressing of sol-

gel-derived nanopowders, Ceram. Int. 41 (2015) 13366–13371, https://doi.org/10.1016/j.ceramint.2015.07.123.

[60] A. Talimian, V. Pouchly, H.F. El-Maghraby, K. Maca, D. Galusek, Transparent magnesium aluminate spinel: effect of critical temperature in two-stage spark plasma sintering, J. Eur. Ceram. Soc. 40 (6) (2020) 2417–2425, https://doi.org/10.1016/j.jeurceramsoc.2020.02.012.

[61] B. Chu, X. Zhou, K. Ren, B. Neese, M. Lin, Q. Wang, et al., A dielectric polymer with high electric energy density and fast discharge speed, Science 313 (5785) (2006) 334–336, https://doi.org/10.1126/science.1127798.

[62] L.B. Kong, T.S. Zhang, J. Ma, F. Boey, Progress in synthesis of ferroelectric ceramic materials via high-energy mechanochemical technique, Prog. Mater. Sci. 53 (2008) 207–322, https://doi.org/10.1016/j.pmatsci.2007.05.001.

[63] B. Qu, H. Du, Z. Yang, Lead-free relaxor ferroelectric ceramics with high optical transparency and energy storage ability, J. Mater. Chem. C 4 (2016) 1795–1803, https://doi.org/10.1039/C5TC04005A.

[64] X. Zeng, X.Y. He, W.X. Cheng, P.S. Qiu, B. Xia, Effect of Dy substitution on ferroelectric, optical and electro-optic properties of transparent $Pb_{0.90}La_{0.10}(Zr_{0.65}Ti_{0.35})O_3$ ceramics, Ceram. Int. 40 (2014) 6197–6202, https://doi.org/10.1016/j.ceramint.2013.11.074.

[65] G.H. Hertling, Improved hot-pressed electrooptic ceramics in the (Pb,La)(Zr,Ti)O3 system, J. Am. Ceram. Soc. 54 (1971) 303–309, https://doi.org/10.1111/j.1151-2916.1971.tb12296.x.

[66] C. Huang, J.M. Xu, Z. Fang, D. Ai, W. Zhou, L. Zhao, et al., Effect of preparation process on properties of PLZT (9/65/35) transparent ceramics, J. Alloys Compd. 723 (2017) 602–610, https://doi.org/10.1016/j.jallcom.2017.06.271.

[67] Y.J. Wu, J. Li, R. Kimura, N. Uekawa, K. Kakegawa, Effects of preparation conditions on the structural and optical properties of spark plasma-sintered PLZT (8/65/35) ceramics, J. Am. Ceram. Soc. 88 (2005) 3327–3331, https://doi.org/10.1111/j.1551-2916.2005.00601.x.

[68] Y. Abe, K. Kakegawa, H. Ushijima, Y. Watanabe, Y. Sasaki, Fabrication of optically transparent lead lanthanum zirconate titanate ($(Pb,La)(Zr,Ti)O_3$) ceramics by a three-stage-atmosphere-sintering technique, J. Am. Ceram. Soc. 85 (2002) 473–475, https://doi.org/10.1111/j.1151-2916.2002.tb00113.x.

[69] G. Huangfu, H. Xiao, L. Guan, H. Zhong, C. Hu, Z. Shi, et al., Visible or near-infrared light self-powered photodetectors based on transparent ferroelectric ceramics, ACS Appl. Mater. Interfaces 12 (30) (2020) 33950–33959, https://doi.org/10.1021/acsami.0c09991.

[70] Y. Zhang, Z. Song, M. Lv, B. Yang, L. Wang, C. Chen, et al., Comparison of PMN–PT transparent ceramics processed by three different sintering methods, J. Mater. Sci. Mater. Electron. 28 (2017) 15612–15617, https://doi.org/10.1007/s10854-017-7448-7.

[71] Z. Ma, Y. Zhang, C. Lu, Y. Qin, Z. Lv, S. Lu, Synthesis and properties of La-doped PMN–PT transparent ferroelectric ceramics, J. Mater. Sci. Mater. Electron. 29 (2018) 6985–6990, https://doi.org/10.1007/s10854-018-8685-0.

[72] X. Wu, S.B. Lu, K.W. Kwok, Photoluminescence, electro-optic response and piezoelectric properties in pressureless-sintered Er-doped KNN-based transparent ceramics, J. Alloys Compd. 695 (2017) 3573–3578, https://doi.org/10.1016/j.jallcom.2016.11.409.

[73] Z.J. Wang, G.H. Zhou, J. Zhang, X.P. Qin, S.W. Wang, Luminescence properties of Eu^{3+}-doped Lanthanum gadolinium hafnates transparent ceramics, Opt. Mater. 71 (2017) 5–8, https://doi.org/10.1016/j.optmat.2016.05.046.

[74] P. Fu, Z.Y. Wang, Z.D. Lin, Y.Q. Liu, V.A.L. Roy, The microwave dielectric properties of transparent $ZnAl_2O_4$ ceramics fabricated by spark plasma sintering, J. Mater. Sci. Mater. Electron. 28 (2017) 9589–9595, https://doi.org/10.1007/s10854-017-6707-y.

[75] S. Alahrache, K.A. Saghir, S. Chenu, E. Veron, D.D.S. Meneses, A.I. Becerro, et al., Perfectly transparent $Sr_3Al_2O_6$ polycrystalline ceramic elaborated from glass crystallization, Chem. Mater. 25 (2013) 4017–4024, https://doi.org/10.1021/cm401953d.

[76] A.J.F. Carrion, K.A. Saghir, E. Veron, A.I. Becerro, F. Porcher, W. Wisniewsld, et al., Local disorder and tunable luminescence in $Sr_{1-x/2}Al_{2-x}Si_xO_4$ ($0.2 \leq x \leq 0.5$) transparent ceramics, Inorg. Chem. 56 (2017) 14446–14458, https://doi.org/10.1021/acs.inorgchem.7b01881.

[77] M. Boyer, X.Y. Yang, A.J.F. Carrion, Q.C. Wang, E. Veron, C. Genevois, et al., First transparent oxide ion conducting ceramics synthesized by full crystallization from glass, J. Mater. Chem. A 6 (2018) 5276–5289, https://doi.org/10.1039/C7TA07621E.

[78] H. Schneider, J. Schreuer, B. Hildmann, Structure and properties of mullite—a review, J. Eur. Ceram. Soc. 28 (2008) 329–344, https://doi.org/10.1016/j.jeurceramsoc.2007.03.017.

[79] K.S. Mazdiyasni, L.M. Brown, Synthesis and mechanical properties of stoichiometric aluminum silicate (mullite), J. Am. Ceram. Soc. 55 (1972) 548–552, https://doi.org/10.1111/j.1151-2916.1972.tb13434.x.

[80] L. Ren, Z.Y. Fu, Y.C. Wang, F. Zhang, J.Y. Zhang, W.M. Wang, et al., Fabrication of transparent mullite ceramic by spark plasma sintering from powders synthesized via sol-gel process combined with pulse current heating, Mater. Des. 83 (2015) 753–759, https://doi.org/10.1016/j.matdes.2015.06.046.

[81] Q. Liu, N. Jiang, J. Li, K. Sun, Y.B. Pan, J.K. Guo, Highly transparent AlON ceramics sintered from powder synthesized by carbothermal reduction nitridation, Ceram. Int. 42 (2016) 8290–8295, https://doi.org/10.1016/j.ceramint.2016.02.041.

[82] B.Y. Ma, Y.Z. Wang, W. Zhang, Q.Y. Chen, Pressureless sintering and fabrication of highly transparent MgAlON ceramic from the carbothermal powder, J. Alloys Compd. 745 (2018) 617–623, https://doi.org/10.1016/j.jallcom.2018.02.254.

[83] R.S. Zhang, Y.Z. Wang, M. Tian, H. Wang, Highly transparent LiAlON ceramic prepared by reaction sintering and post hot isostatic pressing, J. Eur. Ceram. Soc. 38 (2018) 5252–5256, https://doi.org/10.1016/j.jeurceramsoc.2018.07.053.

[84] M. Xiang, Y.F. Zhou, W.T. Xu, X.Q. Li, K. Wang, W. Pan, Transparent AlN ceramics sintered from nanopowders produced by the wet chemical method, J. Ceram. Soc. Jpn. 126 (2018) 241–245, https://doi.org/10.2109/jcersj2.17271.

[85] S.E. Hatch, R.J. Weagley, W.F. Parsons, Hot-pressed polycrystalline CaF_2:Dy^{2+} laser, Appl. Phys. Lett. 5 (1964) 153–154, https://doi.org/10.1063/1.1754094.

[86] X.Q. Chen, Y.Q. Wu, High concentration Ce^{3+} doped BaF_2 transparent ceramics, J. Alloys Compd. 817 (2020) 153075, https://doi.org/10.1016/j.jallcom.2019.153075.

[87] A. Ikesue, Y.L. Aung, Synthesis and performance of advanced ceramic lasers, J. Am. Ceram. Soc. 89 (2006) 1936–1944, https://doi.org/10.1111/j.1551-2916.2006.01043.x.

[88] M. Huang, L. Li, Y. Feng, X. Zhao, W. Pan, C. Wan, Y_3NbO_7 transparent ceramic series for high refractive index optical lenses, J. Am. Ceram. Soc. 104 (2021) 5776–5783, https://doi.org/10.1111/jace.17953.

[89] K.R. Senthil, P. Biswas, R. Johnson, Y.R. Mahajan, Transparent ceramics for ballistic armor applications, in: Y.R. Mahajan, R. Johnson (Eds.), Handbook of Advanced Ceramics and Composites, Springer, Cham, 2020, https://doi.org/10.1007/978-3-030-16347-1_12.

[90] J.A. Salem, Transparent armor ceramics as spacecraft windows, J. Am. Ceram. Soc. 96 (2013) 281–289, https://doi.org/10.1111/jace.12089.

[91] S. Bodhak, V.K. Balla, S. Bose, A. Bandyopadhyay, U. Kashalikar, S.K. Jha, et al., In vitro biological and tribological properties of transparent magnesium aluminate (spinel) and aluminum oxynitride (ALON®), J. Mater. Sci. Mater. Med. 22 (2011) 1511–1519, https://doi.org/10.1007/s10856-011-4332-5.

[92] M. Krishnan, B. Tiwari, S. Seema, N. Kalra, P. Biswas, K. Rajeswari, et al., Transparent magnesium aluminate spinel: a prospective biomaterial for esthetic orthodontic brackets, J. Mater. Sci. Mater. Med. 25 (2014) 2591–2599, https://doi.org/10.1007/s10856-014-5268-3.

Transparent metal oxides in OLED devices

3

Narinder Singh[a] *and Manish Taunk*[b]
[a]Department of Physics, Sardar Patel University, Mandi, Himachal Pradesh, India,
[b]Department of Physics, Maharaja Agrasen University, Baddi, Himachal Pradesh, India

Chapter outline

1 Introduction

Organic light-emitting diodes (OLEDs) have recently emerged as futuristic and advanced optoelectronic devices, attracting the attention of industry and academia worldwide due to their remarkable and intriguing properties such as excellent mechanical flexibility, high color gamut, low-energy consumption, rapid response

Metal Oxides for Next-Generation Optoelectronic, Photonic, and Photovoltaic Applications
https://doi.org/10.1016/B978-0-323-99143-8.00006-7

time, low fabricating cost, unprecedented contrast, no need for backlighting, low operating voltage (3–4 V), lightweight, and uniform flat illumination, which find potential applications in displays (TVs, smartphones), flat panels for lighting applications, automotive, communication, medicine, agriculture, and wearable devices [1]. The OLED has appealing characteristics such as high resolution, quick time response, a wider viewing angle (up to 160 degrees), enhanced color contrast, color tunability, transparency, and the ability to bend and roll up like a poster [2]. OLEDs outperform other optoelectronic devices such as liquid crystal display (LCD), light-emitting diode (LED), or point-shaped light sources due to their larger flexible luminescent display area [3,4].

However, new challenges are emerging in OLED materials and the architecture of flexible OLED devices. The main issues that require innovations and strategies to improve overall performance are poor power conversion efficiency, low light output, reliance on less abundant materials such as indium, the short lifetime of matrix and colors (particularly blue light), and chemical and thermal stability of flexible substrates [5]. In other words, the full commercial potential of OLEDs can be realized by exploring multilayered structures made of suitable materials, optimizing device architectures, choosing appropriate electron and hole injection materials and work functions for electrodes, and forming ultrathin films [6,7].

The anode side of the OLED emits light, which must have the required optoelectronic properties such as the lowest optical absorption in the visible region to enhance light emission, minimal sheet resistance for effective and uniform luminance, minimal surface roughness to reduce current leakage, and low fabrication temperatures (200°C) suitable for flexible substrates such as naphthalate and polyethylene terephthalate [8].

One electrode of an OLED's various layers and two electrodes must be transparent in the visible region and have high electrical conductivity. In general, high electrical conductivity and optical transparency ($E_g \sim 3.3\,eV$) are self-contradictory properties in a material. However, transparent conducting electrodes (TCOs) have both properties. Rupperecht reported the first TCO, indium tin oxide (ITO, In_2O_3:Sn), in 1954, followed by SnO_2, CdO, ZnO, MoO_3, TiO_2, WO_3, and Ga_2O_3 [9]. These TCOs are n-type semiconductors with poor optoelectronic properties for efficient device fabrication. The optoelectronic properties of these TCOs can be improved by doping with appropriate materials. In comparison to n-type TCOs, there are few reports on the synthesis of p-type TCOs, which limits the development of TCO-based p-n junctions. Only a few p-type semiconductors have been reported to be in development, including ZnO, Cu_2SrO_2, and $CuAlO_2$. Zinc oxide is also an n-type semiconductor due to stoichiometry deviations caused by zinc interstitials and oxygen vacancies. ZnO can be doped with both p-type and n-type impurities; however, its p-type behavior is not fully understood [10].

Transparent conductive electrodes (TCEs) with high electrical conductivity, high transparency in the visible region, remarkable mechanical flexibility, and matching work function with charge carrier transport layers are some of the main characteristics of flexible optoelectronic devices. Transparent electrodes fall into several categories, including ultrathin metal films [11], metal nanowires [12], metal grids [13], metal oxides [14], and organic conducting materials (graphene, carbon nanotubes). Hybrid composites and conducting polymers, particularly metal oxides, have proven to be benchmark materials as transparent anodes for use in OLED devices [15]. Doping pristine metal oxides with Al, Ga, F, In, W, Ta, Sn, and Zn, for example, can improve the optical, chemical, and electrical properties of TCOs [16–22].

2 Structure and working principle of OLED

A single film consisting of electroluminescent organic conjugated material constituting an emissive layer (EML) is deposited between the anode, generally indium tin oxide (ITO), and metallic cathodes such as lithium fluoride-aluminum (LiF-Al), magnesium, or calcium in the most basic and simplest OLED single-layer architecture (see Fig. 1A). The organic layer serves as both a charge carrier transport layer and a charge carrier (holes and electrons) recombination region, causing light to be emitted. When a forward-biased voltage is applied to the electrodes, holes from the anode and electrons from the cathode are injected into the organic layer, where they recombine to form excitons, as shown in the energy level diagram (Fig. 1C) [23]. The formation and decay of excitons is the fundamental principle of light emission in OLED. The exciton is a high-energy, bound, uncharged molecular state formed by electron-hole recombination [24]. Following the lifetime of excitons, light with the wavelength corresponding to the energy of excitons is emitted. As a result, color tuning with the molecular design of the color center becomes controllable, which has potential applications in OLED displays [10].

The performance of the fabricated OLED device is determined by the charge carrier concentration, injection, excitons generation, and actual radiative recombination. To improve the device's efficiency, the single-layered structure was further engineered to build a multilayered structure. Each layer of a multilayered OLED architecture comprises different materials that each performs a different function [25]. As shown in Fig. 1B, the number of layers deposited between the anode and cathode has been increased from seven to nine, namely, hole-injection layer (HIL), hole transport layer (HTL), emissive layer (EML), hole blocking layer (HBL), electron transport layer (ETL), and electron injection layer (EIL). These layers are inserted to accelerate charge carrier injection, lowering injection barriers, promoting exciton formation, and then radiative emission from the emissive layer [26]. To reduce the possibility of excitons quenching at the electrodes, the emissive layer is placed in the middle region. CuPc (copper phthalocyanine), NPB (*N*,*N*′-bis (naphthalen-1-yl)-*N*,*N*′-bis(phenyl) benzidine), and NPD (*N*-(1-naphthyl)-*N*-(phenyl) are the most commonly used for HIL and HTL. The most commonly used electron transport material is (tris-(8-hydroxyquinoline) aluminum), which is combined with fluorescent dyes to produce different colored light [27].

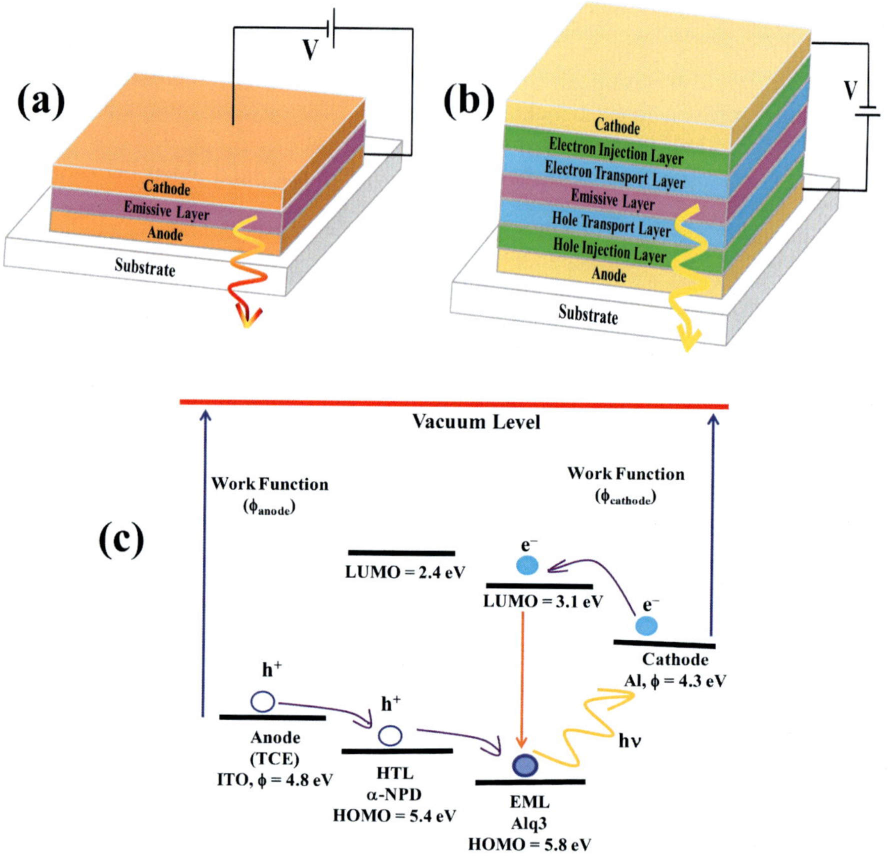

Fig. 1 Schematic illustration of (A) single-layered OLED, (B) multilayered OLED and (C) working principle of multilayered OLED.
(A and B) Adapted from reference G. Hong, X. Gan, C. Leonhardt, Z. Zhang, J. Seibert, J.M. Busch, S. Bräse, A brief history of OLEDs—emitter development and industry milestones, Adv. Mater. 33 (2021) 2005630. (C) Adapted from references A.I. Hofmann, E. Cloutet, G. Hadziioannou, Materials for transparent electrodes: from metal oxides to organic alternatives, Adv. Electron. Mater. 4 (2018) 1700412.

3 Generations and types of OLEDs

Although basic research begun in the early 1960s indicates thermally activated delayed fluorescence from eosin and electroluminescence in anthracene single crystal, it was Tang et al. who reported the first efficient low-voltage OLED in 1987, which sparked renewed interest in the development of OLED technology among academia and industry [28].

The evolution of OLEDs can be divided into four "generations" based on the properties of emissive light. The first generation relies on fluorescent emitters (Fig. 2A),

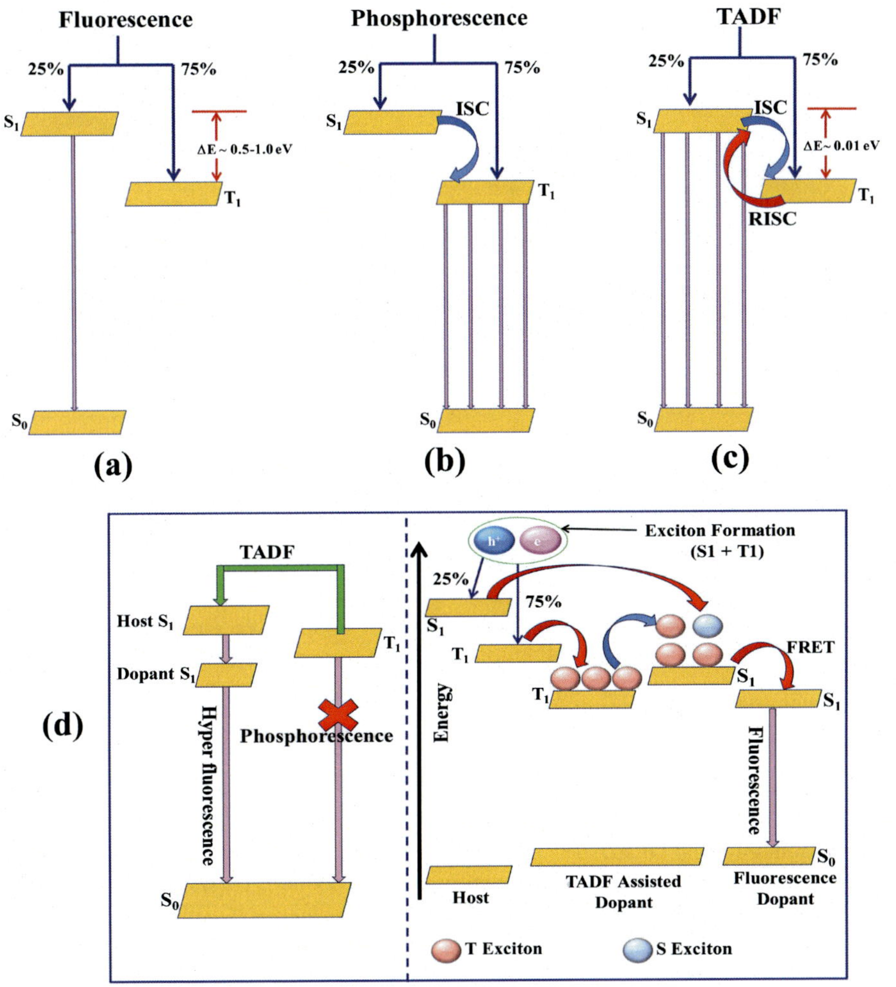

Fig. 2 Emission principles used in OLED devices include (A) fluorescence, (B) phosphorescence, (C) TADF, and (D) hyperfluorescence emission and its mechanism. Adapted from reference G. Hong, X. Gan, C. Leonhardt, Z. Zhang, J. Seibert, J.M. Busch, S. Bräse, A brief history of OLEDs—emitter development and industry milestones, Adv. Mater. 33 (2021) 2005630.

the second generation on phosphorescent emitters (Fig. 2B), and the third generation on thermally activated delayed fluorescence (TADF) as light-emitting pathways (Fig. 2C). These three generations are further subdivided into green, orange-red, and blue emitters, with these colors being critical for color saturation [5,29]. The next emerging and evolving "fourth-generation" is based on hyperfluorescence that employs a fluorescence emitter and TADF assists dopant in the electron emissive layer (Fig. 2D).

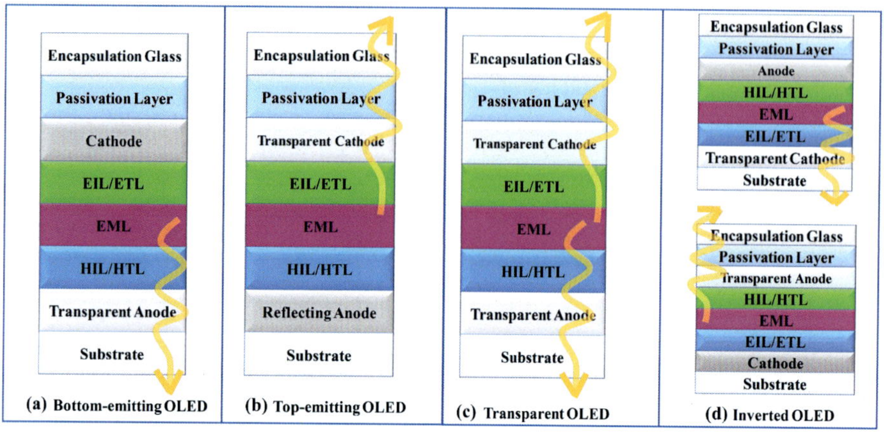

Fig. 3 Schematic illustration of (A) bottom-emitting OLED, (B) top-emitting OLEDs, (C) transparent OLED, emitting light from both sides, and (D) inverted bottom-emitting and topside-emitting OLEDs.
Adapted from reference R. Pode, Organic light emitting diode devices: an energy efficient solid state lighting for applications, Renew. Sust. Energ. Rev. 133 (2020) 110043.

Depending on the position of the transparent conductive electrode (TCE), the OLED device architecture can be classified into four categories (see Fig. 3) [30,31]:

(i) Bottom-emitting OLED
(ii) Top-emitting OLED
(iii) Transparent OLED
(iv) Inverted OLED

(i) **Bottom-emitting OLED:** The transparent electrode is deposited on the transparent substrate in this configuration (usually glass or flexible polymer). The transparent electrode is then covered with an organic layer. Finally, on the organic layer, a reflective electrode is deposited (see Fig. 3A). Light passes through the bottom transparent electrode and the transparent substrate in this case.

(ii) **Top-emitting OLED:** This configuration is similar to the bottom-emitting OLED configuration, except that the positions of the transparent electrode and the reflective electrode (which may be opaque) are reversed (see Fig. 3B). In this case, light passes through the top transparent electrode.

(iii) **Transparent OLED:** As shown in Fig. 3C, the bottom and top electrodes in this architecture are transparent or semitransparent, and light passes through both of them.

(iv) **Inverted OLED:** The light is emitted either through the top anode or semitransparent bottom cathode (see Fig. 3D).

4 Deposition techniques

As shown in Fig. 4, film deposition methods are broadly classified into two categories: physical vapor deposition (PVD) and chemical vapor deposition (CVD). Vapor

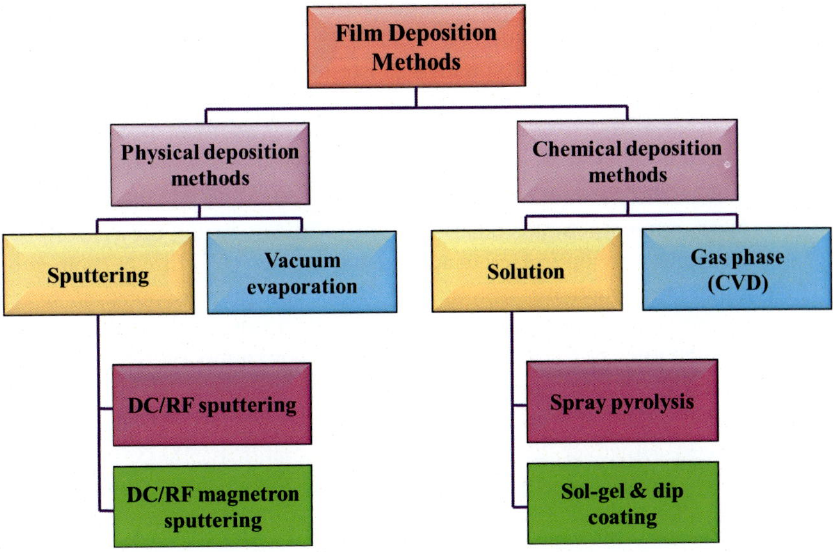

Fig. 4 Different physical and chemical deposition techniques for thin film deposition on the glass substrate.

deposition entails the deposition of thin films on solid surfaces via the transfer of vapors generated either by purely physical methods from a solid/molten source or by chemical reactions in the gas/solution phase. For thin film deposition in PVD or CVD, there are three main steps: (i) The formation of appropriate ionic, atomic, or molecular species in the vapor; (ii) the transport of these species to the substrate via vacuum, and (iii) the vapor condensation on the substrate.

TCO deposition techniques, as well as controlling parameters such as annealing temperature, geometry, substrate nature, film thickness, substrate temperature, and gas pressure play an important role in determining the optoelectronic properties of TCO [32]. RF/DC magnetic sputtering, CVD, thermal or electron beam evaporation, molecular beam epitaxy (MBE), spray pyrolysis, pulsed laser deposition (PLD), atomic-layered deposition (ALD), and sol-gel and dip-coating solution methods are the most commonly used deposition techniques. These deposition methods are optimized or modified to be less expensive, simpler, easier to use, safer, use less power, and be more environmentally friendly.

4.1 Magnetron sputtering

This is one of the most captivating and widely used industrial deposition techniques. This method has several advantages, including better reproducibility and a high deposition rate on a large area. Sputtering is a process that extracts particles (molecules or atoms) from a target material by bombarding it with highly energetic ions or particles made of inert gases or plasma. A magnetic sputtering machine generates a magnetic field, which causes plasma and highly energetic sputtered atoms to form [33]. This is further subdivided into two types based on the power source: radio frequency (RF) and direct current (DC) magnetic sputtering techniques [34]. To obtain a thin film with high transparency and conductivity, a high substrate temperature (200°C) is required in the magnetic sputtering route, which imposes challenges during lift-off fabrication technology. The preheating process removes contaminants from the organic layer, while the annealing process improves crystal growth. High-temperature conditions (pre- and postdeposition) are harmful to both the organic layer and the substrate with a low glass transition temperature, such as plastic [35]. As a result, the deposition processes and parameters are optimized to achieve the desired TCO film at room temperature. Although RF magnetic sputtering allows for film deposition at room temperature, it results in poor structural and optoelectronic properties, which may be attributed to the poor re-crystallization of ITO at room temperature. Although difficult, DC magnetic sputtering can result in film deposition at room temperature. However, the electrode film becomes amorphous, and its electrical conductivity is significantly reduced. However, at high substrate temperatures (300–350°C), both RF and DC magnetic sputtering produced thin electrode films with the same electrical properties, namely electrical conductivity, bandgap, and carrier concentration [32,36–39].

4.2 Pulsed laser deposition

In this technique, a highly powered pulse laser beam is incident on the target material kept in the vacuum chamber. This results in the formation of a laser plume that is directed toward a nearby substrate [40]. Mayer et al. reported that AZO had shown average transmittance of ~73% and current and power efficiencies greater than 43 cd A^{-1} and 27 lm W^{-1}. The device has shown very low current density (3×10^{-5} mA cm^2) at a reverse bias of 6 V [41].

4.3 Spray pyrolysis method

This method involves spraying solution (metals and liquids) onto a preheated substrate (at about 600°C) using a spray nozzle and a compressed carrier gas. The concentration of the precursor solution, the temperature of the substrate, the thickness, the nature of the metal and liquid, the airflow rate, the distance of the spray nozzle from the substrate, the airflow rate, and other key parameters all have a significant impact on the optoelectronic properties of the TCO film [40]. The substrate temperature is a key parameter to ensure the overall performance of the deposited thin film. The very high substrate temperature hinders the film's absorption on the substrate, while crystallization of the thin film degraded at too low substrate temperature. This technique is well known for its adaptability and low cost, as no vacuum and gas protection devices are required [42]. This deposition method also has some drawbacks, such as a complex precursor delivery system and a lack of standardization in terms of liquid delivery, substrate heating, and atmospheric conditions [43].

4.4 Chemical vapor deposition

Because of the fine control over composition, uniformity in film, large-area growth, and better coverage of different steps, this method is preferred for the large-scale production of thin films [44]. Precursors are heated and vaporized in the CVD method before being transported to reaction sites via a piping system that is also heated to avoid the condensation of vaporized precursors. Aerosol-assisted chemical vapor deposition (AACVD) is a more promising method than CVD. This is a safer, more cost-effective, and easier-to-manage method that does not require volatile precursors or fine control of the gas, which makes the CVVD method unreliable [45]. Furthermore, it enables simple doping and fine control over the solvent-dependent morphology [46].

4.5 Sol-gel and dip-coating method

PVD or CVD deposition processes are complex, expensive, and necessitate high-temperature and vacuum conditions. These processes are extremely expensive due to the complicated equipment requirements, extreme physical conditions, vacuum equipment, heavy machinery, and power consumption [47]. The researchers have endeavored to replace these methods with solution-processed methods. The sol-gel deposition process is gaining popularity due to its low cost and ease of use in synthesizing large-scale thin films at room temperature. This method provides high homogeneity, superior compositional control, and a low crystallization temperature [48]. Dip coating is frequently used in conjunction with the sol-gel method. The pre-synthesized sol-gel is placed in the center of the rotating substrate at a fixed revolution per minute. Centrifugal forces result in a uniform substrate coating, which is then dried at room temperature to evaporate the solvent before being annealed at high temperatures in the 300–600°C range used to fabricate thin films. Tsang et al. used a laser-irradiated sol-gel spin coating method to create an AZO electrode at a low substrate temperature (300°C). In the visible region, the AZO film had a low electrical resistivity (44×10^{-3} Ω-cm) and high transmittance (∼90%) in the visible region. They stated that, while the electrical resistivity is one order higher than that of ITO, the condition of low substrate temperature (200°C) can be obtained through further experimental optimization, which holds promise for depositing TCO on the plastic substrate [49–53].

5 Optoelectronic properties of TCEs

The properties of the electrode material are the primary determinants of the device's overall performance. In addition to work function, TCO adhesion, surface roughness, mechanical properties, and processibility of the electrode material, the key optoelectronic parameters are high electrical conductivity and optical transparency. In the case of "top-emitting" OLEDs, the organic layer is vulnerable to such electrode material properties and bombardment of ions or atoms during deposition processes, making it difficult to protect the vulnerable organic layer. The surface roughness of the single-layered TCO increases with increasing film thickness. The irregular surface of the TCO causes an uneven interface with the organic layer, lowering overall OLED performance [54]. The performance of a TCE is evaluated based on two important characteristics [55]: low sheet resistance, R_s (measured in $\Omega\,\text{sq.}^{-1}$), and high transmittance, T (in.%), in the visible region, which, thin film, are related by the following relation [56]:

$$T = \left(1 + \frac{188.5\sigma_{\text{opt.}}}{R_s\sigma_{\text{dc}}}\right)^{-2} \tag{1}$$

where $\sigma_{\text{opt.}}$ Represents the optical conductivity and σ_{dc} is the electrical direct current conductivity. In the case TCO/metal/TCO multilayered structure, each layer is considered a parallel combination of three resistances. Therefore, the total resistance of the structure can be obtained using the following equation [57]:

$$\frac{1}{R} = \frac{1}{R_{\text{TCO(top)}}} + \frac{1}{R_{\text{metal}}} + \frac{1}{R_{\text{TCO(bottom)}}} \tag{2}$$

Generally, the thickness of the bottom and top thickness is taken as equal, which gives the same sheet resistance for both layers. Therefore, Eq. (2) is rewritten as follows:

$$\frac{1}{R} = \frac{2}{R_{\text{TCO(top)}}} + \frac{1}{R_{\text{metal}}} \tag{3}$$

According to Ohm's law, the current density is given as follows:

$$\vec{j} = \sigma_{\text{dc}}\vec{E} \tag{4}$$

Here, $\vec{E}$ ($\omega = 0$) is the static applied electric field and σ_{dc} is frequency-independent electrical conductivity. However, on changing applied electromagnetic field, $\vec{E}$ (ω # 0), $\sigma_{\text{opt.}}$ (ω) becomes frequency dependent, which is related to the film transmission (Lambert-Beer's law).

The term "figure of merit" (FoM) can be defined as the ratio of electrical dc conductivity and optical conductivity, which is used to evaluate and compare the optoelectronic properties of the TCEs. Using Eq. (1) and Eq. (4), Dressel and Gruner proposed to determine FoM as shown in the following expression [58]:

$$\text{FoM} = \frac{\sigma_{\text{dc}}}{\sigma_{\text{opt.}}} = \frac{188.5}{\left(\frac{1}{\sqrt{T}} - 1\right)R_s} \tag{5}$$

The FoM can be maximized if sheet resistance (R_s) is decreased and transmittance (T) is enhanced. FoM > 1 Ω^{-1} is considered good for a TCE to be used in applications. A smaller value of R_s indicates low operating voltage, uniform emission as aging of TEC. The high transmittance value can be ascribed to no absorption of light resulting in increased external quantum efficiency (EQE). Besides expression (5), Haacke proposed another expression to determine the FoM (φ_{TC}) for different TCO thicknesses, which is calculated using the following [59]:

$$\varphi_{TC} = \frac{T_{av}^{10}}{R_s} \tag{6}$$

where R_s is the sheet resistance and T_{av} is the average optical transmittance in the visible range, which can be estimated as follows:

$$T_{av} = \frac{\int V(\lambda)T(\lambda)d\lambda}{\int V(\lambda)d\lambda} \tag{7}$$

where $V(\lambda)$ is the photopic luminous efficiency function defining the standard observer for the photometry and $T(\lambda)$ is the transmittance.

Furthermore, because the thickness of the films is in the range of a few hundred nanometric domains, the smaller surface roughness of the films prevents current leakage and shorting. Moreover, there are two other properties, namely the absorption coefficient (α) and electrical conductivity (σ), that characterize the performance of a TCE, which are given as follows [56]:

$$\sigma = \frac{1}{t\,R_S} = ne\mu \tag{8}$$

$$\alpha = \frac{A}{t} = \frac{-\log T}{t} \tag{9}$$

where t is the thickness of the film and A represents the absorbance; n, e, and μ represent charge carrier density, electronic charge, and charge carrier mobility, respectively. From Eq. (4), it is evident that conductivity can be increased if both charge carrier density and charge carrier mobility are enhanced. The electrical conductivity of TCEs can be increased by doping, while the mobility depends on the microstructural and nanostructural properties of the film [56,60].

Another important property of TCO is the work function, which should be in close alignment with the organic material's highest occupied molecular orbital (HOMO), allowing holes to be injected effectively. As shown in Fig. 1C, the lower energy barrier at the TCO/organic interface facilitates hole injection. The HOMO of an organic emissive layer is typically around 6 eV, which is too deep to inject holes from TCOs with work function values ranging from 4.4 to 4.8 eV. As a result, researchers are looking for TCOs with high transmittance, conductivity, and work function [56].

Furthermore, microstructural parameters such as crystallinity or amorphous nature, crystallographic orientation, morphology, grain or crystallite size, stress, ductility, and stress influence TCO mechanical and thermal stability [61,62].

6 Important TCOs

6.1 Indium tin oxide (ITO)

Indium tin oxide (ITO, In_2O_3:SnO_2) is a ternary ionic bound degenerate n-type TCO, which is obtained by doping indium oxide (In_2O_3 90 wt%) with SnO_2 (10 wt%). The tin (Sn) is an n-type semiconductor that replaces In^{3+} from the indium oxide cubic bixbyite lattice. When Sn impurities are oxidized, electrons are added to indium tin oxide's conduction band, increasing its electrical conductivity [23,63]. The conductivity of ITO is accounted for by the free electrons contributed by Sn^{2+} donors and doubly charged oxygen vacancy donors. The substrate temperature, stoichiometry, sputtering power, sputtering gas pressure, and gas mixture affects the number of oxygen vacancies. Excessive oxidation of the ITO surface by plasma oxygen reduces the number of oxygen vacancies, lowering its conductivity. ITO is regarded as a benchmark material to be used as a transparent conducting anode in OLEDs due to its superior optoelectronic properties like high electrical conductivity ($\sim 10^4$ S cm^{-1}), high transmittance (80%–95%) in the visible region ascribed to a wide optical band gap (3.0–4.6 eV), low sheet resistance (4×10^{-4} Ω $sq.^{-1}$), high carrier concentration (10^{20}–10^{21} cm^{-3}), high Hall mobility (10–70 cm^2 V^{-1} s^{-1}), more stability, easy patterning ability, comparatively high work function, and desired surface morphological properties [40]. ITO's optoelectronic properties are altered by various surface treatments [64,65]. The work function values are changed to match those of HTL, allowing holes to be injected efficiently from ITO to HTL. Gu et al. successfully fabricated the first flexible OLED made of small molecules with ITO deposited on a flexible substrate in 1997. In the commercial world, ITO is used as a single-layer TCE on a flexible substrate. The work function values are changed to match those of HTL, allowing holes to be injected efficiently from ITO to HTL [27].

Without a doubt, ITO has remarkable optoelectronic properties such as high electrical conductivity, high optical transmittance, and environmental stability that can be used to fabricate transparent electrodes for use in optoelectronic devices (OLEDs and OPVs) [66]. However, several unavoidable limitations limit its widespread industrial use. The major drawbacks of ITO as a transparent electrode are excessive use but low abundance due to scarcity of indium on Earth's crust, making it more expensive, toxicity, mechanical brittleness, and low charge injection due to work function (4.3–4.8 eV) nonalignment with electron/hole transport layer energy levels at the interface of TCEs and p-type semiconductor, environmentally hazardous, complex, and costly processing, high surface roughness leading to significant leakage currents and limited device lifetime due to diffusion of oxygen and tin ions from ITO to the active layer, resulting in its oxidation and degradation [56]. Furthermore, deposition techniques and surface modification processes necessitating high temperature (400°C) and high vacuum conditions make ITO more expensive than TCE. The high-energy atoms or ions, radiation, and heat produced during deposition techniques cause damage to the underlying organic layer, especially in "top-emitting" OLEDs. Next-generation flexible, portable, and lightweight OLEDs must be manufactured on flexible substrates rather than glasses, which are not feasible due to the high-temperature-dependent deposition techniques and brittle nature of ITO. According to reports, ITO deposited on glass substrates has higher electrical conductivity and transparency than flexible substrates such as PES (polyethersulphone), PEN (polyethylene naphthalate), PET (polyethylene terephthalate), PPA (polypropylene adipate), and PC (polycarbonate) [67].

However, ITO is deposited and annealed at higher temperatures (>300°C), causing damage to the underlying layer of organic material. Therefore, unlike the bottom electrode, the top electrode in top-emission OLEDs has very limited applications [68]. As a result, it became critical to protect the organic layer from high-temperature conditions without compromising the OLED device's performance. Buffer layers such as organic protective layers, ultrathin metal layers, or transition metal oxide layers are used for it. The buffer layer should be chemically inert and must exhibit high conductivity and transparency and promote charge carrier injection. However, it became challenging to find all such properties in a single material. It was also attempted to use a low-energy magnetron sputtering technique to reduce damage to the underlying organic layer. This method not only resulted in slower deposition processes but also had a negative impact on the optoelectronic properties of the OLEDs [69]. Lee et al. used 1,4,5,8,9,11-hexaazatriphenylenehexacarbonitrile (HATCN), a discoid organic molecule, as an organic layer to protect the underlying emissive layer during sputter deposition of the ITO top electrode. They observed that the average transmittance (81%) and hole injection performance of inverted OLED was not interfered with by the buffer layer [70]. There is still room for improvement in its optoelectronic properties using the low-energy sputtering method.

One of the major issues that prevent the ITO layer from being used in flexible modules is its brittle nature. Microcracks formed as a result of strain or extensive mechanical bending reduce the conductivity of the film [71]. Flexibility is quickly becoming one of the most desired features of optoelectronic devices, as it allows them to be both cost-effective and portable. To overcome such constraints, it became critical to develop indium-free transparent electrodes containing little or no indium at all [21].

6.2 Fluorinated tin oxide (FTO)

Many researchers are investigating fluorine-doped tin oxide (FTO) as a transparent electrode under various reaction conditions as an alternative to costly, toxic, and scarce indium in ITO [72–75]. FTO is becoming popular due to its low cost, easy market availability, lower sheet resistance, and higher thermal and chemical stability [76]. Koo et al. reported that the FTO film prepared by a combustion-activated ultrasonic spray pyrolysis deposition route showed $R_s \sim 6.6\ \Omega\ \text{sq.}^{-1}$, the figure of merit (FoM) values of $\sim 5.34 \times 10^{-2}\ \Omega^{-1}$, and a transmittance of 90.1% [77]. However, the electrical conductivity of FTO is observed to be one order lesser than ITO. Saikia et al. investigated the effect of the NiO film, a buffer layer, on the FTO surface and observed that a buffer layer of thickness of 10 nm facilitated the hole injection process from anode to HTL, resulting in the maximum current efficiency of 7.32 Cd A^{-1} [78].

6.3 Zinc oxide (ZnO)

Zinc oxide has recently gained attention as a potential alternative to ITO as a TCE for various optoelectronic devices. ZnO is a wide bandgap ($E_g \sim 3.3$ eV) semiconductor with increased optical transmittance, excellent thermal and mechanical stability, lower toxicity, natural abundance, and higher excitonic binding energy (60 meV) [58]. The n-type semiconducting properties of the ZnO film can be attributed to intrinsic defects in the form of Zn antisites and Zn interstitials. There are some limitations, such as lower charge carrier concentration and thus lower and unstable electrical conductivity, as well as sensitivity to heat and humidity, which limit its use as TCEs in optoelectronic devices. Furthermore, the work function of untreated ZnO is poorly producible and has been measured to be around 4.3 eV, resulting in an injection barrier for the majority of conventional semiconducting materials [79]. The work function can be altered by using a buffer layer or by passivation of the ZnO surface with a self-assembled monolayer of polar molecules. Using a buffer layer of ethoxylated polyethyleneimine (PEI) or polyethyleneimine, the work function of ZnO can be reduced [80,81]. However, the etchability of ZnO in an acidic environment makes surface modification of ZnO more difficult. Lange et al. successfully modified sol-gel processed ZnO with benzyl phosphonic acid (BPA), a self-assembled molecule, and varied its work function by more than 2.3 eV, allowing it to be used for charge carrier extraction [82].

Untreated ZnO's optoelectronic properties are also influenced by native defects such as zinc vacancy (V_{Zn}), zinc interstitials (Z_{ni}), antisites, and incorporated hydrogen. Doping ZnO with group III elements improves its optoelectronic properties by increasing the free charge carrier concentration and optical band gap [83]. The hydrogen behaves like an amphoteric impurity, acting as a donor (H^+) in p-type semiconductors and an acceptor (H^-) in n-type semiconductors. Gaspar et al. mixed H_2 with argon during ZnO film growth using RF magnetic sputtering without substrate heating or postdeposition annealing. The n-type ZnO:H thin film showed an electrical resistivity of 3×10^{-3} Ω-cm, which may be due to the shallow donor effect of hydrogen. They also concluded that high mobility, $\mu_H \sim 47$ cm^2 V^{-1} s^{-1}, and charge carrier concentration, $n \sim 4.4 \times 10^{19}$ cm^{-3}, might be attributed to passivation of grain boundaries by incorporation of H_2 and formation of Zn interstitials, respectively [84].

ZnO doped with aluminum (ZnO:Al), gallium (ZnO:Ga), and/or tungsten (ZnO:W) has lower sheet resistance and transparency comparable to ITO. When deposited on flexible electrodes, aluminum-doped zinc oxide (AZO) thin film produced more promising results. Park et al. deposited an AZO thin film (thickness ~200 nm) on the flexible polyethersulphone substrate using a magnetron sputtering technique and reported that the values of electrical conductivity and optical transmittance were found to be 2000 S cm^{-1} and 90% (in the temperature range from 20°C to 200°C), respectively, indicating better results when compared to the ITO/PET substrate [85]. It was observed that ITO/AZO had exhibited better optoelectronic properties like electrical resistivity (8.4×10^{-4} vs 1.1×10^{-3} Ω cm), FOM (1.0×10^4 vs 7.4×10^3 Ω^{-1}cm^{-1}), transparency (88.3% vs 87.3%), and sheet resistance (32.9 vs 42.3 Ω sq.$^{-1}$) as compared to ITO/ZnO. The values of sheet resistance and electrical resistivity were reported to decrease as the thickness of ZnO as well as the AZO film was increased [86]. However, AZO has drawbacks such as poor stability, limited flexibility, high reactivity with oxygen, easy diffusion of zinc ions from the electrode into the emissive organic layer, and etching of chemically unstable films in acidic environments. Ga-doped ZnO, which is less expensive, has a higher transparency, and is more resistant to oxidation, outperforms AZO in OLED device performance. Das et al. conducted a comparison study to compare the performance of ZnO:Ga and SnO_2:F as TCO anodes in OLED. They reported that commercially established SnO_2:F, as well as ITO, can be replaced by ZnO:Ga TCO due to its very low sheet resistance $<5.6\,\Omega\,sq.^{-1}$, a smaller resistivity value of $9.6 \times 10^{-5}\,\Omega\,cm$, and a higher transparency of >90% (FOM: 6.22×10^{18}) [87].

6.4 Cadmium oxide (CdO)

Cadmium oxide is a highly degenerate n-type semiconductor that Bädeker reported as a TCO in 1907. Its chemical inertness and ease of adaptability for flexible electrodes, as well as its low production cost, make it an appealing material for TCO [88]. In terms of cost, the potential materials follow the order In > Sn > Ti > Zn > Cd. Various attempts were made to improve its optoelectronic properties using doping with indium (In), tin (Sn), cerium (Ce), copper (Cu), chromium (Cr), fluorine (F), lanthanum (La), and gadolinium (Gd). Sakthivel et al. reported that Gd-doped CdO showed high conductivity ($\sigma = 13{,}882\,S\,cm^{-1}$), low resistivity ($\rho = 7.203 \times 10^{-5}\,\Omega\,cm$), good transparency (93% at 900 nm), and small roughness (1.69 nm) [89]. However, the toxic nature of Cd and its narrow optical bandgap make it a less desirable material for TCO use [90].

6.5 Tin oxide (SnO_2)

Tin oxide, the most stable oxide of Sn, is an n-type semiconductor having promising potential as a TCO due to its better compatibility with different deposition techniques, low sheet resistance, wide optical bandgap (3.6–4.0 eV), and high transmittance. The antimony-doped SnO_2 (ATO) electrode has been investigated extensively due to its lower work function (3.9–4.46 eV), flexibility, and similar surface roughness when compared to the commercially available ITO anode [91]. The carrier mobility and free-electron concentration depend on the nonstoichiometry and antimony doping weight percentage in ATO [92]. Montero et al. sputtered Sb-doped SiO_2 on the glass substrate using oxygen-reactive DC magnetron sputtering at room temperature, which was annealed in a nitrogen atmosphere at 350°C. They investigated the effect of (O_2/O_2 + Ar) in the sputter flow gas on electrical resistivity and optical transmittance of an Sb-doped SiO_2 thin film [93]. Elangoven et al. observed that fluorine-doped SnO_2 showed higher transparency as compared to pristine SnO_2 as well as Sb-doped SnO_2 [94]. Ramrajan et al. deposited Ta-doped SnO_2 films using a more economical spray pyrolysis method and explored their electrical properties in the high-temperature regime. The film showed a transmittance of ~85%, a sheet resistance of $\sim 18\,\Omega\,sq.^{-1}$, and a resistivity value of $4.36 \times 10^{-4}\,\Omega\,cm$ [95].

6.6 TCO/metal/TCO multilayered structures

So far, ITO exhibits excellent conductivity $\sim 10^4\,S\,cm^{-1}$ and high transparency (>90%) in the visible region with the film thickness in the range of 100–300 nm. The ITO films are annealed at a temperature >250°C during or postdeposition processes, which ensure transparent and conducting films by increasing crystallization and decreasing crystalline defects [96]. There is still a scope to reduce the sheet resistance, and surface roughness and increase the mechanical strength of transparent electrodes. To achieve it, TCO/metal/TCO multilayered structure comprising of either ITO/metal/ITO or indium-free electrodes like SnO_2/metal/SnO_2, ZnO/metal/ZnO, IZO/metal/IZO, AZO/metal/AZO and MoO_x/metal/MoO_x [97] or metal coated with metal oxide (Ag/ZnO and Ag/SnO_x) are investigated which not only resulted in the decrease in sheet resistance and surface roughness but also impart mechanical strength, enhance chemical and mechanical stability against corrosion, increases the current efficiency and optical transparency (shown in Table 1). The conductivity order of different metals is as follows: Ag > Cu > Au > Al > Mg > W > Mo > Zn > Ni > In > Pt > Pd > Sn > Cr > Ta > Ti [107]. Silver (Ag) is the most favored metal among these metals due to its high conductivity, followed by copper (Cu) and gold (Au). Silver-based alloys with small amounts of copper, palladium, and gold are also used to provide moisture and thermal stability to the anode. The TCO acts as a conducting medium between the interlayer metallic film and any point on the surface in a TCO/metal/TCO configuration. However, its primary function is to maximize metal film transmittance in the visible region while minimizing reflectance from the metal surface. The thickness of thin interlayer metal or its alloy has a significant impact on sheet resistance, electrical conductivity, and TCO transmittance. Multilayered transparent electrodes are more compatible with plastic electrodes and require less heat to fabricate. The ductility of the interlayer metal results in the desired conductivity even after TCOs have reached their elastic limit. Ji et al. created a highly flexible ZnO/Ag/CuSCN transparent electrode with CuSCN as a hole transport antireflecting film, with a sheet resistance of $9.7\,\Omega\,sq.^{-1}$ that remained unchanged after 10,000 mechanical bendings, an average transmittance of 94%, a high current efficiency of $23.4\,Cd\,A^{-1}$, and a stability of 400 days at 68% humidity [108]. The thickness and surface structure of the layered structure has a significant impact on its optoelectronic properties. The electrical conductivity is increased by the smooth and continuous thin films, while the optical transmittance is improved by the perforated surface [109].

Sibin et al. reported the minimum electrical resistivity ($\sim 7.0 \times 10^{-5}\,\Omega$-cm) and sheet resistance ($\sim 4.8\,\Omega\,sq.^{-1}$), when the thickness of the Ag layer was increased to 25 nm in the ITO/Ag/ITO anode. However, at very low thickness, Ag ~ 5 nm, the values of electrical conductivity were found to be $\sim 5.37 \times 10^{-4}\,\Omega$-cm and $\sim 42.9\,\Omega\,sq^{-1}$, respectively [57]. This behavior could be attributed to the formation of an agglomerated island-like Ag thin film, which could rupture or block electron-conducting channels [110]. Ekmekcioglu et al. deposited ZTO/Ag/ZTO electrodes on three transparent polymer substrates, SLG, PC, and PET using the DC magnetron sputtering method at room temperature. The sheet resistance values for ZTO/Ag/ZTO deposited on SLG, PC, and PET were estimated to be 8.6, 6.7, and $11.2\,\Omega\,sq.^{-1}$, respectively. The micrographs of all electrodes deposited on polymer substrates show

Table 1 List of key optoelectronic parameters of TCOs for use in OLED devices.

Multilayer structure	Deposition method	Transmittance (%)	R_s (Ωsq^{-1})	FoM Ω^{-1}	Substrate	Ref
ITO(60 nm)/Ag(13 nm)/ITO (60 nm)	DC magnetron sputtering	90.2% at 550 nm	6.90	5×10^{-3}	FEP	[57]
ITO(43 nm)/Ag(16 nm)/ITO (43 nm)	RF magnetron sputtering	79.4% at 550 nm	8.90	–	glass	[98]
ITO(50 nm)/Ag(17 nm)/ITO (50 nm)	RF magnetron sputtering	83.2% at 550 nm	6.70	–	PET	[99]
ITO(50 nm)/Ag(15 nm)/ITO (50 nm)	PVD	86% at 550 nm	3.00	–	glass	[100]
TiO_x(40 nm)/Ag(15 nm)/ WO_3(40 nm)	Electron beam evaporation	93.5% at 550 nm	3.97	1.2×10^{-1}	PES	[101]
MoO_3(45 nm)/Ag(10 nm)/ ITO(45 nm)	Thermal evaporation	85% at 530 nm	3.00	–	glass	[102]
TiO_2(40 nm)/Ag(18 nm)/ ITO (40 nm)	RF magnetron sputtering	93.1% at 560 nm	5.80	72.8×10^{-3}	PET	[103]
TiO_2(35 nm)/Ag(13 nm)/ TiO_2(35 nm)	RF magnetron sputtering	74% at 550 nm	3.74	13.2×10^{-3}	glass	[104]
IZO(30 nm)/Ag(12 nm)/IZO (30 nm)	DC magnetron sputtering	84.8% at 550 nm	6.90	28×10^{-3}	PET	[105]
ZITO(100 nm)/Ag/(8 nm) ZITO(42 nm)	RF/DC magnetron sputtering	92% at 550 nm	9.40	27.5×10^{-3}	glass	[106]

very smooth and flat morphologies, indicating the absence of defects that would block shunts or leakage current [111]. Kim et al. observed that island-like growth of Ag is reduced when the thickness is increased from 8 to 22 nm. The electron carrier concentration (N_e) and Hall mobility (μ_H) increase linearly with the thickness of interlayer Ag in the In-Ga-ZnO/Ag/In-Ga-ZnO-based multilayered electrode, which results in an increase in electrical conductivity [112]. The very thin thickness of the film limits its continuity, whereas the large thickness of the metal film reduces optical transmittance, which could be due to the interaction of photons and electrons of the metal [2]. Lee et al. reported the fabrication of asymmetric ITO/Ag/ZTO and ZTO/Ag/ITO electrodes using the roll-to-roll sputtering method. Both electrodes possessed similar optoelectronic and mechanical properties. However, in flexible OLED devices, the ITO/Ag/ZTO anode has shown superior characteristics than the ZTO/Ag/ITO anode. This behavior is attributed to the larger work function of the top TCO (ϕ_{ITO}=4.7–4.8 eV) in the ITO/Ag/ZTO anode as compared to top TCO (ϕ_{ITO}=3.81 eV) in the ZTO/Ag/ITO anode. The increased work function injects holes more efficiently, improving the overall performance of the OLED device [113].

Gold (Au) is another important metal interlayer because it not only has higher conductivity but also has excellent optical transmittance due to the interaction of incident light with surface plasmon polaritons at the metal-dielectric interface [114]. Kim et al. discovered that the Lewis-acid base reaction between MoO_x and Au could reduce the agglomeration of the Au thin film in an MoO_x/Au/MoO_x (MAM) electrode, resulting in a continuous and low-resistant thin film. They concluded that MAM-based devices operate at lower voltage, higher current density (20 mA cm^{-2}), maximum current efficiency (11.46 Cd A^{-1}), and higher luminance density than ITO-based devices, which may be attributed to the larger work function of MoO_x (~5.14 eV) and optimized thickness of the MAM structure [97]. Song et al. deposited IGZO (50 nm)/Ni (15 nm)/IGZO (50 nm) (IGZO=In_2O_3:Ga_2O_3:ZnO) on the polycarbonate substrate by RF and DC magnetron sputtering, showed FoM=$2.01 \times 10^{-4}\,\Omega^{-1}$, lower $\rho=4.74 \times 10^{-4}\,\Omega$-cm, $n=4.48 \times 10^{20}$ cm^{-3}, and $\mu=29.4$ cm^2 V^{-1} s^{-1}, which were observed to be far better than that of single-layered IGZO [115].

The TCO/metal/TCO hybrid structures have exceptional optoelectronic and mechanical properties due to the constituents' synergy. Metals' superior conductivity, a well-matched refractive index of TCO with a metal layer, and metal ductility are responsible for the high electrical conductivity, high optical transmittance, and high mechanical properties, respectively. However, due to the metal's polycrystalline nature, such structures are mechanically weak. When deposited between oxide layers, the metal layer crystallizes easily despite its thin thickness (10 nm). Furthermore, the bottom TCO layer is amorphous when deposited on the glass substrate, whereas the top TCO layer becomes polycrystalline when deposited on the polycrystalline metal layer. Mechanically, the resulting TCO (polycrystalline)/metal/TCO (amorphous) configuration is unstable. When an ITO/Ag/ITO structure was prepared in the presence of 0.2% hydrogen gas flow, Park et al. demonstrated that hydrogen passivation of ITO significantly improved mechanical stability. The presence of hydrogen stabilized the amorphous structure by reducing TCO subgap defects and, as a result, residual stress in the structure [116].

6.7 Multicomponent-based TCOs

Another approach to reducing material toxicity and cost while achieving desired optoelectronic properties is to use multicomponent-based TCOs as an alternative to ITO. Researchers have investigated several promising TCOs, including Zn-In-O (ZIO), Ga-In-O (GIO), Ga-In-Sn-O (GITO), and Zn-In-Sn-O (ZITO). The electrical conductivities of GITO ($3280\,S\,cm^{-1}$) and GITO ($2290\,S\,cm^{-1}$) are more than ZIO ($1030\,S\,cm^{-1}$) and GIO ($700\,S\,cm^{-1}$) and comparable to ITO ($3500\,S\,cm^{-1}$). They also exhibit high transmittance (~90%) and low sheet resistance. More important is the large values of work functions of such TCOs (5.2–6.1 eV vs ~4.7 eV for ITO), which offers perfect alignment with the HOMO of the HTL layer [56,117,118].

7 Surface treatment of TCOs

The efficiency of charge carrier injection and transport, as well as the high transparency and long lifetime of an OLED, determines its performance. Surface electronic properties, work function values, and surface morphologies have a strong influence on the interfacial interaction at the TCO-organic layer. Imperfect bonding at the interface can be caused by surface carbon or hydroxyl contaminants or islands- or hillocks-like morphology, which causes problems such as current leakage, short circuit, and diffusion of indium or zinc into the organic layer at the interface, degrading the overall device performance and lifetime. As a result, the TCO surface is modified or treated in order to tune the work function values of the TCO with the organic layer for effective hole injection and transport [119]. Ultrasonic cleaning, mechanical polishing, surface cleaning, oxygen or argon plasma treatment, self-assembled monolayer (SAM), acids and electrochemical treatments, and other methods are used for surface treatment [120,121]. The presence of hydrogen, for example, in the RF magnetron technique increases the oxygen vacancies in the ITO film, resulting in higher charge carrier concentration. The nonstoichiometric ITO film is caused by the number of oxygen vacancies, which is determined by the deposition conditions [122]. The ITO film prepared in the presence of oxygen, on the other hand, reduces carrier concentration, which could be attributed to oxygen vacancy dissipation [123].

The reduction in surface roughness raises the work function of ITO, resulting in a lower energy barrier at the ITO-HTL interface and an efficient hole-injection rate. This causes variations in contact resistance, lowers turn-on voltage, improves luminous efficiency, and raises the device's current density and efficiency. Increased surface oxygen content, interfacial dipole, Fermi level shift toward the middle of the band gap, tin oxide oxidation, and other factors can be attributed to an increase in the ITO work function. Yahya et al. investigated the optoelectronic properties of an ITO electrode and an OLED device with a glass/ITO/PEDOT:PSS/Alq_3/Al structure after argon plasma treatment. They discovered that surface treatment not only improves the ITO anode's surface smoothness and transmittance but also lowers the turn-on voltage and increases the current density of the OLED. At the ITO-HTL interface, argon plasma improves adhesiveness and lowers the energy barrier [54].

8 TCOs on flexible substrates

The flexibility of an OLED device is rapidly becoming the most desirable feature, as it allows devices to be stretched, rolled, and folded. Flexible OLEDs are lighter, thinner, and suitable for roll-to-roll manufacturing [124]. Heeger et al. deposited conducting polymers on PET, a flexible substrate, in 1992 [125]. For transparent and bottom-emitting OLEDs, the flexible substrate should have the following characteristics: smooth surface, lightweight, remarkable mechanical deformation, improved thermal stability, resistance to oxygen and moisture, and high transparency. Except for thermal durability, plastic-based flexible substrates such as PET, PES, and PEN owe these desirable properties [126]. When ITO is annealed at high temperatures (~300°C), its transparency enhances, increasing OLED light efficiency and brightness. Because of its brittle nature, it is incompatible with flexible substrates. The plastic substrate, on the other hand, has a low glass transition temperature and a smooth surface. The conductivity of the TCOs (particularly ITO and IZO) anode deposited on the plastic substrate was low. When they are annealed at high temperatures, which would damage the plastic substrate, the conductivity and transmittance increase [127]. Jou et al. precoated PES flexible substrate with SiO_2 and deposited ITO using RF magnetron sputtering at 200°C to increase the conductivity and adhesion of ITO with a plastic substrate. They proposed a structure of OLED, showing a maximum external quantum efficiency (EQE) of ~3.4% and a power efficiency (PE) of ~6.5 lm W^{-1}. The high efficiency of the device could be attributed to the presence of the SiO_2 layer between ITO and flexible substrate, which reduces the surface roughness and increases the electrical conductivity [128]. Morales et al. reported that OLED fabricated with ZTO/Ag/ZTO and ZTO/metal (Mo/Al/Mo) grid anodes deposited at low temperature (60°C) on a temperature-sensitive flexible electrode exhibit better performance as compared to ITO and ITO/Ag grid anodes, respectively, when deposited at low-temperature (150°C) [8]. Kwon and his co-workers prepared Al- and Mg-doped ZnO (AMZO) to fabricate AMZO/Ag/AMZO structure and utilized the synergetic effect of mechanical flexible conductive Ag and moisture-resistant AMZO layer to protect the organic layer against the ambient atmosphere. They observed water vapor transmission rate $\sim 10^{-5}\, g^{-2}\, day^{-1}$ and $R_s \sim 5.6\,\Omega\, sq.^{-1}$, FOM value of $60.4 \times 10^{-3}\,\Omega$ and average optical transmittance ~90% when thickness of MAM layer ~110 nm [129].

9 Color tuning with graded ITO thickness

Color tuning in OLED is accomplished through the use of a microcavity structure, which consists of a stack of organic layers inserted between a top cathode and a bottom bilayer reflective anode. The thickness of a single organic layer or TCO is varied to produce a multicolor image, in which the color tuning is achieved by emitting light of different wavelengths, as illustrated in Fig. 5. For hole injection in top-emitting OLEDs, a bilayer reflective electrode composed of metal/ITO (metal: silver,

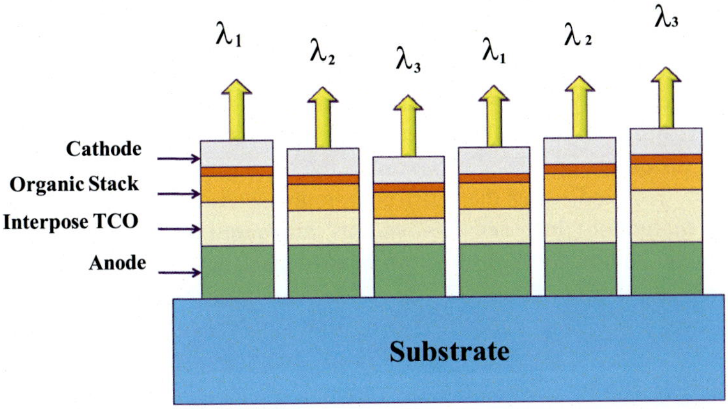

Fig. 5 Schematic illustration (cross-section view) of a multicolor or full-color pixel display with micro-cavities on a substrate.
Adapted from reference F. Zhu, in: Z. Li, H. Meng (Eds). Transparent Electrode for OLEDs. Organic Light-Emitting Materials and Devices, 2017.

chromium, or its alloy) with a graded thickness of ITO can be used. The thickness variation causes constructive microcavity interference. Increased metal layer thickness improves microcavity but decreases optical transmittance. Optical loss may occur due to surface plasmon polariton, light absorption, and the presence of waveguide and substrate modes at the TCO/organic interface and the substrate/air interface, respectively, in addition to the thickness of the metal layer. Furthermore, the metal film causes spectral shifts and color distortion [10]. According to Choi et al., scattering from SnO_x nanocones can improve light extraction efficiency and viewing angle characteristics by reducing total internal reflection at the substrate/air interface. The metal layer serves as a mirror, reflecting light to the cathode. This reflective layer has no effect on the current-voltage characteristics, so the hole injection properties at the TCO/HTL interface remain unchanged. Because metals are opaque, their transmittance, which is typically in the 30%–60% range, must be optimized. One strategy is to deposit the dielectric capping layer (CL) on top of the metal layer. Chung et al. investigated the effect of variation of CL thickness on the transparent OLED having device structure: ITO (150 nm)/PEDOT:PSS (25 nm)/MoO_3 (10 nm)/4,4′-bis (carbazol-9-yl) biphenyl(CBP) (35 nm)/CBP doped with tris(2-phenylpyridine) iridium(III) ($Ir(ppy)_3$, 8 wt%)(15 nm)/4,7-diphenyl-1, 10-phenanthroline(Bphen)/LiF (1 nm)/Al (1 nm)/Ag (15 nm)/ZnS. After considering the effect of surface plasmon polariton loss, the values of transmittance and quantum efficiency were estimated to be 80% and 15%, respectively [130].

Choi et al. used a perforated WO_3 layer to improve the out-coupling efficiency of transparent OLED without causing optical haze [131]. The thickness of ITO has a significant impact on light transmission while having no effect on the light emission efficiencies of the top and bottom sides. As a result, conductivity and transmittance are carefully balanced to maximize device efficiency [132].

10 Conclusions

The indium-free multilayered transparent metal oxide electrodes perform admirably, with high conductivity, high transmittance, and low sheet resistance. However, the multilayered structure raises the cost of electrode production, limiting the commercialization of flexible OLEDs. Furthermore, the most effective use of TCO materials can never be realized using the most commonly used deposition techniques. New strategies, methodologies, and innovations are required to prepare continuous ultrathin metal layers through cost-effective deposition techniques. According to reports, only about 30% of the ITO material reaches the substrates. As a result, recovering left-over ITO from deposition processes and recycling ITO from scrapped OLED devices may be prudent approaches to lowering costs. The cost issue related to the materials can be addressed using indium-free metal oxides.

It has been discovered that ITO is the best-performing TCE among metal oxides, and as a result, it is still used on a commercial scale in optoelectronic devices. Due to the high cost, toxicity, and scarcity of indium, other metal oxides were investigated as potential replacements for ITO. To date, it appears difficult to find a single TCO to replace ITO. The earth's abundant materials, when combined, can be explored further due to their prudent trade-off of conductivity, optical transparency, and flexibility. Other n-type semiconductors (MoO_3, WO_3, Ga_2O_3, TiO_2, and V_2O_5) are also investigated for their optoelectronic properties, which indicated their potential use in OLEDs. These oxides have also shown low optical absorption, high compatibility with, high refractive index, and high stability toward energetic particles bombardment in plasma-based deposition methods.

Similarly, it is desired to replace expensive deposition techniques such as chemical vapor deposition (CVD), magnetron sputtering, molecular beam epitaxy (MBE), thermal evaporation, atomic-layered deposition (ALD), and pulsed laser deposition (PLD) with less expensive and simpler solution-processed techniques that do not compromise the required optoelectronic properties of TCOs.

Acknowledgments

Narinder Singh is thankful to Sardar Patel University, Mandi, for extending their valuable cooperation and motivation. Manish Taunk is thankful to Maharaja Agrasen University for rendering precious support in compiling this chapter.

References

[1] N. Singh, Polypyrrole-based emerging and futuristic hybrid nanocomposites, Polym. Bull. 79 (2022) 6929–7007.

[2] K.-J. Ko, S.-R. Shin, H. Lee, E. Jeong, Y. Yoo, H. Kim, Y. Song, J. Yun, J.-W. Kang, Fabrication of an oxide/metal/oxide structured electrode integrated with antireflective film to enhance performance in flexible organic light-emitting diodes, Mater. Today Energy 20 (2021) 100704.

[3] B. Geffroy, P. Le Roy, C. Prat, Organic light-emitting diode (OLED) technology: materials, devices and display technologies, Polym. Int. 55 (2006) 572–582.
[4] H.-W. Chen, J.-H. Lee, B.-Y. Lin, S. Chen, S.-T. Wu, Liquid crystal display and organic light-emitting diode display: present status and future perspectives, Light: Sci. Appl. 7 (2018) 17168.
[5] G. Hong, X. Gan, C. Leonhardt, Z. Zhang, J. Seibert, J.M. Busch, S. Bräse, A brief history of OLEDs—emitter development and industry milestones, Adv. Mater. 33 (2021) 2005630.
[6] H. Sasabe, J. Kido, Multifunctional materials in high-performance OLEDs: challenges for solid-state lighting, Chem. Mater. 23 (2011) 621–630.
[7] A. Chauhan, P. Jha, D. Aswal, J. Yakhmi, Organic devices: fabrication, applications, and challenges, J. Electron. Mater. 51 (2022) 447–485.
[8] M. Morales-Masis, F. Dauzou, Q. Jeangros, A. Dabirian, H. Lifka, R. Gierth, M. Ruske, D. Moet, A. Hessler-Wyser, C. Ballif, An indium-free anode for large-area flexible OLEDs: defect-free transparent conductive zinc tin oxide, Adv. Funct. Mater. 26 (2016) 384–392.
[9] A. Facchetti, T. Marks, Transparent Electronics: From Synthesis to Applications, John Wiley & Sons, 2010.
[10] F. Zhu, in: Z. Li, H. Meng (Eds.), Transparent Electrode for OLEDs. Organic Light-Emitting Materials and Devices, CRC Press, Boca Raton, 2017. Chapter 6. ISBN: 9781315216775, https://doi.org/10.1201/b18540.
[11] P. Spinelli, R. Fuentes Pineda, M. Scigaj, T. Ahmad, K. Wojciechowski, Transparent conductive electrodes based on co-sputtered ultra-thin metal layers for semi-transparent perovskites solar cells, Appl. Phys. Lett. 118 (2021), 241110.
[12] R. Xu, K. Yang, Y. Zang, ZnO/Ag/ZnO multilayer transparent electrode for highly-efficient ITO-free polymer solar cells, Curr. Appl. Phys. 20 (2020) 425–430.
[13] X. Chen, W. Guo, L. Xie, C. Wei, J. Zhuang, W. Su, Z. Cui, Embedded Ag/Ni metal-mesh with low surface roughness as transparent conductive electrode for optoelectronic applications, ACS Appl. Mater. Interfaces 9 (2017) 37048–37054.
[14] X. Yang, P. Gao, Z. Yang, J. Zhu, F. Huang, J. Ye, Optimizing ultrathin Ag films for high performance oxide-metal-oxide flexible transparent electrodes through surface energy modulation and template-stripping procedures, Sci. Rep. 7 (2017) 1–9.
[15] H.J. Yun, S.J. Kim, J.H. Hwang, Y.S. Shim, S.-G. Jung, Y.W. Park, B.-K. Ju, Silver nanowire-IZO-conducting polymer hybrids for flexible and transparent conductive electrodes for organic light-emitting diodes, Sci. Rep. 6 (2016) 1–12.
[16] V. Sharma, P. Kumar, A. Kumar, K. Asokan, K. Sachdev, High-performance radiation stable ZnO/Ag/ZnO multilayer transparent conductive electrode, Sol. Energy Mater. Sol. Cells 169 (2017) 122–131.
[17] L. Cattin, J. Bernède, M. Morsli, Toward indium-free optoelectronic devices: dielectric/metal/dielectric alternative transparent conductive electrode in organic photovoltaic cells, Phys. Status Solidi A 210 (2013) 1047–1061.
[18] B.-R. Koo, D.-H. Oh, D.-H. Riu, H.-J. Ahn, Improvement of transparent conducting performance on oxygen-activated fluorine-doped tin oxide electrodes formed by horizontal ultrasonic spray pyrolysis deposition, ACS Appl. Mater. Interfaces 9 (2017) 44584–44592.
[19] H.J. Park, J. Kim, J.H. Won, K.S. Choi, Y.T. Lim, J.S. Shin, J.-U. Park, Tin-doped indium oxide films for highly flexible transparent conducting electrodes, Thin Solid Films 615 (2016) 8–12.

[20] R. Qian, J. Liao, G. Luo, H. Wu, ITO-free organic solar cells with oxide/metal/oxide multilayer structure cathode, Org. Electron. 108 (2022), 106614.
[21] S. Sharma, S. Shriwastava, S. Kumar, K. Bhatt, C.C. Tripathi, Alternative transparent conducting electrode materials for flexible optoelectronic devices, Opto-Electron. Rev. 26 (2018) 223–235.
[22] R. Ramarajan, M. Kovendhan, K. Thangaraju, D.P. Joseph, Indium-free large area Nb-doped SnO2 thin film as an alternative transparent conducting electrode, Ceram. Int. 46 (2020) 12224–12231.
[23] S. Sohn, Y.S. Han, Transparent conductive oxide (TCO) films for organic light emissive devices (OLEDS), in: Organic Light Emitting Diode-Material, Process and Devices, Intech Open Publisher, 2011, pp. 233–273. Chapter 9. ISBN: 978-953-51-4475-5, https://doi.org/10.5772/18545.
[24] N. Singh, Quantum dot sensitized solar cells (QDSSCs): a review, Int. J. Res. Anal. Rev. (IJRAR) 9 (2022) 509–529.
[25] S. Ho, S. Liu, Y. Chen, F. So, Review of recent progress in multilayer solution-processed organic light-emitting diodes, J. Photon. Energy 5 (2015), 057611.
[26] N.C. Giebink, S. Forrest, Quantum efficiency roll-off at high brightness in fluorescent and phosphorescent organic light emitting diodes, Phys. Rev. B 77 (2008), 235215.
[27] G. Gu, P. Burrows, S. Venkatesh, S. Forrest, M. Thompson, Vacuum-deposited, non-polymeric flexible organic light-emitting devices, Opt. Lett. 22 (1997) 172–174.
[28] T. Tsujimura, OLED Display Fundamentals and Applications, John Wiley & Sons, 2017.
[29] Z. Zhou, R. Chen, P. Jin, J. Hao, W. Wu, B. Yin, C. Zhang, J. Yao, Interplay between singlet and triplet excited states in interface exciplex OLEDs with fluorescence, phosphorescence, and TADF emitters, Adv. Funct. Mater. (2022) 2211059.
[30] R. Pode, Organic light emitting diode devices: an energy efficient solid state lighting for applications, Renew. Sust. Energ. Rev. 133 (2020) 110043.
[31] W. Cao, J. Li, H. Chen, J. Xue, Transparent electrodes for organic optoelectronic devices: a review, J. Photon. Energy 4 (2014) 040990.
[32] O. Tuna, Y. Selamet, G. Aygun, L. Ozyuzer, High quality ITO thin films grown by dc and RF sputtering without oxygen, J. Phys. D. Appl. Phys. 43 (2010) 055402.
[33] P.J. Kelly, R.D. Arnell, Magnetron sputtering: a review of recent developments and applications, Vacuum 56 (2000) 159–172.
[34] A.K. Sigdel, P.F. Ndione, J.D. Perkins, T. Gennett, M.F. van Hest, S.E. Shaheen, D.S. Ginley, J.J. Berry, Radio-frequency superimposed direct current magnetron sputtered Ga: ZnO transparent conducting thin films, J. Appl. Phys. 111 (2012) 093718.
[35] H. Li, Y.-J. Gao, S.-H. Yuan, D.-S. Wuu, W.-Y. Wu, S. Zhang, Improvement in the figure of merit of ITO-metal-ITO sandwiched films on poly substrate by high-power impulse magnetron sputtering, Coatings 11 (2021) 144.
[36] F. Kurdesau, G. Khripunov, A. Da Cunha, M. Kaelin, A. Tiwari, Comparative study of ITO layers deposited by DC and RF magnetron sputtering at room temperature, J. Non-Cryst. Solids 352 (2006) 1466–1470.
[37] M. Shiravand, N. Ghobadih, E.G. Hatam, Influence of Ar Gas Pressure on the Structural and Optical Properties and Surface Topography of Al-Doped ZnO Thin Films Sputtered by DC-Magnetron Sputtering Method, 2021, https://doi.org/10.21203/rs.3.rs-691578/v1.
[38] F. Lekoui, R. Amrani, W. Filali, E. Garoudja, L. Sebih, I.E. Bakouk, H. Akkari, S. Hassani, N. Saoula, S. Oussalah, Investigation of the effects of thermal annealing on the structural, morphological and optical properties of nanostructured Mn doped ZnO thin films, Opt. Mater. 118 (2021), 111236.

[39] E. Rucavado, Q. Jeangros, D.F. Urban, J. Holovský, Z. Remes, M. Duchamp, F. Landucci, R.E. Dunin-Borkowski, W. Körner, C. Elsässer, Enhancing the optoelectronic properties of amorphous zinc tin oxide by subgap defect passivation: a theoretical and experimental demonstration, Phys. Rev. B 95 (2017) 245204.

[40] B. Janarthanan, C. Thirunavukkarasu, S. Maruthamuthu, M.A. Manthrammel, M. Shkir, S. AlFaify, M. Selvakumar, V.R.M. Reddy, C. Park, Basic deposition methods of thin films, J. Mol. Struct. 1241 (2021), 130606.

[41] J. Meyer, P. Görrn, S. Hamwi, H.-H. Johannes, T. Riedl, W. Kowalsky, Indium-free transparent organic light emitting diodes with Al doped ZnO electrodes grown by atomic layer and pulsed laser deposition, Appl. Phys. Lett. 93 (2008) 306.

[42] D.S. Jung, Y.N. Ko, Y.C. Kang, S.B. Park, Recent progress in electrode materials produced by spray pyrolysis for next-generation lithium ion batteries, Adv. Powder Technol. 25 (2014) 18–31.

[43] A. Zhussupbekova, D. Caffrey, K. Zhussupbekov, C.M. Smith, I.V. Shvets, K. Fleischer, Low-cost, high-performance spray pyrolysis-grown amorphous zinc tin oxide: the challenge of a complex growth process, ACS Appl. Mater. Interfaces 12 (2020) 46892–46899.

[44] C.E. Knapp, G. Hyett, I.P. Parkin, C.J. Carmalt, Aerosol-assisted chemical vapor deposition of transparent conductive gallium-indium-oxide films, Chem. Mater. 23 (2011) 1719–1726.

[45] C. Xu, M.J. Hampden-Smith, T.T. Kodas, Aerosol-assisted chemical vapor deposition (AACVD) of silver, palladium and metal alloy ($Ag_{1-x}Pd_x$, $Ag_{1-x}Cu_x$ and $Pd_{1-x}Cu_x$) films, Adv. Mater. 6 (1994) 746–748.

[46] P.I. Filho, C.J. Carmalt, P. Angeli, E.S. Fraga, Mathematical modeling for the design and scale-up of a large industrial aerosol-assisted chemical vapor deposition process under uncertainty, Ind. Eng. Chem. Res. 59 (2020) 1249–1260.

[47] N. Abid, A.M. Khan, S. Shujait, K. Chaudhary, M. Ikram, M. Imran, J. Haider, M. Khan, Q. Khan, M. Maqbool, Synthesis of nanomaterials using various top-down and bottom-up approaches, influencing factors, advantages, and disadvantages: a review, Adv. Colloid Interf. Sci. 102597 (2021).

[48] A. Sharmin, S. Tabassum, M. Bashar, Z.H. Mahmood, Depositions and characterization of sol–gel processed Al-doped ZnO (AZO) as transparent conducting oxide (TCO) for solar cell application, J. Theor. Appl. Phys. 13 (2019) 123–132.

[49] W.M. Tsang, F.L. Wong, M.K. Fung, J. Chang, C.S. Lee, S.T. Lee, Transparent conducting aluminum-doped zinc oxide thin film prepared by sol–gel process followed by laser irradiation treatment, Thin Solid Films 517 (2008) 891–895.

[50] S. Cai, Y. Li, X. Chen, Y. Ma, X. Liu, Y. He, Optical and electrical properties of Ta-doped $ZnSnO_3$ transparent conducting films by sol–gel, J. Mater. Sci. Mater. Electron. 27 (2016) 6166–6174.

[51] H.J. Kim, M.-J. Maeng, J. Park, M.G. Kang, C.Y. Kang, Y. Park, Y.J. Chang, Chemical and structural analysis of low-temperature excimer-laser annealing in indium-tin oxide sol-gel films, Curr. Appl. Phys. 19 (2019) 168–173.

[52] B.W.N.H. Hemasiri, J.-K. Kim, J.-M. Lee, Fabrication of highly conductive graphene/ITO transparent bi-film through CVD and organic additives-free sol-gel techniques, Sci. Rep. 7 (2017) 1–12.

[53] F.J. Serrao, K. Sandeep, S. Dharmaprakash, Annealing-induced modifications in sol–gel spin-coated Ga: ZnO thin films, J. Sol-Gel Sci. Technol. 78 (2016) 438–445.

[54] M. Yahya, M. Fadavieslam, The effects of argon plasma treatment on ITO properties and the performance of OLED devices, Opt. Mater. 120 (2021), 111400.

[55] P.-C. Hsu, S. Wang, H. Wu, V.K. Narasimhan, D. Kong, H. Ryoung Lee, Y. Cui, Performance enhancement of metal nanowire transparent conducting electrodes by mesoscale metal wires, Nat. Commun. 4 (2013) 1–7.
[56] A.I. Hofmann, E. Cloutet, G. Hadziioannou, Materials for transparent electrodes: from metal oxides to organic alternatives, Adv. Electron. Mater. 4 (2018) 1700412.
[57] K. Sibin, G. Srinivas, H. Shashikala, A. Dey, N. Sridhara, A.K. Sharma, H.C. Barshilia, Highly transparent and conducting ITO/Ag/ITO multilayer thin films on FEP substrates for flexible electronics applications, Sol. Energy Mater. Sol. Cells 172 (2017) 277–284.
[58] S.K. Kang, D.Y. Kang, J.W. Park, K.R. Son, T.G. Kim, Work function-tunable ZnO/Ag/ZnO film as an effective hole injection electrode prepared via nickel doping for thermally activated delayed fluorescence-based flexible blue organic light-emitting diodes, Appl. Surf. Sci. 538 (2021), 148202.
[59] J.-H. Kim, D.-H. Kim, S.-K. Kim, D. Bae, Y.-Z. Yoo, T.-Y. Seong, Control of refractive index by annealing to achieve high figure of merit for TiO2/Ag/TiO2 multilayer films, Ceram. Int. 42 (2016) 14071–14076.
[60] W. Gaynor, S. Hofmann, M.G. Christoforo, C. Sachse, S. Mehra, A. Salleo, M.D. McGehee, M.C. Gather, B. Lüssem, L. Müller-Meskamp, Color in the corners: ITO-free white OLEDs with angular color stability, Adv. Mater. 25 (2013) 4006–4013.
[61] Y. Leterrier, L. Medico, F. Demarco, J.-A. Månson, U. Betz, M. Escola, M.K. Olsson, F. Atamny, Mechanical integrity of transparent conductive oxide films for flexible polymer-based displays, Thin Solid Films 460 (2004) 156–166.
[62] S. Ozbay, N. Erdogan, F. Erden, M. Ekmekcioglu, B. Rakop, M. Ozdemir, G. Aygun, L. Ozyuzer, Surface free energy and wettability properties of transparent conducting oxide-based films with Ag interlayer, Appl. Surf. Sci. 567 (2021) 150901.
[63] J. Ederth, P. Heszler, A. Hultåker, G. Niklasson, C. Granqvist, Indium tin oxide films made from nanoparticles: models for the optical and electrical properties, Thin Solid Films 445 (2003) 199–206.
[64] R. Mahdiyar, M. Fadavieslam, The effects of chemical treatment on ITO properties and performance of OLED devices, Opt. Quant. Electron. 52 (2020) 1–12.
[65] S.-Y. Yu, J.-H. Chang, P.-S. Wang, C.-I. Wu, Y.-T. Tao, Effect of ITO surface modification on the OLED device lifetime, Langmuir 30 (2014) 7369–7376.
[66] W.G. Haines, R.H. Bube, Effects of heat treatment on the optical and electrical properties of indium–tin oxide films, J. Appl. Phys. 49 (1978) 304–307.
[67] C. Guillén, J. Herrero, Comparison study of ITO thin films deposited by sputtering at room temperature onto polymer and glass substrates, Thin Solid Films 480 (2005) 129–132.
[68] T. Minami, Substitution of transparent conducting oxide thin films for indium tin oxide transparent electrode applications, Thin Solid Films 516 (2008) 1314–1321.
[69] S. Prakash, P-224: damage-free cathode coating process for OLEDs, in: SID Symposium Digest of Technical Papers, Wiley Online Library, 2008, pp. 2046–2048.
[70] J.-H. Lee, S. Lee, J.-B. Kim, J. Jang, J.-J. Kim, A high performance transparent inverted organic light emitting diode with 1, 4, 5, 8, 9, 11-hexaazatriphenylenehexacarbonitrile as an organic buffer layer, J. Mater. Chem. 22 (2012) 15262–15266.
[71] E. Kim, J. Kwon, C. Kim, T.-S. Kim, K.C. Choi, S. Yoo, Design of ultrathin OLEDs having oxide-based transparent electrodes and encapsulation with sub-mm bending radius, Org. Electron. 82 (2020), 105704.
[72] Y.-L. Wang, B.-J. Li, S.-S. Li, H. Li, L.-J. Huang, N.-F. Ren, Parameter optimization in femtosecond pulsed laser etching of fluorine-doped tin oxide films, Opt. Laser Technol. 116 (2019) 162–170.

[73] D. Saikia, R. Sarma, Performance improvement of organic light emitting diode using 4,4′-N,N′-dicarbazole-biphenyl (CBP) layer over fluorine-doped tin oxide (FTO) surface with doped light emitting region, Pramana 91 (2018) 1–9.
[74] D. Saikia, R. Sarma, Organic light-emitting diodes with a perylene interlayer between the electrode–organic interface, J. Electron. Mater. 47 (2018) 737–743.
[75] B. Suer, M. Ozenbas, Conducting fluorine doped tin dioxide (FTO) coatings by ultrasonic spray pyrolysis for heating applications, Ceram. Int. 47 (2021) 17245–17254.
[76] A. Way, J. Luke, A.D. Evans, Z. Li, J.-S. Kim, J.R. Durrant, H.K. Hin Lee, W.C. Tsoi, Fluorine doped tin oxide as an alternative of indium tin oxide for bottom electrode of semi-transparent organic photovoltaic devices, AIP Adv. 9 (2019), 085220.
[77] B.-R. Koo, J.-W. Bae, H.-J. Ahn, Optoelectronic multifunctionality of combustion-activated fluorine-doped tin oxide films with high optical transparency, Ceram. Int. 45 (2019) 10260–10268.
[78] D. Saikia, R. Sarma, A comparative study of the influence of nickel oxide layer on the FTO surface of organic light emitting diode, Indian J. Phys. 92 (2018) 307–313.
[79] K. Ellmer, A. Klein, B. Rech, Transparent Conductive Zinc Oxide: Basics and Applications in Thin Film Solar Cells, Springer Berlin, Heidelberg, 2007. ISBN: 978-3-540-73612-7, https://doi.org/10.1007/978-3-540-73612-7.
[80] S. Höfle, A. Schienle, M. Bruns, U. Lemmer, A. Colsmann, Enhanced electron injection into inverted polymer light-emitting diodes by combined solution-processed zinc oxide/polyethylenimine interlayers, Adv. Mater. 26 (2014) 2750–2754.
[81] X. Jia, N. Wu, J. Wei, L. Zhang, Q. Luo, Z. Bao, Y.-Q. Li, Y. Yang, X. Liu, C.-Q. Ma, A low-cost and low-temperature processable zinc oxide-polyethylenimine (ZnO: PEI) nano-composite as cathode buffer layer for organic and perovskite solar cells, Org. Electron. 38 (2016) 150–157.
[82] I. Lange, S. Reiter, J. Kniepert, F. Piersimoni, M. Pätzel, J. Hildebrandt, T. Brenner, S. Hecht, D. Neher, Zinc oxide modified with benzylphosphonic acids as transparent electrodes in regular and inverted organic solar cell structures, Appl. Phys. Lett. 106 (2015) 3231.
[83] A.u.H.S. Rana, A. Shahid, J.Y. Lee, H.S. Kim, High-power microwave-assisted Ga doping, an effective method to tailor n-ZnO/p-Si heterostructure optoelectronic characteristics, Phys. Status Solidi A 215 (2018) 1700763.
[84] D. Gaspar, L. Pereira, K. Gehrke, B. Galler, E. Fortunato, R. Martins, High mobility hydrogenated zinc oxide thin films, Sol. Energy Mater. Sol. Cells 163 (2017) 255–262.
[85] J.H. Park, Y.C. Cho, J.M. Shin, S. Cha, C.R. Cho, H.S. Kim, S.J. Yoon, S. Jeong, S.E. Park, A. Lim, A study of transparent conductive aluminum-doped zinc oxide fabricated on a flexible polyethersulphone (PES) substrate, J-Korean Phys. Soc. 51 (2007) 1968.
[86] M.N. Rezaie, N. Manavizadeh, E.M.N. Abadi, E. Nadimi, F.A. Boroumand, Comparison study of transparent RF-sputtered ITO/AZO and ITO/ZnO bilayers for near UV-OLED applications, Appl. Surf. Sci. 392 (2017) 549–556.
[87] H.S. Das, R. Das, P.K. Nandi, S. Biring, S.K. Maity, Influence of Ga-doped transparent conducting ZnO thin film for efficiency enhancement in organic light-emitting diode applications, Appl. Phys. A 127 (2021) 1–7.
[88] S.M. Rozati, S.A.M. Ziabari, A review of various single layer, bilayer, and multilayer TCO materials and their applications, Mater. Chem. Phys. (2022) 126789.
[89] P. Sakthivel, S. Asaithambi, M. Karuppaiah, R. Yuvakkumar, Y. Hayakawa, G. Ravi, Improved optoelectronic properties of Gd doped cadmium oxide thin films through optimized film thickness for alternative TCO applications, J. Alloys Compd. 820 (2020), 153188.

[90] G. Genchi, M.S. Sinicropi, G. Lauria, A. Carocci, A. Catalano, The effects of cadmium toxicity, Int. J. Environ. Res. Public Health 17 (2020) 3782.
[91] V. Fauzia, M. Yusnidar, L.H. Lalasari, A. Subhan, A.A. Umar, High figure of merit transparent conducting Sb-doped SnO2 thin films prepared via ultrasonic spray pyrolysis, J. Alloys Compd. 720 (2017) 79–85.
[92] G. Dalapati, H. Sharma, A. Guchhait, P. Bamola, S. Zhuk, Q.N. Liu, S. Gopalan, A.M.S. Krishna, S. Mukhopadhyay, A. Dey, Critical review on SnO2 for transparent conductor and Electron transport layer: impact on dopants and functionalization of SnO2 on photovoltaic and energy storage devices, J. Mater. Chem. A (2021).
[93] J. Montero, J. Herrero, C. Guillén, Preparation of reactively sputtered Sb-doped SnO2 thin films: structural, electrical and optical properties, Sol. Energy Mater. Sol. Cells 94 (2010) 612–616.
[94] E. Elangovan, M. Singh, M. Dharmaprakash, K. Ramamurthi, Some physical properties of spray deposited SnO2 thin films, J. Optoelectron. Adv. Mater. 6 (2004) 197–203.
[95] R. Ramarajan, J.M. Fernandes, M. Kovendhan, G. Dasi, N.P. Reddy, K. Thangaraju, D.P. Joseph, Boltzmann conductivity approach for charge transport in spray-deposited transparent Ta-doped SnO2 thin films, J. Alloys Compd. 897 (2022), 163159.
[96] Y.G. Bi, Y.F. Liu, X.L. Zhang, D. Yin, W.Q. Wang, J. Feng, H.B. Sun, Ultrathin metal films as the transparent electrode in ITO-free organic optoelectronic devices, Adv. Opt. Mater. 7 (2019) 1800778.
[97] M. Kim, C. Lim, D. Jeong, H.-S. Nam, J. Kim, J. Lee, Design of a MoOx/Au/MoOx transparent electrode for high-performance OLEDs, Org. Electron. 36 (2016) 61–67.
[98] T.H. Kim, C.H. Kim, S.K. Kim, Y.S. Lee, L.S. Park, High quality transparent conductive ITO/Ag/ITO multilayer films deposited on glass substrate at room temperature, Mol. Cryst. Liq. Cryst. 532 (2010). 112/[528]–118/[534].
[99] T.H. Kim, B.H. Choi, J.S. Park, S.M. Lee, Y.S. Lee, L.S. Park, Transparent conductive ITO/Ag/ITO multilayer films prepared by low temperature process and physical properties, Mol. Cryst. Liq. Cryst. 520 (2010). 209/[485]–214/[490].
[100] C.W. Joo, J. Lee, W.J. Sung, J. Moon, N.S. Cho, H.Y. Chu, J.-I. Lee, ITO/metal/ITO anode for efficient transparent white organic light-emitting diodes, Jpn. J. Appl. Phys. 54 (2015) 02BC04.
[101] C.-H. Peng, P.S. Chen, J.W. Lo, T.W. Lin, S. Lee, Indium-free transparent TiOx/Ag/WO3 stacked composite electrode with improved moisture resistance, J. Mater. Sci. Mater. Electron. 27 (2016) 12060–12066.
[102] M.G. Varnamkhasti, H.R. Fallah, M. Mostajaboddavati, A. Hassanzadeh, Influence of Ag thickness on electrical, optical and structural properties of nanocrystalline MoO3/Ag/ITO multilayer for optoelectronic applications, Vacuum 86 (2012) 1318–1322.
[103] D.-H. Kim, J.H. Kim, H.-K. Lee, J.-Y. Na, S.-K. Kim, J.H. Lee, S.-W. Kim, Y.-Z. Yoo, T.-Y. Seong, Flexible and transparent TiO2/Ag/ITO multilayer electrodes on PET substrates for organic photonic devices, J. Mater. Res. 30 (2015) 1593–1598.
[104] Z. Zhao, T. Alford, The optimal TiO2/Ag/TiO2 electrode for organic solar cell application with high device-specific Haacke figure of merit, Sol. Energy Mater. Sol. Cells 157 (2016) 599–603.
[105] S.-W. Cho, J.-A. Jeong, J.-H. Bae, J.-M. Moon, K.-H. Choi, S.W. Jeong, N.-J. Park, J.-J. Kim, S.H. Lee, J.-W. Kang, Highly flexible, transparent, and low resistance indium zinc oxide–Ag–indium zinc oxide multilayer anode on polyethylene terephthalate substrate for flexible organic light light-emitting diodes, Thin Solid Films 516 (2008) 7881–7885.

[106] E.M. Kim, I.-S. Choi, J.-P. Oh, Y.-B. Kim, J.-H. Lee, Y.-S. Choi, J.-D. Cho, Y.-B. Kim, G.-S. Heo, Transparent conductive ZnInSnO–Ag–ZnInSnO multilayer films for polymer dispersed liquid-crystal based smart windows, Jpn. J. Appl. Phys. 53 (2014), 095505.
[107] M. Girtan, Comparison of ITO/metal/ITO and ZnO/metal/ZnO characteristics as transparent electrodes for third generation solar cells, Sol. Energy Mater. Sol. Cells 100 (2012) 153–161.
[108] Y. Ji, J. Yang, W. Luo, L. Tang, X. Bai, C. Leng, C. Ma, X. Wei, J. Wang, J. Shen, Ultraflexible and high-performance multilayer transparent electrode based on ZnO/Ag/CuSCN, ACS Appl. Mater. Interfaces 10 (2018) 9571–9578.
[109] G. Kumar, S. Biring, Y.-N. Lin, S.-W. Liu, C.-H. Chang, Highly efficient ITO-free organic light-emitting diodes employing a roughened ultra-thin silver electrode, Org. Electron. 42 (2017) 52–58.
[110] J.-A. Jeong, H.-K. Kim, Low resistance and highly transparent ITO–Ag–ITO multilayer electrode using surface plasmon resonance of Ag layer for bulk-heterojunction organic solar cells, Sol. Energy Mater. Sol. Cells 93 (2009) 1801–1809.
[111] M. Ekmekcioglu, N. Erdogan, A.T. Astarlioglu, S. Yigen, G. Aygun, L. Ozyuzer, M. Ozdemir, High transparent, low surface resistance ZTO/Ag/ZTO multilayer thin film electrodes on glass and polymer substrates, Vacuum 187 (2021), 110100.
[112] D.H. Kim, S.Y. Lee, Variation of optical and electrical properties of amorphous In–Ga–Zn–O/Ag/amorphous In–Ga–Zn–O depending on Ag thickness, Thin Solid Films 536 (2013) 327–329.
[113] S.-M. Lee, H.-W. Koo, T.-W. Kim, H.-K. Kim, Asymmetric ITO/Ag/ZTO and ZTO/Ag/ITO anodes prepared by roll-to-roll sputtering for flexible organic light-emitting diodes, Surf. Coat. Technol. 343 (2018) 115–120.
[114] Y. Kim, J. Park, D. Choi, H. Jang, J. Lee, H. Park, J. Choi, D. Ju, J. Lee, D. Kim, ITO/Au/ITO multilayer thin films for transparent conducting electrode applications, Appl. Surf. Sci. 254 (2007) 1524–1527.
[115] Y.-H. Song, T.-Y. Eom, S.-B. Heo, J.-Y. Cheon, B.-C. Cha, D. Kim, Characteristics of IGZO/Ni/IGZO tri-layer films deposited by DC and RF magnetron sputtering, Mater. Lett. 205 (2017) 122–125.
[116] S. Park, J. Yoon, S. Kim, P. Song, Hydrogen-driven dramatically improved mechanical properties of amorphized ITO–Ag–ITO thin films, RSC Adv. 11 (2021) 3439–3444.
[117] J. Cui, A. Wang, N.L. Edleman, J. Ni, P. Lee, N.R. Armstrong, T.J. Marks, Indium tin oxide alternatives—high work function transparent conducting oxides as anodes for organic light-emitting diodes, Adv. Mater. 13 (2001) 1476–1480.
[118] M.G. Kang, L.J. Guo, Nanoimprinted semitransparent metal electrodes and their application in organic light-emitting diodes, Adv. Mater. 19 (2007) 1391–1396.
[119] I.Y. Ahmet, F.F. Abdi, R. van de Krol, Chemical treatment of Sn-containing transparent conducting oxides for the enhanced adhesion and thermal stability of electroplated metals, Adv. Mater. Interfaces 9 (2022) 2201617.
[120] S.A. Paniagua, A.J. Giordano, O.N.L. Smith, S. Barlow, H. Li, N.R. Armstrong, J.E. Pemberton, J.-L. Brédas, D. Ginger, S.R. Marder, Phosphonic acids for interfacial engineering of transparent conductive oxides, Chem. Rev. 116 (2016) 7117–7158.
[121] K.N. Manjunatha, S. Paul, Stability study: transparent conducting oxides in chemically reactive plasmas, Appl. Surf. Sci. 424 (2017) 316–323.
[122] L.-J. Meng, A. Macarico, R. Martins, Study of annealed indium tin oxide films prepared by rf reactive magnetron sputtering, Vacuum 46 (1995) 673–680.

[123] S. Honda, M. Watamori, K. Oura, The effects of oxygen content on electrical and optical properties of indium tin oxide films fabricated by reactive sputtering, Thin Solid Films 281 (1996) 206–208.
[124] E.G. Jeong, J.H. Kwon, K.S. Kang, S.Y. Jeong, K.C. Choi, A review of highly reliable flexible encapsulation technologies towards rollable and foldable OLEDs, J. Inform. Display 21 (2020) 19–32.
[125] G. Gustafsson, Y. Cao, G. Treacy, F. Klavetter, N. Colaneri, A. Heeger, Flexible light-emitting diodes made from soluble conducting polymers, Nature 357 (1992) 477–479.
[126] E. Kim, H. Cho, K. Kim, T.W. Koh, J. Chung, J. Lee, Y. Park, S. Yoo, A facile route to efficient, low-cost flexible organic light-emitting diodes: utilizing the high refractive index and built-in scattering properties of industrial-grade PEN substrates, Adv. Mater. 27 (2015) 1624–1631.
[127] D.-H. Kim, M.-R. Park, G.-H. Lee, Preparation of high quality ITO films on a plastic substrate using RF magnetron sputtering, Surf. Coat. Technol. 201 (2006) 927–931.
[128] J.-H. Jou, C.-P. Wang, M.-H. Wu, H.-W. Lin, H.C. Pan, B.-H. Liu, High-efficiency flexible white organic light-emitting diodes, J. Mater. Chem. 20 (2010) 6626–6629.
[129] J.H. Kwon, Y. Jeon, K.C. Choi, Robust transparent and conductive gas diffusion multibarrier based on Mg-and Al-doped ZnO as indium tin oxide-free electrodes for organic electronics, ACS Appl. Mater. Interfaces 10 (2018) 32387–32396.
[130] J. Chung, H. Cho, T.-W. Koh, J. Lee, E. Kim, J. Lee, J.-I. Lee, S. Yoo, Towards highly efficient and highly transparent OLEDs: advanced considerations for emission zone coupled with capping layer design, Opt. Express 23 (2015) 27306–27314.
[131] C.S. Choi, D.Y. Kim, S.M. Lee, M.S. Lim, K.C. Choi, H. Cho, T.W. Koh, S. Yoo, Blur-free outcoupling enhancement in transparent organic light emitting diodes: a nanostructure extracting surface plasmon modes, Adv. Opt. Mater. 1 (2013) 687–691.
[132] M.S. Farhan, E. Zalnezhad, A.R. Bushroa, A.A.D. Sarhan, Electrical and optical properties of indium-tin oxide (ITO) films by ion-assisted deposition (IAD) at room temperature, Int. J. Precis. Eng. Manuf. 14 (2013) 1465–1469.

Section B

Metal oxide-based phosphors and their applications

Metal oxide-based nanophosphors for next generation optoelectronic and display applications

4

Pooja Yadav and P. Abdul Azeem
Department of Physics, National Institute of Technology, Warangal, TS, India

Chapter outline

1 Introduction

Display technology has been around for a long time, with widespread applications in the production of visual-based devices, the most popular of which are televisions, computer monitors, and mobile phone screens [1]. However, this technology is still evolving, and display device development must continue, with a focus on improving efficiency and picture quality, reducing weight and power consumption, selecting environmentally friendly materials, and developing more advanced futuristic devices such as three-dimensional (3D), flexible, and transparent displays. Because of the constant need for improved display devices, new and previously unknown materials are being investigated as substitutes for traditional materials such as transparent conductive oxides and semiconductors, as well as materials with unique crystal structures and properties [2]. A display device is primarily an output device for the visual presentation of information. Based on the work principles of electronic display devices, it is divided into several categories: cathode ray tube (CRT), liquid-crystal display (LCD), organic light-emitting diode (OLED), and micro-LED [3]. The CRT, which consists of one or more electron guns and a phosphorescent screen to display images, was first

Metal Oxides for Next-Generation Optoelectronic, Photonic, and Photovoltaic Applications
https://doi.org/10.1016/B978-0-323-99143-8.00014-6

proposed in early 1907. A CRT produces an image in a phosphor-pixelated screen based on the principle of electron beam excitation of a fluorescent screen. In 1950, the first color CRT television was created [4]. This is utilized in classic display technology and CRT TV has dominated the display industry for decades because of its great properties, such as superb visual depth of field and high response rate. Following the CRT, the liquid crystal display (LCD) and the plasma display panel (PDP) entered into the market [5]. These screens gained popularity because they were more portable and consumed less power than CRT monitors, and because of the constant focus put into improving LCD technology, the plasma display panel has become uncompetitive [6]. LCDs made it feasible to create gadgets such as digital watches, mobile phones, laptops, and other small-screen electronics. LCDs were originally designed for handheld and portable devices, but they have now moved into sectors where CRTs formerly reigned supreme, such as computer displays and TVs. The LCD display device, on the other hand, has several drawbacks, including a slow response time, low efficiency, and limited color saturation. A variety of improvement approaches have been explored to solve these issues [7]. Consumers continue to be interested in the modern LCD display device. After CRT and LCD, OLED entered the market, prompting a flurry of research into how to increase the OLED's performance [8]. Wide viewing angle, infinite contrast ratio, low power consumption, adaptability, and response time a thousand times greater than those of LCD are the major benefits of OLED. As a result, the commercialization of OLED began in 2000 as a result of its advantages. OLED, like LCD, has drawbacks, including "burn-in" issues and constraints in theoretical work. Following this progress, the demand for smaller devices led to the creation of micro-LED and mini-LED devices. With its high contrast ratio, low power consumption, great brightness, and, most significantly, long life span, micro-LEDs may outperform OLEDs [9]. Mini-LEDs are generally utilized as the backlight of LCDs for high resolution and wide color spectrum [10]. It has received a lot of attention since JIANG's group at Texas Technique University reported the first fabrication of a micro-LED chip with a diameter of 12 m in 2000.

Electronic displays are classified as either emissive display or nonemissive display, with phosphors used to generate visible light in both types of display [11]. To be used in displays, phosphors must have high quantum efficiency, high lifetime, high reflectivity to visible light, required afterglow decay or persistence, proper color coordinates, and temperature as specified by NTSC/PAL standards. When manufacturing display devices, important physical requirements such as particle distribution, particle shape, particle size, surface charge, and body color must be recognized [12]. When Nakamura invented the efficient blue LED at the end of the last century, he made a breakthrough in lighting technology. In LCD backlight panels, compact trichromatic lamp phosphors such as BAM (blue), $LaPO_4$:Ce, Tb (green), and Y_2O_3:Eu (red) are used. Furthermore, various attempts are being made to design LCD backlights with solid-state lighting (SSL) modules in order to improve efficiency, longer life, and stability at high temperatures. To convert a portion of blue light from LEDs to yellow light, a Ce-activated YAG-based phosphor is currently being used. One can generate white light by combining yellow and blue phosphors. UV LEDs and red, green, and blue (RGB) phosphors in solid-state light modules will be efficient backlight for large-area LCD monitors [13]. This breakthrough has allowed LEDs to generate white light in two ways: (i) by combining blue, green, and red light from different LED chips and (ii) by using phosphor conversion technology. In this chapter, silicate-based phosphor materials are discussed in detail. These materials have various sophisticated applications in optoelectronics like waveguides and photoluminescence devices [14]. These materials are also used in microelectronics because of low-k dielectrics, diffusion sources, sensitive membranes, and multilayer metallization planarization [15]. Beyond the technological restrictions of existing lighting technologies, white light-emitting diodes (LEDs) require reliable and efficient phosphor systems in order to be used in solid-state lighting [16]. As a result, inorganic solid-state conversion phosphors must be carefully chosen and evaluated in terms of their unique material properties and optical synergies [17].

Recently, for red-emitting nanophosphors, $Y_2O_3:Eu^{3+}$ and Li^+-activated $Y_2O_3:Eu^{3+}$ were successfully synthesized using coprecipitation technique, and it was observed that enhancement in the emission intensity by the co-doping of Li^+ ion in the host material [18]. $Ca_{0.5}Y_{1.90-x}O_3:xHo^{3+}$ ($0.5 \leq x \leq 3$ mol%) nanophosphor was synthesized using combustion method, and this phosphor exhibits good thermal stability and high color purity. It can be used for solid-state lighting and optoelectronic applications [19]. In recent studies, $Y_2Zr_2O_7:Er^{3+}$ nanophosphor co-doped with Li^+ ion, and as the Li^+ concentration increases, PL intensity also increases. This material shows the best optical properties and can be used for solar cell and high-power LED applications [20]. $CaTiO_3:Gd_2O_3:Eu^{3+}$ perovskite material can also be derived for W-LED devices [21]. $BaLa_2ZnO_5:Dy^{3+}$ nanomaterial is a new white light emitter. This material is a good contender for creating WLEDs, lasers, and sensors [22]. Sm^{3+}-doped $Bi_4MgO_4(PO_4)_2$ is also a good contender for solid-state lighting. It exhibits bright orange color and good structural and luminescence properties [23]. Ca^{2+}-doped Eu: Y_2O_3 nanophosphors recently synthesized using combustion method and it has a promising thermal stability with a high quantum yield of 81% [24]. Novel $SrGdAlO_4:Er^{3+}$ material is an energy-efficient phosphor and the best candidate for PC-WLEDs as well as other optoelectronic device [25]. Eu^{3+}-doped $MGdAl_3O_7$ (M=Mg, Ca, Sr, and Ba) exhibits sharp red emission and after examination, it is concluded that this is a brilliant material for applications in white light-emitting diodes, fluorescent panels, and modern display devices [26].

Apart from LEDs, various displays such as flat panel displays (FPD), plasma display panels (PDP), electroluminescence displays (ELD), and liquid crystal displays (LCD) have a competitive advantage in display applications. The large-screen cathode ray tubes (CRT) used in high-definition television (HDTV) sets are too bulky and take up too much space for widespread acceptance. As a result, these displays are in high demand in the market for large-screen displays [27]. The vacuum ultraviolet (VUV) radiation from an inert gas plasma excites three primary color phosphors in PDPs, imposing a unique requirement on the phosphor [28]. Tm^{3+}-activated lanthanum phosphate (LPTM) phosphor was used for PDPs with good color saturation and better stability [29]. $BaMgAl_{10}O_{17}:Eu^{2+}$ (BAM), an important blue phosphor of plasma display panel (PDP) has been reported [30]. Europium-activated yttrium, gadolinium borate [$(Y,Gd)BO_3:Eu^{3+}$] is an efficient red-emitting phosphor that is currently used in PDPs due to its high QE persistence characteristics [31].

The alkaline earth metal rare-earth doped silicates, also known as long persistent luminescence phosphors (LPL), develop as luminescence phosphors continuously over a long period of time and are currently being studied extensively [32]. Their applications, such as traffic signs, decoration, and textile printing, are constantly expanding in the growing market due to these unique characteristics [33]. Silicate phosphors have advantages over sulfide phosphors in terms of long-term durability and brighter luminescence emission, as well as abundance, ease of preparation, and cost effectiveness [34]. Calcium magnesium silicate CMS-based phosphors have luminescent properties as a result of being co-doped with various rare-earth ions [35]. CMS has been extensively investigated as a host for LPL and PDPs applications. After the stimulation is removed, LPL can last for a long time in the dark. This process has piqued people's interest. As a result, it is used in a variety of applications, including emergency lights, safety indicators, road signals, glossy paints, graphic art pieces, and so on. Full color exposure also necessitates the use of blue, green, and red LPLs, as well as white LPL [36].

As energy-saving technologies, phosphor-converted white light-emitting diodes (LEDs) have emerged as a vital solid-state light source for the next-generation lighting sector, with higher luminous efficiency, a longer lifetime, and adjustable optical performance [37]. Silicate phosphor has become a popular topic all over the world, with the number of publications on the subject growing at an exponential rate over time. Because of silicate's wide range of future applications in luminescence phosphorus in fluorescent ceramics, white light-emitting diodes, and a variety of other optical devices, as well as its emerging ultrawide-bandgap semiconductor material, it has recently attracted a lot of scientific and technological attention [38].

Nanoparticles are usually dispersed in a continuous phase [39]. The size of the phosphor particles has a big impact on how well phosphor screens work [40]. The top two monolayers of nanoparticles contain nearly half of their atoms, making the phosphor optical properties highly sensitive to surface morphology. Nanophosphors have been studied extensively over the last decade due to their potential for use in a variety of high-performance displays and devices. Almost all displays use these as a strategic component [41]. The semiconductor nanophosphor exhibits a wide range of unique optical, electronic, and chemical properties, owing to two main factors: quantum confinement effects and a large surface-to-volume ratio [42]. Due to increased luminescence efficiency and a reduction in radiative lifetime by orders of magnitude from milliseconds to nanoseconds, luminescent nanocrystals, also known as nanophosphors, have attracted considerable interest in the recent decades when compared to bulk crystal [43]. To summarize, LED-based display techniques have a wide range of intriguing properties, including flexibility, printable, wider viewing angle, wide operating temperature, color-temperature tenability, high resolution, better contrast ratio, lower power consumption, transparency, and lightweight, all of which have made them very convincing components in the small and large display field [44].

Silicate phosphors have been designed and developed in the recent years to meet the needs of light-emitting diode applications. Eu^{3+}-doped $Ca_2MgSi_2O_7$ has recently emerged as a novel phosphor for WLED applications [45]. The single-particle diagnosis approach was used to synthesize new nitride silicate phosphates $Ba_5Si_{11}Al_7N_{25}:Eu^{2+}$ and $BaSi_4Al_3N_9:Eu^{2+}$ for solid-state lighting, having a quantum efficiency of 36% [46]. With a quantum yield of 60%, barium yttrium silicate ($Ba_9Y_2Si_6O_{24}:Ce^{3+}$) was discovered to be an effective blue-green phosphor for use in solid-state lighting [47]. Recently, oxo nitride silicate with an appetite structure was successfully synthesized. The obtained $Y_5Si_3O_{12}N:Ce^{3+}/Dy^{3+}$ phosphor is a promising single-phase phosphor for use in white light LEDs [48]. Solid-state reaction was used to create a new blue-emitting zirconium silicate phosphor, $K_2ZrSi_2O_7:Eu^{2+}$, which proved to be a better phosphor for white-light LEDs and field emission displays. $Na_3ScSi_2O_7:Ce^{3+}$ (NSSO:Ce^{3+}) was successfully synthesized using a solid-state method at low temperature with a quantum efficiency of 50.9%, making it a good candidate for thermal stability in warm WLEDs and field emission displays [49]. Europium-doped silicate-phosphor layer was used to boost the solar cell's impressive efficiency to 18.77% [50]. This trend is expected to continue for the foreseeable future as research funding for opto-electronic device development rises day by day [51].

2 Phosphor and luminescence mechanism

The word phosphor comes from the Greek language and it means light bearer. Phosphors are the light emissive material that consists of a host material and a very less quantity of an activator [52]. The impurity concentrations in general are relatively low because of the fact that at higher concentrations, the efficiency of the luminescence process usually decreases due to concentration quenching effects. Activator material is either rare-earth ions or transition metal ions (3d or 4f metal). These materials are having a capability of converting the low wavelength light to high wavelength light due to the electronic transition between the valence band and conduction band of the materials [53]. Phosphor are having a different composition such as (i) optically inactive host material that is doped by a suitable amount of dopant (rare-earth ion or transition metal ion) (ii) chemically stable crystalline host material that inherently contains the luminescent ion, and (iii) defect related (oxygen vacancies) luminescent material. The selection of host material is an important task because it only decides the dopant-ligand coordination, which effects the luminescence properties such as quantum efficiency, emission color, and reduce the nonradiative losses [54]. Usually, the wide bandgap materials such as oxides, nitrites, and silicates are used as a host material due to its wide bandgap which helps in improving the luminescence properties due to its low lattice phonon frequencies. Joining the host lattice and dopant ion are contributes for the luminescence properties. In order to create the emission, the ground state electron needs to be excited in the higher energy level. The relaxation of energetic electron from higher energy state to lower energy state gives the luminescence (Fig. 1A and B).

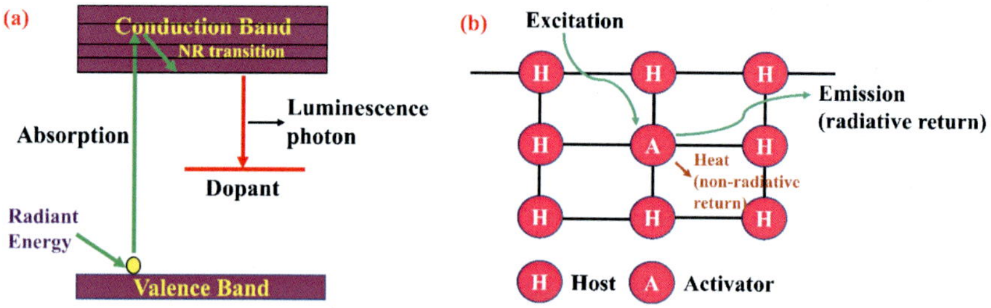

Fig. 1 (A) Luminescence mechanism happen in a material contains dopant ions (B) Luminescence ion A and its host lattice H.

There are different methods of providing the excitation based on that luminescence classified in various categories such as photoluminescence, electroluminescence, thermoluminescence, chemiluminescence, etc. [55]. When the material is excited by the absorbance of electromagnetic radiation of the photon and causes the radiation of a photon, this process is known as photoluminescence (PL). When the electric potential is applied and there is a conversion of electric energy into light emission, this process is known as electroluminescence (EL) and the device that produces the EL is known as light emitting diode (LED) [56]. In LED, p-n semiconductor junction is situated between the two electrodes and at least one electrode should be transparent when a situation where cells of low electrical resistance contacts without blocking light. It is based on the principle of recombination of electrons and holes at the junction when current is passed through the junction and during recombination, energy is released in the form of light (Fig. 2A and B). The color of the LEDs is determined by the material used in the semiconductor element. Usually in industrial applications,

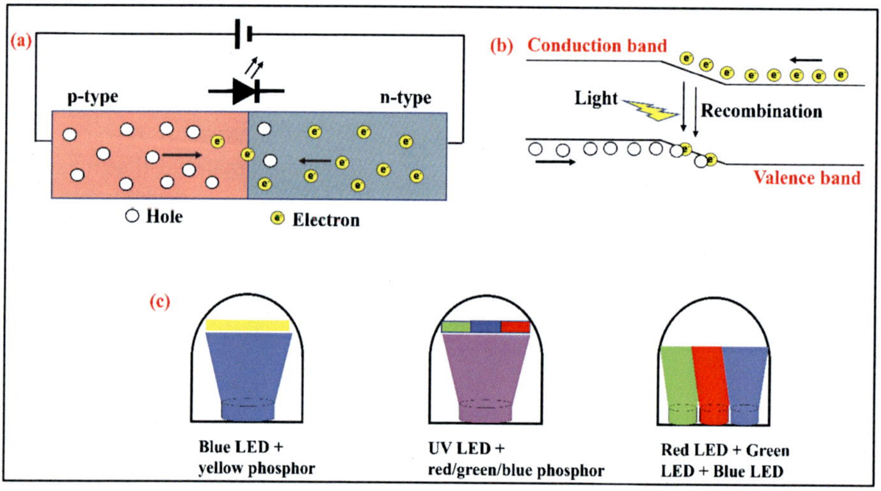

Fig. 2 (A) The forward biased p-n junction (B) Recombination of electrons and holes in the junction of p-n semiconductors (C) Illustration of white light generation by different mode.

white color LEDs are highly in demand as a light source and as backlighting for electronic displays [57]. There are three common methods for producing the white light LED as shown in Fig. 2C (i) combination of three different LEDs red, green, and blue, (ii) mixing of different color luminescent material irradiation by UV light to generate white color, and (iii) irradiation of blue light LED on the yellow color phosphor.

3 Silicate phosphor for LED applications

Energy conservation and environmental protection are gaining worldwide attention as a result of modern society and digitalization. In the recent decades, luminescence materials have played an essential role in the advancement of display technology [58]. The host material and activator are used to make phosphors. Activator is a luminescent center that is also referred to as a doping element. Phosphors are efficient luminescent materials and irreplaceable components of light-emitting devices like cathode ray tubes (CRTs), plasma display panels (PDPs), and field emission displays (FEDs) [59]. Many new and promising LED phosphors have been developed, modified, or invented, and their suitability in LED devices has been demonstrated. In the development of new LED phosphors, the silicate compound family is absolutely essential [60]. Fig. 3 show the various properties of silicate materials. The exploration and characterization of silicate-based luminescent materials has piqued the interest of material scientists because of the following characteristics: Silicate is a low-cost material and has a high stability and moisture-resistant properties. The rare-earth doped silicate materials possess high luminescence efficiency and reflect some remarkable optical, electrical, and magnetic properties.

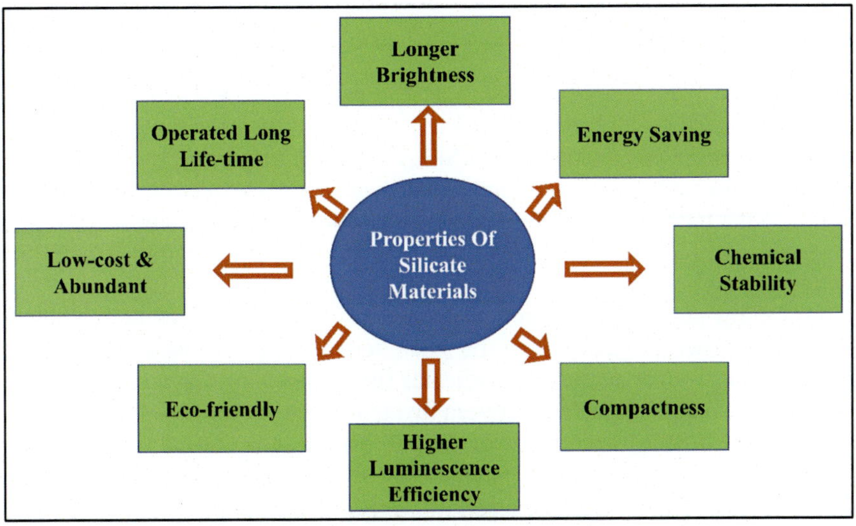

Fig. 3 Various properties exhibit by the silicate-based phosphor materials.

The opto-magnetic characteristics of rare-earth ions doped into a silicate host are altered due to the doped ion's occupation of metal ion positions in the host lattice [61]. As a result, spectroscopic features of doped ions should be considered while designing and developing improved silicate phosphor materials, particularly in white light-emitting diodes [62]. The development of white LEDs necessitates the use of highly efficient and stable silicate materials with a low color-related temperature and a high color rendering index [63]. The green silicate phosphor is used in the packaging of high-power positive white LEDs, and it has good thermal stability, long life, high brightness, uniform particle size, and high conversion efficiency [64]. It also has a low attenuation for long-term use in high temperature and high humidity environments, and can be used to create a temperature and humidity resistant LED lighting device. These LED phosphors can assist LCD manufacturers in producing displays with a wide color gamut, providing materials that match high-performance color filters and giving the screen unprecedented color realism [65]. To package warm white or special light LEDs, the series of silicate phosphors can be combined with other yellow or red phosphors.

The silicates phosphor category is important for long-lasting luminescent materials. For these applications, silicate phosphors are classified as binary silicates or ternary silicates. Orthosilicates and metasilicates are the two subcategories of the binary system. The disilicates and metasilicates containing Mg are the main focus of research in the ternary silicates system. As early as 1949, Smith [66] was reported the luminescent properties of $CaMgSi_2O_6:Eu^{2+}$, but after that the silicates system's persistent luminescent properties were not reported. Until 1975, the persistent luminescence time of $Zn_2SiO_4:Mn^{2+}$, As^{3+} could reach up to 30 min, when Chiba Institute of Technology and Dental University in Japan found it, it was the first time the orthosilicates system's persistent luminescence was identified [67].

4 Basics of silicate

Silicates are the main mineral components of the earth's crust and mantle, and they are composed primarily of Si and O. The $(SiO_4)^{4-}$ tetrahedron is the fundamental unit of all silicate's phosphor materials. The $(SiO_4)^{4-}$ tetrahedra form a variety of silicate minerals by linking to self-similar units that share one, two, three, or all four corner oxygens of the tetrahedron. Orthosilicates or isolated tetrahedral silicates are also a type of silicates, for example olivine, in which tetrahedra share no oxygen atom [68]. Double tetrahedral silicates, also known as sorosilicate, are silicates in which two tetrahedral pairs share oxygen. The various types of silicates are depicted in Fig. 4. When two oxygen atoms on each tetrahedron link to other tetrahedra (cyclosilicates), we get single-chain silicates (inosilicates) or ring silicates.

When some oxygen is shared between two tetrahedra and some between three tetrahedra, double-chain silicates, also known as inosilicates, are formed. Sheet silicates, also known as layered silicates or phyllosilicates, are formed when three oxygen on each tetrahedron link to other tetrahedra to form tetrahedral planes, and framework silicates are formed when all oxygen is shared between tetrahedra (network silicate

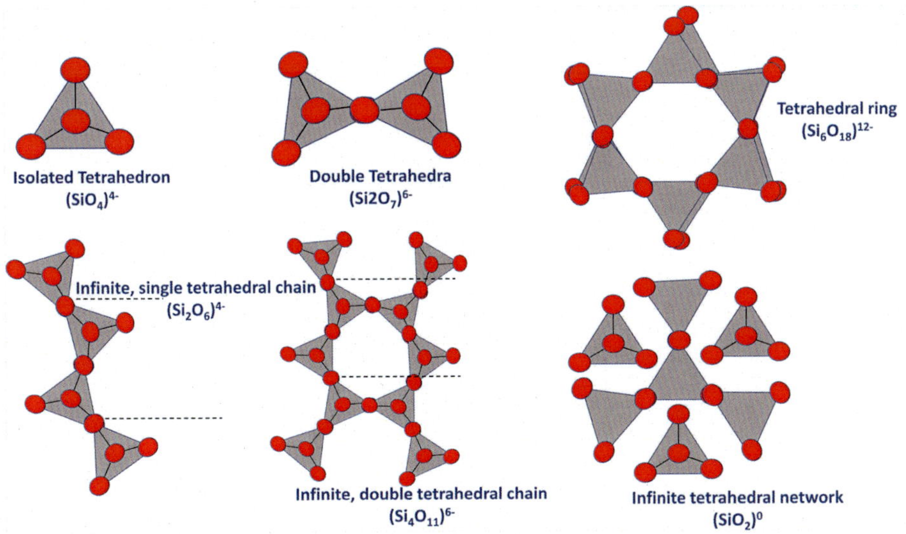

Fig. 4 Various type of silicate structure made from smallest tetrahedron unit.

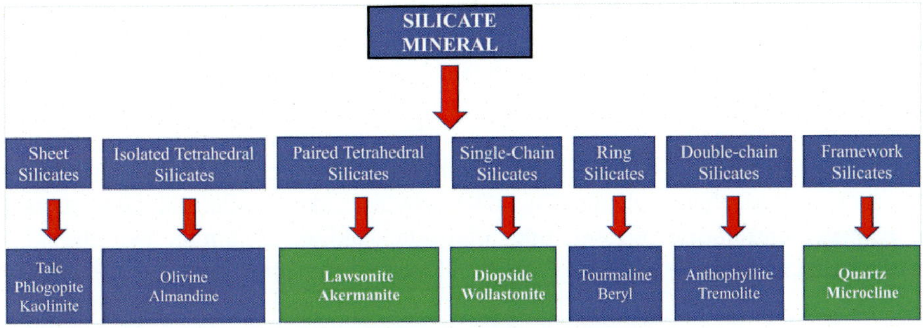

Fig. 5 Classification of silicate minerals based on their structure.

or tectosilicates). Fig. 5 shows the different silicate mineral classifications, paired or double tetrahedral silicates, single-chain silicates, and framework silicates in more detail. The Si:O ratio tells us about the silicate subclass because different ratios result from different amounts of oxygen sharing [69]. Basically, silicate bioceramics, such as wollastonite ($CaSiO_3$), diopside ($CaMgSi_2O_6$), akermanite ($Ca_2MgSi_2O_7$), and merwinite ($Ca_3MgSi_2O_8$) have been developed in the past 10 years [70].

In general, the chemistry of silicates is related to the subclass to which they belong (Table 1). This relationship reflects Si:O ratios as well as how silica polymerization regulates the number and nature of cation sites between anions. Minerals rich in Fe^{2+} and Mg^{2+} can be found in isolated tetrahedral silicates and chain silicates, but not in framework silicates. In framework silicates' three-dimensional polymerization low sufficient anionic charge and the small crystallographic sites required for small, highly

Table 1 Silicate mineral subclass.

Silicate mineral structures					
Silicate subclass	**Minerals examples**	**Mineral formulas**	**Cation coordinations**	**Oxygen shared by tetrahedra**	**$(Si,Al^{IV}):O$**
Single-chain silicates	Wollastonite Diopside	$CaSiO_3$ $CaMgSi_2O_6$	Ca^{VI}, Si^{IV} Ca^{VIII}, Mg^{VI}, Si^{IV}	2	2:6 or 1:3
Paired tetrahedral silicate	Akermanite	$Ca_2MgSi_2O_7$	Ca^{VIII}, Mg^{IV}, Si^{IV}	1	1:3.5
Framework silicates	Quartz Microcline	SiO_2 $KAlSi_3O_8$	Si^{IV} K^X, Al^{IV}, Si^{IV}	4	1:2

charged cations. Even if Al^{3+} replaces Si^{4+}, there is not enough charge left over to allow other cations to existing. Framework silicates are devoid of small, highly charged cations. Because of the large sites Na^+ and K^+ easily accommodate monovalent cations, Na^+ and K^+ have highly polymerized structures for opposite reasons. The alkalis are rare in chain silicates and absent in island silicates [71].

Wollastonite is a calcium-rich silicate that forms rocks in the upper crust [72]. Calcium silicate refers to a group of compounds that contain calcium, silicon, and oxygen and are all hydrated [73]. Fig. 6A depicts a wollastonite crystal [74]. Calcium silicate has the general formula $CaSiO_3$ and is found in two forms: dicalcium silicate (Ca_2SiO_4) and tricalcium silicate (Ca_3SiO_5). Afwillite, akermanite, and radite, as well as calcite, centrallasite, crestmoreite, diopside, eaklite, grammite, gyrolite, hillebrandite, larnite, and wollastonite, are all calcium silicate minerals [75]. Calcium silicate is a white to off-white color in its pure form. It has the ability to absorb up to two-and-a-half its own weight in water. In this state, the hydrated powder retains its ability to flow freely and gel is formed when a mineral acid, such as hydrochloric acid, is added. It is divided into two polymorphs (α- and β-wollastonite), with each polymorph having a number of polytypes [76]. Wollastonite is a phosphor chain-silicate material. Made from a mixture of oxides, including lime (calcium oxide), sand (silicon dioxide), and aluminum, iron, and magnesium oxides, calcium silicate is one of the most common building materials used today in portland cement [77]. When these ingredients are combined with water, a complex set of chemical reactions occurs, with calcium silicates being one of the main products. Portland cement products are known for their high strength, which is due to these calcium silicates [78]. The white color, acicular (needle-like) crystal shape, and alkaline pH of wollastonite set it apart from other nonmetallic industrial minerals. Wollastonite is high purity commercially grades material because it is refined using wet processing and high-intensity magnetic separation to remove trace amounts of impurities. The most common accessory minerals found with wollastonite are quartz, calcium carbonate, diopside ($CaMgSi_2O_6$), garnet,

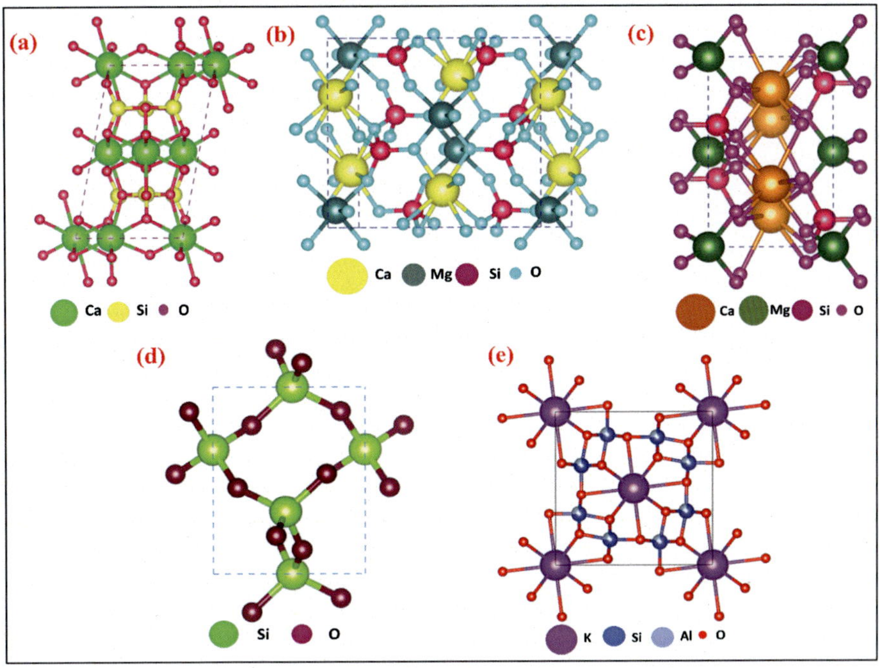

Fig. 6 Ball and stick diagram of (A) wollastonite (B) diopside (C) akermanite (D) quartz (E) microcline.

and prehnite. These impurities are a natural result of the contact metamorphism of silica and limestone that produces wollastonite, which follows the general formula below [79].

$$CaCO_3 + SiO_2 = CaSiO_3 + CO_2$$

Diopside is a single-chain silicate mineral [80]. It belongs to the Pyroxene mineral family. Fig. 6B shows the crystal structure of diopside, which has the general formula $CaMgSi_2O_6$, in which silicon atoms are surrounded by four oxygen atoms, and two oxygen atoms of the tetrahedral group are shared with neighboring groups in a three to one oxygen to silicon atoms ratio [81]. The tetrahedra are linked together by shared oxygen atoms to form endless chains parallel to the crystal's perpendicular axis; they are held together by calcium and magnesium atoms. The crystal's three cleavages (110), (100), and (0l0) are all parallel to these chains, indicating their relative strength [82]. Pyroxene minerals are closely related to amphibole minerals, which contain a variety of fibrous crystals, including asbestos. It is therefore fascinating to find these chains of silicon and oxygen atoms forming a "grain" in the structure of typical pyroxene, as this feature could be found in a variety of minerals [83].

In akermanite, the tetragonal space group P421m crystallizes ($Ca_2MgSi_2O_7$) [84]. It belongs to the paired tetrahedral silicates family, in which the calcium ions are eightfold coordinated and both the Mg^{2+} and Si^{4+} ions are tetrahedrally surrounded by oxygen anions. Despite the fact that in this structure type, two $(SiO_4)^{4-}$ tetrahedra share a corner. Akermanite is a paired tetrahedral silicate that belongs to the sorosilicate family. The structure in Fig. 6C is made up of interleaved tetrahedral and Ca cation layers that are parallel to the [001] plane (also known as $[CaO_8]$ polyhedral layer). Large calcium cations in eight-coordinated sites act as bridges between the tetrahedral layers, which are made up of $[Si_2O_7]$ dimmers connected by $[MgO_4]$ tetrahedrons in the form of five-membered rings [85]. The $Ca_2MgSi_2O_7$ is also known for being a host for persistent phosphors, which are usually activated with rare-earth ions like Eu^{2+}, Nd^{3+}, Dy^{3+}, or Mn^{2+}. When Eu^{2+} and Mn^{2+} are used as dopant ions, the phosphors have the potential to be used in LEDs to generate white luminescence [86]. In recent years, it has also been investigated for biological and medical applications. $Ca_2MgSi_2O_7$ phosphors doped with Eu^{2+} are known to be efficient and stable [87].

Quartz is a crystalline silica-based mineral that is extremely hard (silicon dioxide). It belongs to the framework silicates group. As shown in Fig. 6D, the atoms are connected in a continuous framework of SiO_4 silicon-oxygen tetrahedra, with each oxygen shared between two tetrahedra, yielding a chemical formula of SiO_2 [88]. Silica nanoparticles have a wide range of applications in medicine, batteries, optical fibers, and concrete materials due to their tunable physical, optical, chemical, and mechanical properties. There are two types of silica: crystalline and amorphous. Silica comes in two forms: crystalline and amorphous [89].

$$\text{Quartz} \xleftrightarrow{870^\circ C} \text{Tridymite} \xleftrightarrow{1470^\circ C} \text{Cristoblite}$$

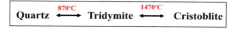

As previously stated, SiO_2 is formed by four oxygen atoms arranged at the corners of a tetrahedron around a central silicon atom through strong, directional covalent bonds with a well-defined local structure. If the orientation is random, an amorphous structure can form. There will only be one silicon atom bonded to some oxygen atoms (nonbridging atoms) [90]. The ratio of bridging to nonbridging determines the "quality" of the oxide. A regular crystal structure called quartz is formed when all oxygen atoms bridge. As a result, much more effort has been put into improving the efficiency of silica's UV and visible emission in order to create an active light-emitting element in next-generation miniature optoelectronic devices [91,92]. Furthermore, because silica is less expensive than most phosphor materials (which frequently use expensive rare-earth elements as activators and/or host components), core-shell phosphor materials will cost less per unit mass than pure phosphor materials [93]. Microcline ($KAlSi_3O_8$) is a potassium aluminum silicate ($KAlSi_3O_8$). The microcline has a triclinic structure. Together with orthoclase and sanidine, microcline makes up the K-feldspar group. The physical properties of these three minerals are nearly identical. Their crystal structure is the only difference between them. Orthoclase and sanidine are monoclinic, whereas microcline are triclinic [94]. The crystalline structure of the microcline is depicted in Fig. 6E. The oxygen atoms are showed by red balls, the silicon, and aluminum atoms are showed by blue balls in the $[SiO_4]^4$ or

$[AlO_4]^5$ tetrahedral, and the potassium cations are represented by purple balls in the tetrahedral void [95]. The network of TO_4 tetrahedra, in which T is Si^{4+} or Al^{3+} and all four atoms are shared with the other tetrahedra, forms the structure of framework silicates.

5 Method of synthesis of silicate phosphors

Controlling the size and surface area of phosphor materials requires the use of synthesis methods [96]. A variety of synthesis methods are available, some of which are described in the following section. The various synthesis techniques used for material synthesis in bulk and nano form are solid-state reaction, hydrothermal, sol-gel, combustion, coprecipitation, modified emulsion precipitation, aerosol, etc. The word hydrothermal means "Hydro" + "Thermal." In the hydrothermal method, a high vapor pressure level and a relatively low temperature are usually required [97]. Depending on the vapor pressure of the main composition in the reaction, either low-pressure or high-pressure conditions can be used to control the morphology of the phosphor materials [98]. The treating temperature is relatively low, ranging between 80°C and 150°C, with the most common range being 95°C–100°C [99]. A strong container is required for a hydrothermal process known as a hydrothermal bomb or "Autoclave," and fill it with a solution [100]. This approach does not require calcination and uses a low-temperature, direct route to oxide powder with a compact size distribution [101]. Chemical equilibrium, chemical kinetics, and thermodynamic features of aqueous systems are all studied in this method [102]. Metal salts, oxides, or hydroxides are heated in a liquid suspension for about 20 h at a controlled temperature and pressure before being applied to ceramic powders. The temperature in a hydrothermal process normally lies between the boiling point of water and the critical temperature with the pressure is more than 100 kPa ($T_c = 374°C$).

Hydrolysis and condensation are the two processes to which the sol-gel method is dependent [103]. It is a very low-cost synthesis technique because metal oxide synthesis requires a lower temperature than solid-state reactions. Making a sol from a homogeneously mixed solution, converting it into a gel via a polycondensation process, and then heat-treating the product to the desired material are all portion of the sol-gel process [104]. Sol refers to a colloidal dispersion of solid in liquid, while gel refers to a colloidal suspension of liquid in solid. If we leave the sol alone for a while or cool it down, it will turn into a gel. In the same way, heating the gel causes it to turn into a sol. Salts are used to introduce acetates and other components for which no alkoxides are available. During the hydrolysis process, the temperature, pH, and concentration of alkoxides, as well as the addition of water and alcohol, are all monitored [105]. At last the final step of sol-gel method is heat treatment, after this final step we get to know that the formation of crystalline materials such as nanoparticles or thin films, as well as noncrystalline materials such as ceramics or glasses [106]. The sol-gel technique is based on two classic chemical reactions of nucleophilic substitutions [107]: hydrolysis and condensation.

The hydrolysis is reproduced by the following reaction:

$$\text{M-OR} + H_2O = \text{M-OH} + \text{ROH}$$

The condensation is given by the reactions below:

$$\text{M-OR} + \text{HO-M} = \text{M-O-M} + \text{R-OH}$$

$$\text{M-OH} + \text{HO} - \text{M} = \text{M-O-M} + \text{HO-H}$$

where, M represents a metal cation such as silicon, titanium or zirconium and the organic R alkyl group. As reported, Eu-doped calcium silicate is synthesized using sol-gel method to obtained the less particle size [108]; and Dy-doped calcium magnesium silicate also synthesis with this technique to increase the luminescence efficiency as performance of material completely depend up to the synthesis process [109].

Combustion is a universal technology that provides the majority of the world's primary energy. Combustion synthesis is a new powder processing technique that can be used to make ceramic pigments [110]. This method is a low-temperature synthesis technique that uses metal nitrates and organic fuel to produce ceramic pigments in a highly exothermic redox reaction. These processes are characterized by high temperatures, rapid heating rates, and a short reaction time. The use of relatively simple equipment, the formation of high-purity products, the stabilization of metastable phases, and the creation of products of virtually any size and shape are all advantages of combustion synthesis [111]. A self-sustaining reaction in a homogeneous solution of different oxidizers is involved in this process. Depending on the type of precursors and the conditions used for process organization, solution combustion synthesis can take place in either volume or layer-by-layer propagating combustion modes. In a single step, this method creates nanoscale oxide materials while also allowing for the uniform (homogeneous) doping of trace amounts of rare-earth impurity ions [112]. $Ba_3MgSi_2O_8{:}Eu^{2+}$, Mn^{2+} phosphor material is synthesized using combustion method and found that it is a promising candidate for white light LED application [113]. Likewise, many papers reported $CaSiO_3{:}Eu^{3+}$ fabricated using the same technique and a good candidate for red phosphor [114–116].

The simultaneous precipitation of a normally soluble component and a macrocomponent from the same solution is a result of the formation of mixed crystals, adsorption, occlusion, or mechanical entrapment. Coprecipitation (CPT), is the process of precipitate transporting substances that are normally soluble in the conditions used. Coprecipitation, on the other hand, is the formation of an antigen-antibody complex from an unbound "antigen in medicine" [117]. Sample preparation for residue analysis has three main goals: isolate the analytes of interest from as many interfering compounds as possible, dissolve the analytes in a suitable solvent, and preconcentrate the analytes. Sample preparation is followed by a separation and detection procedure in an analytical method. Sample preparation is depending up to the following things: (1) The analyte(s), (2) the analyte concentration level(s), (3) the sample matrix, (4) the instrumental measurement technique, and (5) the required sample size all influence the sample preparation method chosen [118]. $(Ba_{1-x}Sr_x)_2SiO_4:Eu^{2+}(0 \leq x \leq 1)$ is synthesized using the coprecipitation method and emits green-yellow phosphor [119]. This approach has the advantage of reducing particle agglomeration formed in individual bubbles. The capacity to synthesize well-dispersed, uniform size nanoparticles, the water-in-oil (w/o) (micro) emulsion precipitation approach has gained special attention among the different procedures established for the creation of nanoparticle materials [120]. By employing the interior of water in oil (micro) emulsion droplets as nanoreactors for nanoparticle creation, the shape, size, and size distribution of the particles may be precisely controlled. To date, (micro)emulsion precipitation has been used to effectively synthesize a variety of nanoparticles, including metal oxide nanoparticles [121]. As a result, the following processing processes can be carried out at very low temperatures. To get the most out of the approach, multicomponent oxide precipitation pathways must be constructed in such a way that an intimate combination of atoms arises during precipitation and chemical homogeneity is maintained throughout subsequent processing. Because emulsion coprecipitations are frequently performed with sample precursors that tend to precipitate at different speeds, resulting in partial phase segregation, there are a number of issues to consider. This method necessitates the addition of appropriate quantities of surfactants to a water-oil mixture in order to create thermally stable emulsion systems [122]. The emulsion system has a modest number of atoms per droplet. The production of a stable precipitate necessitates the exchange of reactive species between droplets. According to the Einstein-Smoluchowski equation, the normal rate of particle development is faster than the comparable rate of droplet exchange [123]. As a result, nucleation and development in emulsions is slower than in homogeneous solution, which prevents the formation of big particles. Multi surfactants are used in the formation of a thermally stable emulsion as well as the control of droplet size [124]. After the water has been removed, other additives act as steric particle stabilizers. The emulsions were made by mixing the oil phase (Cyclohexane or n-heptane) with tergitol surfactants and octanol as a cosurfactant before particle dispersion by filtration or decantation of the organic phase [125]. In stoichiometric proportions, water was added to the system, then violently agitated until a translucent emulsion formed. The emulsion was dropped into alkoxide alcohol solutions one drop at a time and agitated for several hours. After the solvents in the dispersion were removed, the residue was taken up in acetone to eliminate the micelles. Following the decantation of the organic phase, the solid product was dried and calcined into nanocrystalline spinals [126].

Other synthesis method is aerosol method of synthesis also known as a gas phase technique. In large-scale industrial manufacture of multicomponent materials, it is regarded to be convenient and cost-effective. Aerosols are small solid or liquid particles suspended in a gaseous suspension. Ultrafine particles can be made in two ways using aerosol processes. Producing a supersaturated vapor from a reactant is the first step, followed by homogeneous nucleation (gas-to-particle conversion). The second step involves forming liquid droplets, which are then heated to solidify (liquid to particle conversion) [127]. The latter is used to make multicomponent materials. Spray drying and spray pyrolysis are the most common liquid-to-solid conversion methods. The first step is to make a metal precursor (sol) solution, which is then atomized into droplets and placed in a furnace. Inside the furnace, solvent evaporation, drying, precipitation, gas phase reactions, and pyrolysis are used to produce the final product. Spray drying and spray pyrolysis share the same colloidal dispersion particles and precursors [128]. With this technology, uniformly spherical granules with diameters ranging from submicron to millimeter can be produced. Nanoparticles are present in the granules if colloidal nanoparticles (primary particles) are present in the suspension, resulting in nanostructure-granulated powders [129]. As a result, spray drying nanoparticles into submicron spherical granulates that can be compacted into microscopic shapes may be a realistic solution. The composition of final particles is determined by the ratio of reactants dissolved in the starting solution. The size of the atomized droplets and the precursor concentrations in the starting solution dictate the average size and size distribution [130]. Particle shape and agglomeration will be influenced by precursor characteristics, carrier gas flow rate, and temperature [131]. So, for different product requirement, different synthesis procedure can be adopted to enhance the purity of the material.

6 Comparative study of rare-earth/transition metal ion-doped silicate phosphor, synthesis method, characterization, and luminescence studies

The rare-earth ions and transition metal ions are widely used as a luminescent center in the variety of host lattices [132]. The rare-earth elements are 17 elements consisting of the lanthanides from Lanthanum (La) to Lutetium (Lu), Scandium (Sc), and Yttrium (Y). The transition in rare-earth ions is happen in 4f shell. All the rare-earth ions exhibit +3 oxidation state. For the rare-earth element, the +2 and +4 oxidation state is stable because this oxidation state corresponds to the fully empty, partial filled, and completely filled 4f shell [133]. As Eu^{2+} is the most stable state and it is having half filled ($4f^7$) electronic configuration, similarly Ce^{4+} is the most stable state because of fully empty 4f shell in the outer electronic configuration. The rare-earth elements that exhibit +3 oxidation state are showing strong magnetic behavior due to unpaired number of 4f electron that provide net magnetic moment, that may be oriented by the use of external magnetic field. Transition metal ions first series begins with Titanium (Ti) and ends with Zinc (Zn). The second transition series includes the elements from Yttrium (Y) to Cadmium (Cd) and the three series includes the elements from Lanthanum (La) to Mercury (Hg) [134]. These three series include total of 30 elements and known as d-block elements. Most of the transition metal exhibits multiple oxidation states (+1, +2, +3, +4, +5, +6, +7) because it is easy to lose electron from transition metal atoms. The transition metal shows different oxidation states due to unpaired d-electrons undergoes electronic transition from one d-orbital to another and bind to a variety of ligands to form coordination complexes that are often colored.

6.1 Rare-earth/transition metal doped calcium silicate ($CaSiO_3$)

Silicate phosphor are the promising candidate for the solid-state lighting application due to its high radiation intensity, quantum yield up to 70%, water resistance and highly thermal and chemical stability. This type of feature arises from the weak interaction between the host and the activator ion. Besides from above mention properties, these materials are having a wide band gap and absorb in the far-UV region [135]. Earlier sulfide host material are used but now- a-days silicate host in demand due to better characteristics such as longer and brighter luminescence emission, easier preparation, and lower cost. Among several silicate materials, wollastonite ($CaSiO_3$) and its derivatives are known for the excellent host material for luminescence. There are various literature available on doped calcium silicate as shown in Table 2.

6.2 Rare-earth/transition metal doped diopside ($CaMgSi_2O_6$)

$CaMgSi_2O_6$ is a pyroxene mineral and known as a good bioactive material. Recently, alkaline earth silicates have attracted much interest in the field of luminescence since they are suitable hosts due to their high luminescence efficiency and chemical stability. Although this type of material showing the good dielectric properties with low dielectric constant. Jiang et al. prepared a long-lasting phosphor using the alkaline earth silicate ($CaMgSi_2O_6$) doped by different activators Eu^{2+}, Nd^{3+}, and Dy^{3+}. He showed that the prepared luminescent material are having a good blue emission under ultraviolet illumination [146]. There are various literature available for the synthesis of diopside material as shown in Table 3.

6.3 Rare-earth/transition metal doped akermanite ($Ca_2MgSi_2O_7$)

$Ca_2MgSi_2O_7$ is a member of melilite group and it generally crystallize in tetragonal crystal structure. Because of interesting acoustic and nonlinear characteristics due to the absence of inversion center, minerals with melilite type structure have been synthesized so far with the addition of luminescent ions. Due to its tetragonal structure without an inversion symmetry, the transition metal and lanthanides are easily incorporated as an activator in the melilite host [152]. Due to its comfort of incorporation, melilites group are recently used as a host material for rare-earth dopant and transition metal atom to prepare luminescent material, which can be used in variety of application such as light emitting diodes, solar cell, etc. Table 4 shows the reported data available for the synthesis of akermanite material using variety of synthesis technique.

From the above literature survey, it was observed that calcium silicate and calcium silicate-based phosphors like akermanite and diopside are very promising candidates for the display applications. Rare-earth doped phosphors exhibit a broad emission band, and therefore, they are tested as potential materials for white light-emitting diodes (WLEDs). These devices demonstrate their commitment to being extremely energy efficient and thus environmentally friendly. Above-reported tables mention the synthesis technique, optical properties of the silicate phosphor materials. The main disadvantage is that all the material prepared using high-cost precursor materials.

Table 2 Various synthesis techniques for rare-earth and transition metal ion doped in calcium silicate material for display application.

Ref. No.	Authors	Rare-earth	Synthesis method	Study	Application	Remarks
[136]	Basavaraj et al.	Pr^{3+}	Bio-inspired combustion route	Temperature = 950°C Excitation = 606 nm Emission = 467 nm	Cold white-LED	Obtained superstructures resemble the naturally existing materials using *Mimosa pudica* plant extract
[137]	Kang et al.	Eu^{3+}/Tb^{3+}	Simple template route	Excitation = 250 nm Emission = 615 nm Temperature = 600°C	Drug delivery and disease therapy	This luminescent property has the potential to be a useful drug carrier
[138]	Kaur Chhina et al.	Sm^{3+}	Solid-state reaction method	Temperature = 1250°C Emission = 608 nm Excitation = 402 nm	It can be used in a LED device	Agro-food wastes have the potential to be used as phosphor synthesis source materials
[139]	Nagabhushana et al.	Eu^{3+}	Combustion method	Temperature = 900°C Excitation = 254 nm Emission = 614 nm, 593 nm	Display applications	The used combustion route is simple and quick, and the starting materials are inexpensive and readily available
[140]	Kaur et al.	Sm^{3+}	Sol-Gel method	Temperature = 1100°C Excitation = 402 nm Emission = 602 nm CIE coordinates = (0.579, 0.419)	White LED and other photonic device applications	Even at 423 K and 463 K, the luminescence intensity is up to 85.4% and 79.6% of that at room temperature
[141]	Singh et al.	Er^{3+}	Combustion method	Temperature = 1000°C Excitation = 440 nm Emission = 665 nm	LED conversion phosphor	Phosphor has an effective orange red color and great color stability, indicating that it is suitable for display applications

[142]	Manohara et al.	Cr^{3+}	Combustion technique	Temperature = 900°C Excitation = 323 nm Emission = 707 nm	WLEDs and solid-state display applications	The photoluminescence emission spectra demonstrate that the intensity of the emission increases as the Cr^{3+} concentration rises, and quenching occurs beyond 5 mol % concentration
[139]	Sunitha et al.	Eu^{3+}	Low temperature combustion method	Temperature = 950°C Excitation = 613 nm Emission = 254 nm	White-LED	The phosphor is appropriate for white LEDs since the chromaticity coordinates of Eu^{3+} activated samples are situated in white light
[143]	Venkataravanappa et al.	Dy^{3+}	Combustion route	Excitation = 574 nm Emission = 350 nm Temperature = 950°C	Forensic and solid-state lighting applications	The current phosphor proved to be very useful in the manufacture of artificial illumination devices
[144]	Seong Jang et al.	Ce^{3+}, Li^{+}	Solid-state reaction method	Temperature = 1500°C Emission = 559 nm Excitation = 450 nm	Solid-state lighting devices	By combining this phosphor with a blue LED, excellent white light with a high color rendering index of 86 was produced, thanks to the phosphor's wide emission bandwidth
[145]	Verma et al.	Er^{3+}	Combustion synthesis	Temperature = 800°C Excitation = 440 nm (Er^{3+}) Emission = 665 nm (Er^{3+})	Near ultraviolet LED conversion	This phosphor emits green light efficiently and has excellent color stability

Table 3 Various synthesis techniques for rare-earth and transition metal ion doped in diopside material for display application.

Ref. No.	Authors	Rare-earth	Synthesis method	Study	Application	Remarks
[147]	Rosticher et al.	Eu^{2+}, Mn^{2+}, Pr^{3+}	Sol-Gel method	Temperature = 1150°C Emission = 597 nm, 614 nm, 630 nm, 680 nm Excitation = 250 nm, 275 nm, 450 nm	Persistent luminescence	Multi emission peaks due to Eu^{2+}, Mn^{2+}, Pr^{3+} Improved the persistent luminescence
[148]	Lecointre et al.	Mn^{2+}	Sol-Gel method	Temperature = 1100°C	Optical imaging	$CaMgSi_2O_6$:Mn′s long-lasting phosphorescence is investigated for bioimaging applications
[149]	Singh et al.	Pb^{2+}	Sol-Gel method	Temperature = 1100°C Emission = 284 nm Excitation = 247 nm	Assumed to be a viable candidate for a broadband UV application	The optimal Pb^{2+} concentration in phosphor was discovered to be 0.03 mol, which results in the best luminescence
[150]	Ribas de Morais et al.	Dy^{3+}	Sol-Gel method	Temperature = 1000°C Excitation =350, 390, 442 and 452 nm Emission =479 nm, 577 nm CIE coordinates = (x = 0.309, y = 0.363)	White-LED	Light blue-green emission
[151]	Fang et al.	Cr^{3+}	Solid-state reaction method	Temperature = 1250°C Excitation = 460 nm Emission = 638 nm 655 nm	Broadband NIR pc-LED	1% Cr^{3+} and 4% Cr^{3+} doped samples fabricated LED

Table 4 Variety of synthesis techniques for rare-earth and transition metal ion doped in akermanite material for display application.

Ref. No.	Author	Rare-earth	Synthesis method	Study	Application	Remarks
[45]	Sahu et al.	Eu^{3+}	Solid-state reaction method	Emission = 593 nm Excitation = 395 nm Temperature = 1200°C CIE coordinates = (X = 0.5554, Y = 0.4397)	This orange–red-emitting phosphor is a good candidate for the white light-emitting diode (LED)	It shows the efficient orange-red emission and good color stability
[153]	Birkel et al.	Eu^{2+}	Microwave assisted technique	Emission = 541 nm Excitation = 395 nm Temperature = 1200°C CIE coordinates = (X = 0.30, Y = 0.41) Luminescence efficiency = 6 lm/W QY_{77K} = 40% FWHM = 76 nm	LED device Electroluminescence spectrum obtained using LED device	Greenish-white color light appears from the LED device under a forward current of 20 mA
[154]	Sohn et al.	Eu^{2+}	Pulsed laser deposition	Emission = 525 nm Excitation = 254 nm Temperature = 1500°C CIE coordinates = (x = 0.145, y = 0.077)	For display applications, thin-film phosphors were deposited on quartz glass using a pulsed laser deposition technique	By optimizing the processing parameters, a developed thin film with large area homogeneity and minor phases minimized could be produced
[155]	Zhou et al.	Bi^{3+} Eu^{3+}	Orthodox solid phase procedure	Emission = 287 nm (Bi^{3+}), 393 nm (Eu^{3+}) Excitation = 350 nm (Bi^{3+}), 613 nm (Eu^{3+}) Temperature = 1350°C	Optical thermometric device	Future development of $Ca_2MgSi_2O_7$-based LIR technology sensing materials

Continued

Table 4 Continued

Ref. No.	Author	Rare-earth	Synthesis method	Study	Application	Remarks
[156]	He et al.	Eu^{2+} Dy^{3+}	Solid-state reaction method	Emission = 474 nm (Eu^{2+}), 533 nm (Eu^{2+}/Dy^{3+}) Excitation = 350 nm (Eu^{2+}), 375 nm (Eu^{2+}/Dy^{3+}) Temperature = 1300°C	$Ca_2MgSi_2O_7$:Eu^{2+} in white LEDs	The phosphors $Ca_2MgSi_2O_7$:Eu^{2+} and $Ca_2MgSi_2O_7$:Eu^{2+}, Dy^{3+} have a lot of potential as long-lasting phosphors because of their afterglow properties
[157]	Sahu et al.	Eu^{2+}, Dy^{3+}	Solid-state reaction method	Excitation = 200–425 nm Emission = 465 nm and 535 nm Temperature = 1050°C	340-nm pulse nano LED light was used to excite the samples	The ML intensity increased linearly with the impact velocity of the moving piston, according to ML measurements
[158]	Igashira et al.	Eu^{2+} and Eu^{3+}	Solid-state reaction method	Emission = 530 nm (Eu^{2+}), 610 nm (Eu^{3+}) Excitation = 405 nm, 395 nm	Eu-doped CMSM have comparable *LY* (~10,700 ph/MeV)	*QY* increases with Eu concentration
[159]	Zhang et al.	Eu^{2+}	Solid-state reaction method	Emission = 535 nm Excitation = 375 nm Temperature = 1300°C	Fabricated Blue LED has an excellent stability	On UV chips emitting at 395 nm, the phosphors were precoated
[160]	Sahu et al.	Eu^{2+}	Solid-state reaction method	Emission = 530 nm Excitation = 291 nm, 329 nm Temperature = 1250°C CIE coordinates = (x = 0.2791, y = 0.6726)	Mechanoluminescence tested	This type of phosphors can be used to detect an object's stress as self-powered biosensors and stress sensors

[161]	Mondal et al.	Eu^{3+}/Dy^{3+}	Solid-state reaction method	Emission = 617 nm (Eu^{3+}), 576 nm (Dy^{3+}) Excitation = 393 nm (Eu^{3+}), 349 nm (Dy^{3+}) Temperature = 1200°C CIE coordinates = (0.34, 0.33) for Dy^{3+} (0.33, 0.33) for Eu^{3+}	White light LED	Novel white-emitting phosphors for W-LED applications
[162]	Zhou et al.	Ce^{3+} Eu^{2+}	Solid-state reaction method	Emission = 388 nm (Ce^{3+}), 530 nm (Eu^{2+}) Excitation = 342 nm (Ce^{3+}), 420 nm (Eu^{2+}) Temperature = 1350°C	Potential candidates for optical thermometer materials	According to detailed characterizations, the $Ca_{2-x}Sr_xMgSi_2O_7$:Ce^{3+}, Eu^{2+} phosphor series has the potential to be used in optical thermometers with a maximum sensitivity of 2.43% K-1
[87]	Karacaoglu et al.	$Eu^{2+/3+}$, Dy^{3+}, Ce^{3+}, Sm^{3+} and $Pr^{3+/4+}$	Solid-state reaction method	Temperature = 1100–1400°C Excitation = 350 nm (Eu), 350 nm (Dy), 341 nm (Ce), 477 nm (Pr) Emission = 577 nm (Eu), 486 nm and 575 nm (Dy), 395 nm (Ce), 600 nm (Pr)	Photoluminescence investigations for W-LED	The Akermanite type phosphor produced visible white light emission after Ce^{3+} doping, and this study shows that the rare-earth ions (REI 3+/4+) alter the emission properties of the samples
[163]	Sahu et al.	($R^+ = Li^+$, Na^+ and K^+) co-doped with Eu^{3+}	Solid-state reaction method	Temperature = 1200°C Excitation = 242 nm, 396 nm Emission = 614 nm, 595 nm CIE coordinates = (x = 0.568, y = 0.380)	LED application	$Ca_2MgSi_2O_7$:Eu^{3+} and $Ca_2MgSi_2O_7$:Eu^{3+}, R^+ phosphors exhibit orange-red emission, indicating that it has favorable properties for

Continued

Table 4 Continued

Ref. No.	Author	Rare-earth	Synthesis method	Study	Application	Remarks
						application as near ultraviolet LED conversion phosphor
[164]	Onani et al.	Tb^{3+} or Eu^{3+}	Combustion method	Temperature = 600°C Excitation = 335 nm (Tb^{3+}), 497 nm (Eu^{3+}) Emission = 544 nm (Tb^{3+}), 620 nm (Eu^{3+})	LED applications	The combustion synthesis of urea and ammonium nitrate is a straightforward, time-saving, and cost-effective technique for synthesis
[165]	Pandey et al.	Dy^{3+} and Eu^{2+}	Solid-state and combustion method	Temperature = 1200–1400°C for solid-state and 600°C for combustion	White long afterglow phosphor	Various silicate compounds have been created with a white prolonged afterglow after irradiation with a mercury lamp
[166]	Htun et al.	Cu^{2+}	High-energy wet planetary ball-milling route	Temperature = 1200°C	Bone replacement materials in biomedical applications	The creation of materials with a lower densifying sintering temperature may be advantageous in terms of saving time and energy

By using agro waste, food waste, industrial waste materials, low cost and good luminescence properties can be achieved. Food waste like eggshell, peanut shell, sugarcane bagasse, coconut husk; agricultural waste like rice husk, wheat strew and many more which can be used for the formation of phosphor materials. Our research group reported calcium silicate derived from food waste and agro waste [167] for tissue engineering applications, and Sm-doped calcium silicate synthesized using eggshell and rice husk by sol-gel method [168] for display applications. Calcium silicate from rice husk ash and shell of snail *Pomacea Canaliculata* by solid-state reaction also reported [169]. Using bamboo leaf and meretix shell, wollastonite is been synthesized to reduce cost [170]. For intense red emission light, europium-doped dicalcium silicate is been synthesized using sugarcane baggage ash and chicken shell [171]. So, from this literature review it is concluded that, we can make our display device using a phosphor material, which is derived from the recycling material to reduce the environmental problem. The use of food waste to produce luminescent material could be a low-cost approach while at the same time controlling environmental pollution.

7 Conclusion

Luminescence investigations and possible uses of rare-earth (RE) and transition metal ion activated phosphors are presented in this chapter. The overview starts with a brief history of silicate materials, then moves on to the creation of RE-doped and transition metal ion-doped akermanite, diopside, calcium silicate phosphors, and the nanophosphor method of preparation. In terms of the future, it is clear that there is a lot of scope for more structural and dynamical research to improve our current understanding of these phosphors. Another promising future prospect stems from the observation that a more efficient method of synthesizing these phosphors can be highlighted, lowering the cost of the initial precursor and lowering the calcination temperature.

Calcium silicate-based material have been shown to have good luminescence efficiency, long lifetime, and longer brightness. However, a significant challenge is to attain minimum cost of the initial precursor. Silicate has proven to be a promising candidate not only in one field, but also in a variety of research and human applications. Silicate nanomaterials are producing outstanding results and are being used to make optical LED devices all over the world. These devices demonstrate their commitment to being extremely energy efficient and thus environmentally friendly. In this chapter, an attempt has been made to better describe the various characteristic properties of silicate nanophosphors.

As a result, future research on calcium silicate-based materials should focus on two key aspects: (i) development of innovative fabrication processes to lower calcination temperatures and (ii) synthesis of silicate phosphor materials employing waste product extract from the food, agricultural, and manufacturing industries.

References

[1] P.G.K. Martynas Beresna, M. Gecevicius, Adv. Opt. Photon. 6 (2014) 293–339.
[2] E. Feizi, A.K. Ray, J. Disp. Technol. 12 (2016) 451.
[3] L. Ma, Y.F. Shao, J. Cent. South Univ. 27 (2020) 1624–1644.
[4] R.L. Barbin, A.S. Poulos, Encyclopedia of Imaging Science and Technology, John Wiley & Sons, Ltd, 2002.
[5] C. Hilsum, Philos. Trans. R. Soc. A Math. Phys. Eng. Sci. 368 (2010) 1027–1082.
[6] M.L. Socolof, J.G. Overly, J.R. Geibig, J. Clean. Prod. 13 (2005) 1281–1294.
[7] Y. Wang, H. Nie, J. Han, Y. An, Y.M. Zhang, S.X.A. Zhang, Light Sci. Appl. 10 (2021).
[8] M.R. Fernández, E.Z. Casanova, I.G. Alonso, Sustainability 7 (2015) 10854–10875.
[9] Y. Huang, E.L. Hsiang, M.Y. Deng, S.T. Wu, Light Sci. Appl. 9 (2020).
[10] H.E. Lee, J.H. Shin, J.H. Park, S.K. Hong, S.H. Park, S.H. Lee, J.H. Lee, I.S. Kang, K.J. Lee, Adv. Funct. Mater. 29 (2019).
[11] J.-F. Tremblay, C&EN Glob. Enterp. 94 (2016) 30–34.
[12] H. Menkara, R.A. Gilstrap, T. Morris, M. Minkara, B.K. Wagner, C.J. Summers, Opt. Express 19 (2011) A972.
[13] K.H. Lee, S.W.R. Lee, Proc. Electron. Packag. Technol. Conf. EPTC (2006) 379–384.
[14] W.X. Yang, H.L. Zhou, D. Su, Z.R. Yang, Y.J. Song, X.Y. Zhang, T. Zhang, J. Mater. Chem. C 10 (2022) 7352–7367.
[15] N.N. Toan, M.A. Tuan, T.T. Huong, N.T. Huong, C. Bathou, T.K. Anh, L.Q. Minh, N.D. Chien, Conf. Optoelectron. Microelectron. Mater. Devices, Proceedings, COMMAD 2000-Janua, 2000, pp. 304–307.
[16] H.A. Höppe, Angew. Chem. Int. Ed. 48 (2009) 3572–3582.
[17] E.F. Schubert, J.K. Kim, Science 308 (2005) 1274–1278.
[18] S. Limbu, L.R. Singh, J. Solid State Chem. (2022), 122929.
[19] A. Dwivedi, M. Srivastava, S.K. Srivastava, J. Mol. Struct. 1251 (2022), 132061.
[20] A. Kumar, J. Manam, Ceram. Int. 48 (2022) 13615–13625.
[21] S. Sasidharan, G. Jyothi, K.G. Gopchandran, J. Sci.: Adv. Mater. Devices 7 (2022), 100400.
[22] P. Sehrawat, R.K. Malik, R. Punia, S. Maken, N. Kumari, Chem. Phys. Lett. (2022), 139399.
[23] P. Phogat, S.P. Khatkar, V.B. Taxak, R.K. Malik, Mater. Chem. Phys. 276 (2022), 125389.
[24] A. Dwivedi, M. Srivastava, A. Dwivedi, A. Srivastava, A. Mishra, S.K. Srivastava, J. Rare Earths 40 (2021) 1187–1198.
[25] P. Sehrawat, A. Khatkar, A. Hooda, M. Kumar, R. Kumar, R.K. Malik, S.P. Khatkar, V. B. Taxak, Ceram. Int. 45 (2019) 24104–24114.
[26] S. Kadyan, S. Singh, S. Sheoran, A. Samantilleke, B. Mari, D. Singh, Trans. Ind. Ceram. Soc. 78 (2019) 219–226.
[27] T. Kojima, IEICE Trans. E71-E (1988) 1050–1055.
[28] R.G. Kaufman, IEEE Trans. Electron Devices 24 (1977) 884–890.
[29] R.P. Rao, J. Lumin. 113 (2005) 271–278.
[30] K. Yokota, S.X. Zhang, K. Kimura, A. Sakamoto, J. Lumin. 92 (2001) 223–227.
[31] R.P. Rao, S.I.D. Symp, Dig. Tech. Pap. 34 (2003) 1219–1221.
[32] Q. Fei, C. Chang, D. Mao, J. Alloys Compd. 390 (2005) 133–137.
[33] P. Chandrakar, D.P. Bisen, R.N. Baghel, B.P. Chandra, J. Electron. Mater. 44 (2015) 3450–3457.

[34] Y. Chen, X. Cheng, M. Liu, Z. Qi, C. Shi, J. Lumin. 129 (2009) 531–535.
[35] L. Jiang, C. Chang, D. Mao, ChemInform 34 (2003).
[36] I.P. Sahu, D.P. Bisen, N. Brahme, J. Radiat. Res. Appl. Sci. 8 (2015) 381–388.
[37] Z. Xia, Z. Xu, M. Chen, Q. Liu, Dalton Trans. 45 (2016) 11214.
[38] M.H.M. Zaid, H.A.A. Sidek, R. El-Mallawany, K.A. Almasri, K.A. Matori, J. Mater. Res. Technol. 9 (2020) 13153–13160.
[39] R. Yevale, N. Khan, S. Bhadane, J. Drug Deliv. Ther. 9 (2019) 181–184.
[40] R. Kubrin, KONA Powder Part. J. 31 (2014) 22–52.
[41] H. Chander, Mater. Sci. Eng. R. Rep. 49 (2005) 113–155.
[42] T. Edvinsson, R. Soc, Open Sci. 5 (2018).
[43] R. Chatterjee, G. Chandra Das, K.K. Chattopadhyay, Mater. Res. Express 7 (2020).
[44] K. Zhang, T. Han, W.K. Cho, H.S. Kwok, Z. Liu, Nanomaterials 11 (2021) 1–10.
[45] I.P. Sahu, D.P. Bisen, N. Brahme, I.P. Sahu, D.P. Bisen, N. Brahme, J. Radiat. Res. Appl. Sci. 8 (2015) 381–388.
[46] N. Hirosaki, T. Takeda, S. Funahashi, R.J. Xie, Chem. Mater. 26 (2014) 4280–4288.
[47] J. Brgoch, C.K.H. Borg, K.A. Denault, A. Mikhailovsky, S.P. Denbaars, R. Seshadri, Inorg. Chem. 52 (2013) 8010–8016.
[48] S.A. Khan, N.Z. Khan, I. Mehmood, M. Rauf, B. Dong, M. Kiani, J. Ahmed, S.M. Alshehri, J. Zhu, S. Agathopoulos, J. Lumin. 229 (2021), 117687.
[49] Q. Zhang, X. Wang, Z. Tang, Y. Wang, Mater. Chem. Front. 3 (2019) 2120–2127.
[50] W.J. Ho, G.C. Yang, Y.T. Shen, Y.J. Deng, Appl. Surf. Sci. 365 (2016) 120–124.
[51] D. Guo, Q. Guo, Z. Chen, Z. Wu, P. Li, W. Tang, Mater. Today Phys. 11 (2019), 100157.
[52] C.C. Lin, W. Chen, R.S. Liu, Handbook of Advanced Lighting Technology, Springer Nature, 2014.
[53] Y.C. Lin, M. Karlsson, M. Bettinelli, Top. Curr. Chem. 374 (2016) 374–421.
[54] P. Kumari, D. Banerjee, Defects in Rare-Earth-Doped Inorganic Materials, Elsevier, 2022.
[55] K.V.R. Murthy, H.S. Virk, Defect Diffus. Forum 347 (2014) 1–34.
[56] M. Sindhu, N. Ahlawat, S. Sanghi, A. Agarwal, R. Dahiya, N. Ahlawat, Curr. Appl. Phys. 12 (2012) 1429–1435.
[57] T. Walther, E.S. Fry, Optics in Remote Sensing, Springer Nature, 2016.
[58] S.-S. Wang, W.-T. Chen, Y. Li, J. Wang, H.-S. Sheu, R.-S. Liu, J. Am. Chem. Soc. 135 (2013) 45.
[59] P. Psuja, D. Hreniak, W. Strek, J. Nanomater. 2007 (2007) 1–7.
[60] N.C. George, K.A. Denault, R. Seshadri, Annu. Rev. Mater. Res. 43 (2013) 481–501.
[61] M.H. Rahman, M.Z. Rahaman, E.H. Chowdhury, M. Motalab, A.K.M.A. Hossain, M. Roknuzzaman, Mol. Syst. Des. Eng. (2022) 1516–1528.
[62] R. Reddappa, L.L. Devi, C. Basavapoornima, S.R. Depuru, J. Kaewkhao, W. Pecharapa, C.K. Jayasankar, Optik (Stuttg.) 264 (2022), 169360.
[63] D. Singh, S. Sheoran, V. Tanwar, Rev. Artic. 2017 (2017) 656–672.
[64] O. Hyeon Kwon, J. Sik Kim, J. Woo Jang, H. Yang, Y. Soo Cho, S.E. Brinkley, N. Pfaff, K.A. Denault, Z. Zhang, H.T. Hintzen, R. Seshadri, S. Nakamura, S.P. Denbaars, J. Am. Ceram. Soc. 1 (2005) 874–879.
[65] Z. Liu, C.H. Lin, B.R. Hyun, C.W. Sher, Z. Lv, B. Luo, F. Jiang, T. Wu, C.H. Ho, H.C. Kuo, J.H. He, Light Sci. Appl. 91 (9) (2020) 1–23.
[66] A.L. Smith, J. Electrochem. Soc. 96 (1949) 287–296, https://doi.org/10.1149/1.2776791.
[67] Y. Liu, B. Lei, Phosphors, Up Conversion Nano Particles, Quantum Dots and Their Applications, vol. 2, Springer Nature, 2016, pp. 167–214.
[68] V. Labet, P. Colomban, J. Non-Cryst. Solids 370 (2013) 10–17.

[69] M.C. Day, F.C. Hawthorne, Mineral. Mag. 84 (2020) 165–244.
[70] C. Wu, J. Chang, Biomed. Mater. 8 (2013).
[71] F.C. Hawthorne, Y.A. Uvarova, E. Sokolova, Mineral. Mag. 83 (2019) 3–55.
[72] W.A. Deer, R.A. Howie, J. Zussman, An introduction to the rock-forming minerals, third ed., The Mineralogical Society, London, 2013.
[73] R.G. Ribas, T.M.B. Campos, V.M. Schatkoski, B.R.C. de Menezes, T.L.D.A. Montanheiro, G.P. Thim, Ceram. Int. 46 (2020) 6575–6580.
[74] S.J. Edrees, M.M. Shukur, M.M. Obeid, Comput. Condens. Matter 14 (2018) 20–26.
[75] E.P. Flint, H.F. Mcmurdie, L.S. Wells, J. Res. Natl. Bur. Stand. 21 (1938).
[76] L.A. Núñez-Rodríguez, M.A. Encinas-Romero, A. Gómez-Álvarez, J.L. Valenzuela-García, G.C. Tiburcio-Munive, J. Biomater. Nanobiotechnol. 9 (2018) 263–276.
[77] M.A. Saghiri, J. Orangi, A. Asatourian, J.L. Gutmann, F. Garcia-Godoy, M. Lotfi, N. Sheibani, Dent. Mater. J. 36 (2017) 8.
[78] M.A. Mahdy, S.H. Kenawy, I.K. El Zawawi, E.M.A. Hamzawy, G.T. El-Bassyouni, Ceram. Int. 46 (2020) 6581–6593.
[79] W. Joswig, E.F. Paulus, B. Winkler, V. Milman, Zeitschrift Fur Krist. 218 (2003) 811–818.
[80] J.J. Papike, Rev. Geophys. 25 (1987) 1483–1526.
[81] E.T. Rodriguez, L.M. Anovitz, C.D. Clement, A.J. Rondinone, M.C. Cheshire, Sci. Rep. 81 (8) (2018) 1–7.
[82] T. Nonami, S. Tsutsumi, J. Mater. Sci. Mater. Med. 108 (10) (1999) 475–479.
[83] B. Warren, W.L. Bragg, Zeitschrift Für Krist.—Cryst. Mater. 69 (1929) 168–193.
[84] C. Wu, J. Chang, J. Biomater. Appl. 21 (2006) 119–129.
[85] Y. Yang, H. Lu, C. Yu, J.M. Chen, Comput. Mater. Sci. 47 (2009) 35–40.
[86] S. Ye, F. Xiao, Y.X. Pan, Y.Y. Ma, Q.Y. Zhang, Mater. Sci. Eng. R. Rep. 71 (2010) 1–34.
[87] E. Karacaoglu, B. Karasu, Indian J. Chem. 54 (2015) 1394–1401.
[88] C. Büchner, M. Heyde, Prog. Surf. Sci. 92 (2017) 341–374.
[89] S. Prabha, D. Durgalakshmi, S. Rajendran, E. Lichtfouse, Environ. Chem. Lett. 19 (2020) 1667–1691.
[90] E. Pernicka, G.A. Wagner, Int. At. Energy Agency 10 (1979) 79–85.
[91] G.G. Ross, D. Barba, F. Martin, Int. J. Nanotechnol. 5 (2008) 984–1017.
[92] I. Romanov, F. Komarov, O. Milchanin, L. Vlasukova, I. Parkhomenko, M. Makhavikou, E. Wendler, A. Mudryi, A. Togambayeva, J. Nanomater. 2019 (2019).
[93] P.Y. Jia, X.M. Liu, M. Yu, Y. Luo, J. Fang, J. Lin, Chem. Phys. Lett. 424 (2006) 358–363.
[94] J.M. Kalita, M.L. Chithambo, J. Lumin. 224 (2020), 117320.
[95] Q. Zhao, X. Li, Q. Wu, Y. Liu, Y. Lyu, J. Sol-Gel Sci. Technol. 94 (2020) 3–10.
[96] N. Baig, I. Kammakakam, W. Falath, I. Kammakakam, Mater. Adv. 2 (2021) 1821–1871.
[97] X. Wu, G.Q. Lu, L. Wang, Energy Environ. Sci. 4 (2011) 3565–3572.
[98] Y.X. Gan, A.H. Jayatissa, Z. Yu, X. Chen, M. Li, J. Nanomater. 2020 (2020).
[99] S. Sadjadi, Encapsulated Catalysts, Elsevier, 2017, pp. 443–476.
[100] A. Chen, P. Holt-Hindle, Chem. Rev. 110 (2010) 3767–3804.
[101] M. Zawadzki, J. Wrzyszcz, Mater. Res. Bull. 35 (2000) 109–114.
[102] B.N. Tiwari, S. Chandra Kishore, N. Marina, S. Bellucci, Condens. Matter 3 (2018) 1–29.
[103] J. Livage, C. Sanchez, J. Non-Cryst. Solids 145 (1992) 11–19.
[104] D. Levy, M. Zayat, Wiley Online Library, 2015.
[105] M. Zayat, D. Levy, Chem. Mater. 12 (2000) 2763–2769.
[106] T.M. Lopez, D. Aunir, M. Aegerter, Emerging Fields in Sol-Gel Science and Technology, Springer Nature, 2003.
[107] T.K. Tseng, Y.S. Lin, Y.J. Chen, H. Chu, Int. J. Mol. Sci. 11 (2010) 2336–2361.
[108] Q. Meng, J. Lin, L. Fu, H. Zhang, S. Wang, Y. Zhou, J. Mater. Chem. 11 (2001) 3382–3386.

[109] I.P. Sahu, P. Chandrakar, R.N. Baghel, D.P. Bisen, N. Brahme, R.K. Tamrakar, J. Alloys Compd. 649 (2015) 1329–1338.
[110] A. Varma, A.S. Mukasyan, A.S. Rogachev, K.V. Manukyan, Chem. Rev. 116 (2016) 14493–14586.
[111] S.B. Bhaduri, Ceram. Compos. Process. Methods (2012) 391–413.
[112] S.T. Aruna, A.S. Mukasyan, Curr. Opin. Solid State Mater. Sci. 12 (2008) 44–50.
[113] J.S. Kim, S.W. Mho, Y.H. Park, J.C. Choi, H.L. Park, G.S. Kim, Solid State Commun. 136 (2009) 504–507.
[114] H. Nagabhushana, B.M. Nagabhushana, M. Madesh Kumar, K.V.R. Chikkahanumantharayappa, C. Murthy, R.P.S.C. Shivakumara, Spectrochim. Acta. A. Mol. Biomol. Spectrosc. 78 (2011) 64–69.
[115] R. Shukla, A.K. Tyagi, Springer Nature (2021) 51–78.
[116] E. Novitskaya, J.P. Kelly, S. Bhaduri, O.A. Graeve, Int. Mater. Rev. (2020) 188–214.
[117] W.S. Peternele, V. Monge Fuentes, M.L. Fascineli, J. Rodrigues Da Silva, R.C. Silva, C. M. Lucci, R. Bentes De Azevedo, J. Nanomater. 2014 (2014).
[118] H. Dahiya, A. Siwach, M. Dahiya, A. Singh, S. Nain, M. Dalal, S.P. Khatkar, V.B. Taxak, D. Kumar, Chem. Phys. Lett. 754 (2020), 137657.
[119] J.K. Han, M.E. Hannah, A. Piquette, J.B. Talbot, K.C. Mishra, J. McKittrick, ECS J. Solid State Sci. Technol. 1 (2012) R98–R102.
[120] J.H. Fendler, Chem. Rev. 87 (2002) 877–899.
[121] J. Shi, H. Verweij, Langmuir 21 (2005) 5570–5575.
[122] M. Zembyla, B.S. Murray, A. Sarkar, Trends Food Sci. Technol. 104 (2020) 49–59.
[123] M.A. Islam, Phys. Scr. 70 (2004) 120–125.
[124] T. Sharma, G.S. Kumar, B.H. Chon, J.S. Sangwai, J. Ind. Eng. Chem. 22 (2015) 324–334.
[125] M.H. Lee, S.G. Oh, S.K. Moon, S.Y. Bae, J. Colloid Interface Sci. 240 (2006) 83–89.
[126] R. Srivastava, Sage 4 (2012) 17–27.
[127] A.G. Nasibulin, A. Moisala, D.P. Brown, H. Jiang, E.I. Kauppinen, Chem. Phys. Lett. 402 (2005) 227–232.
[128] M. Eslamian, N. Ashgriz, Springer Nature, 2011, pp. 755–773.
[129] D. Guo, G. Xie, J. Luo, J. Phys. D. Appl. Phys. 47 (2013), 013001.
[130] K. Okuyama, I.W. Lenggoro, Chem. Eng. Sci. 58 (2003) 537–547.
[131] R.C. Flagan, Nanostruct. Mater. (1998) 15–30.
[132] P.M. Leong, T.Y. Eeu, T.Q. Leow, R. Hussin, Z. Ibrahim, AIP Conf. Proc. 1528 (2013) 310–315.
[133] W.H. Wells, V.L. Wells, Patty's Toxicol. 1 (2012) 817–840.
[134] Y. Li, S. Qi, P. Li, Z. Wang, RSC Adv. 7 (2017) 38318–38334.
[135] L.S. Ping, C.-H. Huang, T.-S. Chan, T.-M. Chen, ACS Appl. Mater. Interfaces 6 (2014) 7260–7267.
[136] R.B. Basavaraj, H. Nagabhushana, B.D. Prasad, S.C. Sharma, K.N. Venkatachalaiah, J. Alloys Compd. 690 (2017) 730–740.
[137] X. Kang, S. Huang, P. Yang, P. Ma, D. Yang, J. Lin, Dalton Trans. 40 (2011) 1873–1879.
[138] M. Kaur Chhina, K. Singh, Ceram. Int. 47 (2021) 21588–21598.
[139] D.V. Sunitha, H. Nagabhushana, S.C. Sharma, B.M. Nagabhushana, R.P.S. Chakradhar, J. Alloys Compd. 575 (2013) 434–443.
[140] H. Kaur, M. Jayasimhadri, M.K. Sahu, P.K. Rao, N.S. Reddy, Ceram. Int. 46 (2020) 26434–26439.
[141] N.K. Singh, J. Dyn. Control Syst. 10 (2018) 110–116.
[142] B.M. Manohara, H. Nagabhushana, K. Thyagarajan, B. Daruka Prasad, S.C. Prashantha, S.C. Sharma, B.M. Nagabhushana, J. Lumin. 161 (2015) 247–256.

[143] M. Venkataravanappa, R.B. Basavaraj, G.P. Darshan, B. Daruka Prasad, S.C. Sharma, P. Hema Prabha, S. Ramani, H. Nagabhushana, J. Rare Earths 36 (2018) 690–702.
[144] H. Seong Jang, H. You Kim, Y.-S. Kim, H. Mo Lee, D. Young Jeon, Opt. Express 20 (3) (2012) 2761–2771.
[145] S. Verma, A. Mishra, M. Bhuie, N.K. Singh, Int. J. Comput. Sci. Eng. Open Access Res. Pap. 6 (2018) 108–114.
[146] L. Jiang, C. Chang, D. Mao, J. Alloys Compd. 360 (2003) 193–197.
[147] C. Rosticher, C. Chaneac, A. Bos, B. Viana, Oxide-Based Materials and Devices VII, 9749, SPIE, 2016.
[148] A. Lecointre, A. Bessière, K.R. Priolkar, D. Gourier, G. Wallez, B. Viana, Mater. Res. Bull. 48 (2013) 1898–1905.
[149] V. Singh, M.K. Tiwari, Optik (Stuttg.) 202 (2020), 163542.
[150] V. Ribas De Morais, D. De Rezende Leme, C. Yamagata, Int. Conf. Ceram. 62 (2018).
[151] L. Fang, Z. Hao, L. Zhang, H. Wu, H. Wu, G. Pan, J. Zhang, Mater. Res. Bull. 149 (2022), 111725.
[152] S. Ogawa, Y. Ogino, M. Yonemura, T. Fukunaga, H. Kiuchi, K. Nakayama, R. Ishikawa, Y. Ikuhara, Y. Doi, K. Suzuki, M. Saito, T. Motohashi, Chem. Mater. 32 (2020) 6847–6854.
[153] A. Birkel, L.E. Darago, A. Morrison, L. Lory, N.C. George, A.A. Mikhailovsky, C.S. Birkel, R. Seshadri, Solid State Sci. 14 (2012) 739–745.
[154] K.S. Sohn, S.H. Cho, C. Kulshreshtha, N. Shin, Y.R. Do, Electrochem. Solid-State Lett. 10 (2007).
[155] L. Zhou, R. Ye, F. Liu, J. Lin, Y. Chen, J. Fu, W. Guo, D. Deng, S. Xu, Opt. Mater. (Amst.) 116 (2021), 111076.
[156] H. He, R. Fu, X. Song, R. Li, Z. Pan, X. Zhao, Z. Deng, Y. Cao, J. Electrochem. Soc. 157 (2010) 69–73.
[157] I.P. Sahu, D.P. Bisen, N. Brahme, R. Sharma, Res. Chem. Intermed. 41 (2015) 6649–6664.
[158] K. Igashira, D. Nakauchi, T. Ogawa, T. Kato, N. Kawaguchi, T. Yanagida, Mater. Res. Bull. 135 (2021).
[159] M. Zhang, J. Wang, W. Ding, Q. Zhang, Q. Su, Opt. Mater. (Amst.) 30 (2007) 571–578.
[160] I.P. Sahu, D.P. Bisen, N. Brahme, Luminescence 30 (2015) 1125–1132.
[161] K. Mondal, J. Manam, J. Mol. Struct. 1125 (2016) 503–513.
[162] L. Zhou, F. Liu, R. Ye, L. Lei, W. Guo, L. Chen, D. Deng, S. Xu, J. Lumin. 240 (2021), 118417.
[163] I.P. Sahu, D.P. Bisen, K.V.R. Murthy, R.K. Tamrakar, Int. J. Lumin. Appl. 6 (2017).
[164] M.O. Onani, F.B. Dejene, Phys. B Condens. Matter 439 (2014) 137–140.
[165] M.D. Pandey, R. Sharma, N. Brahme, Int. J. Adv. Res. 4 (2016) 1639–1646.
[166] M.M. Htun, A.F.M. Noor, M. Kawashita, Y.M.B. Ismail, IOP Conf. Ser. Mater. Sci. Eng. 943 (2020).
[167] S. Palakurthy, K.V.G. Reddy, R.K. Samudrala, A. Azeem, Mater. Sci. Eng. C 98 (2019) 109–117.
[168] M. Krishnam Raju, R. Prasada Rao, N. Vijayan, P. Abdul Azeem, Ceram. Int. 47 (2021) 26704–26711.
[169] R. Phuttawong, N. Chantaramee, P. Pookmanee, R. Puntharod, Adv. Mater. Res. 1103 (2015) 1–7.
[170] D. Mardina, D. Asmi, M. Badaruddin, A.Z. Syahrial, Mater. Sci. Forum 1029 (2021) 167–173.
[171] R. Priya, I. Khurana, O.P. Pandey, J. Mater. Sci. Mater. Electron. 31 (2020) 1912–1928.

Metal oxide-based phosphors for white light-emitting diodes

5

M.Y.A. Yagoub[a], *Irfan Ayoub*[b], *Vijay Kumar*[a,b], *Hendrik C. Swart*[a], *and E. Coetsee*[a]
[a]Department of Physics, University of the Free State, Bloemfontein, Free State, South Africa,
[b]Department of Physics, National Institute of Technology Srinagar, Hazratbal, Srinagar, Jammu and Kashmir, India

Chapter outline

Metal oxide-based phosphors have increasingly become vitally important in various practical applications. This chapter describes a brief overview on the development of metal oxide-based phosphor for white light-emitting diodes (WLEDs). The mechanism of generating WLEDs is typically classified into two categories: Direct white light generation and discrete color mixing WLEDs. The requirements of metal oxides-based phosphor for WLEDs will also be provided.

1 Introduction

The increasing demand for energy has been growing fast in every aspect in the past few decades. Illumination, however, consumes more than 20% of global energy, which is the second largest energy use [1]. The search for high-efficiency light sources has therefore emerged to minimize energy consumption and meet the rapid increase in

Metal Oxides for Next-Generation Optoelectronic, Photonic, and Photovoltaic Applications
https://doi.org/10.1016/B978-0-323-99143-8.00007-9

energy demand, as well as concerns about climate change and global warming [2–4]. In recent years, extensive research has been conducted to develop new efficient materials for energy-saving lighting systems [5–8]. A great effort was made to replace mercury discharge fluorescent lighting and conventional tungsten lamps. In this regard, various approaches have been established to accomplish white light emission [6,9]. The important key in choosing a proper white light device is the development of materials that produce highly efficient white light emission. Apart from energy efficiency, environmentally friendly materials should also be a choice for lighting technology. In this regard, solid-state lighting has been considered to fulfill all these conditions [1].

White light-emitting diodes (WLEDs) have brought much attention for displays and illumination owing to their small size, high luminous efficiency, low power consumption, environmental friendliness, long life, and high reliability [10–14]. The common strategy for producing commercial WLEDs is a blue-emitting GaN light-emitting diode chip coated with a yellow-emitting phosphor [15–18]. This strategy showed white light with a poor color rendering index (CRI) [19–21]. In order to achieve efficient solid-state lighting, electroluminescence from a multicolor semiconductor in red, green, and red must replace the phosphors in the WLED devices [22,23]. Another strategy was to employ rare earth doped wide bandgap semiconductors for producing stable, narrow, and sharp emission spectra against injection current and ambient temperature. The rare earth doped wide bandgap semiconductors were therefore considered ideal candidates for optoelectronic devices [24–27].

Among the semiconductor WLEDs, there are metal oxide semiconductors. These metal oxide semiconductors have obtained much attention over the last few decades especially in terms of low power consumption, operational lifetime, luminous efficiency, environment friendliness, and durability [28,29]. WLEDs made from wide bandgap metal oxides, including zinc oxide (ZnO), zirconium oxide (ZrO_2) and gallium oxide (Ga_2O_3), have received considerable attention up to date [30–35]. These materials are capable of producing intrinsic emission in the visible region (blue-red) and can also be doped with rare earth ions [30,31]. The majority of metal oxides are widely utilized as phosphor materials in various applications, including vacuum fluorescent displays, due to their emission stability [36]. Furthermore, metal oxide nanostructures can possibly grow on any p-type substrate and hence make high-quality optoelectronic pn-heterojunctions [37]. However, over the last few decades, a vast improvement has been achieved in WLEDs using rare-earth doped oxide materials [8]. It has been reported that white light emission can easily be produced from metal oxides co-doped with various rare-earth ions [24]. Therefore, currently WLEDs are in the spotlight of extensive research using metal oxide nanocrystals. Based on recent research, white light emission associated with metal oxides as phosphor can be generated using these principles: (i) direct white light generation and (ii) discrete color mixing (Fig. 1). Based on these two classifications, this chapter discusses the recent developments in metal oxides for phosphor-based WLEDs. It is accompanied by a brief outline of the base materials for phosphors. The ideal white light properties of nanocrystals as phosphors are also outlined in this chapter.

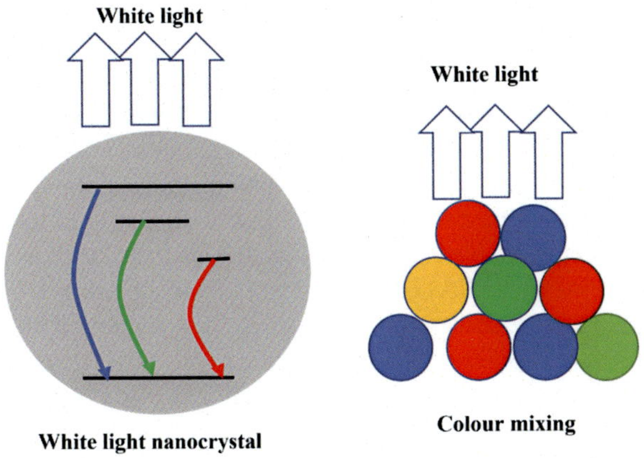

Fig. 1 Schematic diagram of the common approaches used for white light generation from metal oxide nanocrystal-based phosphors for WLEDs. Left: Direct white light generation: White light is generated from a single nanocrystal from an emission layer. Right: Color mixing: The electroluminescent emissions of *blue*, *green*, and *red* are optically mixed to form white light.

2 Phosphors and quantum dots

Phosphors refer to luminescence materials that emit light in conventional down-conversion or upconversion processes when excited with external energy. There are two main classes of absorption and radiative emission from phosphors that can occur, including lattice host sensitization or activators acting as emitting centers. The two processes are schematically illustrated in Fig. 2. In Fig. 1A, the host band to band photon absorption produces electron-hole pairs from where the photon can

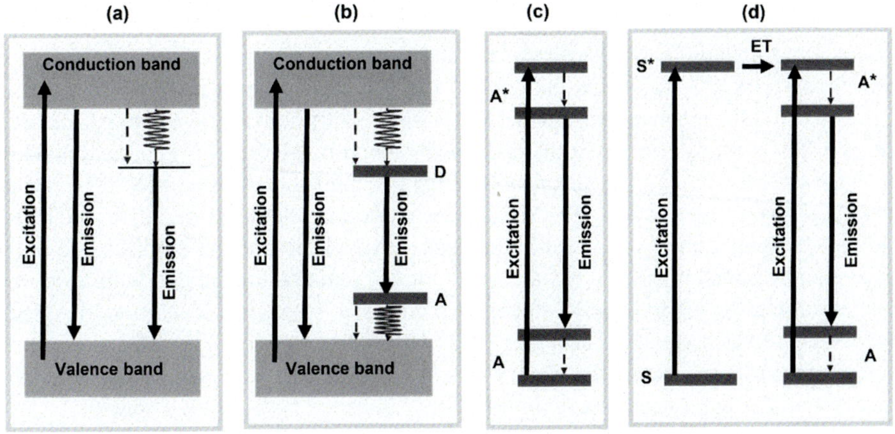

Fig. 2 (A) Host lattice sensitization, (B) host sensitization doped with an activator, (C) an activator acting as luminescence center, and (D) a sensitizer S excited to state S* and the energy transferred to a nearby activator that emits radiatively. Dashed lines represent nonradiative relaxation.

be generated when the electron relaxes from the conduction band to the valance band radiatively. Some of the host materials possess intrinsic defects, and excited electrons can relax nonradiatively to the defect levels and then recombine with a hole in the valance band radiatively. Examples of such materials are quantum dot (QD) semiconductors. QD refers to semiconductor nanocrystals whose absorption and emission bands are size dependent [38]. QDs exhibit a blue shift in wavelength with size reduction. Large-particle quantum dot nanocrystals produce a longer lifetime.

The color of an emitted quantum dot can also be changed with the help of an activator as illustrated in Fig. 2B–D. In these cases, the band to band photon relaxes to a donor level acting as an emission center (labeled D) from where the electron can relax radiatively to the ground state levels (labeled A) of the acceptor. Two activators can both be doped into the host lattice where energy transfer between the sensitizer and acceptor can occur (Fig. 2C). Hence, the emission by sensitization by the host lattice is produced from the sensitizer–acceptor pair. A direct donor excitation is also possible. The simplest situation happens when the sensitizer element (S) gets excited and transfers its energy to another activator dopant element activator (A) from where radiative emission occurs through downconversion processes (Fig. 2B). A tunable fluorescence color output can therefore be produced through energy transfer.

Energy transfer is a vital physical phenomenon for generating white light that occurs in metal oxide phosphors. In the process of energy transfer, the excitation energy is transferred from the luminescent centers (S) to the accepter ions, which in turn release the emission photons.

$$\mathrm{S} + h\nu \rightarrow \mathrm{S}^*$$

$$\mathrm{S}^* + \mathrm{A} \rightarrow \mathrm{S} + \mathrm{A}^*$$

$$\mathrm{A}^* \rightarrow \mathrm{A} + h\nu'.$$

where S and A stand for sensitizer and acceptor, respectively. $h\nu$ is the excited energy, and the asterisk (*) represents the excited ion. The energy transfer processes occur through different mechanisms, including radiative and nonradiative energy transfer [39,40]. The radiative energy transfer requires spectral overlap between the emission of the sensitizer ion and the absorption of the accepter ion. A large spectral overlap results in strong interaction. The significant difference between radiative and nonradiative resonant energy transfer is the decay time of the donor's emission. In radiative energy transfer, the decay time of the donor remains unchanged with acceptor concentration. Resonant energy transfer requires a resonance condition to occur between a sensitizer and an acceptor [41]. The radiative energy transfer process is independent of the distance between the sensitizer and the acceptor. Thus, radiative energy transfer can occur at a relatively larger distance between two species.

Nonradiative energy transfer process is the energy transferred from the sensitizer to accepter ions without photon emission from the sensitizer. Thus, in nonradiative energy transfer, the interaction between the sensitizer and acceptor ions is required. The interactions between the sensitizer and the acceptor can occur through exchange interaction or electrostatic interaction (Coulombic interaction) [42]. Both nonradiative energy transfer processes strongly depend on the distance between the sensitizer and acceptor. The nonradiative energy transfer process probability is inversely proportional to the distance (*d*) between the sensitizer and the acceptor [43]. Both radiative and nonradiative energy transfer have been demonstrated between metal oxide nanocrystals (sensitizer) and lanthanide ions (acceptor) [44,45].

The local symmetry of the crystalline host can influence the luminescence properties of the activator ions. The common activator ions used in WLEDs are transitional metals and rare earth ions [8,24,46]. The crystal field that is produced due to the electrostatic interaction between the activator and ligands splits the d orbital of the luminescent center (both transitional metal and rare-earth ions).

Lanthanide ions are normally utilized as activators in phosphors. The series of lanthanide elements that are used for WLEDs are characterized by an incomplete 4f electron shell. The shielded 4f orbital of the lanthanides produces sharp emission fingerprints of the ions. For example, Tm^{3+}, Tb^{3+}, and Er^{3+} emit blue, green, and red emission, respectively. Therefore, a combination of these three ions can produce white light. This way of producing white light is commonly used in WLED devices [8,41]. The 4f-4f transitions of these ions are mainly responsible for the photon conversion processes. Therefore, they are known as the narrowband conversion process ions.

3 Structure of quantum light-emitting diodes (QLEDs)

There exist three types of charge-transporting materials that are being used in the QLEDs. These three types are all organic (Type I), all inorganic (Type II), and hybrid structures (Type III), and the architecture is illustrated in Fig. 3. All the inorganic QLEDs are often a euphemism as neither the electron nor the hole-transporting materials, including the widely utilized ZnO nanoparticles, are truly inorganic. They

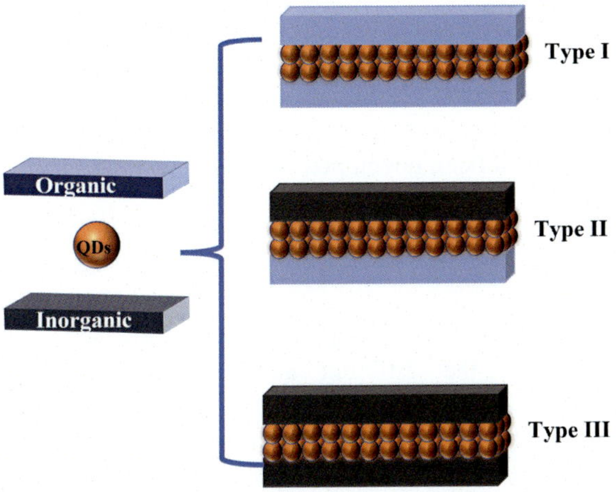

Fig. 3 Three basic structures of QLEDs, Type I all organic, Type II hybrid, and Type III all inorganic [47].

generally include the organic ligands; despite this, the phrase "all inorganic QLED" is often used [47]. It was Colvin and his associates in 1994 who confirmed the fabrication of the first QLED [4]. This fabricated QLED was found to possess an exceptionally low external quantum efficiency (EQE) of the order of 0.01%. This device consisted of a polymer electroluminescent layer sandwiched between the anode and the cathode. Indium tin oxide was the most commonly used anode material. While analyzing the cause of low EQE, it has been found that the metal electrode caused significant quenching impact in the QD emission and was therefore responsible for limiting the efficiency. To minimize the deleterious quenching effect, Colvin et al. [4] build a tri-layer QLED. They have incorporated the monolayer QD in between the organic transport layers as depicted in Fig. 3 (Type I).

The addition of the organic electron transport layers (ETL) in the QD emission played a significant role in decreasing the quenching effect due to the metal electrode. By the incorporation of these changes, an EQE of 0.5% has been attained. However, by considering the remarkable excellent biocompatibility of the inorganic materials, Muller et al. [48] have synthesized a QLED that was found to possess an EQE of the order of 1% in 2005. The fabricated QLED possessed the structural arrangement of Au/p-GaN/CdSe/n-GaN/In. With the progress in science and technology, hybrid structures proved to be the most prominent architect for designing the QLEDs. Due to the remarkable brightness and reliability of the red, blue, and green devices, the hybrid structures have attracted great attention since 2011 [49–56]. The integration of the n-type ZnO nanoparticles into the hybrid QLEDs as an ETL was a key development in the QLED sector. This incorporation has displayed a high brightness of the order of 68,000 cd/m^2 (*green*) along with good stability [54]. Since then, significant improvements have been made both for the EQE that reached the theoretical limit of 20% and for the brightness that is 356,000, 168,000, and 626,000 cd/m^2 for the red, green, and blue devices [50,53,57–65]. It is also noteworthy to mention that the operational lifetime corresponded to the red, green, and blue has been found to be 226,000, 1,760,000, and 32,650 h, respectively. In addition, tethered QLEDs, which combine two or more light-emitting units connected by an interlinked layer, have been developed for further improving the efficiency [57,66–68]. The tethered structures, therefore, have been used for creating high performance white QLEDs. Most commonly, the top-emitting layer, through which the light emanates to the upper transparent electrode, is being used because of its characteristic features such as a large aperture ratio and effective light-out coupling [69,70]. For widening of the applicative part, color tunable and transparent devices have also been developed [66,71,72].

4 Spectroscopy of phosphors materials

The fundamental and applied domains of nanophosphor spectroscopy are now attracting enormous attention due to the different advancements made in nanofabrication. Large numbers of experimental observations have indicated different modifications in the general characteristics of the lanthanide activated phosphors that were enabled due to nanoscale enlargements. Different fabrication techniques such as sol-gel, combustion, microemulsion, and different other template-based approaches are being employed for regulating the size and shape, chemical characteristics, phase, doping quantity, and functionalization in the production of nanophosphors [73–76]. While studying the effect of Mn doping on the size of ZnS nanocrystals, a significant enhancement in the photoluminescence quantum efficiency with an increment in the luminescence lifetime, of nearly five orders of magnitude more than that of the counter bulk part, was observed [77]. Emission amplification with tuning ability was also observed in the 20-nm yttrium oxide (Y_2O_3) nanoparticles co-doped with Yb and Er ions [78]. The simultaneous action of lanthanides' energy levels, along with dopant-induced symmetry disruption of the host lattices incentivized by the surface reduced coordination, resulted in increasing emission efficiencies [79]. Modifying the morphologies of the hexagonal and cubic nanocrystals like $NaREF_4$ (RE = rare earth elements), as well as the $NaYF_4$ nanoparticles co-doped with the Yb, Er, and Tm, significantly increased the upconversion transitions [80]. This increase was attributed to the transition of the shapes that were polyhedra-, rod- and plate-like, etc. [80]. The distinctive emission lines corresponding to the rear-earth elements are directly connected to their anomalous electrical configuration, where the 4f valence electrons are veiled by the outer shells, therefore protecting them from interacting with the crystal lattice and the ligands. On the other hand, in the case of transition metals, the unoccupied outermost protracted and delocalized d-orbitals tend to cause the splitting of the atomic energy levels in the host crystal field. As a result of this configuration, the energy levels in the transition metal dopants are inevitably impacted by the host crystal lattice. But it has been observed that spatial incarceration possesses very little impact on the distinctive emissions resulting from intraconfiguration f-f transitions [81]. However, the emission characteristics of the rear-earths are also caused by the d-f transitions, where host-dependent shifts in energy are being observed in the rear-earth emissions. As a consequence of the L-S coupling and the presence of angular momentum (J) in the excited state of rear-earths, a multiplet represented as $^{2S+1}L_J$ reveals the information that these states are sensitive to the host crystal and thereby generate the host-dependent emission of the rear-earth centers [81]. As a result of the changed coupling sites with regard to the lower energy levels, the separation between the dopant and the ligand along with the site symmetry has a significant impact on the emission acquired from the excited states of the rear-earths in the nanophosphor regime. Furthermore, the dopant-related symmetry breakdown in the host materials tends to produce variations in the emission efficiency. This phenomenon can be more effective due to the site distortion favored by the dopant and the surface reduced coordination [81].

5 Transition metal ions and their role in LED phosphors

The activator ions present in the solid host lattices are generally exposed to the crystalline atmosphere, which possesses a significant impact on the luminescence characteristics of the activator ion. The activator ions present in the crystalline host serve as a luminescence center. Either the rear earth or the transition metal can serve as the activator ion in the host and is usually coordinated by the ligands [8]. The crystal field coupling between the activator ion and the ligand is the essence of the interaction between the oppositely charged activator ion and the nonbonding electrons of the ligands. As a consequence of this interaction, there occurs splitting in the d orbitals of the luminescence centers. The transition metals are more prone to changes in the crystal field surroundings as compared to the rare earths. Thus, the effect of the crystal field is more evident in the transition metal ions in the rear-earth ions. Among the transition metal ions, Mn^{4+} is the most studied element in the domain of the phosphors for LEDs [8]. In the spectral region of 600–750nm, Mn^{4+} is well known for generating efficacious, bright red luminescence [82,83]. While visualizing its configuration, it is being observed that it possesses three electrons in its empty 3d shell. Owing to the simple adequacy of raw manganese ore and low manufacturing costs, Mn^{4+}-doped phosphors have received a lot of interest. Mn^{4+} phosphors have various special properties that allow them to match the specifications of an optimal red-emitting phosphor for LEDs. Due to the fact that Mn^{4+}-based phosphors possess a large effective positive charge, they exhibit significant crystal field effects. Strong crystal fields are usually present in the Mn^{4+} ions. The strength of the crystal fields (weak or strong) is being distinguished by analyzing the emission spectra of the crystal [84]. It has been observed that under the weak crystal field the spin allowed $^2T_{2g}$-$^4A_{2g}$ transition results in wide emission band while under the strong crystal field the spin-forbidden transition ^2E_g-$^4A_{2g}$ observed at high emission peak's position. In particular, it has been observed that the sharp lines associated with the spin-forbidden ^2E_g-$^4A_{2g}$ transition overwhelm the Mn^{4+} emission spectrum. $^4A_2$ corresponds to the ground state of the Mn^{4+} ions and is a spin quartet orbital singlet. The spin-allowed transitions that are $^4A_{2g}$-$^4T_{2g}$ and $^4A_{2g}$-$^4T_{1g}$ arising from the ^4F levels of the Mn^{4+} are readily visible in its excitation spectrum, whereas the third spin-allowed transition that is $^4A_{2g}$-$^4T_{1g}$ arising from the ^4P level generally remains obscured among the host-lattice absorption or charge transfer transition spectra [85]. Due to the presence of the excitation bands occurring in the UV to blue range, the Mn^{4+}-doped phosphor shows efficient absorption in this range and thereby tends to generate string emission lines in the red region. The mentioned aspect provides confirmation that it is a potential candidate for being used in the LED phosphors excited by NUV or blue LEDs. The fabricated phosphors also hold the capability to get readily mixed with yellow or green phosphors with very little reabsorption impact. Significant variation in the energy corresponding to the ^2E_g-$^4A_{2g}$ transition of Mn_{4+} has been observed when doped in different hosts. It has been observed that the transition occurs at 16181 cm^{-1} when doped in Na_2SiF_6 [86], 15,360 cm^{-1} when doped in $Sr_4Al_{14}O_{25}$ [87], 15,220 cm^{-1} when doped in Mg_2TiO_4 [88], etc. This fluctuation in energy corresponding to the transition ^2E_g-$^4A_{2g}$ cannot be attributed to the crystal field variation as the electronic configuration (d^3) of the transition is altogether independent from the crystal field interactions [85]. This energy variation actually relies on the covalency or iconicity of the Mn^{4+}-ligand bonding in the host. The energy belonging to the transition governs the iconicity or covalency of the host compound. If the energy belonging to the transition is more than 15,000 cm^{-1}, then the host is regarded as ionic, and if it is less than 15,000 cm^{-1}, then the host is deemed as covalent. This property of Mn^{4+} permits it to be used as a reliable diagnostic for determining a host's ionic or covalent character. Mn^{4+} is one of the most widely used dopants in oxides and fluorides for developing the red-emitting LED phosphor. While analyzing the stability of the Mn^{4+}-doped hosts, it has been observed that the oxide hosts are thermally and chemically much more stable than that of the fluoride ones. It has been found that Mn^{4+} tends to reside only in the octahedral position of the host lattice [89]. Under certain circumstances, the Mn^{4+} octahedral site's closest six atoms create a deformed octahedron around the Mn^{4+} ions. Manganese in its +2 state also exhibits photoluminescence characteristics. The Mn^{2+} ions possess five electrons in the outermost 3d shell. The orbital singlet $^6A_1$ forms the ground state of the atom and is not affected by the impact of the crystal field. But under the implications of the high crystal field, the ground state is replaced by the spin doublet state $^2T_2$. Mn^{2+} intends to occupy the octahedral or tetrahedral site in the substituted host. At both mentioned sites, Mn^{2+} exhibits the spin-forbidden $^4T_1$-$^6A_1$ transition. The emission color corresponding to d-d transitions in the Mn^{2+} has been discovered to fluctuate contingent on the crystal field intensity. When the crystal field intensity is higher, the Mn^{2+} emission gets shifted to the orange-red regions of the electromagnetic spectrum. On the contrary, in the case of weak crystal field strength, the emission gets shifted toward the green region. Besides manganese, there are other various transition metals that are in different hosts for generating the phenomenon of the luminescence [8]. Chromium ion (Cr^{3+}) is the most prominent for generating luminescence in the red or near-infrared regions of the electromagnetic spectrum [90,91]. Besides being applied in laser technology, Cr^{3+} is being readily explored as a possible luminous ion for designing phosphor in the NIR region [92]. Cr^{3+} holds a similar electronic configuration to that of Mn^{4+}, that is, d^3 configuration. Under the impact of the crystal field, the ^4F levels of the Cr^{3+} get split into three different states, that is, $^4A_2$, $^4T_2$, and $^4T_1$. Among these states, $^4A_2$ corresponds to the ground state of the Cr^{3+} ion. However, the crystal field converts the ^4P level into the $^4T_1$ state rather than splitting it. The spin-allowed transitions $^4A_2$-$^4T_2$ and $^4A_2$-$^4T_1$ residing in the visible region of the electromagnetic spectrum produce two wide absorption bands [93]. Besides the aforementioned, various nonrear-earth-stimulated phosphors have also been identified, which have attained a lot of attention in the field of energy-efficient LED technology [94].

6 WLEDs' requirements

The ideal WLEDs should produce a white light spectrum like natural sunlight and must be most importantly convenient to the human eye. Therefore, WLEDs characterized by high quality white light, cost competitiveness, and long lifetimes are in high demand. Several parameters have been introduced to evaluate the white light quality emitted from WLEDs, including Commission Internationale de l'Eclairage (CIE) chromaticity coordinates, luminous efficacy (LE), correlated color temperature (CCT), and color rendering index (CRI) [6]. For example, the accurate emission color can be measured using the CIE chromaticity coordinates (x, y). The ideal CIE value for white light is (0.33, 0.33). The capability of a light source to show the accurate colors of objects in contrast to a natural light source is known as CRI. The CRI value ranges between 0 and 100. A high CRI is desirable for light sources in color-critical applications. A value greater than 80 is the recommended value of the CRI for WLEDs. The CCT measures the color appearance of a light source, which is determined by comparing the light source chromaticity to that of a black-body radiator. Ideal CCT values for WLEDs are characterized by CCT values in the range from 2500 to 4500K (warm white) to 4500–6500K (cold white). Luminous efficiency (LE) defines the ratio between the total luminous flux and the total lamp power input or power radiantly emitted, which is expressed in lumens/watt. The WLEDs with a luminous efficiency of $200\,\mathrm{L\,mW^{-1}}$ are ideal. The metal oxide materials used for WLEDs should therefore exhibit the desired properties. Thus, the specific requirements for materials for ideal WLED performance are as follows:

- Emission and excitation properties. The primary requirement for a suitable phosphor for WLED's applications is that it should possess a high-absorption cross-section for near-UV or blue LED photons. The material therefore should be characterized with a wide bandgap that could emit in blue, green, and yellow-red, or it can be doped with rare earth ions for white light emission properties. The ideal material for WLEDs should not suffer from self-absorption. In other words, the material should have a large Stokes shift.
- High quantum efficiency. The fluorescence quantum yield defines the ratio between the emitted and absorbed photons. The quantum efficiency is generally measured/calculated from the photoluminescence measurements. An excellent material that can be used as a phosphor for WLEDs depends on the quantum efficiency. Both external and internal quantum efficiency are important factors for WLED materials.
- Environmentally stable. The WLEDs should be stable in different environments. A humid environment was found to influence the stability of some of the quantum dots/phosphor materials [95].
- Thermal quenching: The operation of the semiconductor-based WLEDs sometimes requires high power. At this high power, the electrical input power (approximately 60%) is converted into heat. The local the heat of LED can reach up to 400K, which can increase the entire device's temperature. The thermal quenching (TQ) property of the emitted phosphor is therefore vitally important for WLED devices. The emission peak of the phosphor decreases and broadens with increasing temperature.

Metal oxide semiconductors, however, are characterized by wide band gaps, approximately larger than 3.0eV (Table 1), which can be excited with UV/blue photons. Due to their outstanding properties, they have been utilized in various applications,

Table 1 Bandgap energies of some of the metal oxide nanostructures.

Material	Bandgap energy (eV)	Reference
ZnO	3.27	[24]
Ga_2O_3	4.8	[96]
ZrO_2	5.8	[97]
SnO_2	3.08	[98]
WO_3	3.24	[99]

including energy conversion, microelectronics, optoelectronics, LED, and biosensing [24,48–51]. In addition to that, some of the metal oxide nanostructures possess several intrinsic defects that can emit photons of a wide range in the visible region (*violet* to *red color*) [24]. This property makes metal oxides attractive materials for luminescent applications. Furthermore, metal oxide nanostructures are easy to grow and process in different forms. Metal oxide-based phosphors have been utilized in two different ways: nanostructured metal oxide-based WLEDs, which depend on (i) the broad defect emission from violet to the yellow-red regions and (ii) the rare earth ions doped in the metal oxides.

7 Tuning and role of dopant

The nanophosphor technology is intended for developing simpler device architecture for lightning applications. The main focus of the technology is to achieve broad-range color emission tunability along with multicolored emission by single wavelength stimulations. For the implementation of this basic design, rear earth elements are considered essential because of their complex energy level structure [7]. Proper selection of the dopant, balance of the dopant species, and the dopant-dopant interaction are the important parameters for achieving the multipeak emission from the phosphor. These parameters are necessarily taken into consideration as they are strongly affected by the lattice vibrations and crystal fields of the dopants. By properly adjusting the ratio of intensity among two distinct emission colors, single dopant acting as an emitter will be able to produce the white-light emission [100,101]. However, also by adopting the balanced spectrum and comparative ratio of distinct emission colors, distinct dopant-host interactions can produce diverse color emission. The symmetry possessed by the dopants in diverse host lattices as well as in the dopant-dopant interaction at several lattice sites also intends the phase-dependent luminescence [79]. The effect of the dopant-dopant interaction can be better understood by analyzing the emission spectrum of the hexagonal phase of β-YF_3 nanocrystal under an excitation wavelength of 976 nm. The mentioned nanocrystal generally reveals the f-f transitions corresponding to the energy levels of the co-dopants, which are Er^{3+}, Tm^{3+}, and Yb^{3+}. The observed emission has been visualized to be a mixture of blue, green, and red light emitted by the Yb^{3+}/Tm^{3+}, Yb^{3+}/Er^{3+}, and Tm^{3+}/Er^{3+} pairs, respectively [102].

The amount of the dopants present in the hosts also played an important role in the emission profile of the host. Nowadays the dependency of the output light intensity on the dopant concentration is being used to control multicolor as well as the comparative emission intensity. For example, in the case of $NaYF_4$ nanocrystals doped with the Yb, Tm, and Er, for the fixed Yb^{3+}/Tm^{3+} concentration ratio of 20/0.2 mol%, the color output can be altered from the blue to white by increasing the concentration of Er^{3+} from 0.2 to 1.5 mol% [103]. In case if the $LiYF_4$ nanocrystal is codoped with the Yb^{3+}, Tm^{3+}, and Er^{3+}, it emits blue and green light when excited by a wavelength of the order of 980 nm. It has been observed that the intensity of the green to blue photoluminescence can be altered as it strongly depends on the concentration of Er^{3+} [104]. In this host, the concentration of Er^{3+} is varied from 0.1 to 0.5 mol% for visualizing the shift in the emission color regime. Similarly change in the red-green emission ratio with particle size has been observed in the yttrium oxide nanoparticles codoped with Er^{3+}, Tm^{3+}, and Ho^{3+} [105]. Also it has been observed that the output color emission can be strongly adjusted from blue to red by altering the dopant concentration. The amount of dopant present in the nanophosphors is critical not only for the multicolor tuning but also plays a significant role in the luminescence quenching phenomenon. In bulk phosphors when the concentration of the localized centers exceeds a particular threshold host-based value, the phenomenon of crossrelaxation and energy transfer comes into play and thereby causes luminescence quenching [39,100]. The effect is known as the concentration quenching phenomenon, as it happens at dopant concentrations of approximately 1 mol%. The luminescence steadily keeps on lowering up to different magnitudes till it disappears, starting from this lower limit up to 10–20 mol%. The phenomenon of quenching is also being observed in the nanophosphors with the exception that the concentration at which the phenomenon can become evident is slightly higher for the nanophosphors compared to their bulk equivalent [106]. In particular cases of nanoparticles such as Y_2O_3, rather than decreasing, an increase in the luminescence has been observed while increasing the dopant concentration [107]. The decreased likelihood of the ion-ion energy transfer as a consequence of the resonance along with the lack of phonon modes capable of participating in the energy transfer processes has been recognized as the possible cause for this unique experimental observation [108]. For resolving the quenching phenomenon in the nanocrystals, the incarceration of the lattice dopants has been envisaged, which includes the allocation of the Yb^{3+} ions with high content organized in well defined structures in orthorhombic lattices, which were observed to produce a four-photon-promoted violet light, which is an extremely protracted emission given by Er^{3+} ions [109].

8 Metal oxide-based phosphors for WLEDs

8.1 Direct white light generation

A direct generation of white light is where the white light photoluminescence is produced from a single nanocrystal. The design of such devices normally relies on the UV LED source to excite the nanocrystal. The UV or blue photons from the LED would then interact with the phosphor that is coated on the plastic cover. The phosphor therefore absorbs the UV/blue photons and emits the blue to green and red photons. This is the common approach of generating white light where a blue LED chip and a yellow phosphor are combined [110]. The metal oxide nanocrystals are capable of producing different colors; direct white light generation has been demonstrated from different metal oxide nanocrystals [9,98,99]. White light from metal oxide nanocrystals is produced as a combination of blue, green, and yellow-red emission originated from different emission centers inside the metal oxides. Among those metal oxides, ZnO has brought considerable attention owing to its direct bandgap, low cost, and low temperature synthesis processability [32]. Above all, the various intrinsic defects (deep level emission) in ZnO that produce the visible emission (deep level) are the most outstanding for WLEDs. The deep-level emission of ZnO was found to be dependent on the crystal size, which can be maximized in nanosize particles [111–114]. Fig. 3 illustrates a schematic diagram for possible emissions from a ZnO nanocrystal.

From Fig. 4, the combination of emitted light from a ZnO nanocrystal exhibited brilliant white light [115]. In direct white generation methods, ZnO nanocrystals can therefore be used as the central light emitting center in WLED devices. The ZnO nanocrystals, however, are utilized as phosphors using blue or UV LED excitation or an intrinsic light-emitting layer [116,117]. Making such a device is therefore dependent on the synthesis of ZnO nanostructures of a desired size. Different synthesis methods have been reported for this purpose, including the sol-gel and hydrothermal techniques. The disadvantage of this method is the instability of the broad emission spectra due to the bandgap and deep-level dependence on the device temperature [31].

The second strategy for WLEDs is the employment of lanthanide ions as dopants in wide bandgap semiconductors. The spectra generated in this case are characterized by narrow and sharp emission bands. Lanthanide ion emission could combine with metal oxide semiconductor emission to make WLEDs. Another possibility is producing white light from different lanthanide ions co-doped with metal oxides. Emission spectra produced this way make the emission ultrastable against the surrounding temperature. Therefore, lanthanide ion-doped wide-bandgap metal oxide semiconductors

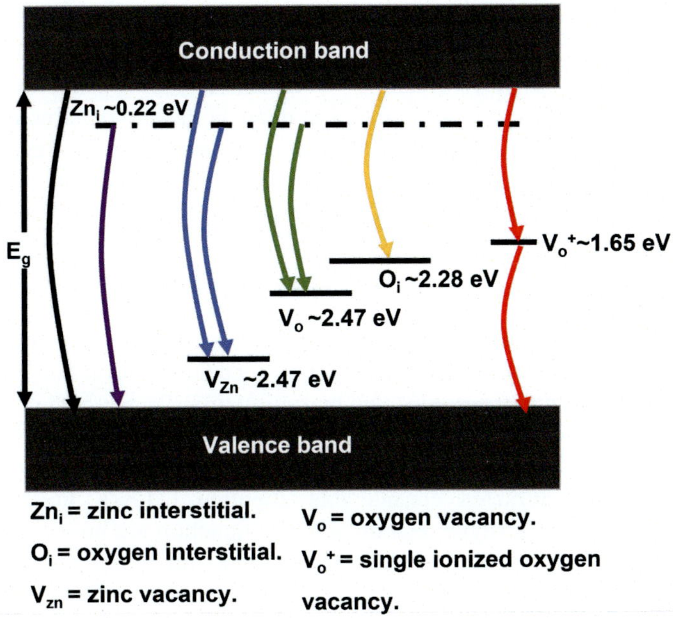

Fig. 4 Schematic diagram of the possible emission in ZnO including deep level emission. Adapted from reference N.H. Alvi, K. ul Hasan, O. Nur, M. Willander, The origin of the red emission in n-ZnO nanotubes/p-GaN white light emitting diodes, Nanoscale Res. Lett. 6 (2011) 1–7. https://doi.org/10.1186/1556-276X-6-130.

have been considered ideal candidates for LED devices [118]. Over the last two decades, substantial investigations have therefore been done in WLEDs using lanthanide ion-doped metal oxide nanocrystals [8,24,97–99]. Most of the studies on lanthanide ion-doped wide-bandgap metal oxide for WLEDs have focused on Tm^{3+}, Tb^{3+}, Eu^{3+}, and Er^{3+} for producing blue, green, and yellow to red emission, respectively. White light emission from metal oxide nanocrystals doped with Eu^{3+} ions and different lanthanide ions has been reported by various groups [24,44,119,120]. Prashant et al. [119] reported white light from Eu^{3+}- and Tb^{3+}-doped Ga_2O_3 nanocrystals. In this system, the absorption of UV photons resulted in blue, green, and red emissions originated from the Ga_2O_3 nanocrystal, Tb^{3+}, and Eu^{3+}, respectively. This type of WLEDs reported ideal CIE chromaticity coordinates of (0.33, 0.33). Different combinations of lanthanide ions, including Eu-, Tb-, Dy-, and Er-doped ZnO, ZrO_2, WO_3, and SnO_2 nanocrystals, have also been reported for white light emission [33,44,45,99,121]. In all these systems, the metal oxides are excited electrically or with photons with energy bigger than their bandgaps from where the energy transfer occurs to the lanthanide ions, producing blue, green, and red emissions. Therefore, substantial studies have been devoted to demonstrating the energy transfer between metal oxide nanocrystals and lanthanide ions [45,122].

White light from lanthanide ions doped in large metal oxide bandgap materials has also been demonstrated [30,123]. To achieve the multipeak tunable emission from large bandgap metal oxide nanophosphors, it requires the right choice of dopants and dopant–dopant interaction. An example of generating a white emission spectrum between dopants is given by ZrO_2 nanocrystals codoped with suitable Tm^{3+}, Tb^{3+}, and Er^{3+} under an UV excitation source [123]. The resulting emission is blue, green, and red originating from Tm^{3+}, Tb^{3+}, and Er^{3+}, respectively. Codopants integrated in a single metal oxide nanocrystal were found to degrade the crystal structure and thus influence the luminous efficiency. Furthermore, crossrelaxation between lanthanide ions affects the color output and leads to white light with low efficiency. Therefore, considerable approaches have been invented to overcome such problems [9,30].

8.2 Homojunction and heterojunction WLEDs

Although the UV/blue LED excitation chip approach is effective and simple, it is considered less efficient as it is based on an absorption and downconversion/downshifting process, which is inevitably accompanied by energy loss. To overcome this problem, various approaches have been invented and considered as solutions [9]. Among these approaches, homojunction and heterojunction WLEDs using metal oxide nanostructures have attracted considerable attention [31,37]. In homojunction WLEDs, both p-type and n-type regions of the pn-junction are built using the same metal oxide nanostructures. The white light emission is therefore generated through electroluminescence, which avoids the absorption processes. ZnO homojunction WLED devices therefore consist of p-type ZnO and n-type ZnO nanostructures [31]. The p-type ZnO region is normally produced with metallic elements doped in the ZnO such as antimony (Sb). Figure 5 illustrates the origin of the white light emission along with a possible radiative transition from the schematic energy bands of

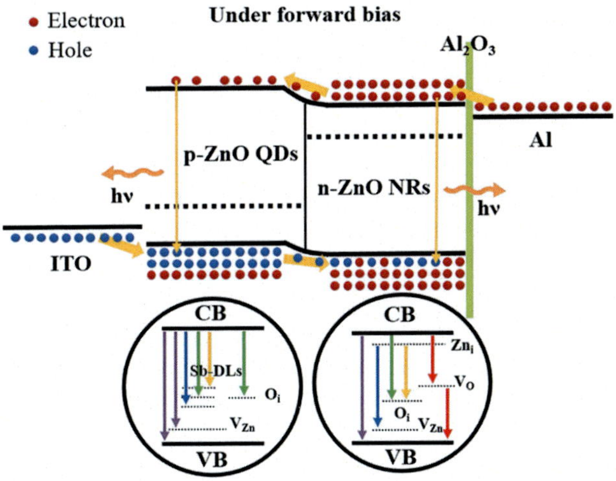

Fig. 5 Schematic energy band diagram illustrating the possible emission from n-ZnO nanorods/Sb-doped ZnO quantum dot homojunction WLEDs [31].

WLEDs. In the homojunction, under forward bias, normally the carriers diffuse toward the adjacent side of the junction; therefore, recombination of the electron and hole occurs at both the p-type and n-type sides simultaneously, resulting in ideal white light with high CIE, CCT, and efficiency [31]. The results showed that homojunction WLED devices could lead to high electrical charge transport, low device resistance, and a long device lifetime. The performance of the homojunction device depends on the availability of the p-type and n-type metal oxide materials. Due to the difficulty of controlling the p-type materials, especially p-type ZnO, metal oxide WLEDs were initially produced with different heterojunction structures.

In heterojunction WLEDs, the metal oxide nanostructures can be grown on a p-type substrate for making the pn-junction [9]. Therefore, in heterojunction WLEDs, the white light is produced from only the n-type metal oxide nanostructures. The most popular heterojunction WLEDs have been the combination of metal oxide semiconductors with group III-nitrides [9]. Alvi et al. [117,124] reported different ZnO nanostructures that were grown on a blue GaN substrate. They found that ZnO nanowall structures produced ideal white light with a CRI of 95 and CCT of 6518 K.

8.3 Discrete color mixing WLEDs

In this approach, the white light is produced by optically mixing the appropriate colors of the electroluminescent emissions from several semiconductor nanocrystals in a suitable ratio. For example, red-, green-, and blue-emitting oxide nanocrystals should be combined to generate WLEDs. By this way, white light emission was generated by mixing a ZnO core wrapped in a shell of a single layer of graphene quantum dots [125]. ZnO exhibited blue and green emissions, while the red emission was obtained through crossrelaxation from the graphene HOMO levels to the ZnO valence band. However, Ga_2O_3 nanocrystals can emit broad blue emission. The mixing of the blue emission of the Ga_2O_3 nanocrystal and Eu^{3+} and Tb^{3+} doped Tb_2O_3 grown on a Si substrate was also reported by the Ling et al. group [33]. The role of the lanthanide ions was to generate green and red emissions.

The application of WLEDs requires emissive material along the entire visible spectrum, which is robust, nonselfabsorptive, and highly efficient [126]. The direct mixing approach therefore has several disadvantages:

- It is difficult to control the proportion of the individual color components for ideal white light [127]. This makes this approach complicated for producing white light with the same quality.
- Self-absorption due to the different absorption centers is high in this approach, which severely affects the overall quantum efficiency. For example, the blue emission from nanocrystals could be absorbed by other red- and green-emitting nanocrystals nearby.

Recently, in order to overcome these difficulties, an electrically driven strategy and phosphor free WLEDs have been suggested [30]. This approach is based on alternative multilayers of single materials of metal oxides doped with lanthanide ions. Huang et al. [30] reported multilayers of $Ga_2O_3:Tm^{3+}$, $Ga_2O_3:Eu^{3+}$, and $Ga_2O_3:Er^{3+}$ films, which were utilized as blue, red, and green, respectively. The film layers were vertically grown on a single GaAs substrate. This strategy of WLEDs produced bright white light color CIE coordinates of (0.3329, 0.3335) and a CCT of 5479 K. The CIE coordinates of this strategy strongly depend on the thickness of the multilayers without altering the dopant concentration.

9 Conclusion

In this chapter, the recent developments in metal oxide nanocrystal-based phosphors for WLEDs were highlighted in terms of the two principles, which include direct white-light generation and discrete color mixing. The approaches were designed to improve the performance of WLEDs. The approaches therefore utilized the ability of the materials for converting photons in the visible region. A noticeable improvement in WLEDs has been observed with lanthanide ion dopants, but solving the issue of emission quenching between lanthanide ions can still lead to enhancement. The homojunction and heterojunction WLEDs based on metal oxide semiconductors were developed to overcome the absorption and downconversion/shifting problem of WLEDs. The homojunction and heterojunction WLEDs therefore generate white light through electroluminescence of blue, green, and red color. Interestingly it appeared that better solutions for WLEDs depend on the homojunction direct electroluminescence of white light. Such innovative approaches have shown promising results irrespective of the LED technology.

The popularity of ZnO nanocrystals stems from brilliant white light properties in the visible range. Furthermore, ZnO nanocrystals can easily be prepared using different simple preparation methods. It appears that so far ZnO heterojunction LEDs are not good homojunction LEDs due to the difficulty of making good p-type ZnO.

Electroluminescence from multilayers of lanthanide doped wide bandgap metal oxides could lead to a noticeable enhancement in the discrete color mixing approach. In this approach, it has been shown that the white light emission characteristics depend on the thickness of the multilayers instead of dopants' concentration. The available results indicate that the white light produced was from the color mixing that depends on the lanthanide ions used. Further investigations and exploration will certainly be fruitful in order to produce affordable WLEDs in order to reduce or maintain production costs and enhance the efficiency.

Acknowledgments

This work is based on the research supported by the South African Research Chairs Initiative of the Department of Science and Technology and National Research Foundation of South Africa (grand 84415). The financial assistance of the National Research Foundation (NRF) and the University of the Free State toward this research is hereby acknowledged.

References

[1] M.E. Coltrin, J.Y. Tsao, Solid-State Lighting Technology Perspective, SANDIA REPORT, 2006. *https://www.osti.gov/servlets/purl/889939/*.

[2] N. Tessler, V. Medvedev, M. Kazes, S.H. Kan, U. Banin, Efficient near-infrared polymer nanocrystal light-emitting diodes, Science 295 (2002) 1506–1508, https://doi.org/10.1126/science.1068153.

[3] S. Coe, W.K. Woo, M. Bawendi, V. Bulović, Electroluminescence from single monolayers of nanocrystals in molecular organic devices, Nature 420 (2002) 800–803, https://doi.org/10.1038/nature01217.

[4] V.L. Colvin, M.C. Schlamp, A.P. Alivisatos, Light-emitting diodes made from cadmium selenide nanocrystals and a semiconducting polymer, Nature 370 (1994) 354–357, https://doi.org/10.1038/370354a0.

[5] A. Baltakesmez, S. Tekmen, S. Tüzemen, ZnO homojunction white light-emitting diodes, J. Appl. Phys. 110 (2011), 054502, https://doi.org/10.1063/1.3627247.

[6] Q. Dai, C.E. Duty, M.Z. Hu, Semiconductor-nanocrystals-based white light-emitting diodes, Small 6 (2010) 1577–1588, https://doi.org/10.1002/SMLL.201000144.

[7] M. Cesaria, B. Di Bartolo, Nanophosphors-based white light sources, Nanomaterials 9 (2019) 1048, https://doi.org/10.3390/nano9071048.

[8] G.B. Nair, H.C. Swart, S.J. Dhoble, A review on the advancements in phosphor-converted light emitting diodes (pc-LEDs): phosphor synthesis, device fabrication and characterization, Prog. Mater. Sci. 109 (2020), 100622, https://doi.org/10.1016/j.pmatsci.2019.100622.

[9] F. Rahman, Zinc oxide light-emitting diodes: a review, Opt. Eng. 58 (2019) 1, https://doi.org/10.1117/1.OE.58.1.010901.

[10] W. Ullah Khan, L. Zhou, X. Li, W. Zhou, D. Khan, S.I. Niaz, M. Wu, Single phase white LED phosphor Ca3YAl3B4O15:Ce3+,Tb3+,Sm3+ with superior performance: color-tunable and energy transfer study, Chem. Eng. J. 410 (2021) 128455, https://doi.org/10.1016/J.CEJ.2021.128455.

[11] J. Chen, J. Wang, X. Xu, J. Li, J. Song, S. Lan, S. Liu, B. Cai, B. Han, J.T. Precht, D. Ginger, H. Zeng, Efficient and bright white light-emitting diodes based on single-layer

heterophase halide perovskites, Nat. Photonics 15 (2020) 238–244, https://doi.org/10.1038/s41566-020-00743-1.

[12] Q. Ma, Q. Li, G. Zhang, S. Han, Q. Zhang, Q. Chen, B. Ma, Color tunable Eu-doped CaSrNb2O7 phosphors for white light-emitting diodes and optical thermometers, J. Lumin. 227 (2020), 117582, https://doi.org/10.1016/J.JLUMIN.2020.117582.

[13] H. Chen, L. Zhu, C. Xue, P. Liu, X. Du, K. Wen, H. Zhang, L. Xu, C. Xiang, C. Lin, M. Qin, J. Zhang, T. Jiang, C. Yi, L. Cheng, C. Zhang, P. Yang, M. Niu, W. Xu, J. Lai, Y. Cao, J. Chang, H. Tian, Y. Jin, X. Lu, L. Jiang, N. Wang, W. Huang, J. Wang, Efficient and bright warm-white electroluminescence from lead-free metal halides, Nat. Commun. 12 (2021) 1–7, https://doi.org/10.1038/s41467-021-21638-x.

[14] P. Balakrishnan, M. Jayachandiran, S.M.M. Kennedy, Color tunable emission from single-phased $Ba_2CaZn_2Si_6O_{17}$: Bi^{3+}, Eu^{3+} phosphors with good energy transfer efficiency for white light emitting diodes, J. Lumin. 215 (2019) 116649, https://doi.org/10.1016/J.JLUMIN.2019.116649.

[15] S. Wang, Q. Sun, B. Devakumar, J. Liang, L. Sun, X. Huang, Novel high color-purity Eu^{3+}-activated $Ba_3Lu_4O_9$ red-emitting phosphors with high quantum efficiency and good thermal stability for warm white LEDs, J. Lumin. 209 (2019) 156–162, https://doi.org/10.1016/J.JLUMIN.2019.01.050.

[16] W. Sun, H. Li, B. Zheng, R. Pang, L. Jiang, S. Zhang, C. Li, Electronic structure and photoluminescence properties of a novel single-phased color tunable phosphor $KAlGeO_4$:Bi^{3+},Eu^{3+} for WLEDs, J. Alloys Compd. 774 (2019) 477–486, https://doi.org/10.1016/J.JALLCOM.2018.10.087.

[17] M. Cui, J. Wang, M. Shang, J. Li, Q. Wei, P. Dang, H.S. Jang, J. Lin, Full visible light emission in Eu^{2+},Mn^{2+}-doped $Ca_9LiY_{0.667}(PO_4)_7$ phosphors based on multiple crystal lattice substitution and energy transfer for warm white LEDs with high colour-rendering, J. Mater. Chem. C. 7 (2019) 3644–3655, https://doi.org/10.1039/C9TC00109C.

[18] H.C. Chen, K.J. Chen, C.C. Lin, C.H. Wang, H.V. Han, H.H. Tsai, H.T. Kuo, S.H. Chien, M.H. Shih, H.C. Kuo, Improvement in uniformity of emission by ZrO_2 nano-particles for white LEDs, Nanotechnology 23 (2012), https://doi.org/10.1088/0957-4484/23/26/265201.

[19] S.A. Khan, A. Jalil, Q. Ullah Khan, R.M. Irfan, I. Mehmood, K. Khan, M. Kiani, B. Dong, N.Z. Khan, J.L. Yu, L. Zhu, S. Agathopoulos, New physical insight into crystal structure, luminescence and optical properties of Y_PO_4:Dy^{3+}\$Eu^{3+}$\$Tb^{3+}$ single-phase white-light-emitting phosphors, J. Alloys Compd. 817 (2020) 152687, https://doi.org/10.1016/J.JALLCOM.2019.152687.

[20] S.A. Khan, N.Z. Khan, Z. Hao, W.W. Ji, H. Abadikhah, L. Hao, X. Xu, S. Agathopoulos, Influence of substitution of Al-O for Si-N on improvement of photoluminescence properties and thermal stability of $Ba_2Si_5N_8$:Eu^{2+} red emitting phosphors, J. Alloys Compd. 730 (2018) 249–254, https://doi.org/10.1016/j.jallcom.2017.09.335.

[21] N.Z. Khan, S.A. Khan, L. Zhan, A. Jalil, J. Ahmed, M.A.M. Khan, M.T. Abbas, F. Wang, X. Xu, Synthesis, structure and photoluminescence properties of Ca_2YTaO_6:Bi^{3+}\Eu^{3+} double perovskite white light emitting phosphors, J. Alloys Compd. 868 (2021) 159257, https://doi.org/10.1016/J.JALLCOM.2021.159257.

[22] M. Auf Der Maur, A. Pecchia, G. Penazzi, W. Rodrigues, A. Di Carlo, Efficiency drop in green InGaN/GaN light emitting diodes: the role of random alloy fluctuations, Phys. Rev. Lett. 116 (2016) 027401, https://doi.org/10.1103/PHYSREVLETT.116.027401.

[23] J.Y. Tsao, M.H. Crawford, M.E. Coltrin, A.J. Fischer, D.D. Koleske, G.S. Subramania, G.T. Wang, J.J. Wierer, R.F. Karlicek, Toward smart and ultra-efficient solid-state lighting, Adv. Opt. Mater. 2 (2014) 809–836, https://doi.org/10.1002/ADOM.201400131.

[24] V. Kumar, O.M. Ntwaeaborwa, T. Soga, V. Dutta, H.C. Swart, Rare earth doped zinc oxide nanophosphor powder: A future material for solid state lighting and solar cells, ACS Photonics. 4 (2017) 2613–2637, https://doi.org/10.1021/acsphotonics.7b00777.
[25] S. Senapati, K.K. Nanda, Designing dual emissions via co-doping or physical mixing of individually doped ZnO and their implications in optical thermometry, ACS Appl. Mater. Interfaces 9 (2017) 16305–16312, https://doi.org/10.1021/ACSAMI.7B00587.
[26] G. Deng, Y. Huang, Z. Chen, C. Pan, K. Saito, T. Tanaka, Q. Guo, Yellow emission from vertically integrated Ga_2O_3 doped with Er and Eu electroluminescent film, J. Lumin. 235 (2021) 118051, https://doi.org/10.1016/J.JLUMIN.2021.118051.
[27] S. Ichikawa, N. Yoshioka, J. Tatebayashi, Y. Fujiwara, Room-temperature operation of near-infrared light-emitting diode based on Tm-doped GaN with ultra-stable emission wavelength, J. Appl. Phys. 127 (2020) 113103, https://doi.org/10.1063/1.5140715.
[28] A. Zainelabdin, O. Nur, G. Amin, S. Zaman, Metal oxide nanostructures and white light emission, in: Oxide-Based Mater, Devices III, SPIE, 2012, p. 82630N, https://doi.org/10.1117/12.909342.
[29] I. Şerban, A. Enesca, Metal oxides-based semiconductors for biosensors applications, Front. Chem. 8 (2020) 354, https://doi.org/10.3389/FCHEM.2020.00354/BIBTEX.
[30] Y. Huang, K. Saito, T. Tanaka, Q. Guo, Strategy toward white LEDs based on vertically integrated rare earth doped Ga_2O_3 films, Appl. Phys. Lett. 119 (2021) 062107, https://doi.org/10.1063/5.0060066.
[31] S.D. Baek, Y.C. Kim, J.M. Myoung, Sb-doped p-ZnO quantum dots: templates for ZnO nanorods homojunction white light-emitting diodes by low-temperature solution process, Appl. Surf. Sci. 480 (2019) 122–130, https://doi.org/10.1016/J.APSUSC.2019.02.209.
[32] M. Willander, O. Nur, J.R. Sadaf, M.I. Qadir, S. Zaman, A. Zainelabdin, N. Bano, I. Hussain, Luminescence from zinc oxide nanostructures and polymers and their hybrid devices, Materials (Basel) 3 (2010) 2643–2667, https://doi.org/10.3390/ma3042643.
[33] L. Li, S. Wang, G. Mu, X. Yin, L. Yi, Multicolor light-emitting devices with Tb_2O_3 on silicon, Sci. Rep. 7 (2017) 1–4, https://doi.org/10.1038/srep42479.
[34] L. Rao, Y. Tang, Z. Li, X. Ding, J. Li, S. Yu, C. Yan, H. Lu, Effect of ZnO nanostructures on the optical properties of white light-emitting diodes, Opt. Express 25 (2017) A432, https://doi.org/10.1364/oe.25.00a432.
[35] A. Báez-Rodríguez, D. Albarrán-Arreguín, A.C. García-Velasco, O. Álvarez-Fregoso, M. García-Hipólito, M.A. Álvarez-Pérez, L. Zamora-Peredo, C. Falcony, White and yellow light emission from ZrO_2:Dy^{3+} nanocrystals synthesized by a facile chemical technique, J. Mater. Sci. Mater. Electron. 29 (2018) 15502–15511, https://doi.org/10.1007/S10854-018-9105-1.
[36] M.J. Weber, A.V. Dotsenko, L.B. Glebov, V.A. Tsekhomsky, Handbook of phosphors, Opt. Mater. (Amst) 23 (2003) 1584. *http://link.aip.org/link/OPEGAR/v36/i6/p1584/s1&Agg=doi.*
[37] M. Willander, O. Nur, Q.X. Zhao, L.L. Yang, M. Lorenz, B.Q. Cao, J. Z̃iga Pérez, C. Czekalla, G. Zimmermann, M. Grundmann, A. Bakin, A. Behrends, M. Al-Suleiman, A. El-Shaer, A.C. Mofor, B. Postels, A. Waag, N. Boukos, A. Travlos, H.S. Kwack, J. Guinard, D. Le Si Dang, Zinc oxide nanorod based photonic devices: recent progress in growth, light emitting diodesand lasers, Nanotechnology 20 (2009) 332001, https://doi.org/10.1088/0957-4484/20/33/332001.
[38] J. McKittrick, L.E. Shea-Rohwer, Review: down conversion materials for solid-state lighting, J. Am. Ceram. Soc. 97 (2014) 1327–1352, https://doi.org/10.1111/JACE.12943.
[39] F. Auzel, Upconversion and anti-stokes processes with f and d ions in solids, Chem. Rev. 104 (2003) 139–173, https://doi.org/10.1021/CR020357G.

[40] M.Y.A. Yagoub, H.C. Swart, E. Coetsee, Energy transfer study between Ce^{3+} and Tb^{3+} ions in a calcium fluoride crystal for solar cell applications, J. Lumin. 187 (2017) 96–101, https://doi.org/10.1016/J.JLUMIN.2017.02.066.

[41] A.L. Rogach, T.A. Klar, J.M. Lupton, A. Meijerink, J. Feldmann, Energy transfer with semiconductor nanocrystals, J. Mater. Chem. 19 (2009) 1208–1221, https://doi.org/10.1039/B812884G.

[42] B. Valeur, Handbook of analytical techniques single-molecule detection in solution, Meth. Appl. (2001).

[43] I. Pelant, J. Valenta, Luminescence Spectroscopy of Semiconductors, Oxford University Press, 2012.

[44] V. Mangalam, K. Pita, C. Couteau, Study of energy transfer mechanism from ZnO nanocrystals to Eu^{3+} ions, Nanoscale Res. Lett. 11 (2016) 1–9, https://doi.org/10.1186/S11671-016-1282-3.

[45] V. Mangalam, K. Pita, Energy transfer efficiency from ZnO-nanocrystals to Eu3+ ions embedded in SiO2 film for emission at 614 nm, Materials (Basel) 10 (2017) 930, https://doi.org/10.3390/ma10080930.

[46] Y. Wang, G. Zhu, S. Xin, Q. Wang, Y. Li, Q. Wu, C. Wang, X. Wang, X. Ding, W. Geng, Recent development in rare earth doped phosphors for white light emitting diodes, J. Rare Earths 33 (2015) 1–12, https://doi.org/10.1016/S1002-0721(14)60375-6.

[47] Q. Yuan, T. Wang, P. Yu, H. Zhang, H. Zhang, W. Ji, A review on the electroluminescence properties of quantum-dot light-emitting diodes, Org. Electron. 90 (2021), 106086, https://doi.org/10.1016/j.orgel.2021.106086.

[48] A.H. Mueller, M.A. Petruska, M. Achermann, D.J. Werder, E.A. Akhadov, D.D. Koleske, M.A. Hoffbauer, V.I. Klimov, Multicolor light-emitting diodes based on semiconductor nanocrystals encapsulated in GaN charge injection layers, Nano Lett. 5 (2005) 1039–1044, https://doi.org/10.1021/NL050384X.

[49] W.K. Bae, Y.S. Park, J. Lim, D. Lee, L.A. Padilha, H. McDaniel, I. Robel, C. Lee, J.M. Pietryga, V.I. Klimov, Controlling the influence of auger recombination on the performance of quantum-dot light-emitting diodes, Nat. Commun. 4 (2013), https://doi.org/10.1038/NCOMMS3661.

[50] X. Dai, Z. Zhang, Y. Jin, Y. Niu, H. Cao, X. Liang, L. Chen, J. Wang, X. Peng, Solution-processed, high-performance light-emitting diodes based on quantum dots, Nature 515 (2014) 96–99, https://doi.org/10.1038/nature13829.

[51] T.H. Kim, K.S. Cho, E.K. Lee, S.J. Lee, J. Chae, J.W. Kim, D.H. Kim, J.Y. Kwon, G. Amaratunga, S.Y. Lee, B.L. Choi, Y. Kuk, J.M. Kim, K. Kim, Full-colour quantum dot displays fabricated by transfer printing, Nat. Photonics 5 (2011) 176–182, https://doi.org/10.1038/nphoton.2011.12.

[52] J. Kwak, W.K. Bae, D. Lee, I. Park, J. Lim, M. Park, H. Cho, H. Woo, D.Y. Yoon, K. Char, S. Lee, C. Lee, Bright and efficient full-color colloidal quantum dot light-emitting diodes using an inverted device structure, Nano Lett. 12 (2012) 2362–2366, https://doi.org/10.1021/NL3003254.

[53] B.S. Mashford, M. Stevenson, Z. Popovic, C. Hamilton, Z. Zhou, C. Breen, J. Steckel, V. Bulovic, M. Bawendi, S. Coe-Sullivan, P.T. Kazlas, High-efficiency quantum-dot light-emitting devices with enhanced charge injection, Nat. Photonics 7 (2013) 407–412, https://doi.org/10.1038/nphoton.2013.70.

[54] L. Qian, Y. Zheng, J. Xue, P.H. Holloway, Stable and efficient quantum-dot light-emitting diodes based on solution-processed multilayer structures, Nat. Photonics 5 (2011) 543–548, https://doi.org/10.1038/nphoton.2011.171.

[55] Y. Yang, Y. Zheng, W. Cao, A. Titov, J. Hyvonen, J.R. Manders, J. Xue, P.H. Holloway, L. Qian, High-efficiency light-emitting devices based on quantum dots with tailored nanostructures, Nat. Photonics 9 (2015) 259–266, https://doi.org/10.1038/nphoton.2015.36.
[56] Z. Zhang, Y. Ye, C. Pu, Y. Deng, X. Dai, X. Chen, D. Chen, X. Zheng, Y. Gao, W. Fang, X. Peng, Y. Jin, High-performance, solution-processed, and insulating-layer-free light-emitting diodes based on colloidal quantum dots, Adv. Mater. 30 (2018), https://doi.org/10.1002/ADMA.201801387.
[57] H. Zhang, X. Sun, S. Chen, Over 100 cd A−1 efficient quantum dot light-emitting diodes with inverted tandem structure, Adv. Funct. Mater. 27 (2017), https://doi.org/10.1002/ADFM.201700610.
[58] Y.-H. Won, O. Cho, T. Kim, D.-Y. Chung, T. Kim, H. Chung, H. Jang, J. Lee, D. Kim, E. Jang, Highly efficient and stable InP/ZnSe/ZnS quantum dot light-emitting diodes, Nature 575 (2019) 634–638, https://doi.org/10.1038/s41586-019-1771-5.
[59] H. Shen, Q. Gao, Y. Zhang, Y. Lin, Q. Lin, Z. Li, L. Chen, Z. Zeng, X. Li, Y. Jia, S. Wang, Z. Du, L.S. Li, Z. Zhang, Visible quantum dot light-emitting diodes with simultaneous high brightness and efficiency, Nat. Photonics 13 (2019) 192–197, https://doi.org/10.1038/s41566-019-0364-z.
[60] Y. Sun, Q. Su, H. Zhang, F. Wang, S. Zhang, S. Chen, Investigation on thermally induced efficiency roll-off: toward efficient and ultrabright quantum-dot light-emitting diodes, ACS Nano 13 (2019) 11433–11442, https://doi.org/10.1021/ACSNANO.9B04879.
[61] Y. Fu, W. Jiang, D. Kim, W. Lee, H. Chae, Highly efficient and fully solution-processed inverted light-emitting diodes with charge control interlayers, ACS Appl. Mater. Interfaces 10 (2018) 17295–17300, https://doi.org/10.1021/ACSAMI.8B05092.
[62] J. Song, O. Wang, H. Shen, Q. Lin, Z. Li, L. Wang, X. Zhang, L.S. Li, Over 30% external quantum efficiency light-emitting diodes by engineering quantum dot-assisted energy level match for hole transport layer, Adv. Funct. Mater. 29 (2019), https://doi.org/10.1002/ADFM.201808377.
[63] L. Wang, J. Lin, Y. Hu, X. Guo, Y. Lv, Z. Tang, J. Zhao, Y. Fan, N. Zhang, Y. Wang, X. Liu, Blue quantum dot light-emitting diodes with high electroluminescent efficiency, ACS Appl. Mater. Interfaces 9 (2017) 38755–38760, https://doi.org/10.1021/ACSAMI.7B10785.
[64] P. Shen, F. Cao, H. Wang, B. Wei, F. Wang, X.W. Sun, X. Yang, Solution-processed double-junction quantum-dot light-emitting diodes with an EQE of over 40%, ACS Appl. Mater. Interfaces 11 (2019) 1065–1070, https://doi.org/10.1021/ACSAMI.8B18940.
[65] H. Zhang, Q. Su, S. Chen, Suppressing Förster resonance energy transfer in close-packed quantum-dot thin film: toward efficient quantum-dot light-emitting diodes with external quantum efficiency over 21.6%, Adv. Opt. Mater. 8 (2020), https://doi.org/10.1002/ADOM.201902092.
[66] H. Zhang, Q. Su, S. Chen, Quantum-dot and organic hybrid tandem light-emitting diodes with multi-functionality of full-color-tunability and white-light-emission, Nat. Commun. 11 (2020), https://doi.org/10.1038/S41467-020-16659-X.
[67] C. Jiang, J. Zou, Y. Liu, C. Song, Z. He, Z. Zhong, J. Wang, H.L. Yip, J. Peng, Y. Cao, Fully solution-processed tandem white quantum-dot light-emitting diode with an external quantum efficiency exceeding 25%, ACS Nano 12 (2018) 6040–6049, https://doi.org/10.1021/ACSNANO.8B02289.
[68] H. Zhang, Y. Feng, S. Chen, Improved efficiency and enhanced color quality of light-emitting diodes with quantum dot and organic hybrid tandem structure, ACS Appl. Mater. Interfaces 8 (2016) 26982–26988, https://doi.org/10.1021/ACSAMI.6B07303.

[69] X. Yang, E. Mutlugun, C. Dang, K. Dev, Y. Gao, S.T. Tan, X.W. Sun, H.V. Demir, Highly flexible, electrically driven, top-emitting, quantum dot light-emitting stickers, ACS Nano 8 (2014) 8224–8231, https://doi.org/10.1021/NN502588K.
[70] T. Lee, D. Hahm, K. Kim, W.K. Bae, C. Lee, J. Kwak, Highly efficient and bright inverted top-emitting InP quantum dot light-emitting diodes introducing a hole-suppressing interlayer, Small 15 (2019), https://doi.org/10.1002/SMLL.201905162.
[71] L. Yao, X. Fang, W. Gu, W. Zhai, Y. Wan, X. Xie, W. Xu, X. Pi, G. Ran, G. Qin, Fully transparent quantum dot light-emitting diode with a laminated top graphene anode, ACS Appl. Mater. Interfaces 9 (2017) 24005–24010, https://doi.org/10.1021/ACSAMI.7B02026.
[72] H.M. Kim, A.R. Bin Mohd Yusoff, T.W. Kim, Y.G. Seol, H.P. Kim, J. Jang, Semi-transparent quantum-dot light emitting diodes with an inverted structure, J. Mater. Chem. C 2 (2014) 2259–2265, https://doi.org/10.1039/C3TC31932F.
[73] F. Wang, Y. Han, C.S. Lim, Y. Lu, J. Wang, J. Xu, H. Chen, C. Zhang, M. Hong, X. Liu, Simultaneous phase and size control of upconversion nanocrystals through lanthanide doping, Nature 463 (2010) 1061–1065, https://doi.org/10.1038/nature08777.
[74] X. Qin, X. Liu, W. Huang, M. Bettinelli, X. Liu, Lanthanide-activated phosphors based on 4f-5d optical transitions: theoretical and experimental aspects, Chem. Rev. 117 (2017) 4488–4527, https://doi.org/10.1021/acs.chemrev.6b00691.
[75] S. Gai, C. Li, P. Yang, J. Lin, Recent progress in rare earth micro/nanocrystals: soft chemical synthesis, luminescent properties, and biomedical applications, Chem. Rev. 114 (2014) 2343–2389, https://doi.org/10.1021/CR4001594.
[76] H. Chander, Development of nanophosphors—a review, Mater. Sci. Eng. R Rep. 49 (2005) 113–155, https://doi.org/10.1016/J.MSER.2005.06.001.
[77] R.N. Bhargava, D. Gallagher, X. Hong, A. Nurmikko, Optical properties of manganese-doped nanocrystals of ZnS, Phys. Rev. Lett. 72 (1994) 416, https://doi.org/10.1103/PhysRevLett.72.416.
[78] F. Vetrone, J.C. Boyer, J.A. Capobianco, A. Speghini, M. Bettinelli, Significance of Yb3+ concentration on the upconversion mechanisms in codoped Y_2O_3:Er^{3+}, Yb^{3+} nanocrystals, J. Appl. Phys. 96 (2004) 661, https://doi.org/10.1063/1.1739523.
[79] D. Tu, Y. Liu, H. Zhu, R. Li, L. Liu, X. Chen, Breakdown of crystallographic site symmetry in lanthanide-doped $NaYF_4$ crystals, Angew. Chem. Int. Ed. 52 (2013) 1128–1133, https://doi.org/10.1002/ANIE.201208218.
[80] Z. Li, Y. Zhang, An efficient and user-friendly method for the synthesis of hexagonal-phase$NaYF_4$:Yb, Er/Tm nanocrystals with controllable shape and upconversion fluorescence, Nanotechnology 19 (2008), 345606, https://doi.org/10.1088/0957-4484/19/34/345606.
[81] M. Cesaria, B. Di Bartolo, Chapter 3 nanophosphors: from rare earth activated multicolor-tuning to new efficient white light sources, NATO Sci. Peace Secur. Ser. B Phys. Biophys. (2018) 27–77, https://doi.org/10.1007/978-94-024-1544-5_3.
[82] R. Cao, W. Wang, J. Zhang, S. Jiang, Z. Chen, W. Li, X. Yu, Synthesis and luminescence properties of Li_2SnO_3:Mn^{4+} red-emitting phosphor for solid-state lighting, J. Alloys Compd. 704 (2017) 124–130, https://doi.org/10.1016/j.jallcom.2017.02.079.
[83] T. Lesniewski, S. Mahlik, M. Grinberg, R.S. Liu, Temperature effect on the emission spectra of narrow band Mn^{4+} phosphors for application in LEDs, Phys. Chem. Chem. Phys. 19 (2017) 32505–32513, https://doi.org/10.1039/C7CP06548E.
[84] Q. Zhou, L. Dolgov, A.M. Srivastava, L. Zhou, Z. Wang, J. Shi, M.D. Dramićanin, M.G. Brik, M. Wu, Mn^{2+} and Mn^{4+} red phosphors: synthesis, luminescence and applications in WLEDs. A review, J. Mater. Chem. C 6 (2018) 2652–2671, https://doi.org/10.1039/c8tc00251g.

[85] M.G. Brik, A.M. Srivastava, On the optical properties of the Mn^{4+} ion in solids, J. Lumin. 133 (2013) 69–72, https://doi.org/10.1016/j.jlumin.2011.08.047.
[86] H.D. Nguyen, C.C. Lin, M.H. Fang, R.S. Liu, Synthesis of $Na_2SiF_6:Mn^{4+}$ red phosphors for white LED applications by co-precipitation, J. Mater. Chem. C 2 (2014) 10268–10272, https://doi.org/10.1039/C4TC02062F.
[87] Y.D. Xu, D. Wang, L. Wang, N. Ding, M. Shi, J.G. Zhong, S. Qi, Preparation and luminescent properties of a new red phosphor ($Sr_4Al_{14}O_{25}:Mn^{4+}$) for white LEDs, J. Alloys Compd. 550 (2013) 226–230, https://doi.org/10.1016/j.jallcom.2012.09.139.
[88] Z. Qiu, T. Luo, J. Zhang, W. Zhou, L. Yu, S. Lian, Effectively enhancing blue excitation of red phosphor $Mg_2TiO_4:Mn^{4+}$ by Bi^{3+} sensitization, J. Lumin. 158 (2015) 130–135, https://doi.org/10.1016/j.jlumin.2014.09.032.
[89] Y. Li, S. Qi, P. Li, Z. Wang, Research progress of Mn doped phosphors, RSC Adv. 7 (2017) 38318–38334, https://doi.org/10.1039/c7ra06026b.
[90] R. Zhong, J. Zhang, Red photoluminescence due to energy transfer from Eu^{2+} to Cr^{3+} in $BaAl_{12}O_{19}$, J. Lumin. 130 (2010) 206–210, https://doi.org/10.1016/j.jlumin.2009.09.021.
[91] B. Malysa, A. Meijerink, T. Jüstel, Temperature dependent Cr^{3+} photoluminescence in garnets of the type $X_3Sc_2Ga_3O_{12}$ (X = Lu, Y, Gd, La), J. Lumin. 202 (2018) 523–531, https://doi.org/10.1016/j.jlumin.2018.05.076.
[92] R.E. Samad, G.E.C. Nogueira, S.L. Baldochi, N.D. Vieira, Development of a flashlamp-pumped Cr:LiSAF laser operating at 30 Hz, Appl. Opt. 45 (2006) 3356–3360, https://doi.org/10.1364/AO.45.003356.
[93] Q. Shao, H. Ding, L. Yao, J. Xu, C. Liang, J. Jiang, Photoluminescence properties of a $ScBO3:Cr^{3+}$ phosphor and its applications for broadband near-infrared LEDs, RSC Adv. 8 (2018) 12035–12042, https://doi.org/10.1039/c8ra01084f.
[94] M. Li, L. Wang, W. Ran, Q. Liu, C. Ren, H. Jiang, J. Shi, Broadly tunable emission from $Ca_2Al_2SiO_7$:Bi phosphors based on crystal field modulation around Bi ions, New J. Chem. 40 (2016) 9579–9585, https://doi.org/10.1039/c6nj01755j.
[95] R. Karlicek, C.C. Sun, G. Zissis, R. Ma, Handbook of advanced lighting technology, Handb. Adv. Light. Technol. (2017) 1–1185, https://doi.org/10.1007/978-3-319-00176-0.
[96] M. Mohamed, I. Unger, C. Janowitz, R. Manzke, Z. Galazka, R. Uecker, R. Fornari, The surface band structure of β-Ga_2O_3, J. Phys. Conf. Ser. 286 (2011), 012027, https://doi.org/10.1088/1742-6596/286/1/012027.
[97] J. Li, S. Meng, J. Niu, H. Lu, Electronic structures and optical properties of monoclinic ZrO_2 studied by first-principles local density approximation + U approach, J. Adv. Ceram. 6 (2017) 43–49, https://doi.org/10.1007/S40145-016-0216-Y.
[98] V. Vasanthi, M. Kottaisamy, K. Anitha, V. Ramakrishnan, Yellow emitting Cd doped SnO2 nanophosphor for phosphor converted white LED applications, Mater. Sci. Semicond. Process. 85 (2018) 141–149, https://doi.org/10.1016/J.MSSP.2018.06.001.
[99] V.S. Kavitha, R. Reshmi Krishnan, R. Sreeja Sreedharan, K. Suresh, C.K. Jayasankar, V. P. Mahadevan Pillai, Tb^{3+}-doped WO3 thin films: a potential candidate in white light emitting devices, J. Alloys Compd. 788 (2019) 429–445, https://doi.org/10.1016/J.JALLCOM.2019.02.222.
[100] S. Do Han, S.P. Khatkar, V.B. Taxak, G. Sharma, D. Kumar, Synthesis, luminescence and effect of heat treatment on the properties of Dy^{3+}-doped YVO_4 phosphor, Mater. Sci. Eng. B 129 (2006) 126–130, https://doi.org/10.1016/J.MSEB.2006.01.002.
[101] H. Lai, A. Bao, Y. Yang, W. Xu, Y. Tao, H. Yang, Preparation and luminescence property of Dy^{3+}-doped YPO4 phosphors, J. Lumin. 128 (2008) 521–524, https://doi.org/10.1016/J.JLUMIN.2007.09.027.

[102] D. Chen, Y. Wang, K. Zheng, T. Guo, Y. Yu, P. Huang, Bright upconversion white light emission in transparent glass ceramic embedding $Tm^{3+}/Er^{3+}/Yb^{3+}$:β-YF_3 nanocrystals, Appl. Phys. Lett. 91 (2007), 251903, https://doi.org/10.1063/1.2825285.
[103] F. Wang, X. Liu, Upconversion multicolor fine-tuning: visible to near-infrared emission from lanthanide-doped $NaYF_4$ nanoparticles, J. Am. Chem. Soc. 130 (2008) 5642–5643, https://doi.org/10.1021/JA800868A.
[104] A.R. Hong, S.Y. Kim, S.H. Cho, K. Lee, H.S. Jang, Facile synthesis of multicolor tunable ultrasmall $LiYF_4$:Yb,Tm,Er/$LiGdF_4$ core/shell upconversion nanophosphors with sub-10 nm size, Dyes Pigments 139 (2017) 831–838, https://doi.org/10.1016/J.DYEPIG.2016.12.048.
[105] X. Qin, T. Yokomori, Y. Ju, Flame synthesis and characterization of rare-earth (Er^{3+}, Ho^{3+}, and Tm^{3+}) doped upconversion nanophosphors, Appl. Phys. Lett. 90 (2007), 073104, https://doi.org/10.1063/1.2561079.
[106] W. Zhang, P. Xie, C. Duan, K. Yan, M. Yin, L. Lou, S. Xia, J.C. Krupa, Preparation and size effect on concentration quenching of nanocrystalline Y_2SiO_5:Eu, Chem. Phys. Lett. 292 (1998) 133–136, https://doi.org/10.1016/S0009-2614(98)00656-3.
[107] M.A. Flores-Gonzalez, G. Ledoux, S. Roux, K. Lebbou, P. Perriat, O. Tillement, Preparing nanometer scaled Tb-doped Y_2O_3 luminescent powders by the polyol method, J. Solid State Chem. 178 (2005) 989–997, https://doi.org/10.1016/J.JSSC.2004.10.029.
[108] J. Collins, Non-radiative processes in crystals and in nanocrystals, ECS J. Solid State Sci. Technol. 5 (2015) R3170, https://doi.org/10.1149/2.0221601JSS.
[109] J. Wang, R. Deng, M.A. Macdonald, B. Chen, J. Yuan, F. Wang, D. Chi, T.S. Andy Hor, P. Zhang, G. Liu, Y. Han, X. Liu, Enhancing multiphoton upconversion through energy clustering at sublattice level, Nat. Mater. 13 (2013) 157–162, https://doi.org/10.1038/nmat3804.
[110] K. Li, C. Shen, White light LED based on YAG:Ce^{3+} and YAG:Ce^{3+},Gd^{3+} phosphor, Mater. Sci. 7658 (2010) 797–802, https://doi.org/10.1117/12.865938.
[111] H.Q. Shi, W.N. Li, L.W. Sun, Y. Liu, H.M. Xiao, S.Y. Fu, Synthesis of silane surface modified ZnO quantum dots with ultrastable, strong and tunable luminescence, Chem. Commun. 47 (2011) 11921–11923, https://doi.org/10.1039/C1CC15411G.
[112] A. Asok, A.R. Kulkarni, M.N. Gandhi, Defect rich seed mediated growth: a novel synthesis method to enhance defect emission in nanocrystals, J. Mater. Chem. C 2 (2014) 1691–1697, https://doi.org/10.1039/C3TC32107J.
[113] A. Asok, M.N. Gandhi, A.R. Kulkarni, Enhanced visible photoluminescence in ZnO quantum dots by promotion of oxygen vacancy formation, Nanoscale 4 (2012) 4943–4946, https://doi.org/10.1039/C2NR31044A.
[114] Y.S. Fu, X.W. Du, S.A. Kulinich, J.S. Qiu, W.J. Qin, R. Li, J. Sun, J. Liu, Stable aqueous dispersion of ZnO quantum dots with strong blue emission via simple solution route, J. Am. Chem. Soc. 129 (2007) 16029–16033, https://doi.org/10.1021/JA075604I.
[115] N.H. Alvi, K. ul Hasan, O. Nur, M. Willander, The origin of the red emission in n-ZnO nanotubes/p-GaN white light emitting diodes, Nanoscale Res. Lett. 6 (2011) 1–7, https://doi.org/10.1186/1556-276X-6-130.
[116] B. Sundarakannan, M. Kottaisamy, Synthesis of blue light excitable white light emitting ZnO for luminescent converted light emitting diodes (LUCOLEDs), Mater. Lett. 165 (2016) 153–155, https://doi.org/10.1016/J.MATLET.2015.11.091.
[117] N.H. Alvi, S.M. Usman Ali, S. Hussain, O. Nur, M. Willander, Fabrication and comparative optical characterization of n-ZnO nanostructures (nanowalls, nanorods, nanoflowers and nanotubes)/p-GaN white-light-emitting diodes, Scr. Mater. 64 (2011) 697–700, https://doi.org/10.1016/J.SCRIPTAMAT.2010.11.046.

[118] P.N. Favennec, H. L'haridon, M. Salvi, D. Moutonnet, Y. le Guillou, Luminescence of erbium implanted in various semiconductors: IV, III-V and II-VI materials, Electron. Lett. 25 (1989) 718–719, https://doi.org/10.1049/EL:19890486.

[119] P. Patil, J. Park, S.Y. Lee, J.-K. Park, S.-H. Cho, White light generation from single gallium oxide nanoparticles co-doped with rare-earth metals, Appl. Sci. Converg. Technol. 23 (2014) 296–300, https://doi.org/10.5757/ASCT.2014.23.5.277.

[120] L. Luo, F.Y. Huang, G.S. Dong, Y.H. Wang, Z.F. Hu, J. Chen, White light emission and luminescence dynamics in Eu3+/Dy3+ codoped ZnO nanocrystals, J. Nanosci. Nanotechnol. 16 (2016) 619–625, https://doi.org/10.1166/JNN.2016.10812.

[121] K.J. Chen, H.V. Han, H.C. Chen, C.C. Lin, S.H. Chien, C.C. Huang, T.M. Chen, M.H. Shih, H.C. Kuo, White light emitting diodes with enhanced CCT uniformity and luminous flux using ZrO_2 nanoparticles, Nanoscale 6 (2014) 5378–5383, https://doi.org/10.1039/C3NR06894C.

[122] N. Jin, H. Li, F. Liu, Y.H. Xie, Microstructure and luminescence properties of Tb^{3+} doped ZnO quantum dots, J. Nanosci. Nanotechnol. 16 (2016) 3592–3596, https://doi.org/10.1166/JNN.2016.11796.

[123] L.X. Lovisa, V.D. Araújo, R.L. Tranquilin, E. Longo, M.S. Li, C.A. Paskocimas, M.R.D. Bomio, F.V. Motta, White photoluminescence emission from ZrO_2 co-doped with Eu^{3+}, Tb^{3+} and Tm^{3+}, J. Alloys Compd. 674 (2016) 245–251, https://doi.org/10.1016/J.JALLCOM.2016.03.037.

[124] N.H. Alvi, M. Riaz, G. Tzamalis, O. Nur, M. Willander, Fabrication and characterization of high-brightness light emitting diodes based on n-ZnO nanorods grown by a low-temperature chemical method on p-4H-SiC and p-GaN, Semicond. Sci. Technol. 25 (2010), 065004, https://doi.org/10.1088/0268-1242/25/6/065004.

[125] D.I. Son, B.W. Kwon, D.H. Park, W.S. Seo, Y. Yi, B. Angadi, C.L. Lee, W.K. Choi, Emissive ZnO-graphene quantum dots for white-light-emitting diodes, Nat. Nanotechnol. 7 (2012) 465–471, https://doi.org/10.1038/nnano.2012.71.

[126] M.J. Bowers, J.R. McBride, S.J. Rosenthal, White-light emission from magic-sized cadmium selenide nanocrystals, J. Am. Chem. Soc. 127 (2005) 15378–15379, https://doi.org/10.1021/JA055470D.

[127] A. Nag, D.D. Sarma, White light from Mn^{2+}-doped CdS nanocrystals: a new approach, J. Phys. Chem. C 111 (2007) 13641–13644, https://doi.org/10.1021/JP074703F.

Thermographic phosphors for remote temperature sensing

6

Shriya Sinha[a,b] *and Manoj Kumar Mahata*[c]
[a]Department of Physics, Indian Institute of Technology (Indian School of Mines), Dhanbad, Jharkhand, India, [b]Department of Physics, Shahid Chandrashekhar Azad Govt. P. G. College, Jhabua, Madhya Pradesh, India, [c]Third Institute of Physics, Georg-August-Universität Göttingen, Göttingen, Germany

Chapter outline

1 Introduction

Upconversion (UC) phosphors are a type of luminescent material that can convert lower-energy photons to higher-energy photons [1–6]. This process is known as upconversion, and it is achieved by exciting certain electron transitions within the phosphor material. The most common upconversion process is the two-photon process, in which two lower-energy photons are absorbed and converted into a single higher-energy photon. Upconversion phosphors have a wide range of applications, including biomedical imaging, spectroscopy, and solar energy conversion. In biomedical imaging, upconversion phosphors are used as contrast agents to enhance the visibility of certain tissues or cells. In spectroscopy, upconversion phosphors can be used to detect and analyze trace amounts of certain elements or molecules. In solar energy conversion, upconversion phosphors can be used to increase the efficiency of solar cells by converting lower-energy photons to higher-energy photons that can be used to generate electricity.

UC phosphors can be made from a variety of materials, including rare-earth (RE) elements such as ytterbium and erbium. These materials are typically synthesized

Metal Oxides for Next-Generation Optoelectronic, Photonic, and Photovoltaic Applications
https://doi.org/10.1016/B978-0-323-99143-8.00009-2

using methods such as solid-state reactions and hydrothermal synthesis. The efficiency of upconversion is generally low, but recent advances in material design and synthesis have led to significant improvements in upconversion efficiency. Upconversion phosphors have also been used in other fields like sensing, biolabeling, security, anticounterfeiting, and laser-pumping. One of the main advantages of upconversion phosphors is their ability to convert infrared light, which is invisible to the human eye, into visible light, making them useful in applications where infrared light is used but visible light is desired.

The use of phosphors doped with rare-earth ions that exhibit UC luminescence has gained significant interest due to its wide range of applications in areas such as lighting and displays, security printing, laser technology, pH and temperature sensing, and biomedical field. Noncontact optical thermometry techniques have gained popularity as they offer advantages over traditional contact thermometry methods, such as thermocouples, in terms of spatial resolution, sensitivity, and accuracy of temperature detection [7–13]. The most widely used technique for noncontact measurement is the fluorescence intensity ratio (FIR) between two thermally coupled levels (TCLs) of RE ions. The FIR technique is expected to provide a more accurate and higher spatial resolution of temperature sensing [14–22]. The RE ions possess abundant energy levels, and emissions arising from the intra-4f transitions provide unique spectroscopic features such as narrow bandwidth emissions, high UC luminescence lifetimes, and large anti-Stokes shifts [23–29]. UC luminescence can be achieved through the use of low-power continuous wave lasers and is more efficient than other techniques such as simultaneous two-photon absorption and second harmonic generation [30,31]. Previous studies have classified the mechanisms of upconversion processes in RE-based materials into five categories: excited state absorption, energy transfer upconversion, cooperative upconversion, energy migration upconversion, and photon avalanche. These have been discussed in more detail in previous works [29–39]. Ground-state absorption (GSA) is a process in which a ground-state molecule absorbs a photon to reach an excited state. This is the starting point for many other photon absorption processes. Excited-state absorption (ESA) is a mechanism where a molecule in an excited state takes in another photon, leading to further energy transfer or chemical reactions. Sequential absorption is a process where a molecule absorbs multiple photons in sequence, leading to the production of highly excited molecules. Energy-transfer upconversion (ETU) is a process where multiple low-frequency photons are absorbed by a sensitizer molecule, which then transfers its energy to an emitter molecule, leading to the overall upconversion of the absorbed light. Photon avalanche (PA) is a process where a photon absorbed by a molecule leads to the rapid production of many more photons through a chain reaction. These processes can be used to achieve different objectives depending on the desired outcome and the properties of the system used. To produce high-quality upconverting nanoparticles (UCNPs), there are three widespread methods of synthesis: thermal decomposition, hydrothermal synthesis, and ionic liquids-based synthesis.

The most utilized RE activators for UC thermometry are Er^{3+}, Ho^{3+}, Tm^{3+}, Pr^{3+}, Eu^{3+}, Dy^{3+}, and Nd^{3+} [40–43] because of their TCLs are found in the visible and NIR spectral ranges. Besides, host materials used for optical thermometry generally involve RE-based inorganic oxide and fluoride compounds. Oxide materials such as molybdate, niobate, titanate, tungstate, phosphate, and vanadate have been proven to be efficient hosts in the development of novel optical thermometers [44–52] because they have good physical and chemical stability, as well as low phonon threshold energy.

In this chapter, we present a brief summary of recent advancements in FIR-based temperature sensors of UC phosphors and discuss the fundamental principles of various RE-doped inorganic phosphors for optical thermometry.

2 Optical temperature sensing

Optical thermometry is a method of measuring temperature that is based on the temperature-sensitive properties of luminescent materials. These properties include the intensity of absolute luminescence, emission intensity ratio, fluorescence lifetime, and so on. The materials used in optical thermometry comprise hosts and luminescence centers. The hosts provide an appropriate crystal or coordination field for the RE to generate luminescence, while the activators and sensitizers affect the local crystal field and enhance the probability of RE emission. The efficiency of luminescence is largely influenced by the selection of host materials and activators, particularly when utilizing sensitizers. Choosing the right luminescence centers, typically trivalent RE, is essential for optical thermometry because of their abundance of energy levels in 4f configurations [9,31]. The temperature change can greatly affect the emission intensity of RE ions, as it influences the coupling state between the host's lattice field and the luminescence center. Additionally, the nonradiative relaxation probability of RE ions in the excited state increases with thermal agitation, resulting in a decrease of the luminescence intensity.

2.1 Basic principle of fluorescence intensity ratio-based temperature sensing

The UC luminescent materials have received great attention for noncontact optical temperature sensing utilizing the temperature dependent FIR of two closely spaced energy levels [9,12–14]. FIR technique has the advantage over conventional temperature measurements in thermally or electromagnetically harsh environments, thanks to its fast response, real-time temperature monitoring, and high spatial resolution. Activator ions such as Er^{3+}, Ho^{3+}, Tm^{3+}, and Nd^{3+} have been shown to possess several TCLs when excited by NIR light, and the most accurate results are achieved when the energy gap separation is between 200 and 2000 cm^{-1} [4]. The mechanism of temperature sensing has been discussed in many reports by researchers [3,4,12–14]. A simplified energy level diagram used to measure the FIR from two TCLs is represented in Fig. 1, and the relative population of such TCLs satisfies the Boltzmann-type population distribution [4]. According to the Boltzmann distribution law, the FIR from two TCLs can be expressed as follows:

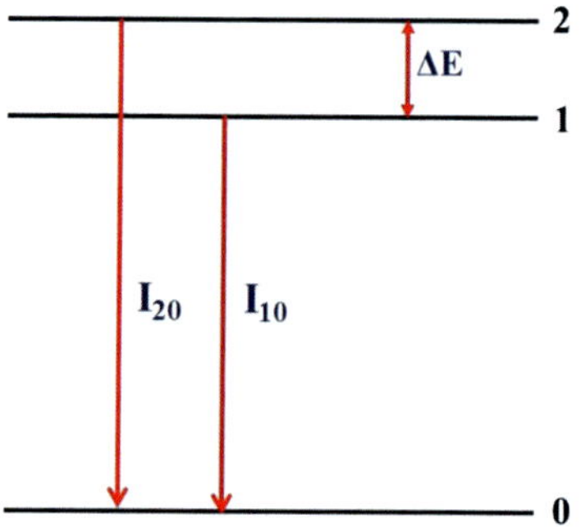

Fig. 1 Schematic diagram of fluorescence intensity ratio (FIR)-based temperature sensing.

$$\mathrm{FIR} = \frac{\mathrm{I}_{20}}{\mathrm{I}_{10}} = \frac{g_2\sigma_{20}\omega_{20}}{g_1\sigma_{10}\omega_{10}} \exp\left(-\frac{\Delta E}{KT}\right) = B\exp\left(-\frac{\Delta E}{KT}\right) \tag{1}$$

where I_{20} and I_{10} denote the integrated fluorescence intensities of transitions from the upper and lower thermalizing energy levels to a ground energy level, respectively. g, σ, and ω represent the degeneracy of the energy level, emission cross-section, and the transition's angular frequency. ΔE is the energy gap between the upper and lower levels. K is the Boltzmann constant and T is the absolute temperature. Taking natural logarithm on both sides of Eq. (1) gives ln(FIR) versus inverse absolute temperature ($1/T$) relation as follows [35]:

$$\ln(\mathrm{FIR}) = \ln(B) - \frac{\Delta E}{KT} \tag{2}$$

For application purposes, it is essential to know the thermometric capability of the phosphor as a sensor. Based on the FIR, the absolute (S_a) and relative (S_r) sensor sensitivities can be calculated by using the following formulae [18]:

$$S_a = \left|\frac{d(\mathrm{FIR})}{dT}\right| = \mathrm{FIR}\frac{\Delta E}{KT^2} \tag{3}$$

and

$$S_r = \left|\frac{1}{\mathrm{FIR}}\frac{d(\mathrm{FIR})}{dT}\right| = \frac{\Delta E}{KT^2} \tag{4}$$

The above equations suggest that thermal sensitivities of phosphors are related to the energy gap separation between the TCLs and the value of FIR. Generally, larger energy differences contribute to higher sensitivity. However, the energy gap difference between the two levels is restricted, increasing the sensitivity of an activator further is difficult. As an alternative method to improve the temperature sensitivity has been proposed in recent years, which is based on the FIR of non-TCLs of the activators

[3,12,14]. The two non-TCLs are usually generated from one or two RE emitting centers and are no longer limited by the energy gap, offering higher thermal sensitivity. The FIR thermometry based on two TCLs or non-TCLs of RE's of Er^{3+}, Ho^{3+}, Tm^{3+}, and Nd^{3+} is briefly discussed in the next sections.

2.2 Optical thermometry based on Er^{3+} emission

The most studied activator for FIR-based temperature sensing is the Er^{3+} ion in the spectroscopic range from the ultraviolet to NIR region. It emits prominent green light ($^2H_{11/2}$, $^4S_{3/2} \rightarrow {}^4I_{15/2}$), along with relatively weak red ($^4F_{9/2} \rightarrow {}^4I_{15/2}$), UV ($^4D_{7/2}$,$^4G_{9/2} \rightarrow {}^4I_{15/2}$ and $^4G_{11/2}$, $^2H_{9/2} \rightarrow {}^4I_{15/2}$), and NIR ($^4I_{13/2} \rightarrow {}^4I_{15/2}$) emissions [23]. The optical thermometers have been extensively explored in Er^{3+}/Yb^{3+} incorporated UC phosphors based on couples of adjacent TCLs and non-TCLs of Er^{3+} ions [44–52]. Previously, we developed highly sensitive thermographic phosphors based on several types of titanates, molybdate, and vanadate materials using two thermally coupled green ($^2H_{11/2}/^4S_{3/2} \rightarrow {}^4I_{15/2}$) levels ($\Delta E \sim 700$–$800\,cm^{-1}$) of Er^{3+} [17,19,35]. For example, a temperature sensitivity of $0.0105\,K^{-1}$ at 450 K was reported by our group for $Gd_2Mo_3O_9$:Er^{3+}/Yb^{3+} phosphor based on $^2H_{11/2} \rightarrow {}^4I_{15/2}$ (538 nm) and $^4S_{3/2} \rightarrow {}^4I_{15/2}$ (549 nm) transitions [35]. Fig. 2A–D shows the temperature-dependent UC spectra of $Gd_2Mo_3O_9$:Er/Yb irradiated by 980 nm light for various Er^{3+} concentrations (0.3–3 mol%). The plot of FIR (I_{524}/I_{549}) versus absolute temperature is

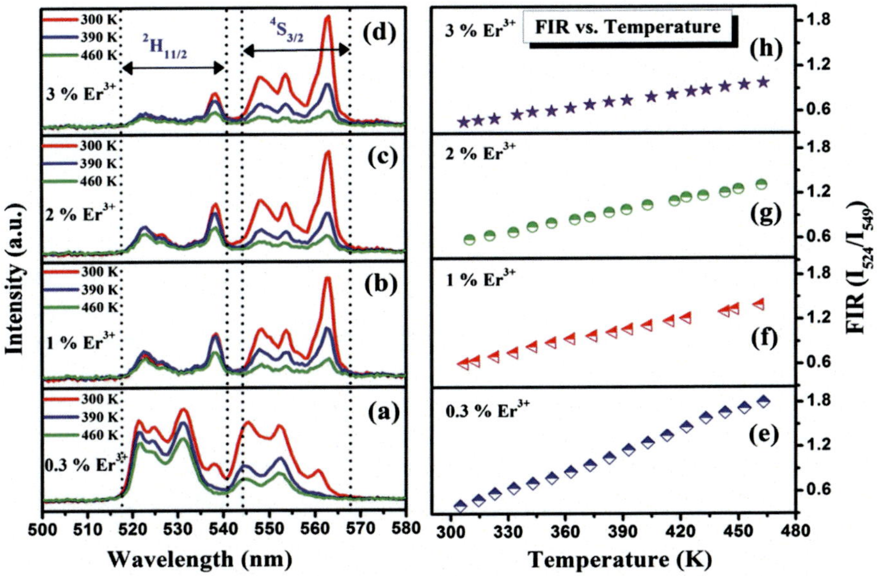

Fig. 2 (A–D) Upconversion emission spectra of $Gd_2Mo_3O_9$:Er^{3+}/Yb^{3+} phosphor with Er^{3+} concentration, recorded at different temperatures under 980 nm excitation; (E–H) temperature dependence of fluorescence intensity ratios of I_{524} and I_{549} with Er^{3+} concentration [35].

presented in Fig. 2E–H. The obtained absolute sensitivity S_a in the investigated temperature range with different Er^{3+} concentrations was also investigated. Moreover, a better behavior as a thermographic phosphor has been obtained in $CaMoO_4:Er^{3+}/Yb^{3+}$ phosphor (~0.0216 K^{-1} at 490 K) via K^+/Na^+ ions co-doping [36], and it is due to the tailoring effect of the crystal field around Er^{3+} ions in calcium molybdate. Zhang et al. [45] investigated the temperature sensing performance of $Ba_3Y_4O_9:Er^{3+}/Yb^{3+}$ phosphor based on thermally coupled green levels and red Stark sublevels of Er^{3+} over a range from 83 to 563 K. The result concluded that green emissions are suitable for high temperature with a maximum S_a of 0.00248 K^{-1} at 563 K, and the red emissions are appropriate for low temperatures with a maximum S_a of 0.00371 K^{-1} at 143 K. Similar findings have been reported in the recent works [46–48] based on Stark sublevels of Er^{3+} via the FIR technique. Apart from this, temperature sensor based on non-TCLs pairs of Er^{3+} was reported in some literature [49–52]. Since sensitivity is restricted to ΔE of the TCLs, utilizing non-TCLs with a larger ΔE could be a valid approach to enhance the sensitivity. For instance, the temperature-dependent FIR between green ($^2H_{11/2} \rightarrow {}^4I_{15/2}$) and red emission ($^4F_{9/2} \rightarrow {}^4I_{15/2}$) bands from a pair of non-TCLs was studied in $La_2MoO_6:Er^{3+}/Yb^{3+}$ by Chen's group [49]. It was observed that with the raising of temperature from 30 to 350°C, the green emission (523 nm) increases, while the red emission (673 nm) decreases. The maximum value of S_r was estimated to be 1.6% K^{-1} at room temperature. Moreover, by using the FIR technique, temperature sensing behavior based on $^4D_{7/2}$ (256 nm)/$^4G_{9/2}$ (276 nm) $\rightarrow {}^4I_{15/2}$ transitions of Er^{3+} in the UV region was first time reported by Qin et al. [50] and found S_a of 0.0052 K^{-1} at 303 K. In Table 1, we summarize the temperature sensing range, sensitivity, and involved transition of Er^{3+}/Yb^{3+}-based phosphors.

2.3 Optical thermometry based on Ho^{3+} emission

Ho^{3+} is known for its excellent UC luminescence over a wide range from blue to NIR, showing intense green UC emission around 537 and 550 nm ($^5F_4$, $^5S_2/ \rightarrow {}^5I_8$) with faint blue emissions at 465 and 489 nm ($^3K_8$, $^5F_3 \rightarrow {}^5I_8$), red emission around 665 nm ($^5F_5 \rightarrow {}^5I_8$), and NIR at 757 nm ($^5F_4/^5S_2 \rightarrow {}^5I_7$). The Ho^{3+}/Yb^{3+}doped phosphor is a good option for measuring temperature using NIR due to its ideal energy level layout and the $^5I_7$ level that can store population for the UC process [54]. The optical thermometry based on TCLs and non-TCLs of Ho^{3+} has been explored and reported by researchers in different host materials [15,54–62]. Our group has recently used the $BaTiO_3:Ho^{3+}/Yb^{3+}$ nanocrystals for temperature sensing at the cryogenic temperature range [16]. It shows a maximum sensitivity of 0.0095 K^{-1} (at 12 K) by analyzing temperature-dependent UC from Ho^{3+}'s $^5F_4$ and $^5S_2$ states.

In another work, Rai's group [55] reported optical thermometry of $Gd_2O_3:Ho^{3+}/Yb^{3+}/Li^+$ phosphor by using thermally coupled $^5F_4 \rightarrow {}^5I_8$ (538 nm) and $^5S_2 \rightarrow {}^5I_8$ (549 nm) levels corresponding to green emissions ($\Delta E \sim 305\ cm^{-1}$). The FIR of I_{538} and I_{549} are temperature dependent and showed S_a of 0.0092 K^{-1} at 505 K by 980 nm laser excitation. In addition, very recently, this group reported the maximum value of S_r (0.0013 at 300 K) and S_a (0.0014 K^{-1} at 600 K) for $ZnGa_2O_4:Ho^{3+}/Yb^{3+}/Bi^{3+}$ phosphor via $^5F_4/^5S_2 \rightarrow {}^5I_8$ transitions [56].

Table 1 Comparison of temperature sensitivity in Er^{3+}/Yb^{3+} co-doped phosphors.

Materials	Transitions	Range (K)	Sensitivity (K^{-1})	Ref.
$Gd_2Mo_3O_9$:Er/Yb	$^2H_{11/2}/^4S_{3/2} \rightarrow {}^4I_{15/2}$	300–480	0.0105 (450 K)	[35]
$CaMoO_4$:Er/Yb	$^2H_{11/2}/^4S_{3/2} \rightarrow {}^4I_{15/2}$	306–513	0.0216 (490 K)	[36]
$Na_2Gd_2Ti_3O_{10}$:Er/Yb	$^2H_{11/2}/^4S_{3/2} \rightarrow {}^4I_{15/2}$	290–490	0.0058 (490 K)	[44]
$Na_2La_2Ti_3O_{10}$:Er/Yb	$^2H_{11/2}/^4S_{3/2} \rightarrow {}^4I_{15/2}$	290–490	0.0061 (470 K)	[44]
$Ba_3Y_4O_9$:Er/Yb	$^4F_{9/2(1,2)} \rightarrow {}^4I_{15/2}$	83–563	0.0037 (143 K)	[45]
$Gd_2(WO_4)_3$:Er/Yb	$^2H_{11/2}/^4S_{3/2(2)} \rightarrow {}^4I_{15/2}$	296–620	0.0165 (395 K)	[46]
	$^4S_{3/2(1,2)} \rightarrow {}^4I_{15/2}$		0.0013 (296) K	
	$^4F_{9/2(1,2)} \rightarrow {}^4I_{15/2}$		0.0012 (296 K)	
$Ba_5Gd_8Zn_4O_{21}$:Er/Yb	$^2H_{11/2}/^4S_{3/2} \rightarrow {}^4I_{15/2}$	200–490	0.0032 (490 K)	[47]
	$^4F_{9/2(1,2)} \rightarrow {}^4I_{15/2}$		0.0029 (200 K)	
ZrO_2:Er/Yb/Li	$^2H_{11/2}/^4S_{3/2} \rightarrow {}^4I_{15/2}$	323–673	0.0035 (643 K)	[48]
	$^4F_{9/2(1,2)} \rightarrow {}^4I_{15/2}$		0.0026 (110 K)	
$K_3Y(PO_4)_2$:Er/Yb	$^2H_{11/2}/^4S_{3/2} \rightarrow {}^4I_{15/2}$	293–553	0.0030 (553 K)	[51]
	$^2H_{11/2}/^4F_{9/2} \rightarrow {}^4I_{15/2}$		0.44 (293 K)	
$Ca_2Gd_8(SiO_4)_6O_2$:Er/Yb	$^2H_{11/2}/^4S_{3/2} \rightarrow {}^4I_{15/2}$	293–553	0.0037 (553 K)	[52]
	$^2H_{11/2}/^4F_{9/2} \rightarrow {}^4I_{15/2}$		1.12 (293 K)	
$CaWO_4$:Er/Yb	$^4G_{11/2}/^2H_{9/2} \rightarrow {}^4I_{15/2}$	303–873	0.0073 (873 K)	[53]

The temperature-dependent UC emissions in the blue region at 465 and 491 nm from the thermally coupled transitions $^3K_8 \rightarrow {}^5I_8$ and $^5F_3 \rightarrow {}^5I_8$ ($\Delta E \sim 750\,cm^{-1}$) have been investigated by Pandey et al. in $Y_2O_3:Ho^{3+}/Yb^{3+}/Zn^{2+}$ [57]. The maximum S_a was determined to be $0.0030\,K^{-1}$ at 673 K, indicating its potentiality as a high thermal probe. A similar study has been carried out by Dey et al. [58] in $CaMoO_4:Ho^{3+}/Yb^{3+}/Mg^{2+}$ phosphor with a sensor sensitivity of $0.0066\,K^{-1}$ at 353 K. The temperature-dependent blue emissions from $^5F_{2,3}/^3K_8 \rightarrow {}^5I_8$ and $^5G_6/^5F_1 \rightarrow {}^5I_8$ transitions ($\Delta E \sim 1300\,cm^{-1}$) of $CaWO_4:Ho^{3+}/Yb^{3+}$ phosphor have been investigated from 303 to 923 K by Cao's group [54]. An excellent temperature sensitivity of $0.005\,K^{-1}$ at 923 K was obtained with good accuracy and high resolution.

The temperature-dependent UC emissions in the red region at 650 and 660 nm of $KLu(WO_4)_2:Ho^{3+}/Yb^{3+}$ were investigated by Diaz et al. [15]. Using two thermally coupled $^5F_{5(1,2)} \rightarrow {}^5I_8$ Stark sublevels of the Ho^{3+}, a sensitivity of $0.00385\,K^{-1}$ at 297 K was observed, and the phosphor has been demonstrated to have great potential for use in optical thermometry [15,59].

Furthermore, temperature-dependent UC emissions are realized using non-TCLs, green, and NIR emissions corresponding to $^5F_4/^5S_2 \rightarrow {}^51_8$, $^5I_7$ transitions. Martin et al. [60] introduced a four-level system scheme to explain the thermalization process between green ($^5F_4/^5S_2 \rightarrow {}^51_8$) and NIR ($^5F_4/^5S_2 \rightarrow {}^51_7$) levels. Later, Lojpur et al. [61] investigated temperature dependent intensity ratio of bands in the green region to the NIR region in $Y_2O_3:Yb^{3+}/Ho^{3+}$ phosphor within 10–300 K range. A remarkable sensitivity of $0.097\,K^{-1}$ at 85 K corresponding to FIR (I_{536}/I_{772}) was detected. Further, an optical thermometer based on NIR-green FIR (I_{757}/I_{540}) and red-green FIR

($I_{641},_{665}/I_{540},_{549}$) for $ZnWO_4:Ho^{3+}/Yb^{3+}$ phosphor was studied by Yao's group [62] from 83 to 503 K. The FIR (I_{757}/I_{540}) corresponding to $^5F_4/^5S_2 \rightarrow {}^5I_8$, $^5I_7$ transitions based on a four-level system showed the highest sensitivity as 0.0064 K^{-1} at low temperatures. Fig. 3A exhibits the UC of $ZnWO_4:Ho^{3+}/Yb^{3+}$ phosphor at different temperatures under 980 nm excitation. Fig. 3B shows that Ho^{3+} has a simple 4-level energy system, which makes it a good option for optical thermometry. The temperature-dependent variation of FIR (I_{757}/I_{540}) and sensitivity are presented in Fig. 3C and D. In Table 2, the thermometric performances of Ho^{3+}/Yb^{3+}-based phosphors have been compared.

2.4 Optical thermometry based on Tm^{3+} emission

Tm^{3+} ions are known for their ability to produce emissions at 470 nm, 800 nm, 650 nm, and 700 nm. The most intense band is at 470 nm due to the $^1G_4 \rightarrow {}^3H_6$ transition, and NIR emission at 800 nm ($^3H_4 \rightarrow {}^3H_6$), along with weak red UC emissions at 650 nm ($^1G_4 \rightarrow {}^3F_4$) and 700 nm ($^3F_{2,3} \rightarrow {}^3H_6$). The $^3H_4$ and $^3F_{2,3}$ levels of Tm^{3+} are TCLs, and the energy gap difference is much larger ($\Delta E \sim 1500$–$2000\ cm^{-1}$), so it is favorable to acquire good thermal sensitivity and measurement resolution. Up to now, several groups have investigated the temperature-sensing behavior in different phosphors doped with Tm^{3+}/Yb^{3+} ions [41,63–70]. For instance, Sheng et al. [59] demonstrated optical thermometry in $BiPO_4:Yb^{3+}/Tm^{3+}$ phosphor based on two pairs of TCLs ($^3F_{2,3}$ and $^3H_4$) and ($^3H_{4(1)}$ and $^3H_{4(2)}$). The developed phosphor showed a higher value of $S_a \sim 0.00083\ K^{-1}$ at lower temperature using FIR (I_{794}/I_{804}:$^3H_{4(1)}/^3H_{4(2)} \rightarrow {}^3H_6$) while its value using FIR (I_{700}/I_{804}:$^3F_{2,3}/^3H_4 \rightarrow {}^3H_6$) was calculated 0.00022 K^{-1} at higher temperature. Lis's group [41] presented a method for monitoring very high-temperature values (up to ~1000 K) by using $YVO_4:Tm^3/Yb^{3+}$ (Fig. 4A). Luminescence thermometry was performed by using Tm^{3+}'s TCLs and non-TCLs, corresponding to ($^3F_{2,3} \rightarrow {}^3H_6/^3H_4 \rightarrow {}^3H_6$; 700 nm/800 nm) and ($^2F_{5/2} \rightarrow {}^2F_{7/2}/{}^3H_4 \rightarrow {}^3H_6$; 940 nm/800 nm), respectively (Fig. 4B and C). At ~1000 K for TCLs, it was noticed that sensitivity and temperature resolution reduced significantly, whereas, for non-TCLs, a very high relative sensitivity ($S_r \sim 2.1\%\ K^{-1}$) and good temperature resolution (1.4 K) was achieved (Fig. 5A–D) [41]. It was inferred that the developed optical thermometer could be useful in industrial fields.

Furthermore, Soni et al. [66] reported the temperature sensing performance in $Na_2Y_2B_2O_7:Tm^{3+}/Yb^{3+}$ phosphor using the Stark subcomponents of the Tm^{3+}'s $^1G_4$ state. The maximum value of S_a was found as 0.00045 K^{-1} at 300 K. The two thermally coupled $^1G_{4(1)}$ (477 nm) and $^1G_{4(2)}$ (488 nm) sublevels of the $^1G_4$ with a small energy gap ($\Delta E \sim 473\ cm^{-1}$) were treated as in thermal equilibrium. Wei et al. [67] also showed a similar technique by using $^1G_{4(1)} \rightarrow {}^3H_6$ and $^1G_{4(2)} \rightarrow {}^3H_6$ in $BaGd_2ZnO_5:Tm^{3+}/Yb^{3+}$ phosphor and reported S_a of 0.0055 K^{-1} at room temperature. In Table 3, we compare the thermometric performances of Tm^{3+}/Yb^{3+} based phosphors.

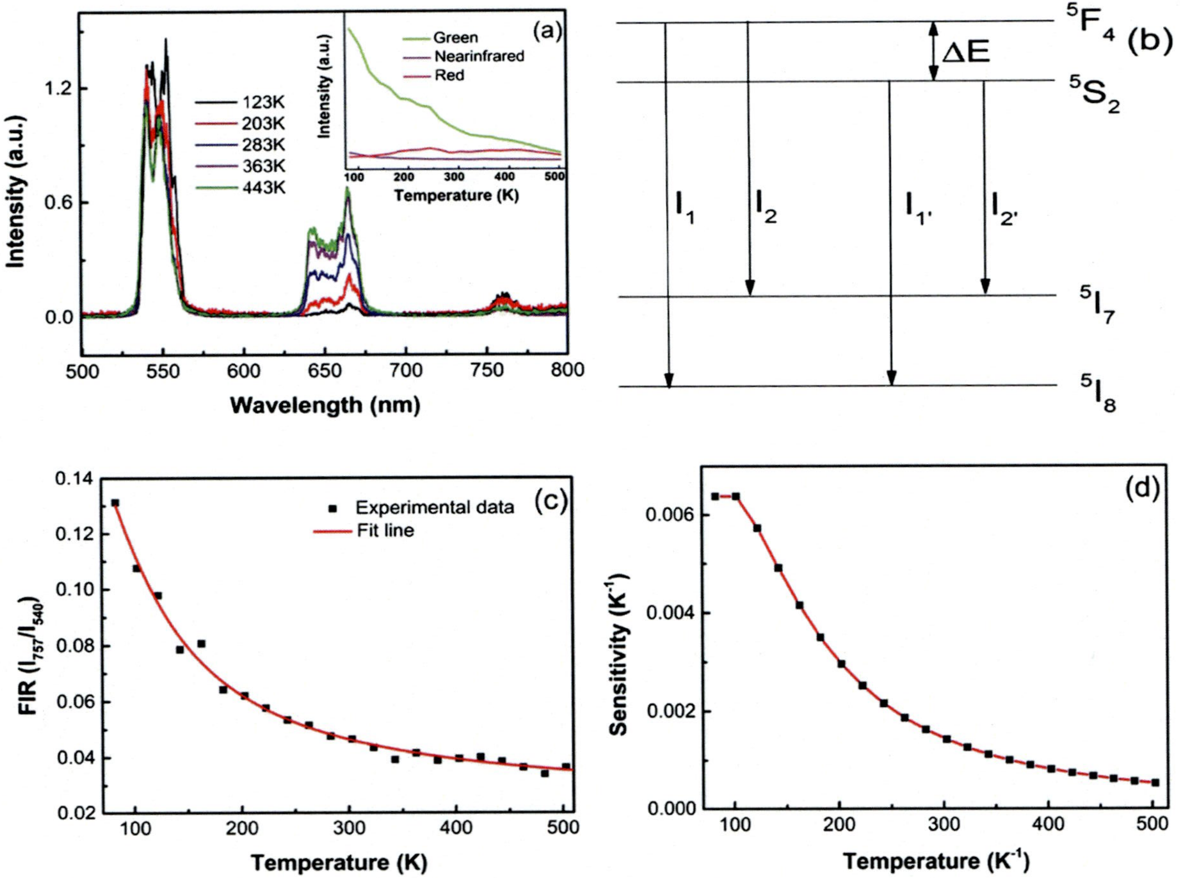

Fig. 3 (A) Temperature dependent UC of $ZnWO_4$:Ho^{3+}/Yb^{3+} (inset figure shows temperature-dependent upconversion), (B) simplified energy diagram, (C) ratio: I_{757}/I_{540} with temperature, and (D) sensitivity of I_{757}/I_{540} with temperature [62].

Table 2 Comparison of the maximum values of sensitivity in Ho^{3+}/Yb^{3+} co-doped phosphors using the FIR technique.

Materials	Transitions	Range (K)	Sensitivity (K^{-1})	Ref.
Gd_2O_3:Ho/Yb/Li	$^5F_4/^5S_2 \rightarrow {}^5I_8$	300–673	0.0092 (505 K)	[55]
$ZnGa_2O_4$:Ho/Yb/Bi	$^5F_4/^5S_2 \rightarrow {}^5I_8$	300–600	0.0013 (300 K)	[56]
Y_2O_3:Ho/Yb/Zn	$^3K_8/^5F_3 \rightarrow {}^5I_8$	299–673	0.0030 (673 K)	[57]
$CaMoO_4$:Ho/Yb/Mg	$^3K_8/^5F_3 \rightarrow {}^5I_8$	303–543	0.0066 (353 K)	[58]
$CaWO_4$:Ho/Yb	$^5F_{2,3}/^3K_8 \rightarrow {}^5I_8$ and $^5G_6/^5F_1 \rightarrow {}^5I_8$	303–923	0.005 (923 K)	[54]
$KLu(WO_4)_2$:Ho/Yb	$^5F_{5(1,2)} \rightarrow {}^5I_8$	297–673	0.00385 (297 K)	[15]
$BiPO_4$:Ho/Yb	$^5F_{5(1,2)} \rightarrow {}^5I_8$	313–573	0.00079 (333*K*)	[59]
Y_2O_3:Yb/Ho	$^5F_4/^5S_2 \rightarrow {}^5I_8, {}^5I_7$	10–300	0.097 (85 K)	[61]
$ZnWO_4$:Ho/Yb	$^5F_4/^5S_2 \rightarrow {}^5I_8, {}^5I_7$	83–503	0.0064 (83 K)	[62]
$NaY_9(SiO_4)_6O_2$:Ho/Yb	$^5F_5/(^5F_4/^5S_2) \rightarrow {}^5I_8$	298–523	1.53% (298 K)	[63]

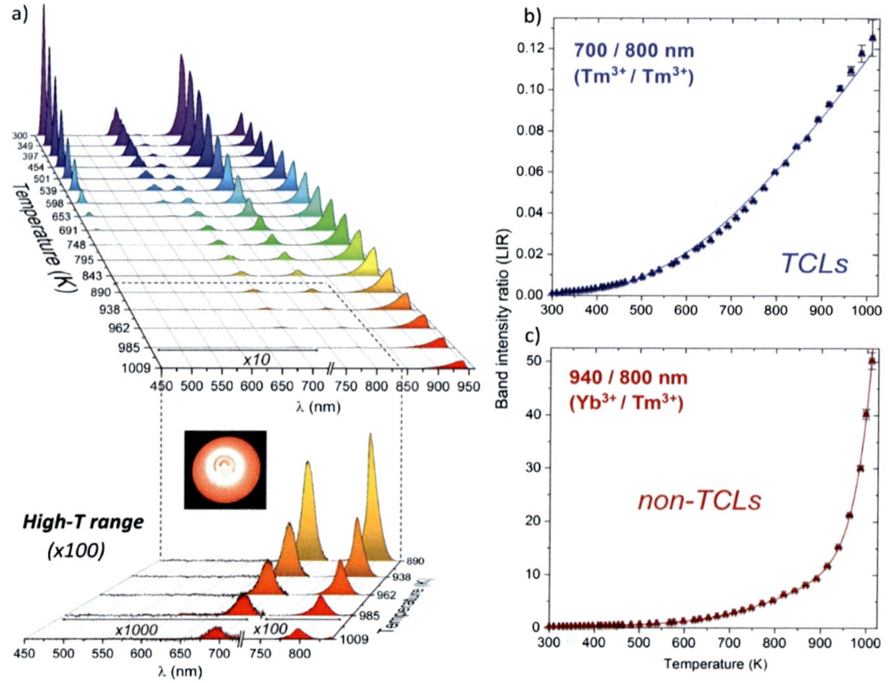

Fig. 4 (A) Temperature-dependent UC of Yb^{3+}-Tm^{3+} doped YVO_4 under 975 nm excitation; (B, C) temperature-dependent FIR for (B) TCLs and (C) non-TCLs; inset (A) photograph of the specimen at ~1000 K upon NIR irradiation, [41].

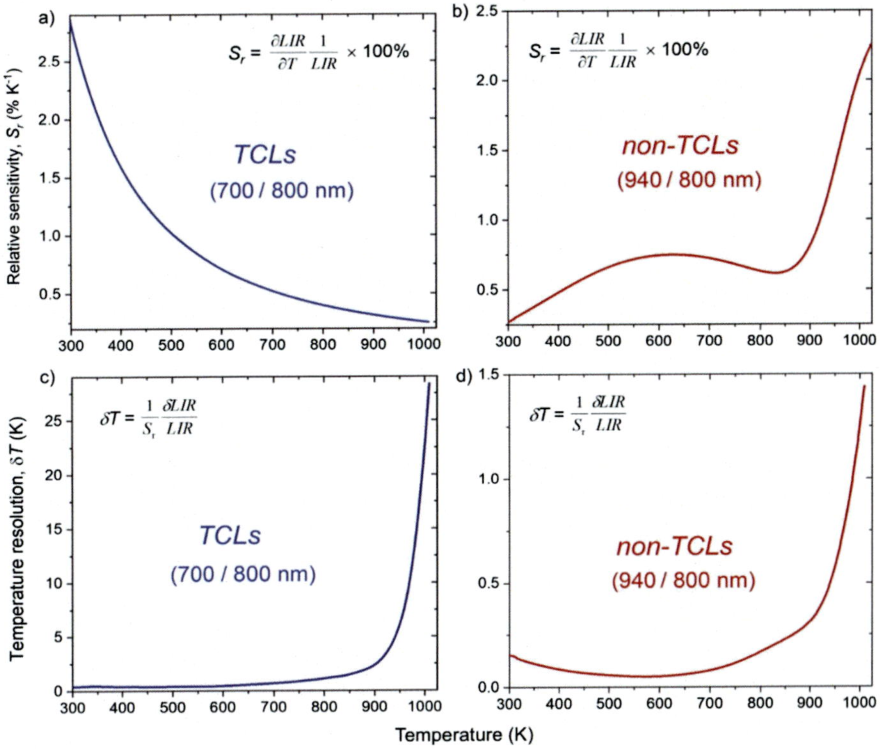

Fig. 5 (A and B) Temperature-dependent sensitivity for TCLs and non-TCLs of $YVO_4:Yb^{3+}/Tm^{3+}$ (C and D) temperature resolutions (δT) for TCLs and non-TCLs [41].

Table 3 Comparative list of sensitivity in Tm^{3+}/Yb^{3+} co-doped phosphors

Materials	Transitions	Range (K)	Sensitivity (K^{-1})	Ref.
$NaY_9(SiO_4)_6O_2$:Tm/Yb	$^3F_{2,3}/^3H_4 \rightarrow {}^3H_6$	298–548	0.00034 (548 K)	[63]
$Bi_2Ti_2O_7$:Tm/Yb	$^3F_{2,3}/^3H_4 \rightarrow {}^3H_6$	300–505	0.024 (300 K)	[64]
$BiPO_4$:Tm/Yb	$^3F_{2,3}/^3H_4 \rightarrow {}^3H_6$ $^3H_{4(1)}/\ ^3H_{4(2)} \rightarrow {}^3H_6$	313–573	0.00022 (548 K) 0.00083 (313 K)	[59]
YVO_4:Tm/Yb	$^3F_{2,3} \rightarrow {}^3H_6/^3H_4 \rightarrow {}^3H_6$ $^2F_{5/2} \rightarrow {}^2F_{7/2}/$ $^3H_4 \rightarrow {}^3H_6$	300–1009	0.25% (300 K) 2.13% (1009 K)	[41]
$Na_2Y_2B_2O_7$:Tm/Yb	$^1G_{4(1,2)} \rightarrow {}^3H_6$	300–623	0.0045 (300 K)	[66]
$BaGd_2ZnO_5$:Tm/Yb	$^1G_{4(1,2)} \rightarrow {}^3H_6$	313–573	0.0055 (323 K)	[67]
$ZnWO_4$:Tm/Yb/Mg	$^1G_{4(1,2)} \rightarrow {}^3H_6$	300–600	0.0034 (300 K)	[68]
Gd_2ZnTiO_6:Tm/Yb	$^1G_4 \rightarrow {}^3F_4/^3H_4 \rightarrow {}^3H_6$	290–473	1.7351 (473 K)	[69]
$Y_{4.67}Si_3O_{13}$:Tm/Yb	$^1G_4 \rightarrow {}^3F_4/^3H_4 \rightarrow {}^3H_6$ $^1G_{4(1,2)} \rightarrow {}^3H_6$	273–553	0.00027 (553 K) 0.0041 (273 K)	[70]

2.5 Optical thermometry based on Nd^{3+} emission

Nd^{3+} is another widely reported RE ions for FIR thermometry. It has excitation and emission in the first (700–980nm) and the second (1000–1400nm) biological windows (BWs) [43,71–76]. Several researchers have reported optical thermometry based on the Stark sublevels of $^4F_{3/2}$ level and TCLs of Nd^{3+} [73–79]. For example, Yin et al. [74] reported the temperature-sensing behavior of $La_2O_2S:Nd^{3+}$ phosphor in the 30–600K temperature range. Temperature sensing has been realized by monitoring the FIR between the transitions $^4F_{5/2} \rightarrow {}^4I_{9/2}$ (818nm) and $^4F_{3/2} \rightarrow {}^4I_{9/2}$ (897nm) and the two thermally coupled Stark levels of $^4F_{3/2}$, i.e., between the $^4F_{3/2} \rightarrow {}^4I_{9/2(1)}$ (891nm) and $^4F_{3/2} \rightarrow {}^4I_{9/2(2)}$ (897nm) of Nd^{3+}. The phosphor exhibited excellent sensitivity from 17.5% K^{-1} to 0.04% K^{-1} in a larger temperature range from 30 to 600K via FIR of two thermally coupled Stark sublevels. Wei et al. [77] reported Nd^{3+}- and Yb^{3+}-sensitized $Ba_4La_2Ti_4Nb_6O_{30}$ for temperature sensing study based on the TCLs of the Nd^{3+} ions composed of TCL1 ($^4F_{7/2}/^4S_{3/2}$ and $^4F_{5/2}/^2H_{9/2}$), TCL2 ($^4F_{7/2}/^4S_{3/2}$ and $^4F_{3/2}$), and TCL3 ($^4F_{5/2}/^2H_{9/2}$ and $^4F_{3/2}$). Fig. 6A shows the UC in the NIR region (700–1000nm) increased remarkably with increasing the temperature from 303 to 573K, and it is a result due to the increase of the transfer of energy from Yb^{3+} to Nd^{3+}. Fig. 6B shows the plots of the FIR between the TCL1, TCL2, and TCL3 with temperature. The variation of sensitivities (S_a and S_r) based on different TCLs with temperature is presented in Fig. 6C. The maximum value of S_a was noticed as 0.031 K^{-1} at 573K, corresponding to TCL3. Furthermore, Wu et al. [78] reported a thermal sensitivity of 0.0071 K^{-1} in Li^+ co-doped $CaWO_4:Nd^{3+}/Yb^{3+}$ phosphor. Song et al. [79] reported the maximum sensitivity of 0.02857 K^{-1} at 453K by using FIR (I_{801}/I_{869}) corresponding to $^4F_{5/2}/^4F_{3/2} \rightarrow {}^4I_{9/2}$ TCL for $SrWO_4:Nd^{3+}/Yb^{3+}$ phosphor [79]. In Table 4, we compare the thermometric performances of Nd^{3+}/Yb^{3+}-based phosphors.

3 Lifetime-based thermometry

Time-resolved spectroscopy is a technique that can be used to measure temperature by observing how the lifetime of energy levels changes over time. This can be done by monitoring the decay of excited states in a material as its temperature changes. The technique is commonly used to study the properties of materials and their behavior at different temperatures. Time-resolved spectroscopy can be applied to a wide range of phosphor and semiconductor materials and is particularly useful for studying temperature-sensitive processes. One of the main advantages of the technique is its high temporal resolution, which allows for the measurement of fast temperature changes.

In practice, time-resolved spectroscopy is done using a laser, a detector, and a timing system. The material is first excited by a laser pulse and the subsequent lifetime at the excited state is recorded. The data is then analyzed to extract information about the temperature of the sample. There are several different types of time-resolved spectroscopy, each with its own unique advantages and applications.

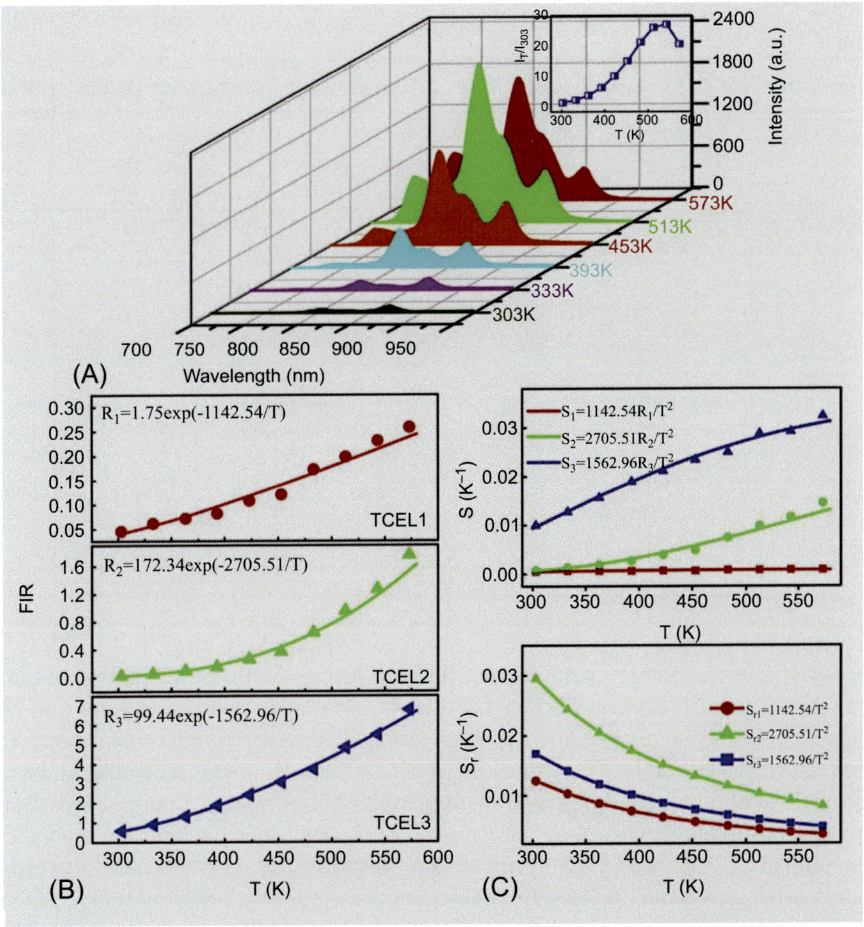

Fig. 6 (A) Upconversion spectra of $Ba_4La_2Ti_4Nb_6O_{30}$: Nd^{3+}/Yb^{3+} at different temperatures under 980 nm light; Inset shows a comparison of the intensity ratio of overall near-infrared emission band with that at 303 K at various temperatures. (B) Variation of fluorescence intensity ratio for TCL1, TCL2, and TCL3 with temperature (303–573 K). (C) Absolute and relative sensitivities variation for TCLs of Nd^{3+} with temperature [77].

Time-resolved spectroscopy utilizes the absorption of IR light by phosphor materials to determine the temperature. The absorption of IR light by a material depends on the energy levels of the energy acceptor and can change with temperature. By measuring the lifetime at different temperatures, it is possible to know the temperature of the sample. Additionally, it has the capability to measure material temperatures in demanding conditions characterized by high pressure in industrial processes.

Therefore, time-resolved spectroscopy is a versatile technique that can provide valuable information about the temperature of a wide range of materials. It can be used to measure temperature on a microsecond range, study temperature-sensitive

Table 4 Comparison of the maximum sensitivity values in Nd^{3+}/Yb^{3+} co-doped phosphors using the FIR technique.

Materials	Transitions	Range (K)	Sensitivity (K^{-1})	Ref.
La_2O_2S:Nd	$^4F_{3/2(1,\,2)} \rightarrow {}^4I_{9/2}$	30–600	17.5% (30*K*)	[74]
Gd_2O_3:Nd	$^4F_{5/2}/^4F_{3/2} \rightarrow {}^4I_{9/2}$	288–323	1.75% (288 K)	[75]
La_2O_3:Nd/Yb	$^4F_{7/2}/^4F_{3/2} \rightarrow {}^4I_{9/2}$	290–1230	0.0003 (693 K)	[76]
$Ba_4La_2Ti_4Nb_6O_{30}$: Nd/Yb	$^4F_{7/2}/^4S_{3/2} \rightarrow {}^4I_{9/2}$ and $^4F_{5/2}/^2H_{9/2} \rightarrow {}^4I_{9/2}$ $^4F_{7/2}/^4S_{3/2} \rightarrow {}^4I_{9/2}$ and $^4F_{3/2} \rightarrow {}^4I_{9/2}$ $^4F_{5/2}/^2H_{9/2} \rightarrow {}^4I_{9/2}$ and $^4F_{3/2} \rightarrow {}^4I_{9/2}$	303–573	0.0008 (573 K) 0.013 (573 K) 0.031 (573 K)	[77]
$CaWO_4$:Nd/Yb/Li	$^4F_{7/2}$ and $^4F_{5/2}$, $^4F_{3/2} \rightarrow {}^4I_{9/2}$	303–773	0.0071 (769 K)	[78]
$SrWO_4$:Nd/Yb	$^4F_{7/2}/^4S_{3/2} \rightarrow {}^4I_{9/2}$ $^4F_{7/2}/^4F_{5/2} \rightarrow {}^4I_{9/2}$ $^4F_{5/2}/^4F_{3/2} \rightarrow {}^4I_{9/2}$	313–453	0.01263 (453 K) 0.00164 (453 K) 0.02857 (453 K)	[79]

processes, and measure temperature distribution. Additionally, it can be used to measure temperature in nonequilibrium conditions, providing unique insights into the dynamics of these systems. Another application of time-resolved spectroscopy for temperature sensing is in the field of thermal imaging. By using a detector array, it is possible to obtain images of the temperature distribution within a sample. Additionally, time-resolved spectroscopy can be used in the field of noncontact temperature sensing. This can be achieved by measuring the temperature-dependent properties of a material such as its lifetime. By measuring the lifetime at different temperatures, it is possible to extract information about the temperature of the sample in a noncontact way. This can be useful in situations where direct contact with the sample is not possible or desired.

Thus, time-resolved spectroscopy is a powerful technique that can be used for temperature sensing in a wide range of applications. It can provide detailed information about the temperature distribution within a sample and can be used to study temperature-sensitive processes. Phosphor's rare-earth lifetime-based temperature sensing is a method of measuring temperature by observing the decay of excited states in rare-earth phosphors. The lifetime of the excited state in phosphors is temperature dependent. To measure the temperature using this method, the lifetime of the excited state can be determined by fitting the decay curve to an exponential mathematical model. The temperature then can be calculated from the lifetime using a calibration curve.

This temperature sensing method has several advantages, such as its high sensitivity, fast response time, and the ability to measure temperature noninvasively. These sensors are developed to use in medicine, biology, and materials science to measure temperature in living cells, biological tissues, and other temperature-sensitive environments. Phosphor's rare-earth lifetime-based temperature sensing is also used in industrial processes where high temperature is present and it's not possible to measure temperature with contact methods.

In addition to its high sensitivity and fast response time, phosphor rare-earth lifetime-based temperature sensing also has a high spatial resolution, allowing temperature measurements to be made at specific locations within a sample. Phosphors can be made that are sensitive to different ranges of temperature, from cryogenic temperatures to high temperatures above 1000°C.

Furthermore, it can also be used in combination with other techniques to measure temperature in challenging conditions, for example, by using a phosphor layer to coat a sensor, which can be used to measure temperature in harsh environments. Finally, it is worth noting that the accuracy of this method can be improved by using a more complex mathematical model to fit the decay curve and by using more accurate calibration curves.

As an example of this technique, the lifetime of Ho^{3+} ions in $YVO_4{:}Ho^{3+}/Yb^{3+}$ phosphor was studied with temperature [42]. The material had been found to have both downconversion and upconversion properties, meaning it can convert light from a higher-energy state to a lower energy state (downconversion) and vice versa (upconversion). The time-resolved spectroscopic measurements were conducted within a temperature range of 12–300 K. It was found that the material is highly suitable for temperature sensing below room temperature. We determined the temperature of the nanocrystals using the decay time and rise time of vanadate and Ho^{3+} energy levels. The material showed a maximum relative sensor sensitivity of 1.35% K^{-1} using the rise time of the Ho^{3+} energy level. The change in the decay time of the VO_4^{3-} emission band as a function of temperature is illustrated in Fig. 7A. The data on the lifetime of the material displays a constant value between 12 and around 150 K, and then it consistently decreases as the temperature increases from 150 to 300 K. The decreasing trend observed above 150 K can be attributed to the suppression of the vanadate emission through nonradiative processes [42].

The decay time and rise time at various (temperatures) of the $^5F_4/^5S_2$ level were fitted according to an Arrhenius equation. It was found that the rise time of $^5F_4/^5S_2$ (Fig. 7B) and decay time of the vanadate emission level (Fig. 7A) were almost constant below 150 K, but within the range of 150–300 K, temperature-dependent behavior was exhibited. Based on this, $^5F_4/^5S_2$ level's rise time and vanadate level's decay time can be employed for temperature measurement within this temperature range. As shown in Fig. 7C, it was also found that $^5F_4$ level's decay time exhibited dependence temperature dependence firmly in the $300 \geq T \geq 12$ K and can be used for temperature sensing in a broader range. We defined the variation of decay time with temperature as $\tau^{-1} = 0.4 + 1.6 * \exp(-449/T)$.

The relative sensitivity of the temperature sensing method using the time-resolved measurements is depicted in Fig. 7D–F. The highest relative sensitivity values using the rise and decay times of $^5F_4/^5S_2$ are 1.35% and 0.33% per K, respectively, at room temperature, whereas the highest sensitivity based on the decay time of the VO_4^{3-} emitting level is 1.22% per K at room temperature [42].

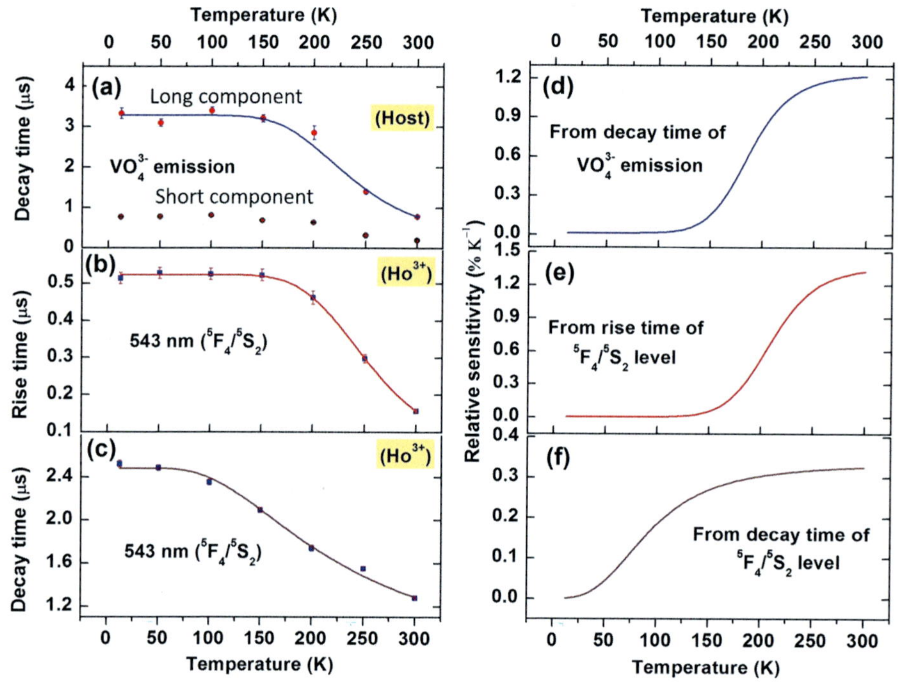

Fig. 7 (A) vanadate decay time, (B) $^5F_4/^5S_2$'s rise time, (C) $^5F_4/^5S_2$'s decay time, (D–F) relative sensitivities at various temperature in Ho^{3+}/Yb^{3+} doped YVO_4 phosphor [42].

Overall, phosphor rare-earth lifetime-based temperature sensing is a versatile and accurate technique that can be used to measure temperature in a wide range of applications, from biology to industry. It is based on the observation of the decay of excited states in rare-earth phosphors, whose lifetime is temperature dependent.

4 Upconverting nanothermometers in biomedical applications

Upconverting nanothermometers are also used to measure temperature in biomedical applications. This technique allows precise temperature measurements in small, localized areas, such as within cells or tissues. Upconverting nanothermometers have been used in various biomedical applications, including in vivo imaging of temperature changes in living organisms, monitoring temperature changes during drug delivery, and monitoring thermal therapy for cancer treatment.

One of the advantages of using upconverting nanothermometers in biomedical applications is their high sensitivity and accuracy. Additionally, they can be functionalized with biomolecules such as antibodies or peptides, allowing them to target specific cells or tissues.

Up to now, considerable research efforts have been made to utilize RE-doped upconverting nanoparticles as thermal sensing agents in ex vivo, in vitro, and in vivo studies [71,80–89]. The RE-doped UC nanoparticles display the advantages of deep tissue penetration depth, minimum photodamage to living organisms, and the absence of autofluorescence background from biological tissues [23,24]. Moreover, RE-doped UC nanoparticles can provide noninvasive real-time control temperature sensing during thermal treatments of cancers in both living organisms and culture cells and minimizing overheating of normal tissues [71,82,86]. Among several RE ions, Nd^{3+}-based luminescent thermometers have attracted significant attention in biological systems due to their excitation and emission of light wavelengths lying in the BWs, which allow deep tissue penetration and noninvasive thermal sensing [71–76]. For example, Benayas et al. [80] investigated the temperature sensing properties of Nd^{3+}:YAG nanoparticles in the first BW using the emission lines from different Stark sublevels of the $^4F_{3/2} \rightarrow {}^4I_{9/2}$ transitions, located at 938 nm and 945 nm under 808 nm laser irradiation. In order to verify their ability as thermal probes in biological tissues, an ex vivo thermal sensing experiment was designed [80]. A chicken breast tissue sample was injected with 100 μL of a 0.1 wt% water dispersion of Nd^{3+}:YAG nanoparticles at a depth of 5 mm followed by the heating of hot air flow (70°C) for 90 s. The subtissue emission and the intensity ratio of the Nd:YAG nanoparticles during the heating/cooling cycles were continuously monitored for a period of more than 8 min. A linear increase of the subtissue temperature with time was observed during the heating process, reaching a maximum of 55°C and then exponentially decreased to room temperature when turning off the hot air flow. In another study, Rodriguez's group reported maximum thermal sensitivity (2.5×10^{-2}°C^{-1}) for a hybrid nanostructure ($NaGdF_4$:Nd^{3+} UCNPs and the semiconductor PbS/CdS/ZnS quantum dots) operating in the second BW for real subtissue thermal sensing in ex vivo experiment [81]. Furthermore, Ximendes et al. [71] reported Nd^{3+}:LaF_3@Yb^{3+}:LaF_3 core@shell nanoparticles as accurate and reliable subcutaneous thermal sensing probes for in vivo applications operating in the second BW under NIR excitation. Moreover, previous works have demonstrated that Nd^{3+}-based UC thermometers at high doping levels are capable of converting light into heat effectively under 808 nm excitation, which could be useful for in vivo and ex vivo photothermal experiments [82–85]. For instance, Carrasco et al. [83] demonstrated in vivo temperature-controlled photothermal treatment of cancer tumors in mice by using heavily Nd^{3+} ion doped LaF_3 nanoparticles. Further, the same research group synthesized ultrasmall $NdVO_4$ stoichiometric (100% constituent Nd^{3+} ions) nanoparticles and showed their potential applications in ex vivo and in vivo photothermal experiments. They found a photothermal conversion efficiency of 72% at an excitation wavelength of 808 nm [84]. In addition, Kolesnikov et al. [85] demonstrated the potential application of YVO_4:Nd^{3+} nanoparticles in real-time temperature-controlled photothermal therapies by performing ex vivo experiments in biological systems.

The fluorescent nanothermometer used for in vitro measurement of living cells such as HeLa cancer cells was first reported by Vetrone et al. [86] using water-dispersible PEI-capped $NaYF_4$:Er^{3+}/Yb^{3+} NPs. The internal temperature of the individual HeLa cells was measured within the physiological range, from 25°C to its thermally induced death at 45°C. In another study, Shi et al. [87] synthesized $NaYF_4$:Yb^{3+},Er^{3+}@$NaYF_4$:Yb^{3+},Nd^{3+}@DMSA nanoparticles for in vitro temperature monitoring of NIH-3T3 cells. Another study by Garcia et al. [88] developed a theranostics nanoplatform integrating UC luminescence imaging and temperature-controlled PTT using $NaYF_4$:Yb^{3+}/Er^{3+} UCNPs decorated with gold nanoparticles. The nanocomposite was used for in vitro studies to track the MCF-7 breast cancer cells. Zhu et al. [89] designed a core-shell nanostructure of carbon-coated UCNPs ($NaLuF_4$:Yb,Er@$NaLuF_4$@Carbon), for real-time temperature monitoring at the microscopic level during photothermal therapy in HeLa cells and in vivo. This study showed that the photothermal effect of core-shell nanostructure under mild apparent temperature can kill the cancer cells without damaging the normal cells.

5 Conclusion and prospects

In this chapter, we have presented temperature sensing of UC phosphors that have most of the required properties including high photostability, low photobleaching or photoblinking, longer lifetime of the excited states, and larger anti-Stokes shifts [90–101]. More importantly, background-free fluorescence detection is helpful for their usage in developing UC-based sensors in biological imaging. The emission profiles can be tuned easily by controlling the type of RE^{3+} dopants and their concentrations. The surface of the UCNPs can also be modified by biomolecules and conjugated to other materials.

Through this chapter, we have compared FIR-based temperature determination performances of UC phosphors and discussed how to improve the performances of temperature sensitivity. For an ideal temperature sensor, two main requirements are high sensitivity and wide detection range, and these properties can be modified by adjusting the distribution of non-RE or RE dopants. Separately, it should be noted that UCNPs can be used as remote optical sensors in the biomedical industry to measure local temperature, and in vivo testing of UCNPs in this regard has already begun successfully. However, there are still some challenges to overcome in the future, such as low quantum yield, cytotoxicity, and UCNP-based technology.

Although upconverting nanothermometers hae the potential to be used in biomedical applications, some challenges still need to be addressed. Some of these are explained below:

Upconversion-based temperature sensing has several limitations, including lower sensitivity than other methods such as thermocouples, making it difficult to detect small temperature changes. The technique relies on measuring the intensity of specific wavelengths emitted by thermographic phosphors, but these wavelengths can overlap with other sources of light and result in inaccurate measurements. Additionally, the technique is only effective within a limited temperature range; it can be costly and complex and may require a high level of expertise to set up and operate correctly. It can also be prone to interference, have a lower resolution, and require regular calibration. The technique is heavy and bulky in nature, has a short lifetime, low signal-to-noise ratio, needs external energy sources, and may require advanced analysis to retrieve useful information.

Furthermore, upconversion-based temperature sensing systems are sensitive to environmental changes and can drift over time, which can result in inaccurate measurements. The phosphors used in these systems can degrade over time, which can reduce their sensitivity and accuracy. The signal-to-noise ratio in upconversion-based temperature sensing is often low, making it difficult to retrieve useful temperature information from the data. The technique also requires an external energy source to excite the thermographic phosphors, which can increase the power consumption of the system and may not be suitable for some applications. The size and weight of the systems can also make them difficult to use in some applications. The data generated by upconversion-based temperature sensing systems can be complex and may require advanced analysis techniques to extract meaningful information. Overall, while upconversion-based temperature sensing has its own unique advantages, it also has several limitations and challenges that need to be considered before implementing it in any application. Overall, upconverting nanothermometers have great potential, but more research and development is needed to overcome the challenges.

References

[1] A. Rapaport, J. Milliez, A. Cassanho, H.J. Jenssen, Review of the properties of up-conversion phosphors for new emissive display, J. Disp. Technol. 2 (2006) 68–78, https://doi.org/10.1109/JDT.2005.863781.

[2] M.K. Mahata, H. Bae, K.T. Lee, Upconversion luminescence sensitized pH-nanoprobes, Molecules 22 (2017) 2064, https://doi.org/10.3390/molecules22122064.

[3] C. Wang, Y. Jin, R. Zhang, Q. Yao, Y. Hu, A review and outlook of ratiometric optical thermometer based on thermally coupled levels and non-thermally coupled levels, J. Alloys Compd. 894 (2022), 162494.

[4] X. Wang, Q. Liu, Y. Bu, C.S. Liu, T. Liu, X. Yan, Optical temperature sensing of rare-earth ion doped phosphors, RSC Adv. 5 (2015) 86219.

[5] R. De, M.K. Mahata, K.T. Kim De, Structure-based varieties of polymeric nanocarriers and influences of their physicochemical properties on drug delivery profiles, Adv. Sci. 9 (2022) 2105373.

[6] M.K. Mahata, R. De, K.T. Lee, Near-infrared-triggered upconverting nanoparticles for biomedicine applications, Biomedicine 9 (2021) 756.

[7] P.R.N. Childs, J.R. Greenwood, C.A. Long, Review of temperature measurement, Rev. Sci. Instrum. 71 (2000) 2959–2978.

[8] C. Abram, B. Fond, F. Beyrau, Temperature measurement techniques for gas and liquid flows using thermographic phosphor tracer particles, Prog. Energy Combust. 64 (2018) 93–156.

[9] C.D.S. Brites, P.P. Lima, N.J.O. Silva, A. Millan, V.S. Amaral, F. Palacio, L.D. Carlos, Thermometry at the nanoscale, Nanoscale 4 (2012) 4799–4829.

[10] D. Teyssieux, L. Thiery, B. Cretin, Near-infrared thermography using a charge-coupled device camera: application to microsystems, Rev. Sci. Instrum. 78 (2007), 034902.

[11] M. Kuball, J.M. Hayes, M.J. Uren, T. Martin, J.C.H. Birbeck, R.S. Balmer, B.T. Hughes, Measurement of temperature in active high-power AlGaN/GaN HFETs using Raman spectroscopy, IEEE Electron Device Lett. 23 (2002) 7–9.

[12] L.H. Fischer, G.S. Harms, O.S. Wolfbies, Upconverting nanoparticles for nanoscale thermometry, Angew. Chem. Int. Ed. 50 (2011) 4546–4551.

[13] D. Jaque, F. Vetrone, Luminescence nanothermometry, Nanoscale 4 (2012) 4301–4326.

[14] S.A. Wade, S.F. Collins, G.W. Baxter, Fluorescence intensity ratio technique for optical fiber point temperature sensing, J. Appl. Phys. 94 (2003) 4743.

[15] O.A. Savchuk, J.J. Carvajal, M.C. Pujol, E.W. Barrera, J. Massons, M. Aguilo, F. Diaz, Ho,Yb:$KLu(WO_4)_2$ nanoparticles: a versatile material for multiple thermal sensing purposes by luminescent thermometry, J. Phys. Chem. C 119 (2015) 18546–18558.

[16] M.K. Mahata, T. Koppe, K. Kumar, H. Hofsäss, U. Vetter, Upconversion photoluminescence of Ho^{3+}-Yb^{3+} doped barium titanate nanocrystallites: optical tools for structural phase detection and temperature probing, Sci. Rep. 10 (2020) 8775.

[17] M.K. Mahata, T. Koppe, T. Mondal, C. Brüsewitz, K. Kumar, V. Kumar Rai, H. Hofsäss, U. Vetter, Incorporation of Zn^{2+} ions into $BaTiO_3$:Er^{3+}/Yb^{3+} nanophosphor: An effective way to enhance upconversion, defect luminescence and temperature sensing, Phys. Chem. Chem. Phys. 17 (32) (2015) 20741–20753.

[18] S. Sinha, K. Kumar, Studies on up/down-conversion emission of Yb^{3+} sensitized Er^{3+} doped $MLa_2(MoO_4)_4$ (M = Ba, Sr and Ca) phosphors for thermometry and optical heating, Opt. Mater. 75 (2018) 770–780.

[19] M.K. Mahata, K. Kumar, V.K. Rai, Er^{3+}/Yb^{3+} doped vanadate nanocrystals: a highly sensitive thermographic phosphor and its optical nanoheater behavior, Sens. Actuators B Chem. 209 (2015) 775–780.

[20] S. Sinha, M.K. Mahata, K. Kumar, Comparative thermometric properties of bi-functional Er^{3+}-Yb^{3+} doped rare earth (RE = Y, Gd and La) molybdates, Mater. Res. Express 5 (2018), 026201.

[21] S. Sinha, M.K. Mahata, K. Kumar, Enhancing the upconversion luminescence properties of Er^{3+}/Yb^{3+} doped yttrium molybdate through Mg^{2+} incorporation: effect of laser excitation power on temperature sensing and heat generation, New J. Chem. 43 (15) (2019) 5960–5971.

[22] S. Sinha, A. Mondal, K. Kumar, H.C. Swart, Enhancement of upconversion emission and temperature sensing of paramagnetic $Gd_2Mo_3O_9$: Er^{3+}/Yb^{3+} phosphor via Li^+/Mg^{2+} co-doping, J. Alloys Compd. 747 (2018) 455–464.

[23] A. Nadort, J. Zhao, E.M. Goldys, Lanthanide upconversion luminescence at the nanoscale: fundamentals and optical properties, Nanoscale 8 (27) (2016) 13099–13130.

[24] M.K. Mahata, K.T. Lee, Development of near-infrared sensitized core-shell-shell upconverting nanoparticles as pH-responsive probes, Nanoscale Adv. 1 (6) (2019) 2372–2381.

[25] S. Sinha, M.K. Mahata, K. Kumar, S.P. Tiwari, V.K. Rai, Dualistic temperature sensing in Er^{3+}/Yb^{3+} doped $CaMoO_4$ upconversion phosphor, Spectrochim. Acta Part A: Mol. Biomol. Spectrosc. 173 (2017) 369–375.

[26] R. De, Y.-H. Song, M.K. Mahata, K.T. Lee, pH-responsive polyelectrolyte complexation on upconversion nanoparticles: a multifunctional nanocarrier for protection, delivery, and 3D-imaging of therapeutic protein, J. Mater. Chem. B 10 (2022) 3420–3433, https://doi.org/10.1039/D2TB00246A.

[27] M.K. Mahata, S.P. Tiwari, S. Mukherjee, K. Kumar, V.K. Rai, YVO_4:Er^{3+}/Yb^{3+} phosphor for multifunctional applications, J. Opt. Soc. Am. B 31 (8) (2014) 1814–1821.

[28] F. Auzel, Upconversion and anti-stokes processes with f and d ions in solids, Chem. Rev. 104 (2004) 139–174.

[29] M.K. Mahata, K. Kumar, V.K. Rai, Structural and optical properties of Er^{3+}/Yb^{3+} doped barium titanate phosphor prepared by co-precipitation method, Spectrochim. Acta Part A: Mol. Biomol. Spectrosc. 124 (2013) 285–291.

[30] A. Kumari, M.K. Mahata, Upconversion nanoparticles for sensing applications, in: Upconversion Nanophosphors, Elsevier, 2022, pp. 311–336.

[31] M.K. Mahata, H. Hofsaess, U. Vetter, Photon-upconverting materials: Advances and prospects for various emerging applications, in: J. Thirumalai (Ed.), Luminescence—An Outlook on the Phenomena and Their Applications, InTech, Croatia, 2016, pp. 109–131.
[32] S. Sinha, M.K. Mahata, What are upconversion nanophosphors: basic concepts and mechanisms, in: Upconversion Nanophosphors, Elsevier, 2022, pp. 19–48.
[33] M.K. Mahata, R. De, K.T. Lee, Upconversion nanoparticles in pH sensing applications, in: Upconverting Nanoparticles: From Fundamentals to Applications, John Wiley & Sons, 2022, pp. 395–416.
[34] K. Kumar, S.K. Maurya, M.K. Mahata, Upconversion hybrid phosphors for biological applications, in: Hybrid Phosphor Materials, Springer, Cham, 2022, pp. 195–222.
[35] S. Sinha, M.K. Mahata, K. Kumar, Up/down-converted green luminescence of Er^{3+}-Yb^{3+} doped paramagnetic gadolinium molybdate: a highly sensitive thermographic phosphor for multifunctional applications, RSC Adv. 6 (92) (2016) 89642–89654.
[36] S. Sinha, M.K. Mahata, H.C. Swart, A. Kumar, K. Kumar, Enhancement of upconversion, temperature sensing and cathodoluminescence in the K^+/Na^+ compensated $CaMoO_4$: Er^{3+}/Yb^{3+} nanophosphor, New J. Chem. 41 (13) (2017) 5362–5372.
[37] P. Sukul, M.K. Mahata, U.K. Ghorai, K. Kumar, Crystal phase induced upconversion enhancement in Er^{3+}/Yb^{3+} doped $SrTiO_3$ ceramic and its temperature sensing studies, Spectrochim. Acta A Mol. Biomol. Spectrosc. 212 (2019) 78–87.
[38] M.K. Mahata, A. Kumari, V.K. Rai, Er3+, Yb3+ doped yttrium oxide phosphor as a temperature sensor, AIP Conf. Proc. 1536 (1) (2013) 1270–1271.
[39] S.P. Tiwari, M.K. Mahata, K. Kumar, V.K. Rai, Enhanced temperature sensing response of upconversion luminescence in ZnO–$CaTiO_3$: Er^{3+}/Yb^{3+} nano-composite phosphor, Spectrochim. Acta A Mol. Biomol. Spectrosc. 150 (2015) 623–630.
[40] S. Sinha, M.K. Mahata, V.K. Rai, K. Kumar, Upconversion emission study of Er^{3+} doped $CaMoO_4$ phosphor, AIP Conf. Proc. 1728 (2016) 020476.
[41] M. Runowski, P. Wony, N. Stopikowska, I.N. Martin, V. Lavin, S. Lis, Luminescent nanothermometer operating at very high temperature—sensing up to 1000 K with upconverting nanoparticles (Yb^{3+}/Tm^{3+}), ACS Appl. Mater. Interfaces 12 (2020) 43933–43941.
[42] M.K. Mahata, T. Koppe, K. Kumar, H. Hofsäss, U. Vetter, Demonstration of temperature dependent energy migration in dual mode YVO_4: Ho^{3+}-Yb^{3+} nanocrystals for low temperature thermometry, Sci. Rep. 6 (2016) 36342.
[43] M.A.H. Rodríguez, A.D.L. Gorrín, I.R. Martín, U.R.R. Mendoza, V. Lavín, Comparison of the sensitivity as optical temperature sensor of nano-perovskite doped with Nd^{3+} ions in the first and second biological windows, Sensor 255 (2018) 970–976.
[44] Z. Zhang, C. Guo, H. Suo, X. Zhao, N. Zhang, T. Li, Thermometry and up-conversion luminescence of Yb^{3+}–Er^{3+} co-doped $Na_2Ln_2Ti_3O_{10}$ (Ln = Gd, La) phosphors, Phys. Chem. Chem. Phys. 18 (2016) 18828–18834.
[45] H. Wu, Z. Hao, L. Zhang, X. Zhang, Y. Xiao, G.H. Pan, H. Wu, Y. Luo, L. Zhang, J. Zhang, Er^{3+}/Yb^{3+} codoped phosphor $Ba_3Y_4O_9$ with intense red upconversion emission and optical temperature sensing behavior, J. Mater. Chem. C 6 (2018) 3459–3467.
[46] H. Lu, R. Meng, H. Hao, Y. Bai, Y. Gao, Y. Song, Y. Wang, X. Zhang, Stark sublevels of Er^{3+}-Yb^{3+} codoped $Gd_2(WO_4)_3$ phosphor for enhancing the sensitivity of a luminescent thermometer, RSC Adv. 6 (2016) 57667.
[47] H. Suo, C. Guo, T. Li, Broad-scope thermometry based on dual-color modulation up-conversion phosphor $Ba_5Gd_8Zn_4O_{21}$:Er^{3+}/Yb^{3+}, J. Phys. Chem. C 120 (2016) 2914–2924.

[48] L. Liu, Y. Wang, X. Zhang, K. Yang, Y. Bai, C. Huang, Y. Song, Optical thermometry through green and red upconversion emissions in $Er^{3+}/Yb^{3+}/Li^{+}$: ZrO_2 nanocrystals, Opt. Commun. 284 (2011) 1876–1879.
[49] F. Huang, T. Yang, S. Wang, L. Lin, T. Hu, D. Chen, Temperature sensitive cross relaxation between Er^{3+} ions in laminated hosts: a novel mechanism for thermochromic upconversion and high-performance thermometry, J. Mater. Chem. C 6 (2018) 12364–12370.
[50] K. Zheng, W. Song, G. He, Z. Yuan, W. Qin, Five-photon UV upconversion emissions of Er3+ for temperature sensing, Opt. Express 23 (2015) 7653–7658.
[51] J. Zhang, Y. Zhang, X. Jiang, Investigations on upconversion luminescence of $K_3Y(PO_4)_2$: Yb^{3+}-$Er^{3+}/Ho^{3+}/Tm^{3+}$ phosphors for optical temperature sensing, J. Alloys Compd. 748 (2018) 438–445.
[52] J. Zhang, G. Chen, Z. Zhai, H. Chen, Y. Zhang, Optical temperature sensing using upconversion luminescence in rare-earth ions doped $Ca_2Gd_8(SiO_4)_6O_2$ phosphors, J. Alloys Compd. 771 (2019) 838–846.
[53] W. Xu, Z. Zhang, W. Cao, Excellent optical thermometry based on short-wavelength upconversion emissions in Er^{3+}/Yb^{3+} codoped $CaWO_4$, Opt. Lett. 37 (2012) 4865.
[54] W. Xu, H. Zhao, Y. Li, L. Zheng, Z. Zhang, W. Cao, Optical temperature sensing through the upconversion luminescence from Ho^{3+}/Yb^{3+} co-doped $CaWO_4$, Sens. Actuators B Chem. 188 (2013) 1096–1100.
[55] P. Singh, P.K. Shahi, A. Rai, A. Bahadur, S.B. Rai, Effect of Li^{3+} ion sensitization and optical temperature sensing in Gd_2O_3: Ho^{3+}/Yb^{3+}, Opt. Mater. 58 (2016) 432–438.
[56] Monika, R.S. Yadav, A. Rai, S.B. Rai, NIR light guided enhanced photoluminescence and temperature sensing in $Ho^{3+}/Yb^{3+}/Bi^{3+}$ co-doped $ZnGa_2O_4$ phosphor, Sci. Rep. 11 (2021) 4148.
[57] A. Pandey, V.K. Rai, Improved luminescence and temperature sensing performance of Ho^{3+}-Yb^{3+}-Zn^{2+}: Y_2O_3 phosphor, Dalton Trans. 42 (2013) 11005.
[58] R. Dey, A. Kumari, A.K. Soni, V.K. Rai, $CaMoO_4$: Ho^{3+}-Yb^{3+}-Mg^{2+} upconverting phosphor for application in lighting devices and optical temperature sensing, Sens. Actuators B Chem. 210 (2015) 581–588.
[59] N. Wang, Z. Fu, Y. Wei, T. Sheng, Investigation for the upconversion luminescence and temperature sensing mechanism based on $BiPO_4$: Yb^{3+}, RE^{3+} (RE^{3+}= Ho^{3+}, Er^{3+} and Tm^{3+}), J. Alloys Compd. 772 (2019) 371–380.
[60] P. Haro-González, S.F. León-Luis, S. González-Pérez, I.R. Martín, Analysis of Er^{3+} and Ho^{3+} codoped fluoroindate glasses as wide range temperature sensor, Mater. Res. Bull. 46 (2011) 1051–1054.
[61] V. Lojpur, M. Nikolic, L. Mancic, O. Milosevic, M.D. Dramicanin, Y_2O_3: Yb,Tm and Y_2O_3:Yb,Ho powders for low-temperature thermometry based on up-conversion fluorescence, Ceram. Int. 39 (2013) 1129–1134.
[62] X. Chai, J. Li, X. Wang, Y. Li, X. Yao, Upconversion luminescence and temperature sensing properties of Ho^{3+}/Yb^{3+}-codoped $ZnWO_4$ phosphors based on fluorescence intensity ratios, RSC Adv. 7 (2017) 40046.
[63] J. Zhang, J. Chen, Y. Zhang, S. An, Yb^{3+}/Tm^{3+} and Yb^{3+}/Ho^{3+} doped $NaY_9(SiO_4)_6O_2$ phosphors: upconversion luminescence processes, temperature-dependent emission spectra and optical temperature-sensing properties, J. Alloys Compd. 860 (2021), 158473.
[64] W. Ge, M. Xu, J. Shi, J. Zhu, Y. Li, Highly temperature-sensitive and blue upconversion luminescence properties of $Bi_2Ti_2O_7$: Tm^{3+}/Yb^{3+} nanofibers by electrospinning, Chem. Eng. J. 391 (2020), 123456.

[65] A.F. Pereira, K.U. Kumar, W.F. Silva, W.Q. Santos, D.J.C. Jacinto, Yb^{3+}/Tm^{3+} co-doped $NaNbO_3$ nanocrystals as three-photon-excited luminescent nanothermometers, Sens. Actuators B Chem. 213 (2015) 65–71.

[66] A.K. Soni, R. Dey, V.K. Rai, Stark sublevels in Tm^{3+}-Yb^{3+} codoped $Na_2Y_2B_2O_7$ nanophosphor for multifunctional applications, RSC Adv. 5 (2015) 34999–35009.

[67] Z. Sun, G. Liu, Z. Fu, X. Zhang, Z. Wu, Y. Wei, High sensitivity thermometry and optical heating Bi-function of Yb^{3+}/Tm^{3+} Co-doped $BaGd_2ZnO_5$ phosphors, Curr. Appl. Phys. 17 (2017) 255–261.

[68] R.S. Yadav, S.J. Dhoble, S.B. Rai, Enhanced photoluminescence in Tm^{3+}, Yb^{3+}, Mg^{2+} tri-doped $ZnWO_4$ phosphor: three photon upconversion, laser induced optical heating and temperature sensing, Sens. Actuators B Chem. 273 (2018) 1425–1434.

[69] Y. Wu, S. Xu, F. Lai, B. Liu, J. Huang, X. Ye, W. You, Intense near-infrared emission, upconversion processes and temperature sensing properties of Tm^{3+} and Yb^{3+} co-doped double perovskite Gd_2ZnTiO_6 phosphors, J. Alloys Compd. 804 (2019) 486–493, https://doi.org/10.1016/j.jallcom.2019.07.036.

[70] G. Chen, J. Zhang, Investigation on optical temperature sensing behaviour for $Y_{4.67}Si_3O_{13}$: Tm^{3+}, Yb^{3+} phosphors based on upconversion luminescence, Opt. Mater. Express 8 (2018) 1841, https://doi.org/10.1364/OME.8.001841.

[71] E.C. Ximendes, W.Q. Santos, U. Rocha, U.K. Kagola, F.S. Rodriguez, N. Fernandez, A. da Silva Gouveia-Neto, D. Bravo, A.M. Domingo, B. del Rosel, C.D.S. Brites, L. D. Carlos, D. Jaque, C. Jacinto, Unveiling in vivo subcutaneous thermal dynamics by infrared luminescent nanothermometers, Nano Lett. 16 (2016) 1695–1703, https://doi.org/10.1021/acs.nanolett.5b04611.

[72] M. Quintanilla, Y. Zhang, L.M.L. Marzan, Subtissue plasmonic heating monitored with CaF_2:Nd^{3+}, Yb^{3+} Nanothermometers in the second biological window, Chem. Mater. 30 (2018) 2819–2828, https://doi.org/10.1021/acs.chemmater.8b00806.

[73] G. Dantelle, M. Matulionyte, D. Testemale, A. Cantarano, A. Ibanez, F. Vetrone, Nd^{3+} doped Gd3Sc2Al3O12 nanoparticles: towards efficient nanoprobes for temperature sensing, Phys. Chem. Chem. Phys. 21 (2019) 11132, https://doi.org/10.1039/C9CP01808E.

[74] G. Jiang, X. Wei, S. Zhou, Y. Chen, C. Duan, M. Yin, Neodymium doped lanthanum oxysulfide as optical temperature sensors, J. Lumin. 152 (2014) 156–159, https://doi.org/10.1016/j.jlumin.2013.10.027.

[75] S. Balabhadra, M.L. Debasu, C.D.S. Brites, L.A.O. Nunes, O.L. Malta, J. Rocha, M. Bettinelli, L.D. Carlos, Boosting the sensitivity of Nd^{3+}-based luminescent nanothermometers, Nanoscale 7 (2015) 17261–17267, https://doi.org/10.1039/C5NR05631D.

[76] G. Gao, D. Busko, S.K. Weiss, A. Turshatov, I. Howard, B.S. Richards, Wide-range non-contact fluorescence intensity ratio thermometer based on Yb^{3+}/Nd^{3+} co-doped La_2O_3 microcrystals operating from 290 to 1230 K, J. Mater. Chem. C 6 (2018) 4163–4170, https://doi.org/10.1039/C8TC00782A.

[77] Y. Shi, F. Yang, C. Zhao, Y. Huang, M. Li, Q. Zhou, Q. Li, Z. Li, J. Liu, T. Wei, Highly sensitive up-conversion thermometric performance in Nd^{3+} and Yb^{3+} sensitized $Ba_4La_2Ti_4Nb_6O_{30}$ based on near-infrared emissions, J. Phys. Chem. Solids 124 (2019) 130–136, https://doi.org/10.1016/j.jpcs.2018.09.013.

[78] W. Xu, Y. Hu, L. Zheng, Z. Zhang, W. Cao, H. Liu, X. Wu, Enhanced NIR-NIR luminescence from $CaWO_4$: Nd^{3+}/Yb^{3+} phosphors by Li^+ co-doping for thermometry and optical heating, J. Lumin. 208 (2019) 415–423, https://doi.org/10.1016/j.jlumin.2019.01.005.

[79] H. Song, Q. Han, X. Tang, X. Zhao, K. Ren, T. Liu, Nd^{3+}/Yb^{3+} codoped $SrWO_4$ for highly sensitive optical thermometry based on the near infrared emission, Opt. Mater. 84 (2018) 263–267, https://doi.org/10.1016/j.optmat.2018.06.054.
[80] A. Benayas, B. del Rosal, A.P. Delgado, K.S. Gomez, D. Jaque, G.A. Hirata, F. Vetrone, Nd:YAG near-infrared luminescent nanothermometers, Adv. Opt. Mater. 3 (2015) 687–694, https://doi.org/10.1002/adom.201400484.
[81] E.N. Ceron, D.H. Ortgies, B. del Rosal, F. Ren, A. Benayas, F. Vetrone, D. Ma, F.S. Rodriguez, J.G. Sole, D. Jaque, E.M. Rodriguez, Hybrid nanostructures for high-sensitivity luminescence nanothermometry in the second biological window, Adv. Mater. 27 (2015) 4781–4787, https://doi.org/10.1002/adma.201501014.
[82] X. Qiu, Q. Zhou, X. Zhu, Z. Wu, W. Feng, F. Li, Ratiometric upconversion nanothermometry with dual emission at the same wavelength decoded via a time-resolved technique, Nat. Commun. 4 (2020) 11, https://doi.org/10.1038/s41467-019-13796-w.
[83] E. Carrasco, B. del Rosal, F.S. Rodriguez, A.J.J. de la Fuente, P.H. Gonzalez, U. Rocha, K.U. Kumar, C. Jacinto, J.G. Sole, D. Jaque, Intratumoral thermal reading during photothermal therapy by multifunctional fluorescent nanoparticles, Adv. Funct. Mater. 25 (2014) 615–626, https://doi.org/10.1002/adfm.201403653.
[84] B. del Rosal, A.P. Delgado, E. Carrasco, D.J. Jovanovic, M.D. Dramicanin, G. Drazic, A. J. de la Fuente, F.S. Rodriguez, D. Jaque, Neodymium-based stoichiometric ultrasmall nanoparticles for multifunctional deep-tissue photothermal therapy, Adv. Opt. Mater. 4 (2016) 782–789, https://doi.org/10.1002/adom.201500726.
[85] I.E. Kolesnikov, E.V. Golyeva, A.A. Kalinichev, M.A. Kurochkin, E. Lahderanta, M.D. Mikhailov, Nd3+ single doped YVO4 nanoparticles for sub-tissue heating and thermal sensing in the second biological window, Sens. Actutors B: Chem. 243 (2017) 338–345, https://doi.org/10.1016/j.snb.2016.12.005.
[86] F. Vetrone, R. Naccache, A. Zamarro, A.J. de la Fuente, F.S. Rodriguez, L.M. Maestro, E.M. Rodriguez, D. Jaque, J.G. Sole, J.A. Capobianco, Temperature sensing using fluorescent nanothermometers, ACS Nano 4 (2010) 3254–3258, https://doi.org/10.1021/nn100244a.
[87] Z. Shi, Y. Duan, X. Zhu, Q. Wang, D.D. Li, K. Hu, W. Feng, F. Li, C. Xu, Dual functional NaYF4:Yb3+, Er3+@NaYF4:Yb3+, Nd3+ Core-shell nanoparticles for cell temperature sensing and imaging, Nanotechnology 29 (2018), 094001, https://doi.org/10.1088/1361-6528/aaa44a.
[88] G.R. Garcia, M.A.H. Colin, E. De la Rosa, T.L. Luke, S.S. Panikar, J.d.J.I. Sanchez, V. Piazz, Theranostic nanocomplex of gold-decorated upconversion nanoparticles for optical imaging and temperature-controlled photothermal therapy, J. Photochem. Photobiol. A 384 (2019) 112053, https://doi.org/10.1016/j.jphotochem.2019.112053.
[89] X. Zhu, W. Feng, J. Chang, Y.W. Tan, J. Li, M. Chen, Y. Sun, F. Li, Temperature-feedback upconversion nanocomposite for accurate photothermal therapy at facile temperature, Nat. Commun. 7 (2016) 10437, https://doi.org/10.1038/ncomms10437.
[90] Y.C. Goh, Y.H. Song, G. Lee, H. Bae, M.K. Mahata, K.T. Lee, Cellular uptake efficiency of nanoparticles investigated by three-dimensional imaging, Phys. Chem. Chem. Phys. 20 (2018) 11359–11368, https://doi.org/10.1039/C8CP00493E.
[91] S. Wen, J. Zhou, K. Zheng, A. Bednarkiewicz, X. Liu, D. Jin, Advances in highly doped upconversion nanoparticles, Nat. Commun. 9 (2018) 1–12, https://doi.org/10.1038/s41467-018-04813-5.

[92] A.K. Soni, M.K. Mahata, Photoluminescence and cathodoluminescence studies of Er^{3+}-activated strontium molybdate for solid-state lighting and display applications, Mater. Res. Express 4 (2017), 126201, https://doi.org/10.1088/2053-1591/aa9a52.
[93] G. Liang, H. Wang, H. Shi, H. Wang, M. Zhu, A. Jing, J. Li, G. Li, Recent progress in the development of upconversion nanomaterials in bioimaging and disease treatment, J. Nanobiotechnol. 18 (1) (2020) 1–22, https://doi.org/10.1186/s12951-020-00713-3.
[94] A.K. Soni, V.K. Rai, M.K. Mahata, Yb^{3+} sensitized $Na_2Y_2B_2O_7$: Er^{3+} phosphors in enhanced frequency upconversion, temperature sensing and field emission display, Mater. Res. Bull. 89 (2017) 116–124, https://doi.org/10.1016/j.materresbull.2017.01.009.
[95] P.P. Sukul, M.K. Mahata, K. Kumar, NIR optimized dual mode photoluminescence in Nd doped Y_2O_3 ceramic phosphor, J. Lumin. 185 (2017) 92–98, https://doi.org/10.1016/j.jlumin.2017.01.008.
[96] M.K. Mahata, S. Sinha, K. Kumar, Frequency upconversion in Er^{3+} and Yb^{3+} co-doped $MgTiO_3$ phosphor, AIP Conf. Proc. 1728 (2016) 020555, https://doi.org/10.1063/1.4946606.
[97] G. Gao, D. Busko, N. Katumo, R. Joseph, E. Madirov, A. Turshatov, I.A. Howard, B.S. Richards, Ratiometric luminescent thermometry with excellent sensitivity over a broad temperature range utilizing thermally-assisted and multiphoton upconversion in triply-doped La_2O_3: $Yb^{3+}/Er^{3+}/Nd^{3+}$, Adv. Opt. Mater. 9 (2021) 2001901, https://doi.org/10.1002/adom.202001901.
[98] M.K. Mahata, T. Koppe, H. Hofsäss, K. Kumar, U. Vetter, Host sensitized luminescence and time-resolved spectroscopy of YVO_4: Ho^{3+} nanocrystals, Phys. Proc. 76 (2015) 125–131, https://doi.org/10.1016/j.phpro.2015.10.023.
[99] S. Sinha, K. Kumar, Hydrothermal synthesis infrared to visible upconversion luminescence of SrMoO4: Er3+/Yb3+ phosphor, AIP Conf. Proc. 1942 (1) (2018), 050027, https://doi.org/10.1063/1.5028658.
[100] S. Sinha, M.K. Mahata, K. Kumar, Synthesis and upconversion emission properties of green emitting Y_4MoO_9: Er^{3+}/Yb^{3+} phosphor, Mater. Focus 5 (2016) 222–226, https://doi.org/10.1166/mat.2016.1318.
[101] M.S. Arai, A.S.S. de Camargo, Exploring the use of upconversion nanoparticles in chemical and biological sensors: from surface modifications to point-of-care devices, Nanosc. Adv. 3 (2021) 5135–5165, https://doi.org/10.1039/D1NA00327E.

Metal oxide-based phosphors for chemical sensors

7

Sibel Oguzlar[a] *and Merve Zeyrek Ongun*[b]
[a]Center for Fabrication and Application of Electronic Materials, Dokuz Eylul University, Izmir, Turkey, [b]Chemistry Technology Program, Izmir Vocational High School, Dokuz Eylul University, Izmir, Turkey

Chapter outline

Abbreviations

CCFLs	cold cathode fluorescent lamp
CCT	correlated color temperature
CIE	International Commission on Illumination
CRI	color rendering index
CRTs	cathode ray tubes
CVD	chemical vapor deposition method
DSSC	dye sensitized solar cell

Metal Oxides for Next-Generation Optoelectronic, Photonic, and Photovoltaic Applications
https://doi.org/10.1016/B978-0-323-99143-8.00020-1

ELs	electroluminescent displays
EQE	external quantum efficiency
FEDs	field emission displays
FWHM	full width at half maximum
IQE	internal quantum efficiency
LCD	liquid crystal display
LEDs	light-emitting diodes
NIR	near infrared region
NPs	nanoparticles
NTSC	National Television System Committee Standard
NUV-wLEDs	near ultraviolet white light-emitting diodes
pc-wLEDs	phosphor-converted white light-emitting diodes
PDP	plasma display panel
PL	photoluminescence
PSL	photo-stimulated luminescence
QE	quantum efficiency
QY	quantum yield
RE	rare earth
RP-wLEDs	remote-packaging white light-emitting diodes
SSFM	self-sum-frequency mixing
UV-LEDs	ultraviolet light-emitting diodes
VUV	vacuum ultraviolet light
wLEDs	white light-emitting diodes
WPE	wall plug efficiency
X-ray	X-radiation

1 Introduction

The quantitative and qualitative determination of molecules is of great importance both in environmental monitoring and in the health-care, automotive, mining, food, and pharmaceutical industries. In this context, the device that enables the determination of the relevant molecule or analyte is also called a "sensor." Sensors can be classified in various ways.

One of these classifications is "chemical" and "physical" sensors, which are based on the working principle of the transducer. In this classification, physical sensors record parameters such as temperature, pressure, and acceleration of the systems; chemical sensors provide information about the presence and concentration of various chemicals in the environment. On the other hand, all chemical sensors can also be subdivided into mass-sensitive, electrochemical, optical, electrical, magnetic, and thermometric devices [1–4] (see Fig. 1).

Except for these, it is possible to classify them as reversible/nonreversible or direct/indirect sensors depending on the application and sizes.

"Chemical sensors" are defined as devices that convert chemical information (from concentration to overall composition) of a particular sample component, such as ion, element, and chemical/biological molecule, into an analytically useful signal [5–7].

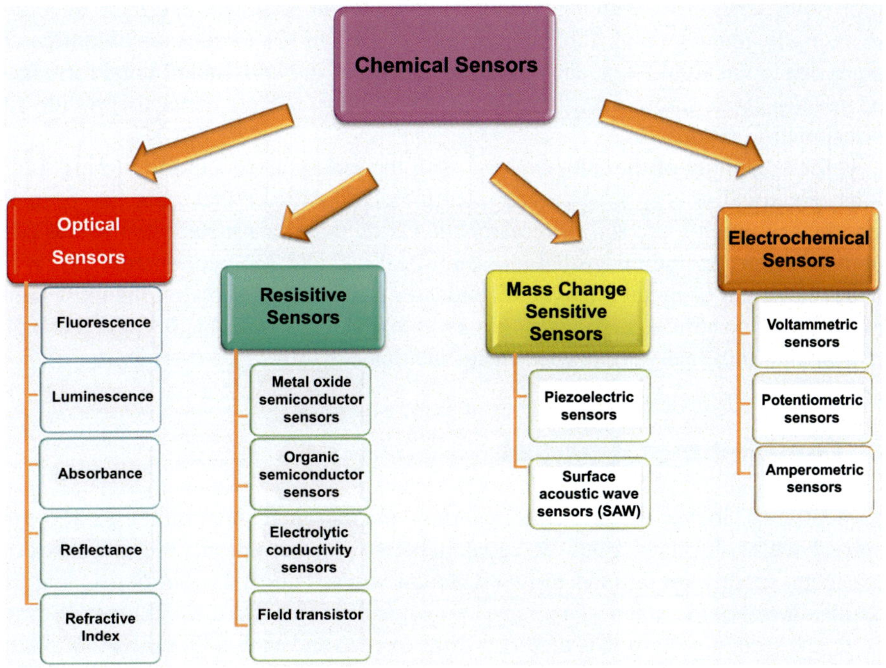

Fig. 1 Classification of chemical sensors.

Modern chemical sensors can also be defined as transducers. Many transduction methods such as optical, electrical, electrochemical, surface acoustic waves, and piezoelectricity have been mentioned in many literature studies.

As the industry has developed over time, there have been quite good developments in the field of chemical sensors in terms of sociable and financial advantages. As technology and processes develop in all these areas, it becomes more difficult to develop parameters such as sensitivity, selectivity, detection limit, reliability, and cost required for sensors. In this study, the use of metal oxide-based phosphors for chemical sensors will be discussed in general. For this reason, first, information on metal oxide-based materials will be given.

2 Metal oxide materials

An oxide material is a crystalline solid and an important chemical compound that occurs in a significant part of the chemical compound group formed by oxygen combining with another element. While simple oxides (binary oxides) contain a single metal combined with oxygen (structures such as AO, A_2O, and A_2O_3), complex oxides contain oxygen and at least two elements in their structure. Most of the Earth's crust consists of oxides because of the oxidation of individual crustal elements by oxygen in air or water. Therefore, with a general definition, it can be stated that oxide materials are naturally abundant and stable compounds in the soil [8]. Oxides are ubiquitous in nature due to the atom's tendency to attract electrons and consequent bond formation and readily form stable chemical bonds with almost all elements to provide the corresponding oxides.

At the beginning of the 20th century, with the industrial revolution turning into a global electrical energy process, the electronic property of materials has been an important research topic for all researchers. Oxide materials include the whole conductivity range, including insulators, semimetallics, and superconductors [9]. With the development of metal oxide semiconductor technology following the implementation of the first solid-state transistor in 1948, oxide materials have been increasingly incorporated into many solid-state electronic devices [10–12].

3 Complex metal oxides

As mentioned above, metal oxide materials are inorganic materials crystalline in nature, have a solid state lattice, and contain both oxygen atoms and metal ions. Complex metal oxides are defined in two different ways. The first definition is that the complex metal oxide material contains oxygen and at least two different metallic elements; the second definition is that it contains oxygen and a metallic element in two or more oxidation states in the lattice. Among metal oxide materials, complex metal oxides display some of the most advantageous properties.

Complex metal oxide materials are important materials with a wide variety of magnetic and electronic properties such as piezoelectricity, ferroelectricity, ferromagnetism, colossal magnetoresistance, and high-temperature superconductivity due to the electrons in the d and f orbitals in their structures.

Although each structure is different, there are specific properties displayed by the complex oxide, which differs from material to material. Therefore, it is of interest to researchers to search for new metal oxide materials with new types of structures and useful functions [13,14].

The use of mixed and complex metal oxides enables the production of materials with excellent properties such as increased sensitivity, improved absorption properties, high catalytic activity, and thermodynamic stability [15–17].

On the other hand, in the application of new material synthesis, many factors such as starting material, chemical component ratio, temperature, reaction time must be carefully selected and optimized [18–20].

However, the realization of synthesis methods with atomic layer precision has revealed even more amazing behavior derived from interfaces between different oxide materials. A few other interesting behaviors have also been found at oxide interfaces. Recently, the interfaces of non-oxide and complex oxide materials have shown thrilling results.

There are three types of metal oxides consisting of alkali and alkaline earth metals: oxides (contains the oxide anion), peroxides (contains the peroxide anion), and superoxides (contains the superoxide anion).

It is known that metal oxides are generally catalytically active, resistant to high temperatures, used as insulators/semiconductors, optically transparent in the visible region, and also have a wide bandgap. The key properties required for chemical sensing applications include both optical and electrical properties as well as chemical reactive possessions. Since the surface of metal oxide is an important factor for interaction with target molecules, studies to improve these surface properties lead to more precise and improved properties [21–23].

4 Nano-structured metal oxides

The advanced technology allows controlling the development of thin rectified film structures such as two-dimensional nanostructures, one-dimensional nanowires, nanotubes, zero-dimensional nanoparticles, and quantum dots with a diameter of a few nanometer. The miniaturization of smaller electronic, optical, and mechanical devices than ever, results in materials with excellent properties compared to bulk materials. When the material particle size is on the order of nanometer, its energy structure is similar to an atom, which can only be excited to discrete energy levels [24,25]. For this reason, nanomaterials are essential in new industrial applications due to their superior chemical, electrical, optical, magnetic, electrical, and thermal properties. The properties of these nanosized materials are largely dependent on their grain size, shape, and crystallinity [26,27].

Research for both the synthesis and characterization of metal oxide nanoparticles and their applications in different areas like photonics [28,29], solar cells [30,31], electronics [32], optics [33], supercapacitors [34], Li-ion batteries [35], energy storage devices [36], ceramics [37], adsorbents [38], semiconductors [39], biochemical sensing [40], and molecule detection [41] are among the latest focuses of nanotechnology (see Fig. 2).

Advances in nanotechnology offer new possibilities in metal oxide applications, as in every field. The main purpose here is to obtain nanostructures that have more specific properties compared to bulk or single-particle types. Nanosized oxide particles can show special physical and chemical possessions that influence basic properties in any material due to their high density of corner/edge surface areas and their limited size. Materials whose grain size is reduced to nanoscale provide new features such as wider bandgap and more advanced photoluminescence (PL) properties. when the surface area increases. The new physical effects occurring in nanostructured metal oxides can be used to develop sensor tools with improved parameters.

Metal oxide materials are among the promising materials used in many application areas. Since oxygen has a high electronegativity, the compounds formed when positive metal and negative oxygen combine have long-term stability. The combination of metal and oxygen atoms in the crystal structure lattice creates between 1 and 10 eV bandgap values, which means that materials are classified as insulators or semiconductors [42]. Considering all these advantages, it is inevitable that these oxide structures play an important role in our lives, and that they are the focus of attention of researchers in different scientific fields. Since metal oxides are widely used in sensor

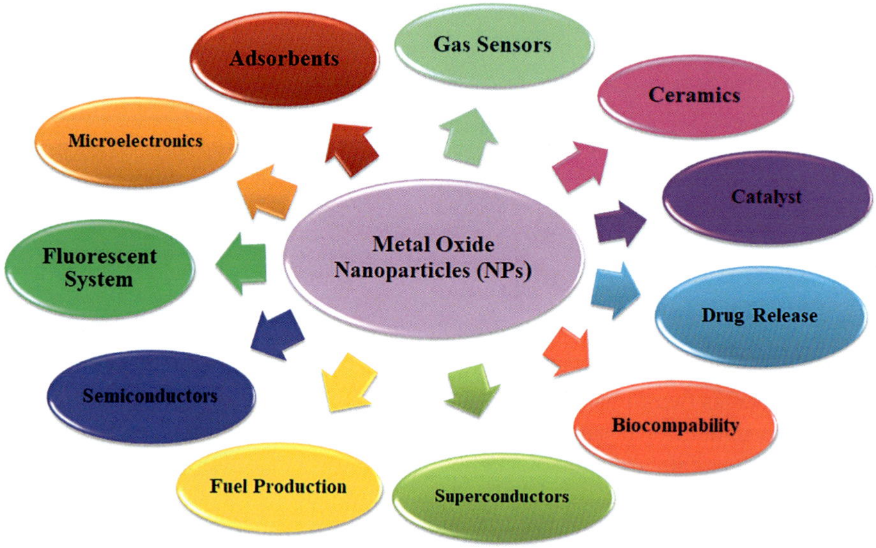

Fig. 2 Applications of metal oxide nanoparticles.

applications, they have great importance in fields such as chemistry, physics, and materials science. Thanks to the physicochemical properties of metal elements, a wide variety of oxide compounds with metallic, semiconductor, and insulating properties are formed, which are used in many fields such as catalysis, gas sensor, ceramic production, energy storage and recycling, biosensors, piezoelectric materials and applications of fuel cells, microelectronics and transistors available in general delivering lower cost and larger device stability [12,43].

In order to improve the material properties, in addition to producing nanosized materials, the use of various dopant materials also changes the properties of the relevant material. For this reason, additives are added to a wide variety of metal oxides to change their properties. In some cases, the aim is to provide the desired features and to improve them, while in some cases, it is to eliminate or reduce the undesirable effects [44–47].

Since many metal oxides are used to form thin films, they can be used as both charge-injecting and charge-transporting layers [48,49]. Additionally, laser diodes [50], light-emitting diodes (LEDs) [51], transparent conductive electrodes [52], photocatalysis [53], photodetectors [54], spintronic devices [55], and solar cells [56] are produced in a form of bulk or thin films using metal oxides.

5 Synthesis of metal oxide structures

Many techniques have been used to synthesize metal oxides with well-defined morphology and particle size. These methods are listed such as sonochemical method [57], solvothermal method [58], co-precipitation method [59], sol-gel method [60], microwave-assisted method [61], microemulsion method [62], chemical vapor deposition method (CVD) [63], reactive magnetron sputtering [64], combustion method

[65], template/surface-mediated synthesis [66], and biological synthesis [67]. Among these mentioned methods, sol-gel and hydrothermal synthesis methods (bottom-up approach) and sputtering techniques (top-down) are the most used techniques in the synthesis of metal oxide materials. As it is known, the synthesis method has a significant effect on the form, grain size, and detection properties of the material to be obtained. Recently, one-dimensional structures such as wire and fiber have drawn considerable attention in terms of application in the detection of the relevant analyte due to the increase in their surface areas [25,68]. However, the electrospinning method has emerged as a convenient method for producing metal oxide-based fiber structures [69]. In addition, molecular imprinting and template-based synthesis methods also have recently been the subject of research to obtain highly ordered metal oxide structures with monodisperse populations [70].

6 Phosphors (or luminescent materials)

Phosphors, known as luminescent materials, are defined as materials that absorb energy in the form of light, heat, electricity, and emit light. The emission of light can also occur in other areas, such as ultraviolet or infrared, that have energy beyond thermal equilibrium. These luminescent phosphors have an important place in the development of new devices and circuits in the electronics, communication and optoelectronic industries. Increasing demand for these advanced luminescent materials has motivated scientific and technological studies to improve existing properties and develop phosphors with more effective properties.

With a general definition, phosphor material can be expressed as a material that absorbs energy in forms such as electricity, light, and heat, and emits it as light.

For example, cathode ray televisions are used as a phosphor that absorbs electrons and emits light, while fluorescent lamps are used as a phosphor material that absorbs ultraviolet rays and emits light. Another example is that the phosphors used in white light-emitting diodes (wLEDs) produce yellow and red light while absorbing the violet-blue light emitted by the light-emitting diodes (LEDs).

A wide variety of phosphors have been designed for various application areas. However, oxide-structured phosphor materials are compeletely used because of having excellent features such as thermal stability, high photoluminescence efficiency, cheap raw materials, and easy synthesis methods. This section will provide information with a greater focus on improving the performance of practical oxide phosphors for use in oxide-based phosphors for display and lighting, three-band fluorescent lamps, plasma displays and wLEDs.

6.1 Oxide type phosphors

It has been found that some oxide-structured minerals are luminescent when excited. In some of these, the activator needs to be defined within the crystal structure. Ruby-Al_2O_3 with Cr^{3+} activator (red); willemite-Zn_2SiO_4 with Mn^{2+} activator (green) and scheelite-$CaWO_4$ without an activator (blue) [71] can be given as example.

All these minerals are produced synthetically in much higher yields than those found naturally. Materials like Zn_2SiO_4, Zn_2BeSiO_4, $Zn_3(BO_2)_3$ and $Cd_3(BO_2)_3$, and $Cd_3(PO_4)_2$ emitted from red to green in the spectrum when activated by manganese. Additionally, blue-emitting ZnS:Ag, green-emitting Zn_2SiO_4:Mn, and red-emitting YVO_4:Eu metal oxide crystals have also been integrated into color television screens to diffuse colors [72].

6.2 Photoluminescence mechanism based on centers, activators, and coactivators

The luminescence phenomenon is a common occurrence in daily life, bearings all types of displays, lights, LEDs, and cathode-ray tubes (CRTs) including excitation/emission and energy transfer events. In the case of luminescence, light-emitting materials are separated into three groups as self-activating matrices, host including activator, and host including sensitizer and activator complex materials. Activator or sensitizer can be used to improve the luminescence properties of the material used. By the way, applications such as mono- and co-doping techniques can be used to enhance luminescence properties [73–76].

In 1890, on phosphor chemistry, physicist Philipp Anton Lenard in Germany is known as the first person to define activator ions dispersed in ZnS and other crystalline structures that act as host crystals. Host crystal ions surround the activators, which form the luminescence centers where the absorption and emission process of phosphor occur. Since the luminescence centers can inactivate each other, they should not be close to each other in the host crystal. To ensure high efficiency, the distribution of activator ions must be as uniform and traceable as possible within the host crystal. Otherwise, in the case of high concentrations, activator ions act to inhibit the luminescence feature. For example, even a very small amount of ions such as iron, cobalt, and nickel inhibits the luminescence property of phosphors.

Self-activating hosts (metal oxides, molybdates, tantalates, zirconates, niobates, titanates, tungstates, vanadates, chromates) do not need activators because they have luminescent centers due to special groups (SiO_4 or WO_4) in their structures and so can emit radiation by absorbing excitation energy.

Due to the spectral shifts exhibited by phosphors activated by lanthanide ions, which have the ability to radiate in the narrow spectral region, it has been clearly shown that the luminescent properties of a center are largely dependent on the symmetry of neighboring ion groups when viewed with respect to the entire phosphor molecule. Even small amounts (2%) of titanium combined into zinc orthosilicate provide a great enhancement in photoluminescence due to this modifying effect on the symmetry of the luminescence centers. Therefore, here Titanium is called a condensing activator because it enhances the luminescence property of the host crystal.

Materials such as sodium chloride, which facilitate the incorporation of activator ions into the structure, are defined as coactivators.

For example, copper (II)-Cu^{2+} ions are used as activator ions of $ZnCl_2$-based phosphors and are presented into the structure in the form of copper (II) or copper.

However, if copper (II)-Cu^{2+} ions are added to the ZnS structure by heating, copper (I) sulfide (Cu_2S) crystals will form in a form that does not match the host crystal structure. This situation causes the formation of very few luminescent centers.

On the other hand, when copper (II) salt is used together with a coactivator material such as sodium chloride (NaCl), CuCl crystals with the same structure as the host crystal is formed by reducing Cu^{2+} ions to Cu^{+} ions.

To summarize, it would be correct to use terms such as activator, coactivator, emission spectrum, persistence, crystalline class and chemical composition, temperature and time of the crystallization process to describe the luminescent phosphor material [77].

6.3 Chemical sensors based on metal oxide-based phosphors

Nowadays, the design of rare-earth ion or transition metal ion-activated phosphors and their synthesis by different methods are very popular in order to improve the luminescence properties of micro and nanoscale metal oxide-based phosphors. The usage in the lighting industries of the metal oxide-based phosphors as a result of developments in lighting technologies has become inevitable due to their high stability, easy integration into technological devices, and low production costs. Fig. 3 shows the chemical sensor-based applications of the metal oxide phosphors.

6.3.1 Metal oxide-based phosphors for three-band fluorescent lamps

Oxide phosphor materials used for tri-band fluorescent lamps include simple metal oxides such as aluminate, phosphate, and borate. Aluminate-based phosphor materials are the fairly common phosphor materials used in tri-band fluorescent lamps because of improved physical and chemical stability and high emission efficiencies. Metal oxide phosphors, consisting of both oxide and oxyacid salts, are excited at 254 nm UV light.

6.3.2 Metal oxide-based phosphors for plasma display panels (PDPs)

Oxide-based luminescent materials can be excited by vacuum ultraviolet (VUV) light and contain simple oxides such as aluminate, silicate, and borate structures for plasma display panels (PDPs). PDP is a kind of display device which is coated with phosphor and consists of small gas discharge cells of pixels. Green-emitting phosphors as Y_2O_3: Eu^{3+}, (Y,Gd)BO_3:Eu^{3+}; red-emitting phosphor materials like $BaAl_{12}O_{19}$:Mn^{2+}, Zn_2SiO_4:Mn^{2+}, and blue-emitting metal oxide-based phosphor as $BaMgAl_{10}O_{17}$: Eu^{2+} are important because of their high luminescence efficiency, compatibility for full-color PDPs, and resistance to VUV light.

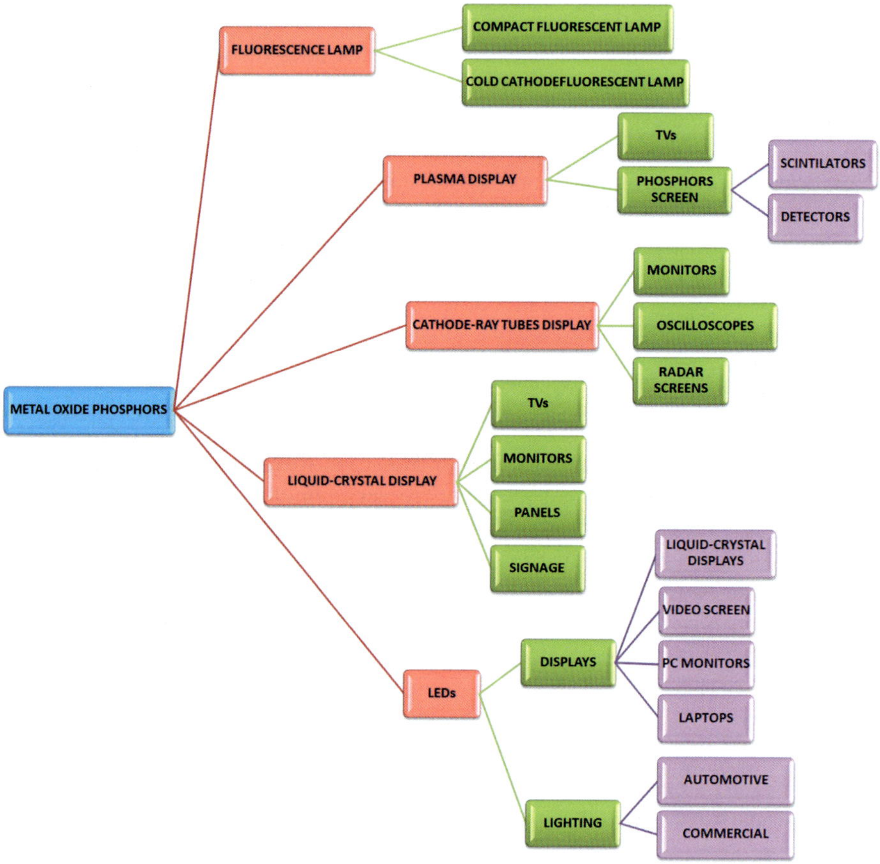

Fig. 3 The various applications range of the metal oxide phosphors.

6.3.3 Metal oxide-based phosphors for white light-emitting diodes (wLEDs)

Metal oxide-based phosphors like silicate and aluminate phosphors are generally preferred for wLEDs, which are generally preferred in lighting due to their superior specifications as energy-saving, environmentally friendly structure, long life, and high efficiency. Today, wLEDs are generally produced by using yellow-emitting phosphors and blue LEDs chips together. Also, the white light thus obtained has a highly correlated color temperature (CCT) and a low color rendering index (CRI) due to the lack of red light. Red-emitting luminescent phosphor materials are added to make up for the deficiency of red emission. In this way, wLEDs produced with the combination of luminescent materials emit blue, green, and red are highly preferred in the production of high-quality white light due to the advantages such as high color strength and superior color rendering.

6.4 Characteristics of phosphors for LEDs applications

6.4.1 Correlated color temperature (CCT)

The CCT, as measured in Kelvin (K), is basically defined as an indication of how blue or yellow, the color of the light emitted from the light source appears. It is most generally found between 2200 and 6500 K. The temperature of an ideal black body radiator that emits light of a similar color to the light source is expressed as the color temperature of the light source. Cool light with a CCT over the 4500 K exhibits bluish-white light; warm white light with a CCT under the 4500 K exhibits a yellowish light [78].

One of the most significant factors determining the efficiency of LEDs is color temperature. Color temperature not only affects luminous efficacy but can also be used in the structure of existing lighting arrangements. LEDs with a high CCT have a higher efficiency than those with a low CCT. To produce white light with LEDs, a semiconductor chip that emits blue light is often used. Some of this light is converted to light by means of a converter mixture or phosphor of longer wavelengths (green, yellow, and red light). Combining all these colors together results in white light. This conversion process involves losses that increase with the wavelength of the converted light. This loss increases as the difference in energy between light at a higher energy level (blue light) and light at a lower energy level (red light) are converted into heat. Minimizing losses requires calling for the exact alignment of the absorber and emission wavelengths of the transducer.

6.4.2 Colorimetry

The color purity and dominant wavelength of a light source are defined with the aid of the International Commission on Illumination (CIE) 1931 diagram. The dominant wavelength calculated for the chromaticity coordinate of the light source is obtained with the help of the equal energy wave of a light source and the line that cuts the chromaticity coordinate. The color purity of a point on the CIE 1931 chart can be figured by Eq. (1).

$$\text{Color purity} = \frac{\sqrt{(x-0.33)^2 + (y-0.33)^2}}{\sqrt{(x_d-0.33)^2 + (y_d-0.33)^2}} \quad (1)$$

where (x, y) and (x_d, y_d) shows the light source and dominant wavelength point chromaticity coordinates, respectively [79].

6.4.3 Color rendering index (CRI)

The CRI describes how colors appear under an artificial light source. In other words, it is the measurement of a light source's ability to faithfully reproduce the colors of objects of interest compared to a reference archetype light source. The CRI index is measured between 0 and 100. Ideally, the CRI value of the light source used for illumination should be above 90. CIE - Color Rendering Index (R_a) is the tool of choice for lighting. True color can never be identified, so the processed chromaticity of the eight tests of reference illuminators with the highest R_a processed under blackbody or daylight radiation is considered essential. The "Test-Color Method" is based on a shift in the color appearance of the chromaticity coordinates when samples of eight relatively saturated test colors are illuminated with test and reference light sources. Fifteen test samples of different colors are taken into account to calculate the CRI value of a lamp. Fifteen test swatches of different colors are taken into account to calculate the CRI of a light bulb. Each test color presents a distinct CRI, represented as R_a ($R_1 - R_{15}$), corresponding to the chromatic difference of each test color sample illuminated by the reference and test lights. The arithmetic averages of the first eight color samples as $R_1 - R_8$ are equal to the test illuminator CRI value, while the other monochrome CRI as $R_9 - R_{15}$ defines the specific CR [80]. As already mentioned regarding color temperature, the composition of the color spectrum based on the selection of a suitable converter has a decisive influence on the efficiency of the LEDs. The converter mix has been developed specifically for different CRIs and is specifically optimized for CRI as well as efficiency. The differences between CRI 70, 80, and 90 are very clearly visible in the rendering of red hues. A high percentage of longwave light is required to reproduce these hues as faithfully as possible.

6.4.4 Quantum efficiency

Besides adjusting the quantum efficiency (QE), the LEDs opto-semiconductor provides a good size or current density that is adjusted. LEDs are typically categorized on the basis of brightness and color for some operating current. In particular grouping conditions, there is a typical efficacy that can be adjusted by the application and the desired level of light efficacy. The semiconductor LEDs performance is generally defined by the external quantum efficiency (EQE), which is specified as the product of its internal quantum efficiency (IQE) multiplied by the extraction efficiency. The first step for LEDs to produce a high luminous flux depends on their high internal quantum efficiency. The internal quantum efficiency (IQE) is the ratio of the number of photons produced in the LEDs to the number of electrons fed into the LEDs. Photon generation appears when an electron in the valence band recombines with an electron in the conduction band. IQE is calculated as following Eq. (2);

$$\eta_{\text{IQE}} = \frac{\int P(\lambda)d\lambda}{\int \{E(\lambda) - R(\lambda)\}d\lambda} \tag{2}$$

The light output power of the LEDs is proportional to the EQE, that is, the ratio of the number of photons emitted from the device to the number of fed electrons. The external quantum efficiency is defined by three components. The EQE is the ratio of the optical power measured outside the LEDs to the total electrical power consumed. EQE is calculated as following Eq. (3);

$$\eta_{\mathrm{EQE}} = \frac{\int P(\lambda)d\lambda}{\int E(\lambda)d\lambda} \tag{3}$$

where η is luminous efficiency, $E(\lambda)/h\nu$-excitation, $R(\lambda)/h\nu$-reflectance or transmittance, and $P(\lambda)/h\nu$-phosphor emission spectrum represent the number of photons. Also, in the field of phosphor research, IQE can also be called quantum yield (QY), with a similar definition of the ratio of the number of emitted photons to the number of photons absorbed [81].

6.4.5 Factors affecting of LEDs efficiency

The LEDs performance depends on various aspects like the emission wavelength, the LEDs chip optical power based on EQE, the emission spectrum overlap with the photopic regime curve, and wall plug efficiency (WPE) values, conversion efficiency, and phosphor absorption. To be used in LEDs chip applications, the phosphor must have the following properties.

1. Ability to provide robust absorption between the 360 and 460 nm.
2. The lifetime value of the activator center should have a short decay time of 10^{-9}–10^{-6} s for blue light-emitting LEDs.
3. Phosphors conversion efficiency, expressed by IQE and external EQE, is as high as possible.
4. In order to provide the best light efficiency and color rendering, the full width at half maximum (FWHM) value of the emission spectra should be close to 50 nm.
5. Temperature independent emission-based spectrum, IQE, and emission state lifetime as LEDs chips can reach high temperatures of up to 150°C.
6. Emission spectrum should provide ideal white light for daylight (CCT $\sim$ 6500 K) and night lighting (CCT $<$ 3500 K).
7. Sub-micro-sized powders to reduce light scattering.
8. Low cost, simplicity, and efficiency.
9. Reliability, photostability, and chemical stability.
10. Environmentally friendly composition.

The following sections provide information on the various classes of phosphors used in LEDs applications, including metal oxide-based phosphors and phosphors activated by lanthanide ions or transition metal. Transition metal-doped phosphors and lanthanide and/or rare-earth (RE) ions come to the fore due to their thermal stability and not containing too many toxic metals in their compositions. RE-doped phosphors have gained a lot of concern in recent years due to the necessity of phosphors showing superior material properties with improved luminescence properties in solid-state lighting applications. The addition of even a small amount of RE ions to the host matrix can significantly alter the photoluminescence properties of the phosphor material. Although RE ions are categorized by the partially filled 4f shell surrounded by 5s2 and 5p6 electrons, they are also considered as luminescence centers in various inorganic host lattices. By the way, these ions are integrated to produce highly quantum efficient phosphors used in solid-state lighting applications [82,83].

7 Types of metal oxide-based phosphors

In this part of the chapter, the luminous properties of aluminate, zincate, phosphate, silicate, borate, gallate, etc. metal oxide-based phosphors in structures, which have attracted a lot of attention in the literature in recent years, either commercially or as a synthesis product, and chemical sensor applications are presented (see Fig. 4). In order to improve the performance of the relevant metal oxide-based phosphors, detailed information is given about the optimization of the compositions and ratios of the additives, the surface modifications made during or after the synthesis, and the improvements made in the synthesis methods.

7.1 Aluminate-based phosphors

In recent years, the focus has been on rare-earth ion-doped aluminates with more complex structures than alkaline earth sulfide hosts [84,85]. Lanthanide-based or doped inorganic aluminate phosphors are used in various applications such as wLEDs [84], lasers [85], bioimaging [86], luminescent solar concentrator [87], solar cell [88], and optical temperature sensors [89]. Table 1 shows the emission color of certain aluminate-based phosphors.

Among aluminates, $BaAl_2O_4$ phosphors have been reported as the most suitable host cage because it is a material with good pyroelectric and dielectric properties, a large solid-soluble region, and a high melting point [100,101]. $ZnAl_2O_4$ phosphor, which is known to have good optical and wide bandgap semiconductor properties, is widely

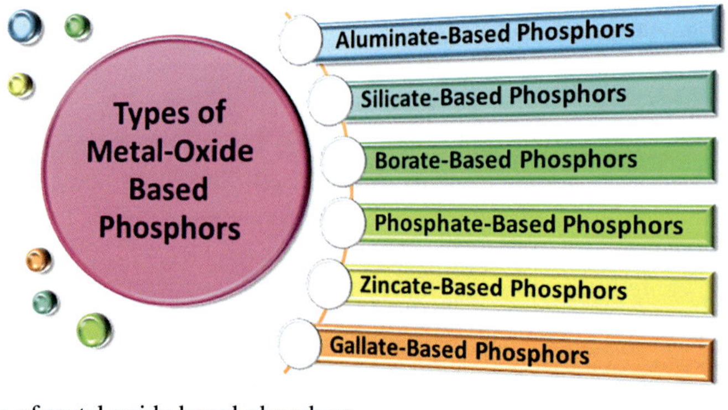

Fig. 4 Types of metal oxide based phosphors.

Table 1 The emission color of certain aluminate-based phosphors.

Aluminate phosphors	Color	λ_{ex} (nm)	λ_{em} (nm)	Reference
$CaAl_2O_4:Eu^{2+},Nd^{3+}$	Blue	250, 296	440	[90]
$SrAl_2O_4$: Eu^{2+} ,Dy^{3+}	Green	320, 370	525	[91]
$BaAl_2O_4:Eu^{2+},Dy^3$	Blue-green	339	499	[92]
$ZnAl_2O_4:Dy^{3+}$	Blue, Yellow	225, 245	480, 575	[93]
$BaMgAl_{10}O_{17}:Eu^{2+}$	Green	147	515	[94]
$BaAl_{12}O_{19}:Mn^{2+}$	Blue	270, 328, 380	450	[95]
$Y_3Al_5O_{12}:Ce^{3+}$	Yellow	465	575	[96]
$CaYHf_2Al_3O_{12}:Ce^{3+},Tb^{3+}$	Green	408	543	[97]
$Ca_2TbZr_2Al_3O_{12}$: Ce^{3+}	Green	395	543	[98]
$CaAl_2O_4:Mn^{4+}$	Red	325	658	[99]

preferred as a host material in thin-film electroluminescent displays, mechanical-optical sensors, and applications in imaging devices [102–107].

Recently, RE-doped $ZnAl_2O_4$ phosphors have attracted great interest due to their enhanced optical, electrical, and wide bandgap properties. Eu^{3+}-doped $ZnAl_2O_4$ phosphors were suggested as a candidate by Satapathy et al. [108] because of their linear dose-response in dosimetric use. It has been reported by Araujo et al. [109] that $ZnAl_{1.9}Eu_{0.05}O_4$ phosphors can be used as a luminescent marker in applications in the field of biomaterials. Kumar et al. [93] reported that Dy^{3+} doped $ZnAl_2O_4$ phosphor has a potential application in wLEDs, since it can emit yellow light under UV excitation. It has been stated that Er^{3+}/Yb^{3+} doped $ZnAl_2O_4$ phosphors function as upconversion (UC) material with high efficiency [110,111].

Plasma display panels contain $BaAl_{12}O_{19}:Mn^{2+}$ green-emitting and $BaMgAl_{10}O_{17}$: Eu^{2+} blue-emitting aluminate phosphors. $BaAl_{12}O_{19}:Mn^{2+}$ phosphor, which exhibits an intense green light emission, is considered a promising green phosphor in PDPs and lamps due to its brightness, colorimetric purity, good stability, and high efficiency under VUV stimulation. The broad emission band of $BaAl_{12}O_{19}:Mn^{2+}$ phosphors observed at 515 nm is caused by the transition of Mn^{2+} ions $3d^5$ ($^4T_{1g}$)-$3d^5$ ($^6A_{1g}$) [112].

The excitation spectrum of blue-emitting $BaMgAl_{10}O_{17}:Eu^{2+}/Mn^{2+}$ phosphor has two broad bands in the UV region in the range of 210–410 nm and plays a very significant role as a fluorescent structure for ultraviolet light-emitting diodes (UV-LEDs). wLEDs have been produced by coating the synthesized $BaMgAl_{10}O_{17}:Eu^{2+}/Mn^{2+}$ phosphors with the addition of AlF_3 and BaF_2 on a UV-LEDs chip that offers a high CRI value (R_a = 95) by Ke et al. [113].

$MAl_2O_4:Eu^{2+}$ (M = Sr, Ca, Ba) phosphors, which are preferred in warning and emergency signs, lighting, radiation sensors, fluorescent lamps, LEDs, and new generation display applications, have attracted great interest recently due to their strong photoluminescence properties in the visible region and chemical stabilities [114–119].

In Ce^{3+} and Dy^{3+} doped $BaAl_2O_4:Eu^{2+}$ phosphors, rare-earth ions cause a decrease in the number of traps due to the reduction in the number of oxygen vacancies around Eu^{2+} ions during the co-doping process. As a result of reducing the number of oxygen vacancies, the luminescence efficiency of Eu^{2+}-doped phosphors can be increased. As a result, Eu^{2+} and Dy^{3+} co-doped $BaAl_2O_4$ phosphors can exhibit good photoluminescence properties such as high luminescence intensity, stable chemical stability, favorable color diffusion, and long-term persistence [120–126].

It has been reported by Luo et al. that the dye-sensitized solar cell (DSSC) light-collection efficiency increased as a result of the increase in the power conversion efficiency and short-circuit current density of the blue light-emitting $CaAl_2O_4:Eu^{2+}$, Nd^{3+} phosphor by adding $\leq 5\%$ by weight phosphor particles to the TiO_2 matrix [127].

The CIE chromaticity coordinates of $CaAl_2O_4:Mn^{4+}$ phosphors, which gave an emission maxima approximately at 658 nm due to the $^2E \rightarrow {}^4A_2$ transition of the Mn^{2+} ion, were found to be (0.7181, 0.2813). These properties show that Mn^{4+}-doped $CaAl_2O_4$ phosphor can be used as a high-efficiency material for wLEDs [99].

As a result of efficient energy transfer from Ce^{3+} to Tb^{3+} further to $^5D_4$-$^7F_5$ transitions, $Ca_2TbZr_2Al_3O_{12}:Ce^{3+}$ phosphor, which emits green, shows a 9.74 nm narrow FWHM value and high IQE (QY = 60%). The wLEDs performance using $Ca_2TbZr_2Al_3O_{12}:Ce^{3+}$ phosphor shows a high CRI value of 90.1%, a low CCT of 3686 K, and a wide color gamut of 61% of the National Television System Committee standard (NTSC) at the CIE. By the way, the CIE 1931 color space offers great potential for violet-based wLEDs [98].

After the excitation of Eu^{2+} from $^4f_7$ to $^4f_6^5d_1$, a trap level released by Eu^{2+} is captured by Dy^{3+} and as a result of escaping of electrons from the trap level, Eu^{2+}/Dy^{3+}-doped strontium aluminate ($SrAl_2O_4$) phosphorescent crystals show high afterglow and brightness properties [128].

Rare-earth doped blue-emitted $SrAl_2O_4:Eu^{2+}$,Dy^{3+} and green-emitted $BaAl_2O_4$: Eu^{2+},Dy^{3+} phosphors are used in many applied fields such as in vivo medical imaging, home decoration, and emergency signs because of their optical properties as long lifetime, afterglow, and photo-resistance properties [129–131].

Yellow-emitting rare-earth ion-doped yttrium aluminium garnet ($Y_3Al_5O_{12}$) phosphors have gained increasing attention. As they have high luminescence efficiencies and excellent thermal stability, their usage in high-power LEDs has been limited. Xia and co-workers reported that the novel synthesized $Y_3Al_5O_{12}:Ce^{3+}$- embedded $CaBi_2B_4O_{10}$ glass-ceramics, encourage applicants for using in warm and high-power wLEDs because of the optimal optical parameters, such as the luminous efficiency of 105.3 lm/W, correlated color temperature of 3940 K, and color rendering index of 70.1 [96].

Chen et al. reported the C_3N_4-coated $Y_3Al_5O_{12}$ phosphor as a white luminescence device with the values of bright white light with chromatic coordinates (0.368, 0.337), high color rendering index (R_a = 89), and a color temperature of 4056 K [132].

7.2 Silicate-based phosphors

The silicate-based phosphor materials used for applications of wLEDs refer to cerium (Ce^{3+}) activated garnet type silicates and europium (Eu^{2+}) activated

Table 2 The emission color of certain silicate-based phosphors.

Silicate phosphors	Color	λ_{ex} (nm)	λ_{em} (nm)	Reference
$Ca_3Sc_2Si_3O_{12}:Ce^{3+}$	Green	450	502	[133]
$Sr_3SiO_5:Eu^{2+}$	Yellow-orange	395	560	[134]
$CaMgSi_2O_6:Eu^{2+},Mn^{2+}$	Red	453	680	[135]
$CaMgSi_2O_6:Eu^{2+}$	Blue	365	446	[136]
$Y_2SiO_5:Sm^{3+}$	Orange-red	401	602	[137]
$Ba_2CaZn_2Si_6O_{17}:Sm^{3+}$	Orange-red	405	600	[138]
$Zn_2SiO_4:Mn^{2+}$	Green	420	525	[139]
$Ca_3SiO_4Cl_2:Eu^{2+}, Mn^{2+}$	Orange-green	395	512, 570	[140]
$Lu_2SrAl_4SiO_{12}:Ce^{3+}$	Green	448	514	[141]
$RbLi(Li_3SiO_4)_2:Eu^{2+}$	Green	460	530	[142]

metasilicates and orthosilicates. Table 2 shows the emission color of certain silicate-based phosphors.

In the studies, it has been proved that oxyorthosilicate phosphors are activated with cerium ions such as $Y_2SiO_5:Ce^{3+}$, $Gd_2SiO_5:Ce^{3+}$, and $Lu_2SiO_5:Ce^{3+}$ are very good scintillation phosphors because of their properties like fast decay time, high luminescence intensity, and good light output [143]. Europium-activated silicate phosphors for PDPs refer to $CaMgSi_2O_6:Eu^{2+}$, $Sr_3SiO_5:Eu^{2+}$, and $Sr_2SiO_4:Eu^{2+}$ phosphors and are still preferred due to their high luminous efficiencies, cheap cost raw materials, and low synthesis temperature. Recently, in their study, Shen and Kacharu [144] proposed Y_2SiO_5: Eu^{2+} phosphor material for practical developments and introduced a new process for parallel data storage with the use of coherent time domain optical memory. The photo-stimulated luminescence (PSL) in silicate-based materials can be used in the development of X-radiation (X-ray) storage phosphors. In literature studies, it has been shown that $Ba_5SiO_4Br_6:Eu^{2+}$ [145] and $Y_2SiO_5:Ce^{3+},Sm^{3+}$ [146] phosphors are highly-efficient X-ray storage phosphors due to their characteristic features.

In addition, $Bi_{12}SiO_{20}$ single crystals with photorefractive properties are of increasing interest in the field of optoelectronics. These optoelectronic fields can be specified as holographic information storage, spatio-time light modulators, and many other application areas in nonlinear optics [147].

$Ca_3Sc_2Si_3O_{12}:Ce^{3+}$ phosphor, which is produced by the sol-gel technique, can be used in the production of monochromatic classic LEDs and high-power LEDs, as it absorbs blue light and converts it to green light [133].

The energy transfer resulting from the dipole-dipole interaction is due to the $^4f_6^5d_1 \rightarrow {}^4f_7$ transitions of Eu^{2+} in two different crystallographic regions in Sr_3SiO_5 and exhibits an orange-yellow emission under blue light excitation. Due to this feature, it has been thought that it can be used as potential materials for applications in the development of Near Ultraviolet White Light-Emitting Diodes (NUV-wLEDs) [134].

Red-emitting $CaMgSi_2O_6:Eu^{2+}$, Mn^{2+} phosphor materials have been used to enhance the color rendering properties of remote-packaging white LEDs (RP-wLEDs) due to their higher luminous efficiency and better color uniformity [135].

$CaMgSi_2O_6:Eu^{2+}$ shows a strong blue emission at 450 nm when excited under the NUV region of the electromagnetic spectrum. $CaMgSi_2O_6:Eu^{2+}$ is often used for plasma display panels because of its low synthesis temperature, high thermal stability, and good optical properties [136].

Sm^{3+} -doped Y_2SiO_5 phosphors have three emission peaks in the orange-red region at ~570, 602, and 656 nm, which are attributed to $^4G_{5/2} \rightarrow {}^6H_J$ (J = 5/2, 7/2, and 9/2) transitions in Sm^{3+} ions. This produced phosphors has excellent luminescence and wLEDs are used in important areas such as dosimetry and fingerprint applications in forensic science [137].

$Ba_2CaZn_2Si_6O_{17}:Sm^{3+}$ phosphors with four emission bands arise at 562, 600, 647, and 708 nm corresponding to $^4G_{5/2} \rightarrow {}^6H_J$ (J = 5/2, 7/2, 9/2, and 11) transitions of Sm^{3+}, respectively are seen as the matrix that shows great potential for pc-wLED applications that can be pumped with a near-UV LED chip [138].

$(Lu_{0.2}Gd_{0.78}Ce_{0.01}Sm_{0.01})_2SiO_5$ and $(Lu_{0.2}Gd_{0.78}Ce_{0.01}Dy_{0.01})_2SiO_5$ contributing to the excitation spectra of the Sm^{3+} and Dy^{3+} luminescence of the band associated with the 4f-5d transition of Ce^{3+} to Ce^{3+}-Sm^{3+} and Ce^{3+}-Dy^{3+} enable energy transfers to occur. Thanks to the broadband luminescence from 400 to 650 nm of Ce^{3+}-doped Lu_2SiO_5 and Gd_2SiO_5 hosts, these phosphors are considered as a potential matrix for the design of UV-LEDs-excited light sources [148].

7.3 Borate-based phosphors

Rare-earth borates and oxyborates are host materials that have found wide application areas for plasma display panel (PDP) phosphor materials becuase of their large bandgap, excellent transparency in the vacuum ultraviolet (VUV) region, and high stability. Since most RE borates are optically transparent down to 140–180 nm, VUV light can directly activate the impurity activator in these hosts. Its advantages are high photoluminescence efficiency and superior stability, while its disadvantages are relatively nonadjustable color coordinates and low color purity. In general, Gd^{3+}, Eu^{3+}, and Tb^{3+} ions are utilized as sensitizing ions to increase the excitation of VUV light and luminescence efficiency. Table 3 shows the emission color of certain borate - based phosphors.

As a result of the studies, it is known that borate materials such as lithium triborate (LiB_3O_5) [159], barium metaborate (BaB_2O_4) [160], cesium lithium borate ($CsLiB_6O_{10}$) [161], and potassium aluminum borate ($K_2Al_2B_2O_7$) [162] have excellent nonlinear optical crystalline properties. Due to their excellent optical properties, rare earth-doped gadolinium calcium oxoborate ($Ca_4GdO(BO_3)_3$) [163–166] and yttrium aluminum borate ($YAl_3(BO_3)_4$) [167] compounds are a very good candidates for a new solid-state laser materials and has great attention in many applications. However, while lithium borate ($Li_2B_4O_7$) is used for VUV laser applications [168], Cu^{2+}-doped lithium borate ($Li_2B_4O_7:Cu$) [169] and Dy^{3+}-doped magnesium borate ($MgB_4O_7:Dy$) [170] materials are also used in commercial dosimeter systems.

Eu^{2+}-doped alkaline haloborates ($M_2B_5O_9X:Eu^{2+}$ (M:Ca, Sr, Ba; X:Cl, Br)) [171] and strontium borophosphate ($SrBPO_5$) [172] compounds can be used for the neutron

Table 3 The emission color of certain borate-based phosphors.

Borate phosphor	Color	λ_{ex} (nm)	λ_{em} (nm)	Reference
$GdBO_3:Ce^{3+},Tb^{3+},Eu^{3+}$	Red	356	592, 612, 624	[149]
$Ba_2Ca(BO_3)_2:Ce^{3+}/Mn^{2+}$	Blue/red	345/410	420/630	[150]
$Ca_3Y(GaO)_3(BO_3)_4$: Ce^{3+}, Tb^{3+},Mn^{2+}	White	365	409, 488, 544, 581, 620	[151]
$K_2Al_2B_2O_7:Eu^{2+}$	Blue	325	450	[152]
$Ca_4GdO(BO_3)_3$: Nd^{3+}	Red	530	812	[153]
$YAl_3(BO_3)_4:Nd^{3+}$	Infra-red	592	1062, 1337	[154]
$LiCaBO_3:Ce^{3+},Mn^{2+}$	Yellow-orange	407	572	[155]
$Sr_3B_2O_6:Ce^{3+},Eu^{2+}$	Blue-yellow	351	434, 574	[156]
$SrMgB_6O_{11}:Eu^{2+},Mn^{2+}$	Blue-red	380	460, 630	[157]
$NaMgBO_3:Ce^{3+}$	Bluish green	370	480, 520	[158]

radiography and X-ray imaging applications by using photo-stimulated luminescence (PSL), respectively. Lanthanum pentaborate phosphors are used in lamp applications [173].

Eu^{2+}-doped strontium tetraborate (SrB_4O_7) phosphor finds application in commercial sun tanning lamps [174].

Tb^{3+}-doped indium orthoborate ($InBO_3:Tb^{3+}$) cathodoluminescent phosphor material is used both as a green-emitting phosphor in color projection TV and in neutrino detection applications [175]. However, Tb^{3+}-doped (Y, Gd)BO_3 is used as green phosphors in PDP applications [176].

For the wLED application, Dy^{3+}/Pr^{3+} doped lithium borate glasses with emission very close to white light have been studied [177].

Lithium Lead Alumino Borate (LiPbAlB) glasses with 0.5 mol% Dy^{3+} doped have been reported to show the best values of CCT temperatures, CIE coordinates, quantum efficiency, emission cross sections, and confocal images so that they can be used for wLEDs and lasers [178].

Two distinct emission peaks in the photoluminescence spectra of Dy^{3+} ions-doped magnesium borate glasses indicated that they could donate to the development of photonic devices and the use of solid-state laser applications [179].

Eu^{3+} doped-Na_2O-CaO-P_2O_5-B_2O_3-ZrO_2 glass systems have been proposed by Yao et al as advantageous materials for solid-state lighting, wLED, and display device applications [180].

Rico et al. reported that a continuous-wave efficient simultaneous oscillation, which can be performed in the two ${}^4F_{3/2} \rightarrow {}^4I_{11/2}$ and ${}^4F_{3/2} \rightarrow {}^4I_{3/2}$ infrared (IR) laser channels, at 1062 and 1338 nm in $YAl_3(BO_3)_4$: Nd^{3+} and they observed 592 nm laser light generation Type I self-sum-frequency mixing (SSFM) for yellow laser light generation [154].

When excited at 325 nm, a novel $K_2Al_2B_2O_7:Eu^{2+}$ phosphor exhibited a blue emission band centered at 450 nm, which can be assigned to the $4f_65d_1 \rightarrow 4f_7Eu^{2+}$ ions transitions, and it was suggested as a potential near-UV blue-emitting phosphor for wLEDs applications by Xiao and co-workers [152].

Ce^{3+}/Eu^{2+}-codoped $Sr_3B_2O_6$ exhibited two broad asymmetric emission bands centering at 434 and 574 nm under UV of 351 and 377 nm excitation, which is assigned to the $^5d_1 \rightarrow ^4f_1$ Ce^{3+} transition and the typical $4f_65d_1 \rightarrow 4f_7(8S_{7/2})$ transition of Eu^{2+}, respectively. The energy transfer from Ce^{3+} to Eu^{2+} in $Sr_3B_2O_6:Ce^{3+}$, Eu^{2+} caused the tunable color hues from blue through white and finally to yellow-orange [156].

The luminescence of $SrMgB_6O_{11}:Eu^{2+}$, Mn^{2+} shows emission bands as a broad blue at 420 nm and a red at 650 nm from Eu^{2+} and Mn^{2+}, respectively. It has been observed that the energy transfer from Eu^{2+} to Mn^{2+} is very effective in $SrMgB_6O_{11}$. The optimal composition of $SrMgB_6O_{11}:0.02Eu^{2+}$, $0.12Mn^{2+}$ produces white light, indicating that the resulting glass would be a good material for wLEDs applications [157].

7.4 Phosphate-based phosphors

Phosphate-based phosphor materials are significant photoluminescence hosts, have excitation in the VUV region between 100 and 200 nm, high brightness, excellent thermal stability, good stability, glow after a long time, as well as low-temperature synthesis, cheap cost. Table 4 shows the emission color of certain phosphate-based phosphor materials.

Table 4 The emission color of certain phosphate-based phosphors.

Phosphate phosphors	Color	λ_{ex} (nm)	λ_{em} (nm)	Reference
$(SrCa)_2La(PO_4)_3O:Eu^{2+}$	Blue	365	460	[181]
$Ca_9Y(PO_4)_7:Eu^{2+},Mn^{2+}$	Blue-orange	365	486, 638	[182]
$Ba_3Ce(PO_4)_3:Tb^{3+}/Mn^{2+}/Sm^{3+}$	Green/red/red	377/387/387	551/606/606	[183]
$Na_3Sc_2(PO_4)_3:Eu^{2+}/Tb^{3+}/Mn^{2+}$	Blue/green/red	360	458, 541, 610	[184]
$Ca_{9.75}Mg_{0.75}(PO_4)_7:Eu^{2+}$	Blue-red	365	418, 465, 630	[185]
$Ca_9Gd(PO_4)_7:Eu^{2+}$, Mn^{2+}	Green-red	380	490, 652	[186]
$Ca_9LiY_{0.667}(PO_4)_7:Eu^{2+}$, Mn^{2+}	Green-red	365	490, 655	[187]
$Na_3Al_2(PO_4)_3:Ce^{3+}/Eu^{3+}/Mn^{2+}$	Green-red	328	515, 615	[188]
$K_3Al_2(PO_4)_3$: Dy^{3+}/Eu^{3+}	Blue, green, yellow	351, 385	484, 576	[189]
$NaCaPO_4:Eu^{2+}$, Tb^{3+}	Green	350	504	[190]

Phosphate compounds are remarkable materials as host in luminescent particles because of their stable structure, inexpensive raw material needs, and easy synthesis. In particular, for tri-band lamps (La, Ce, Tb)PO_4 green-emitting phosphate phosphor can be used [191]. For tri-band lamps, $Sr_5(PO_4)_3Cl:Eu^{2+}$ blue-emitting phosphor is used to obtain bright white light [192].

Phosphate phosphors get great attention from researchers due to their wide usage in display and lighting application areas. Solid-state lighting, which utilize LEDs and phosphor to obtain white light, has many advantages like less power consumption, more efficient light performance, and a long lifetime. It is highly preferred in the lighting industry due to its tricolor phosphate phosphors, which may be excited by NUV and UV light, have gained great interest because of important applications in solid-state wLEDs. Orthophosphate materials that have the chemical formula as $A_IB_{II}PO_4$ (A_I: monovalent cation, B_{II}: divalent cation) are receiving increasing attention as luminescence hosts.

The morphology, composition, thermal stability, bandgap of the host material, photoluminescence properties, and CIE chromaticity index of mixed orthophosphates including $KBaPO_4:Eu^{3+}$, $LiSrPO_4:Tb^{3+}$, and $LiSrPO_4:Sm^{3+}$ were investigated by Lin and co-workers for the development of appropriate host properties for application in wLEDs [193].

Two strong emission peaks observed in the PL emission spectrum of Dy^{3+} doped $X_6AlP_5O_{20}$ (X = Sr, Ba, Ca, and Mg) phosphors when excited near UV light at 350 nm suggest that these materials are potential candidates for UV-excited LEDs [189].

Considering the brightness and post-radiation decay time (6–9 ms) caused by the green-emitting $YPO_4:Tb^{3+}$ phosphor, it is preferred in display applications, especially when used with a blue absorber filter in televisions, since it has a single peak at 543 nm [194].

When excited at 325 nm, green-emitting Ce^{3+}/Tb^{3+} co-doped $Ca_3Y(PO_4)_3$ phosphors has emission bands as blue arising from the 5d-4f Ce^{3+} transition and green originating from the $^5D_4$-^7F_J Tb^{3+} transition. Thanks to the high efficiency energy transfer between $Ce^{3+} \rightarrow Tb^{3+}$ ions, it was predicted that this phosphor could be used in wLEDs [195].

Research on materials doped with lanthanide ions in two different oxidation states is important for obtaining wLEDs. When doping a matrix with Eu^{2+} and Eu^{3+} at the same time, phosphors can emit white light between 400 and 720 nm, covering almost the entire visible region, due to the emission of phosphors, the blue-green luminescence of Eu^{2+}, and the red emission of Eu^{3+}. After reduction, an intense broad band attributed to d-f transitions in Eu^{2+} and a narrow band at 619 nm attributed to Eu^{3+} ion were observed in the emission spectrum of europium and silicon ions-doped sodium calcium phosphate phosphors by Górecka et al. [196].

Meanwhile, due to its good color purity and unique thermal resistance, the potential blue-emitting $KBaPO_4$ phosphor Eu^{2+} was activated and tested for thermal stability and moisture resistance for plasma display panels, cold cathode fluorescent lamps (CCFLs), and phosphor-converted white light-emitting diodes (pc-wLEDs) [197].

Various activator ions like Eu^{3+}, Ce^{3+}, Tb^{3+}, and Pr^{3+} can be added to halophosphate phosphors, which have formulas such as $Ca_5(PO_4)_3(F, Cl)$. Added activator ions increased the effectiveness of the phosphors color centers and lumen maintenance in halophosphate phosphors [198].

The fact that Eu^{2+}/Mn^{2+} doped $Ca_5(PO_4)_3Cl$ phosphor can be excited efficiently with NUV light as a result of energy transfer between Eu^{2+} and Mn^{2+} shows that this phosphor is a potential material for LEDs applications [199].

The photoluminescent properties of Dy^{3+}/Li^+ co-doped $KSrPO_4$ phosphors were researched under NUV excitation, and a significant increase in photoluminescence of $KSrPO_4$:Dy^{3+} was observed with the addition of Li^+ ions. This increase in photoluminescence was attributed to improved crystallinity and charge compensation by Gao et al. Temperature dependence of photoluminescence for $KSrPO_4$:0.01Dy^{3+}, 0.06Li^+ showed that it can be applied for high power LEDs [200]. $Ca_5(PO_4)_3F$:Dy^{3+}, Eu^{3+} phosphor when excited under near ultraviolet excitation, blue (486 nm) and yellow (579 nm) emissions corresponding to the $^4F_{9/2} \rightarrow {}^6H_{15/2}$ and $^4F_{9/2} \rightarrow {}^6H_{13/2}$ transitions of the Dy^{3+} ion and the red emission (631 nm) of Eu^{3+} ion were observed. These results show that $Ca_5(PO_4)_3F$:Dy^{3+}, Eu^{3+} phosphor materials have potential applications as a single component phosphor in warm wLEDs [201].

When Dy^{3+}-doped $NaCa_3Bi(PO_4)_3F$ phosphors excited at 349 nm, three emission peaks were observed at 480 nm ($^4F_{9/2}$-$^6H_{15/2}$), 577 nm ($^4F_{9/2}$-$^6H_{13/2}$), and 662 nm ($^4F_{9/2}$-$^6H_{11/2}$). Phosphate phosphors is a yellow-emitting fabricated apatite type, revealing its potential due to its high activation energy ($E_a = 0.32$ eV), excellent thermal stability, well CCT of 5083 K, high CRI ($R_a = 92$). Its applicability for wLEDs was confirmed as yellow-emitting $NaCa_3Bi(PO_4)_3F$:Dy^{3+} fluorophosphate phosphor [202].

In the emission spectrum of $Na_3Al_2(PO_4)_3$:Ce^{3+} phosphor at 350 nm, a band corresponding to $^2D_{3/2} \rightarrow {}^2F_{5/2}$ and $^2D_{3/2} \rightarrow {}^2F_{7/2}$ transitions is observed, which indicates that the relevant phosphor can be used in scintillators. Also, the PL emission spectrum of $Na_3Al_2(PO_4)_3$:Eu^{3+} contains two emission bands at 615 and 593 nm, attributable to the Eu^{3+} transitions ($^5D_0 \rightarrow {}^7F_2$ and $^5D_0 \rightarrow {}^7F_1$). $Na_3Al_2(PO_4)_3$:Mn^{2+} phosphor materials can be shifted to orange-red from the green emission peak (515 nm) because of the $^4T_1 \rightarrow {}^6A_1$ transition of the Mn^{2+} ion. $Na_3A_{12}(PO_4)_3$:Eu^{3+} and $Na_3Al_2(PO_4)_3$:Mn^{2+} can be utlized in fluorescent mercury lamps, PDP, and solid-state lighting devices [188].

7.5 Zincate-based phosphors

Metal oxide-based zincate phosphors are materials that attract more attention than other types of phosphors due to their easy synthesis and their structures suitable for doping with various rare-earth ions. Zinc-based metal oxide phosphors have optical, magnetic, superconducting, dielectric properties as well as excellent thermal, chemical, and physical stability [203–205].

Because of these properties, zincate phosphors have a broad range of applications, including field emission displays (FEDs), thermographic sensors, plasma display devices, electroluminescent displays (ELs), wLEDs, solar cells, cathode ray tubes

Table 5 The emission color of certain zincate-based phosphors.

Zincate phosphors	Color	λ_{ex} (nm)	λ_{em} (nm)	Reference
$Ca_3Al_4ZnO_{10}:Sm^{3+}$	Orange	401	601	[208]
$BaYAlZn_3O_7:Dy^{3+}$	Blue-orange	355	485, 579	[209]
$Sr_{0.85}Ca_{0.15}ZnO_2:Eu^{3+}$	Red	374	614	[210]
$SrZnO_2:Sm^{3+}$	Red	380, 413	607, 655	[211]
$BaYAlZn_3O_7:Sm^{3+}$	Red	413	708	[212]
$BaLa_2ZnO_5:Sm^{3+}$	Red	413	708	[213]
$CaZnO_5:Mn^{2+}$	Red	530	615	[214]
$BaLa_2ZnO_5:Dy^{3+}$	Blueyellow	355	485, 578	[215]
$BaLa_2ZnO_5:Tb^{3+}$	Green	278	544	[216]
$BaGd_2ZnO_5:Er^{3+}$	Green-red	453	530, 553, 667	[217]

(CRTs), and high performance imaging devices in the biomedical fields [206,207]. Table 5 shows the emission color of certain zincate-based phosphors.

$Ca_3Al_4ZnO_{10}:Sm^{3+}$ phosphor synthesized in the solid state reaction acts as an efficient orange emitting phosphor for display, white light emitting, and solid state lighting devices in terms of high emission intensity due to $^4G_{5/2}\rightarrow{}^6H_{7/2}$ transitions at 601 nm, color-purity and appropriate color temperature [208].

Dy^{3+}-activated $BaYAlZn_3O_7$ phosphor has been considered suitable for cool wLEDs, digital signage, horticulture, bio-imaging, solar cells, and lasers due to its crystalline properties, color brightness, wide bandgap, and favorable colorimetric parameters [209].

A broad emission band and sensitivity characteristics of the host cage can be observed at $Sr_{0.85}Ca_{0.15}ZnO_2$:0.02 Eu^{3+} phosphors, indicating that energy transfer from the host to Eu^{3+} occurs, resulting in tunable red-emitting. The noval phosphor with 90.05% CP may be a poteantial material as a red-emitting phosphor for wLEDs [210].

$SrZnO_2:Sm^{3+}$ phosphor, a promising candidate for wLEDs, is excited in a wide region in the range of 350–420 nm and shows a red emission band between 600 and 730 nm [211].

$BaYAlZn_3O_7:Sm^{3+}$ phosphors, which have hexagonal phase crystallographic properties, high photoluminescence brightness, wide bandgap, decay time (2.4876 $\pm$0.0311 ms), intrinsic lifetime (3.4380 ms), quantum efficiency (72.36%), and relaxation rate (0.1111 ms^{-1}) may be used as a potential material in ultraviolet-triggered wLEDs, digital signages, lasers, bioimaging, solar cells, and security labels with low CCT values [212].

Quadrilateral type $BaLa_2ZnO_5:Sm^{3+}$ powder produced as a phosphor material with a dark orange-red luminosity subjected to the $^4G_{5/2} \rightarrow {}^6H_{11/2}$ transition has a bandgap energy of 4.34 eV, the quantum efficiency of 86.13%, an intrinsic lifetime of 1.0242 ms, color purity of 71.02%, and a high-performance hot-wLEDs due to having CIE color coordinates (0.59, 0.42) have been noted as a promising possibility for applications in signage, security labels, bioimaging, sensors, solar cells, and lasers [213].

Green-emitting $BaGd_{1.94}Er_{0.06}ZnO_5$ phosphors are suitable for producing green light for the application of blue/green/red (B/G/R) white light-emitting diodes (wLED), cold in X-ray imaging, digital signages, solar cells, data transmission, sensors, and laser because of having good color purity (87.86%), color temperature (6916K), energy bandgaps (Eg = 3.21 and 3.77 eV), and suitable chromaticity coordinates (0.2531, 0.6106), NTSC (0.30, 0.60) [218].

Under 355 nm excitation, $BaLa_2ZnO_5:Dy^{3+}$ nanomaterials showed white luminosity due to two bands showing blue at 485 nm and yellow at 578 nm. $BaLa_2ZnO_5:Dy^{3+}$ phosphors can be used in photovoltaic cells, high-performance cold-wLEDs, indoor cultivation, lasers, and sensors due to their high quantum efficiency (74.95%), favorable color temperature (9290K), and high purity white light coordinates (0.2693, 0.3278) [215].

Zhang et al. investigated the ultraviolet luminescence of $BaGd_2ZnO_5:Er^{3+}$/YAG (8/2 ratio) phosphors and recommended it for interior lighting use due to its excellent white light performance, UV intensity (84.55 mW/cm^2), and color temperature (5977K) [217]. $BaLa_2ZnO_5:Tb^{3+}$ phosphors with an emission band at 544 nm, corresponding to the $^5D_4 \rightarrow {}^7F_5$ transition, were considered suitable candidates for green-emitting ultraviolet (UV) convertible phosphors and UV-excited wLEDs, as they exhibit tunable CIE coordinates that shift from blue to green [216].

7.6 Gallate-based phosphors

Recently, Cr^{3+} activator-doped gallate phosphors, which are self-illuminating from ultraviolet rays to visible light, have a long photoluminescence excitation range of 250–680 nm, and show persistent luminescence emitted in the near-infrared region, have promising applications in bioimaging and night vision surveillance. In addition, various gallate phosphors obtained by doping Cr^{3+}, Eu^{3+}, Mn^{2+}, Co^{3+}, Yb^{3+}, Tm^{3+}, Er^{3+}, etc. give emission bands in red, green, and blue colors. As a result of white emission by blending each phosphor, these phosphors can be used in plasma display panels, solid-state lasers, UV photodetectors, white light-emitting diodes, cold cathode fluorescent lamps, plasma display panels, lasers, electroluminescent displays, and vacuum fluorescent displays. Table 6 shows the emission color of certain gallate-based phosphors.

Table 6 The emission color of certain gallate-based phosphors.

Gallate phosphors	Color	λ_{ex} (nm)	λ_{em} (nm)	Reference
$ZnGa_2O_4$:Ge/Sn	Red	364	693	[219]
$La_3Ga_5GeO_{14}:Cr^{3+},Nd^{3+}$	Infra-red	808	1064	[220]
$LiGa_5O_8:Cr^{3+}$	Red	400	716	[221]
$Ca_3Ga_2Ge_3O_{12}:Cr^{3+},Yb^{3+},Tm^{3+}$	Red	458	750	[222]
$Zn(Ga1\text{-}xAlx)_2O_4:Cr^{3+},Bi^{3+}$	Red	350	695	[223]
$SrGa_{12}O_{19}:Cr^{3+},Nd^{3}$	Infra-red	450	900, 1050	[224]
$Zn_3Ga_2SnO_8:Cr^{3+}, Yb^{3+}, Er^{3+}$	Red	279	696	[225]
$Ga_2O_3:Cr^{3+}$	Red	300	720	[226]
$ZnGa_2O_4:Mn^{2+}$	Green	290	505	[227]
$ZnGa_2O_4:Cr^{3+}$	Red	254	695	[228]

The $ZnGa_2O_4:Mn^{2+}$ phosphor excited by electron beam or ultraviolet light exhibits a bright green color emission and can be used in thin-film electroluminescent devices, field emission displays (FEDs), and vacuum fluorescent displays (VFDs) [227].

$La_3Ga_5GeO_{14}:Cr^{3+}$ phosphors containing different dopant ions such as Sc^{3+}, Ba^{2+}, and Sn^{3+} have been proposed as promising candidates for the near infrared region (NIR)-LED applications [229].

$La_3Ga_5GeO_{14}:Cr^{3+},Nd^{3+}$ phosphor exhibits persistent broadband luminescence at 700 nm. In particular, it exhibits NIR emission signals under excitation at 808 nm with the addition of Nd^{3+}, and it has been observed that the residence time also increases due to the increase in trap content. Therefore, $La_3Ga_5GeO_{14}:Cr^{3+},Nd^{3+}$ biodetection material is suggested as a candidate for optical imaging of structural processes in tissues and cells [220].

Cr^{3+}activator-doped $SrGa_{12}O_{19}$ phosphors with excitation between 250 and 700 nm has broadband emission between 600 and 950 nm. $SrGa_{12}O_{19}:Cr^{3+},Nd^{3+}$ phosphors are promising as materials to be used as spectral converters on photovoltaic solar cells, as the emission band of the $SrGa_{12}O_{19}:Cr^{3+},Nd^{3+}$ broadens the NIR range [224].

8 Conclusion and future remarks

This chapter provides a comprehensive overview of metal oxide-based phosphors and their use in a variety of chemical-based sensor applications. Basic information about chemical sensors, basic principles of luminescence, properties and applications of metal oxides, classification of metal oxide-based phosphors, preparation techniques of phosphors by sol-gel and other methods, optical properties, and their use in industry are discussed in detail. It also defines solid-state lighting phosphors used in today's phosphor technology, plasma display phosphors, bioluminescent phosphors, phosphors used in liquid crystal display (LCD), cathode ray tubes, lasers, LEDs, lamps, X-rays, and ionizing radiation detection.

References

[1] D. Dey, T. Goswami, Optical biosensors: a revolution towards quantum nanoscale electronics device fabrication, J. Biomed. Biotechnol. 2011 (2011).

[2] A.A. Tomchenko, Printed chemical sensors: from screen-printing to microprinting*, Encycl. Sens. 10 (2006) 279–290.

[3] M.N. Velasco-Garcia, T. Mottram, Biosensor technology addressing agricultural problems, Biosyst. Eng. 84 (1) (2003) 1–12.

[4] C. Cristea, V. Hârceagă, R. Săndulescu, Electrochemical sensor and biosensors, in: Environmental Analysis by Electrochemical Sensors and Biosensors, Springer, 2014, pp. 155–165.

[5] N. Ullah, M. Mansha, I. Khan, A. Qurashi, Nanomaterial-based optical chemical sensors for the detection of heavy metals in water: recent advances and challenges, TrAC Trends Anal. Chem. 100 (2018) 155–166.

[6] V. Naresh, N. Lee, A review on biosensors and recent development of nanostructured materials-enabled biosensors, Sensors 21 (4) (2021) 1109.
[7] A. Lobnik, Optical chemical sensors and personal protection, in: S. Jayaraman, P. Kiekens, A.M. Grancaric (Eds.), Intelligent Textiles for Personal Protection and Safety, vol. 3, IOS Press, 2006, pp. 107–131.
[8] N.N. Greenwood, A. Earnshaw, Chemistry of the Elements, Elsevier, 2012.
[9] N. Tsuda, K. Nasu, A. Fujimori, K. Siratori, Electronic Conduction in Oxides, vol. 94, Springer Science & Business Media, 2000.
[10] Y. Liang, C. Zhao, H. Yuan, Y. Chen, W. Zhang, J. Huang, et al., A review of rechargeable batteries for portable electronic devices, Info Mat. 1 (1) (2019) 6–32.
[11] S. Basu, P. Bhattacharyya, Recent developments on graphene and graphene oxide based solid state gas sensors, Sensors Actuators B Chem. 173 (2012) 1–21.
[12] A. Pérez-Tomás, A. Mingorance, D. Tanenbaum, M. Lira-Cantú, Metal oxides in photovoltaics: all-oxide, ferroic, and perovskite solar cells, in: The Future of Semiconductor Oxides in Next-Generation Solar Cells, Elsevier, 2018, pp. 267–356.
[13] A. Goldman, Applications and functions of ferrites, in: A. Goldman (Ed.), Modern Ferrite Technology, Springer, New York, NY, 2006, pp. 217–226.
[14] T. Ishihara, Perovskite Oxide for Solid Oxide Fuel Cells, Springer Science & Business Media, 2009.
[15] R.V. Bovhyra, S.I. Mudry, D.I. Popovych, S.S. Savka, A.S. Serednytski, Y.I. Venhryn, Photoluminescent properties of complex metal oxide nanopowders for gas sensing, Appl. Nanosci. 9 (5) (2019) 775–780.
[16] R. Shi, G.I.N. Waterhouse, T. Zhang, Recent progress in photocatalytic CO2 reduction over perovskite oxides, Solar RRL 1 (11) (2017) 1700126.
[17] S. Stankic, S. Suman, F. Haque, J. Vidic, Pure and multi metal oxide nanoparticles: synthesis, antibacterial and cytotoxic properties, J. Nanobiotechnol. 14 (1) (2016) 1–20.
[18] N. Tsunoji, D. Shimono, K. Tsuchiya, M. Sadakane, T. Sano, Formation pathway of AEI zeolites as a basis for a streamlined synthesis, Chem. Mater. 32 (1) (2019) 60–74.
[19] Z. Shao, W. Zhou, Z. Zhu, Advanced synthesis of materials for intermediate-temperature solid oxide fuel cells, Prog. Mater. Sci. 57 (4) (2012) 804–874.
[20] J. Li, C. Lin, Y. Min, Y. Yuan, G. Li, S. Yang, et al., Discovery of complex metal oxide materials by rapid phase identification and structure determination, J. Am. Chem. Soc. 141 (12) (2019) 4990–4996.
[21] M.E. Franke, T.J. Koplin, U. Simon, Metal and metal oxide nanoparticles in chemiresistors: does the nanoscale matter? Small 2 (1) (2006) 36–50.
[22] M.J. Limo, A. Sola-Rabada, E. Boix, V. Thota, Z.C. Westcott, V. Puddu, et al., Interactions between metal oxides and biomolecules: from fundamental understanding to applications, Chem. Rev. 118 (22) (2018) 11118–11193.
[23] P. Cao, Z. Yang, S.T. Navale, S. Han, X. Liu, W. Liu, et al., Ethanol sensing behavior of Pd-nanoparticles decorated ZnO-nanorod based chemiresistive gas sensors, Sensors Actuators B Chem. 298 (2019), 126850.
[24] J. Zheng, C. Zhou, M. Yu, J. Liu, Different sized luminescent gold nanoparticles, Nanoscale 4 (14) (2012) 4073–4083.
[25] E. Comini, C. Baratto, G. Faglia, M. Ferroni, A. Vomiero, G. Sberveglieri, Quasi-one dimensional metal oxide semiconductors: preparation, characterization and application as chemical sensors, Prog. Mater. Sci. 54 (1) (2009) 1–67.
[26] K.Y. Kim, P.S. Bin, Preparation and property control of nano-sized indium tin oxide particle, Mater. Chem. Phys. 86 (1) (2004) 210–221.

[27] J.P. Zuniga, M. Abdou, S.K. Gupta, Y. Mao, Molten-salt synthesis of complex metal oxide nanoparticles, J. Vis. Exp. 140 (2018), e58482.
[28] F. Diao, Y. Wang, Transition metal oxide nanostructures: premeditated fabrication and applications in electronic and photonic devices, J. Mater. Sci. 53 (6) (2018) 4334–4359.
[29] R. Schneider, R. Schneider, E.A. de Campos, J.B.S. Mendes, J.F. Felix, P.A. Santa-Cruz, Lead–germanate glasses: an easy growth process for silver nanoparticles and their promising applications in photonics and catalysis, RSC Adv. 7 (66) (2017) 41479–41485.
[30] K. Valadi, S. Gharibi, R. Taheri-Ledari, S. Akin, A. Maleki, A.E. Shalan, Metal oxide electron transport materials for perovskite solar cells: a review, Environ. Chem. Lett. 19 (3) (2021) 2185–2207.
[31] D.K. Kumar, J. Kříž, N. Bennett, B. Chen, H. Upadhayaya, K.R. Reddy, et al., Functionalized metal oxide nanoparticles for efficient dye-sensitized solar cells (DSSCs): a review, Mater. Sci. Energy Technol. 3 (2020) 472–481.
[32] B.Y. Zhang, K. Xu, Q. Yao, A. Jannat, G. Ren, M.R. Field, et al., Hexagonal metal oxide monolayers derived from the metal–gas interface, Nat. Mater. 20 (8) (2021) 1073–1078.
[33] J. Huang, H. Wang, Z. Qi, P. Lu, D. Zhang, B. Zhang, et al., Multifunctional metal–oxide nanocomposite thin film with plasmonic Au nanopillars embedded in magnetic $La_{0.67}Sr_{0.33}MnO_3$ matrix, Nano Lett. 21 (2) (2021) 1032–1039.
[34] S.A. Delbari, L.S. Ghadimi, R. Hadi, S. Farhoudian, M. Nedaei, A. Babapoor, et al., Transition metal oxide-based electrode materials for flexible supercapacitors: a review, J. Alloys Compd. 857 (2021), 158281.
[35] Q. Li, H. Li, Q. Xia, Z. Hu, Y. Zhu, S. Yan, et al., Extra storage capacity in transition metal oxide lithium-ion batteries revealed by in situ magnetometry, Nat. Mater. 20 (1) (2021) 76–83.
[36] N. Li, Y. Li, Q. Li, Y. Zhao, C.-S. Liu, H. Pang, NiO nanoparticles decorated hexagonal nickel-based metal-organic framework: self-template synthesis and its application in electrochemical energy storage, J. Colloid Interface Sci. 581 (2021) 709–718.
[37] N. Barati, M.M. Husein, J. Azaiez, Modifying ceramic membranes with in situ grown iron oxide nanoparticles and their use for oily water treatment, J. Membr. Sci. 617 (2021), 118641.
[38] B. Fatima, S.I. Siddiqui, R.K. Nirala, K. Vikrant, K.-H. Kim, R. Ahmad, et al., Facile green synthesis of ZnO–$CdWO_4$ nanoparticles and their potential as adsorbents to remove organic dye, Environ. Pollut. 271 (2021), 116401.
[39] H. Liu, H. Fu, Y. Liu, X. Chen, K. Yu, L. Wang, Synthesis, characterization and utilization of oxygen vacancy contained metal oxide semiconductors for energy and environmental catalysis, Chemosphere 272 (2021), 129534.
[40] K. Kannan, D. Radhika, K.R. Reddy, A.V. Raghu, K.K. Sadasivuni, G. Palani, et al., Gd^{3+} and Y^{3+} co-doped mixed metal oxide nanohybrids for photocatalytic and antibacterial applications, Nano Express 2 (1) (2021) 10014.
[41] H. Zeng, G. Zhang, K. Nagashima, T. Takahashi, T. Hosomi, T. Yanagida, Metal–oxide nanowire molecular sensors and their promises, Chemosensors 9 (2) (2021) 41.
[42] J. Portier, H.S. Hilal, I. Saadeddin, S.J. Hwang, M.A. Subramanian, G. Campet, Thermodynamic correlations and band gap calculations in metal oxides, Prog. Solid State Chem. 32 (3–4) (2004) 207–217.
[43] S. Laurent, S. Boutry, R.N. Muller, Metal oxide particles and their prospects for applications, in: Iron Oxide Nanoparticles for Biomedical Applications, Elsevier, 2018, pp. 3–42.
[44] Z. Xia, A. Meijerink, Ce^{3+}-doped garnet phosphors: composition modification, luminescence properties and applications, Chem. Soc. Rev. 46 (1) (2017) 275–299.

[45] S.R. Anishia, M.T. Jose, O. Annalakshmi, V. Ramasamy, Thermoluminescence properties of rare earth doped lithium magnesium borate phosphors, J. Lumin. 131 (12) (2011) 2492–2498, https://doi.org/10.1016/j.jlumin.2011.06.019.
[46] B. Basar, A. Tezcaner, D. Keskin, Z. Evis, Improvements in microstructural, mechanical, and biocompatibility properties of nano-sized hydroxyapatites doped with yttrium and fluoride, Ceram. Int. 36 (5) (2010) 1633–1643, https://doi.org/10.1016/j.ceramint.2010.02.033.
[47] Y.L. Shang, Y.L. Jia, F.H. Liao, J.R. Li, M.X. Li, J. Wang, et al., Preparation, microstructure and electrorheological property of nano-sized TiO_2 particle materials doped with metal oxides, J. Mater. Sci. 42 (8) (2007) 2586–2590.
[48] Z. Yin, Q. Zheng, S.C. Chen, D. Cai, L. Zhou, J. Zhang, Bandgap tunable $Zn_{1-x}Mg_xO$ thin films as highly transparent cathode buffer layers for high-performance inverted polymer solar cells, Adv. Energy Mater. 4 (7) (2014) 1–6.
[49] M.G. Helander, Z.B. Wang, Z.H. Lu, Chlorinated indium tin oxide as a charge injecting electrode for admittance spectroscopy, Org. Electron. 12 (9) (2011) 1576–1579.
[50] L. Znaidi, Sol-gel-deposited ZnO thin films: a review, Mater. Sci. Eng. B Solid State Mater. Adv. Technol. 174 (1–3) (2010) 18–30.
[51] Y. Hu, W. Zhuang, H. Ye, D. Wang, S. Zhang, X. Huang, A novel red phosphor for white light emitting diodes, J. Alloys Compd. 390 (1–2) (2005) 226–229.
[52] K.S. Lee, J.W. Lim, H.K. Kim, T.L. Alford, G.E. Jabbour, Transparent conductive electrodes of mixed TiO_{2-x}-indium tin oxide for organic photovoltaics, Appl. Phys. Lett. 100 (21) (2012) 1–4.
[53] G. Zhuang, J. Yan, Y. Wen, Z. Zhuang, Y. Yu, Two-dimensional transition metal oxides and chalcogenides for advanced photocatalysis: progress, challenges, and opportunities, Solar RRL 5 (6) (2021) 1–50.
[54] M.I. Chen, A.K. Singh, J.L. Chiang, R.H. Horng, D.S. Wuu, Zinc gallium oxide—a review from synthesis to applications, Nanomaterials 10 (11) (2020) 1–37.
[55] D. Guo, Q. Guo, Z. Chen, Z. Wu, P. Li, W. Tang, Review of Ga_2O_3-based optoelectronic devices, Mater. Today Phys. (2019) 11.
[56] L.G. Gerling, C. Voz, R. Alcubilla, J. Puigdollers, Origin of passivation in hole-selective transition metal oxides for crystalline silicon heterojunction solar cells, J. Mater. Res. 32 (2) (2017) 260–268.
[57] R.S. Yadav, P. Mishra, A.C. Pandey, Growth mechanism and optical property of ZnO nanoparticles synthesized by sonochemical method, Ultrason. Sonochem. 15 (5) (2008) 863–868.
[58] X. Li, H. Liu, J. Wang, H. Cui, S. Yang, I.R. Boughton, Solvothermal synthesis and luminescent properties of YAG:Tb nano-sized phosphors, J. Phys. Chem. Solids 66 (1) (2005) 201–205.
[59] R. Srinivasan, N.R. Yogamalar, J. Elanchezhiyan, R.J. Joseyphus, A.C. Bose, Structural and optical properties of europium doped yttrium oxide nanoparticles for phosphor applications, J. Alloys Compd. 496 (1–2) (2010) 472–477, https://doi.org/10.1016/j.jallcom.2010.02.083.
[60] H. Xiao, P. Li, F. Jia, L. Zhang, General nonaqueous sol-gel synthesis of nanostructured Sm_2O_3, Gd_2O_3, Dy_2O_3, and $Gd_2O_3:Eu^{3+}$ phosphor, Society (2009) 0–7.
[61] F.W. Liu, C.H. Hsu, F.S. Chen, C.H. Lu, Microwave-assisted solvothermal preparation and photoluminescence properties of $Y_2O_3:Eu^{3+}$ phosphors, Ceram. Int. 38 (2) (2012) 1577–1584, https://doi.org/10.1016/j.ceramint.2011.09.044.
[62] R.E. Rojas-Hernandez, F. Rubio-Marcos, M.Á. Rodriguez, J.F. Fernandez, Long lasting phosphors: $SrAl_2O_4$:Eu, Dy as the most studied material, Renew. Sust. Energ. Rev. 2018 (81) (2016) 2759–2770, https://doi.org/10.1016/j.rser.2017.06.081.

[63] C.S. Cheng, M. Serizawa, H. Sakata, T. Hirayama, Electrical conductivity of Co_3O_4 films prepared by chemical vapour deposition, Mater. Chem. Phys. 53 (3) (1998) 225–230.
[64] M. Verma, V. Kumar, A. Katoch, Synthesis of ZrO_2 nanoparticles using reactive magnetron sputtering and their structural, morphological and thermal studies, Mater. Chem. Phys. 212 (2018) 268–273.
[65] S. Samantaray, B.G. Mishra, D.K. Pradhan, G. Hota, Solution combustion synthesis and physicochemical characterization of ZrO_2–MoO_3 nanocomposite oxides prepared using different fuels, Ceram. Int. 37 (8) (2011) 3101–3108.
[66] J.S. Beck, J.C. Vartuli, W.J. Roth, M.E. Leonowicz, C.T. Kresge, K.D. Schmitt, et al., A new family of mesoporous molecular sieves prepared with liquid crystal templates, J. Am. Chem. Soc. 114 (27) (1992) 10834–10843.
[67] A. Rajan, E. Cherian, G. Baskar, Biosynthesis of zinc oxide nanoparticles using Aspergillus fumigatus JCF and its antibacterial activity, Int. J. Mod. Sci. Technol. 1 (2) (2016) 52–57.
[68] X. Lu, C. Wang, Y. Wei, One-dimensional composite nanomaterials: synthesis by electrospinning and their applications, Small 5 (21) (2009) 2349–2370.
[69] S. Santangelo, F. Pantò, C. Triolo, S. Stelitano, P. Frontera, F. Fernández-Carretero, et al., Evaluation of the electrochemical performance of electrospun transition metal oxide-based electrode nanomaterials for water CDI applications, Electrochim. Acta 309 (2019) 125–139.
[70] A. Katz, M.E. Davis, Molecular imprinting of bulk, microporous silica, Nature 403 (6767) (2000) 286–289.
[71] P.J. Modreski, R. Aumente-Modreski, Fluorescent minerals, Rocks Miner. 71 (1) (1996) 14–22.
[72] M. Singh, Luminescent materials for communication devices: an overview, Int. J. Eng. Technol. Manag. Res. 5 (2) (2020) 96–100, https://doi.org/10.29121/ijetmr.v5.i2.2018.619.
[73] M. Shang, C. Li, J. Lin, How to produce white light in a single-phase host? Chem. Soc. Rev. 43 (5) (2014) 1372–1386.
[74] D. Pan, J. Zhang, Z. Li, M. Wu, Hydrothermal route for cutting graphene sheets into blue-luminescent graphene quantum dots, Adv. Mater. 22 (6) (2010) 734–738.
[75] Ü. Özgür, Y.I. Alivov, C. Liu, A. Teke, M.A. Reshchikov, S. Doğan, et al., A comprehensive review of ZnO materials and devices, J. Appl. Phys. 98 (4) (2005) 1–103.
[76] G. Blasse, B.C. Grabmaier, A general introduction to luminescent materials, in: G. Blasse, B.C. Grabmaier (Eds.), Luminescent Materials, Springer, Berlin, Heidelberg, 1994, pp. 1–9.
[77] I. Gupta, S. Singh, S. Bhagwan, D. Singh, Rare earth (RE) doped phosphors and their emerging applications: a review, Ceram. Int. 47 (14) (2021) 19282–19303, https://doi.org/10.1016/j.ceramint.2021.03.308.
[78] A.K.R. Choudhury, Principles of Colour and Appearance Measurement: Object Appearance, Colour Perception and Instrumental Measurement, Elsevier, 2014.
[79] M. Mariyappan, S. Arunkumar, K. Marimuthu, White light emission and spectroscopic properties of Dy^{3+} ions doped bismuth sodiumfluoroborate glasses for photonic applications, J. Alloys Compd. 723 (2017) 100–114, https://doi.org/10.1016/j.jallcom.2017.06.244.
[80] X. Guo, K.W. Houser, A review of colour rendering indices and their application to commercial light sources, Light. Res. Technol. 36 (3) (2004) 183–199.
[81] J. Xu, X. Liu, J. Li, Solid-state lighting, in: A. Ikesue (Ed.), Processing of Ceramics: Breakthroughs in Optical Materials, Wiley, 2021, pp. 187–274.

[82] V. Kumar, S. Som, S. Dutta, H.C. Swart, Novel Zincate Phosphors: A New Red Emitting Phosphor for LED Applications, South African Institute of Physics Pretoria, 2015.

[83] A.G. Bispo-Jr, L.F. Saraiva, S.A.M. Lima, A.M. Pires, M.R. Davolos, Recent prospects on phosphor-converted LEDs for lighting, displays, phototherapy, and indoor farming, J. Lumin. 237 (December 2020) (2021), 118167, https://doi.org/10.1016/j.jlumin.2021.118167.

[84] J. Liao, S. Liu, H.-R. Wen, L. Nie, L. Zhong, Luminescence properties of ZrW2O8: Eu3+ nanophosphors for white light emitting diodes, Mater. Res. Bull. 70 (2015) 7–12.

[85] M. Zhang, A. Yang, Y. Peng, B. Zhang, H. Ren, W. Guo, et al., Dy^{3+}-doped Ga-Sb-S chalcogenide glasses for mid-infrared lasers, Mater. Res. Bull. 70 (2015) 55–59, https://doi.org/10.1016/j.materresbull.2015.04.019.

[86] Y. Liu, Z. Wang, G. Cheng, J. Zhang, G. Hong, J. Ni, Synthesis and characterization of Fe_3O_4@ YPO_4:Eu^{3+} multifunctional microspheres, Mater. Lett. 152 (2015) 224–227.

[87] O.M. Ten Kate, K.W. Krämer, E. Van Der Kolk, Efficient luminescent solar concentrators based on self-absorption free, Tm^{2+} doped halides, Sol. Energy Mater. Sol. Cells 140 (2015) 115–120, https://doi.org/10.1016/j.solmat.2015.04.002.

[88] F. Steudel, S. Loos, B. Ahrens, S. Schweizer, Luminescent borate glass for efficiency enhancement of CdTe solar cells, J. Lumin. 164 (2015) 76–80, https://doi.org/10.1016/j.jlumin.2015.03.022.

[89] D. Chen, Z. Wan, Y. Zhou, P. Huang, J. Zhong, M. Ding, et al., Bulk glass ceramics containing Yb^{3+}/Er^{3+}:β-$NaGdF_4$ nanocrystals: phase-separation-controlled crystallization, optical spectroscopy and upconverted temperature sensing behavior, J. Alloys Compd. 638 (2015) 21–28, https://doi.org/10.1016/j.jallcom.2015.02.170.

[90] Y. Yuan, W. Jidong, Z. Cunyi, S. Xiaolei, L. Jing, G. Mingqiao, Preparation and characterization of $CaAl_2O_4$:Eu^{2+},Nd^{3+} luminous phosphor synthesized by coprecipitation-sol-gel method, Mater. Res. Express 5 (8) (2018).

[91] B. Lu, M. Shi, Z. Pang, Y. Zhu, Y. Li, Study on the optical performance of red-emitting phosphor: $SrAl_2O_4$:Eu^{2+}, Dy^{3+}/$Sr_2MgSi_2O_7$:Eu^{2+}, Dy^{3+}/light conversion agent for long-lasting luminous fibers, J. Mater. Sci. Mater. Electron. 32 (13) (2021) 17382–17394, https://doi.org/10.1007/s10854-021-06270-1.

[92] R. Ianoş, R. Lazəu, R.C. Boruntea, Solution combustion synthesis of bluish-green $BaAl_2O_4$: Eu^{2+}, Dy^{3+} phosphors, Ceram. Int. 41 (2) (2015) 3186–3190.

[93] M. Kumar, S.K. Gupta, R.M. Kadam, Near white light emitting $ZnAl_2O_4$:Dy^{3+} nanocrystals: sol-gel synthesis and luminescence studies, Mater. Res. Bull. 74 (2016) 182–187, https://doi.org/10.1016/j.materresbull.2015.10.029.

[94] B. Liu, Y. Wang, Y. Wen, F. Zhang, G. Zhu, J. Zhang, Photoluminescence properties of S-doped $BaAl_{12}O_{19}$:Mn^{2+} phosphors for plasma display panels, Mater. Lett. 75 (2012) 137–139, https://doi.org/10.1016/j.matlet.2012.02.011.

[95] R.S. Yadav, S.K. Pandey, A.C. Pandey, Blue-shift and enhanced photoluminescence in $BaMgAl_{10}O_{17}$:Eu^{2+} nanophosphor under VUV excitation for PDPs application, Mater. Sci. Appl. 01 (01) (2010) 25–31.

[96] L. Xia, Q. Xiao, X. Ye, W. You, T. Liang, Erosion behavior and luminescence properties of $Y_3Al_5O_{12}$:Ce^{3+}-embedded calcium bismuth borate glass-ceramics for wLEDs, J. Am. Ceram. Soc. 102 (4) (2019) 2053–2065.

[97] S. Wang, B. Devakumar, Q. Sun, J. Liang, L. Sun, X. Huang, Highly efficient near-UV-excitable $Ca_2YHf_2Al_3O_{12}$:Ce^{3+},Tb^{3+} green-emitting garnet phosphors with potential application in high color rendering warm-white LEDs, J. Mater. Chem. C 8 (13) (2020) 4408–4420.

[98] T. Zhang, N. Li, Z. Yu, P. Du, B. Tian, Z. Li, et al., Violet-light-excitable super-narrow band green emitting phosphor for high-quality white LEDs, J. Lumin. 225 (February) (2020).
[99] R. Cao, F. Zhang, C. Cao, X. Yu, A. Liang, S. Guo, et al., Synthesis and luminescence properties of $CaAl_2O_4$:Mn^{4+} phosphor, Opt. Mater. 38 (2014) 53–56, https://doi.org/10.1016/j.optmat.2014.10.002.
[100] S.Y. Huang, R. Von Der Mühll, J. Ravez, P. Hagenmuller, Structural, ferroelectric and pyroelectric properties of nonstoichiometric ceramics based on $BaAl_2O_4$, J. Phys. Chem. Solids 55 (1) (1994) 119–124.
[101] M.M. Ali, S.K. Agarwal, S. Agarwal, S.K. Handoo, Kinetics and diffusion studies in $BaAl_2O_4$ formation, Cem. Concr. Res. 25 (1) (1995) 86–90.
[102] W. Strek, P. Dereń, A. Bednarkiewicz, M. Zawadzki, J. Wrzyszcz, Emission properties of nanostructured Eu^{3+} doped zinc aluminate spinels, J. Alloys Compd. 300 (2000) 456–458.
[103] H. Matsui, C.N. Xu, H. Tateyama, Stress-stimulated luminescence from $ZnAl_2O_4$:Mn, Appl. Phys. Lett. 78 (8) (2001) 1068–1070.
[104] M. Zawadzki, J. Wrzyszcz, W. Strek, D. Hreniak, Preparation and optical properties of nanocrystalline and nanoporous Tb doped alumina and zinc aluminate, J. Alloys Compd. 323–324 (2001) 279–282.
[105] M. García-Hipólito, C.D. Hernández-Pérez, O. Alvarez-Fregoso, E. Martínez, J. Guzmán-Mendoza, C. Falcony, Characterization of europium doped zinc aluminate luminescent coatings synthesized by ultrasonic spray pyrolysis process, Opt. Mater. 22 (4) (2003) 345–351.
[106] S.K. Sampath, D.G. Kanhere, R. Pandey, Electronic structure of spinel oxides: zinc aluminate and zinc gallate, J. Phys. Condens. Matter 11 (18) (1999) 3635–3644.
[107] S.K. Sampath, J.F. Cordaro, Optical properties of zinc aluminate, zinc gallate, and zinc aluminogallate spinels, J. Am. Ceram. Soc. 81 (3) (1998) 649–654.
[108] K.K. Satapathy, G.C. Mishra, F. Khan, $ZnAl_2O_4$:Eu novel phosphor: SEM and mechanoluminescence characterization synthesized by solution combustion technique, Luminescence 30 (5) (2015) 564–567.
[109] P.M.A.G. Araújo, P.T.A. Santos, P.T.A. Santos, F.N. Silva, A.C.F.M. Costa, E.M. Araújo, Obtaining of chitosan/znal1,9EU0,05O4 film for application as biomaterial, Mater. Sci. Forum 805 (2015) 65–70.
[110] R.J. Wiglusz, A. Watras, M. Malecka, P.J. Deren, R. Pazik, Structure evolution and up-conversion studies of ZnX_2O_4:Er^{3+}/Yb^{3+} (X = Al^{3+}, Ga^{3+}, In^{3+}) nanoparticles, Eur. J. Inorg. Chem. 6 (2014) 1090–1101.
[111] I. Kamińska, K. Fronc, B. Sikora, K. Koper, R. Minikayev, W. Paszkowicz, et al., Synthesis of $ZnAl_2O_4$:(Er^{3+},Yb^{3+}) spinel-type nanocrystalline upconverting luminescent marker in HeLa carcinoma cells, using a combustion aerosol method route, RSC Adv. 4 (100) (2014) 56596–56604.
[112] R.S. Yadav, S.K. Pandey, A.C. Pandey, $BaAl_{12}O_{19}$:Mn^{2+} green emitting nanophosphor for PDP application synthesized by solution combustion method and its vacuum ultra-violet photoluminescence characteristics, J. Lumin. 131 (9) (2011) 1998–2003, https://doi.org/10.1016/j.jlumin.2011.04.022.
[113] W.-C. Ke, C.C. Lin, R.-S. Liu, M.-C. Kuo, Energy transfer and significant improvement moist stability of $BaMgAl_{10}O_{17}$:Eu^{2+},Mn^{2+} as a phosphor for white light-emitting diodes, J. Electrochem. Soc. 157 (8) (2010) J307. Available from: https://iopscience.iop.org/article/10.1149/1.3454755.

[114] D. Ravichandran, S.T. Johnson, S. Erdei, R. Roy, W.B. White, Crystal chemistry and luminescence of the Eu^{2+}-activated alkaline earth aluminate phosphors, Displays 19 (4) (1999) 197–203.

[115] C. Li, C.N. Xu, L. Zhang, H. Yamada, Y. Imai, W.X. Wang, Dynamic visualization of stress distribution by mechanoluminescence image, Key Eng. Mater. 388 (August 1999) (2009) 265–268.

[116] J.S. Kim, P.E. Jeonny, J.C. Choi, H.L. Park, S.I. Mho, G.C. Kim, Warm-white-light emitting diode utilizing a single-phase full-color $Ba_3MgSi_2O_8$:Eu^{2+}, Mn^{2+} phosphor, Appl. Phys. Lett. 84 (15) (2004) 2931–2933.

[117] Z. Qiu, Y. Zhou, M. Lü, A. Zhang, Q. Ma, Combustion synthesis of long-persistent luminescent MAl_2O_4: Eu^{2+}, R^{3+} (M = Sr, Ba, Ca, R = Dy, Nd and La) nanoparticles and luminescence mechanism research, Acta Mater. 55 (8) (2007) 2615–2620.

[118] H. Yamamoto, S. Okamoto, H. Kobayashi, Luminescence of rare-earth ions in perovskite-type oxides: from basic research to applications, J. Lumin. 100 (1–4) (2002) 325–332.

[119] K. Van den Eeckhout, P.F. Smet, D. Poelman, Persistent luminescence in Eu^{2+}-doped compounds: a review, Materials 3 (4) (2010) 2536–2566.

[120] Y. Ding, Y. Zhang, Z. Wang, W. Li, D. Mao, H. Han, et al., Photoluminescence of Eu single doped and Eu/Dy codoped $Sr_2Al_2SiO_7$ phosphors with long persistence, J. Lumin. 129 (3) (2009) 294–299.

[121] R. Sakai, T. Katsumata, S. Komuro, T. Morikawa, Effect of composition on the phosphorescence from $BaAl_2O_4$:Eu^{2+},Dy^{3+} crystals, J. Lumin. 85 (1–3) (1999) 149–154.

[122] T. Katsumata, R. Sakai, S. Komuro, T. Morikawa, H. Kimura, Growth and characteristics of long duration phosphor crystals, J. Cryst. Growth 198–199 (PART I) (1999) 869–871.

[123] Y. Lin, Z. Zhang, Z. Tang, J. Zhang, Z. Zheng, X. Lu, Characterization and mechanism of long afterglow in alkaline earth aluminates phosphors co-doped by Eu_2O_3 and Dy_2O_3, Mater. Chem. Phys. 70 (2) (2001) 156–159.

[124] K.T. Lee, P.B. Aswath, Synthesis of hexacelsian barium aluminosilicate by a solid-state process, J. Am. Ceram. Soc. 83 (12) (2000) 2907–2912.

[125] C. Zhang, L. Wang, L. Cui, Y. Zhu, A novel method for the synthesis of nano-sized BaAl2O4 with thermal stability, J. Cryst. Growth 255 (3–4) (2003) 317–323.

[126] L. Xingdong, M. Zhong, R. Wang, Roles of Eu^{2+}, Dy^{3+} ions in persistent luminescence of strontium aluminates phosphors, J. Wuhan Univ. Technol. Mater. Sci. Ed. 23 (5) (2008) 652–657.

[127] X. Luo, J.Y. Ahn, S.H. Kim, Aerosol synthesis and luminescent properties of $CaAl_2O_4$: Eu^{2+}, Nd^{3+} down-conversion phosphor particles for enhanced light harvesting of dye-sensitized solar cells, Sol. Energy 178 (July 2018) (2019) 173–180, https://doi.org/10.1016/j.solener.2018.12.029.

[128] J. Kaur, R. Shrivastava, B. Jaykumar, N.S. Suryanarayana, Studies on the persistent luminescence of Eu^{2+} and Dy^{3+}-doped $SrAl_2O_4$ phosphors: A review, Res. Chem. Intermed. 40 (1) (2014) 317–343.

[129] X. Zhang, J. Tang, H. Li, Y. Wang, X. Wang, Y. Wang, et al., Red light emitting nano-PVP fibers that hybrid with Ag@SiO_2@Eu(tta)3phen-NPs by electrostatic spinning method, Opt. Mater. 78 (2018) 220–225.

[130] Y. Zhu, M. Ge, Effect of y/x ratio on luminescence properties of $xSrO{\cdot}yAl_2O_3$:Eu^{2+},Dy^{3+} luminous fiber, J. Rare Earths 32 (7) (2014) 598–602, https://doi.org/10.1016/S1002-0721(14)60114-9.

[131] J. Li, J. Wang, Y. Yu, Y. Zhu, M. Ge, Preparation and luminescence properties of rare-earth doped fiber with spectral blue-shift: $SrAl_2O_4$:Eu^{2+},Dy^{3+} phosphors/

triarylsulfonium hexafluoroantimonate based on polypropylene substrate, J. Rare Earths 35 (6) (2017) 530–535, https://doi.org/10.1016/S1002-0721(17)60944-X.
[132] H.R. Chen, C. Cai, Z.W. Zhang, L. Zhang, H.P. Lu, X. Xu, et al., Enhancing the luminescent efficiency of $Y_3Al_5O_{12}$:Ce^{3+} by coating graphitic carbon nitride: toward white light-emitting diodes, J. Alloys Compd. 801 (2019) 10–18, https://doi.org/10.1016/j.jallcom.2019.06.122.
[133] Y. Chen, J. Li, S. Zeng, H. Fan, J. Feng, L. Tan, Use of LiF flux in the preparation of $Ca_3Sc_2Si_3O_{12}$:Ce^{3+} phosphor by sol-combustion method, Opt. Mater. 37 (2014) 464–469.
[134] H.K. Yang, H.M. Noh, B.K. Moon, J.H. Jeong, S.S. Yi, Luminescence investigations of Sr_3SiO_5:Eu^{2+} orange–yellow phosphor for UV-based white LED, Ceram. Int. 40 (8) (2014) 12503–12508.
[135] G.-F. Luo, N.T.P. Loan, N.D.Q. Anh, H.-Y. Lee, Enhancement of color quality and luminous flux for remote-phosphor LEDs with red-emitting CaMgSiO: Eu, Mn, Mater. Sci.-Pol. 38 (3) (2020) 409–415.
[136] Y. Li, W. Liu, X. Wang, G. Zhu, C. Wang, Y. Wang, A double substitution induced Ca $(Mg_{0.8},Al_{0.2})(Si_{1.8},Al_{0.2})O_6$:$Eu^{2+}$ phosphor for w-LEDs: synthesis, structure, and luminescence properties, Dalton Trans. 44 (29) (2015) 13196–13203.
[137] M.M. Gowri, G.P. Darshan, Y.V. Naik, H.B. Premkumar, D. Kavyashree, S.C. Sharma, et al., Phase dependent photoluminescence and thermoluminescence properties of Y_2SiO_5: Sm^{3+} nanophosphors and its advanced forensic applications, Opt. Mater. 96 (2019), 109282.
[138] G. Annadurai, S.M.M. Kennedy, V. Sivakumar, Photoluminescence properties of a novel orange-red emitting $Ba_2CaZn_2Si_6O_{17}$:Sm^{3+} phosphor, J. Rare Earths 34 (6) (2016) 576–582.
[139] C. Wang, J. Wang, J. Jiang, S. Xin, G. Zhu, Redesign and manually control the commercial plasma green Zn_2SiO_4:Mn^{2+} phosphor with high quantum efficiency for white light emitting diodes, J. Alloys Compd. 814 (2020), 152340.
[140] W. Ding, J. Wang, Z. Liu, M. Zhang, Q. Su, J. Tang, An intense green/yellow dual-chromatic calcium chlorosilicate phosphor $Ca_3SiO_4Cl_2$:Eu^{2+}–Mn^{2+} for yellow and white LED, J. Electrochem. Soc. 155 (5) (2008) J122.
[141] Y. Xiao, W. Xiao, L. Zhang, Z. Hao, G.-H. Pan, Y. Yang, et al., A highly efficient and thermally stable green phosphor ($Lu_2SrAl_4SiO_{12}$ Ce^{3+}) for full-spectrum white LEDs, J. Mater. Chem. C 6 (45) (2018) 12159–12163.
[142] M. Zhao, H. Liao, L. Ning, Q. Zhang, Q. Liu, Z. Xia, Next-generation narrow-band green-emitting RbLi $(Li_3SiO_4)_2$:Eu^{2+} phosphor for backlight display application, Adv. Mater. 30 (38) (2018) 1802489.
[143] V.B. Bhatkar, S.K. Omanwar, S.V. Moharil, Combustion synthesis of silicate phosphors, Opt. Mater. 29 (8) (2007) 1066–1070.
[144] X.A. Shen, R. Kachru, High-speed holographic recording of 500 images in a rare earth doped solid, J. Alloys Compd. 250 (1–2) (1997) 435–438.
[145] A. Meijerink, G. Blasse, L. Struye, A new photostimulable phosphor: Eu^{2+}-activated bariumbromosilicate ($Ba_5SiO_4Br_6$), Mater. Chem. Phys. 21 (3) (1989) 261–270.
[146] A. Meijerink, W.J. Schipper, G. Blasse, Photostimulated luminescence and thermally stimulated luminescence of Y_2SiO_5-Ce, Sm, J. Phys. D. Appl. Phys. 24 (6) (1991) 997–1002.
[147] V. Marinova, M. Veleva, D. Petrova, I.M. Kourmoulis, D.G. Papazoglou, A.G. Apostolidis, et al., Optical properties of $Bi_{12}SiO_{20}$ single crystals doped with 4d and 5d transition elements, J. Appl. Phys. 89 (5) (2001) 2686–2689.

[148] W. Ryba-Romanowski, B. Macalik, M. Berkowski, Down-and up-conversion of femtosecond light pulse excitation into visible luminescence in cerium-doped Lu_2SiO_5–Gd_2SiO_5 solid solution crystals co-doped with Sm^{3+} or Dy^{3+}, Opt. Express 23 (4) (2015) 4552–4562.
[149] X. Zhang, L. Zhou, Q. Pang, M. Gong, A broadband-excited and narrow-line $GdBO_3$: Ce^{3+}, Tb^{3+}, Eu^{3+} red phosphor with efficient Ce^{3+}→(Tb^{3+}) n→ Eu^{3+} energy transfer for NUV LEDs, Opt. Mater. 36 (7) (2014) 1112–1118.
[150] C. Guo, L. Luan, Y. Xu, F. Gao, L. Liang, White light–generation phosphor $Ba_2Ca(BO_3)_2$:Ce^{3+},Mn^{2+} for light-emitting diodes, J. Electrochem. Soc. 155 (11) (2008) J310.
[151] C.-H. Huang, T.-M. Chen, A novel single-composition trichromatic white-light $Ca_3Y(GaO)_3(BO_3)_4$: Ce^{3+}, Mn^{2+}, Tb^{3+} phosphor for UV-light emitting diodes, J. Phys. Chem. C 115 (5) (2011) 2349–2355.
[152] W. Xiao, X. Zhang, Z. Hao, G.-H. Pan, Y. Luo, L. Zhang, et al., Blue-emitting $K_2Al_2B_2O_7$: Eu^{2+} phosphor with high thermal stability and high color purity for near-UV-pumped white light-emitting diodes, Inorg. Chem. 54 (7) (2015) 3189–3195.
[153] F. Mougel, F. Augé, G. Aka, A. Kahn-Harari, D. Vivien, F. Balembois, et al., New green self-frequency-doubling diode-pumped Nd: $Ca_4GdO(BO_3)_3$ laser, Appl. Phys. B 67 (5) (1998) 533–535.
[154] M.L. Rico, J.L. Valdés, J. Martínez-Pastor, J. Capmany, Continuous-wave dual-wavelength operation at 1062 and 1338 nm in Nd^{3+}: $YAl_3(BO_3)_4$ and observation of yellow laser light generation at 592 nm by their self-sum-frequency-mixing, Opt. Commun. 282 (8) (2009) 1619–1621.
[155] C. Guo, J. Yu, X. Ding, M. Li, Z. Ren, J. Bai, A dual-emission phosphor $LiCaBO_3$:Ce^{3+}, Mn2+ with energy transfer for near-UV LEDs, J. Electrochem. Soc. 158 (2) (2010) J42.
[156] C.-K. Chang, T.-M. Chen, $Sr_3B_2O_6$:Ce^{3+}, Eu^{2+}: a potential single-phased white-emitting borate phosphor for ultraviolet light-emitting diodes, Appl. Phys. Lett. 91 (8) (2007) 81902.
[157] Z.-F. Pan, C.-J. Zhu, C.-K. Duan, J. Xu, W.-H. Liu, L.-L. Wang, Luminescence and energy transfer of $SrMgB_6O_{11}$: Eu^{2+}, Mn^{2+} for white light emitting diodes, J. Electrochem. Soc. 159 (1) (2011) P18.
[158] J. Zhong, Y. Zhuo, S. Hariyani, W. Zhao, J. Wen, J. Brgoch, Closing the cyan gap toward full-spectrum LED lighting with $NaMgBO_3$:Ce^{3+}, Chem. Mater. 32 (2) (2019) 882–888.
[159] J.W. Kim, C.S. Yoon, H.G. Gallagher, Dielectric properties of lithium triborate single crystals, Appl. Phys. Lett. 71 (22) (1997) 3212–3214.
[160] Z. Guoqing, X. Jun, C. Xingda, Z. Heyu, W. Siting, X. Ke, et al., Growth and spectrum of a novel birefringent α-BaB_2O_4 crystal, J. Cryst. Growth 191 (3) (1998) 517–519.
[161] J.-M. Tu, D.A. Keszler, $CsLiB_6O_{10}$: A noncentrosymmetric polyborate, Mater. Res. Bull. 30 (2) (1995) 209–215.
[162] Z.-G. Hu, T. Higashiyama, M. Yoshimura, Y. Mori, T. Sasaki, Flux growth of the new nonlinear optical crystal: $K_2Al_2B_2O_7$, J. Cryst. Growth 212 (1–2) (2000) 368–371.
[163] C.S.K. Mak, P.A. Tanner, Z. Zhuo, Absorption spectroscopy of Er: GdCOB and Yb: GdCOB crystals, J. Alloys Compd. 323 (2001) 292–296.
[164] M. Malinowski, M. Kowalska, R. Piramidowicz, T. Lukasiewicz, M. Swirkowicz, A. Majchrowski, Optical transitions of Pr^{3+} ions in $Ca_4GdO(BO_3)_3$ crystals, J. Alloys Compd. 323 (2001) 214–217.
[165] G. Aka, F. Mougel, F. Auge, A. Kahn-Harari, D. Vivien, J.M. Benitez, et al., Overview of the laser and non-linear optical properties of calcium-gadolinium-oxo-borate $Ca_4GdO(BO_3)_3$, J. Alloys Compd. 303 (2000) 401–408.

[166] G. Dominiak-Dzik, W. Ryba-Romanowski, S. Golab, A. Pajaczkowska, Thulium-doped $Ca_4GdO(BO_3)_3$ crystals. An investigation of radiative and non-radiative processes, J. Phys. Condens. Matter 12 (25) (2000) 5495.

[167] D.P. Shumov, V.S. Nikolov, D.D. Nihtianova, J.J. Macicek, J.K. Georgieva, A.T. Nenov, High-temperature solutions suitable for the growth of $NdAl_3(BO_3)_4$ crystals, J. Cryst. Growth 172 (3–4) (1997) 478–485.

[168] V. Petrov, F. Rotermund, F. Noack, R. Komatsu, T. Sugawara, S. Uda, Vacuum ultraviolet application of $Li_2B_4O_7$ crystals: generation of 100 fs pulses down to 170 nm, J. Appl. Phys. 84 (11) (1998) 5887–5892.

[169] Y. Kutomi, M.H. Kharita, S.A. Durrani, Characteristics of TL and PTTL glow curves of gamma irradiated pure $Li_2B_4O_7$ single crystals, Radiat. Meas. 24 (4) (1995) 407–410.

[170] M. Prokić, Development of highly sensitive $CaSO_4$: Dy/Tm and MgB_4O_7: Dy/Tm sintered thermoluminescent dosimeters, Nucl. Inst. Methods 175 (1) (1980) 83–86.

[171] M.J. Knitel, V.R. Bom, P. Dorenbos, C.W.E. Van Eijk, I. Berezovskaya, V. Dotsenko, The feasibility of boron containing phosphors in thermal neutron image plates, in particular the systems $M_2B_5O_9X$: Eu^{2+} (M= Ca, Sr, Ba; X= Cl, Br). Part I: simulation of the energy deposition process, Nucl. Instrum. Methods Phys. Res. Sect. A 449 (3) (2000) 578–594.

[172] A. Karthikeyani, R. Jagannathan, Eu^{2+} luminescence in stillwellite-type $SrBPO_5$—a new potential X-ray storage phosphor, J. Lumin. 86 (1) (2000) 79–85.

[173] A.M. Srivastava, D.A. Doughty, W.W. Beers, Photon cascade luminescence of Pr^{3+} in $LaMgB_5O_{10}$, J. Electrochem. Soc. 143 (12) (1996) 4113.

[174] K. Machida, G. Adachi, J. Shiokawa, Luminescence properties of Eu (II)-borates and Eu2 +-activated Sr-borates, J. Lumin. 21 (1) (1979) 101–110.

[175] J.P. Chaminade, A. Garcia, M. Pouchard, C. Fouassier, B. Jacquier, D. Perret-Gallix, et al., Crystal growth and characterization of $InBO_3$:Tb^{3+}, J. Cryst. Growth 99 (1–4) (1990) 799–804.

[176] I.-E. Kwon, B.-Y. Yu, H. Bae, Y.-J. Hwang, T.-W. Kwon, C.-H. Kim, et al., Luminescence properties of borate phosphors in the UV/VUV region, J. Lumin. 87 (2000) 1039–1041.

[177] P.P. Pawar, S.R. Munishwar, R.S. Gedam, Physical and optical properties of Dy^{3+}/Pr^{3+} Co-doped lithium borate glasses for W-LED, J. Alloys Compd. 660 (2016) 347–355.

[178] N. Deopa, A.S. Rao, Photoluminescence and energy transfer studies of Dy^{3+} ions doped lithium lead alumino borate glasses for w-LED and laser applications, J. Lumin. 192 (2017) 832–841.

[179] A. Ichoja, S. Hashim, S.K. Ghoshal, I.H. Hashim, R.S. Omar, Physical, structural and optical studies on magnesium borate glasses doped with dysprosium ion, J. Rare Earths 36 (12) (2018) 1264–1271.

[180] L. Yao, G. Chen, S. Cui, H. Zhong, C. Wen, Fluorescence and optical properties of Eu^{3+}-doped borate glasses, J. Non-Cryst. Solids 444 (2016) 38–42.

[181] H. Wang, X. Tang, X. Bian, C. Ma, M. Sardar, Z. Zhai, et al., Synthesis and spectroscopic investigation of blue-emitting $(SrCa)_2La(PO_4)_3O$:Eu^{2+} phosphors, Opt. Mater. 107 (2020), 110104.

[182] C.-H. Huang, T.-M. Chen, W.-R. Liu, Y.-C. Chiu, Y.-T. Yeh, S.-M. Jang, A single-phased emission-tunable phosphor $Ca_9Y(PO_4)_7$: Eu^{2+}, Mn^{2+} with efficient energy transfer for white-light-emitting diodes, ACS Appl. Mater. Interfaces 2 (1) (2010) 259–264.

[183] S. Xu, Z. Wang, P. Li, T. Li, Q. Bai, J. Sun, et al., Single-phase white-emitting phosphors $Ba_3Ce_{(1-x-y)}(PO_4)_3$: xTb^{3+}, yMn^{2+} and $B_{a3}Ce_{(1-x-z)}(PO_4)_3$: xTb^{3+}, zSm^{3+}: structure,

luminescence, energy transfer and thermal stability, RSC Adv. 7 (32) (2017) 19584–19592.

[184] D. Liu, Y. Jin, Y. Lv, G. Ju, C. Wang, L. Chen, et al., A single-phase full-color emitting phosphor $Na_3Sc_2(PO_4)_3$: $Eu^{2+}/Tb^{3+}/Mn^{2+}$ with near-zero thermal quenching and high quantum yield for near-UV converted warm w-LEDs, J. Am. Ceram. Soc. 101 (12) (2018) 5627–5639.

[185] Z. Leng, R. Li, L. Li, D. Xue, D. Zhang, G. Li, et al., Preferential neighboring substitution-triggered full visible spectrum emission in single-phased $Ca_{10.5-x}$ Mg_x $(PO_4)_7$: Eu^{2+} phosphors for high color-rendering white LEDs, ACS Appl. Mater. Interfaces 10 (39) (2018) 33322–33334.

[186] C.-H. Huang, W.-R. Liu, T.-M. Chen, Single-phased white-light phosphors Ca_9Gd $(PO_4)_7$: Eu^{2+}, Mn^{2+} under near-ultraviolet excitation, J. Phys. Chem. C 114 (43) (2010) 18698–18701.

[187] M. Cui, J. Wang, M. Shang, J. Li, Q. Wei, P. Dang, et al., Full visible light emission in Eu^{2+}, Mn^{2+}-doped $Ca_9LiY_{0.667}(PO_4)_7$ phosphors based on multiple crystal lattice substitution and energy transfer for warm white LEDs with high colour-rendering, J. Mater. Chem. C 7 (12) (2019) 3644–3655.

[188] I.M. Nagpure, K.N. Shinde, V. Kumar, O.M. Ntwaeaborwa, S.J. Dhoble, H.C. Swart, Combustion synthesis and luminescence investigation of $Na_3Al_2(PO_4)_3$: RE (RE= Ce^{3+}, Eu^{3+} and Mn^{2+}) phosphor, J. Alloys Compd. 492 (1–2) (2010) 384–388.

[189] K.N. Shinde, S.J. Dhoble, Luminescence in Dy^{3+} and Eu^{3+} activated $K_3Al_2(PO_4)_3$, J. Fluoresc. 21 (6) (2011) 2053–2056.

[190] Y. Wang, M.G. Brik, P. Dorenbos, Y. Huang, Y. Tao, H. Liang, Enhanced green emission of Eu2+ by energy transfer from the 5D3 level of Tb^{3+} in $NaCaPO_4$, J. Phys. Chem. C 118 (13) (2014) 7002–7009.

[191] K.V.R. Murthy, D.R. Joshi, M. Parmar, Y.S. Patil, K.G. Chaudhari, Photoluminescence characteristics of $LaPO_4$:Tb, in: Proceedings of National Conference on Luminescence and Its Applications, 2005.

[192] J. Zheng, Q. Cheng, S. Wu, Z. Guo, Y. Zhuang, Y. Lu, et al., An efficient blue-emitting Sr_5 $(PO_4)_3Cl:Eu^{2+}$ phosphor for application in near-UV white light-emitting diodes, J. Mater. Chem. C 3 (42) (2015) 11219–11227.

[193] C.C. Lin, Z.R. Xiao, G.-Y. Guo, T.-S. Chan, R.-S. Liu, Versatile phosphate phosphors $ABPO_4$ in white light-emitting diodes: collocated characteristic analysis and theoretical calculations, J. Am. Chem. Soc. 132 (9) (2010) 3020–3028.

[194] R.P. Rao, D.J. Devine, RE-activated lanthanide phosphate phosphors for PDP applications, J. Lumin. 87 (2000) 1260–1263.

[195] N. Feng, S. Bai, C. Wang, G. Wu, G. Zhang, J. Yang, Energy transfer and thermal stability of novel green-emitting $Ca_3Y(PO_4)_3$:Ce^{3+}, Tb^{3+} phosphors for white LEDs, Opt. Mater. 96 (2019), 109317.

[196] N. Górecka, K. Szczodrowski, A. Lazarowska, M. Grinberg, The influence of Si^{4+} co-doping on the spectroscopic properties of β-$NaCaPO_4$:Eu^{2+}/Eu^{3+}, New J. Chem. 43 (8) (2019) 3409–3418.

[197] I.W. Bin, H.S. Yoo, S. Vaidyanathan, K.H. Kwon, H.J. Park, Y.-I. Kim, et al., A novel blue-emitting silica-coated $KBaPO_4$: Eu^{2+} phosphor under vacuum ultraviolet and ultraviolet excitation, Mater. Chem. Phys. 115 (1) (2009) 161–164.

[198] K.N. Shinde, S.J. Dhoble, H.C. Swart, K. Park, Phosphate Phosphors for Solid-State Lighting, Springer Science & Business Media, 2012.

[199] J. Yu, C. Guo, Z. Ren, J. Bai, Photoluminescence of double-color-emitting phosphor $Ca_5(PO_4)_3Cl:Eu^{2+},Mn^{2+}$ for near-UV LED, Opt. Laser Technol. 43 (4) (2011) 762–766.

[200] J. Gou, J. Wang, B. Yu, D. Duan, S. Ye, S. Liu, Li doping effect on the photoluminescence behaviors of $KSrPO_4:Dy^{3+}$ phosphors for WLED light, Mater. Res. Bull. 64 (2015) 364–369.

[201] M. Yu, W. Zhang, S. Qin, J. Li, K. Qiu, Synthesis and luminescence properties of single-component $Ca_5(PO_4)_3F$: Dy^{3+}, Eu^{3+} white-emitting phosphors, J. Am. Ceram. Soc. 101 (10) (2018) 4582–4590.

[202] Y. Wang, G. Lu, Y. Qiu, W. Sun, S. Qin, Y. Lin, et al., Synthesis and optical properties of novel apatite-type $NaCa_3Bi(PO_4)_3F:Dy^{3+}$ yellow-emitting fluorophosphate phosphors for white LEDs, J. Rare Earths 40 (12) (2021) 1827–1836.

[203] L. Li, C. Guo, S. Jiang, D.K. Agrawal, T. Li, Green up-conversion luminescence of Yb^{3+}-Er^{3+} co-doped $CaLa_2ZnO_5$ for optically temperature sensing, RSC Adv. 4 (13) (2014) 6391–6396.

[204] D. Xu, D. Haranath, H. He, S. Mishra, I. Bharti, D. Yadav, et al., Studies on phase stability, mechanical, optical and electronic properties of a new Gd_2CaZnO_5 phosphor system for LEDs, CrystEngComm 16 (9) (2014) 1652–1658.

[205] F. Li, L. Li, C. Guo, T. Li, H.M. Noh, J.H. Jeong, Up-conversion luminescence properties of Yb^{3+}–Ho^{3+} co-doped $CaLa_2ZnO_5$, Ceram. Int. 40 (5) (2014) 7363–7366.

[206] H.-R. Shih, Y.-Y. Tsai, K.-T. Liu, Y.-Z. Liao, Y.-S. Chang, The luminescent properties of Pr^{3+} ion-doped BaY_2ZnO_5 phosphor under blue light irradiation, Opt. Mater. 35 (12) (2013) 2654–2657.

[207] M. Dalal, V.B. Taxak, S. Chahar, J. Dalal, A. Khatkar, S.P. Khatkar, Judd-Ofelt and structural analysis of colour tunable BaY_2ZnO_5: Eu^{3+} nanocrystals for single-phased white LEDs, J. Alloys Compd. 686 (2016) 366–374.

[208] S. Kaur, A.S. Rao, M. Jayasimhadri, Spectroscopic and photoluminescence characteristics of Sm^{3+} doped calcium aluminozincate phosphor for applications in w-LED, Ceram. Int. 43 (10) (2017) 7401–7407.

[209] P. Sehrawat, R.K. Malik, R. Punia, S.P. Khatkar, V.B. Taxak, Augmenting the photoluminescence efficiency via enhanced energy-relocation of new white-emanating $BaYAlZn_3O_7:Dy^{3+}$ nano-crystalline phosphors for WLEDs, J. Alloys Compd. 879 (2021), 160371.

[210] X. Zhu, Y. Pu, X. Pu, X. Li, J. Sun, S. Zhang, et al., Tuning of red-emitting luminescence through host-sensitized luminescence effect in Eu^{3+} doped $Sr_{1-x}Ca_xZnO_2$ phosphors, Mater. Lett. 237 (2019) 176–179.

[211] L. Yang, X. Yu, S. Yang, C. Zhou, P. Zhou, W. Gao, et al., Preparation and luminescence properties of LED conversion novel phosphors $SrZnO_2$:Sm, Mater. Lett. 62 (6–7) (2008) 907–910.

[212] P. Sehrawat, R.K. Malik, R. Punia, M. Sheoran, N. Kumari, S.P. Khatkar, et al., Luminescence tuning and structural analysis of new $BaYAlZn_3O_7$: Sm^{3+} nanomaterials with excellent performance for advanced optoelectronic appliances, J. Mater. Sci. Mater. Electron. 32 (12) (2021) 15930–15943.

[213] P. Sehrawat, R.K. Malik, R. Punia, S.P. Khatkar, V.B. Taxak, Probing into multifunctional deep orange-red emitting Sm^{3+}-activated zincate based nanomaterials for wLED applications, Chem. Phys. Lett. 777 (2021), 138743.

[214] J.-C. Zhang, L.-Z. Zhao, Y.-Z. Long, H.-D. Zhang, B. Sun, W.-P. Han, et al., Color manipulation of intense multiluminescence from $CaZnOS:Mn^{2+}$ by Mn^{2+} concentration effect, Chem. Mater. 27 (21) (2015) 7481–7489.

[215] P. Sehrawat, R.K. Malik, R. Punia, S. Maken, N. Kumari, Ecofriendly synthesis and white light-emitting properties of $BaLa_2ZnO_5$: Dy^{3+} nanomaterials for lighting application in NUV-WLEDs and solar cells, Chem. Phys. Lett. 139399 (2022).

[216] V. Singh, G. Lakshminarayana, A. Wagh, Photoluminescence investigation on green-emitting Tb^{3+}-doped $BaLa_2ZnO_5$ nanophosphors, J. Electron. Mater. 49 (1) (2020) 510–517.

[217] Y. Zhang, Q. Cui, Z. Wang, G. Liu, T. Tian, J. Xu, Up-converted ultraviolet luminescence of Er^{3+}: $BaGd_2ZnO_5$ phosphors for healthy illumination, Front. Mater. Sci. 10 (3) (2016) 328–333.

[218] P. Sehrawat, R.K. Malik, R. Punia, N. Kumari, Design of bright-green radiating Er^{3+}-singly activated zincate-based nanomaterials for high-performance optoelectronic devices, J. Electron. Mater. 51 (1) (2022) 391–402.

[219] M. Allix, S. Chenu, E. Véron, T. Poumeyrol, E.A. Kouadri-Boudjelthia, S. Alahrache, et al., Considerable improvement of long-persistent luminescence in germanium and tin substituted $ZnGa_2O_4$, Chem. Mater. 25 (9) (2013) 1600–1606.

[220] Y. Wu, Y. Li, X. Qin, R. Chen, D. Wu, S. Liu, et al., Dual mode NIR long persistent phosphorescence and NIR-to-NIR Stokes luminescence in $La_3Ga_5GeO_{14}$:Cr^{3+}, Nd^{3+} phosphor, J. Alloys Compd. 649 (2015) 62–66.

[221] F. Liu, W. Yan, Y.-J. Chuang, Z. Zhen, J. Xie, Z. Pan, Photostimulated near-infrared persistent luminescence as a new optical read-out from Cr^{3+}-doped $LiGa_5O_8$, Sci. Rep. 3 (1) (2013) 1–9.

[222] D. Chen, Y. Chen, H. Lu, Z. Ji, A bifunctional Cr/Yb/Tm: $Ca_3Ga_2Ge_3O_{12}$ phosphor with near-infrared long-lasting phosphorescence and upconversion luminescence, Inorg. Chem. 53 (16) (2014) 8638–8645.

[223] Y. Zhuang, J. Ueda, S. Tanabe, Tunable trap depth in Zn $(Ga_{1-x}Al_x)_2O_4$: Cr, Bi red persistent phosphors: considerations of high-temperature persistent luminescence and photostimulated persistent luminescence, J. Mater. Chem. C 1 (47) (2013) 7849–7855.

[224] V. Anselm, T. Jüstel, On the photoluminescence and energy transfer of $SrGa_{12}O_{19}$: Cr^{3+}, Nd^{3+} microscale NIR phosphors, J. Mater. Res. Technol. 11 (2021) 785–791.

[225] P. Ge, K. Sun, H. Li, F. Yang, S. Ren, Y. Ding, Near-infrared up-converted persistent luminescence in $Zn_3Ga_2SnO_8$: Cr^{3+}, Yb^{3+}, Er^{3+} nano phosphor for imaging, Optik 218 (2020), 164944.

[226] Y.-Y. Lu, F. Liu, Z. Gu, Z. Pan, Long-lasting near-infrared persistent luminescence from β-Ga2O3: Cr3+ nanowire assemblies, J. Lumin. 131 (12) (2011) 2784–2787.

[227] A. Luchechko, Y. Zhydachevskyy, S. Ubizskii, O. Kravets, A.I. Popov, U. Rogulis, et al., Afterglow, TL and OSL properties of Mn^{2+}-doped ZnGa2O4 phosphor, Sci. Rep. 9 (1) (2019) 1–8.

[228] J.S. Kim, J.S. Kim, H.L. Park, Optical and structural properties of nanosized $ZnGa_2O_4$: Cr^{3+} phosphor, Solid State Commun. 131 (12) (2004) 735–738.

[229] G.N.A.D. Guzman, V. Rajendran, Z. Bao, M.H. Fang, W.K. Pang, S. Mahlik, R.S. Liu, Multi-site cation control of ultra-broadband near-infrared phosphors for application in light-emitting diodes, Inorg. Chem. 59 (2020) 15101–15110.

Advancing biosensing with photon upconverting nanoparticles

8

Anita Kumari[a], *Ranjit De*[b], *and Manoj Kumar Mahata*[c]
[a]Department of Physics, Indian Institute of Technology (Indian School of Mines), Dhanbad, India, [b]Department of Materials Science and Engineering, Pohang University of Science and Technology, Pohang, South Korea, [c]Third Institute of Physics, Georg-August-Universität Göttingen, Göttingen, Germany

Chapter outline

1 Introduction

The upconverting nanoparticles (UCNP) are well known for excellent sensing capability and have a large number of unique properties, making UCNP-based nanoprobes convenient for biosensing applications [1–10]. These materials are becoming a topic of attraction for biomolecules because they can be decorated with an effective surface with high biological activity [11–15]. Additionally, these materials show narrow spectral lines, zero autofluorescence, enhanced signal-to-noise ratio, and very low photobleaching, which are promising features for designing luminescent sensors [16–21]. By tuning the emissions, UCNPs can be used in sensitive and efficient multiplexed biodetection [22]. Moreover, UCNPs exhibit various properties such as biocompatibility, nontoxicity, excellent optical, chemical, and mechanical properties, along with high penetration depth and high detection sensitivity [23–31]. These properties of the UCNPs facilitate the biomolecules, a suitable environment that is very important for biosensing. The conventional bioprobes were mostly based on organic dyes and semiconductor quantum dots. Organic dyes, despite possessing numerous advantageous properties, typically exhibit excitation by visible or ultraviolet light, which can be detrimental to

Metal Oxides for Next-Generation Optoelectronic, Photonic, and Photovoltaic Applications
https://doi.org/10.1016/B978-0-323-99143-8.00015-8

biological entities. Furthermore, in spite of their comparably low size-tunable light emission, intense photoluminescence, fair photostability, wide UV excitation, and somewhat fine emission bands, the quantum dots are not suitable for biotechnology due to their inherent toxicity, chemical instability, hydrophobic characteristics, and cost implications [32–39]. Therefore, UCNPs are excellent nanoparticles for biosensing applications and have better features than organic dyes and quantum dots, mostly because of their capability of low-to high-energy light conversion. These nanoparticles are excited by near-infrared (NIR) radiation and exhibit outstanding luminescent properties, characterized by their nontoxic nature, excellent photostability, chemical stability, and narrow absorption and emission spectra [40,41].

UCNPs exhibit a wide array of potential applications in the field of biosensing, which involves the detection and measurement of biological molecules or processes. Some of the main applications of UCNPs in biosensing include (i) biomarker detection: UCNPs have been employed to detect specific biomarkers, such as cancer markers; (ii) cell imaging: UCNPs can be used to label cells, which can then be visualized using a microscope; (iii) biomolecule sensors: UCNPs can be used to design biosensors; (iv) Point-of-care diagnostics: UCNPs can be used in the development of portable diagnostic devices; (v) in vivo imaging: these nanoparticles are also employed for bioimaging living organisms or tissues. Therefore, UCNPs have numerous advantages in biosensing applications due to their unique optical properties, high sensitivity and specificity, and ability to be functionalized with various biomolecules.

For biodetection, several studies on Forster resonance energy transfer (FRET) or luminescence resonance energy transfer (LRET) UCNP-based nanosensors have been developed [42–48]. In such a system, luminescence donor, acceptor, and various biomolecular entities are conjugated. The donor's emission can be actively absorbed by the acceptor in which energy is transmitted via intermolecular dipole-dipole coupling from the donor to the acceptor [42]. For example, Hazara et al. have studied the 3,5-dinitrobenzoic acid-coated UCNPs for melamine detection. In this work, to detect melamine, the use of UCNPs has been done for the first time [49]. For the purpose of detecting antihuman immunoglobulin (IgG) in goat, Wang et al. reported a FRET-based nanosensor employing donor UCNPs and acceptor gold NPs [50]. An oligonucleotide detection system on the basis of LRET was studied to know the sickle cell disease by Zhang et al. [51]. In this study, *N,N,N′,N′*-tetramethyl-6-carboxyrhodamine (TAMRA) was utilized as an acceptor, whereas UCNPs were served as donors.

Within this chapter, our attention is directed toward the utilization of UCNPs as nanoprobes in biosensing applications, accompanied by an exploration of their background and synthesis. An overview of recent advancements in the employment of various UCNP-based phosphors for biosensing purposes, along with their underlying mechanisms, is presented.

2 Background of UCNPs and their synthesis

Usually, UCNPs are combinations of a host and the luminescent centers (activators and sensitizers) [52–57]. The activators and sensitizers are rare-earth ions. Sensitizers are employed to enhance the optical absorption of activators, which absorb energy, followed by energy migration to the other activator ions. The transfer of energy takes place in this process through luminescent materials when the absorption of the activator ions is weak. Basically, in the upconversion process, UCNPs are excited by near-infrared (NIR) light, which has a long wavelength, leading to the emission of shorter-wavelength light. The emission intensity and the color emitted from the UCNP materials can be controlled by selecting the appropriate dopant ions, as well as by varying the concentration of the dopants [7,12,58–60]. The host lattice can also cause optical absorption and emission through the transfer of energy to activators or sensitizers. The underlying mechanisms of conversion phenomena are ground-state absorption (GSA), excited-state absorption (ESA), energy-transfer upconversion (ETU), and photon avalanche (PA) [61].

Ground-state absorption (GSA): In this process, a ground-state molecule absorbs a photon to reach an excited state. It is the most basic phenomenon of absorption, and this straightforward phenomenon serves as the foundation for various other photon absorption processes. GSA constitutes the initial step in the absorption of photons and sets the stage for the exploration of more multifaceted absorption phenomena.

Excited-state absorption (ESA): In this mechanism, a molecule in an excited state takes another photon, leading to further energy transfer or chemical reactions. ESA can be used to generate a population of highly excited molecules.

Energy-transfer upconversion (ETU): This is a process in which multiple low-frequency photons are absorbed by a sensitizer molecule, which then transfers its energy to an emitter molecule. Afterward, the emitter molecule can release a high-energy photon, leading to the overall upconversion of the absorbed light.

Photon avalanche (PA): This is a process in which a photon absorbed by a molecule leads to the rapid production of many more photons through a chain reaction. PA can be used to generate intense light.

Overall, these different processes are used to achieve different objectives, and the selection of the appropriate process will depend on the desired outcome and the properties of the system used. To produce high-quality UCNPs, there are three widespread methods of synthesis: (a) thermal decomposition, (b) hydrothermal synthesis, and (c) ionic liquid-based synthesis.

2.1 Thermal decomposition technique

The thermal decomposition method is a widely used technique for producing well-shaped and precisely sized particles. It involves dissolving organic precursors, such as trifluoroacetate compounds, utilizing surfactants composed of hydrocarbon chains such as oleylamine, oleic acid, and 1-octadecence and polar capping groups in conjunction with high-boiling organic solvents. The method requires high temperatures (around 250–330 °C) in an inert gas environment, and the use of organic solvents. However, the presence of surfactants used to stabilize the nanomaterials can pose challenges for biological applications. [62,63]. One of the advantages of thermal decomposition is that it is a relatively simple and easy-to-use method compared to other methods of UCNP synthesis. Additionally, it allows UCNPs' synthesis with large particle sizes, which can be beneficial for certain applications. Thermal decomposition also allows for the synthesis of UCNPs with high crystallinity, which can lead to improved optical properties. However, thermal decomposition also has some limitations. For example, the high temperatures used in the synthesis can cause damage to some of the precursors. Additionally, thermal decomposition is not compatible with aqueous solutions, and it can be difficult to scale up for mass production.

2.2 Hydrothermal synthesis

Hydrothermal synthesis is a method used to synthesize UCNPs by using high-temperature and high-pressure in aqueous reaction medium. In this method, the precursors of the UCNPs, such as rare-earth ions, are dissolved and then heated to high temperatures (~300 °C) and high pressures (typically around 10–20 bar) in a sealed reactor. The high temperature and pressure conditions facilitate the formation of UCNPs through processes such as nucleation and crystal growth. This method is effective and suitable for developing inorganic UCNPs with controllable morphologies [64,65]. One of the advantages of hydrothermal synthesis is that it enables precise regulation of the particle size and shape, which can affect the optical properties and performance of the UCNPs. Hydrothermal synthesis also allows for the synthesis of UCNPs with uniform size and shape, which is difficult to achieve with other methods. Additionally, hydrothermal synthesis can be conducted in aqueous solutions, which makes it suitable for biological applications.

2.3 Ionic liquid-based synthesis

UCNP synthesis can be achieved through the use of an ionic liquid-based method by using ionic liquids as the reaction medium. In ionic liquid-based synthesis, the precursors of the UCNPs, such as rare-earth ions, are dissolved in the ionic liquid and then heated to a certain temperature, typically around 150–300 °C. The ionic liquid can act as both solvent and reactant, and the unique properties of ionic liquids can also help to improve the nucleation and crystal growth processes of the UCNPs. It allows for a milder reaction condition compared to other methods such as thermal decomposition, making it more suitable for certain precursors. However, ionic liquid-based synthesis also has some limitations. For example, the cost of ionic liquids can be relatively high, and the reaction conditions can be difficult. In this method, organic solvents are not needed to develop the UCNPs; also, it takes relatively less time and a low temperature to complete the reaction. Though the synthesis procedure is environment friendly, the produced nanoparticles have a wide range of size sequences, less dispersity, and non-uniform particle size compared to those of the two methods mentioned earlier. This is the reason that the hydrothermal and thermal decomposition methods are considered the best techniques for developing nanoparticles with the requisite shape and size.

3 Application of UCNP-based biosensors

A potential application of UCNPs in biology is in the field of biosensing. Biosensing refers to the detection and measurement of biological molecules or processes, and UCNPs, having been shown to be an effective tool for this purpose, have proven to be a powerful resource. UCNPs can be functionalized with various biomolecules, such as antibodies, aptamers, or peptides, to target specific molecules. UCNPs have been used for the detection of disease markers, such as cancer biomarkers, by using the UCNP to capture the biomarker and then using a fluorescent probe to detect the presence of the biomarker. This method can be used to detect the presence of specific molecules such as nucleic acids or proteins with high sensitivity and specificity. UCNPs have also been used to create biosensors.

Currently, UCNPs have been used in various biosensing applications. Even so, their potential has yet to be fully realized, as certain challenges remain to be overcome. One of the main challenges is to enhance the precision and specificity of biosensors based on UCNPs, which will require the progress of novel UCNP-based materials, and the optimization of the sensing mechanism. Another challenge is to make UCNPs more biocompatible, which will necessitate the creation of novel techniques for surface modification.

Modifying the surface of UCNP nanoprobes is crucial for specific detection, as the surface plays a critical role in numerous physical and chemical processes. Following are the promising methodologies for surface modification: (i) synthetic routes for a hydrophilic surface, (ii) ligand exchange, (iii) ligand oxidation, and interaction (iv) silica encapsulation [66–71]. UCNP nanoprobes exhibit high response and sensitivity to detection targets when paired with suitable recognition entities.

Fig. 1A shows the mechanism of LRET for upconversion [72]. Generally, this is the case within 1–10 nm distance. In this process, UC emission intensity of UCNPs decreases, while the emission of the acceptor increases. The donor's lifetime is shortened as a result of energy transfer.

In Fig. 1B–D, the mechanism for the inner filter effect (IFE) has been shown [73]. In this case, the lifetime of the upconversion donor is constant. The inner filter effect normally involves emission photons and their reabsorption. In addition to interacting with the additional analyte via chemical reaction, there must also be a significant change in the probe's absorption band after interaction with the analyte, leading to an alteration in the intensity of UC emission from the UCNPs.

3.1 Applications of UCNPs as biosensors based on FRET/LRET process

The Förster resonance energy transfer (FRET) or luminescence resonance energy transfer (LRET) process is one of the most widely used mechanisms for UCNP-based biosensing. This process involves the transfer of energy from a donor molecule (UCNP) to an acceptor molecule (sensor molecule) when they are in close proximity. The FRET/LRET process is highly sensitive and selective, making it ideal for biosensing applications.

Wang et al. developed a sensor using $NaYF_4$:Yb/Er UCNPs and gold nanoparticles that utilizes FRET. With the addition of avidin to the mixture of biotin-modified gold nanoparticles and $NaYF_4$:Yb/Er UCNPs, FRET process occurs, and hence the energy transfers from the donor (UCNPs) to that of the acceptors (biotin-modified gold

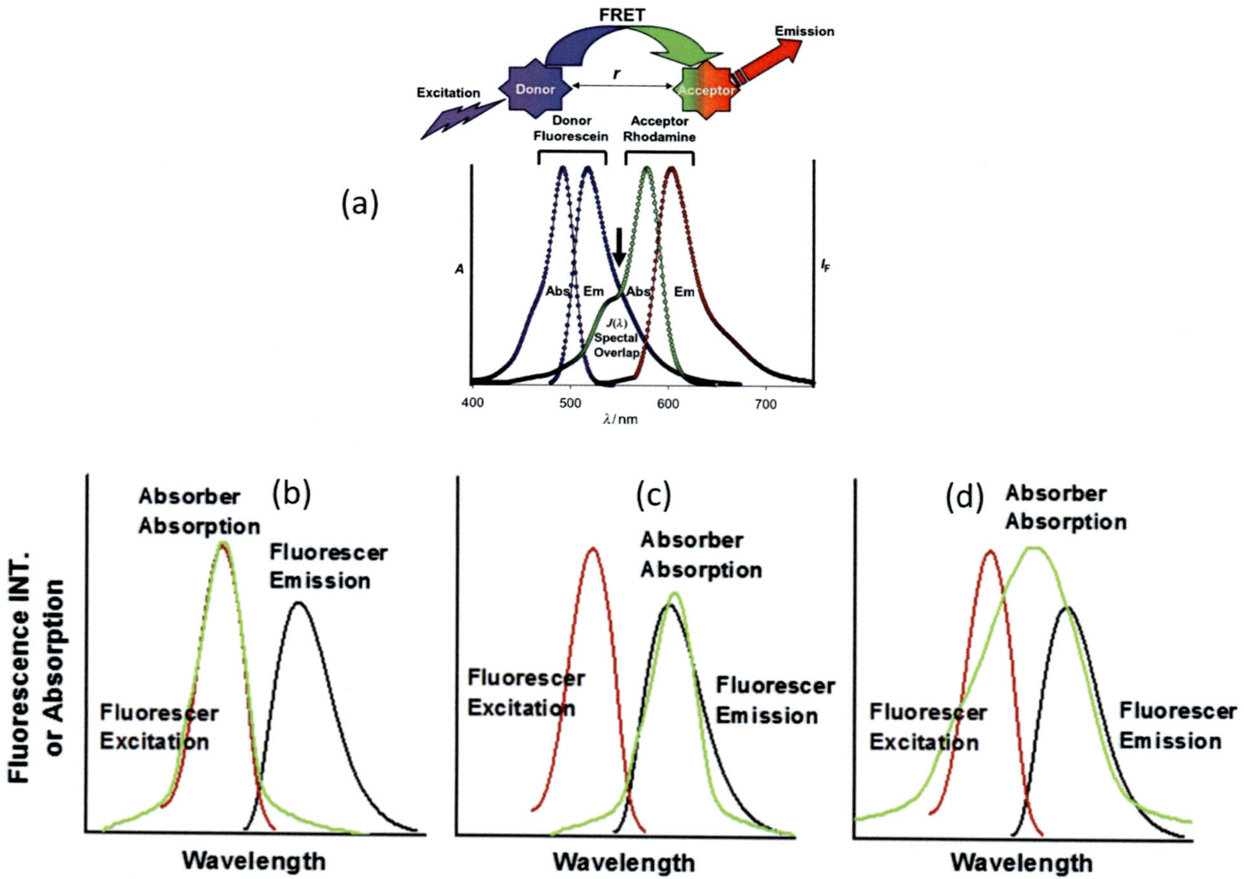

Fig. 1 (A) Schematic description of FRET process [72]; conditions for inner filter effect: The absorption of fluorescence should overlap with the spectrum of excitation (B), emission (C), and both excitation and emission (D) [73].

nanoparticles), resulting in a decrease in the emission intensity of the UCNPs. This allowed for the detection of avidin through a linear correlation of the relative luminescence intensity with the avidin concentration [74].

Another study by Kim et al. [75] used LRET to detect homogenous glycated hemoglobin (HbA1c) using the donor $NaYF_4:Yb^{3+},Er^{3+}$ and the acceptor HbA1c. HbA1c's absorption in the green region around 541 nm matches with UCNPs' emission. Upon laser irradiation at 980 nm, the specific binding between the anti-HbA1c monoclonal antibody-functionalized UCNPs and HbA1c triggered LRET, causing a reduction in the upconversion emission intensity when HbA1c is present (Fig. 2) [75].

Liu and his team employed a spectral overlap approach to detect cyanide ions (CN^-) in water using the LRET mechanism [76]. They utilized $NaYF_4$:20% Yb,1.6%Er,0.4%Tm UCNPs in combination with a CN^--responsive chromophoric Ir(III) complex (Ir1) to achieve this detection. They detected the CN^- by using the Tm^{3+} ion's 800-nm emission (reference signal) with a 0.18-μM detection limit.

The chemical reaction that occurs between CN^- and the a,b-unsaturated carbonyl moiety resulted in the observation of a weak absorption band in around ~400 to 600 nm visible range in the presence of CN^-. Due to this reduction in the overlap of the spectral band between the UCNPs' emission and Ir1 absorption, LRET was expected to be suppressed (Fig. 3) [76].

LRET-based UCNP probe for Cu^{2+} detection is a highly specific method for detecting copper ions.

Shi's group developed a multifunctional UCNP sensor based on LRET for detecting Cu^{2+} ions (Fig. 4A). The UCNPs were coated with 1,2-distearoyl-sn-glycero-3-

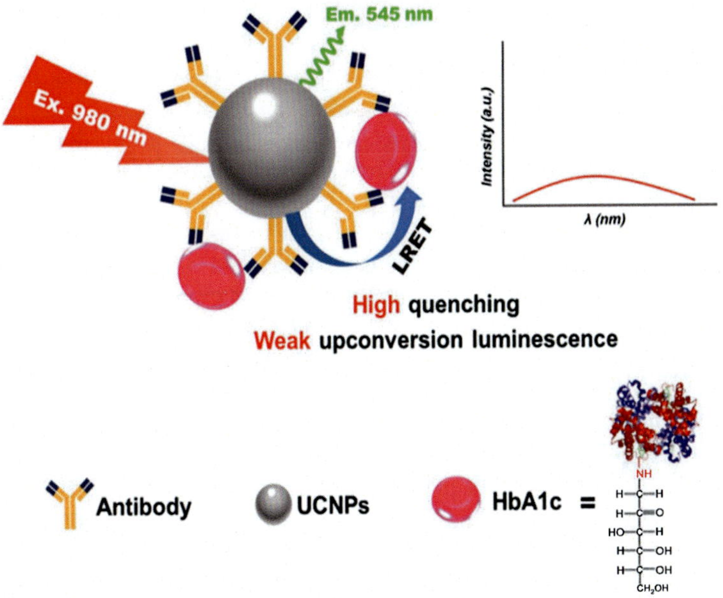

Fig. 2 LRET-based upconversion immunosensor [75].

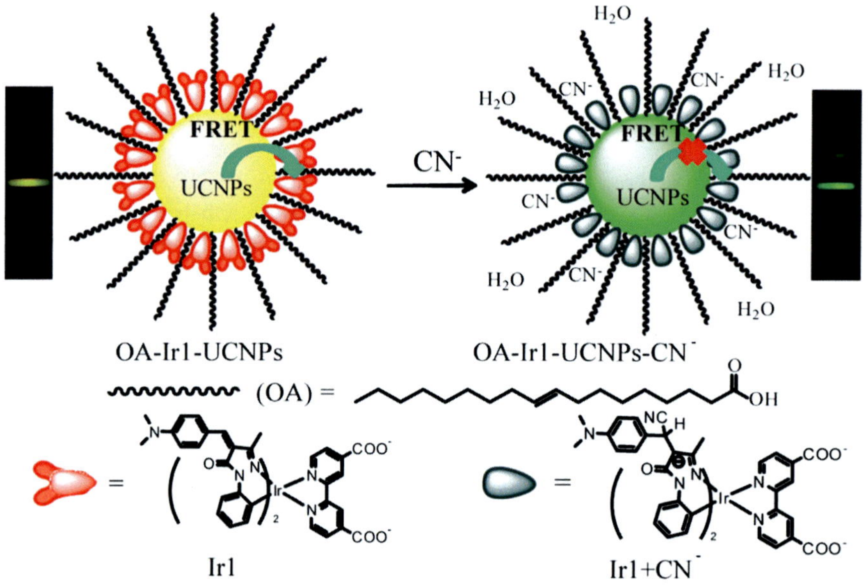

Fig. 3 (A) Detection of CN^- employing $NaYF_4$:20%Yb,Er,Tm, before and after the treatment of CN^-.
Adapted with permission from Liu, J., Liu, Y., Liu, Q., Li, C., Sun, L., Li, F., J. Am. Chem. Soc. 133 (2011), 15276–15279.

phosphoethanolamine-*N*-[maleimide (polyethylene glycol)-2000] (DSPE-PEG2000-NH_2) and 8-hydroxyquinoline-2-carboxylic acid (HQC). The developed composite is also useful for the treatment of Alzheimer's disease [77]. However, further research and development are needed to overcome current challenges and improve its performance.

In addition, to detect H_2S in living cells, Huang and his group demonstrated an upconversion-based nanoprobe [78]. This probe has been designed as a hybrid core-shell (core: $NaYF_4$:Yb,Er,Tm, and shell: $mSiO_2$) incorporating a merocyanine (MC)-based dye that is sensitive to H_2S, namely MC, as shown in Fig. 4B. The UCNPs exhibited an emission around 514–560 nm, which had a strong overlap with the absorption of the MC-based H_2S-sensitive dye at 548 nm. This enabled efficient LRET with 980 nm excitation. When H_2S was introduced, the absorbance at 548 nm reduced, which decreased the efficiency of LRET, but UCNPs' green emission remained unchanged.

Li et al. have reported findings utilizing self-assembled chiroplasmonic nanopyramids made of gold and upconversion nanoparticles for detecting microRNA in live cells [79]. The Au-UCNP pyramids have strong circular dichroism (CD) and luminescence properties, which can be used to detect miRNA. The study found that the CD signal was much more sensitive to miRNA concentration than the luminescence signal, owing to the robust circular dichroism intensity generated by the photon interaction with chiral nanostructures and the amplification of the inherent chirality of

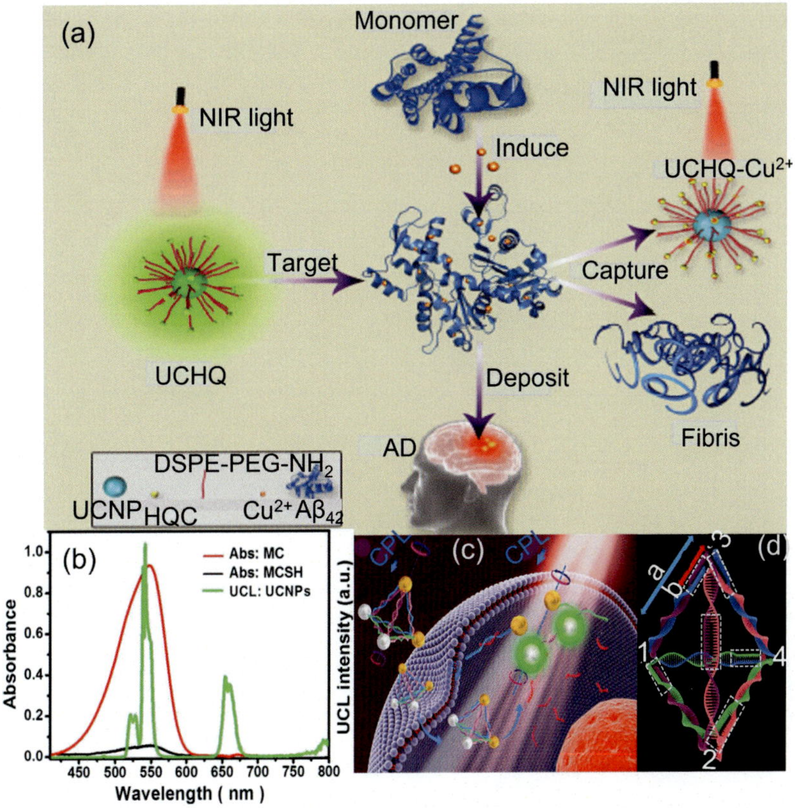

Fig. 4 (A) LRET-based UCNP nanoprobe for Cu^{2+} sensing [77], (B) absorption and upconversion spectra of merocyanine (MC), the HS^- adduct of MC (MCSH), and UCNPs ($NaYF_4$:Yb,Er,Tm) [78], (C) the use of Au-UCNP pyramids for detecting miRNA [79], and (D) the pyramid's nucleic acid structure used in the detection process [79].

DNA molecules within the pyramid. The miRNA recognition sequences, represented by part a in Fig. 4D, are located on both sides of the DNA frame. Additionally, non-complementary sequences, represented by part b and indicated by white dashed lines in Fig. 4D, also appear on both sides of the DNA frame. The separation of gold nanoparticles and UCNPs within the pyramid leads to a significant improvement in the recovery of emission from UCNPs, as seen in Fig. 4C and D [79].

Peng and co-authors have reported a new method utilizing a luminescent method to evaluate hepatotoxicity in living organisms, using UCNPs that are conjugated with chromophores, which emit "turn-on" luminescence, as illustrated in Fig. 5 [80]. This system can detect $ONOO^-$ with a very low limit and quickly, with a response time of less than a second. Additionally, the distinct optical characteristics of these nanoparticles have allowed for the detection of reactive nitrogen species (RNS) in living animals with hepatotoxicity caused by a certain drug. This sensing platform could be used as a simple test kit for identifying hepatotoxicity in new drugs.

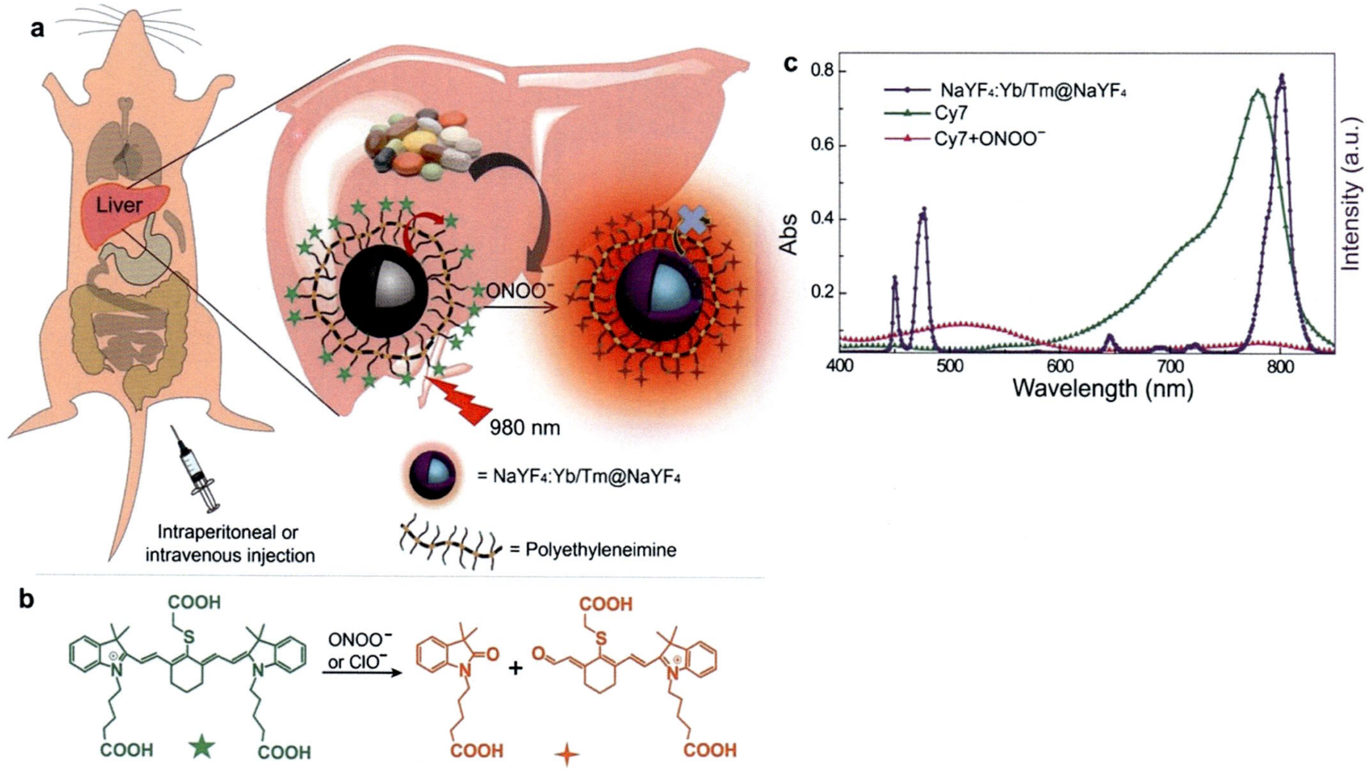

Fig. 5 (A) Chromophore-assembled UCNPs for nitrosative hepatotoxicity detection in vivo. (B) A proposed mechanism for activating luminescence involving the degradation of Cy7 (energy acceptor, represented by a *green star*) through $ONOO^-$ or ClO^- oxidation. (C) The absorption of the chromophore when not present (*green line*) and when present (*red line*) $ONOO^-$, as well as the upconversion spectrum when excited by 980 nm light (*purple line*) [80].

They presented a novel platform of UCNPs for the detection of $ONOO^-$ with the lowest concentration that could be detected, which was 0.08 μM. In this work, donor $NaYF_4$:Yb/Tm@$NaYF_4$@PEI@cyanine (Cy7) was the nanoprobe where UCNPs were selected as the donor and NIR-absorbing chromophore Cy7 was the acceptor. UCNPs cannot absorb Cy7's luminescence in the presence of reactive nitrogen species, resulting in a suppressed LRET process [80].

The detection of oligonucleotides has been presented in Fig. 6A [81]. In this UCNP-based sensing technique, two short oligonucleotides can be combined to form the longer target nucleotide, which is attached to a fluorophore and the upconverting

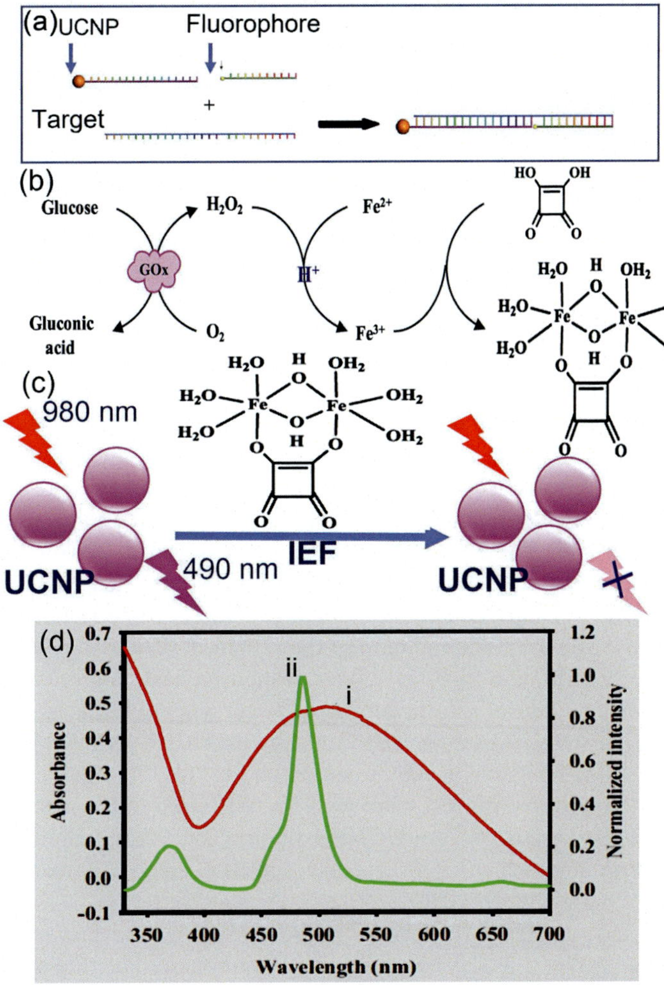

Fig. 6 (A) Design of nucleotide sensor [81], (B,C) glucose detection mechanism [82]; (D) absorption spectrum of SQA-iron (III) (curve i) overlaps with UC of $NaYF_4$:Yb,Tm (curve ii) [82].

nanoparticle. In this situation, the UCNPs served as the source of energy and the fluorophore acted as the recipient of energy. The fluorophore was brought close to the UCNPs when the luminescent nanoparticle was excited and a FRET process occurred between the upconverting nanoparticle and the fluorophore. Therefore, the oligonucleotide can be identified by observing the intensity of the upconversion emission. Thus, a new nucleotide sensor that utilizes photon-upconverting particles as energy sources and a suitable fluorophore as energy receptors has been developed. This design demonstrates high sensitivity, specificity, and self-calibration abilities. By using Er^{3+}–doped $NaYF_4$ particles as energy donors and fluorophore-TAMRA as energy receptors, they were able to quantitatively detect perfectly matched target DNA with a limit of detection of 1.3 nM and distinguish targets with single nucleotide variations using a 50-mW excitation energy. This type of nucleotide sensor is expected to be useful for DNA/RNA detection and protein-DNA/RNA interaction studies due to its low background noise, high sensitivity, and minimal reliance on the optical configuration.

3.2 Application of UCNPs as biosensor based on IFE process

A novel fluorescent probe was developed by Li and his group based on the method of IFE to detect Cys in human serum with a detection limit of 20 μM in a water-based solution [83]. In the study, a combination of donor, $NaYF_4$:Yb,Tm UCNPs, and acceptor, 5(6)-carboxyfluorescein-O,Ó-diacrylate, was used.

The detection of fluoride ion is very important because its excessive intake can be harmful and it can result in a significant health risk. Therefore, the detection of the presence of fluoride's accurate quantity is needed. Liu and his group have reported a sensor based on IFE with high efficiency using UCNPs to detect fluoride ion [84]. In this report, Yb^{3+}-, Er^{3+}-, and Tm^{3+}-doped $NaYF_4$ UCNPs were mixed with the recognition element curcumin to make a nanosystem. With the addition of fluoride ion, bathochromic shift has been observed in the absorption peak of curcumin, resulting in quenching UC emission around 546 nm and 657 nm, while the upconversion at 758 nm and 812 nm did not change. The developed nanosensor could detect F^- ions based on the color and ratio of fluorescence, with high precision, a low detection limit of 25 μM, and high selectivity, responsiveness, and sensitivity.

A UCNP-based nanosensor was developed to detect glucose levels by utilizing the IFE between UCNPs and SQA-Iron (III). The $NaYF_4$:Yb,Tm UCNPs served as the donor and SQA-Iron(III) as the acceptor, shown in Fig. 6B–D [82]. The overlap of the UCNP's upconversion emission band with the SQA-Iron (III)'s absorption band led to a substantial decrease in the UCNP's fluorescence. This approach is cost-effective, sensitive, and selective for glucose detection, with a detection limit of 2.3 μM.

It is crucial and necessary to have a simple, sensitive, and fast technique for detecting tyrosine because it plays a vital role in hormone production in the human body. A photoinduced UCNP biosensor, utilizing the PET mechanism between $NaYF_4$:Yb,Tm UCNPs and melanin-like polymers, was developed to detect tyrosine [85].

3.3 Other biosensing applications

Apart from these, some physiological conditions, such as temperature and pH, can be measured by UCNPs [1,5,6,13]. In addition, UCNPs can be used to improve diagnostic techniques. UCNPs can be utilized to detect specific biomarkers, such as cancer markers. The biomarker is captured by the UCNP, and its presence is detected using a fluorescent probe [86].

Furthermore, UCNPs have been used in in vivo imaging of small animals, and their small size and high brightness make them ideal for deep tissue imaging. This can be used to study the progression of diseases, monitor the effectiveness of treatments, and study the distribution of specific molecules in the organism [87].

Another potential application of UCNPs in biology is in the field of bioimaging. UCNPs have been used in bioimaging due to their ability to convert low-energy NIR light into higher-energy visible light; this allows them to be excited with NIR light and be detected with high sensitivity and specificity. This property makes them suitable for in vivo bioimaging as NIR light has the ability to penetrate deep into tissues without significant absorption or scattering. This allows for real-time, noninvasive, and high-resolution imaging of biological systems. UCNPs have also been used in superresolution imaging [88–90]. Superresolution imaging is an advanced imaging technique that allows for the visualization of structures at a higher resolution than the diffraction limit of light. UCNPs have been used as a contrast agent in superresolution imaging to increase the signal-to-noise ratio and improve the resolution of the image. Additionally, UCNPs have been used in photoacoustic imaging [91,92], a technique that combines the advantages of optical imaging and ultrasonic imaging. In photoacoustic imaging, UCNPs are used as a contrast agent that absorbs light and generates heat; this generates an ultrasonic wave that can be detected and used to generate an image. This technique has the potential to be used for the early diagnosis of cancer and other diseases. Thus, UCNPs have great potential for bioimaging applications due to their unique optical properties, high sensitivity, and specificity. Their ability to convert light makes them suitable for in vivo bioimaging, superresolution imaging, and photoacoustic imaging. These techniques have the potential to be used for the early diagnosis of cancer and other diseases and to study biological systems with high resolution and noninvasively [93–96].

Therefore, UCNPs are a versatile and powerful tool for biological applications and possess diverse potential uses in imaging, drug delivery, and point-of-care diagnostics [97]. Their unique optical properties, high sensitivity and specificity, and ability to be functionalized with various biomolecules make them an attractive option for future biological research and applications.

UCNPs have been explored for pH-sensing applications [13]. pH is a measure of the acidity or basicity of a solution, and it is an important parameter in many biological and chemical processes. The pH sensing ability of UCNPs is based on the fact that the luminescence of the UCNPs changes with changes in the pH of the solution. One of the ways in which UCNPs have been used for pH sensing is by incorporating pH-sensitive ligands onto the surface of the UCNPs. These ligands can be designed to respond to changes in pH by altering their conformation, which in turn changes the luminescence of the UCNPs. This allows for the sensitive and specific detection of pH changes over a wide range of pH values [6,98–100].

UCNPs have also been used for pH sensing through the pH-dependent quenching of the UCNP's luminescence. This happens when the UCNP is near a quenching molecule, such as a heavy metal ion, and the quenching efficiency changes with changes in pH, enabling detection of pH changes across a wide range of pH values. They can be functionalized with pH-sensitive ligands or by using pH-dependent quenching of the UCNP's luminescence, which enables the detection of pH changes in biological systems. Intracellular pH and temperature reflect cell heath status, and to understand the pathological and physiological processes, these two parameters are important to know. We have recently shown the versatility of using $NaYF_4:Yb^{3+}/Tm^{3+}$ $@NaYF_4:Yb^{3+}@NaYF_4:Yb^{3+}/Nd^{3+}$ UCNPs with fluorescein-5-isothiocyanate (FITC) dye molecules for pH sensing upon NIR laser light excitation [13]. The mechanism of sensing relies on changes in the Tm^{3+} upconversion luminescence bands at different pHs, which is caused by the absorption of blue-upconverted light by FITC molecules. The red or near-infrared bands serve as a reference signal.

The intensity of the upconverted light emitted by the UCNPs is highly dependent on the temperature of the surrounding environment, making UCNPs an attractive option for temperature sensing. UCNPs can be used for temperature sensing by incorporating them into a sensing system and measuring the intensity of the upconverted light emitted by the nanoparticles. The intensity (or intensity ratio) of the light is directly dependent on the temperature of the UCNPs, making it possible to accurately measure the temperature of the surrounding environment. Fluorescence intensity ratio (FIR)-based temperature sensing is a method of measuring temperature using the fluorescence emitted by nanophosphor materials. It is a highly sensitive and selective method that relies on the changes in the fluorescence intensity ratio of upconversion bands upon temperature changes.

One of the main advantages of using UCNPs for temperature sensing is that they can operate in the near-infrared (NIR) region of the electromagnetic spectrum, which allows for noninvasive and remote sensing. Additionally, UCNPs are highly stable and can be functionalized for targeted sensing applications. UCNP temperature sensing applications are being developed for a wide range of fields, such as biomedical, industrial, and environmental monitoring. For example, in biomedical applications, UCNPs can be used to monitor the temperature of specific areas of the body, such as tumors or deep tissues, to help diagnose and monitor disease progression [101].

The intensity of the upconverted light emitted by the UCNPs is highly dependent on the temperature of the surrounding environment. This is because the temperature affects the population of energy levels of the rare-earth ions in the UCNPs, which affects the efficiency of the energy transfer process and therefore the intensity of the upconverted light. Another advantage of UCNPs for temperature sensing is that they can be integrated into different platforms such as fibers, films, and microspheres, which can further enhance their performance and versatility.

UCNPs also have several other advantages that make them an attractive option for temperature sensing applications. For example, they can be made with a variety of different materials, such as silica, gold, and polymers, which allows for flexibility in the design of the sensing system. Additionally, the size and shape of the UCNPs can be tailored to optimize the upconversion process and the intensity of the emitted light, which can further improve the temperature sensing [56].

In terms of research and development, scientists are currently working on developing new UCNP materials and designs that can further improve the sensitivity and accuracy of temperature sensing. For example, some researchers are exploring the use of UCNPs made with new host materials, which have unique electronic and optical properties that may enhance the upconversion process. Other researchers are working on developing new methods for functionalizing UCNPs to target specific areas or environments, which could improve the sensing. Thus, UCNPs are a promising technology for temperature sensing applications [24,25]. Research and development is ongoing in the field to further improve the temperature sensing applications.

Upconversion lifetime thermometry is a method of measuring temperature that uses the lifetime of the upconverted photons emitted by UCNPs. The lifetime of the upconverted photons is highly dependent on the temperature of the UCNPs. In upconversion lifetime thermometry, UCNPs are incorporated into a sensing system, and the lifetime of the upconverted photons emitted by the nanoparticles is measured using a technique called time-resolved fluorescence spectroscopy.

It's also worth mentioning that the development of new and advanced UCNPs is an active area of research. Researchers are continuously working on developing new UCNP materials, designs, and synthesis methods to improve the performance of UCNPs for temperature sensing and other applications. In summary, the use of UCNPs for temperature sensing is a promising technology that has several advantages, such as noninvasive, high sensitivity, and specificity.

4 Conclusions

UCNPs have shown promise in many different types of applications. These include their capability of converting low-energy photons to higher energy ones, which allows for the detection of signals at lower light intensities. Additionally, UCNPs have high photostability, chemical stability, and biocompatibility, which makes them suitable for use in bioimaging, biosensing, and biomedical applications. They have been used in a variety of research areas, including bioimaging, drug delivery, and the sensing of biomolecules, due to their ability to emit light in the NIR region, which can penetrate deep into biological tissues. With the ongoing research in this direction, the prospects of UCNPs are bright, and it is expected that they will have a significant impact on the fields of biotechnology and medicine in the future.

Because of the distinctive characteristics of UCNPs, several variables can be focused, such as UCNP emission profile, the recognition element, integration of the recognition element with the nanoparticle's surface, the sensing technique, etc. The utilization of cutting-edge techniques in the optimization of the biosensor's selectivity and sensitivity has led to the development of highly sophisticated devices capable of detecting trace amounts of analytes in complex matrices with unparalleled precision.

However, despite the realization of UCNPs in biosensing, there are still some major challenges to deal with.

(i) The low quantum yield of UCNPs restricts their use in ultra-high-resolution imaging. This is due to the fact that the energy transfer process between the different energy levels of the UCNPs is not very efficient.

(ii) The biosensing techniques are based on the dark and bright states of the FRET or LRET processes. This technique has limited signal to background noise due to the FRET mechanism, and thus, developing a method to solve this issue is a constraint.

(iii) The potential clinical application and biocompatibility of these nanoparticles are the important concerns. Though cytotoxicity of UCNPs is usually measured prior to its biological application, the exposure duration of the nanoparticles is not clear. It takes from several hours to several weeks to degrade, and thus, a longer exposure may interact with the immune and nervous systems of the body.

There are several other challenges and problems that researchers face in the field of upconversion research using UCNPs. Some of these challenges include the following:

- It is difficult to control the size and shape of UCNPs; this can affect their optical properties and limit their applications.
- The synthesis of UCNPs can be costly and time-consuming, which can make it difficult to scale up their production for commercial applications.
- UCNPs can be sensitive to environmental factors such as temperature and humidity, which can affect their stability and limit their applications.
- The compatibility of UCNPs with different materials and systems is a challenge, as the UCNPs need to be stable and retain their optical properties in these different environments.
- The UCNP's excitation and emission spectra's overlap and other fluorophores can also be a problem as it can lead to interference and false signals.
- Some of the materials used in the synthesis of UCNPs can be toxic, which can be a concern for their application in biomedical fields.
- UCNPs have potential applications in the biomedical field, such as imaging, drug delivery, and biosensing. Research should be on delivering UCNPs to specific cells or tissues and optimizing their performance in these applications.
- UCNPs could be used in energy applications, such as solar energy conversion and photocatalysis. An area of emphasis in research could be the cultivation and advancement of methodologies to improve the efficiency of UCNPs in these applications.
- UCNPs could be used for environmental monitoring, such as detecting pollutants in water or air. The area of investigation could focus on the creation and advancement of innovative techniques to make UCNPs more sensitive to specific pollutants or more stable in different environmental conditions.
- Research could focus on developing new methods to combine UCNPs with other technologies, such as quantum dots or graphene, to create new hybrid materials with unique optical properties.
- The focus of research could be directed toward the exploration and implementation of novel strategies to multiplex UCNPs, for example, by using different colors, to allow for multiplexed detection of multiple analytes.

Overall, these challenges highlight the need for further research and development to improve the properties of UCNPs and make them more suitable for various applications. In summary, deep penetrating and noninvasive NIR light-driven UCNPs are promising for biological applications, but a detailed procedure and benchmark for cytotoxicity assessment should be developed for the clinical translation for UCNP biosensing applications. Therefore, UCNPs have great potential for biosensing applications due to their unique optical properties, high sensitivity and specificity, and ability to be functionalized with various biomolecules. They can be used to detect disease markers, create biosensors, and monitor the presence of specific biomolecules in real time. These applications have the potential to be used for the early diagnosis of diseases and to study biomolecular interactions in depth.

References

[1] M.K. Mahata, T. Koppe, K. Kumar, H. Hofsäss, U. Vetter, Upconversion photoluminescence of Ho^{3+}-Yb^{3+} doped barium titanate nanocrystallites: optical tools for structural phase detection and temperature probing, Sci. Rep. 10 (2020) 8775.
[2] M. Bettinelli, L. Carlos, X. Liu, Lanthanide-doped upconversion nanoparticles, Phys. Today 68 (2015) 38–44.
[3] A. Kumari, L. Mukhopadhyay, V.K. Rai, $Er^{3+}/Yb^{3+}/Li^{+}/Zn^{2+}$: $Gd_2(MoO_4)_3$ upconverting nanophosphors in optical thermometry, J. Rare Earths 37 (2019) 242–247.
[4] M.K. Mahata, K. Kumar, V.K. Rai, Er^{3+}/Yb^{3+} doped vanadate nanocrystals: a highly sensitive thermographic phosphor and its optical nanoheater behavior, Sens. Actuators B Chem. 209 (2015) 775–780.
[5] M.K. Mahata, R. De, K.T. Lee, Near-infrared-triggered upconverting nanoparticles for biomedicine applications, Biomedicines 9 (2021) 756.
[6] M.K. Mahata, R. De, K.T. Lee, Upconversion nanoparticles in pH sensing applications, in: Upconverting Nanoparticles: From Fundamentals to Applications, 2022, pp. 395–416.
[7] A. Kumari, A.K. Soni, V.K. Rai, Near infrared to blue upconverting $Tm^{3+}/Yb^{3+}/Li^{+}$: $Gd_2(MoO_4)_3$ phosphors for light emitting display devices, Infrared Phys. Technol. 81 (2017) 313–319.
[8] K. Kumar, S.K. Maurya, M.K. Mahata, Upconversion hybrid phosphors for biological applications, in: Hybrid Phosphor Materials, Springer, Cham, 2022, pp. 195–222.
[9] A. Kumari, L. Mukhopadhyay, V.K. Rai, Energy transfer and dipole–dipole interaction in $Er^{3+}/Eu^{3+}/Yb^{3+}$:$Gd_2(MoO_4)_3$ upconverting nanophosphors, New J. Chem. 43 (2019) 6249–6256.
[10] M.K. Mahata, S. Sinha, K. Kumar, Frequency upconversion in Er^{3+} and Yb^{3+} co-doped $MgTiO_3$ phosphor, AIP Conf. Proc. 1728 (2016) 020555.
[11] D. Zhang, Y. Ni, H. Liu, W. Yao, J.S. Bu, Magnesium silicide nanoparticles as a deoxygenation agent for cancer starvation therapy, Nat. Nanotechnol. 12 (2017) 378–386.
[12] A. Kumari, M. Mondal, V.K. Rai, S.N. Singh, Photoluminescence study in $Ho^{3+}/Tm^{3+}/Yb^{3+}/Li^{+}$:$Gd_2(MoO_4)_3$ nanophosphors for near white light emitting diode and security ink applications, Methods Appl. Fluoresc. 6 (2018) 15003.
[13] M.K. Mahata, K.T. Lee, Development of near-infrared sensitized core-shell-shell upconverting nanoparticles as pH-responsive probes, Nanoscale Adv. 1 (6) (2019) 2372–2381.
[14] S. Sinha, M.K. Mahata, K. Kumar, S.P. Tiwari, V.K. Rai, Dualistic temperature sensing in Er^{3+}/Yb^{3+} doped $CaMoO_4$ upconversion phosphor, Spectrochim. Acta Part A Mol. Biomol. Spectrosc. 173 (2017) 369–375.

[15] R. De, Y.-H. Song, M.K. Mahata, K.T. Lee, pH-responsive polyelectrolyte complexation on upconversion nanoparticles: a multifunctional nanocarrier for protection, delivery, and 3D-imaging of therapeutic protein, J. Mater. Chem. B 10 (2022) 3420–3433.
[16] J. Zhou, Q. Liu, W. Feng, Y. Sun, F. Li, Upconversion luminescent materials: advances and applications, Chem. Rev. 115 (2015) 395–465.
[17] F. Wang, D. Banerjee, Y.S. Liu, X.Y. Chen, X.G. Liu, Upconversion nanoparticles in biological labeling, imaging, and therapy, Analyst 135 (2010) 1839–1854.
[18] S. Sinha, M.K. Mahata, What are upconversion nanophosphors: Basic concepts and mechanisms, in: Upconversion Nanophosphors, Elsevier, 2022, pp. 19–48.
[19] J. Zhou, Z. Liu, F.Y. Li, Upconversion nanophosphors for small-animal imaging, Chem. Soc. Rev. 41 (2012) 1323–1349.
[20] Y.C. Goh, Y.H. Song, G. Lee, H. Bae, M.K. Mahata, K.T. Lee, Cellular uptake efficiency of nanoparticles investigated by three-dimensional imaging, Phys. Chem. Chem. Phys. 20 (2018) 11359–11368.
[21] Q. Su, W. Feng, D. Yang, F. Li, Resonance energy transfer in upconversion nanoplatforms for selective biodetection, Acc. Chem. Res. 50 (2017) 32–40.
[22] M.K. Mahata, A. Kumari, V.K. Rai, Er^{3+}, Yb^{3+} doped yttrium oxide phosphor as a temperature sensor, AIP Conf. Proc. 1536 (1) (2013) 1270–1271.
[23] R. De, M.K. Mahata, K.T. Kim, Structure-based varieties of polymeric nanocarriers and influences of their physicochemical properties on drug delivery profiles, Adv. Sci. 9 (2022) 2105373.
[24] M.K. Mahata, S.P. Tiwari, S. Mukherjee, K. Kumar, V.K. Rai, YVO_4: Er^{3+}/Yb^{3+} phosphor for multifunctional applications, J. Opt. Soc. Am. B 31 (8) (2014) 1814–1821.
[25] M.K. Mahata, T. Koppe, K. Kumar, H. Hofsäss, U. Vetter, Demonstration of temperature dependent energy migration in Dual Mode YVO_4: Ho^{3+}-Yb^{3+} nanocrystals for low temperature thermometry, Sci. Rep. 6 (2016) 36342.
[26] P. Sukul, M.K. Mahata, U.K. Ghorai, K. Kumar, Crystal phase induced upconversion enhancement in Er^{3+}/Yb^{3+} doped $SrTiO_3$ ceramic and its temperature sensing studies, Spectrochim. Acta Part A Mol. Biomol. Spectrosc. 212 (2019) 78–87.
[27] T. Pulli, T. Donsberg, T. Poikonen, F. Manoocheri, P. Karha, E. Ikonen, Advantages of white LED lamps and new detector technology in photometry, Light Sci. Appl. 4 (2015) e332.
[28] M.K. Mahata, H. Bae, K.T. Lee, Upconversion luminescence sensitized pH-nanoprobes, Molecules 22 (2017) 2064.
[29] A. Kumari, M.K. Mahata, Upconversion nanoparticles for sensing applications, in: Upconversion Nanophosphors, Elsevier, 2022, pp. 311–336.
[30] M. Eleftheriadou, G. Pyrgiotakis, P. Demokritou, Nanotechnology to the rescue: using nano-enabled approaches in microbiological food safety and quality, Curr. Opin. Biotechnol. 44 (2017) 87–93.
[31] H. Kang, L. Wang, M. O'Donoghue, Y.C. Cao, W., Tan, in: F.S. Ligler, C.R. Taitt (Eds.), Optical Biosensors, second ed., Elsevier, Amsterdam, 2008, pp. 583–621.
[32] Z. Farka, T. Juřík, D. Kovář, L. Trnková, P. Skládal, Nanoparticle-based immunochemical biosensors and assays: recent advances and challenges, Chem. Rev. 117 (2017) 9973–10042.
[33] S. Sinha, M.K. Mahata, K. Kumar, Up/down-converted green luminescence of Er^{3+}-Yb^{3+} doped paramagnetic gadolinium molybdate: a highly sensitive thermographic phosphor for multifunctional applications, RSC Adv. 6 (92) (2016) 89642–89654.
[34] Z. Zhang, S. Shikha, J. Liu, J. Zhang, Q. Mei, Y. Zhang, Upconversion nanoprobes: recent advances in sensing applications, Anal. Chem. 91 (2019) 548–568.

[35] M.K. Mahata, K. Kumar, V.K. Rai, Structural and optical properties of Er^{3+}/Yb^{3+} doped barium titanate phosphor prepared by co-precipitation method, Spectrochim. Acta Part A Mol. Biomol. Spectrosc. 124 (2013) 285–291.

[36] A.K. Soni, V.K. Rai, M.K. Mahata, Yb^{3+} sensitized $Na_2Y_2B_2O_7$: Er^{3+} phosphors in enhanced frequency upconversion, temperature sensing and field emission display, Mater. Res. Bull. 89 (2017) 116–124.

[37] R. Hardman, A toxicologic review of quantum dots: toxicity depends on physicochemical and environmental factors, Environ. Health Perspect. (2006) 165–172.

[38] S. Sinha, M.K. Mahata, V.K. Rai, K. Kumar, Upconversion emission study of Er^{3+} doped $CaMoO_4$ phosphor, AIP Conf. Proc. 1728 (2016) 020476.

[39] J. Shen, L.D. Sun, C.H. Yan, Luminescent rare earth nanomaterials for bioprobe applications, Dalton Trans. (2008) 5687–5697.

[40] A. Kumari, A.K. Soni, V.K. Rai, Upconversion emission study in Tm^{3+} and Ho^{3+} doped $Gd_2(MoO_4)_3$ phosphor, AIP Conf. Proc. 2018 (1953) 030035.

[41] I.L. Medintz, N. Hildebrandt (Eds.), FRET-Förster Resonance Energy Transfer, Wiley-VCH Verlag GmbH & Co, Weinheim, Germany, 2014.

[42] X. Zhou, Y. Liu, Q. Liu, L. Yan, M. Xue, W. Yuan, M. Shi, W. Feng, C. Xu, F. Li, Point-of-care ratiometric fluorescence imaging of tissue for the diagnosis of ovarian cancer, Theranostics 9 (2019) 4597–4607.

[43] N. Hildebrandt, C.M. Spillmann, W. Russ Algar, T. Pons, M.H. Stewart, E. Oh, K. Susumu, S.A. Díaz, J.B. Delehanty, I.L. Medintz, Energy transfer with semiconductor quantum dot bioconjugates: a versatile platform for biosensing, energy harvesting, and other developing applications, Chem. Rev. 117 (2017) 536–711.

[44] M. Senarisoy, C. Barette, F. Lacroix, S. De Bonis, M. Stelter, F. Hans, J.P. Kleman, M.O. Fauvarque, J. Timmins, Förster resonance energy transfer based biosensor for targeting the hNTH1–YB1 interface as a potential anticancer drug target, ACS Chem. Biol. 15 (2020) 990–1003.

[45] K. Nie, B. Dong, H. Shi, L. Chao, M. Long, H. Xu, Z. Liu, B. Liang, Facile construction of AIE-based FRET nanoprobe for ratiometric imaging of hypochlorite in live cells, J. Lumin. 220 (2020) 117018.

[46] X. Han, Z. Zhai, X. Yang, D. Zhang, J. Tang, J. Zhu, X. Zhu, Y. Ye, A FRET-based ratiometric fluorescent probe to detect cysteine metabolism in mitochondria, Org. Biomol. Chem. 18 (2020) 1487–1492.

[47] Q. Wan, T. Truongvo, H.E. Steele, A. Ozcelikkale, B. Han, Y. Wang, J. Oh, H. Yokota, S. Na, Subcellular domain-dependent molecular hierarchy of SFK and FAK in mechanotransduction and cytokine signaling, Sci. Rep. 7 (2017) 1–15.

[48] Y. Konagaya, K. Terai, Y. Hirao, K. Takakura, M. Imajo, Y. Kamioka, N. Sasaoka, A. Kakizuka, K. Sumiyama, T. Asano, M. Matsuda, A highly sensitive FRET biosensor for AMPK exhibits heterogeneous AMPK responses among **cells** and organs, Cell Rep. 21 (2017) 2628–2638, https://doi.org/10.1016/j. celrep.2017.10.113.

[49] C. Hazra, V.N.K.B. Adusumalli, V. Mahalingam, 3,5-Dinitrobenzoic acid-capped upconverting nanocrystals for the selective detection of melamine, ACS Appl Mater. Interfaces 6 (2014) 7833–7839.

[50] W. Yue, L. Menghua, C. Zehan, D. Zilong, L. Ningtao, F. Jianwei, Z. Weixian, A fluorescence resonance energy transfer probe based on DNA-modified upconversion and gold nanoparticles for detection of lead ions, Front. Chem. 8 (2020) 238.

[51] M. Kumar, P. Zhang, Synthesis, characterization and biosensing application of photon upconverting nanoparticles, Proc. SPIE 7188 (2009) 71880F–1.

[52] S. Sinha, M.K. Mahata, K. Kumar, Enhancing the upconversion luminescence properties of Er^{3+}/Yb^{3+} doped yttrium molybdate through Mg^{2+} incorporation: effect of laser excitation power on temperature sensing and heat generation, New J. Chem. 43 (15) (2019) 5960–5971.
[53] J.G. Bunzli, C. Piguet, Taking advantage of luminescent lanthanide ions, Chem. Soc. Rev. 34 (2005) 1048–1077.
[54] S. Sinha, M.K. Mahata, K. Kumar, Comparative thermometric properties of bi-functional Er^{3+}-Yb^{3+} doped rare earth (RE = Y, Gd and La) molybdates, Mater. Res. Express 5 (2018) 026201.
[55] J. Zhang, Z. Zhang, Z. Tang, Y. Tao, X. Long, Luminescent properties of the $BaMgAl_{10}O_{17}$:Eu^{2+},M^{3+} (M = Nd, Er) phosphor in the VUV region, Chem. Mater. 14 (2002) 3005–3008.
[56] M.K. Mahata, T. Koppe, T. Mondal, C. Brüsewitz, K. Kumar, V. Kumar Rai, H. Hofsäss, U. Vetter, Incorporation of Zn^{2+} ions into $BaTiO_3$:Er^{3+}/Yb^{3+} nanophosphor: an effective way to enhance upconversion, defect luminescence and temperature sensing, Phys. Chem. Chem. Phys. 17 (32) (2015) 20741–20753.
[57] A.K. Soni, M.K. Mahata, Photoluminescence and cathodoluminescence studies of Er^{3+}-activated strontium molybdate for solid-state lighting and display applications, Mater. Res. Exp. 4 (2017) 126201.
[58] G. Chen, H. Qiu, P.N. Prasad, X. Chen, Upconversion nanoparticles: design, nanochemistry, and applications in theranostics, Chem. Rev. 114 (2014) 5161.
[59] P.P. Sukul, M.K. Mahata, K. Kumar, NIR optimized dual mode photoluminescence in Nd doped Y_2O_3 ceramic phosphor, J. Lumin. 185 (2017) 92–98.
[60] M.K. Mahata, H. Hofsaess, U. Vetter, Photon-upconverting materials: advances and prospects for various emerging applications, in: J. Thirumalai (Ed.), Luminescence—An Outlook on the Phenomena and Their Applications, InTech, Croatia, 2016, pp. 109–131.
[61] J. Chen, J.X. Zhao, Upconversion nanomaterials: synthesis, mechanism, and applications in sensing, Sensors 12 (2012) 2414–2435.
[62] M.K. Mahata, T. Koppe, H. Hofsäss, K. Kumar, U. Vetter, Host sensitized luminescence and time-resolved spectroscopy of YVO4: Ho3+ nanocrystals, Phys. Proc. 76 (2015) 125–131. Elsevier B.V., Poland.
[63] S. Sinha, M.K. Mahata, H.C. Swart, A. Kumar, K. Kumar, Enhancement of upconversion, temperature sensing and cathodoluminescence in the K^+/Na^+ compensated $CaMoO_4$: Er^{3+}/Yb^{3+} nanophosphor, New J. Chem. 41 (13) (2017) 5362–5372.
[64] C.X. Li, J. Yang, P.P. Yang, G. Lian, J. Lin, Hydrothermal synthesis of lanthanide fluorides LnF_3 (Ln = La to Lu) nano-/microcrystals with multiform structures and morphologies, Chem. Mater. 20 (13) (2008) 4317–4326.
[65] G. Yi, B. Sun, F. Yang, D. Chen, Y. Zhou, Synthesis and characterization of high-efficiency nanocrystal up-conversion phosphors: Ytterbium and erbium codoped lanthanum molybdate, J. Cheng Chem. Mater. 14 (2002) 2910–2914.
[66] X. Fan, D. Pi, F. Wang, J. Qiu, M. Wang, Hydrothermal synthesis and luminescence behavior of lanthanide-doped GdF/sub 3/nanoparticles, IEEE Trans. Nanotechnol. 5 (2006) 123–128.
[67] J. Peng, W. Xu, C.L. Teoh, S. Han, B. Kim, A. Samanta, J.C. Er, L. Wang, L. Yuan, X. Liu, Y.-T. Chang, High-efficiency in vitro and in vivo detection of Zn^{2+} by dye-assembled upconversion nanoparticles, J. Am. Chem. Soc. 137 (2015) 2336–2342.
[68] Y. Liu, M. Chen, T. Cao, Y. Sun, C. Li, Q. Liu, T. Yang, L. Yao, W. Feng, F. Li, A cyanine-modified nanosystem for in vivo upconversion luminescence bioimaging of methylmercury, J. Am. Chem. Soc. 135 (2013) 9869–9876.

[69] L. Yao, J. Zhou, J. Liu, W. Feng, F. Li, Iridium-complex-modified upconversion nanophosphors for effective LRET detection of cyanide anions in pure water, Adv. Funct. Mater. 22 (2012) 2667–2672.

[70] H. Hu, L. Xiong, J. Zhou, F. Li, Multimodal-luminescence core–shell nanocomposites for targeted imaging of tumor cells, Chem. Eur. J. 15 (2009) 3577–3584.

[71] J. Lu, Y. Chen, D. Liu, W. Ren, Y. Lu, Y. Shi, J. Piper, I. Paulsen, D. Jin, One-step protein conjugation to upconversion nanoparticles, Anal. Chem. 87 (2015) 10406–10413.

[72] K.E. Sapsford, L. Berti, I.L. Medintz, Materials for fluorescence resonance energy transfer analysis: beyond traditional donor–acceptor combinations, Angew. Chem. Int. Ed. 45 (2006) 4562–4588.

[73] S. Chen, Y.L. Yu, J.H. Wang, Inner filter effect-based fluorescent sensing systems: a review, Anal. Chim. Acta 999 (2018) 13–26.

[74] L.Y. Wang, R.X. Yan, Z.Y. Hao, L. Wang, J.H. Zeng, J. Bao, X. Wang, Q. Peng, Y.D. Li, Fluorescence resonant energy transfer biosensor based on upconversion-luminescent nanoparticles, Angew. Chem. Int. Ed. 44 (2005) 6054–6057.

[75] E.J. Jo, H. Mun, M.-G. Kim, Homogeneous immunosensor based on luminescence resonance energy transfer for glycated hemoglobin detection using upconversion nanoparticles, Anal. Chem. 88 (2016) 2742–2746.

[76] J. Liu, Y. Liu, Q. Liu, C. Li, L. Sun, F. Li, Iridium (III) complex-coated nanosystem for ratiometric upconversion luminescence bioimaging of cyanide anions, J. Am. Chem. Soc. 133 (2011) 15276–15279.

[77] Z. Cui, W. Bu, W. Fan, J. Zhang, D. Ni, Y. Liu, J. Wang, J. Liu, Z. Yao, J. Shi, Sensitive imaging and effective capture of Cu^{2+}: towards highly efficient theranostics of Alzheimer's disease, Biomaterials 104 (2016) 158–167.

[78] S. Liu, L. Zhang, T. Yang, H. Yang, K.Y. Zhang, X. Zhao, W. Lv, Q. Yu, X. Zhang, Q. Zhao, X. Liu, W. Huang, Development of upconversion luminescent probe for ratiometric sensing and bioimaging of hydrogen sulfide, ACS Appl. Mater. Interfaces 6 (2014) 11013–11017.

[79] S. Li, L. Xu, W. Ma, X. Wu, M. Sun, H. Kuang, L. Wang, N.A. Kotov, C. Xu, Dual-mode ultrasensitive quantification of microRNA in living cells by chiroplasmonic nanopyramids self-assembled from gold and upconversion nanoparticles, J. Am. Chem. Soc. 138 (2016) 306–312.

[80] J. Peng, A. Samanta, X. Zeng, S. Han, L. Wang, D. Su, D.T. Loong, N.Y. Kang, S.J. Park, A.H. All, W. Jiang, L. Yuan, X. Liu, Y.T. Chang, Real-time in vivo hepatotoxicity monitoring through chromophore-conjugated photon-upconverting nanoprobes, Angew. Chem. Int. Ed. 56 (2017) 4165–4169.

[81] P. Zhang, S. Rogelj, K. Nguyen, D. Wheeler, Design of a highly sensitive and specific nucleotide sensor based on photon upconverting particles, J. Am. Chem. Soc. 128 (12) (2006) 410–411.

[82] H. Chen, A. Fang, H. Li, Y. Zhang, S. Yao, Sensitive fluorescent detection of H_2O_2 and glucose in human serum based on inner filter effect of squaric acid-iron(III) on the fluorescence of upconversion nanoparticle, Talanta 164 (2017) 580–587.

[83] Y. Guan, S. Qu, B. Li, L. Zhang, H. Ma, L. Zhang, Ratiometric fluorescent nanosensors for selective detecting cysteine with upconversion luminescence, Biosens. Bioelectron 77 (2016) 124–130.

[84] Y. Liu, Q. Ouyang, H. Li, Z. Zhang, Q. Chen, Development of an inner filter effects-based upconversion nanoparticles-curcumin nanosystem for the sensitive sensing of fluoride ion, ACS Appl. Mater. Interfaces 9 (2017) 18314–18321.

[85] S. Zhu, J. Zhang, X.E. Zhao, H. Wang, G. Xu, J. You, Electrochemical behavior and voltammetric determination of L-tryptophan and L-tyrosine using a glassy carbon

electrode modified with single-walled carbon nanohorns, Mikrochim. Acta 181 (2014) 445–451.
[86] A. Hlaváček, Z. Farka, M.J. Mickert, U. Kostiv, J.C. Brandmeier, D. Horák, P. Kládal, F. Foret, H.H. Gorris, Bioconjugates of photon-upconversion nanoparticles for cancer biomarker detection and imaging, Nat. Protocols 17 (4) (2022) 1028–1072.
[87] N. Akhtar, P.W. Wu, C.L. Chen, W.Y. Chang, R.S. Liu, C.T. Wu, A. Girigoswami, S. Chattopadhyay, Radiolabeled Human protein functionalized upconversion nanoparticles for multimodal cancer imaging, ACS Appl. Nano Mater. 5 (5) (2022) 7051–7062.
[88] H. Dong, L.D. Sun, C.H. Yan, Lanthanide-doped upconversion nanoparticles for super-resolution microscopy, Front. Chem. 8 (2021) 619377.
[89] R. Xu, H. Cao, D. Lin, B. Yu, J. Qu, Lanthanide-doped upconversion nanoparticles for biological super-resolution fluorescence imaging, Cell Rep. Phys. Sci. 3 (6) (2022) 100922.
[90] S. De Camillis, P. Ren, Y. Cao, M. Plöschner, D. Denkova, X. Zheng, Y. Lu, J.A. Piper, Controlling the non-linear emission of upconversion nanoparticles to enhance super-resolution imaging performance, Nanoscale 12 (39) (2020) 20347–20355.
[91] R. Lv, D. Wang, L. Xiao, G. Chen, J. Xia, P.N. Prasad, Stable ICG-loaded upconversion nanoparticles: silica core/shell theranostic nanoplatform for dual-modal upconversion and photoacoustic imaging together with photothermal therapy, Sci. Rep. 7 (1) (2017) 1–11.
[92] S.K. Maji, S. Sreejith, J. Joseph, M. Lin, T. He, Y. Tong, H. Sun, S. Yu, Y. Zhao, Upconversion nanoparticles as a contrast agent for photoacoustic imaging in live mice, Adv. Mater. 26 (32) (2014) 5633–5638.
[93] E.M. Mettenbrink, W. Yang, S. Wilhelm, Bioimaging with upconversion nanoparticles, Adv. Photon. Res. 3 (12) (2022) 200098.
[94] H. Chen, B. Ding, J. Lin, Recent progress in upconversion nanomaterials for emerging optical biological applications, Adv. Drug Deliv. Rev. 118 (2022) 114414.
[95] G.-R. Tan, M. Wang, C.-Y. Hsu, N. Chen, Y. Zhang, Small upconverting fluorescent nanoparticles for biosensing and bioimaging, Adv. Opt. Mater. 4 (7) (2016) 984–997.
[96] C. Song, S. Zhang, Q. Zhou, H. Hai, D. Zhao, Y. Hui, Upconversion nanoparticles for bioimaging, Nanotechnol. Rev. 6 (2) (2017) 233–242.
[97] A. Qu, M. Sun, L. Xu, C. Hao, X. Wu, C. Xu, N.A. Kotov, H. Kuang Quan, Quantitative zeptomolar imaging of miRNA cancer markers with nanoparticle assemblies, Proc. Natl. Acad. Sci. 116 (9) (2019) 3391–3400.
[98] E.S. Tsai, S.F. Himmelstoß, L.M. Wiesholler, T. Hirsch, E.A.H. Hall, Upconversion nanoparticles for sensing pH, Analyst 144 (18) (2019) 5547–5557.
[99] R. Arppe, T. Näreoja, S. Nylund, L. Mattsson, S. Koho, J.M. Rosenholm, T. Soukka, M. Schäferling, Photon upconversion sensitized nanoprobes for sensing and imaging of pH, Nanoscale 6 (12) (2014) 6837–6843.
[100] S. Radunz, E. Andresen, C. Würth, A. Koerdt, H.R. Tschiche, U. Resch-Genger, Simple self-referenced luminescent pH sensors based on upconversion nanocrystals and pH-sensitive fluorescent BODIPY dyes, Anal. Chem. 91 (12) (2019) 7756–7764.
[101] D. Jaque, B.D. Rosal, E.M. Rodríguez, L.M. Maestro, P. Haro-Gonzalez, J.G. Solé, Fluorescent nanothermometers for intracellular thermal sensing, Nanomedicine 9 (7) (2014) 1047–1062.

Section C

Metal oxides for photonic and optoelectronic applications

Metal oxide-based LEDs and lasers

9

Harjot Kaur and Samarjeet Singh Siwal
Department of Chemistry, M.M. Engineering College, Maharishi Markandeshwar (Deemed to be University), Mullana-Ambala, Haryana, India

Chapter outline

1 Introduction

Nanotechnology is a multidisciplinary area of physics, chemistry, and materials science encompassing the design and fabrication of nanomaterials (NMs) and their usages. It is suggested to comprehend the essential physical and biological effects and the spectacle of NMs and nanocomposites. The science of NMs has developed as a front investigation field because of the new implementation of NMs [1–3]. Feynman indicated the importance of nanoscience at the annual conference of the American Physical Society in 1959, in a lecture entitled "There's Plenty of Room at the Bottom." Over the intervening years, numerous findings have been made and designs constructed in nanoscience regarding manufacturing novel materials and employing them toward different usages. Different methods with respective and select NMs incorporation have been found [4].

Numerous analysis steps with various synthesis strategies have enabled production of different NMs, generally categorized within three classes based on their dimensionality: zero dimensional (0D), one dimensional (1D), and two and three dimensional (2,3D) [5]. Quantum dots (QDs) and the individual molecules fall under the 0D networks where the nanoparticles are separated from others [6–8]. 1D nanostructures, where at least one of the structures dimensions exists within the of nanoscale range, are widely employed and has many uses in the nanodevice applications. Thin nanofilms lie on the 2D assemblies and are examined broadly to utilize nanodevice

Metal Oxides for Next-Generation Optoelectronic, Photonic, and Photovoltaic Applications
https://doi.org/10.1016/B978-0-323-99143-8.00005-5

applications [9–11]. 3D NMs possess deposits, fibrous, multilayer and polycrystalline materials where the 0D, 1D and 2D physical essentials are in neighboring interaction and making boundaries. An essential 3D nanostructured constituent is a dense (bulk) polycrystal with nanosized ounces whose total capacity is occupied with those nanograins [12].

In this chapter, we provide a general overview of metal oxides, their synthesis techniques, and important properties of metal oxides. Furthermore, the most advanced applications of metal oxides in LEDs and lasers are addressed along with concluding remarks and future prospects.

2 General overview of metal oxides

The electronic configuration of MO_x discrepancies shows metallic, semiconductor, or insulator textures. Metal oxide (MO_x) offers a wide range of physiochemical and electronic properties and can be considered as either a semiconductor or a conductor [13]. The synthesis of these substances can be achieved via new preparation methods, categorized as physical and chemical approaches. Generally, two methods have been employed to synthesize these MO_x nanocomposites: top-down and bottom-up incorporation methods. These methods involve liquid-solid or gas-solid modifications [14–16]. Among the large adaptable classes of semiconductor nanostructures, MO_x nanocomposites are one of the utmost typical, miscellaneous and presumably most affluent classes of substances owing to their vast operational, physical and biochemical effects and functionalities. Recently, MO_x has been at the core of numerous surprising advances within materials science. These substances show the most engaging and comprehensive scope of effects [17,18].

Due to the distinctive and tunable characteristics of these MO_xs, for example, optical, optoelectronic, captivating, electric, automated, thermal, catalytic, photochemical, etc., they represent ideal candidates for a range of high-performance technical applications [19]. For example, fuel cells [20,21], subordinate battery ingredients [22], ceramic ware, biosensors [23], different types of energy storage devices [24–26], optical tools, lasers, waveguides, infrared (IR), and insulators within dynamics arbitrary contact memories. Therefore, MO_x nanostructure substances have been vigorously investigated, providing a more comprehensive understanding among researchers. Thus, examining its interpretation in incredible detail in its preparation, characteristics and applications is crucial. Therefore, this chapter is intended to examine zinc oxide (ZnO), doped ZnO, and nickel oxide (NiO) MO_x nanostructures [27]. In this chapter, we will explore the synthesis of the metal oxide nanostructures noted above via two approaches: hydrothermal and thermal evaporation. Several MO_x nanostructures have been prepared, such as ZnO [28,29], copper oxide (CuO) [30,31], NiO [32,33], gallium oxide (Ga_2O_3) [34], etc. Various MO_x structures, for example, nanorods, nanowires, and nanotubes, have been developed. These nanostructures have been synthesized through various techniques, including thermal evaporation [35], chemical vapor deposition (CVD) [36], chemical preparation [37], etc.

3 Synthesis of metal oxides

Synthesis of NMs with select surfaces and structures is the utmost thought-provoking analysis in the nanoscience and nano field. In recent years, the amalgamation of MO_x nanocomposites has received significant attention because of their unique effects that facilitate the manufacture of efficient miniaturized devices for application in various nanoelectronics and photonics contexts. Consequently, different incorporation methods have been analyzed within the literature for synthesizing these MO_x nanostructures, for example, thermal evaporation [38], metalorganic (MO) compunds and CVD [39], hydrothermal amalgamation [40], template-dependent amalgamation [41], etc. Different crystallinity methods have been investigated to regulate the width feature proportion, for example, thermal loss, pulse laser deposition (PLD), MOCVD, a sputtering process, thermal CVD, cyclic CVD (CFCVD), etc.

Vapor phase deposition is the ultimate multipurpose method for developing a universal group of nanostructures. In the vapor phase preparation of nanoparticles, circumstances are arranged so that the vapor phase combination is thermodynamically irregular compared to that of the solid substance to be synthesized in a nanoparticulate state [42,43]. It includes the specific condition of highly dense gas. It also possesses the third method, "biochemical highly viscous," that is thermodynamically promising for the vapor stage particles to react chemically to form a condensed component. Uncertainty, the crack of supersaturation, is adequate, and the reaction/condensation kinetics access, atoms will nucleate consistently. While nucleation occurs, remaining supersaturation may be facilitated with condensation or reaction of the vapor-phase substance upon the consequent atoms, and atom consequence will occur preferably than further nucleation. Thus, to design a small constituent part, one desires to increase supersaturation by generating a high diffusion density and instantly assuages the design by releasing the origin of supersaturation or slowing the kinetics, which the atoms do not produce. It happens quickly (milliseconds to seconds) within a comparatively disorderly manner and progresses the issue to a restarted or quasi-incessant method.

The building methodologies of various copper selenide (CuSe) specimens can be derived. For the RT-CuSe, because of the low response temperature, insufficient energy delivered limited the expression of RT-CuSe, thus creating the combination of CuSe with each unstructured and hexagonal surface (as illustrated in Fig. 1) [44].

The thermal evaporation method boosts different substance's nanostructures. In this approach, there is a necessity for a high-temperature thermal oven employed for evaporating the reference material and decreasing the nanostructures' deposit at relatively low temperatures [45]. In this method, vapor species of the origin substances are first developed by physical or chemical processes, and then condensed under temperature, pressure, and silicon support conditions [46]. This approach has developed significant NMs that range from fundamental nanowires to various semiconductor substances [47]. The thermal evaporation method typically employs a flat quartz cylinder oven, including a rotational pump and gas source in the design.

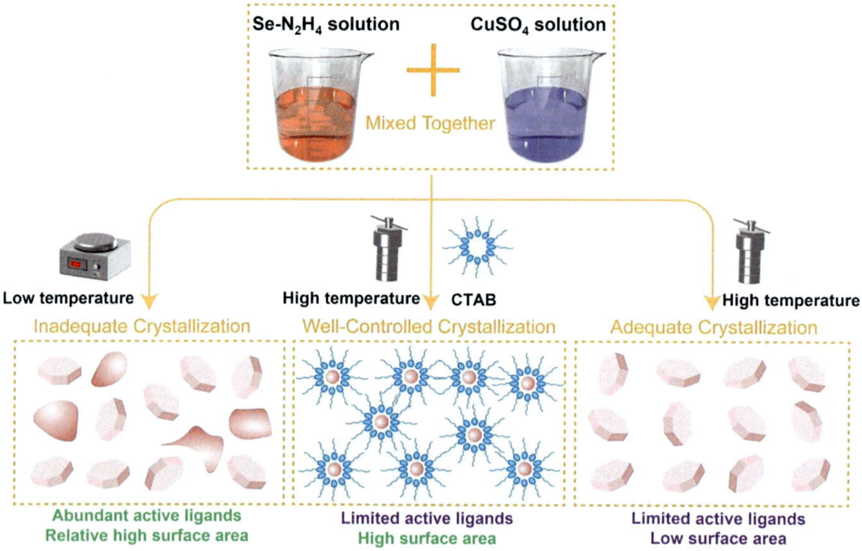

Fig. 1 Schematic design of the development of RT-CuSe, HT-CuSe, and CT-CuSe.
Reused with permission from H. Li, W. Zheng, W. Qu, L. Leng, Y. Feng, F. Xin, G. Sheng, Z. Yang, K. Shih, J. Yang, Facile pathway towards crystallinity adjustment and performance enhancement of copper selenide for vapor-phase elemental mercury sequestration, Chem. Eng. J. 430 (2022) 132811.

The interchange of laser radioactivity, including solid substances, is mainly defined by the character of the ingredients and the effects of the laser beam. Using a pulsed laser to generate the stoichiometric transmission of substantial from a reliable origin to support manufacturing before flash evaporation techniques were documented initially in 1965 [48].

A pulsed laser ray shows a quick material reduction from a compact mark and the formation of an active plasma spiral that shrinks on the support material. In difference to the transparency of the method, the instruments in PLD with extirpation, plasma appearance, plume propagation, and nucleation and development are relatively complicated. In the laser ablation method, the photons are transformed first within electronic irritations, then within the thermal, chemical, and automated power [49], and subsequently within the quick elimination of substances from an exterior. This approach has been investigated extensively because of its significance in laser machining. Boiler speeds are as advanced as 1011 Ks^{-1}, and immediate gas pressures around 10–500 atm are observed on the board surface.

PLD has been utilized to grow high-temperature curates and multiple other complex oxides, such as materials that cannot be obtained via equilibrium. It has been demonstrated that the materials development processes through PLD plumes are distinct from those seen in thermal disappearing [50]. This process has been effective in the film preparation of Y-type magneto plumbite (including a c-axis matrix limit of 43.5 Å) [51] and garnets by 160 particles per unit cell [52]. Some other contemporary illustrations prepare magnetic oxide nanoparticles (NPs), titania NPs, and hydrogenated silicon NPs [53].

A final example of vaporizing a significant is sputtering, including a ray of stationary gas ions. NPs of a dozen different metals have been demonstrated utilizing magnetron sputtering of metallic targets [54]. They linked the NPs and incorporated them as nanostructured layers on silicon supports. This technique should be carried out at moderately low pressures, making the additional processing of the NPs in aerosol form complicated. Due to collision, it is primarily generated by speed interaction between ions and atoms in the material. Surface scattering is generally utilized to produce nanoscale islands or rods using the sputtering method. Recently, this approach has been employed to synthesize types of nanostructures such as ZnO, W, Si, B, CN, etc. [55].

The metalorganic chemical vapor deposition (MOCVD) method is also known as metalorganic vapor phase epitaxy (MOVPE) and is widely utilized to prepare epitaxial systems incorporating atoms on wafer supports [56]. This approach has been extensively utilized for developing different thin films,. The method is straightforward. The desired atoms protected with a specific compound organic gas-particle are handed over a warm semiconductor wafer toward apparent crystal growth. Due to the temperature, the composite organic particles split up and deposit the required atoms on the supporting surface. The unwanted components are destroyed or deposited on the surface of the reactor. The crystal properties at the near-atomic scale can be modified by altering the gas composition. Utilizing this method, layers of exactly regulated thickness may be achieved, which is essential for producing specific optical and electrical characteristics. With MOCVD, it is feasible to produce the semiconductor qualities required for photodetectors and lasers. Similarly, researchers have recently focused on growing nanostructures with this approach and achieving thin-film development. This approach has synthesized different semiconductor nanostructures, as described in the literature.

4 Properties of metal oxides

NPs of MO_xs have drawn growing technical and industrial interest due to their unique optical, magnetic, electrical, and catalytic characteristics, which provide improved physical effects in terms of resilience or chemical inactivity. Numerous physical properties of NPs vary drastically from a single crystal of the same chemical composition. Due to the confinement impact of electronic states and many surface atoms at nanoscale sizes, their bulk phases affect multiple physical properties. MO_x NPs have technical applications in catalysts, passive electronic components, and high-performance ceramics. These substances play an essential part in the surface transformation of various supports in coatings. Compared to bulk MO_x phase composite, MO_x includes a significant particle of fundamental particles as surface particles that contribute to the free energy and result in considerable differences in the thermodynamic effects (melting point hollow solid-solid phase growth peak). Furthermore, the quantum size effect transfers the intrinsic characteristics of MO_x, which varies in the optic and electrical products, with dimensions arising because of the modification within the thickness of electronic energy stages. Table 1 shows the important physical characteristics of some significant MO_x semiconductors [57].

Table 1 Important physical characteristics of some significant MO_x semiconductors.

MO_x	Crystal structure	Conductivity	Band gap
Zinc oxide	Hexagonal	n-type	3.37
Tin oxide	Tetragonal	n-type	3.6
Cupric oxide	Cubic	n-type	2.1.7
Gallium (III) trioxide	Monoclinic	n-type	4.2
Iron (III) oxide	Rhombohedral	n-type	2.1
Indium (III) oxide	Cubic	n-type	3.6
Cadmium oxide	Cubic	n-type	0.55
Cerium (IV) oxide	Cubic	n-type	3.2

Reused with permission from T. Zhai, X. Fang, M. Liao, X. Xu, H. Zeng, B. Yoshio, D. Golberg, A comprehensive review of one-dimensional metal-oxide nanostructure photodetectors, Sensors 9 (2009).

5 Application of metal oxides in LEDs and lasers

The possible applications for MO_x NMs include paint dyes, cosmetics, medicines, catalysis and substrates, medicinal diagnostics, magnetic and optical tools, flat-panel arrays, batteries and fuel cells, electronic and magnetic tools, biomaterials, designed substances, and protective coatings. Oxide NMs are achieving a wide range of applications in various areas due to their unique properties, such as high reactivity owing to their high specific surface area, regular size, and dispersal. Considerably, attention has been given to synthesized NM oxides due to their implementations into optic and electrical properties. They are sufficiently shown for a broad scope of applications.

Among the most commonly employed photocatalysts is nanostructured MO_x. Nevertheless, their current usage is primarily in solution techniques, which give some difficulties regarding the separation of the solid via removal of the solvent and the removal of the support and the additive. The photogenerated couple (e^-/h^+) can facilitate and/or oxidize a complex adsorbed upon the photocatalyst exterior. A graphic illustration of these methods is shown in Fig. 2 [58].

5.1 Application of metal oxides in LEDs

Metal oxides exhibits characteristics such as flexible morphology, high electrical conductivities, and charge injection, and are suitable for deposition on large areas through cost-effective methods [59]. Recent studies reported various metal oxides in LED; for instance, on top of indium tin oxide (ITO) [60], n-type metal oxides like TiO_2, ZrO_2, and ZnO_2 are used as the lowest electron injection while MoO_3 is utilized head cavity injection. For the operation of all optoelectronic and electrical tools, electric contacts are very important. ZnO possesses n-type conductivity, having a bandgap of 3.4 eV, and can be doped on p-type materials for different types of LEDs. ZnO fulfills the required conditions of LEDs, i.e., band gap above 3 eV, and naturally have oxygen vacancies, which results in additional energy states and emission of different wavelengths [61]. During 2001/2002, heterostructure ZnO LEDs were reported by Japan.

In brief, the role of transparent cathode was taken by a layer of Sn-doped indium oxide; then ZnO was deposited; and finally, p-type strontium copper oxide acts as

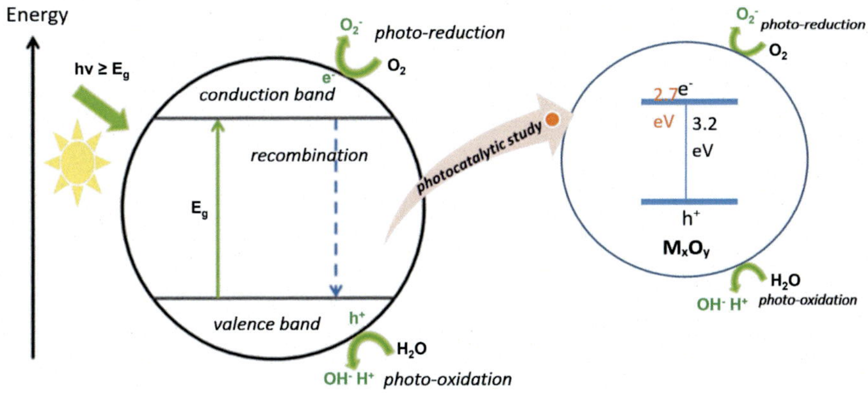

Fig. 2 Photocatalytic performance of nanostructure MO_x.
Reused with permission from C. Diaz, M.L. Valenzuela, M.Á. Laguna-Bercero, Solid-state preparation of metal and metal oxides nanostructures and their application in environmental remediation, Int. J. Mol. Sci. 23 (2022).

p-type transparent conductive oxide. The top layer, which acts as the anode, was Ni-electrode [62]. With higher than 80% optical transparency indium tin oxide (ITO) and fluorine-doped tin oxide (FTO) are considered to be among the best semiconductors and that's why metal oxides have various applications in the field of display and opto-electronic techniques. MoS_2-MoO_3 NMs were prepared by Yin et al. [63] by a thermal annealing-crystallization method. When this NM was utilized as the active layer in LEDs, electroluminescent emission of various wavelengths was emitted. The as synthesized material acted as a p-type conductor and was used with n-type SiC. MoO_3-based NM (MoO_3/Ag/MoO_3) was also utilized in LEDs by another team of researchers [64]. The authors observed the performance of prepared NM with SiO_2 NM on NOA63/3M elastomer, and enhanced performance was observed, with current efficiency and external quantum efficiency of 82.4 cd/A and 22.3%, respectively. Table 2 shows the device structure and its considerable significance.

Metal oxides have been utilized for the fabrication of charge transport layers in LEDs. A general requirement for transparent conductors in electro-optical devices is average transmittance in the visible range (more than 80%) and resistivity must be lower than $10^{-3}\,\Omega$, i.e., energy band gap of more than 3.0 eV and carrier concentration of at least $10^{20}\,cm^{-3}$ [73]. Several metal oxides fulfill this requirement when utilized with certain dopants. A variety of characteristics, such as high chemical stability and charge mobility, low toxicity, and cost effectiveness, makes metal oxides important in the field of light-emission devices. There are a large number of techniques to deposit thin layers of metal oxide in LEDs [74]. Some metal oxides utilized in LEDs, along with their performance and synthesis methods, are described in the following sections.

5.1.1 Metal oxides in quantum dot LEDs (QD-LEDs)

To improve luminance and quality of LEDs, QDs have been utilized as color down-converters [75]. The importance of QDs for display devices lies in characteristics such as pure color, tunable wavelength, higher luminance, longer lifetime, and compatibility

Table 2 The device structure and significant importance.

Device structure	Efficiency (cdA^{-1})	Performance	Turn on voltage (V)	Thickness (nm)	Max. brightness (cdm^{-2})	Reference
Transparent ZnO/AgOx/ZnO OMO TCEs	12.3	Up to 6mm radius it retains 100% of its initial luminance	2.8 ± 0.1	180	9400	[65]
Molybdenum-trioxide/gold/molybdenum-trioxide	82.4	External quantum efficiency of ~22.3% was achieved	–	MoO_3—15nm Au—14nm MoO_3—5nm	–	[64]
NiO_x NPLs	7.60	High transmittance, i.e., 90%	3.4	30	68,646	[66]
TiO_x/Ag/Al:ZnO	5.74	Shows higher current efficiency of 30% and 260% on glass and PET, respectively	4.5	27	100,000	[67]
ITO/Gr	16.4	Significantly higher photoluminescence lifetime of perovskite film	5	10	–	[68]
Al-doped ZnO NPs	59.7	Shows 14.1% external quantum efficiency	12	–	577,200	[69]
Li and Mg Co-doped Zinc Oxide	69.1	Shows power efficiency (PEmax) of 73.8lm/W	–	30	–	[70]
ITO/GO/SY/LiF/Al	19.1	Power efficiency of 11.0lm/W was obtained at 4.4V	1.8	4.3	39,000	[71]
ITO/NiO:K_2CO_3	31.3	External quantum efficiency surged from 7.1% to 9.76%	3.9	1	36,714	[72]

OMO, Oxide/metal/oxide; *TCE*, transparent conducting electrode; *NPLs*, nanoporous layers; *PET*, polyethylene terephthalate; *ITO/Gr*, indium tin oxide/graphene; *PEN*, polyethylene naphthalate.

with solution processes [76]. Consequently, QD-LEDs attract much attention from researchers as potential candidates for display applications. To ensure the high performance of QD-LEDs, some criteria, such as efficient charge transport layer and reduced electric field, must be taken into account when designing them [77]. Metal, metal alloys, and metal oxides can be utilized in QD-LEDs as charge transfer layers. Understanding the structural characteristics of metal oxides-based materials helps in the systematic fabrication of LEDs with increased efficiency and multicolor emissions [78].

ZnO nanomaterial was utilized as charge transport layer, where thermally polymerized polymer was used as a hole transfer layer. The obtained LED displayed a maximum luminance level of 4200 cd/cm^2 for blue light, 68,000 cd/cm^2 for green light, and 31,000 cd/cm^2 for orange-red light. In addition to this, ZnO NM requires a lower driving voltage to generate high power efficiency. Luminance efficiencies of 0.17 lm/W, 8.2 lm/W, and 3.8 lm/W at 2.4 V, 1.8 V, and 1.7 V were obtained for blue, green, and orange-red lights, respectively. The authors used a solution-precipitation method to fabricate ZnO NMs and the structure of ZnO NM-based QD-LED (ZnO/QD-LED). At an operating voltage of 3 V, the full-width at high maximum for green, orange-red, and blue devices was 38 nm, 39 nm, and 28 nm, respectively, as demonstrated in normalized electro-brightness spectra of ZnO/QD-LED. Furthermore, the electrochemical spectra consisted of (i) current density and brightness v/s driving voltage and (ii) luminance power efficiency and external quantum efficiency v/s luminance of ZnO/QD-LED [79]. These high luminance efficiencies indicate the great potential of incorporation of ZnO NMs as transport layers for QLEDs. Kim et al. [80] fabricated Li-doped-ZnO as an electron transport layer to study the performance of green QLED. Maximum luminance and charge efficiencies of 14.3 lm/W and 16.2 cd/A were observed, respectively.

The device performances for R-, G-, and B-QLEDs with stacked ETL are shown in Fig. 3. Fig. 3A–C shows the normalized EL spectra, EQE versus current density, and CIE coordinates of the inverted R-, G-, and B-QLEDs, respectively. The EL peaks are at 643, 523, and 444 nm at the maximum EQEs, with FWHMs of 32, 34, and 24 nm, respectively. Fig. 3D and E shows the histograms of current and power efficiencies of 30 devices from seven batches, yielding average current and power efficiencies of 33.1 cd A^{-1} and 29.1 lm/W, respectively. The good reproducibility of the devices demonstrates the feasibility of LZO as a second ETL in inverted QLEDs.

Yang et al. [81] studied double-sided metal oxide NM as a charge transport layer, where WO_3 served as electron transfer layer and ZnO served as hole transfer layer. Briefly, a solution processing method was utilized to subsequently deposit WO_3, poly[*N*,*N*′-bis(4-butylphenyl)-*N*,*N*′-bis(phenyl)benzidine] (poly-TPD), QDs and ZnO NM layers. A schematic representation of the obtained QLED along with energy level diagrams is provided in Fig. 4A and B. Fig. 4C displays atomic force microscopic (AFM) images of double-sided metal oxide NMs-based QLED. Fig. 4D shows an applied lifetime test as a function of time, from which a lifetime of 95 h was observed.

Choi et al. studied V_2O_5 film as hole injection layer for QLEDs. Briefly, a hole injection layer of V_2O_5 was fabricated on a patterned ITO anode, which was then transferred to a N_2-filled compartment to coat poly(9,9-dioctylfluorene-*alt*-N-(4-s-butylphenyl)-diphenylamine) (TFB), QD, and ZnO NM layers. Lastly, under

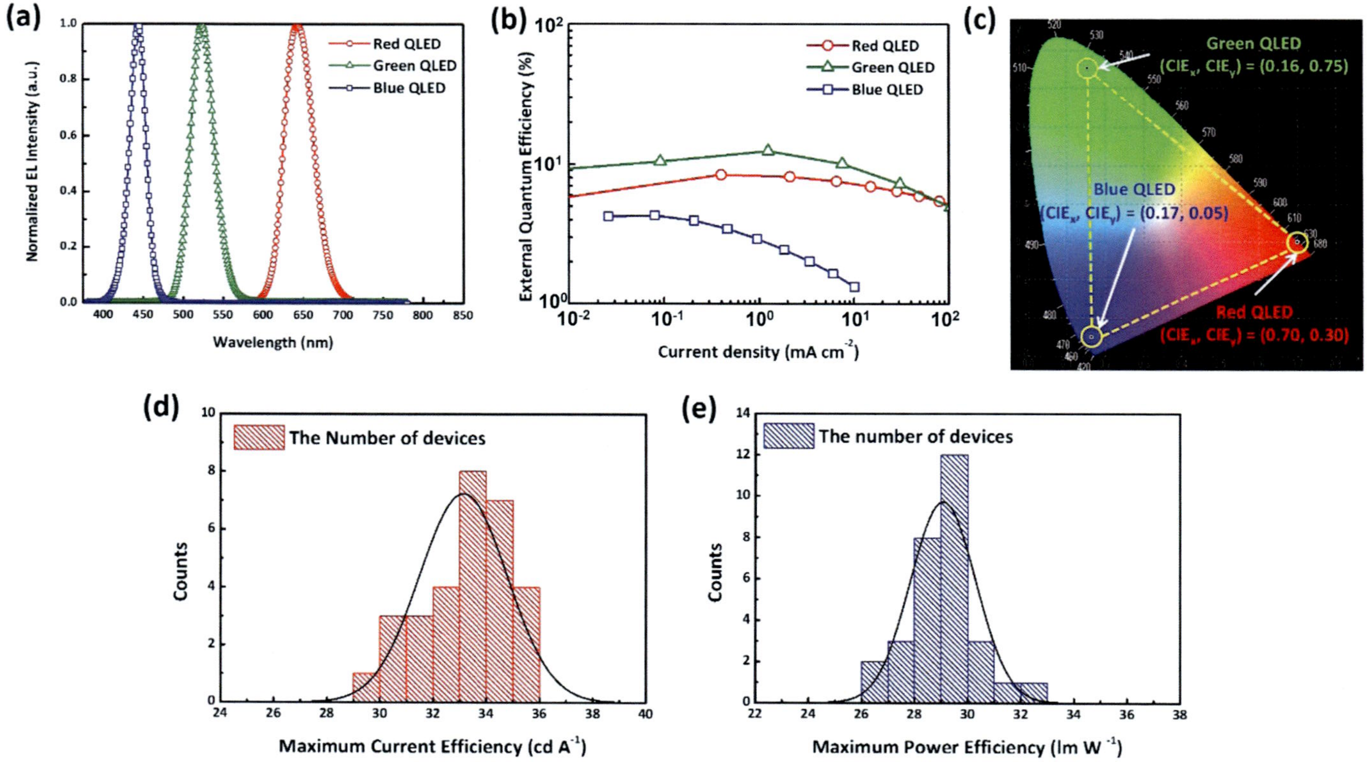

Fig. 3 (A) Normalized EL spectra, (B) EQEs, and (C) CIE coordinates for the fabricated R-, G-, and B-QLEDs with LZO as a 2nd ETL. The device structure of QLEDs with stack ETL; ITO/AZO/LZO/QDs (R, G-, and B-)/TCTA/NPB/HAT-CN/Al. Histogram of maximum (D) current and (E) power efficiencies measured from 30 devices. The device structure of G-QLEDs with stack ETL; ITO/AZO/LZO/G-QDs/TCTA/NPB/HAT-CN/Al. The average current and power efficiencies of G-QLEDs with stack ETL are 33.1 cd A^{-1} and 29.1 lm/W, respectively, and the black line is the Gaussian fitting.

Reused with permission from H.-M. Kim, D. Geng, J. Kim, E. Hwang, J. Jang, Metal-oxide stacked electron transport layer for highly efficient inverted quantum-dot light emitting diodes, ACS Appl. Mater. Interfaces 8 (2016) 28727–28736.

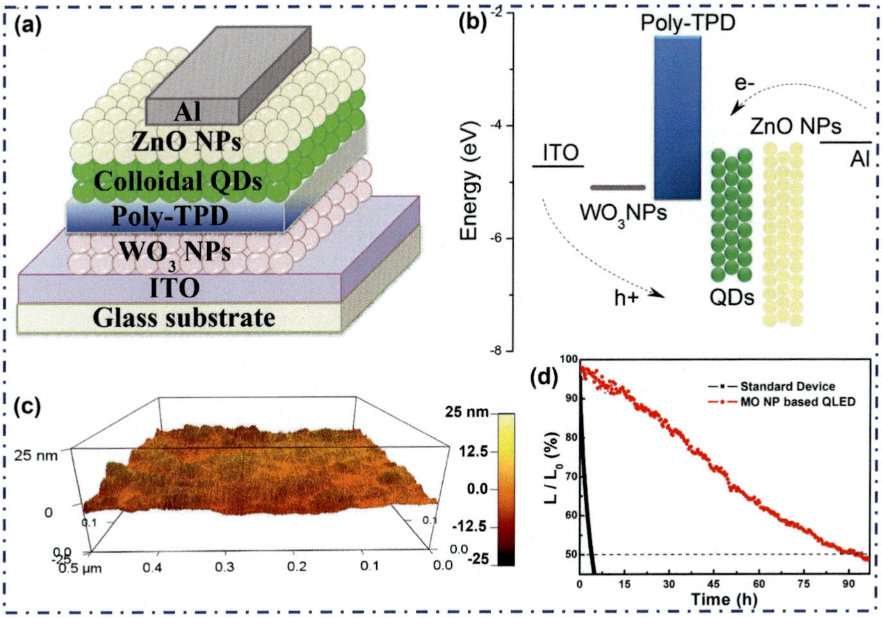

Fig. 4 (A and B) Graphical presentation of different layers, and energy level diagram of double-sided metal oxide based QLED; (C) AFM image of all layers of QLED; (D) lifetime test as a function of time.

Reused with permission from X. Yang, Y. Ma, E. Mutlugun, Y. Zhao, K.S. Leck, S.T. Tan, H.V. Demir, Q. Zhang, H. Du, X.W. Sun, Stable, efficient, and all-solution-processed quantum dot light-emitting diodes with double-sided metal oxide nanoparticle charge transport layers, ACS Appl. Mater. Interfaces 6 (2014) 495–499.

10^{-6} Torr pressure, the Al electrode was evaporated [82]. The maximum luminance of V_2O_5 was 198,542 cd/A and maximum current efficiency was 20.3 cd/A with optimized work function of 4.73 eV and turn-on voltage of 1.7 V. For the synthesis of QLEDs, NiO_x nanoporous layers were utilized as a hole injection layer due to cost effectiveness, high charge mobility, and stability. Maximum brightness of 68,646 cd/cm^2 was achieved with a thickness of 30 nm, and current efficiency of 7.60 cd/A was achieved at low turn-on voltage of 3.4 V [66].

MoS_2@MoO_3 nanowires were fabricated by a hydrothermal technique to be utilized in green QD-LEDs as a hole injection layer. To observe the semiconductive nature of nanowires, optical analysis was performed and electroluminance spectra were recorded in the voltage range from 0.7 to 20 V, where the peak was obtained at 540 nm. Brightness intensity had a linear relationship with the concentration of MoS_2@MoO_3 nanowires. By increasing the concentration of nanowire from 3% to 10%, luminance increased from 44,300 to 68,630 cd/m^2 and efficiency was improved from 1.6% to 2.3% [83]. For green QLEDs, Yb-modified MoO_3/Ag/MoO_3 was fabricated as top cathode of an electron injection layer [84]. Current density of 38 cd/A was achieved with average transmittance of nearly 70% in the visible region at sheet resistance of 12.2 Ω/sq. Furthermore, 9.8% external quantum efficiency was observed with coordinates (0.199, 0.742) of color and improvement in current efficiency was 22.9%.

NiO are p-type semiconductors with wide bandgap, which can be used as a hole transport layer to reduce the hole injection barrier in QLEDs [85]. There are some natural defects, including interstitial oxygen and Ni vacancies from 1:1 stoichiometric ration of Ni and oxygen occurring during fabrication of NiO. Excess oxygen promotes Ni^{3+} formation, which makes it a p-type semiconductor [86,87]. Zhang et al. utilized NiO as hole injection layer in QD-LEDs and observed improved luminance performance from 17,260 to 25,836 cd/m^2 with a current efficiency of 6.05 cd/A and quantum efficiency of 3.68% [88]. Using a solvothermal method, NiO nanocrystals were prepared at 200°C to be utilized as hole injection layer in orange-red QD-LEDs [89]. The obtained device had a longer lifetime, of 11,491 h, which is much better in comparison with poly(3,4-ethylenedioxythiophene) based devices. Current efficiency of 5.38 cd/A was achieved with maximum brightness of 25,580 cd/m^2. NiO film of 85 nm was synthesized by a combustion process and used as hole transport layer in QD-LEDs with a lifetime of 2711 h [90]. The achieved current density was 4.56 cd/A with brightness of 33,852 cd/m^2 at 5.6 V. The estimated bandgap energy of QD-LEDs was nearly 2.38 eV. Cao et al. [91] also utilized NiO as hole injection layer in QD-LEDs. At turn-on voltage of 3 V, maximum luminance of 61,030 cd/m^2 was achieved with current efficiency of 45.7 cd/A.

5.1.2 Metal oxides in polymer LEDs (PLEDs)

PLEDs can be considered potential candidates for solid-state lighting and display technologies due to several characteristics such as easy preparation, quick, cost-effectiveness, and large area [92]. To enhance the performance (lower turn-on voltage and high efficiency) of PLEDs, interlayers have been utilized, which improves their internal structure. Numerous metal oxides, such as Ag, NiO, MoO_3, and WO_3, have been used as electron or hole injection layers in PLEDs, which enhances the performance of PLEDs by proving better charge balance between emitting layers of PLED [93].

Conventionally, poly(3,4-ethylenedioxythiophene) (PEDOT): poly(styrene sulfonate) (PSS) is utilized as hole conductor layer. However, this system has poor stability, and high-performance devices cannot be attained with good stability [94]. Metal oxides can be utilized as charge injection and transport layers in PLEDs as an alternative due to high mechanical and electrical strength, solution processable production, cost effectiveness, controllable film morphology, and stability. For instance, SnO_2 was utilized as electron injection layer in PLED, prepared by a spray deposition technique. The authors observed that, when $Ba(OH)_2$ was coated on a SnO_2 layer of 70 nm thickness, a current density of 20 cd/A and external quantum efficiency of 6.5% was achieved [95]. Chen et al. [96] fabricated Pt/NiO_x material as anode for PLED. The thickness of NiO_x film was nearly 60 nm, which was measured continuously by AFM, and resulted in maintaining 80%–85% mean transmittance in the visible range. Pt nanoparticle was doped on NiO_x film utilizing ion sputter coater to lessen the impact on mean transmission of NiO_x. After coating of Pt, the thickness of material became 62 nm and nearly 2.5% lower mean transmittance was observed. The authors also noted that PEDOT:PSS/ITO has almost same mean transmittance as that of NiO_x but Pt/NiO_x has higher stability and can be considered as a suitable substrate of PLED.

A schematic illustration of different layers of a Pt/NiO_x-based device is shown in Fig. 5A; AFM images for surface morphology are shown in Fig. 5B left side, while the right side shows the current distribution of the proposed PLED. The comparison of L-V and J-V (inset) characteristics of the proposed PLED with conventional PEDOT:PSS/ITO PLED is given in Fig. 5C. Maximum brightness obtained in the case of Pt/NiO_x PLED was 3852 cd/m^2 at 5.5 V, while in the case of conventional PLED, it was 3884 cd/m^2 at 6.4 V. Fig. 5D shows that the maximum current efficiency of proposed PLED was 1.42 cd/A at 233 mA/cm^2 and maximum current efficiency of conventional PLED was 1.21 cd/A at 271 mA/cm^2. It is also noted that the proposed PLED has 1.4 times more average current density as compared to conventional PLED. Electroluminescence spectra at 5.5 V of both the proposed PLED and the conventional PLED is shown in Fig. 5E.

In another study, ZnO and graphene nanorods were incorporated in PLED, which resulted in enhanced light emission at low turn-on voltage. The as-prepared device exhibited efficiency of 4 mA/cm^2, which is 60 times higher in comparison to pristine-based PLED [97]. Bernardo et al. also studied the effect of doping of ZnO on PLED. ZnO doped In_2O_3 was synthesized at room temperature by utilizing an RF magnetron sputtering method. The proposed PLED showed average visible optical transmission of 78% and hall mobility of 44.5 cm^2/Vs with work function of 5.07n and resistivity of $3.44 \times 10^{-4}\,\Omega$cm. This proposed PLED had higher resistance than basic ITO, but still showed higher efficiency of 0.015 cd/A as compared to ITO, which was 0.010 cd/A. This characteristic of ZnO-doped In_2O_3 makes it a suitable candidate for optoelectronic devices [98]. Sol-gel method was utilized by Chen et al. [99] to fabricate Ag-NiO, which was coated on ITO for PLED to use as a hole transport layer. The authors compared the performance of the proposed PLED with typical PEDOT:PSS-based PLED and found that the proposed PLED have better performance; for example, Ag/NiO/ITO PLED has 4.60 times more average current efficiency and 2.55 times higher electro luminance intensity than typical PLED.

5.2 Application of metal oxides in lasers

Nanoparticles of numerous metal oxides, such as CuO, AgO, ZnO, CeO_2, etc., have attracted the attention of researchers due to semiconductive metallic characteristics, which are utilized in the field of optical and electronic technology. Any material that absorbs the laser bandgap of material should be less than or equal to that of laser radiations. Metal oxide nanoparticles can be considered to exhibit the required bandgap [100]. In 2005, a group of researchers studied the laser applications of metal oxide-SiO_2 composite, where the metals were Mg, Zn, and Zr. A low threshold of 85 kW/cm^2 and directional emissions were obtained, as compared to conventional methods. To induce oxygen

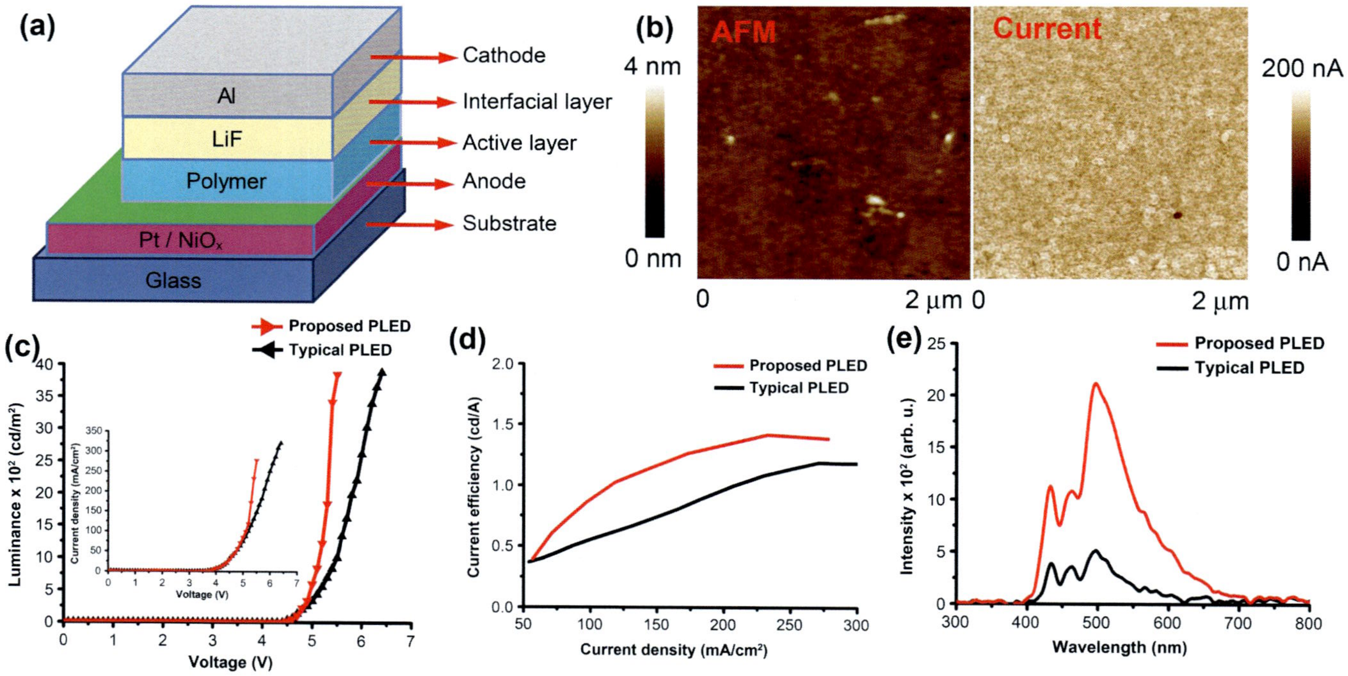

Fig. 5 (A) Schematic diagram of different layers of Pt/NiO$_x$-based PLED; (B) AFM and current distribution of Pt/NiO$_x$ film; (C–E) che comparison of L-V characteristics, current efficiencies, and electroluminescence spectra at 5.5 V of both the proposed PLED and the conventional PLED. Reused with permission from S.-H. Chen, Y.-C. Tu, D.-R. Wang, J.-D. Hwang, P.-C. Kao, Highly-luminous performance of polymer light-emitting devices utilizing platinum/nickelous oxide as the anode material, Synth. Met. 277 (2021a) 116796.

deficiency and structure transformation of several metal oxides, laser irradiation was used [13]. Laser irradiation also induces magnetic switching, excluding extrinsic metal impurities, within a few insulating and semiconducting metal oxides. Rao et al. performed laser irradiation (248 nm) on $SrTiO_3$; field-dependent ferromagnetic characteristics were obtained that remained intact even after 2 months of irradiation. SrTiO3 is a nonmagnetic oxide, and exposure to laser radiation induce oxygen deficiency in it, which plays a critical role in establishing a stable ferromagnetic state [101].

Metal oxides have high optical absorbance in the UV range; therefore, irradiation with laser beams of metal oxide thin films helps in polycrystalline growth of these films. With a laser ablation technique, metal oxide nanoparticles can be synthesized [102].

ZnO nanoparticles are nontoxic, can be easily synthesized at room temperature, and are cost-effective [100]. An ultraviolet lasers with a wavelength of <375 nm is used for fabrication of ZnO thin films. These films can feasibly be tuned with several metallic elements, such as Ag [103] and Li [104], through doping. Al-doped ZnO have been extensively studied; reduction in the resistivity was found from 1×10^{-3} to $5 \times 10^{-4}\,\Omega$cm, and 5% increment from 85% in visible transmittance is also observed (i.e., 90%) [105].

Due to high corrosion and heat resistance NiO can be used as a semiconductor in numerous fields such as LEDs and lasers [72] and sensing [106]. Through laser processing in synthesis of Ni electrodes from NiO nanoparticles, sintering and reduction take place concurrently. Table 3 illustrates the epitaxial growth of metal oxides films via laser.

5.2.1 Metal oxide-based lasers

UV random laser emission was produced by utilizing Ag-ZnO based electrospun cellulose acetate fibers [111]. The properties of random laser and flexibility degree of the as-prepared fiber were characterized by varying the bending radius of curvature from 0.22 to $0.80\,cm^{-1}$, as shown in Fig. 6A–D, and the apparatus used for the evaluation of the prepared material is described in Fig. 6E. In brief, when reducing the bending radius, the threshold energy for the emission of random laser

Table 3 Epitaxial growth of metal oxides films via laser.

Metal oxide	Laser	Substrate	Crystalline form	Reference
$Ce_{0.9}Zr_{0.1}O_{2-y}$	Nd:YAG (266 nm)	Y_2O_3:ZrO_2	Epitaxial	[107]
$LaNiO_3$	Nd:YAG (266 nm)	SrTiO3	Epitaxial	
Tin-doped-ITO	KrF (248 nm)	Glass	Polycrystalline	[108]
WO_3	KrF (248 nm)	TiO_2 (001)	Polycrystalline	[109]
Al-doped ZnO	–	PEN	Polycrystalline	[103]
VO_2	KrF (248 nm)	TiO_2 (001)	Epitaxial	[110]
Li-doped ZnO	He-Cd (325 nm)	Si (100)	Polycrystalline	[104]

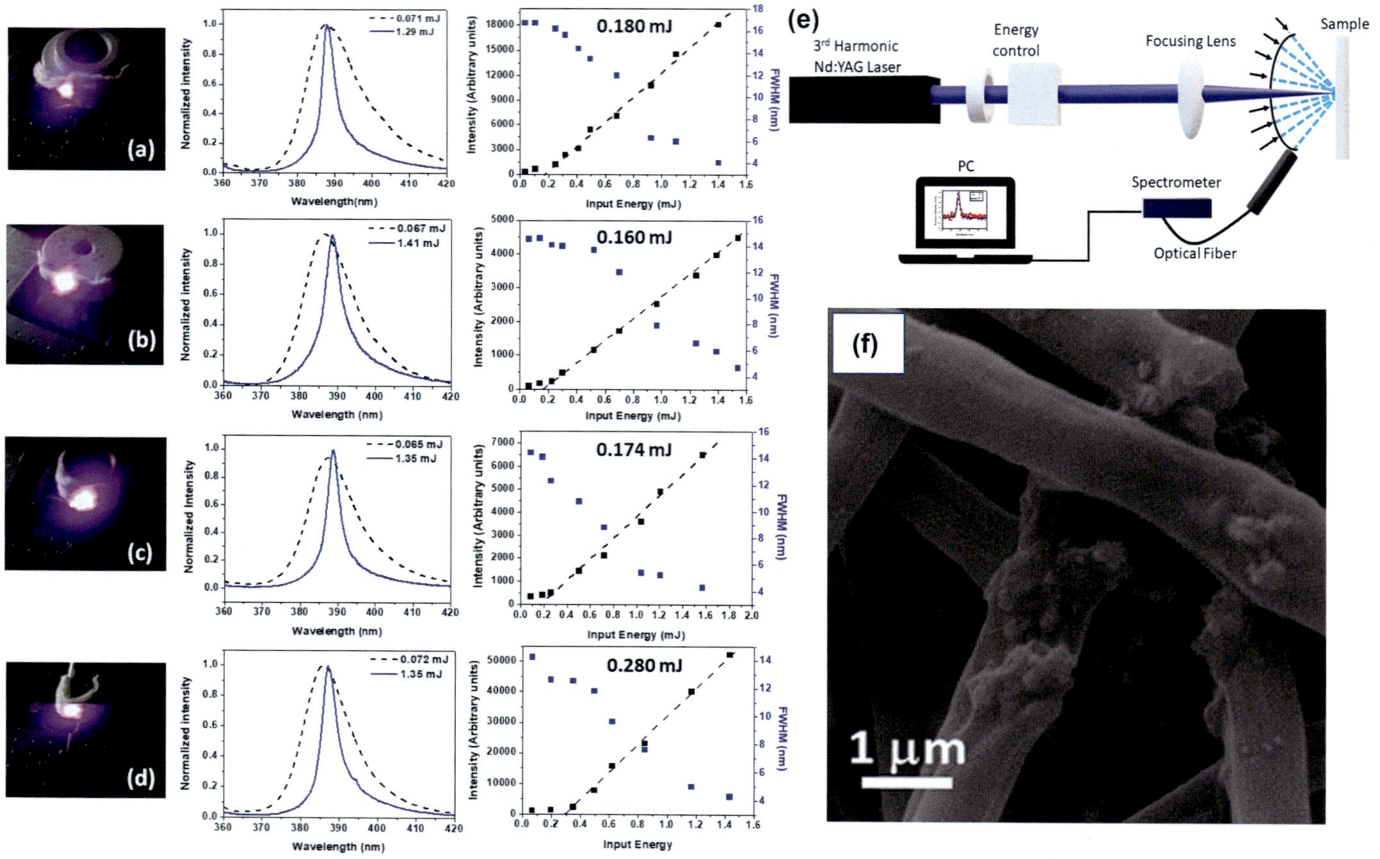

Fig. 6 (A–D) Normalized intensity v/s input energy with different bending radius; (E) pictorial presentation of the apparatus used for the evaluation of the prepared material; (F) SEM image of Ag-ZnO-based electrospun cellulose acetate fibers.

Reused with permission from M.L. Da Silva-Neto, M.C.A. De Oliveira, C.T. Dominguez, R.E.M. Lins, N. Rakov, C.B. De Araújo, L.D.S. Menezes, H. P. De Oliveira, A.S.L. Gomes, UV random laser emission from flexible ZnO-Ag-enriched electrospun cellulose acetate fiber matrix, Sci. Rep. 9 (2019) 11765.

also decreases, which implies that higher performance is provided by the proposed fiber at lower values. An SEM image of Ag-ZnO-based electrospun cellulose acetate fibers is shown in Fig. 6F.

ZnO microwire/MgO/p-GaN was used by Yang et al. [112] to develop electrically pumped lasers. At an injection current of 14 mA, output power of 2.41 μW and threshold of 12.5 mA was achieved. Another study also revealed that ZnO-based nanomaterial enhances the emission of UV random laser [113]. SiO_2-doped ZnO fabricated by chemical bath deposition was utilized to improve the emission performance of UV laser.

Er^{3+}-doped bismuthate glass was investigated by Guo et al. and an emission of 2.7 μm and larger spontaneous transition probability of $65.265\,s^{-1}$ was observed. The obtained results showed that Er^{3+}-doped bismuthate glass has potential applications in laser materials [114]. Rusdi et al. [115] reported Q-switched fiber laser using NiO saturable absorber synthesized by a solution casting method. Maximum pulse energy of 0.58 nJ with pulse width of 3.13 μs was achieved at 3.74% efficiency. Stable pulse was observed at 528 mW.

6 Concluding remarks

Recently, significant intensive investigation steps within nanotechnology have exhibited significant prospects. There have been substantial advances made in preparing specific inorganic nanomaterials (NMs) with various potential applications, as documented within the literature. The result of NMs permits us to move toward a bright and livable subsequently. These materials show specific processes for managing humanity's many unsolved issues and necessities. There are unlimited opportunities for enhanced tools, systems and materials if background may be developed for the assembly at small networks. Nanotechnology has gained the top science level with real and research possibilities within all critical mental sciences. An essential component of nanotechnology is to bridge the gap between the atomic and molecular hierarchies of basic sciences and the microstructural scale of engineering and manufacturing. Therefore, a broad interdisciplinary understanding is to be investigated and related to the suitability of society. It will guide an understanding and combine new high-technical apparatuses. Nanotechnology is not a development but goes within something like a computer that becomes an impact. Metal oxides can be considered as suitable candidates for commercialization, with practical applications in LEDs and lasers. Efforts should be made to establish cost-effective manufacturing methods and to improve the effectiveness of such devices.

Acknowledgment

The authors acknowledge the support from the Department of Chemistry and Research & Development Cell of Maharishi Markandeshwar (Deemed to be University), Mullana, Ambala, Haryana, India.

References

[1] M.N. Eisler, "The ennobling unity of science and technology": materials sciences and engineering, the Department of Energy, and the Nanotechnology Enigma, Minerva 51 (2013) 225–251.

[2] B. Luo, S. Liu, L. Zhi, Chemical approaches toward graphene-based nanomaterials and their applications in energy-related areas, Small 8 (2012) 630–646.

[3] S.S. Siwal, K. Sheoran, K. Mishra, H. Kaur, A.K. Saini, V. Saini, D.-V.N. Vo, H.Y. Nezhad, V.K. Thakur, Novel synthesis methods and applications of Mxene-based nanomaterials (MBNS) for hazardous pollutants degradation: future perspectives, Chemosphere 293 (2022), 133542.

[4] H. Zhan, Z. Xiong, C. Cheng, Q. Liang, J.Z. Liu, D. Li, Solvation-involved nanoionics: new opportunities from 2D nanomaterial laminar membranes, Adv. Mater. 32 (2020) 1904562.

[5] J.N. Tiwari, R.N. Tiwari, K.S. Kim, Zero-dimensional, one-dimensional, two-dimensional and three-dimensional nanostructured materials for advanced electrochemical energy devices, Prog. Mater. Sci. 57 (2012) 724–803.

[6] R.L.G. Lecaros, K.Y. Chua, L.L. Tayo, W.-S. Hung, C.-C. Hu, Q.-F. An, H.-A. Tsai, K.-R. Lee, J.-Y. Lai, The fine-structure characteristics and isopropanol/water dehydration through pervaporation composite membranes improved with graphene quantum dots, Sep. Purif. Technol. 247 (2020), 116956.

[7] A. Yuan, H. Lei, F. Xi, J. Liu, L. Qin, Z. Chen, X. Dong, Graphene quantum dots decorated graphitic carbon nitride nanorods for photocatalytic removal of antibiotics, J. Colloid Interface Sci. 548 (2019) 56–65.

[8] C. Zhang, B. Xie, Y. Zou, D. Zhu, L. Lei, D. Zhao, H. Nie, Zero-dimensional, one-dimensional, two-dimensional and three-dimensional biomaterials for cell fate regulation, Adv. Drug Deliv. Rev. 132 (2018) 33–56.

[9] K. Sakakibara, J.P. Hill, K. Ariga, Thin-film-based nanoarchitectures for soft matter: controlled assemblies into two-dimensional worlds, Small 7 (2011) 1288–1308.

[10] X. Zhang, C. Gong, O.U. Akakuru, Z. Su, A. Wu, G. Wei, The design and biomedical applications of self-assembled two-dimensional organic biomaterials, Chem. Soc. Rev. 48 (2019) 5564–5595.

[11] Z. Zhang, E.S. Penev, B.I. Yakobson, Two-dimensional boron: structures, properties and applications, Chem. Soc. Rev. 46 (2017) 6746–6763.

[12] K. Mishra, N. Devi, S.S. Siwal, Q. Zhang, W.F. Alsanie, F. Scarpa, V.K. Thakur, Ionic liquid-based polymer nanocomposites for sensors, energy, biomedicine, and environmental applications: roadmap to the future, Adv. Sci. 9 (2022) 2202187.

[13] H. Palneedi, J.H. Park, D. Maurya, M. Peddigari, G.-T. Hwang, V. Annapureddy, J.-W. Kim, J.-J. Choi, B.-D. Hahn, S. Priya, K.J. Lee, J. Ryu, Laser irradiation of metal oxide films and nanostructures: applications and advances, Adv. Mater. 30 (2018) 1705148.

[14] C. Århammar, A. Pietzsch, N. Bock, E. Holmström, C.M. Araujo, J. Gråsjö, S. Zhao, S. Green, T. Peery, F. Hennies, S. Amerioun, A. Föhlisch, J. Schlappa, T. Schmitt, V.N. Strocov, G.A. Niklasson, D.C. Wallace, J.-E. Rubensson, B. Johansson, R. Ahuja, Unveiling the complex electronic structure of amorphous metal oxides, Proc. Natl. Acad. Sci. 108 (2011) 6355.

[15] B. Jia, Y. Hu, X. Guan, J. Hao, B. Yan, Y. Zu, G. Liu, Q. Zhang, G.-D. Peng, P. Lu, Atomic structures and electronic properties of different interface types at Al/c-SiO2 interfaces, Appl. Surf. Sci. 578 (2022), 151932.

[16] J. Mayandi, T.G. Finstad, Ø. Dahl, P. Vajeeston, M. Schrade, O.M. Løvvik, S. Diplas, P.A. Carvalho, Thin films made by reactive sputtering of high entropy alloy FeCoNiCuGe: optical, electrical and structural properties, Thin Solid Films 139083 (2022).
[17] M.E. King, M.V. Fonseca Guzman, M.B. Ross, Material strategies for function enhancement in plasmonic architectures, Nanoscale 14 (2022) 602–611.
[18] M. Rahimi, M.H. Abbaspour-Fard, A. Rohani, Synergetic effect of N/O functional groups and microstructures of activated carbon on supercapacitor performance by machine learning, J. Power Sources 521 (2022), 230968.
[19] B. Mondal, P.K. Gogoi, Nanoscale heterostructured materials based on metal oxides for a chemiresistive gas sensor, ACS Appl. Electron. Mater. 4 (2022) 59–86.
[20] K. Mishra, V. Kumar Thakur, S. Singh Siwal, Graphitic carbon nitride based palladium nanoparticles: a homemade anode electrode catalyst for efficient direct methanol fuel cells application, Mater. Today Proc. 56 (2022) 107–111.
[21] S. Siwal, S. Matseke, S. Mpelane, N. Hooda, D. Nandi, K. Mallick, Palladium-polymer nanocomposite: an anode catalyst for the electrochemical oxidation of methanol, Int. J. Hydrog. Energy 42 (2017) 23599–23605.
[22] S.S. Siwal, Q. Zhang, N. Devi, K.V. Thakur, Carbon-based polymer nanocomposite for high-performance energy storage applications, Polymers 12 (2020).
[23] H. Kaur, V.K. Thakur, S.S. Siwal, Recent advancements in graphdiyne-based nanomaterials for biomedical applications, Mater. Today Proc. 56 (2022) 112–120.
[24] S. Karamveer, V.K. Thakur, S.S. Siwal, Synthesis and overview of carbon-based materials for high performance energy storage application: a review, Mater. Today Proc. 56 (2022) 9–17.
[25] H. Lei, X. Li, C. Sun, J. Zeng, S.S. Siwal, Q. Zhang, Galvanic replacement–mediated synthesis of ni-supported pd nanoparticles with strong metal–support interaction for methanol electro-oxidation, Small 15 (2019) 1804722.
[26] S.S. Siwal, Q. Zhang, C. Sun, V.K. Thakur, Graphitic carbon nitride doped copper–manganese alloy as high–performance electrode material in supercapacitor for energy storage, Nanomaterials 10 (2019).
[27] S. Siwal, N. Devi, V. Perla, R. Barik, S. Ghosh, K. Mallick, The influencing role of oxophilicity and surface area of the catalyst for electrochemical methanol oxidation reaction: a case study, Mater. Res. Innov. (2018) 1–8.
[28] T. Gur, I. Meydan, H. Seckin, M. Bekmezci, F. Sen, Green synthesis, characterization and bioactivity of biogenic zinc oxide nanoparticles, Environ. Res. 204 (2022), 111897.
[29] K. Mondal, R. Kumar, B. Isaac, G. Pawar, 5—Metal oxide nanofibers and their applications for biosensing, in: V. Esposito, D. Marani (Eds.), Metal Oxide-Based Nanofibers and Their Applications, Elsevier, 2022.
[30] H.N. Cuong, S. Pansambal, S. Ghotekar, R. Oza, N.T. Thanh Hai, N.M. Viet, V.-H. Nguyen, New frontiers in the plant extract mediated biosynthesis of copper oxide (CuO) nanoparticles and their potential applications: a review, Environ. Res. 203 (2022), 111858.
[31] S.S. Siwal, W. Yang, Q. Zhang, Recent progress of precious-metal-free electrocatalysts for efficient water oxidation in acidic media, J. Energy Chem. 51 (2020) 113–133.
[32] J. Zeng, L. Chen, S. Siwal, Q. Zhang, Solvothermal sulfurization in deep eutectic solvent: a novel route to synthesize Co-doped Ni3S2 nanosheets supported on Ni foam as active materials for ultrahigh-performance pseudocapacitors, Sustain. Energy Fuels 3 (2019) 1957–1965.
[33] J. Zeng, J. Liu, S.S. Siwal, W. Yang, X. Fu, Q. Zhang, Morphological and electronic modification of 3D porous nickel microsphere arrays by cobalt and sulfur dual synergistic

modulation for overall water splitting electrolysis and supercapacitors, Appl. Surf. Sci. 491 (2019) 570–578.
[34] R. Lorenzi, N.V. Golubev, E.S. Ignat'eva, V.N. Sigaev, C. Ferrara, M. Acciarri, G.M. Vanacore, A. Paleari, Defect-assisted photocatalytic activity of glass-embedded gallium oxide nanocrystals, J. Colloid Interface Sci. 608 (2022) 2830–2838.
[35] Y. Tian, X. Liu, S. Xu, J. Li, A. Caratenuto, Y. Mu, Z. Wang, F. Chen, R. Yang, J. Liu, M.L. Minus, Y. Zheng, Recyclable and efficient ocean biomass-derived hydrogel photothermal evaporator for thermally-localized solar desalination, Desalination 523 (2022), 115449.
[36] G. Ahmadpour, M.R. Nilforoushan, B. Shayegh Boroujeny, M. Tayebi, S.M. Jesmani, Effect of substrate surface treatment on the hydrothermal synthesis of zinc oxide nanostructures, Ceram. Int. 48 (2022) 2323–2329.
[37] T.S. Aldeen, H.E. Ahmed Mohamed, M. Maaza, ZnO nanoparticles prepared via a green synthesis approach: physical properties, photocatalytic and antibacterial activity, J. Phys. Chem. Solids 160 (2022), 110313.
[38] C. Lou, G. Lei, X. Liu, J. Xie, Z. Li, W. Zheng, N. Goel, M. Kumar, J. Zhang, Design and optimization strategies of metal oxide semiconductor nanostructures for advanced formaldehyde sensors, Coord. Chem. Rev. 452 (2022), 214280.
[39] J. Yan, T. Liu, X. Liu, Y. Yan, Y. Huang, Metal-organic framework-based materials for flexible supercapacitor application, Coord. Chem. Rev. 452 (2022), 214300.
[40] R. Kumar, R. Thangappan, Electrode material based on reduced graphene oxide (RGO)/transition metal oxide composites for supercapacitor applications: a review, Emerg. Mater. 5 (2022) 1881–1897.
[41] R. Safdar Ali, H. Meng, Z. Li, Zinc-based metal-organic frameworks in drug delivery, cell imaging, and sensing, Molecules 27 (2022).
[42] T. Wasiak, D. Janas, Nanowires as a versatile catalytic platform for facilitating chemical transformations, J. Alloys Compd. 892 (2022), 162158.
[43] W.K. Wong, J.T.Y. Chin, S.A. Khan, F. Pelletier, E.C. Corbos, Robust continuous synthesis and in situ deposition of catalytically active nanoparticles on colloidal support materials in a triphasic flow millireactor, Chem. Eng. J. 430 (2022), 132778.
[44] H. Li, W. Zheng, W. Qu, L. Leng, Y. Feng, F. Xin, G. Sheng, Z. Yang, K. Shih, J. Yang, Facile pathway towards crystallinity adjustment and performance enhancement of copper selenide for vapor-phase elemental mercury sequestration, Chem. Eng. J. 430 (2022), 132811.
[45] S.S. Khudiar, U.M. Nayef, F.A.H. Mutlak, S.K. Abdulridha, Characterization of No2 gas sensing for ZnO nanostructure grown hydrothermally on porous silicon, Optik 249 (2022), 168300.
[46] S.O. Ganiyu, S. Sable, M. Gamal El-Din, Advanced oxidation processes for the degradation of dissolved organics in produced water: a review of process performance, degradation kinetics and pathway, Chem. Eng. J. 429 (2022), 132492.
[47] P. Hamdi-Mohammadabad, T. Tohidi, R. Talebzadeh, R. Mohammad-Rezaei, S. Rahmatallahpur, Evaluation of structural and optical properties of TeO2 nano and micro structures grown on glass and silicon substrates using thermal evaporation method, Mater. Sci. Semicond. Process. 139 (2022), 106363.
[48] H.M. Smith, A.F. Turner, Vacuum deposited thin films using a ruby laser, Appl. Opt. 4 (1965) 147–148.
[49] T. Ikeda, J.-I. Mamiya, Y. Yu, Photomechanics of liquid-crystalline elastomers and other polymers, Angew. Chem. Int. Ed. 46 (2007) 506–528.

[50] H. Sankur, W.J. Gunning, J. Denatale, J.F. Flintoff, High-quality optical and epitaxial Ge films formed by laser evaporation, J. Appl. Phys. 65 (1989) 2475–2478.
[51] I. Ohkubo, Y. Matsumoto, K. Itaka, T. Hasegawa, K. Ueno, M. Ohtani, M. Kawasaki, H. Koinuma, Quick optimization of Y-type magnetoplumbite thin films growth by combinatorial pulsed laser deposition technique, Appl. Surf. Sci. 197–198 (2002) 312–315.
[52] P.R. Willmott, P. Manoravi, K. Holliday, Production and characterization of Nd,Cr:Gsgg thin films on Si(001) grown by pulsed laser ablation, Appl. Phys. A 70 (2000) 425–429.
[53] A. Hossain, P. Bandyopadhyay, A. Karmakar, A.K.M.A. Ullah, R.K. Manavalan, K. Sakthipandi, N. Alhokbany, S.M. Alshehri, J. Ahmed, The Hybrid Halide Perovskite: Synthesis Strategies, Fabrications, and Modern Applications, Ceramics International, 2021.
[54] F.K. Urban, A. Hosseini-Tehrani, P. Griffiths, A. Khabari, Y.W. Kim, I. Petrov, Nanophase films deposited from a high-rate, nanoparticle beam, J. Vac. Sci. Technol. B 20 (2002) 995–999.
[55] L.M. Cao, Z. Zhang, L.L. Sun, C.X. Gao, M. He, Y.Q. Wang, Y.C. Li, X.Y. Zhang, G. Li, J. Zhang, W.K. Wang, Well-aligned boron nanowire arrays, Adv. Mater. 13 (2001) 1701–1704.
[56] C. Hums, J. Bläsing, A. Dadgar, A. Diez, T. Hempel, J. Christen, A. Krost, K. Lorenz, E. Alves, Metal-organic vapor phase epitaxy and properties of AlInN in the whole compositional range, Appl. Phys. Lett. 90 (2007), 022105.
[57] T. Zhai, X. Fang, M. Liao, X. Xu, H. Zeng, B. Yoshio, D. Golberg, A comprehensive review of one-dimensional metal-oxide nanostructure photodetectors, Sensors 9 (2009).
[58] C. Diaz, M.L. Valenzuela, M.Á. Laguna-Bercero, Solid-state preparation of metal and metal oxides nanostructures and their application in environmental remediation, Int. J. Mol. Sci. 23 (2022).
[59] H.J. Bolink, E. Coronado, J. Orozco, M. Sessolo, Efficient polymer light-emitting diode using air-stable metal oxides as electrodes, Adv. Mater. 21 (2009) 79–82.
[60] T.-W. Lee, J. Hwang, S.-Y. Min, Highly efficient hybrid inorganic–organic light-emitting diodes by using air-stable metal oxides and a thick emitting layer, ChemSusChem 3 (2010) 1021–1023.
[61] C.-S. Lee, C.-H. Jeon, B.-T. Lee, S.-H. Jeong, Abrupt conversion of the conductivity and band-gap in the sputter grown Ga-doped ZnO films by a change in growth ambient: Effects of oxygen partial pressure, J. Alloys Compd. 742 (2018) 977–985.
[62] R. Faiz, Zinc oxide light-emitting diodes: a review, Opt. Eng. 58 (2019) 1–20.
[63] Z. Yin, X. Zhang, Y. Cai, J. Chen, J.I. Wong, Y.-Y. Tay, J. Chai, J. Wu, Z. Zeng, B. Zheng, H.Y. Yang, H. Zhang, Preparation of MoS2–MoO3 hybrid nanomaterials for light-emitting diodes, Angew. Chem. Int. Ed. 53 (2014) 12560–12565.
[64] D.K. Choi, D.H. Kim, C.M. Lee, H. Hafeez, S. Sarker, J.S. Yang, H.J. Chae, G.-W. Jeong, D.H. Choi, T.W. Kim, S. Yoo, J. Song, B.S. Ma, T.-S. Kim, C.H. Kim, H.J. Lee, J.W. Lee, D. Kim, T.-S. Bae, S.M. Yu, Y.-C. Kang, J. Park, K.-H. Kim, M. Sujak, M. Song, C.-S. Kim, S.Y. Ryu, Highly efficient, heat dissipating, stretchable organic light-emitting diodes based on a MoO3/Au/MoO3 electrode with encapsulation, Nat. Commun. 12 (2021) 2864.
[65] K.J. Ko, S.R. Shin, H.B. Lee, E. Jeong, Y.J. Yoo, H.M. Kim, Y.M. Song, J. Yun, J.W. Kang, Fabrication of an oxide/metal/oxide structured electrode integrated with antireflective film to enhance performance in flexible organic light-emitting diodes, Mater. Today Energy 20 (2021), 100704.
[66] W.-S. Chen, S.-H. Yang, W.-C. Tseng, W.W.-S. Chen, Y.-C. Lu, Utilization of nanoporous nickel oxide as the hole injection layer for quantum dot light-emitting diodes, ACS Omega 6 (2021) 13447–13455.

[67] L. Kinner, T. Dimopoulos, G. Ligorio, E.J.W. List-Kratochvil, F. Hermerschmidt, High performance organic light-emitting diodes employing Ito-free and flexible TiOx/Ag/Al: ZnO electrodes, RSC Adv. 11 (2021) 17324–17331.
[68] S.-J. Kwon, S. Ahn, J.-M. Heo, D.J. Kim, J. Park, H.-R. Lee, S. Kim, H. Zhou, M.-H. Park, Y.-H. Kim, W. Lee, J.-Y. Sun, B.H. Hong, T.-W. Lee, Chemically robust indium tin oxide/graphene anode for efficient perovskite light-emitting diodes, ACS Appl. Mater. Interfaces 13 (2021) 9074–9080.
[69] Y. Sun, W. Wang, H. Zhang, Q. Su, J. Wei, P. Liu, S. Chen, S. Zhang, High-performance quantum dot light-emitting diodes based on al-doped ZnO nanoparticles electron transport layer, ACS Appl. Mater. Interfaces 10 (2018) 18902–18909.
[70] H.-M. Kim, S. Cho, J. Kim, H. Shin, J. Jang, Li and Mg Co-doped zinc oxide electron transporting layer for highly efficient quantum dot light-emitting diodes, ACS Appl. Mater. Interfaces 10 (2018) 24028–24036.
[71] B.R. Lee, J.-W. Kim, D. Kang, D.W. Lee, S.-J. Ko, H.J. Lee, C.-L. Lee, J.Y. Kim, H.S. Shin, M.H. Song, Highly efficient polymer light-emitting diodes using graphene oxide as a hole transport layer, ACS Nano 6 (2012) 2984–2991.
[72] Malvin, C.-T. Tsai, C.-T. Wang, Y.-Y. Chen, P.-C. Kao, S.-Y. Chu, Improved hole-injection and external quantum efficiency of organic light-emitting diodes using an ultra-thin K-doped NiO buffer layer, J. Alloys Compd. 797 (2019) 159–165.
[73] T. Minami, Transparent conducting oxide semiconductors for transparent electrodes, Semicond. Sci. Technol. 20 (2005) S35–S44.
[74] M. Sessolo, H.J. Bolink, Hybrid organic–inorganic light-emitting diodes, Adv. Mater. 23 (2011) 1829–1845.
[75] Y.-M. Huang, K.J. Singh, A.-C. Liu, C.-C. Lin, Z. Chen, K. Wang, Y. Lin, Z. Liu, T. Wu, H.-C. Kuo, Advances in quantum-dot-based displays, Nanomaterials 10 (2020).
[76] T.-H. Kim, K.-S. Cho, E.K. Lee, S.J. Lee, J. Chae, J.W. Kim, D.H. Kim, J.-Y. Kwon, G. Amaratunga, S.Y. Lee, B.L. Choi, Y. Kuk, J.M. Kim, K. Kim, Full-colour quantum dot displays fabricated by transfer printing, Nat. Photonics 5 (2011) 176–182.
[77] Y.-Q. Liu, D.-D. Zhang, H.-X. Wei, Q.-D. Ou, Y.-Q. Li, J.-X. Tang, Highly efficient quantum-dot light emitting diodes with sol-gel ZnO electron contact, Opt. Mater. Express 7 (2017) 2161–2167.
[78] V. Wood, M.J. Panzer, J.E. Halpert, J.M. Caruge, M.G. Bawendi, V. Bulović, Selection of metal oxide charge transport layers for colloidal quantum dot leds, ACS Nano 3 (2009) 3581–3586.
[79] L. Qian, Y. Zheng, J. Xue, P.H. Holloway, Stable and efficient quantum-dot light-emitting diodes based on solution-processed multilayer structures, Nat. Photonics 5 (2011) 543–548.
[80] H.-M. Kim, D. Geng, J. Kim, E. Hwang, J. Jang, Metal-oxide stacked electron transport layer for highly efficient inverted quantum-dot light emitting diodes, ACS Appl. Mater. Interfaces 8 (2016) 28727–28736.
[81] X. Yang, Y. Ma, E. Mutlugun, Y. Zhao, K.S. Leck, S.T. Tan, H.V. Demir, Q. Zhang, H. Du, X.W. Sun, Stable, efficient, and all-solution-processed quantum dot light-emitting diodes with double-sided metal oxide nanoparticle charge transport layers, ACS Appl. Mater. Interfaces 6 (2014) 495–499.
[82] S.-G. Choi, H.-J. Seok, S. Rhee, D. Hahm, W.K. Bae, H.-K. Kim, Magnetron-sputtered amorphous V2O5 hole injection layer for high performance quantum dot light-emitting diode, J. Alloys Compd. 878 (2021), 160303.
[83] N. Bastami, E. Soheyli, A. Arslan, R. Sahraei, A.F. Yazici, E. Mutlugun, Nanowire-shaped MoS2@MoO3 nanocomposites as a hole injection layer for quantum dot light-emitting diodes, ACS Appl. Electron. Mater. 4 (2022) 3849–3859.

[84] C.-Y. Lee, Y.-M. Chen, Y.-Z. Deng, Y.-P. Kuo, P.-Y. Chen, L. Tsai, M.-Y. Lin, Yb: MoO3/Ag/MoO3 multilayer transparent top cathode for top-emitting green quantum dot light-emitting diodes, Nanomaterials 10 (2020).
[85] X.L. Zhang, H.T. Dai, J.L. Zhao, C. Li, S.G. Wang, X.W. Sun, Effects of the thickness of NiO hole transport layer on the performance of all-inorganic quantum dot light emitting diode, Thin Solid Films 567 (2014) 72–76.
[86] U.T. Nakate, R. Ahmad, P. Patil, Y.T. Yu, Y.-B. Hahn, Ultra thin NiO nanosheets for high performance hydrogen gas sensor device, Appl. Surf. Sci. 506 (2020), 144971.
[87] S. Yang, J. Kim, Y. Choi, H. Kim, D. Lee, J.-S. Bae, S. Park, Annealing environment dependent electrical and chemical state correlation of Li-doped NiO, J. Alloys Compd. 815 (2020), 152343.
[88] Y. Zhang, L. Zhao, H. Jia, P. Li, Study of the electroluminescence performance of NiO-based quantum dot light-emitting diodes: the effect of annealing atmosphere, Appl. Surf. Sci. 526 (2020), 146732.
[89] Y. Zhang, S. Wang, L. Chen, Y. Fang, H. Shen, Z. Du, Solution-processed quantum dot light-emitting diodes based on NiO nanocrystals hole injection layer, Org. Electron. 44 (2017) 189–197.
[90] Y. Zhang, Glycine-assisted fabrication of NiO-based quantum dots light-emitting diodes, J. Lumin. 192 (2017) 1015–1019.
[91] F. Cao, H. Wang, P. Shen, X. Li, Y. Zheng, Y. Shang, J. Zhang, Z. Ning, X. Yang, High-efficiency and stable quantum dot light-emitting diodes enabled by a solution-processed metal-doped nickel oxide hole injection interfacial layer, Adv. Funct. Mater. 27 (2017) 1704278.
[92] Z.B. Wang, M.G. Helander, J. Qiu, Z.W. Liu, M.T. Greiner, Z.H. Lu, Direct hole injection in to 4,4′-N,N′-dicarbazole-biphenyl: a simple pathway to achieve efficient organic light emitting diodes, J. Appl. Phys. 108 (2010), 024510.
[93] Y. Wang, Q. Niu, C. Hu, W. Wang, M. He, Y. Zhang, S. Li, L. Zhao, X. Wang, J. Xu, Q. Zhu, S. Chen, Ultrathin nickel oxide film as a hole buffer layer to enhance the optoelectronic performance of a polymer light-emitting diode, Opt. Lett. 36 (2011) 1521–1523.
[94] A. De Morais, J. Freitas, Metal oxide films as charge transport layers for solution-processed polymer light-emitting diodes, J. Braz. Chem. Soc. (2022) 1–12.
[95] L.P. Lu, C.E. Finlayson, R.H. Friend, A study of tin oxide as an election injection layer in hybrid polymer light-emitting diodes, Semicond. Sci. Technol. 29 (2014), 125002.
[96] S.-H. Chen, Y.-C. Tu, D.-R. Wang, J.-D. Hwang, P.-C. Kao, Highly-luminous performance of polymer light-emitting devices utilizing platinum/nickelous oxide as the anode material, Synth. Met. 277 (2021), 116796.
[97] K. Gautam, I. Singh, R. Bhatnagar, P.K. Bhatnagar, K. Rao Peta, Synergistic effect of graphene and ZnO nanorods in enhancing the performance of MEH-PPV based polymer light emitting diode, Displays 73 (2022), 102170.
[98] G. Bernardo, G. Gonçalves, P. Barquinha, Q. Ferreira, G. Brotas, L. Pereira, A. Charas, J. Morgado, R. Martins, E. Fortunato, Polymer light-emitting diodes with amorphous indium-zinc oxide anodes deposited at room temperature, Synth. Met. 159 (2009) 1112–1115.
[99] S.-H. Chen, Y.-T. Wu, S.-H. Hsiao, C. Tseng, Silver-doped nickel oxide as an efficient hole-transport layer in polymer light-emitting diodes, Microsc. Res. Tech. 85 (2022) 2390–2396.
[100] V.B. Nam, T.T. Giang, S. Koo, J. Rho, D. Lee, Laser digital patterning of conductive electrodes using metal oxide nanomaterials, Nano Converg. 7 (2020) 23.
[101] D.A. Carballo-Córdova, M.T. Ochoa-Lara, S.F. Olive-Méndez, F. Espinosa-Magaña, First-principles calculations and Bader analysis of oxygen-deficient induced magnetism in cubic BaTiO3−x and SrTiO3−x, Philos. Mag. 99 (2019) 181–197.

[102] E. Maurer, S. Barcikowski, B. Gökce, Process chain for the fabrication of nanoparticle polymer composites by laser ablation synthesis, Chem. Eng. Technol. 40 (2017) 1535–1543.
[103] Y. Li, R. Yao, H. Wang, X. Wu, J. Wu, X. Wu, W. Qin, Enhanced performance in Al-doped ZnO based transparent flexible transparent thin-film transistors due to oxygen vacancy in ZnO film with Zn–Al–O interfaces fabricated by atomic layer deposition, ACS Appl. Mater. Interfaces 9 (2017) 11711–11720.
[104] D. Wang, J. Zhou, G. Liu, Effect of Li-doped concentration on the structure, optical and electrical properties of p-type ZnO thin films prepared by sol–gel method, J. Alloys Compd. 481 (2009) 802–805.
[105] S.O. El Hamali, W.M. Cranton, N. Kalfagiannis, X. Hou, R. Ranson, D.C. Koutsogeorgis, Enhanced electrical and optical properties of room temperature deposited aluminium doped zinc oxide (AZO) thin films by excimer laser annealing, Opt. Lasers Eng. 80 (2016) 45–51.
[106] S.M. Majhi, G.K. Naik, H.-J. Lee, H.-G. Song, C.-R. Lee, I.-H. Lee, Y.-T. Yu, Au@NiO core-shell nanoparticles as a p-type gas sensor: novel synthesis, characterization, and their gas sensing properties with sensing mechanism, Sensors Actuators B Chem. 268 (2018) 223–231.
[107] A. Queraltó, A. Pérez Del Pino, M. De La Mata, J. Arbiol, M. Tristany, X. Obradors, T. Puig, Ultrafast epitaxial growth kinetics in functional oxide thin films grown by pulsed laser annealing of chemical solutions, Chem. Mater. 28 (2016) 6136–6145.
[108] T. Tsuchiya, F. Yamaguchi, I. Morimoto, T. Nakajima, T. Kumagai, Microstructure control of low-resistivity tin-doped indium oxide films grown by photoreaction of nanoparticles using a KrF excimer laser at room temperature, Appl. Phys. A 99 (2010) 745–749.
[109] T. Nakajima, T. Kitamura, T. Tsuchiya, Visible light photocatalytic activity enhancement for water purification in Cu(II)-grafted Wo3 thin films grown by photoreaction of nanoparticles, Appl. Catal. B Environ. 108–109 (2011) 47–53.
[110] M. Nishikawa, T. Nakajima, T. Manabe, T. Okutani, T. Tsuchiya, High temperature coefficients of resistance of Vo2 films grown by excimer-laser-assisted metal organic deposition process for bolometer application, Mater. Lett. 64 (2010) 1921–1924.
[111] M.L. Da Silva-Neto, M.C.A. De Oliveira, C.T. Dominguez, R.E.M. Lins, N. Rakov, C.B. De Araújo, L.D.S. Menezes, H.P. De Oliveira, A.S.L. Gomes, UV random laser emission from flexible ZnO-Ag-enriched electrospun cellulose acetate fiber matrix, Sci. Rep. 9 (2019) 11765.
[112] X. Yang, C.-X. Shan, P.-N. Ni, M.-M. Jiang, A.-Q. Chen, H. Zhu, J.-H. Zang, Y.-J. Lu, D.-Z. Shen, Electrically driven lasers from van der Waals heterostructures, Nanoscale 10 (2018) 9602–9607.
[113] A.T. Ali, W. Maryam, Y.-W. Huang, H.C. Hsu, N.M. Ahmed, N. Zainal, M.S. Jameel, SiO2 capped-ZnO nanorods for enhanced random laser emission, Opt. Laser Technol. 147 (2022), 107633.
[114] Y. Guo, Y. Tian, L. Zhang, L. Hu, J. Zhang, Erbium doped heavy metal oxide glasses for mid-infrared laser materials, J. Non-Cryst. Solids 377 (2013) 119–123.
[115] M.F.M. Rusdi, A.A. Latiff, M.T. Ahmad, M.F.A. Rahman, A.H.A. Rosol, S.W. Harun, H. Ahmad, Nickel oxide as a Q-switcher for short pulsed thulium doped fiber laser generation, J. Phys. Conf. Ser. 1151 (2019), 012029.

All metal oxide-based photodetectors

10

Nupur Saxena[a], *Savita Sharma*[b], *and Pragati Kumar*[c]
[a]Organisation for Science Innovations and Research, Bah, Uttar Pradesh, India,
[b]Department of Physics, Kalindi College, University of Delhi, New Delhi, Delhi, India,
[c]Nano Materials and Device Lab, Department of Nanoscience and Materials, Central University of Jammu, Samba, Jammu and Kashmir, India

Chapter outline

1 Introduction

The optoelectronic devices, as the name suggests, deal with the optical, i.e., light and electric or electronic, circuits. This amalgamation of light and electric/electronic circuit can take place in various forms, viz., (i) generation of light via electric/electronic circuit as in light emitting diodes (LEDs), (ii) generation of electric current/voltage via absorption of light as in solar cell/photovoltaic devices, and (iii) the detection of light signals by converting them into electric or electronic signals as in the case of photosensor or photodetector (PD). Particularly, PD is a device which converts incident

Metal Oxides for Next-Generation Optoelectronic, Photonic, and Photovoltaic Applications
https://doi.org/10.1016/B978-0-323-99143-8.00012-2

light into electric signals, either directly via photoelectric effect or indirectly via photothermal effect. In photoelectric effect, photogenerated electron-hole (e-h) pairs are detected as variations in electrical resistance/current in the external circuit like in fast-responding Si and GaAs PDs [1]. On the other hand, in photothermal effect-based PDs, simply the temporal temperature gradient (dT/dt) developed due to light serves as an input to the electric output as in the case of pyroelectric detectors [2].

Other than this, the PDs can be classified in a number of ways. Based on their basic structure and working mechanism, PDs can be of (a) photoconductor type, (b) photodiode type (p-n, avalanche, Schottky), (c) metal-semiconductor-metal (M-S-M) PD, (d) metal-insulator-semiconductor (MIS) PD, and (e) phototransistor type. In the **photoconductor** PDs, light illumination on the photosensitive material generates charge carriers which are separated by the applied external electric field and then collected at each electrode and hence exhibit high external quantum efficiency (EQE) and responsivity (*R*) [3] as well as large gain (*G*) [4]. Whereas in the **photodiodes**, the photogenerated paired charge carriers (e-h) are separated by the built-in potential formed via a p-n/Schottky junction and then transferred to the opposite electrodes with a low-driving voltage [3]. Typically, such PDs have no gain, but based on charge tunneling injection induced by traps, a high gain can also be accomplished [4]. A contemporary practical structure of photodiode PDs comprises a photoactive layer between the electron transfer layer (ETL) and the electron injection layer (EIL) and between the hole injection layer (HIL) and the hole transfer layer (HTL). The other PD based on **M-S-M** geometry is composed of two Schottky junctions in a tandem manner separated by a layer of semiconductor. The charge carriers (e-h) are generated in the semiconducting layer on impinging of light and are collected by the applied electric field during the operation and thus can form a photocurrent. The M-S-M PDs are relatively simpler and exhibit low dark current, noise, and capacitance per unit area along with short response time and high sensitivity [5]. To suppress the high leakage current and to achieve high barrier height in M-S-M PDs, an insulating material layer was inserted between the metal and semiconductor and **MIS** PD came into picture. This PD exhibited improved ratio of photocurrent-to-dark current and UV-to-visible rejection ratio [6]. The conventional **phototransistor** PD, another type of PD and like a transistor, is a 3-terminal device namely source, drain, and gate. The photoactive material behaves as the channel and the gate voltage controls the carrier mobility in the channel consequently suppressing the dark current and enhancing the photocurrent [4]. The presence of insulating layers and gate in the device architecture ensures reduced noise, enhanced electrical signal, and thereby high performance in terms of improved responsivity and gain [7]. Fig. 1 illustrates the basic architectures of the abovementioned PDs with their schematics [8–10].

The performance of any of these PDs is assessed by a certain set of parameters namely photosensitivity (*S*) [11,12], photoresponsivity (*R*) [3,4], external quantum efficiency (EQE) [3,4,12], specific detectivity (D^*) [3,11,12], gain (*G*) [3,11],

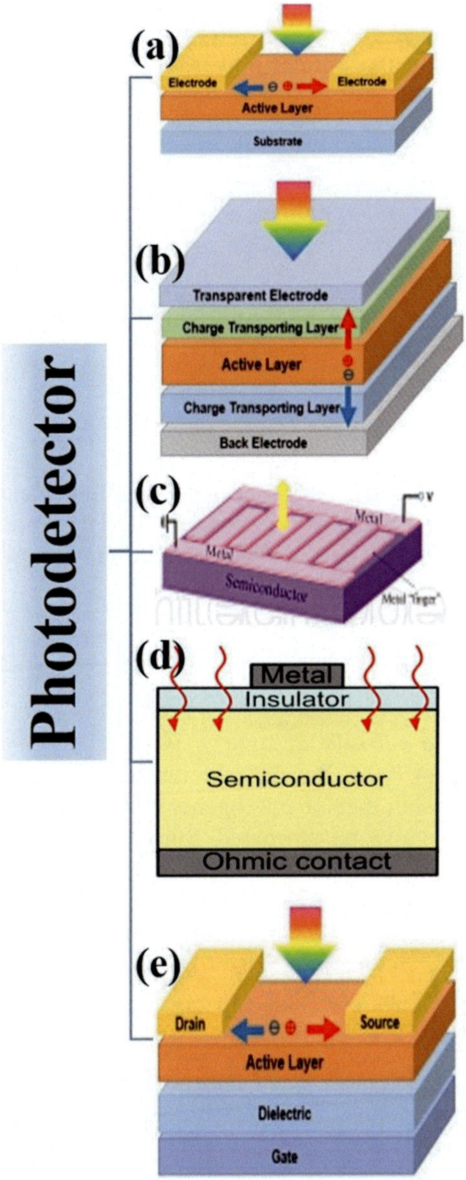

Fig. 1 Schematics of basic architectures of photodetectors used practically. (A) Photoconductor, (B) photodiode, (C) M-S-M, (D) MIS, (E) phototransistor. (A), (B), and (E) Adapted with permission from reference N. Li, Z. Lan, L. Cai, F. Zhu, Advances in solution-processable near-infrared phototransistors, J. Mater. Chem. C 7 (13) (2019) 3711–3729 © 2019 RSC. (C) Adapted from F.F. Masouleh, N. Das, Application of metal-semiconductor-metal photodetector in high-speed optical communication systems, Adv. Opt. Commun. 2014 open access (opentech). (D) Adapted from C.H. Lin, C.W. Liu, Metal-insulator-semiconductor photodetectors, Sensors (Basel) 10 (10) (2010) 8797–8826 open access (MDPI).

response time/speed (τ), and noise equivalent power (NEP) [11,12]. These parameters or figures of merit (FOM) are described here one by one:

The relative change in the photocurrent with respect to dark current is termed as photosensitivity (S) and is given by the expression [11,12]:

$$S = \frac{I_{\mathrm{Ph}} - I_{\mathrm{d}}}{I_{\mathrm{d}}} \tag{1}$$

where, I_d and I_{ph} are the dark and photocurrent, respectively.

"*R*" indicates the ability of optical power conversion into electrical signals. In general, it is the ratio of photocurrent to the optical power and is expressed as [3,4]:

$$R = \frac{I_{Ph}}{P_{in}} \tag{2}$$

where, P_{in} indicates the light intensity of the source.

With better accuracy, spectral responsivity "R_λ" is the ratio of change in the electric current under light illumination to the power of illumination source falling on per unit area of the device and is given by [7,12]:

$$R_\lambda = \frac{I_{Ph} - I_d}{P_{in}} \left(\mathrm{AW}^{-1}\right) \tag{3}$$

EQE is the measurement of the number of collected charge carriers per unit incident photons and is directly proportional to R_λ. Therefore, it is defined as the ratio of the photogenerated carriers (N_c) to the number of incident photons (N_{ip}) and can be expressed as [3,4,12]:

$$\mathrm{EQE}(\eta_{ex}) = \frac{N_c}{N_{ip}} = R_\lambda \frac{hc}{e\lambda} = R_\lambda \frac{1.24}{\lambda(\mathrm{nm})} \times 10^3 \tag{4}$$

where h is the Planck's constant, c is the velocity of light, e is the electronic charge, and λ is the wavelength of incident light. Besides, the internal quantum efficiency (IQE) which is the ratio of the number of photogenerated carriers to the number of absorbed photons (N_{ap}) can also be estimated. Since, under light illumination, a PD absorbs only a portion of photons ($N_{ap} = N_{ip}\eta_a$) depending on the light absorption efficiency (η_a) of photoactive material, therefore *IQE* is given by [3]:

$$\mathrm{IQE}(\eta_{in}) = \frac{N_c}{N_a} = \frac{\mathrm{EQE}}{\eta_a} \text{ or } \frac{\eta_{ex}}{\eta_a} \tag{5}$$

The photosensitivity per unit area of a detector is termed as specific detectivity "D^*" [12]. In other words, it is the minimum detectable power of the illumination which decides the threshold of the PD and makes it easier to compare the characteristics of different PDs [7] and therefore is the most important figure of merit. The dark current, which is sometimes also termed as noise signal, significantly influences the detection of a small amount of light, i.e., smaller dark current, and better and precise detection to weak light. Therefore, "D^*" can be expressed as [3,7]:

$$D^* = \frac{R_\lambda}{I_N}\sqrt{A\Delta f}\left(\text{Jones or cm Hz}^{1/2}\,\mathrm{W}^{-1}\right) \tag{6}$$

where I_N is the noise current, A is the effective area, and Δf is the electrical bandwidth. Generally, the dominating contribution is from I_d in the noise signal, hence the above equation can take the form as [7,12,13]:

$$D^* = R_\lambda \sqrt{\frac{\mathrm{A}}{2qI_d}} \left(\text{Jones or cm Hz}^{1/2}\ \text{W}^{-1}\right) \tag{7}$$

Noise equivalent power (NEP) is the minimum measurable signal's power at the signal-to-noise ratio (SNR) of "1" or "0 dB" in a one-Hertz-output bandwidth. NEP can endorse the minimum illuminating light distinguished from the noise and is given by [3,7]:

$$\text{NEP} = \frac{\sqrt{A\Delta f}}{D^*} = \frac{I_N}{R_\lambda} \tag{8}$$

NEP may also be defined as the incident light power when the effective value of output current becomes equal to the mean square root current value of the noise. It can be estimated as [4]:

$$\text{NEP} = \frac{\sqrt{\overline{I_N^2}}}{R_\lambda} \tag{9}$$

Other important parameter of PDs is gain "G," which represents the number of carriers passing through an external circuit under the illumination of certain amount of light. This is the parameter that describes multiple charge carrier's generation capability by single incident photon. Typically, the multiple charge generation process occurs when τ_l (lifetime of trapped carrier say hole) is longer than τ_t (transit time of the other carrier say electron). In such condition, there is a high probability of multiple circulation of electrons in the channel, which are drifted toward the external circuit subsequently developing the photoconductive gain "G" as given by [3,7]:

$$G = \frac{\tau_l}{\tau_t} = \frac{\tau_l}{L^2}\mu V_{\text{DS}} \tag{10}$$

where L is the channel length, μ is the carrier mobility, and V_{DS} is the applied bias voltage.

Alternative critical performance parameter is "τ" that tells about the switching speed of PD's characteristics under dark-to-illumination conditions. Naturally, "τ" includes the rise time (τ_r) and the fall/decay time (τ_f/τ_d) which are defined as the time taken for the rise and decay of the photocurrent from 10% to 90% and from 90% to 10%, respectively [3,7,11–13]. "τ" may also determine the cutoff

Table 1 Essential properties of some widely used MOx for PD application.

S. no.	Metal oxide	Crystal structure	Conductivity type	Bandgap (eV)	Sensing light
1.	ZnO	Hexagonal	n	3.37	UV
2.	SnO_2	Tetragonal	n	3.6	UV
3.	Cu_2O	Cubic	p	2.17	Visible
4.	β-Ga_2O_3	Monoclinic	n	4.2–4.9	UV
5.	α-Fe_2O_3	Rhombohedral	n	2.1	Visible
6.	In_2O_3	Cubic	n	1.6 (direct) 2.5 (indirect)	UV/ visible
7.	CdO	Cubic	n	2.27 (direct) 0.55 (indirect)	Visible/ IR
8.	CeO_2	Cubic	n	3.2	UV
9.	MoO_3	Orthorhombic	n	2.39–2.90	Visible
10.	WO_3	Monoclinic	n	2.5–2.8	Visible
11.	V_2O_5	Orthorhombic	n	1.7–2.6	Visible
12.	Nb_2O_5	Monoclinic	n	3.1–5.3	UV

Adapted with permission from reference T. Zhai, X. Fang, M. Liao, X. Xu, H. Zeng, B. Yoshio, D. Golberg, A comprehensive review of one-dimensional metal-oxide nanostructure photodetectors, Sensors (Basel) 9 (8) (2009) 6504–6529 copyright MDPI 2009.

frequency or 3dB bandwidth for PDs. In this case, the time taken for the output signal to attain $(1-e^{-1})\approx 63\%$ of its peak steady-state value is "τ". The cutoff frequency or 3dB bandwidth for PD can be estimated as [4]:

$$f_c = \frac{1}{2\pi\tau} \tag{11}$$

There are numerous materials ranging from metals to insulators and ceramics to polymers that have been explored in some sense for the fabrication of PDs. The specific material and architecture has its pros and cons over others. The world of metal oxides (MOx) or better say functional oxides has opened up multiple channels in terms of electronic devices, sensors, optical devices, etc., owing to their eco-friendly nature, cost-effective synthesis, and solution processability. Nonetheless, there are some inherent properties in MOx that make them highly suitable for PDs. Table 1 summarizes a few properties essentially required for PD application of some widely used MOx as an illustration [14].

2 Synthesis of miscellaneous forms of MOx for photodetection

The excellent properties of semiconductor materials (mainly composed of large bulk-state, nano-state, and film-state) are governed by their crystal and band structure, shape and size, surface and bulk defect states and can be controlled by synthesis processes. The various geometrical structures like 0, 1, 2, and 3 dimensional nanostructures (NSs) of semiconductors including MOs have been synthesized via both the top down (TD) and bottom up (BU) approaches using different physical and chemical synthesis routes, respectively. For the synthesis of different types on metal oxide (MOx) nanostructures in the form of thin film, stable colloid and solid powder, several chemical, as well as physical routes have been explored. On the basis of experimental conditions, various forms of nanostructures like particles, wires, tetrapod, rods, helical, spiral, and flower are observed in both the chemical and physical routes. In spite of both of the advantages and shortcomings of different routes, various chemical and physical routes have been explained in this section for the synthesis of different MOx nanostructures. Here a glimpse of synthesis methods that are used frequently to grow the MOs to design PDs is presented.

2.1 Synthesis of MOx QDs

In order to synthesize the MOx nanoparticles using wet chemical method, solution of desired molarity $M(NO_3)_2$ and a solution of NaOH having usually twice the molarity of $M(NO_3)_2$ were used. The generation of metal hydroxide was carried out by dissolving the nitrates of the respective metals in water. This was only possible by the addition of NaOH drop-wise with continuous stirring for >4 h at <100°C. After filtering the reaction mixture, it was let to dry in oven at <100°C. The oxide nanoparticles were successfully formed followed by the calcination of the hydroxides in furnace at >400°C. The oxide nanoparticles thus formed were utilized for the characterization using SEM, EDAX, and XRD facilities.

In the **hydrothermal method**, a Teflon-lined sealed autoclave was used for the reaction of acetate salt of the metal dissolved in methanol with NaOH in the presence of heat under the autogenous pressure at 110°C for 6 h in an oven. The resultant product was then filtered, washed, dried, and then further utilized for characterization.

2.2 Synthesis of 1D MOx

Although a numerous method are available in literature for the growth of 1D nanostructures, however it can mainly be categorized into two classes: — vapor phase (VP) growth and liquid phase (LP) or, solution phase (SP) growth on the basis of synthesis environment. The well-developed vapor phase technique is used for the synthesis of most of the metal oxide nanostructures. The reaction between metal vapor and oxygen gas plays the major role in vapor phase technique. In addition to this, the vapor-solid (VS) and the vapor-liquid (solution)-solid (VLS) and process are the controlling mechanisms. It should be noted that for cost-effective and more flexible synthesis process, solution-phase growth methods are highly appreciable as an alternative method.

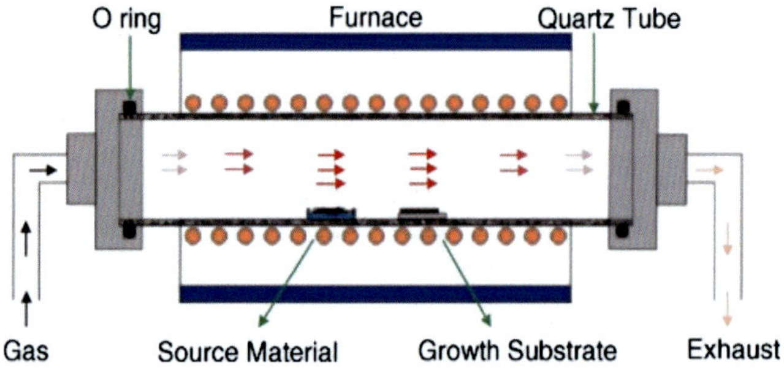

Fig. 2 Schematic of thermal furnace synthesis system that is used in vapor phase growth methods including CVD, thermal evaporation, and PLD.
Adopted with the permission from reference J.G. Lu, P. Chang, Z. Fan, Quasi-one-dimensional metal oxide materials—synthesis, properties and applications, Mater. Sci. Eng.: R: Rep. R52 (1–3) (2006) 49–91 © 2006 Elsevier.

2.2.1 *Vapor phase (VP) growth*

The high temperature vapor phase growth (VPG) is a simple method in which the thermal furnace is used to facilitate the reaction between oxygen gas and the vapor of metal source due to intense heating. Various methods such as pulsed-laser deposition (PLD), thermal chemical vapor deposition (CVD), metal-organic chemical vapor deposition (MOCVD), and direct thermal evaporation have been utilized for controlling crystallinity, aspect ratio, and diameter [15–22]. The synthesis techniques are mainly based on two mechanisms: vapor-liquid-solid (VLS) and vapor-solid (VS) (Fig. 2).

2.2.2 *Solution-phase (SP) growth*

The synthesis of nanoneedles, nanowires, and nanorods in SP growth method has been successfully realized. For the reduction of complexity and to make fabrication cost-effective, the ambient temperature is preferable usually for this method. A number of methods have been implemented by the researchers for the development of strategies that can assist and trap for the growth direction of 1D. Such methods are: template-assisted (TA) method and template-free (TF) method.

2.2.3 *Electrochemical synthesis (ECS)*

An Italian Chemist Luigi V. Brugnatelli invented electrochemistry, which is a technique that utilizes the deposition of a metal/metal oxide layer on the conducting electrode [23]. During the ECS synthesis of ZnO for the first time, thallium oxide was used, which is reported by Izaki and Omi, and Peulon and Lincot [24]. To control the size, shape, composition, and morphology of synthesized nanostructures, cathode potential, current density, deposition temperature, electrolyte composition, and concentration are some key parameters. Variety of conducting electrodes like ITO, FTO, transparent semiconductors metals electrodes like Au, Sn, Pt, Zn, and Cu, AAO, and silicon are used to report the number of papers on ECD [25–28].

2.2.4 Laser ablation on solid liquid interface

The laser ablation in the liquid media is a physical route which is based on solution and is used for the synthesis of nanomaterials. It is a method in which no chemical is used except some surfactants for the prevention of aggregation and agglomeration. For the synthesis of iron oxide nanoparticles, laser ablation in liquid media was used by Partil and coworkers [29] for the first time. Ruby laser with wavelength 694 nm was used for that purpose. It is regarded as the clean and green approach for the synthesis of metal oxide nanostructures. In addition to this, it is chemical contamination free, simple, fast, and efficient method for the synthesis of ultra-fine nanostructures.

2.2.5 Chemical vapor deposition (CVD)

CVD is the chemical technique in which high-purity and high-performance solid materials are produced. Due to this reason, for the production of thin films most of the semiconductor industry use this technique. The substrate (wafer) is exposed to one or more than one volatile metal precursors in a typical CVD method. The decomposition takes place on the wafer surface and results in the formation of metal oxide nanostructures. The flow of gas inside the reaction chamber is helpful to remove the volatile byproducts usually in this method. Nature of substrate, chamber pressure, temperature of both vapor and substrate, molecular weight, and nature of carrier gas are the major parameters to control the morphology of metal oxide nanostructures in CVD method. The reported works for the synthesis of metal oxide nanostructures using CVD method are also mentioned in references in this report.

2.2.6 Physical vapor deposition (PVD)

PVD refers to the general term including the various vacuum deposition techniques in which the thin film is deposited over the surface of various substrates after the condensation of a vaporized form of the material. In contrast to the chemical vapor deposition (CVD), the PVD method involves the coating of materials over the surface of the substrates by means of either high temperature vacuum evaporation or plasma sputter bombardment. The thermal evaporation, pulsed laser deposition, magnetron sputtering, rf-sputtering, and cathodic arc are the major PVD methods that are frequently used for the deposition of metal oxide thin films or nanostructured materials.

3 Designing and performance of MOx photosensing devices

The fabricated all MOx-based PDs are typically responded in three spectral regions; solar blind (SB), ultraviolet (UV), and visible vis.). The MOx materials, strategies, and architectures have been used to fabricate said PDs with varying FOMs. This section deals with the efforts and progress in the three categories of PDs.

3.1 Solar blind photodetectors

A high performance SBPD is developed by homoepitaxial β-Ga_2O_3 thin film. The device with final architecture Ti-Al-Ni/Sn:β-Ga_2O_3/n^+-Ga_2O_3/Ge:β-Ga_2O_3/Pt-Ni-Au shows high *R* in SB region and attends maxima at 230 nm [30]. In subsequent work, they replace Ge:β-Ga_2O_3 by Si:β-Ga_2O_3 in the architecture and observed even high *R* for SB region with slight blue shift (222 nm) in maxima [31]. The comparison of two devices with other commercial devices is illustrated in Fig. 3A. Cui et al. fabricated SBPDs of architecture PET/β-Ga_2O_3/ITO. Here ITO works as a transparent conducting electrode and device works in M-S-M geometry. The developed device shows good *R* for 254 nm illumination and repeatability up to 500 bending cycles (Fig. 3B) [32]. SBPDs were fabricated using InZnSnO/β-Ga_2O_3/InSnO on ITO-coated glass substrate. The fabricated device shows high Schottky barrier height and low-defect interfaces. They compared the photoresponse characteristics of pristine and 600°C postannealed IZTO/β-Ga_2O_3 Schottky PDs under the illumination of 255, 385, and 500 nm wavelengths of light emitting diodes (LEDs) of the power density 1 mW/cm^2. The two devices show good SB *S* and *R* for 255 nm, as the increase in current density under 255 nm is much larger than under 385 and 500 nm (Fig. 3C and D) [33]. The β-Ga_2O_3 MSM SBPDs with graphene electrodes exhibited excellent operating characteristics including high *R* (~29.8 A/W), photo-to-dark current ratio ($\sim 1 \times 10^6$%), and rejection ratio (R_{254nm}/R_{365nm}, $\sim 9.4 \times 10^3$), D^* ($\sim 1 \times 10^{12}$ J) [34]. A high performance and completely flexible SBPDs was developed using amorphous Ga_2O_3 thin film and unconventional amorphous indium-zinc-oxide (a-IZO) conducting electrodes. The device showed high *R* of 43.99 A/W and high EQE of 2.18×10^4% in contrast to conventional Ag electrode devices as illustrated in Fig. 4A, and B [35]. A SBPD of fast photoresponse time (10 ms) along with the highest possible *R* (1.8 A/W) and excellent D^* (2×10^{13} J) developed with Ni/NiO/ZnO/FTO/glass configuration [36]. While self-powered SBPDs (SPSBPDs) with the ultrafast photoresponse time (19 μs), high-photoresponse ratio (1944), and D^* (7.2×10^{11} J) were fabricated in Au/NiO/ZnO/FTO/PET configuration [37]. Liquid-metal exfoliated ZnO nanosheets were used to design SBPDs in M-S-M configurations over Si/SiO_2 with Au electrodes. The device showed fast response (1.6 μs) with *R* (18 A/W) [38].

Highly crystallized single ZnO-Ga_2O_3 core-shell heterostructure (Fig. 5A) was employed to fabricate SPSBPDs with a sharp cutoff wavelength at 266 nm. The device shows an ultrahigh *R* (9.7 mA/W) at 251 nm with a high UV/visible rejection ratio (R_{251nm}/R_{400nm}) of 6.9×10^2 and ultrafast response time (raising time 100 μs and falling time 900 μs) under zero bias [39]. The heterostructure glass/FTO/TiO_2/Co_3O_4/NiO/Ag was used to develop SPSBPDs. The enhancement of both photovoltage and photocurrent of the device from 0.17 to 0.78 V and from 0.54 to 13.5 mA cm^{-2} is noticed as a result of improved carrier lifetime from 1.94 to 9 ms [40]. The heterojunction of ZnO/Co_3O_4 with architecture glass/FTO/ZnO/Co_3O_4/AgNws is also used

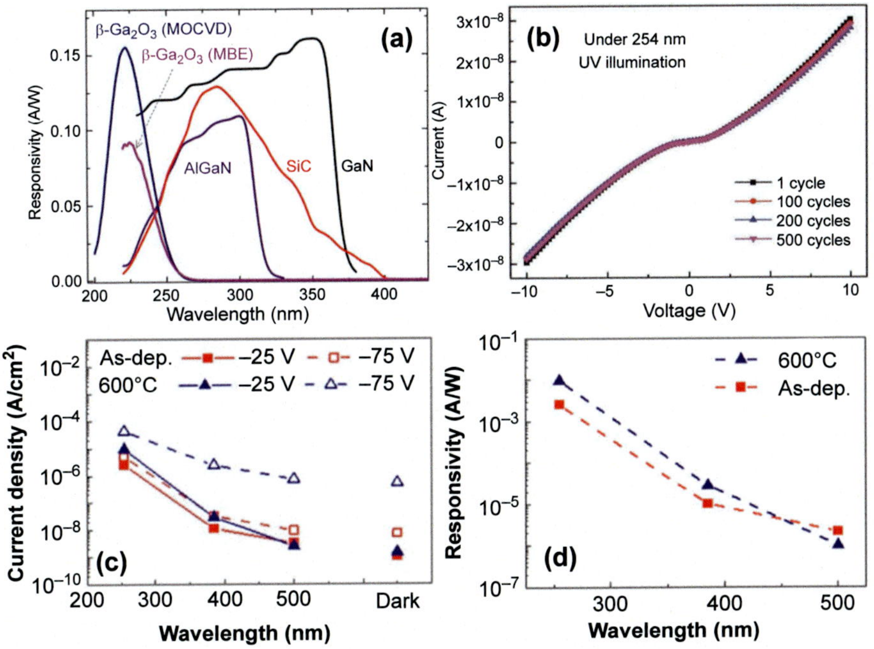

Fig. 3 (A) Comparison of the photoresponsivity of the β-Ga_2O_3:Si photodiode grown using MOCVD (*blue trace*) with commercial devices based on GaN (*black*), SiC (*red*), AlGaN (*purple*), and a MBE grown β-Ga_2O_3:Ge photodiode (*magenta*) wide bandgap semiconductors. (B) Photocurrent-voltage curves of PET/β-Ga_2O_3/ITO device (under 254 nm UV light illumination) under flat and different bending radius. (C) The spectral response at different negative biases (−25 and −75 V) and (D) on-resistance (R_{ON}) changes of as-deposited and 600°C annealed IZTO/β-Ga_2O_3 PDs under illumination.
(A) Adopted with the permission from reference F. Alema, B. Hertog, P. Mukhopadhyay, Y. Zhang, A. Mauze, A. Osinsky, W.V. Schoenfeld, J.S. Speck, T. Vogt, Solar blind Schottky photodiode based on an MOCVD-grown homoepitaxial β-Ga2O3 thin film, APL Mater. 7 (2) (2019) © 2019 AIP publishing group. (B) Adopted with the permission from reference S. Cui, Z. Mei, Y. Zhang, H. Liang, X. Du, Room-temperature fabricated amorphous Ga2O3 high-response-speed solar-blind photodetector on rigid and flexible substrates, Adv. Opt. Mater. 5 (19) (2017) 1700454 © 2017 Wiley-VCH Verlag GmbH & Co. KGaA, Weinheim. (C) Adopted with the permission from reference H. Kim, H.-J. Seok, J.H. Park, K.-B. Chung, S. Kyoung, H.-K. Kim, Y.S. Rim, Fully transparent InZnSnO/β-Ga2O3/InSnO solar-blind photodetectors with high schottky barrier height and low-defect interfaces, J. Alloys Compd. 890 (2022) 161931 © 2021 Elsevier.

to fabricate SPSBPDs that showed significant *S* of 4.57×10^4. Besides, the device showed broad wavelengths (UV-IR) absorbing the capability and exhibited a fast response speed of 81.7 μs and a decay time of 178.8 μs [41]. Alwadi et al. constructed SPSBPDs via synthesis of n-ZnO QDs/p-CuO micropyramid heterostructures over p-Si using Au/Ti and ITO as electrodes. The device showed superior photo-*R* of ~956 mA/W at 244 nm with a faster photoresponse (<80 ms) and 260 nm cutoff.

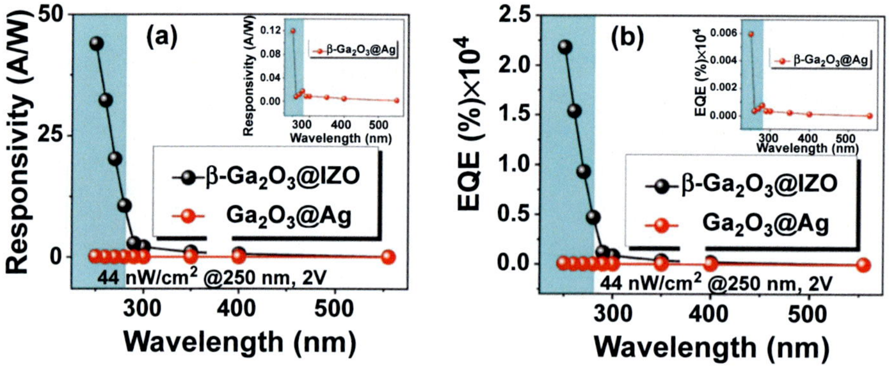

Fig. 4 (A) The photoresponsivity and (B) The EQE as a function of illuminated wavelengths at ambient condition for Ga_2O_3 PD with a-IZO and conventional Ag electrodes, respectively. (Inset) Enlarged view of the same for PD with conventional Ag electrodes.
Adopted with the permission from reference N. Kumar, K. Arora, M. Kumar, High performance, flexible and room temperature grown amorphous Ga_2O_3 solar-blind photodetector with amorphous indium-zinc-oxide transparent conducting electrodes, J. Phys. D. Appl. Phys. 52 (33) (2019) © 2019 IOP Publishing.

The device demonstrated a high photo-*R* of ~29 mA/W under self-bias condition [42]. The hetrojunction of NiO/ZnO over ITO was used to fabricate M-S-M geometry of PD with Al contacts. The devices show excellent UV sensitive visible blind photodetection, with the UV-to-visible rejection ratio ($R_{310\ nm}/R_{450\ nm}$) of 1600 with maximum *R*, D^*, and EQE of 21.8 A/W, 1.6×10^{12} cm Hz$^{1/2}$/W, and 88% respectively at 310 nm [43]. Whereas, an architecture glass/ITO/NiO NPs/$Zn_{1-x}Mg_xO$ NPs/Ti-Au offered tunability in UV photoresponse from 3.24 and 3.49 eV by increasing Mg content (Fig. 5A and B). The device showed maximum *R* and D^* of 0.4 A/W and 2.2×10^{12} cm (Hz)$^{1/2}$/W, respectively [44]. Praveen et al. demonstrated SPSB performance of PDs designed by β-Bi_2O_3/SnO_2 QDs Schottky heterojunction in the configuration of glass/ITO/p-Bi_2O_3/n-SnO_2/Ag. The fabricated device exhibits a maximum *R* of 62.5 μA W^{-1} and D^* of 4.5×10^9 J [45].

3.2 UV photodetectors

The structure of Au/Al:ZnO/Mg, N:$CuCrO_2$ proposed for UV-PDs demonstrated fast responses in the order of subseconds with *S*, *R*, and D^* of ~41,000, 1.6 mA/W, 3.52×10^{12} J, respectively, that are significantly improved in comparison with the Al-doped ZnO photoconductor where these parameters were ~24, 0.655 mA/W and 10×10^9 J [46]. Highly efficient (90%) heterojunction UV-PDs with maximum *R*, and D^* of 0.28 A/W and 6.3×10^{11} J at an applied reverse bias of 1 V was developed using glass/FTO/NiO/ZnO/Ti-Au [47]. Whereas the enhanced parameters with maximum rectification ration, *R*, EQE, and D^* of 5×10^2, 10.2 A/W, 1.8×10^3%, and 6.3×10^{11} J respectively were obtained by glass/ITO/NiO/ZnO/Al heterostructure device [48]. The shelf life of the EQE and detectivity of the device is illustrated

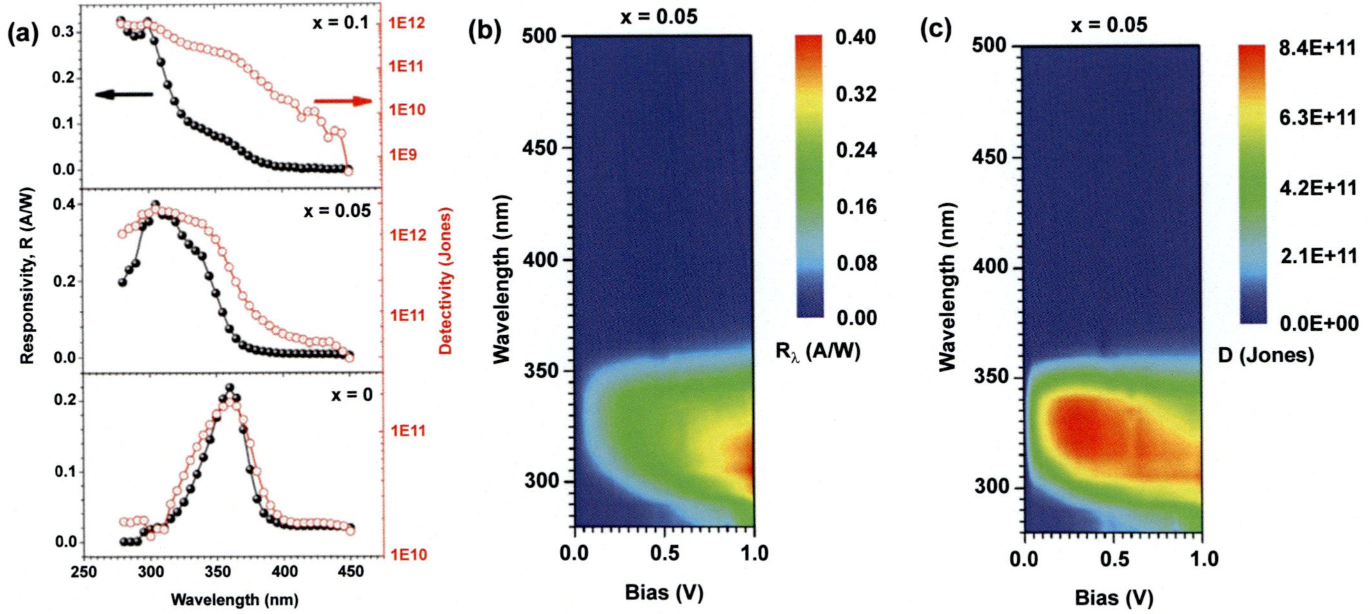

Fig. 5 (A) Responsivity and detectivity of p-Nio/n-$Zn_{1-x}Mg_xO$-based PDs with varying Mg content at 1 V reverse bias, (B) and (C) the device (with 5% Mg content) performance below 1 V reverse bias.
Adopted with the permission from reference T. Xie, G. Liu, B. Wen, J.Y. Ha, N.V. Nguyen, A. Motayed, R. Debnath, Tunable ultraviolet photoresponse in solution-processed p-n junction photodiodes based on transition-metal oxides, ACS Appl. Mater. Interfaces 7 (18) (2015) 9660–9667 © 2015 American Chemical Society.

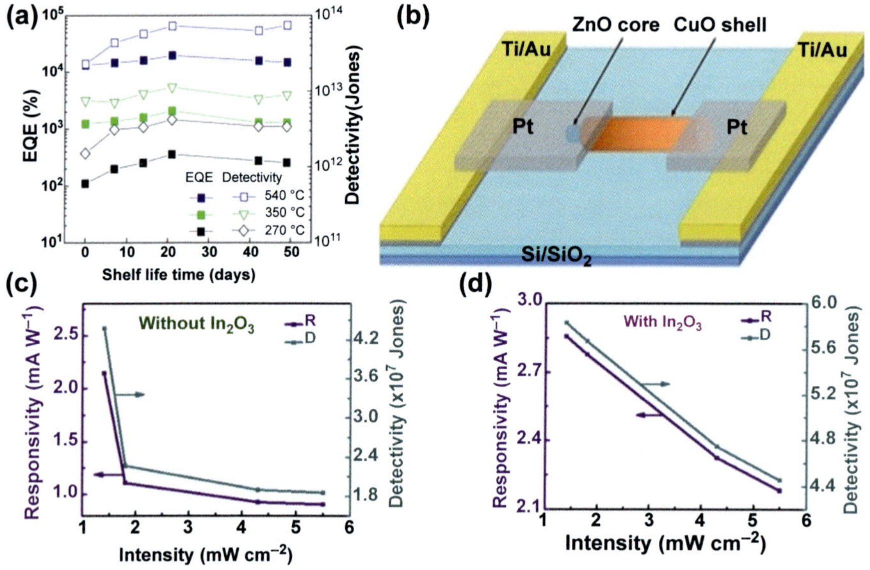

Fig. 6 (A) Stability test of NiO/ZnO heterostructure-based UV PD. (B) Schematic representation of ZnO@Cu_xO core-shell radial heterojunction nanowire-based PD. Responsivity and detectivity of SnO_2/SnS_2 hybrid-based PD (C) without and (D) with In_2O_3 interlayer.

(A) Adopted with the permission from reference D.Y. Kim, J. Ryu, J. Manders, J. Lee, F. So, Air-stable, solution-processed oxide p-n heterojunction ultraviolet photodetector, ACS Appl. Mater. Interfaces 6 (3) (2014) 1370–1374 © 2014 American Chemical Society. (B) Adopted from reference A. Costas, C. Florica, N. Preda, N. Apostol, A. Kuncser, A. Nitescu, I. Enculescu, Radial heterojunction based on single ZnO-CuxO core-shell nanowire for photodetector applications, Sci. Rep. 9 (1) (2019) 5553 open access. (C) and (D) Adopted with the permission from reference S. Abbas, D.-K. Ban, J. Kim, Functional interlayer of In2O3 for transparent SnO2/SnS2 heterojunction photodetector, Sensors Actuators A Phys. 293 (2019) 215–221 © 2019 Elsevier.

in Fig. 6A. UV-PDs were fabricated using single ZnO@Cu_xO core-shell radial heterojunction nanowire contacted by EBL (Fig. 6B). The devices showed slow response of 43 s with maximum *R*, EQE, and D^* of 43.95 A/W, 149.3% and 7.55×10^{10} J, respectively [49].

A UV-SPPD of architecture Ag_xO/TiO_2/ITO/PET showed high *R* (323 mA/W), D^* (4.2×10^8 J), and a noise equivalent power of 2.3×10^{-9} W/$Hz^{1/2}$ [50]. Whereas the UV-SPDD of configuration AgNWs/NiO/TiO_2/FTO/glass showed *R*, D^*, noise equivalent power, and response time of 136 mW/A, 1.1×10^9 J, 9.2×10^{-10} W/Hz, and 7.6 ms, respectively [51] that improved up to 187 mA/W, 5.13×10^9 J, 9.5×10^{-11} W/$Hz^{1/2}$, and ~0.43 ms, respectively, by replacing NiO through Cu_4O_3, i.e., using AgNWs/Cu_4O_3/TiO_2/FTO/glass [52]. The heterostructures of TiO_x/TiO_2 NRs/NiO NFs over FTO-coated glass with Au electrode were also utilized to developed SPPDs with a rectification ratio of 1.45×10^5 and *R* of 5.5 mA/W at bias voltage of 0 V [53]. The hybrid of SnO_2/SnS_2 was used to fabricate UV-SPPDs and study the effect of interlayer of In_2O_3 on photosensing properties. The insertion of In_2O_3 reduced dark current (6.9–2.8 μA) and simultaneously increased the photocurrent (11.6–14.7 μA) along with improvement in rise time (858–59 μs) and fall time (868–79 μs). The device exhibits maximum *R* and *D** of 2.9 mA/W and 5.9×10^7 J respectively after introduction of In_2O_3 interlayer as illustrated in Fig. 6B and C [54]. Whereas the effect of interface layer of Al_2O_3 was studied to investigate the performance of SPPDs fabricated in configuration glass/ITO/Al:ZnO/ZnO@Al_2O_3@NiO/Au. Devices showed that increase in interface layer's thickness reduced both dark current as well as photocurrent, hence thicker interface layer is not suitable for high performance PDs. The fastest device showed response time of 40 ms with *R* and *S* of 1.4 mA/w and 24 respectively for the PD consisting of 1 nm thick interface layer of Al_2O_3 [55]. The heterostructure glass/Ag/ITO/ZnO@$CuCrO_2$ NWs/Ag was employed to fabricate ultrafast (35 μs) SPPDs with high rectification ratio of 1.2×10^4 and *R* of 3.43 mA/W [56].

3.3 Visible MOx photodetectors

The vis.-PDs employing Au/Cu_4O_3 NRs/ZnO NRs/ITO/glass showed broad spectral response (500–700 nm) with response speed of 20 s [57], whereas the NiO/Cu_4O_3/ZnO/ITO/glass showed broad spectral response (450–625 nm) with maximum photoresponse ratio of 546, rise time of 33 ms, and fall time of 89 ms [58]. Single CuO and CuO@ZnO core-shell nanowires were contacted by employing electron beam lithography (EBL) and focused ion beam induced deposition (FIBID) to fabricate vis.-PDs. The estimated *R*, EQE, and *D** values for device were 4.84 A/W, 11.59%, and 2.2×10^{10} J under 520 nm and 8.74 A/W, 26.56%, and 4.05×10^{10} J under 405 nm, respectively with slow response time of 35 s [59]. The mesoporous ZnO/$ZnAl_2O_4$ mixed MOx-based Zn/Al-layered double hydroxide was used as an effective anode material for vis.-PDs. The device showed *R* and *D** values of 11 mA/mW and 10^{11} J respectively with response time of 0.9 s [60].

The vis.-SPPDs fabricated using CuO/ZnO and Cu_2O/ZnO heterojunction show rise times of 35 and 67 ms and fall times of 62 and 154 ms with signal-to-noise ratios 50 and 47, respectively [61]. Whereas PET/ITO/Cu_4O_3/ZnO-based SPPDs showed broad visible spectral response from 400 to 625 nm with high response time of 20 μs [62]. The broad visible spectral (455–620 nm) response was also recorded by the SPPDs based on V_2O_5/ZnO/ITO/PET. The device showed maximum *R* and *D* values of 55 mA/W and 4.13×10^{12} J with response time of ~5–16 ms [63]. The effect of Co_3O_4 shell thickness and a thin buffer layer of the Al_2O_3 between ZnO and Co_3O_4 core shell was investigated, which may inhibit charge recombination, thus boosting performance of vis.-SPPDs (Fig. 7A and B). The photoresponse of the Al_2O_3 inserted PDs at zero bias was six times higher compared to that of glass/FTO/Au/ZnO@Co_3O_4/Au. The *R* and *D** of the best device were 21.80 mA/W and 4.12×10^{12} J, respectively [64].

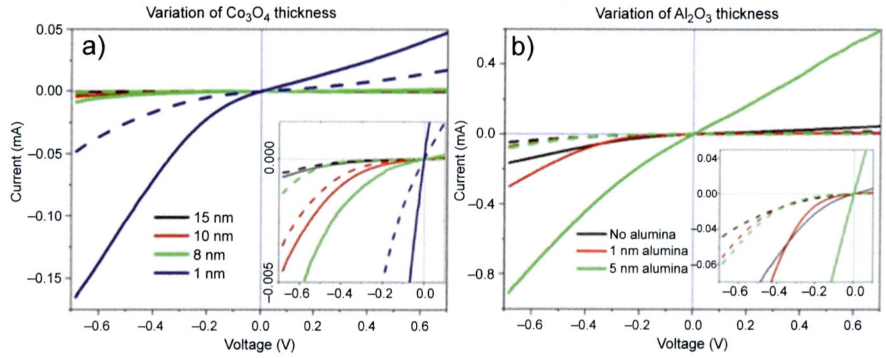

Fig. 7 *I–V* characteristics dark (dashed lines) and illumination (solid lines) of the (A) ZnO-Co_3O_4 core-shell structure with varying shell thicknesses and (B) ZnO-Co_3O_4 (1 nm) with varying thicknesses of the buffer layer (Al_2O_3) between ZnO and Co_3O_4. The insets show the zoom-in of the *I–V* curves.
Adopted with permission from reference P. Ghamgosar, F. Rigoni, M.G. Kohan, S. You, E.A. Morales, R. Mazzaro, V. Morandi, N. Almqvist, I. Concina, A. Vomiero, Self-powered photodetectors based on Core-Shell ZnO-Co3O4 nanowire heterojunctions, ACS Appl. Mater. Interfaces 11 (26) (2019) 23454–23462 © 2019 American Chemical Society.

4 Effect of harsh conditions on performance of MOx photodetectors

MOx PD faces a large number of challenges for the accomplishment of high performance with high tolerance toward harsh environment [65]. Currently, gallium oxide (GaO) having significantly high band gap is appreciably demanded in harsh environment like space exploration and flame prewarning. Similarly, in the various critical environment such as flame prewarning, environmental monitoring, secure communication, and missile tracking, the solar blind photodetectors play a crucial roles. The requirements for the operation of solar blind photodetectors (PD) in harsh environment are: (i) superior chemical and thermal stability and (ii) withstanding nature in radiation hard atmosphere. These PDs have a wide range of applications like sterilization, UV astronomy, water treatment, insect disinfection in crops, and space communications. For the stable operation in space applications, about 100k-Gray tolerance capacity of radiation must have needed for detectors to environment the radiation. The deep effect on materials may perform with the help of gamma irradiation as the PD performance. The long-lived defects like decreased carrier concentration, increased defect concentrations, and carrier lifetime happened due to the radiation matter interaction. Similarly, transient defects like rapid annealing of minority carriers in semiconductors are observed due to such interactions. Hence, the radiation hard materials are preferred for the solar blind PD under harsh environment. The results observed by Tak et al. [66] revealed that the Ga_2O_3 solar blind PD is comparatively less susceptible to radiation environment.

5 Applications of MOx photodetectors

Among the most global technology used nowadays, photodetector is one of them. The main fields in which the photodetectors are used presently are: the devices used for business purposes, research, industry, and in amusement [67]. In addition to this, the photodetection automatically opens supermarket doors, fabrication of photodiodes (PDs) in a fiber optic, in the field of astronomy where it is used for the study of radiations in different parts of the universe to the CCD in a video camera and in the receivers of TV and VCR remote controls [68].

Basically, photodetectors are the devices which are used for various applications. The two main types of applications in which the photodetectors can be divided are: communications and remote sensing. In communications for an encoded signal, it acts as the carrier but in different configuration of remote sensing, radiation is the means of signal used to convey the information about an object or scene. On the basis of registering photons, photodetector works from far infrared up to gamma rays with frequencies radio waves [69,70]. Fig. 8 depicts a schematic of multifarious applications of photodetectors [71].

The photodetectors used in fiber optic communication systems are the most prevalent in market which is observed in the infrared region, receives high speed signals. As the laser drivers provide plenty of radiation to the fiber, these detectors do not need high S. The basic criteria that the photodetector must possess are: high reliability, extremely fast response, and cost effective [72]. In optical communications, the photodiodes based on indium gallium arsenide (InGaAs) act as the work horses. They currently attained high data communication at the rate of 2.5 Gbits/s whose capacity is >200,000 times in comparison to the single copper telephone wire capacity. The detectors are used for the estimation of relative strength of signals that finally determine the position of mouse and orientation in advanced cordless mouse devices used in PCs, which is related to remote control [73,74].

5.1 Safety and security

The commercially cheap and mostly irregular but simplest type of photodetectors are used in remote sensing applications which concern only the identification of the absence or presence of an object or a position for security and safety monitoring purposes. The most common examples of the abovementioned applications are the IR-sensitive motion detectors used in home security systems. In a driver's "blind spot," the most recently developed example of such automotive is helpful to monitor objects and passenger detectors which conclude when to activate the air bags. In factories, the safety detectors execute such tasks and are sensitive to visible or ultraviolet (UV) wavelengths and are regarded as electrical arc detection and cutting off the current automatically happens where arcing occurs [67,68].

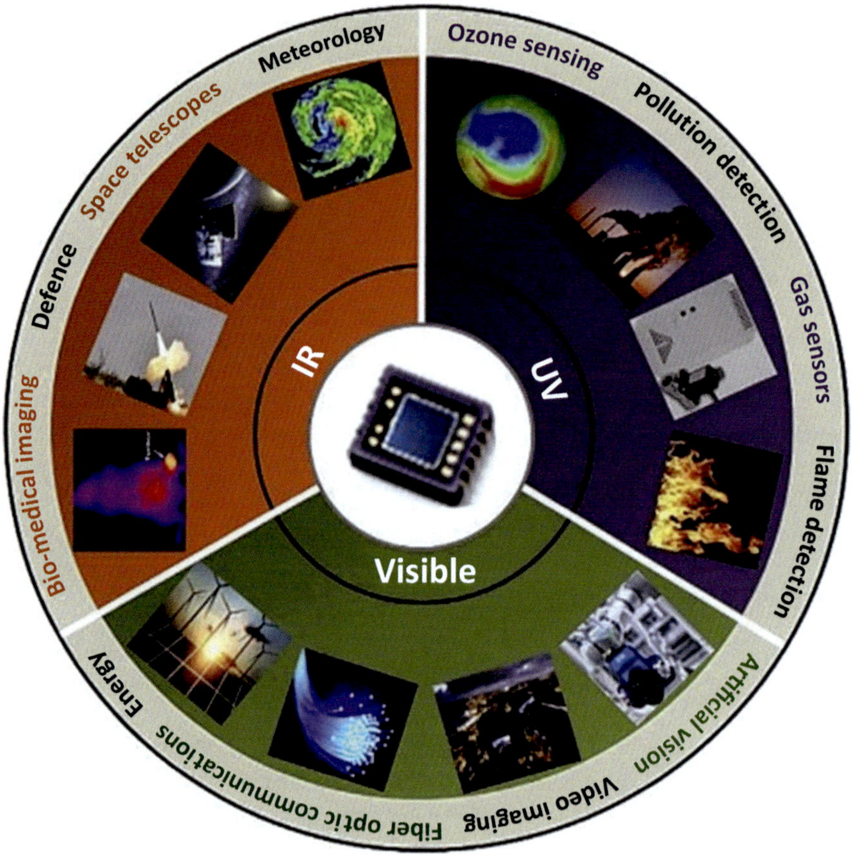

Fig. 8 Schematic diagram of multifarious applications of photodetectors belonging to different regions of electromagnetic spectrum.
Adopted with permission from reference P.V.K. Yadav, B. Ajitha, Y.A. Kumar Reddy, A. Sreedhar, Recent advances in development of nanostructured photodetectors from ultraviolet to infrared region: a review, Chemosphere 279 (2021) 130473 © 2021 Elsevier.

5.2 *Process control*

Another large volume consumer application using photodetectors which comes to use is in the process control applications. These devices in many cases act as simple position sensors to examine whether a work piece is in suitable position or not and also to supply response (feedback) for robotic systems. At different wavelength range, the corresponding values of radiation intensity are needed for the correctly repeatable detectors like photodiodes or phototransistors. The natural fluorescence with optically filtered detectors of certain plastics in recycling plants can be used to sort different materials while different types of glasses that can be differentiated from other spectroscopic tests is the suitable example [74].

5.3 The cutting edge

Measuring the heat radiated by human body in the IR detectors has long been helpful in thermography as a routine diagnostic test. In biomedicine, environmental monitoring and laboratory research for low volume specified fields, the latest detectors tend to be fabricated. To observe rapid biochemical reactions, the recent developments have allowed research biologists for ultrahigh speed CCDs [67,74].

5.4 Environmental sensing

A broad range of photodetectors are used for environmental monitoring today which has range from UV to the IR. Generally, the pollution detection measures the strength of the absorption lines with detectors which depend on UV spectroscopy for the detection of pollutants like ozone, nitrous oxides, and sulfur dioxide. For the monitoring process of solid particulates in air and water, simple techniques are used [75]. In these processes, basically the quantity of light scattered by the particles determines the pollution levels continuously. The photodiodes with their robustness and fast response times can be used as often availability of ample light. The space-based environmental monitoring has great demands on detectors both for tracking pollutants and for weather as well as used for both *S* and for compactness. For the measurement of ozone concentrations from space on satellite-mounted IR spectrometers, the pyroelectric detectors are used for converting heat to electric impulses [67,75].

5.5 Astronomy

In the field of astronomy, the most difficult application of photodetectors comes into play without any confusion. The wavelength range increases from the far IR (at hundreds of microns) to cosmic ray photons with 1020 eV of energy and wavelength of 10–20 μm having the range of 22 orders in magnitude. The European Space Agency at the long wavelength end is planning for the construction of far infrared and submilimeter space telescope. One of the spectral bands will be opened up due to this of the spectral band which is not examined still from 60 to 670 μm. The huge arrays of optical detectors indirectly entering the top of Earth's atmosphere at highest energies respond to UV cosmic rays [67–69].

6 Conclusions

Summarizing, this chapter presents a comprehensive view of MO-based PDs in a systematic way. Starting from categorizing the PDs according to their working principle, the parameters or the figures of merit are discussed here. The synthesis of MOx is elaborated for different nanostructures, viz., QDs, 1Ds, and thin films. Further, in the most important section, only whole MO-based designs of PDs are covered where the architecture includes all MOx (except electrodes in some cases). This section is divided in subsections of solar blind, UV, and visible range PDs and their performances are reviewed along with discussion on the suitability of material for a particular range. The effect of harsh conditions such as space exploration and flame prewarning is included as the applications of PDs range from environmental monitoring to space communication, missile tracking to everyday electronics. The detailed applications of PDs are presented in the last section. Different applications of PDs in safety and security, process control, environment sensing, and astronomy are discussed one by one.

References

[1] F. Zhuge, Z. Zheng, P. Luo, L. Lv, Y. Huang, H. Li, T. Zhai, Nanostructured materials and architectures for advanced infrared photodetection, Adv. Mater. Technol. 2 (8) (2017) 1700005.

[2] K. Song, N. Ma, Y.K. Mishra, R. Adelung, Y. Yang, Achieving light-induced ultrahigh pyroelectric charge density toward self-powered UV light detection, Adv. Electron. Mater. 5 (1) (2019) 1800413.

[3] S. Miao, Y. Cho, Toward green optoelectronics: environmental-friendly colloidal quantum dots photodetectors, Front. Energy Res. 9 (2021) 1–18.

[4] K. Xu, W. Zhou, Z. Ning, Integrated structure and device engineering for high performance and scalable quantum dot infrared photodetectors, Small 16 (47) (2020) e2003397.

[5] Fabrication of ZnO Based MSM Photodetectors, 2021. https://www.inup.cense.iisc.ac.in/msm-photodetectors. (16 October 2021).

[6] S.J. Young, L.W. Ji, S.J. Chang, S.H. Liang, K.T. Lam, T.H. Fang, K.J. Chen, X.L. Du, Q.K. Xue, ZnO-based MIS photodetectors, Sensors Actuators A Phys. 135 (2) (2007) 529–533.

[7] S. Guo, L. Wang, C. Ding, J. Li, K. Chai, W. Li, Y. Xin, B. Zou, R. Liu, Tunable optical loss and multi-band photodetection based on tin doped CdS nanowire, J. Alloys Compd. 835 (2020) 155330.

[8] N. Li, Z. Lan, L. Cai, F. Zhu, Advances in solution-processable near-infrared phototransistors, J. Mater. Chem. C 7 (13) (2019) 3711–3729.

[9] F.F. Masouleh, N. Das, Application of metal-semiconductor-metal photodetector in high-speed optical communication systems, advances in optical, Communication (2014).

[10] C.H. Lin, C.W. Liu, Metal-insulator-semiconductor photodetectors, Sensors (Basel) 10 (10) (2010) 8797–8826.

[11] P. Kumar, N. Saxena, S. Dewan, F. Singh, V. Gupta, Giant UV-sensitivity of ion beam irradiated nanocrystalline CdS thin films, RSC Adv. 6 (5) (2016) 3642–3649.

[12] T. Kalsi, P. Kumar, Cd1-xMgxS CQD thin films for high performance and highly selective NIR photodetection, Dalton Trans. 50 (36) (2021) 12708–12715.

[13] P. Kumar, N. Saxena, F. Singh, V. Gupta, Ion beam assisted fortification of photoconduction and photosensitivity, Sens. Actuator A Phys. 279 (2018) 343–350.

[14] T. Zhai, X. Fang, M. Liao, X. Xu, H. Zeng, B. Yoshio, D. Golberg, A comprehensive review of one-dimensional metal-oxide nanostructure photodetectors, Sensors (Basel) 9 (8) (2009) 6504–6529.

[15] Z.R. Dai, Z.W. Pan, Z.L. Wang, Novel nanostructures of functional oxides synthesized by thermal evaporation, Adv. Funct. Mater. 13 (2003) 9.

[16] D. Zhang, C. Li, S. Han, X. Liu, T. Tang, W. Jin, C. Zhou, Electronic transport studies of single-crystalline In2O3 nanowires, Appl. Phys. Lett. 82 (1) (2003) 112.

[17] Z. Liu, D. Zhang, S. Han, C. Li, T. Tang, W. Jin, X. Liu, B. Lei, C. Zhou, Laser ablation synthesis and electron transport studies of tin oxide nanowires, Adv. Mater. 15 (20) (2003) 1754–1757.
[18] J.G. Lu, P. Chang, Z. Fan, Quasi-one-dimensional metal oxide materials—synthesis, properties and applications, Mater. Sci. Eng.: R: Rep. R52 (1–3) (2006) 49–91.
[19] X. Liu, C. Li, S. Han, J. Han, C. Zhou, Synthesis and electronic transport studies of CdO nanoneedles, Appl. Phys. Lett. 82 (12) (2003) 1950–1952.
[20] G. Malandrino, S.T. Finocchiaro, R. Lo Nigro, C. Bongiorno, C. Spinella, I.L. Fragala, Free-standing copper (II) oxide nanotube arrays through an MOCVD template process, Chem. Mater. 16 (2004) 5559.
[21] W.I. Park, D.H. Kim, S.W. Jung, G.C. Yi, Metalorganic vapor-phase epitaxial growth of vertically well-aligned ZnO nanorods, Appl. Phys. Lett. 80 (2002) 4232.
[22] H.W. Kim, N.H. Kim, Synthesis of βGa2O3 nanowires by an MOCVD approach, Appl. Phy. A 81 (4) (2005) 763–765.
[23] D. Mandler, The future of electrochemical deposition: nanomaterial building blocks, J. Solid State Electrochem. 24 (9) (2020) 2133–2135.
[24] S.C. Singh, J. Singh, R. Gopal, O.N. Srivastava, Zno nanostructures synthesized by laser and thermal evaporation, in: Y.B.H. Ahmad Umar (Ed.), Metal Oxide Nanostructures and their Applications, Elsevier, 2009.
[25] A. Umar, S.H. Kim, Y.S. Lee, K.S. Nahm, Y.B. Hahn, Catalyst-free large-quantity synthesis of ZnO nanorods by a vapor–solid growth mechanism: structural and optical properties, J. Cryst. Growth 282 (1–2) (2005) 131–136.
[26] S.-W. Kim, S. Fujita, H.-K. Park, B. Yang, H.-K. Kim, D.H. Yoon, Growth of ZnO nanostructures in a chemical vapor deposition process, J. Cryst. Growth 292 (2) (2006) 306–310.
[27] N. Kumar, T.T. Nguyen, M. Patel, S. Kim, J. Kim, Transparent and all oxide-based highly responsive n-n heterojunction broadband photodetector, J. Alloys Compd. 898 (2022).
[28] A. Umar, S. Lee, Y.H. Im, Y.B. Hahn, Flower-shaped ZnO nanostructures obtained by cyclic feeding chemical vapour deposition: structural and optical properties, Nanotechnology 16 (10) (2005) 2462–2468.
[29] P.P. Patil, D.M. Phase, S.A. Kulkarni, S.V. Ghaisas, S.K. Kulkarni, S.M. Kanrtkar, S.B. Ogale, Pulsed laser induced reactive quenching at a liquid solid interface: aqueous oxidation of iron, Phys. Rev. Lett. 58 (1987) 238.
[30] F.H. Teherani, D.C. Look, D.J. Rogers, F. Alema, B. Hertog, A.V. Osinsky, P. Mukhopadhyay, M. Toporkov, W.V. Schoenfeld, E. Ahmadi, J. Speck, Vertical solar blind Schottky photodiode based on homoepitaxial Ga2O3 thin film, in: Oxide-Based Materials and Devices VIII, 2017, pp. 101051M1–101051M8. San Francisco, California, United States.
[31] F. Alema, B. Hertog, P. Mukhopadhyay, Y. Zhang, A. Mauze, A. Osinsky, W.V. Schoenfeld, J.S. Speck, T. Vogt, Solar blind Schottky photodiode based on an MOCVD-grown homoepitaxial β-Ga2O3 thin film, APL Mater. 7 (2) (2019).
[32] S. Cui, Z. Mei, Y. Zhang, H. Liang, X. Du, Room-temperature fabricated amorphous Ga2O3 high-response-speed solar-blind photodetector on rigid and flexible substrates, Adv. Opt. Mater. 5 (19) (2017) 1700454.
[33] H. Kim, H.-J. Seok, J.H. Park, K.-B. Chung, S. Kyoung, H.-K. Kim, Y.S. Rim, Fully transparent InZnSnO/β-Ga2O3/InSnO solar-blind photodetectors with high schottky barrier height and low-defect interfaces, J. Alloys Compd. 890 (2022) 161931.

[34] S. Oh, C.-K. Kim, J. Kim, High responsivity β-Ga2O3 metal–semiconductor–metal solar-blind photodetectors with ultraviolet transparent graphene electrodes, ACS Photon. 5 (3) (2017) 1123–1128.
[35] N. Kumar, K. Arora, M. Kumar, High performance, flexible and room temperature grown amorphous Ga2O3 solar-blind photodetector with amorphous indium-zinc-oxide transparent conducting electrodes, J. Phys. D. Appl. Phys. 52 (33) (2019).
[36] M. Patel, H.S. Kim, H.H. Park, J. Kim, Active adoption of void formation in metal-oxide for all transparent super-performing photodetectors, Sci. Rep. 6 (2016) 25461.
[37] M. Patel, J. Kim, Transparent NiO/ZnO heterojunction for ultra-performing zero-bias ultraviolet photodetector on plastic substrate, J. Alloys Compd. 729 (2017) 796–801.
[38] V. Krishnamurthi, T. Ahmed, M. Mohiuddin, A. Zavabeti, N. Pillai, C.F. McConville, N. Mahmood, S. Walia, A visible-blind photodetector and artificial optoelectronic synapse using liquid-metal exfoliated ZnO nanosheets, Adv. Opt. Mater. 9 (16) (2021).
[39] B. Zhao, F. Wang, H. Chen, L. Zheng, L. Su, D. Zhao, X. Fang, An ultrahigh responsivity (9.7 mA/W) self-powered solar-blind photodetector based on individual ZnO–Ga2O3 heterostructures, Adv. Funct. Mater. 27 (17) (2017).
[40] P. Mahala, M. Patel, D.-K. Ban, T.T. Nguyen, J. Yi, J. Kim, High-performing self-driven ultraviolet photodetector by TiO2/Co3O4 photovoltaics, J. Alloys Compd. 827 (2020).
[41] A.K. Rana, M. Patel, T.T. Nguyen, J.-H. Yun, J. Kim, Transparent Co3O4/ZnO photovoltaic broadband photodetector, Mater. Sci. Semicond. Process. 117 (2020).
[42] N. Alwadai, S. Mitra, M.N. Hedhili, H. Alamoudi, B. Xin, N. Alaal, I.S. Roqan, Enhanced-performance self-powered solar-blind UV-C photodetector based on n-ZnO quantum dots functionalized by p-CuO micro-pyramids, ACS Appl. Mater. Interfaces 13 (28) (2021) 33335–33344.
[43] S. Park, S.J. Kim, J.H. Nam, G. Pitner, T.H. Lee, A.L. Ayzner, H. Wang, S.W. Fong, M. Vosgueritchian, Y.J. Park, M.L. Brongersma, Z. Bao, Significant enhancement of infrared photodetector sensitivity using a semiconducting single-walled carbon nanotube/C60 phototransistor, Adv. Mater. 27 (4) (2015) 759–765.
[44] T. Xie, G. Liu, B. Wen, J.Y. Ha, N.V. Nguyen, A. Motayed, R. Debnath, Tunable ultraviolet photoresponse in solution-processed p-n junction photodiodes based on transition-metal oxides, ACS Appl. Mater. Interfaces 7 (18) (2015) 9660–9667.
[45] S. Praveen, S. Veeralingam, S. Badhulika, A flexible self-powered UV photodetector and optical UV filter based on β-Bi2O3/SnO2 quantum dots Schottky heterojunction, Adv. Mater. Interfaces 8 (15) (2021).
[46] M. Ahmadi, M. Abrari, M. Ghanaatshoar, An all-sputtered photovoltaic ultraviolet photodetector based on co-doped CuCrO2 and Al-doped ZnO heterojunction, Sci. Rep. 11 (1) (2021) 18694.
[47] R. Debnath, T. Xie, B. Wen, W. Li, J.Y. Ha, N.F. Sullivan, N.V. Nguyen, A. Motayed, A solution-processed high-efficiency p-NiO/n-ZnO heterojunction photodetector, RSC Adv. 5 (19) (2015) 14646–14652.
[48] D.Y. Kim, J. Ryu, J. Manders, J. Lee, F. So, Air-stable, solution-processed oxide p-n heterojunction ultraviolet photodetector, ACS Appl. Mater. Interfaces 6 (3) (2014) 1370–1374.
[49] A. Costas, C. Florica, N. Preda, N. Apostol, A. Kuncser, A. Nitescu, I. Enculescu, Radial heterojunction based on single ZnO-CuxO core-shell nanowire for photodetector applications, Sci. Rep. 9 (1) (2019) 5553.
[50] S. Abbas, M. Kumar, J. Kim, All metal oxide-based transparent and flexible photodetector, Mater. Sci. Semicond. Process. 88 (2018) 86–92.

[51] S. Abbas, J. Kim, All-metal oxide transparent photodetector for broad responses, Sensors Actuators A Phys. 303 (2020).
[52] S. Abbas, M. Kumar, H.-S. Kim, J. Kim, J.-H. Lee, Silver-nanowire-embedded transparent metal-oxide heterojunction Schottky photodetector, ACS Appl. Mater. Interfaces 10 (17) (2018) 14292–14298.
[53] Y. Gao, J. Xu, S. Shi, H. Dong, Y. Cheng, C. Wei, X. Zhang, S. Yin, L. Li, TiO2 nanorod arrays based self-powered UV photodetector: heterojunction with NiO nanoflakes and enhanced UV photoresponse, ACS Appl. Mater. Interfaces 10 (13) (2018) 11269–11279.
[54] S. Abbas, D.-K. Ban, J. Kim, Functional interlayer of In2O3 for transparent SnO2/SnS2 heterojunction photodetector, Sensors Actuators A Phys. 293 (2019) 215–221.
[55] Z. Chen, B. Li, X. Mo, S. Li, J. Wen, H. Lei, Z. Zhu, G. Yang, P. Gui, F. Yao, G. Fang, Self-powered narrowband p-NiO/n-ZnO nanowire ultraviolet photodetector with interface modification of Al2O3, Appl. Phys. Lett. 110 (12) (2017).
[56] T. Cossuet, J. Resende, L. Rapenne, O. Chaix-Pluchery, C. Jiménez, G. Renou, A.J. Pearson, R.L.Z. Hoye, D. Blanc-Pelissier, N.D. Nguyen, E. Appert, D. Muñoz-Rojas, V. Consonni, J.-L. Deschanvres, ZnO/CuCrO2 core-shell nanowire heterostructures for self-powered UV photodetectors with fast response, Adv. Funct. Mater. 28 (43) (2018).
[57] A. Tahira, R. Mazzaro, F. Rigoni, A. Nafady, S.F. Shaikh, A.A. Alothman, R.A. Alshgari, Z.H. Ibupoto, A simple and efficient visible light photodetector based on Co3O4/ZnO composite, Opt. Quant. Electron. 53 (9) (2021).
[58] H.-S. Kim, M.D. Kumar, W.-H. Park, M. Patel, J. Kim, Cu4O3-based all metal oxides for transparent photodetectors, Sensors Actuators A Phys. 253 (2017) 35–40.
[59] A. Costas, C. Florica, N. Preda, A. Kuncser, I. Enculescu, Photodetecting properties of single CuO-ZnO core-shell nanowires with p-n radial heterojunction, Sci. Rep. 10 (1) (2020) 18690.
[60] E.Y. Salih, M.F. Mohd Sabri, M.H. Eisa, K. Sulaiman, A. Ramizy, M.Z. Hussein, S.M. Said, Mesoporous ZnO/ZnAl2O4 mixed metal oxide-based Zn/Al layered double hydroxide as an effective anode material for visible light photodetector, Mater. Sci. Semicon. Proc. 121 (2021).
[61] H.-S. Kim, M. Patel, P. Yadav, J. Kim, A. Sohn, D.-W. Kim, Optical and electrical properties of Cu-based all oxide semi-transparent photodetector, Appl. Phys. Lett. 109 (10) (2016).
[62] H.-S. Kim, P. Yadav, M. Patel, J. Kim, K. Pandey, D. Lim, C. Jeong, Transparent Cu4O3/ZnO heterojunction photoelectric devices, Superlattice. Microst. 112 (2017) 262–268.
[63] H.-S. Kim, K.R. Chauhan, J. Kim, E.H. Choi, Flexible vanadium oxide film for broadband transparent photodetector, Appl. Phys. Lett. 110 (10) (2017).
[64] P. Ghamgosar, F. Rigoni, M.G. Kohan, S. You, E.A. Morales, R. Mazzaro, V. Morandi, N. Almqvist, I. Concina, A. Vomiero, Self-powered photodetectors based on core-shell ZnO-Co3O4 nanowire heterojunctions, ACS Appl. Mater. Interfaces 11 (26) (2019) 23454–23462.
[65] J.R.D.R. Der-Hsien Lien, J.-J. Ke, C.-F. Kang, J.-H. He, Surface effect in metal oxide-based nanodevices, Nanoscale 7 (2015) 19874–19884.
[66] B.R. Tak, M. Garg, A. Kumar, V. Gupta, R. Singh, Gamma irradiation effect on performance of β-Ga2O3 metal-semiconductor-metal solar-blind photodetectors for space applications, ECS J. Solid State Sci. Technol. 8 (7) (2019) Q3149–Q3153.
[67] R. Hui, Photodetectors, in: S. Merken (Ed.), Introduction to Fiber-Optic Communications, Academic Press, Elsevier, UK, 2020, pp. 125–154.

[68] P. Fay, Photodetectors, in: R.W.C.K.H.J. Buschow, M.C. Flemings, B. Ilschner, E.J. Kramer, S. Mahajan, P. Veyssière (Eds.), Encyclopedia of Materials: Science and Technology, Elsevier, Netherlands, 2001, pp. 6909–6923.
[69] T. Nagatsuma, Photodetectors for microwave photonics, in: B. Nabet (Ed.), Photodetectors, Woodhead Publishing, 2016, pp. 297–314.
[70] G. Konstantatos, Current status and technological prospect of photodetectors based on two-dimensional materials, Nat. Commun. 9 (2018).
[71] P.V.K. Yadav, B. Ajitha, Y.A. Kumar Reddy, A. Sreedhar, Recent advances in development of nanostructured photodetectors from ultraviolet to infrared region: a review, Chemosphere 279 (2021) 130473.
[72] A. Ghatak, K. Thyagarajan, Detectors for optical fiber communication, in: Introduction to Fiber Optics, Cambridge University Press, 1998, pp. 238–248.
[73] T. Hayashi, Multi-core optical fibers, in: I.P. Kaminow, T. Li, A.E. Willner (Eds.), Optical Fiber Telecommunications. Volume VIA: Components and Subsystems, 6th, Academic Press, Elsevier, UK, 2013, pp. 321–352.
[74] M.K. Serkan Kaya, N. Pala, THz detectors, in: B. Nabet (Ed.), Photodetectors, Woodhead Publishing, 2016, pp. 373–414.
[75] H. Chen, K. Liu, L. Hu, A.A. Al-Ghamdi, X. Fang, New concept ultraviolet photodetectors, Mater. Today 18 (9) (2015) 493–502.

Metal oxide charge transport layers for halide perovskite light-emitting diodes

11

Jean Maria Fernandes[a], *D. Paul Joseph*[a], *and M. Kovendhan*[b]
[a]Department of Physics, National Institute of Technology Warangal, Warangal, Telangana, India, [b]Department of Physics and Nanotechnology, SRM Institute of Science and Technology, Kattankulathur, Tamilnadu, India

Chapter outline

Metal Oxides for Next-Generation Optoelectronic, Photonic, and Photovoltaic Applications
https://doi.org/10.1016/B978-0-323-99143-8.00019-5

Abbreviations

LED	light-emitting diode
OLED	organic light-emitting diode
QLED	quantum dot light-emitting diode
PeLED	perovskite light-emitting diode
EQE	external quantum efficiency
HIL	hole injection layer
HTL	hole transport layer
EML	emissive layer
ETL	electron transport layer
EIL	electron injection layer
CTL	charge transport layer
ITO	indium tin oxide
HOMO	highest occupied molecular orbital
LUMO	lowest unoccupied molecular orbital
VBM	valence band maximum
CBM	conduction band minimum
MO	metal oxide
$MAPbI_3$	methyl ammonium lead iodide ($CH_3NH_3PbI_3$)
$CsPbX_3$	cesium lead halide
PEI	polyethylenimine
PEIE	polyethyleneimine ethoxylated
WF	work function
PEDOT:PSS	poly(3,4-ethylenedioxythiophene):polystyrene sulfonate
PVP	poly(vinylpyrrolidone)
PVD	physical vapor deposition
CVD	chemical vapor deposition
PLD	pulsed laser deposition
ALD	atomic layer deposition
SAALD	spatial atmospheric atomic layer deposition
XPS	X-ray photoelectron spectroscopy
UPS	ultraviolet photoelectron spectroscopy
KPFM	kelvin probe force microscopy
TRPL	time-resolved photoluminescence spectroscopy
SCLC	space charge limited current

1 Overview of next-generation halide perovskite light-emitting diodes

In a bid to resolve several critical global issues related to healthcare, energy, climate crisis, etc., there has been a quest for innovative and revolutionary modern technologies which are accessible and sustainable. Over the past two decades, a wide range of newer optoelectronic applications and electronic devices have achieved tremendous progress and breakthroughs, which are defining the future of innovation and transforming the way society functions by efficiently utilizing technology. These include light-emitting diodes (LEDs), photovoltaics, thin film transistors, smart windows, sensors, biomedical devices, and IoT applications [1–8].

One of the ways of resolving the issue of global energy crisis is to replace traditional lighting and display products with modern energy-efficient counterparts. This has driven tremendous interest toward emerging organic/polymer and hybrid materials which are cost-effective, lighter in weight, and flexible; the availability of a very large variety of these materials makes them promising alternatives for several applications [9,10]. An added benefit is that these materials can be processed on a massive scale (except a few materials) by adapting relatively low-temperature methods, solution processing, or even screen-printing technique for device fabrication. Their electronic and optical properties can also be tailored by chemically modifying the molecular structure [9,10].

The past few decades have witnessed a steady rise in the development of organic light-emitting diode (OLED) and quantum dot light-emitting diode (QLED) technologies due to their attractive features and advantages compared to conventional inorganic LEDs [11,12]. In 2014, the latest entrant to this family of emerging technologies has been the metal halide perovskite light-emitting diode (PeLED) with emission at room temperature [13]. This PeLED was based on organometal halide perovskite and showed a very low external quantum efficiency (EQE) of <1%. However, in a short span of time, rapid and drastic growth with exceptional performance of PeLEDs has been witnessed (Fig. 1) by surpassing the EQE milestone of 20% as reported by Cao et al. and Lin et al. in 2018 [14,15]. More recently, He et al. [16] reported achievement of a PeLED with a peak EQE of 32.8% corresponding to near-unity internal quantum efficiency at a low temperature (45 K). Motivated by the revolutionary progress, researchers have focused efforts on the further development of PeLEDs by identifying and strategizing methods to overcome the remaining challenges in this field.

Metal halide perovskites have emerged as powerful game changers [17,18] in the recent past in energy-efficient display and lighting applications due to their excellent

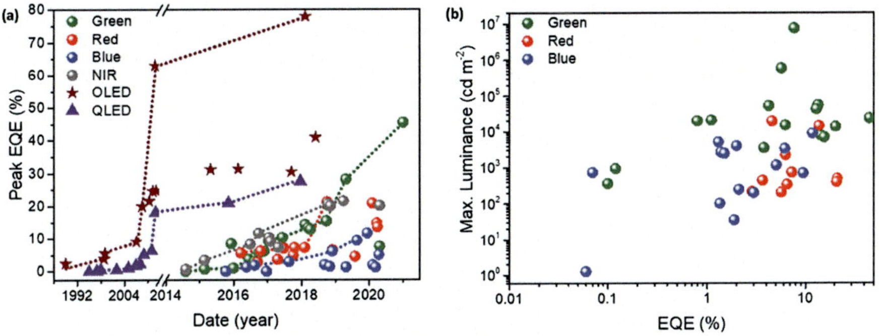

Fig. 1 Yearly growth of peak EQE compared with that of OLED and QLED technologies (A) and maximum luminance as a function of EQE for red, green, and blue PeLEDs (B). Reproduced with permission from S. Kar, N.F. Jamaludin, N. Yantara, S.G. Mhaisalkar, W.L. Leong, Recent advancements and perspectives on light management and high performance in perovskite light-emitting diodes, Nanophotonics, 10 (2020) 2103–2143, https://doi.org/10.1515/nanoph-2021-0033.

optoelectronic properties such as high carrier mobility ($\cong 10\,cm^2\,V^{-1}\,s^{-1}$) [19,20], high transparency to visible light, facile color tunability, narrow emission bandwidth, high color purity, and high photoluminescence quantum efficiency, as well as cost-effective, facile, relatively low-temperature, and solution-based processing techniques [17–20]. A few commercially developed PeLED-based display prototypes and lab-scale devices are shown in Fig. 2A–C.

Ever since the demonstration of the first PeLED (emission at 300 K) in 2014, lead (Pb)-based hybrid organic-inorganic halide perovskites (especially red and green emitting) have shown tremendous progress in a short span of time (EQE of 23.4% as of Jan 2021 [21]) as seen in recent research on emerging display technologies. However, blue PeLEDs still lag far behind their red and green counterparts in terms of color stability and efficiency [EQE $\cong$ 3%–6% (max. $\cong$ 10%)], which is a bottleneck for full color or white

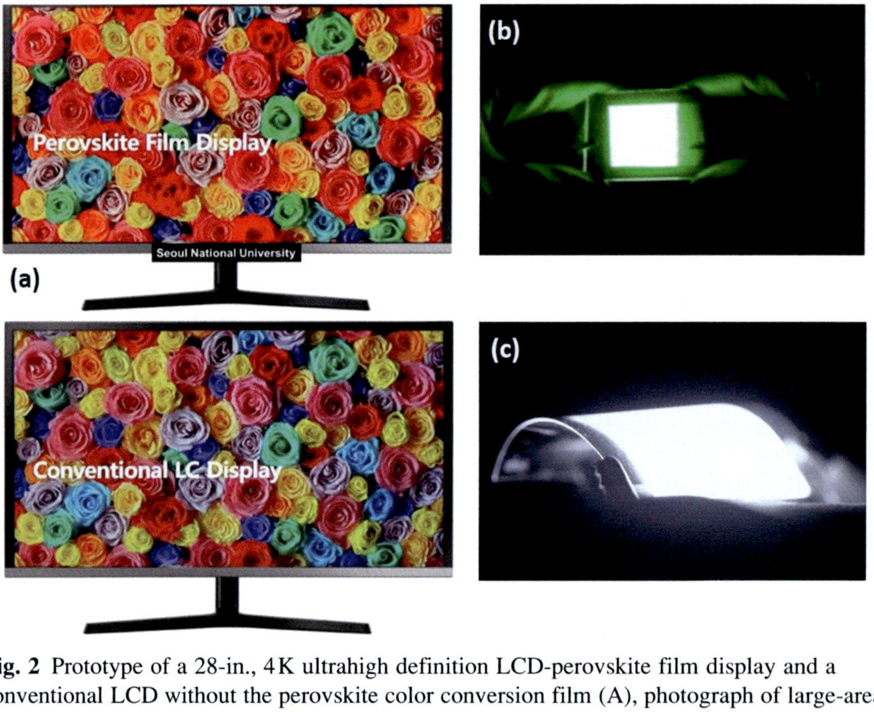

Fig. 2 Prototype of a 28-in., 4K ultrahigh definition LCD-perovskite film display and a conventional LCD without the perovskite color conversion film (A), photograph of large-area green PeLED with EQE 23.4% (B), large-area near-infrared PeLED on flexible PET substrate for medical applications (C).
Source: (A) Reproduced with permission from H. Lee, J. Park, S. Kim, S.-C. Lee, Y.-H. Kim, T.-W. Lee, Perovskite emitters as a platform material for down-conversion applications, Adv. Mater. Technol. 5 (2020) 2000091, https://doi.org/10.1002/admt.202000091; (B) reproduced with permission from Y.H. Kim, S. Kim, A. Kakekhani, et al. Comprehensive defect suppression in perovskite nanocrystals for high-efficiency light-emitting diodes, Nat. Photonics 15 (2021) 148–155, https://doi.org/10.1038/s41566-020-00732-4; (C) reproduced with permission from X. Zhao, Z.-K. Tan, Large-area near-infrared perovskite light-emitting diodes, Nat. Photon. 14 (2020) 215–218, https://doi.org/10.1038/s41566-019-0559-3.

light display and lighting applications [22–24]. Moreover, most of the blue PeLEDs investigated are in the sky-blue range, requiring attention toward the development of pure blue or deep blue colored PeLEDs [25]. In addition, the large-scale commercialization of PeLEDs faces major hurdles, mainly due to their overall instability to factors such as moisture, temperature, etc., as well as short lifetime, poor performance, and Pb-toxicity (for Pb-based). Furthermore, the performance of PeLEDs is also limited by phenomena such as current density-voltage (*I-V*) hysteresis, degradation, nonradiative recombination, etc., generally caused by defect-induced deep traps in perovskites [26–28]. Other disadvantages include ion migration, difficulty in hole injection, color instability due to electric field-induced phase segregation in Br/Cl mixed perovskites, etc. [27]. Organic-inorganic hybrid halides mainly suffer from: (i) structural instability under thermal conditions, (ii) rapid performance degradation when exposed to oxygen and moisture, and (iii) harmful impact on environment and human health due to toxic Pb in their structure [28]. In order to improve the stability and EQE of PeLEDs, innovative strategies can be adapted for developing novel material compositions, device architectures, interface, and defect engineering, etc. [26].

2 Multi-dimensional hybrid organic-inorganic and all-inorganic halide-based diodes

In general, the chemical formula of metal halide perovskites is represented as ABX_3, where "A" and "B" denote cations (A is much larger than B), and "X" is an anion [28]. As the name implies, the A site is often occupied by an organic component (in the case of hybrid perovskites) or inorganic component, B site is substituted by a metal ion (usually Pb^{2+}), and the X site is occupied by an inorganic halide (I^-, Cl^-, Br^-, or their combination) in the metal halide perovskite structure. Based on this, metal halide perovskite structures in diodes can be broadly classified into two main categories as shown in Fig. 3: (i) Hybrid organic-inorganic [e.g., methyl ammonium lead iodide ($CH_3NH_3PbI_3$) (also denoted as $MAPbI_3$)] and (ii) All-inorganic halide (for e.g., Cesium (Cs^+)-based: $CsPbX_3$, where X= Cl, Br, and I) [28,29].

3 Lead-free halide perovskite light-emitting diodes

The potential impact and inherent dangers on human health and environment due to toxicity of lead (Pb) in perovskite structures are of great concern. In order to mitigate the harmful effects of Pb in metal halide PeLEDs, several alternative materials such as Sn, Ge, Bi, Sb, etc. and their compositions have been explored as replacement to Pb [30–33]. Since the ionic radius and chemical properties of Sn^{2+} are similar to that of Pb^{2+}, Sn is regarded as one of the most suitable replacements for Pb [30,31]. However, Sn^{2+} is highly unstable in ambient atmosphere and spontaneously undergoes oxidation to Sn^{4+}, which proves to be a huge challenge in the development of Sn-based perovskite devices. Moreover, all-inorganic halides (Pb-based and Pb-free) suffer from low EQE due to several factors, and this issue is more serious for Pb-free devices.

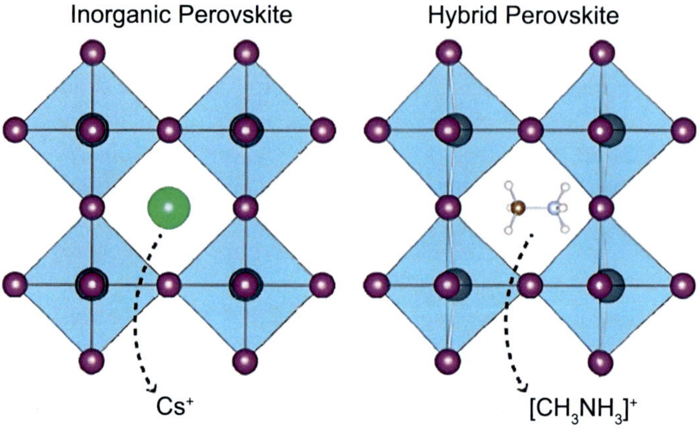

Fig. 3 Schematic representation of inorganic and hybrid perovskite structures.
Reproduced with permission from Q. Chen, N.D. Marco, Y.M. Yang, T.-B. Song, C. C. Chen, H. Zhao, Z. Hong, H. Zhou, Y. Yang, Under the spotlight: The organic—inorganic hybrid halide perovskite for optoelectronic applications, Nanotoday, 10 (2015) 355-396, https://doi.org/10.1016/j.nantod.2015.04.009.

Nevertheless, preliminary investigations on Pb-free metal halide perovskite nanocrystals and quantum dots demonstrate their viability in PeLED applications [32–34]. This leads to further scope for improvement, considering the fact that research efforts in this direction are extremely limited and very recent.

4 Device architectures

A typical multilayer PeLED architecture is quite similar to that of an OLED and constitutes several layers, including anode, hole injection layer (HIL), hole transport layer (HTL), emissive layer (EML), electron transport layer (ETL), electron injection layer (EIL), and cathode as shown in Fig. 4. These devices can be realized via different

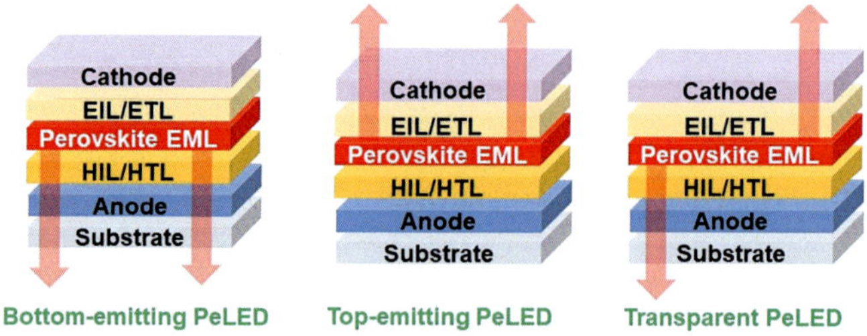

Fig. 4 Typical conventional multilayer PeLED structures (Note that the EIL/ETL and HIL/HTL layers can be interchanged to form inverted structures).
Reproduced with permission from P. Xiao, Y. Yu, J. Cheng, Y. Chen, S. Yuan, J. Chen, J. Yuan, B. Liu, Advances in Perovskite Light-Emitting Diodes Possessing Improved Lifetime, Nanomaterials 11 (1) (2021) 103, https://doi.org/10.3390/nano11010103.

architectures such as conventional (bottom-emitting, top-emitting, transparent) and inverted structures. The most widely used conventional PeLED device architecture is the bottom-emitting structure [35]. Recently, inverted PeLED device architectures have been developed to minimize hysteresis, enhance efficiency, and maximize luminescence [35–37].

In a conventional PeLED, holes from the anode and electrons from the cathode together form excitons, which should ideally decay radiatively in the EML made of metal halide perovskite. However, due to several factors such as imbalance in electron and hole mobilities as well as presence of defects, the recombination may take place in the proximity of the electrodes, resulting in quenching of excitons, which leads to reduction in device efficiency and lifetime [38,39]. In order to mitigate this issue, inclusion of hole/electron blocking layers on adjacent sides of the emissive layer helps in confining the charge carriers within the EML, which significantly improves the functioning of the device [38–40].

Further, the inclusion of HIL and EIL in the device reduces the charge injection barriers at the anode and cathode, respectively. The HTL and ETL, also collectively named as charge transport layers (CTLs), facilitate easy transit for the injected carriers to reach the EML [35,38]. The frequently used anode is indium tin-oxide (ITO) which has excellent transparency in the visible region, owing to its large band gap [38]. The cathode typically constitutes low work function metals like aluminum, etc. or their alloys with silver. The thickness and mobility of these layers significantly influence the PeLED efficiency and require optimization.

The materials are selected considering the energy band alignment at the interfaces between several layers of the device for highly efficient charge injection and transport into the EML. Fig. 5 shows a pictorial representation of the energy level matching in conventional PeLEDs. When an external voltage is administered to the device, there is bending of energy levels (refer Section 6.2). The HTL should maintain high hole mobility and low ionization potential to permit efficient hole transfer from the

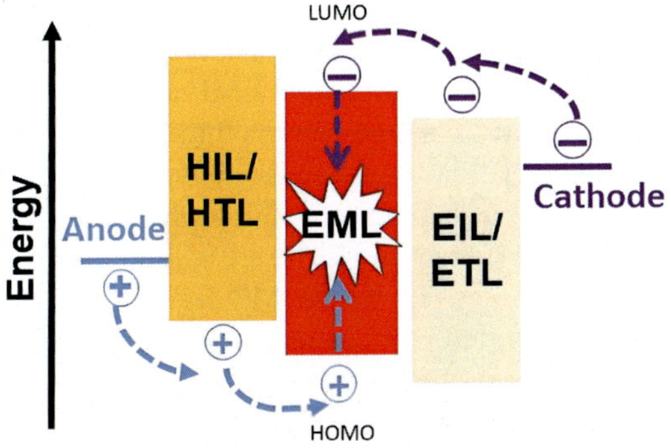

Fig. 5 Schematic representation of PeLED energy-level diagram.
Reproduced with permission from P. Xiao, Y. Yu, J. Cheng, Y. Chen, S. Yuan, J. Chen, J. Yuan, B. Liu, Advances in perovskite light-emitting diodes possessing improved lifetime, Nanomaterials, 11 (1) (2021) 103, https://doi.org/10.3390/nano11010103.

HIL. Similarly, the ETL should possess high electron mobility and high electron affinity so as to match the work function of the cathode and facilitate efficient electron transfer to the EML [35,38].

The highest occupied molecular orbital (HOMO) of organic or hybrid semiconductors is the energy state analogous to the valence band maximum (VBM) in inorganic semiconductors. Likewise, the lowest unoccupied molecular orbital (LUMO) is the energy state analogous to the conduction band minimum (CBM).

The external quantum efficiency (EQE) of a PeLED is the ratio of the number of photons emitted to the number of electrons injected, and is strongly affected by the device architecture. The EQE is a product of four parameters as given in Eq. (1) [41]:

$$\mathrm{EQE} = \gamma \times \eta_{rad,eff} \times \eta_{S/T} \times \eta_{out} \tag{1}$$

where γ is the electrical efficiency quantifying the fraction of injected carriers transformed into excitons, $\eta_{rad,eff}$ signifies radiative recombination efficiency, $\eta_{S/T}$ indicates the proportion of photons generated from the excitons in the EML, and η_{out} is the outcoupling efficiency.

The electrical efficiency γ can be maximized by effectively creating excitons in the EML via balancing of electron and hole injection to the EML [41]. However, the presence of defects or luminescence quenching centers in the device leads to the formation of nonradiative recombination paths which have to be blocked in order to increase the $\eta_{rad,eff}$. The device structure is crucial in deciding the outcoupling efficiency η_{out} which is determined by the effectiveness of the generated photons in the EML escaping out of the device [41]. All these factors can be successfully controlled by device engineering in PeLEDs.

Certain distinct features in PeLEDs set them apart from OLEDs, although both of them possess similar device architectures. One of the most prominent drawbacks in PeLEDs is that they are prone to luminescence quenching at the interfaces adjoining the EML, which can be due to low exciton binding energy and long exciton diffusion length in metal halide perovskites [42]. This leads to the requirement of extreme care while designing the device architecture. Another factor of concern is the serious charge imbalance at the interfaces between the EML and adjacent layers. This issue is caused by the quick transport of charge carriers through the EML due to extremely high carrier mobility in metal halides compared to conventional EMLs such as organic or polymeric materials, thereby leading to buildup of excess charge carriers at the interface [43]. The excess charges in the EML are also known to degrade the metal halide perovskite material. Moreover, the HOMO level of metal halide perovskites is relatively deep compared to that of its organic emitter counterparts. This can lead to migration of the accumulated charges to the adjacent CTLs, causing undesirable emission from the transport layer, thereby reducing the device efficiency [41,43]. For efficient charge carrier injection and transport, the HOMO level (or VBM) of the HIL or HTL should be deeper than that required in conventional OLEDs.

Several studies have concentrated on the optimization of the emissive perovskite layer [39,44,45]. However, a thorough understanding of how the adjacent CTLs affect the perovskite crystallization, film quality, and overall device performance is of critical importance which needs to be addressed for the development of highly efficient PeLEDs. In this regard, subsequent sections of this chapter are dedicated to the discussion on CTLs (particularly metal oxides) in PeLEDs.

5 Charge transport layers in perovskite light-emitting diodes

In order to improve the stability, efficiency, and overall device performance of a PeLED, the development of optimal charge transport layers (CTLs) is extremely crucial [46,47]. The most widely used CTL materials for PeLEDs are classified as follows [36,46,48]:

(1) Metal oxides
(2) p- or n-type organic materials
(3) Organic-inorganic hybrid layers
(4) Ionic materials

Research on PeLEDs initially adapted the know-how of OLED technology, wherein various commonly used organic and inorganic CTLs in OLEDs were also adapted in PeLEDs [44]. However, with rapid progress in PeLED research, there emerged a requirement for more complex and multifunctional roles of the CTLs to resolve perovskite-specific challenges, in addition to charge carrier injection and transport [48,49].

Ideal CTLs contribute to the enhancement of crystallinity and suppression of interfacial defects in the device [50]. This leads to significant reduction in ion migration and phase transformation, thereby facilitating long-term performance stability in PeLEDs. Moreover, due to high thermal conductivity of these CTLs, the Joule heating effect from the perovskite layer of the device can be lowered, leading to reduced degradation of the perovskite material during PeLED operation [51]. In addition, charge carrier balance in the device is crucial in order to maximize recombination rate in the perovskite EML and enhance light outcoupling. This can be achieved by engineering the energy level alignment and blocking opposite charges using suitable CTLs for efficient charge carrier injection and transport in the PeLED [38]. Another crucial factor affecting PeLED performance is the growth mechanism of the perovskite film and the quality of the CTL-perovskite interface, which is influenced by the bottom CTL [52]. Defects at the CTL-perovskite interface cause the formation of deep traps within the band gap, forming a predominant channel for nonradiative recombination [53], and they can be mitigated using suitable interface engineering approaches [52,53].

6 Characteristics of effective metal oxide charge transport layers

Metal oxides have evolved as a unique and prominent class of materials which have revolutionized the field of optoelectronics due to their remarkable structural, optical, and electrical properties [54,55]. They are best suited as CTLs in optoelectronic devices such as LEDs, solar cells, etc. due to their abundance in nature, low cost, long-term stability, high optical transparency (>80%), and high efficiency to enable selective transport of charge carriers [56,57]. Apart from these qualities, metal oxide CTLs (MO-CTLs) also possess high charge carrier mobility ($\cong 10^{-3}$–$10\,cm^2\,V^{-1}\,s^{-1}$ depending on deposition conditions [57]), tunable band gaps, as well as chemical compatibility and energy band alignment with the perovskite layer and electrodes for effective charge transport [53,56,57]. In addition, their structural properties such as crystallinity and surface morphology also play a key role in the functioning of the device [57].

To arrive at the best performance of PeLEDs, all the layers in the device should be morphologically uniform and thermally stable. Most of the MO-CTL materials are capable of forming pinhole-free smooth films at optimal conditions. They also possess high glass transition temperatures, leading to morphological and thermal stability at relatively high operating temperatures [58]. The wide tunable optical band gap (E_g) of MO-CTLs improves the passage of generated visible light through the device. Their interfacial compatibility helps in confining injected carriers within the EML by aligning the energy bands, leading to radiative recombination. This is achieved when the path for nonradiative recombination is reduced by improving the interface between MO-CTL and perovskite layer [52,53]. The quality of the interface can be improved by depositing crystalline metal oxides, which enables the growth of highly crystalline perovskite overlayer [52]. Charge accumulation at the CTL-perovskite interfaces can be avoided as a result of high charge carrier mobility of the MO-CTLs.

6.1 Properties of metal oxide charge transport layers

The surface morphology of MO-CTLs in PeLEDs can be either in the form of conventional thin films or nanoparticles dispersed in thin films [59,60]. Various defects (surface, interfacial, and bulk) in thin film layers of PeLEDs play a critical role in the functioning of the device due to their influence on the charge carrier transport characteristics [61]. These defects induce intrinsic or extrinsic traps which are confined electronic states associated with chemical impurities, structural disarrangement and surface states [61,62]. Extrinsic traps due to impurities can be minimized by repeated sublimation process, whereas intrinsic traps are created due to structural disarrangement caused by varying growth parameters, and mainly influence the functioning of the device. Fig. 6A–F shows the structural, optical, and device properties of some of the MO-CTLs [62–64].

A smooth and homogeneous spin-coated NiO_x film annealed at 400°C and subjected to UV-ozone treatment is shown in Fig. 6A with r.m.s. surface roughness of 0.64 nm. Similarly, Fig. 6B shows a uniform ZnMgO film with roughness of 1.75 nm. The PL intensity of NiO_x films is shown in Fig. 6C. It is reported that the replacement of conventional organic PEDOT: PSS HTL with inorganic NiO_x HTL solves the stability issues caused by the hygroscopic and acidic characteristic of PEDOT: PSS [64]. Moreover, since the $CsPbBr_3$ perovskite used in these PeLEDs is also an inorganic material, its combination with inorganic NiO_x HTL leads to improved operational stability of the devices evaluated under repeated bias scans (Fig. 6D–F) [64].

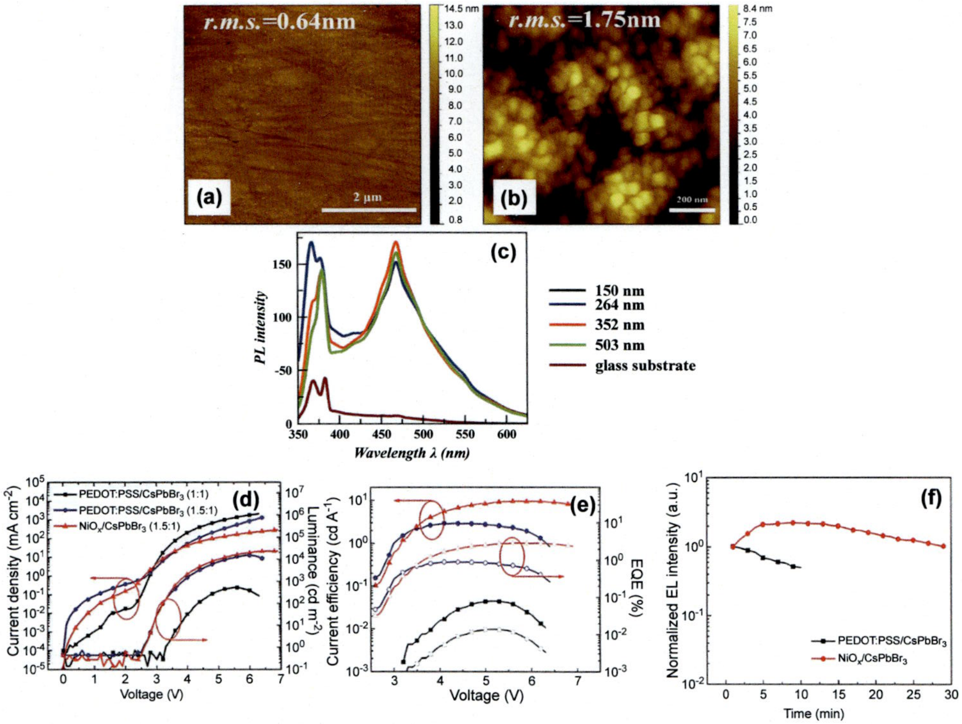

Fig. 6 Two-dimensional AFM images of NiO_x HTL (A) and ZnMgO ETL (B); Photoluminescence emission spectra of NiO_x at different film thicknesses (C), Current density-luminance-voltage characteristics (D), current efficiency-EQE-voltage characteristics (E) and EL intensity vs time (F) of PeLED with NiO_x HTL compared to that with PEDOT:PSS HTL.

Reproduced with permission from L. Liu, Z. Wang, W. Sun, J. Zhang, S. Hu, T. Hayat, A. Alsaedi, Z. Tan, All-solution-processed perovskite light-emitting diodes with all metal oxide transport layers, Chem. Commun. 54 (2018) 13283–13286. https://doi.org/10.1039/C8CC07821A; A.H. Hammad, M.S.A. Wahab, S. Vattamkandathil, A.R. Ansari, Growth and correlation of the physical and structural properties of hexagonal nanocrystalline nickel oxide thin films with film thickness, Coatings 9 (10) (2019) 615. https://doi.org/10.3390/coatings9100615; Y. Gong, S. Zhang, H. Gao, Z. Ma, S. Hu, Z. Tan, Recent advances and comprehensive insights on nickel oxide in emerging optoelectronic devices, Sustain. Energy Fuels 4 (2020) 4415–4458, https://doi.org/10.1039/D0SE00621A.

6.2 Interfacial energetics

Traps are temperature-dependent and can be categorized as deep and shallow traps based on their position occupied in the distribution of energies [61,65]. Shallow traps lie close to the edge of conduction band or valence band, and contribute to a lesser extent toward device degradation. However, deep trap states located within the bandgap of the semiconductor significantly affect the charge transport and device performance. Nevertheless, the dominant defects and their formation energies can vary depending on the stoichiometry of perovskite material, charge transport layer properties, or due to device interfaces [66].

In general, hybrid or metal halide PeLEDs exhibit "defect tolerance," which is the ability to achieve high device performance despite the presence of intrinsic point defects in the system. However, the overall contribution of trapped charge carriers at the defect sites on the surface, grain boundaries and/or bulk leads to nonradiative recombination phenomena, *I-V* hysteresis, instability, and degradation, thereby reducing the carrier lifetime and device efficiency [61,65]. Interfacial engineering is one of the several effective methods to control defects in perovskite films [67], wherein chemically and thermally stable hole and electron transport layers (e.g., metal oxides) and their modified structures having proper energy level alignments and compatibility are incorporated in the device for selective charge extraction and transport.

Fig. 7 shows the energy levels at the interface between $MAPbBr_3$ perovskite and ZnO ETL before and after surface modification with polyethylenimine (PEI) [68]. An energy offset between CBMs (or LUMOs) of ZnO ETL and $MAPbBr_3$ perovskite is observed at the interface, with an electron injection barrier (Φ_{inj}) of 0.71 eV (Fig. 7A). In order to lower this energy barrier and smoothen the surface of the perovskite film, an ultrathin buffer layer of PEI is deposited on top of it as shown in Fig. 7B. This leads to variation in the work function (WF) or Fermi energy (E_F) of the material, and is usually due to bending of the valence and conduction band energy levels (E_{VB} and E_{CB}). In the context of organic semiconductors, this is understood as a shift in the vacuum energy, or as creation of an interfacial dipole [69] with band offset $\Delta = 0.45$ eV, leading to improved electron injection into the perovskite layer.

In another study by Liu et al., it has been observed that incorporation of chemical elements in metal oxides leads to tunable energy bandgap of the MO-CTL, thereby lowering the injection barrier [62]. Here, the relative potential electron energy barrier at the interface between ZnO ETL and CsPbBr perovskite EML is lowered by incorporating Mg into ZnO lattice. The energy bandgap of the resulting ZnMgO becomes larger, causing the CBM level to upshift and lower the energetic barrier at the ETL-EML interface. The ZnMgO layer with deep VBM of 7.51 eV also blocks holes from reaching the Al cathode. Similarly, at the anode side, NiO_x HTL has high CBM level which facilitates blocking of electrons from the perovskite EML, thereby leading to improved charge transport and recombination in the device [62].

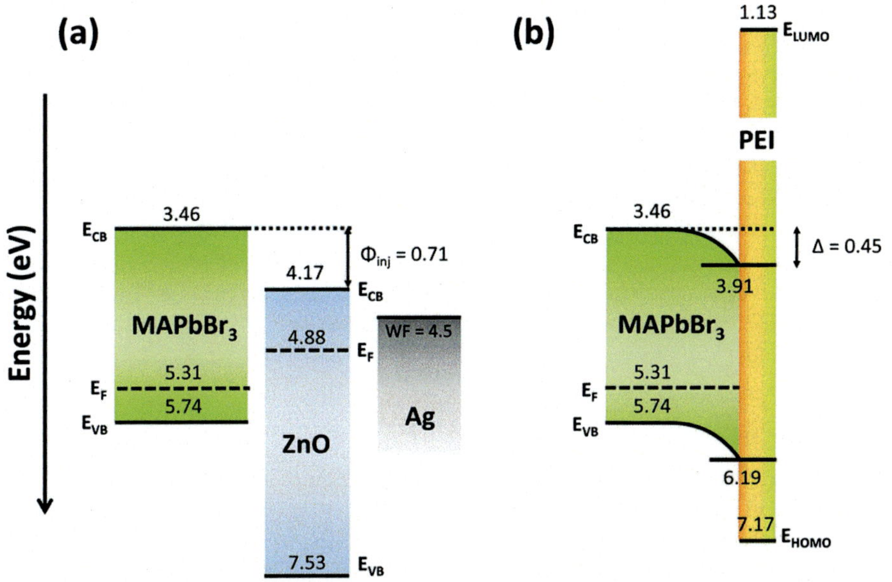

Fig. 7 Charge injection barrier at the perovskite interface before surface modification with PEI (A) and after surface modification (B).
Reproduced with permission from H.-B. Kim, Y. J. Yoon, J. Jeong, J. Heo, H. Jang, J. H. Seo, B. Walker, J.Y. Kim, Peroptronic devices: perovskite-based light-emitting solar cells, Energ. Environ. Sci. 10 (2017) 1950–1957, https://doi.org/10.1039/C7EE01666B.

7 Classification of metal oxides in charge transport layers

Metal oxides can be classified into different types based on their crystal structure, functional properties, etc. In this section, various metal oxides used as CTLs in PeLEDs are discussed. The main groups of metal oxides for charge transport include binary, ternary, and bipolar metal oxides [54,55,70–76].

7.1 Binary and ternary metal oxides

Several inorganic metal oxides such as NiO_x, ZnO, TiO_2, SnO_2, MgO, etc. as CTLs in different PeLED structures are extensively explored due to their noteworthy properties like high carrier mobility, good optical transparency, and intrinsic chemical stability (that protects the device from moisture), which improve the device stability and performance [62,68,70–76]. The most commonly used metal oxides in these CTLs are binary in nature, with very few reports on ternary metal oxide CTLs in PeLEDs.

Although ternary metal oxides (e.g., $BaSnO_3$, ZnMgO, Zn_2SnO_4, etc.) can be synthesized using facile techniques, and they exhibit improved optical and electrical properties due to their unique crystal structures, they have no significant advantages when considering the best performing devices using conventional binary oxides such as SnO_2, ZnO, or TiO_2 [74–76]. Ternary oxides can include metal-doped or modified binary oxides, and mixture of two binary oxides or ternary oxide compounds. These metal oxides can be further classified into electron transport layers (MO-ETLs) and hole transport layers (MO-HTLs) based on the polarity of their charge transfer.

Hybrid MO-CTLs are also used in PeLEDs, wherein the combination of organics and metal oxides forms a hybrid system which exploits the benefits of both the organic and inorganic structures, along with redressal of their disadvantages [77]. Fig. 8 shows the energy level diagram of some common MO-ETLs and MO-HTLs. Table 1 shows the electronic properties of some binary and ternary MO-ETLs employed in optoelectronics.

7.1.1 Metal oxide electron transport layers

Most of the widely used MO-CTLs are n-type in nature, for example, TiO_2, ZnO, SnO_2, etc. and their doped forms; and they are usually employed as ETLs in PeLEDs [74,76,89]. The bandgaps of these materials are usually large with a suitable CBM for electron injection and a deep VBM for hole blocking. Moreover, their oxides in thin film form exhibit high transparency in the wavelength region 400–900nm, thereby minimizing optical energy losses caused by the ETL [90]. Among them, TiO_2 is commonly used as an MO-ETL in high-efficiency PeLEDs, although it requires

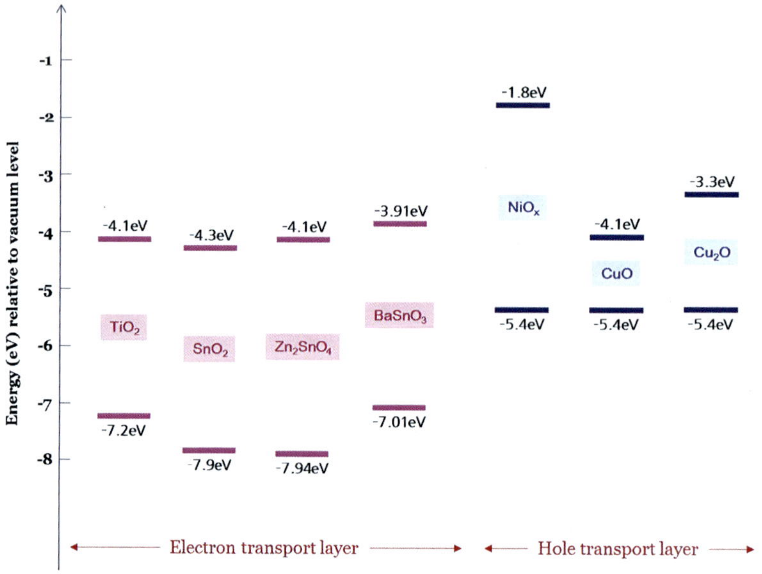

Fig. 8 Energy level diagram of some common MO-ETLs and MO-HTLs.
Reproduced with permission from S.S. Shin, S.J. Lee, S. Il Seok, Metal oxide charge transport layers for efficient and stable perovskite solar cells, Adv. Funct. Mater. 29 (2019) 1900455, https://doi.org/10.1002/adfm.201900455.

Table 1 Electronic parameters of few binary and ternary MO-ETLs used in optoelectronics.

MO-ETL	CBM (eV)	E_g (eV)	Bulk mobility μ ($cm^2 V^{-1}s^{-1}$)	Refractive index	References
TiO_2	−4.1	3.0–3.2	1	2.4–2.5	[78]
SnO_2	−4.3	3.6–4.0	250	2	[56,79]
ZnO	−4.17	3.3	200	2.2	[80,81]
Nb_2O_5	−4.25	3.4	0.2	2.1–2.4	[82]
Zn_2SnO_4	−4.1	3.8	10–30	2.0	[83]
$BaSnO_3$	−3.91	3.1	150	2.07	[84,85]
WO_x	−4.5	2.6-3.1	10–20	1.95	[86]
Cr_2O_3	−3.93	3.5	10^{-5}–1	1.4–2.1	[87,88]

high-processing temperature. However, under UV illumination, the use of TiO_2 leads to reduced device stability. Moreover, TiO_2 also exhibits comparatively low-electron mobility which can be even lesser than that of the perovskite EML [89]. In order to overcome these drawbacks, researchers have been exploring ways for possible replacement of TiO_2.

The prominent alternative MO-ETL materials with higher electron mobilities, good hole blocking effect, moderate deposition temperature, and easy film forming ability are ZnO and SnO_2 [74,76]. Compared to organic ETLs, their electron injection level can be tuned via doping, which enhances the device stability [74,76]. Due to ease of deposition and facile synthesis of ZnO in a variety of nanostructured morphologies using various techniques, it has been extensively explored as ETL in PeLEDs [76,91]. However, devices using ZnO ETL face the issue of significant surface recombination attributed to high concentration of surface defects. This can lead to deterioration in the device performance and efficiency [92]. Another limitation is when the underlying perovskite is made of organic-inorganic hybrid material, for e.g., $CH_3NH_3PbI_3$, there can be deprotonation of CH_3NH_3 in contact with ZnO, depending on its surface polarity, which is facilitated by chemisorbed species on the surface [91]. However, this may not be the case for all-inorganic PeLEDs, wherein the ZnO ETL has been observed to improve the device performance [76]. When compared to ZnO, SnO_2 is known to possess superior chemical stability [93]. Therefore, the overall device performance can be significantly improved by using SnO_2-ZnO bilayer ETL as compared to a single MO-ETL layer. This strategy is widely used in perovskite solar cells and also has a scope for implementation in PeLEDs [94,95].

Presently, bilayer ETLs which are reported in PeLEDs constitute organic materials such as 4,6-Bis(3,5-di-3-pyridylphenyl)-2-methylpyrimidine (B3PYMPM) and 2,2′,2-(1,3,5-Benzinetriyl)-tris(1-phenyl-1-H-benzimidazole) (TPBi) [96], and there are hardly any reports on bilayer MO-ETLs. Nevertheless, the benefits of SnO_2 such as facile synthesis, relatively low deposition temperature, and potential to reduce hysteresis in the devices make it a suitable candidate for replacement of TiO_2 in PeLEDs [93]. However, the performance of the device can vary, depending on several factors such as type of deposition method and processing conditions. In addition to these materials, ZnMgO is a ternary metal oxide which has also been used as ETL in all-solution processed PeLEDs. Here, nanoparticles of ZnMgO are used due to their higher CBM which is tunable for effective energy level alignment with the perovskite layer [62].

7.1.2 Metal oxide hole transport layers

Poly(3,4-ethylenedioxythiophene):polystyrene sulfonate (PEDOT:PSS) is a very common and frequently employed organic HTL in a variety of devices [42,97]. However, the major drawback of PEDOT:PSS is its acidic and hydroscopic nature that is detrimental to long-term device stability. Moreover, device performance is known to be adversely affected due to degradation caused by luminescence quenching at the PEDOT:PSS-perovskite interface [42]. This has motivated researchers to investigate alternative inorganic p-type HTL materials as potential replacement for PEDOT: PSS.

Certain p-type metal oxides such as nickel oxide (NiO_x), molybdenum oxide (MoO_3), and vanadium oxide (V_2O) have been introduced as promising HTLs in PeLEDs due to their excellent air stability, high transparency, and high carrier mobility [58,67,70,98,99]. Ternary p-type metal oxide MgNiO has also been used as HTL in all-inorganic PeLEDs [100]. Hole transporting metal oxide such as MoO_3 has also been used as interface layer between the commonly used PEDOT:PSS and Poly(9-vinylcarbazole) HTL in order to enhance the hole injection in PeLEDs [101]. Here, the interface layer is also known as electric dipole layer which generates electric dipoles due to its deep conduction band level. These electric dipoles produce strong electric fields that exponentially increase the average hopping frequency of the charge carriers. Shi et al. observed that the lifetime and stability of PeLEDs significantly improved when MgNiO was used as HTL instead of PEDOT:PSS [100].

7.1.3 Bipolar metal oxides

Bipolar metal oxide films have the potential to serve as inexpensive and environmentally stable interconnection layers for tandem optoelectronic devices, thereby minimizing the complexities in device design [54,72]. In the year 2021, highly efficient perovskite solar cells have been developed by incorporating low-temperature processed bipolar MO-CTL in their device architecture [72]. The intercalation of p-type NiO with Cs_2CO_3 resulted in a material with bipolar charge transporting ability, leading to improvement in the device performance [72].

Although the use of bipolar MO-CTLs is reported for perovskite solar cells [54], there are hardly any such reports for PeLEDs. Moreover, even in solar cells, the reports exploring this type of strategy are very few, mainly due to challenges in depositing the top metal oxide layer without harming the underlying perovskite layer. Hence, there sometimes arises a requirement of interlayers (generally organic) either between the perovskite and top CTL or between the top CTL and electrode [57]. For example, Zhang et al. demonstrated highly efficient PeLEDs fabricated by depositing a hydrophilic and insulating poly(vinylpyrrolidone) (PVP) over zinc oxide (ZnO)

ETL, where the PVP interlayer is found to suppress nonradiative recombination by compensating defects at the ZnO-perovskite interface [98].

8 Recent progress on device engineering using metal oxide layers

Innovative and effective strategies are employed to improve the efficiency and stability of PeLEDs. Out of the several approaches, device and interface engineering techniques using MO-CTLs have been prominently used in recent years for energy-efficient photon generation and improved light outcoupling in PeLEDs [36,38,102].

In 2017, all-inorganic $CsPbBr_3$-based PeLEDs have been fabricated by Shan et. al with NiO_x and ZnO as HTL and ETL, respectively [103]. These devices showed low turn-on voltage of 2.4 V due to well-aligned energy levels between the MO-CTLs and the EML, causing effective charge carrier injection. The stability of the fabricated PeLEDs using MO-CTLs was investigated under humid conditions and compared with the PeLED stability achieved using conventional organic CTLs (PEDOT:PSS and TPBi). It was observed that the MO-CTL-based all-inorganic PeLEDs showed only 30% reduction in electroluminescence intensity after 1 h 45 min, whereas the PeLEDs with conventional organic CTLs showed up to 70% degradation within 30 min under 65% humidity. Moreover, the sustainable operation of the all-inorganic PeLEDs in water without encapsulation was observed to be around ten times more than that of the organic CTL-based PeLEDs [103].

Not only is ZnO used as ETL in all-inorganic PeLEDs but also in inverted organic-inorganic hybrid PeLEDs [37]. Hoye et al. in 2015 reported PeLEDs using ZnO instead of organic ETL, where the ZnO thin film was directly deposited onto hybrid methylammonium lead tribromide ($CH_3NH_3PbBr_3$) perovskite EML using spatial atmospheric atomic layer deposition (SAALD) at 60°C, unlike spray-pyrolysis requiring temperatures >350°C [104]. There have been efforts by researchers to develop PeLEDs with solution-processed TiO_2 as ETL [74]. However, the EQE of these PeLEDs is very low (≅0.48%). Furthermore, TiO_2 thin film is prepared by sintering at high temperature (>450°C), which is not favorable for the development of low-cost, flexible, and stretchable PeLEDs [74]. Therefore, researchers have developed interest in solution-processed SnO_2 ETL which can be deposited using cost-effective and facile techniques.

Wang et al. in 2018 reported efficient n-i-p-structured PeLEDs for the first time, using solution-processed SnO_2 as ETL, and these devices were compared with the ones using ZnO ETL [74]. Here, the three-dimensional perovskites in the EML showed significantly high chemical compatibility with SnO_2 layer compared to ZnO. Moreover, the EQE of these devices is 7.9%, which is considerably high, indicating the viability of SnO_2 ETL in solution-processed PeLEDs [74].

Reports on MO-HTLs for PeLEDs are very few compared to the research on MO-ETLs. Very recently, in 2020, Zhuang et al. investigated the electroluminescence properties of PeLEDs using optimized NiO HTL [70]. They observed that, although the EQE values of the PeLEDs are low (≅0.08%–0.2%), the optimization of NiO layer by varying deposition parameters such as sputtering power, pressure, gas flow rate, etc. significantly improves the device performance, yielding two times higher EQE compared to the device with unoptimized NiO layer. The low EQE is attributed to the sputtering of NiO films directly on top of the perovskite layer, causing damage to the perovskite surface [70].

9 Metal oxide charge transport layer deposition techniques

Thin film deposition techniques for MO-CTLs are broadly classified as solution-process and vacuum deposition methods [76,83,103,105]. Other auxiliary chemical or mechanical techniques are also used for MO-CTL deposition [54,57,105,106]. It is well known that the thin film morphology, scalability, kinematics of film growth process, etc. are greatly influenced by the deposition method, thereby affecting device properties. The existing properties of a thin film can be modified and new properties can also be introduced via by the choice of deposition techniques. Proper consideration needs to be given to the adjacent layers in the device when depositing a thin film. This section gives an overview of various techniques used for the preparation and deposition of MO-CTLs in PeLEDs.

9.1 Solution-processing methods

Due to ease of deposition, solution-process methods are commonly used for the preparation of various metal oxide layers with nanostructured features [57,62,105]. These techniques are cost-effective, facile, accessible, and do not require high-vacuum and high-temperature conditions [103,105]. Among the solution-process techniques, the most widely employed method is spin-coating process. Other usual methods include sol-gel, chemical precipitation, spray-pyrolysis, chemical combustion, chemical bath deposition, dip-coating, screen printing, etc. [57,105]. Despite their remarkable advantages, solution-processed MO-CTLs face certain challenges such as nonuniform coverage of the film, damage to the perovskite layer due to solvents used, etc. [57]. In order to minimize these obstacles, several approaches such as solvent engineering, surface modification, etc. (discussed in Section 10) have been investigated by researchers [57,62,74,92].

9.2 Vacuum deposition methods

Vacuum deposition techniques can be broadly classified as physical vapor deposition (PVD) and chemical vapor deposition (CVD). The PVD techniques are those in which the particulate form of the material required to be deposited is heated in vacuum at sufficiently high temperature. The sublimation and subsequent condensation of vapors on the substrate positioned at an appropriate distance with relatively low temperature yields a thin film. Some techniques of PVD include thermal evaporation, sputtering, pulsed laser deposition (PLD), etc.

Thin films of organic small molecule CTLs produced by vacuum deposition (e.g., thermal evaporation or resistive heating) are homogenous and free of contaminants, rendering enhanced device properties when compared to solution-processed films [107,108]. This is possible since organic materials can be handled at relatively low temperatures (<200°C) [9]. However, due to high-melting temperatures of several MO-CTLs, it becomes quite challenging to deposit them as thin films via resistive heating [54]. Nevertheless, magnetron sputtering (DC/RF), PLD, and e-beam evaporation methods can be used for their deposition. It is challenging to control the exact stoichiometry of ternary metal oxides; and in order to obtain pure phase of the desired composition, it becomes advantageous to use the PLD method for deposition [109]. However, compared to solution-based processing, vapor deposition techniques such as sputtering show inferior performance and require intense optimization [83,110].

Chemical vapor deposition techniques are widely used for the growth of thin films due to their versatility for depositing several elements and compounds, and easy control of their synthesis parameters at relatively low-deposition temperature [111–113]. Thin films can be deposited with required stoichiometry in the form of vitreous and crystalline layers with high degree of purity. However, both PVD and CVD are considered to be expensive techniques due to use of vacuum, expensive targets/metal oxide precursors, etc. during deposition process [108,111]. There are other variants of CVD, which include plasma-enhanced CVD, metal-organic CVD, low-pressure CVD, etc. [54,57].

In addition to these techniques, atomic layer deposition (ALD) is another route to develop high-quality and highly uniform MO-CTLs with precise control of film thickness at atomic scale [76,114]. Moreover, this technique requires low-deposition temperature, and yields excellent film conformality as well as reproducibility over large-area samples, thereby fulfilling the requirements for preparing uniform pinhole-free MO-CTLs for PeLEDs at relatively low processing temperatures [76]. However, when ALD is used to deposit metal oxide directly on top of the perovskite layer, the metal precursor can degrade the underlying perovskite layer [76]. Only few studies have been devoted to the use of ALD technique for the processing of MO-CTLs in PeLEDs.

In order to overcome the issue of perovskite damage when depositing the top MO-CTL, SAALD technique is used, wherein oxidant H_2O is replaced using oxygen gas [104]. Surface passivation with polyethyleneimine ethoxylated (PEIE) dissolved in chlorobenzene has also been used to facilitate the growth of ALD-deposited ZnO on $CsPbBr_3$ [76]. Here, the hydroxyl groups of PEIE act as surface sites which react with the Zn precursor and minimize damage to the perovskite layer.

9.3 Other deposition methods

Low-dimensional metal oxides can be prepared by mechanochemical process (combination of physical and chemical methods) such as high-energy ball milling or ball grinding [54,115]. In this method, raw materials are ground and mixed by hard spheres in a grinding jar with periodic movements caused by high-speed rotation and vibration of the ball mill [115]. The resulting products are nano-sized compounds formed by induced chemical reactions due to reduced reaction activation energy and refined grains. Moreover, the high speed and high energy of the grinding balls induce significant rise in temperature, thereby changing the material properties and improving the crystallinity of the synthesized samples [115].

The advantage of grinding process is that it is clean, cost-effective, and suitable for mass production. In addition, the resulting metal oxide nanoparticles have fine crystallite size, and their shape and size can be controlled without changing their intrinsic properties [115]. The nanomaterial dispersion of these synthesized metal oxide nanoparticles can be spin-coated to form compact MO-CTLs in PeLEDs. The major benefits of using this technique are that PeLEDs can be fabricated on flexible substrates, and the top MO-CTL deposited via this process on the perovskite layer does not cause damage or thermal degradation [115,116].

Another effective deposition method to prepare metal oxides is electrodeposition technique [117]. This technique is versatile, cost-effective, and scalable, wherein the thickness and chemical composition of deposited films can be easily controlled by varying the deposition current. Ultra-thin films of metal oxides can be deposited using this technique by donating electrons to the ions in a solution.

The most frequently used deposition techniques for preparing MO-CTLs are shown in Fig. 9A and B. Table 2 shows the emission wavelength (λ) and EQE values of PeLEDs with selected MO-CTLs deposited using various techniques. The high EQE (14.64%) of the PeLED using NiO_x HTL can be attributed to the minimal density of traps/defects at the NiO_x/perovskite interface, as well as highly balanced charge carriers in the perovskite layer, leading to enhanced recombination output of the carriers [122].

10 Approaches for optimizing metal oxide charge transport layers

Charge carrier mobility and the overall properties of MO-CTLs in PeLEDs can be significantly improved using several approaches (or their composites) such as doping, surface and interface modification, incorporation of nanostructures, usage of multilayer or composite charge transport layers, etc. [52,54,55,57,59,70,90,116]. This section highlights the most common techniques that are used for the optimization and improvement of performance of MO-CTLs.

10.1 Doping strategy and the use of nanostructures in metal oxide charge transport layers

Doping of metal oxides with suitable elements, organic compounds, etc. is a widely used technique which is found to significantly alter device properties [55,57,123]. Doped metal oxide thin films can be manipulated/altered by incorporating different nanostructured features. Although this method has been widely used in perovskite

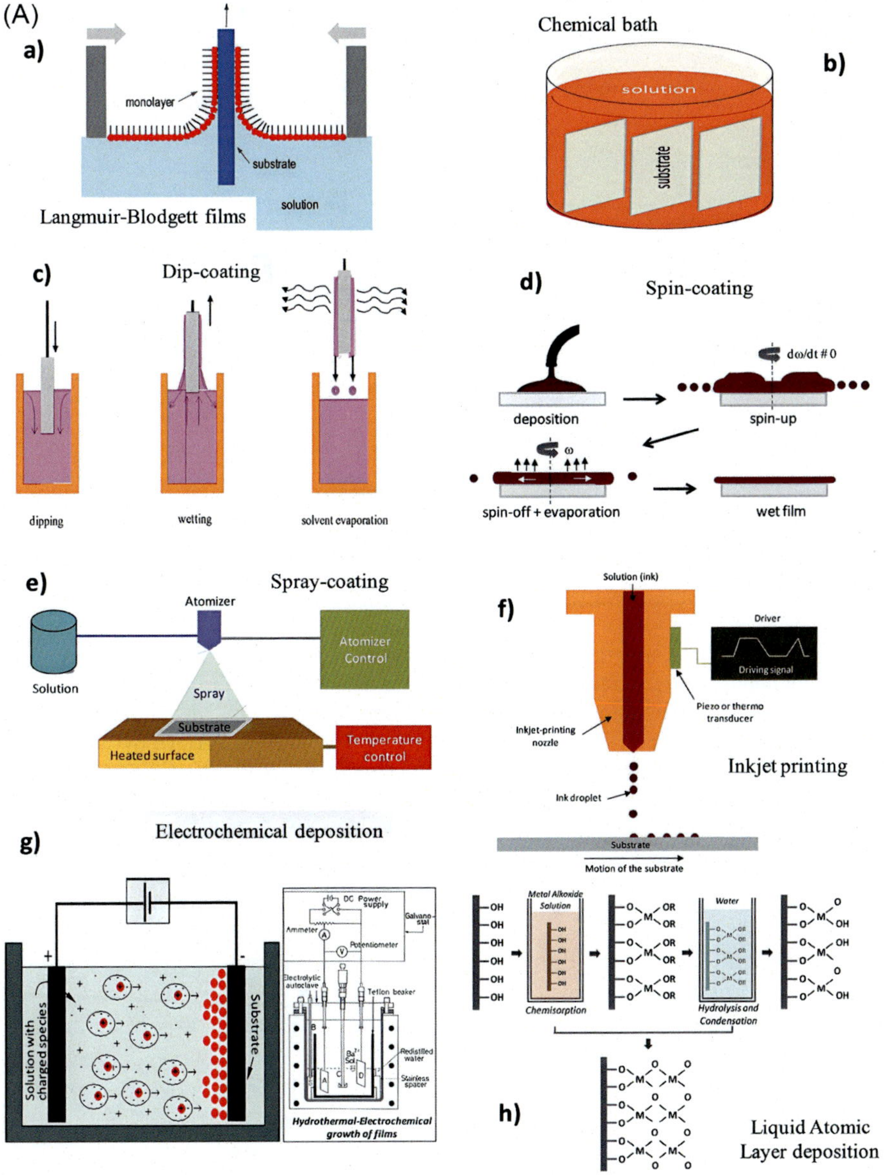

Fig. 9 See figure legend on next page.

Continued

(B)

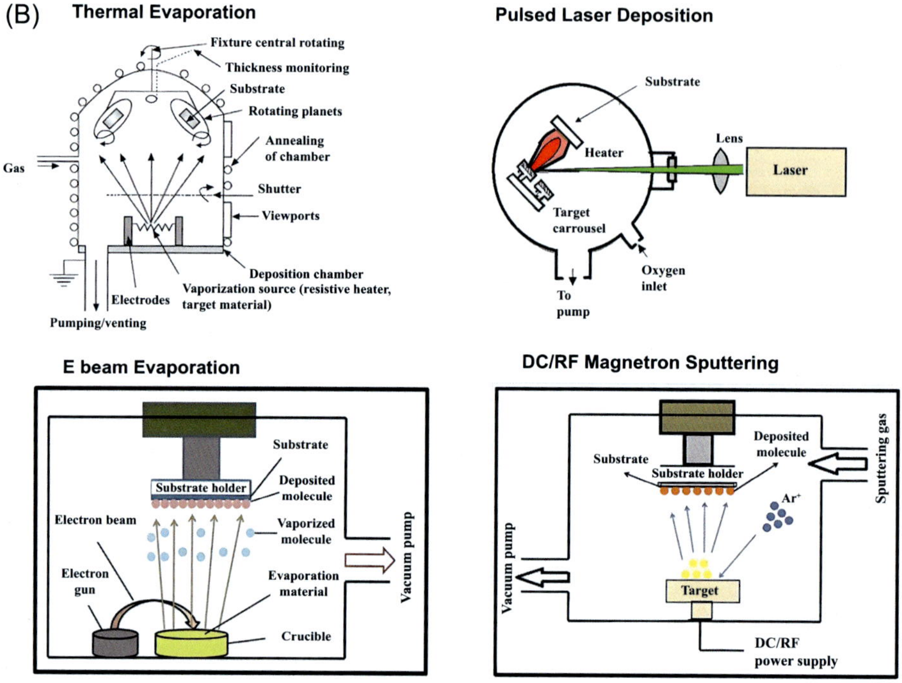

Fig. 9, Cont'd (A) Solution-processing deposition techniques used for preparing MO-CTLs. (B) Vacuum deposition techniques used for preparing MO-CTLs.
(A) Reproduced with permission from I. Bretos, R. Jiménez, J. Ricote, M.L. Calzada, Low-temperature crystallization of solution-derived metal oxide thin films assisted by chemical processes, Chem. Soc. Rev. 47 (2018) 291–308, https://doi.org/10.1039/C6CS00917D;
(B) reproduced with permission from A. Jilani, M.S. Abdel-wahab, A. Hosny Hammad, Advance deposition techniques for thin film and coating, in: Modern Technologies for Creating the Thin-film Systems and Coatings, IntechOpen (2017) https://doi.org/10.5772/65702.

solar cells [55,57], there is limited research available on doped MO-CTLs in PeLEDs. One such study is by Qasim et al., where a specially designed n-type ETL based on Ca-doped ZnO nanoparticles synthesized using sol-gel method was used in high-performance multicolor PeLEDs [123]. They observed significant enhancement in the optoelectronic properties of the PeLEDs, with nearly eight-fold increase in the EQE. More recently, in order to reduce interfacial energy loss and to simplify the device manufacturing process, solution-processed Al-doped ZnO nanoparticle-based ETL has been introduced in deep red PeLED device structure [90]. Improvement in PeLED performance is also reported by Wu et al. where NiO HTL is doped with metals such as Li and Cs in appropriate concentrations [124]. Nearly two-fold increase in the efficiency of PeLEDs with Li/Cs-doped NiO HTL has been observed when compared to that of the device with conventional PEDOT:PSS HTL [124].

Table 2 Performance of PeLEDs for selected MO-CTLs deposited using various techniques.

Functionality	MO-CTL	Deposition method	PeLED device architecture	EQE (%)	λ (nm)	Reference
MO-ETL	ZnO	RF magnetron sputtering	ITO/NiO_x/$CsPbBr_3$ QDs/ZnO/Al	0.11	516	[118]
	Mg-doped ZnO (MZO)	Spin-coating	ITO/MZO/$CsPbBr_3$ NCs/CBP/TCTA/MoO_x/Au (CBP: 4,4′-bis(carbazole-9-yl)biphenyl)	1.1	Green	[119]
	$LiTiO_2$	Hydrothermal method	FTO/$LiTiO_2$/$CsPbBr_3$ QD/p-PD/PEDOT:PSS/Au	2.38	530	[120]
	Amorphous Zn-Si-O (a-ZSO)	RF magnetron sputtering	ITO/a-ZSO/$CsPbBr_3$/NPD/MoO_x/Ag (NPD: *N*,*N*′-Di(1-naphthyl)-*N*,*N*′-diphenyl-(1,1′-biphenyl)-4,4′-diamine)	≅ 10	523	[121]
MO-HTL	NiO	RF magnetron sputtering	ITO/NiO_x/$CsPbBr_3$ QDs/ZnO/Al	0.11	516	[118]
	NiO_x	Spin-coating	ITO/NiO_x/$BA_2FA_2PbBr_{10}$ film/TPBi/LiF/Al	14.64	543	[122]

Metal oxide nanostructures are preferred due to their remarkable electrical, optical, and light outcoupling properties (structural advantage), ease of synthesis at relatively low temperatures, etc. [125–128]. A moth-eye nanostructured ZnO ETL has been reported to enhance hole injection and light outcoupling properties in PeLED device [125]. This moth-eye nanostructure was embedded at the interface between the front electrode and perovskite via soft imprinting method. More recently, the effect of Codoping in ZnO ETL was investigated by Tang et al. in 2021, where they observed significant improvement in electron mobility along with reduced exciton quenching at the ZnO-perovskite interface [126]. This is attributed to passivated oxygen vacancies and reduced electron concentration, resulting from trapped electrons due to deep impurity level induced by Co^{2+}. The size confinement effect from ZnO nanostructures also helps in widening of the bandgap. It was observed that all these factors led to 70% increase in the EQE as well as increase in the maximum luminance from 867 cd m^{-2} (for PeLEDs with undoped ZnO) to 1858 cd m^{-2} (for PeLEDs with Codoped ZnO ETL) [126].

Another recent study showed that a combination of perovskite layer and ZnO nanotubes reduces the turn-on voltage of $CH_3NH_3PbI_3$ PeLEDs (simulated using SILVACO TCAD software) from 13.7 V to 1.1 V [127]. Huang et al. demonstrated an HTL consisting of colloidal NiO_x nanoparticles in $CsPbBr_3$ PeLEDs and compared it with PeLEDs using conventional PEDOT:PSS HTL [59]. It was observed that the colloidal NiO_x nanoparticle-based HTL facilitates injection of holes into the EML by reducing the energy barrier due to the deeper HOMO (or VBM) level (−5.47 eV) of 5 nm-sized NiO_x nanoparticles; and in the process, the transmittance of the HTL is not sacrificed. Similarly, all-inorganic PeLEDs with Cu-doped NiO_x nanoparticle-based HTL have also been fabricated with maximum luminance of 1780 cd m^{-2} [128].

10.2 Surface and interface modification

Defect-engineering using various growth techniques and deposition conditions (e.g., grain size enlargement, etc.) plays a vital role in overcoming phenomena such as nonradiative recombination, hysteresis, etc. in PeLEDs [21,60,65,66,122,129]. Defect passivation is one reliable approach where a surface passivation layer is deposited on the perovskite film to suppress defect reservoirs [129]. As mentioned in the previous sections, the interface between hybrid perovskite layer and top MO-CTL is prone to instabilities such as damage by solvent used for solution-processed MO-CTLs, nonuniform coverage of the MO-CTL, etc. [130–133]. Similarly, the physicochemical properties of the bottom MO-CTL surface can affect the crystallization process of the perovskite layer deposited on top of it. This in turn affects the film quality of the perovskite EML, and eventually the performance of the device [62,67]. These issues can be mitigated by approaches such as surface passivation of MO-CTLs, surface polarity control, defect passivation, etc. [62,67,92,130–133], which can be broadly classified as 1) solvent engineering, 2) interface engineering, and 3) nonsolution processing [38]. The main principle of solvent engineering involves dispersion of target MO nanoparticles with appropriate solvents without significantly affecting the perovskite film quality [62]. Interface engineering involves either surface modification of nanoparticles [92] or the use of bilayer CTL for surface defect passivation and enhancement of its electrical conductivity [130]. Nonsolution processing involves the use of vacuum or mechanical methods.

It has been reported that the photoluminescence of $CsPbBr_3$ quantum dot film can be improved by chemical modification of the surface polarity of nanocrystalline ZnO ETL layer [92]. The ZnO surface treatment employing a self-assembled monolayer of phenethyl trichlorosilane removes hydroxyl groups on the oxide layer and generates a hydrophobic surface, thereby reducing emission quenching in the adjacent $CsPbBr_3$ quantum dot film. Moreover, by varying the concentration of phenethyl trichlorosilane, the surface polarity can be engineered; and it was shown to increase the photoluminescence quantum output by up to 50% and the thermal stability up to 140°C [92].

All-solution-processed $CsPbBr_3$ PeLEDs have been realized with all-metal oxide-CTLs where ZnMgO nanoparticles are dispersed in ethanol and spin-coated on the perovskite EML [62]. In order to protect the perovskite film from damage due to ethanol during deposition, n-butylammonium bromide and polyethylene oxide are added, resulting in significant improvement of the perovskite film quality, thereby paving a way toward realizing an all-solution-processed device structure with minimum turn-on voltage (2.5 V) and a maximum luminance of 17017 cd m^{-2} [62].

Yuan et al. have reported unique interfacial reactions due to modification of ZnO bottom layer with an ultra-thin PEIE layer which promotes perovskite crystallization during growth of the film [131]. Another study demonstrated surface modification of ZnO quantum dot ETL (produced by hydrothermal method) with oleic acid molecules which minimized agglomeration of quantum dots during spin-coating and yielded a smooth and highly transparent ETL layer, leading to enhanced PeLED performance [132].

Similarly, for effective hole injection and transport in PeLEDs, the surface of Cu_2O nanoparticles used as HIL can be modified using thiols and silane ligands, which in turn tailor the conduction and valence band edge energies, leading to suitable band alignment at the perovskite-HIL interface, which enhances the device efficiency [133]. Recently, a novel interfacial method has been demonstrated to enhance the NiO_x-perovskite interface properties [67]. Herein, the NiO_x HTL was coated with sodium dodecyl sulfate and subjected to oxygen plasma treatment up to 4 min. It was observed that this interfacial layer activates the formation of a large surface dipole, owing to the synergistic effect of dipoles due to sodium dodecyl sulfate layer and increase in the N^{3+} species concentration [67]. This leads to substantial improvement in the work function of NiO_x from 4.23 eV to 4.85 eV, and reduced energy offset between NiO_x and the perovskite from 0.69 eV to 0.38 eV, thereby improving hole injection. Moreover, it was observed that the sodium dodecyl sulfate layer subjected to oxygen plasma passivates electronic surface trap states of perovskite films and diminishes exciton quenching by NiO_x layer, resulting in negligible mass nonradiative recombination at the interface between NiO_x and perovskite layers. This was evidenced by the improvement in EQE of the $CsPbBr_3$ PeLEDs from 0.052% to 2.5%, and that of $FAPbBr_3$ nanocrystal PeLED from 5.6% to 7.6% [67]. Fig. 10 summarizes the efficiencies of devices fabricated using various concepts and strategies for the development of high-performance PeLEDs.

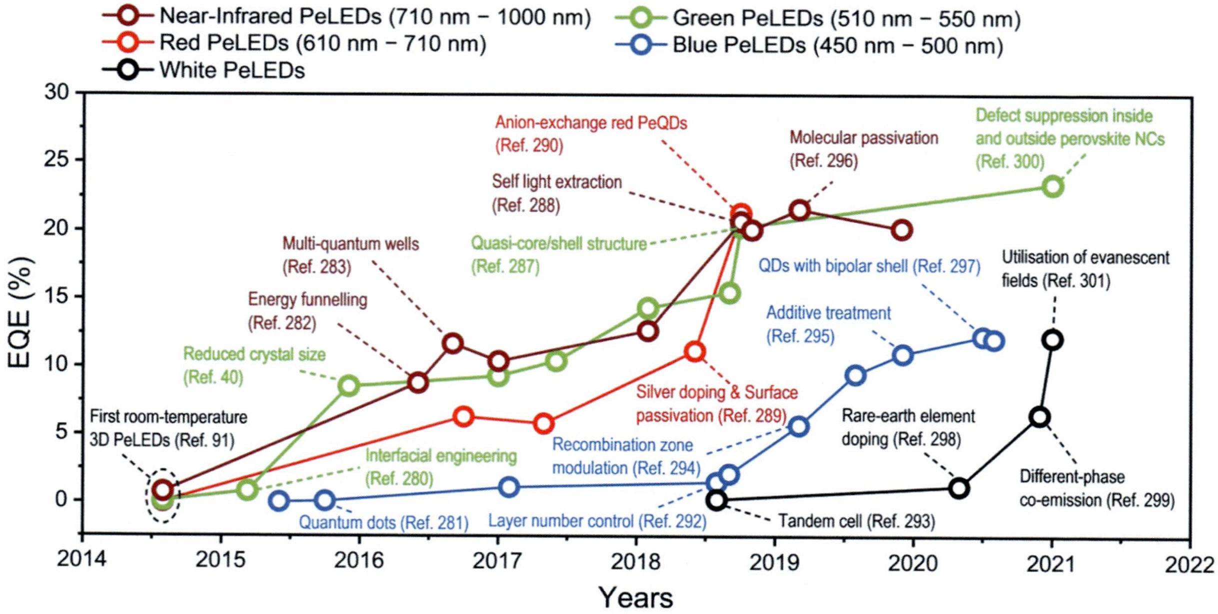

Fig. 10 An overview of PeLED efficiencies for various device engineering concepts and strategies.
Reproduced with permission from Z. Chen, Z. Li, T.R. Hopper, A.A. Bakulin, H-L. Yip, Materials, photophysics and device engineering of perovskite light-emitting diodes, Rep. Prog. Phys. 84 (2021) 046401, https://doi.org/10.1088/1361-6633/abefba.

11 Characterization techniques used for metal oxide charge transport layers

The prominently used characterization techniques for MO-CTLs (in addition to other standard characterization techniques) include [57,134–136]:

(i) X-ray photoelectron spectroscopy (XPS), ultraviolet photoelectron spectroscopy (UPS), etc. to determine the elemental composition, binding energy, charge state, and work function
(ii) Kelvin probe force microscopy (KPFM) to determine surface work function from contact potential difference
(iii) Photoluminescence spectroscopy, transient absorption for analysis of charge transfer dynamics, etc. (optical characterization)
(iv) Hall effect measurement, space charge limited current (SCLC) analysis, Mott-Schottky analysis, impedance analysis, etc. (electrical characterization)

The XPS is a quantitative and surface-sensitive spectroscopic method that can be used to measure the composition and chemical/electronic states of elements existing on the surface of the MO-CTL [57]. UPS provides details about the density of states, work function, and band gap energetics (Fig. 11A) [57,134]. KPFM is a surface-sensitive method mapping the contact potential difference between the sample surface and cantilever, providing information about the surface potential and work function [57]. The lifetime of charge carriers, luminescence decay, information about trap states, etc. can be obtained from time-resolved photoluminescence spectroscopy (TRPL) (Fig. 11B) [135]. Hall effect is a key electrical transport measurement technique to obtain significant information of the MO-CTL; for example, charge carrier type and concentration, mobility, sheet resistance, resistivity, etc. [136]. Another important technique to determine the electrical conduction and trap density at the CTL-perovskite interface is SCLC (Fig. 11C). In this technique, measurement of I-V characteristics is carried out only for electron-only (for ETLs) or hole-only (for HTLs) devices [135].

Based on the SCLC model, at low voltages, a linear ohmic response region exists, followed by a trap filling regime; and finally, the SCLC regime at higher bias voltages [135]. The voltage at which the transition from linear ohmic to the trap-filling regime takes place is known as the trap-filled limit voltage (V_{TFL}) which is used to determine the trap density (N_t) as follows [135,137]:

$$N_t = \left(\frac{2\varepsilon\varepsilon_0}{eL^2}\right)V_{\mathrm{TFL}} \tag{2}$$

where, ε is the dielectric constant of perovskite, ε_0 is the vacuum permittivity, L is the perovskite film thickness, and e is the electronic charge.

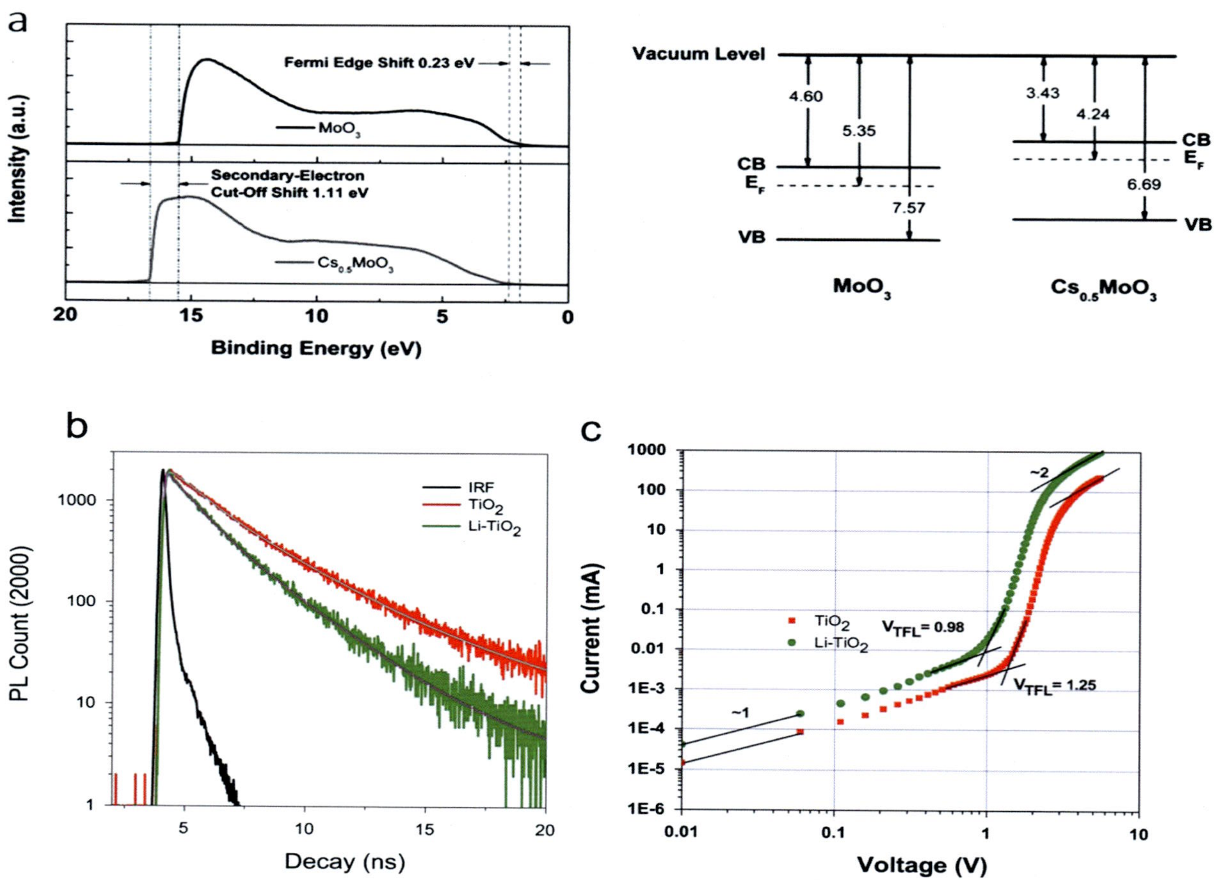

Fig. 11 (A) UPS results of MoO_3 and Cs-doped MoO_3. (B) TRPL decay curves and (C) *I–V* curves of TiO_2 and Li–TiO_2 samples.
(A) Reproduced with permission from Z. Huang, D. Ouyang, C.-J. Shih, B. Yang, W.C.H. Choy, Solution-processed ternary oxides as carrier transport/injection layers in optoelectronics, Adv. Energy. Mater. 10 (2020) 1900903, https://doi.org/10.1002/aenm.201900903; (B, C) reproduced with permission from J.H. Heo, M.S. You, M.H. Chang, W. Yin, T.K. Ahn, S.-J. Lee, S.-J. Sung, D.H. Kim, S.H. Im, Hysteresis-less mesoscopic CH3NH3PbI3 perovskite hybrid solar cells by introduction of Li-treated TiO2 electrode, Nano Energy 15 (2015) 530–539, https://doi.org/10.1016/j.nanoen.2015.05.014.

12 Charge transport dynamics at the metal oxide-perovskite interfaces

The overall PeLED device performance depends on several factors such as charge carrier injection, transport (electron and hole mobilities) and recombination, along with traps, defects, photon generation, and light outcoupling [113,138–141]. For highly efficient radiative recombination leading to the best performance of the PeLED, there should be a good balance between the holes and electrons injected into the perovskite layer [26,36,38].

This section mainly deals with the charge transport dynamics at the interfaces between perovskite EML and MO-HTL/MO-ETL. As mentioned in Eq. (1) (Section 4), the efficiency of charge injection in the PeLED can be quantified by the electrical efficiency γ; and the efficiency of radiative recombination in the presence of traps and defects is $\eta_{rad,eff}$. Considering the case of a high energy barrier at the HTL and a relatively lower barrier at the ETL, a pictorial representation of the charge transport processes at the perovskite-CTL interfaces in a PeLED is shown in Fig. 12.

As observed in Fig. 12A, electrons injected from the cathode are blocked at the HTL-perovskite interface and get accumulated there due to fast injection into the perovskite layer [38]. Meanwhile, due to the presence of energetic barriers between ITO anode and HTL (denoted as ϕ_1), and between the HTL and perovskite (denoted as ϕ_2), the holes get accumulated at these interfaces. Moreover, if the HTL has poor hole mobility, significant number of holes are confined to the HTL without getting injected into the perovskite layer. In such a situation, the accumulated electrons and holes recombine at the HTL-perovskite interface, causing a loss for radiative recombination in the perovskite, thereby causing lower EQE [36,38]. This can be avoided by incorporating an HTL with a deeper HOMO level wherein the energy barrier ϕ_1 is increased and ϕ_2 is decreased as shown in Fig. 12B.

The reduction in ϕ_2 is important because low-hole accumulation at the HTL-perovskite interface suppresses the charge carrier loss and leads to increased injection into the perovskite layer. In order to avoid the relatively high ϕ_1 due to hole aggregation at the ITO-HTL interface, a secondary HTL can be introduced in the device to result in a cascade-type energy level alignment as shown in Fig. 12C, thereby improving its performance [38].

Another way to modulate the charge transport (using different strategies) is by slowing down electron transport in the ETL and/or by accelerating hole transport in the HTL (Fig. 12D), thereby achieving charge carrier balance by way of suppressing the charge carrier loss [38]. The combined effect of energy level alignment and charge transport modulation further improves the PeLED efficiency (Fig. 12E).

The *I-V* characteristics (Fig. 12F) of PeLEDs with poor charge balance show two significant jumps attributed to asynchronous electron and hole injection into the perovskite layer, whereas the PeLEDs with better charge balance show only one jump corresponding to simultaneous electron and hole injection at a given voltage [38].

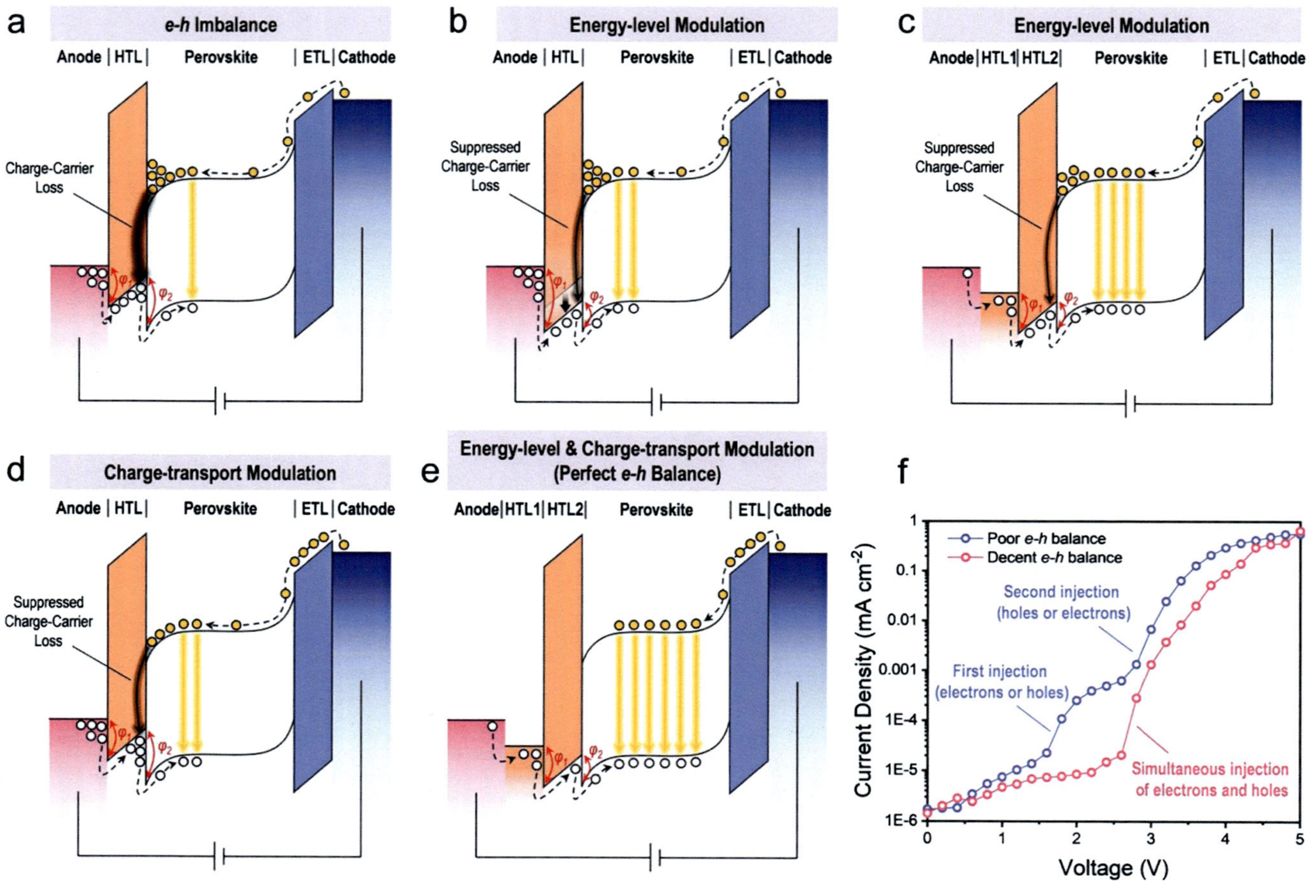

Fig. 12 (A) Energy level representation of a PeLED under operation in the case of charge imbalance. (B) Lowering of HOMO of the HTL to reduce energy barrier. (C) Hole injection improvement using cascade-type HTL. (D) Charge transport modulation of HTL and ETL. (E) Energy diagram of an operational PeLED under ideal charge balance. (F) *I–V* curves of PeLEDs with poor and moderate charge balances.
Reproduced with permission from Z. Chen, Z. Li, T.R. Hopper, A.A. Bakulin, H-L. Yip, Materials, photophysics and device engineering of perovskite light-emitting diodes, Rep. Prog. Phys. 84 (2021) 046401, https://doi.org/10.1088/1361-6633/abefba.

Similarly, the efficiency of radiative recombination $\eta_{rad,eff}$ is affected by parasitic loss of charge carriers due to (i) inefficient blocking of opposite charges by the respective CTLs, (ii) low-coverage perovskite film, and (iii) trap states [36,38,61].

If the injected charge carriers cannot efficiently recombine in the perovskite layer, and instead recombine at the CTLs or electrodes, then this leads to loss of charge carriers, and leakage current in the PeLED increases (Fig. 13A). A perovskite film with pinholes due to nonuniform coverage can lead to direct contact of the ETL with the HTL as shown in Fig. 13B, resulting in leakage current due to direct recombination of charge carriers without getting injected into the perovskite. Fig. 13C shows that charge carriers injected into the CTLs can be trapped by deep energy states present in these layers, and may probably recombine at the interfaces without detrapping, resulting in charge carrier loss [38]. All these parasitic losses can be mitigated or eliminated using the several device engineering approaches discussed in the previous sections.

13 Conclusion, challenges ahead, and perspectives for future work

MO-CTLs are extremely crucial for development of highly efficient and stable PeLEDs due to their remarkable structural, optical, and electrical properties, thereby revolutionizing the field of optoelectronics. Their ease of deposition, characterization, chemical tenability, and excellent interfacial compatibility in PeLEDs make them attractive CTL candidates.

Although MO-CTLs have contributed to remarkable progress in perovskite lighting technologies, there are still several issues which have to be resolved before commercialization of PeLEDs on a large scale. Some of the key challenges of PeLED research and their possible remedies in the foreseeable future are as follows:

(i) The realization of highly efficient blue (particularly deep-blue) PeLEDs for full-color display applications is facing major hurdles due to large carrier injection barriers, inadequate carrier blocking capabilities of CTLs, and comparatively low photoluminescence quantum yield of blue perovskite emitters. These challenges are more severe in deep-blue PeLEDs due to large bandgaps of the underlying perovskites. To solve these issues, newer charge transport layers with energy levels compatible with that of the blue emitters should be developed. Several strategies discussed in this chapter and other novel methods can also be used to manipulate and improve the properties of the existing MO-CTLs.

(ii) The challenges faced by blue PeLEDs directly impact the development of high-quality white PeLEDs, which is a research area still in its infancy. White PeLEDs have great potential in energy-efficient and flexible or stretchable lighting applications; and there is wide scope for research in this direction, once the challenges in blue PeLEDs are tackled.

(iii) The overall device performance is adversely affected by parasitic current losses in PeLEDs. Novel light outcoupling strategies such as incorporation of textured/patterned metal oxide interfaces can be used to address this issue.

(iv) Short device lifetimes, instability, and degradation in PeLEDs are the main obstacles for their commercialization. More systematic studies should be carried out on the metal halide perovskite materials to fully understand their basic ionic characteristics and address the issue of ion transport in PeLEDs.

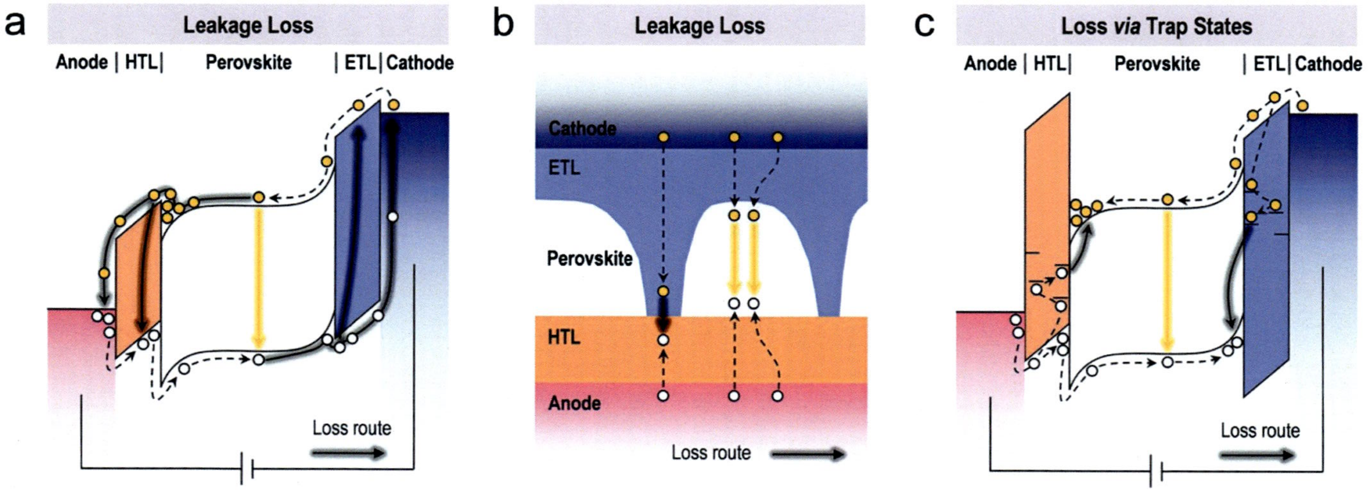

Fig. 13 (A) Leakage loss of charge carriers due to incapable blocking properties of the CTLs, (B) for a nonuniform perovskite film, and (C) loss of charge carriers via trap states in the CTLs.

Reproduced with permission from Z. Chen, Z. Li, T.R. Hopper, A.A. Bakulin, H-L. Yip, Materials, photophysics and device engineering of perovskite light-emitting diodes, Rep. Prog. Phys. 84 (2021) 046401, https://doi.org/10.1088/1361-6633/abefba.

In spite of the several obstacles existing at present in the field of PeLED research, this technology has seen tremendous growth in a short span of time; and possesses great potential for next-generation large-area flexible solid-state display and lighting applications.

References

[1] S.-J. Zou, Y. Shen, F.-M. Xie, J.-D. Chen, Y.-Q. Li, J.-X. Tang, Recent advances in organic light-emitting diodes: toward smart lighting and displays, Mater. Chem. Front. 4 (2020) 788–820, https://doi.org/10.1039/C9QM00716D.

[2] A. Ren, H. Wang, W. Zhang, J. Wu, Z. Wang, R.V. Penty, I.H. White, Emerging light-emitting diodes for next-generation data communications, Nature Electron. 4 (2021) 559–572, https://doi.org/10.1038/s41928-021-00624-7.

[3] M. V Dambhare, B. Butey, S.V. Moharil (2021) Solar photovoltaic technology: A review of different types of solar cells and its future trends, J. Phys.: Conf. Ser. 1913, 012053, https://doi.org/10.1088/1742-6596/1913/1/012053.

[4] E. Fortunato, P. Barquinha, R. Martins, Oxide semiconductor thin-film transistors: a review of recent advances, Adv. Mater. 24 (2012) 2945–2986, https://doi.org/10.1002/adma.201103228.

[5] M. Brzezicki, A systematic review of the most recent concepts in smart windows technologies with a focus on electrochromics, Sustainability 13 (17) (2021) 9604, https://doi.org/10.3390/su13179604.

[6] Z.T. Ardakani, O. Hosu, C. Cristea, M.M. Ardakani, G. Marrazza, Latest trends in electrochemical sensors for neurotransmitters: a review, Sensors 19 (9) (2019) 2037, https://doi.org/10.3390/s19092037.

[7] M. Nasseri, T.P. Attia, B. Joseph, N.M. Gregg, E.S. Nurse, P.F. Viana, G. Worrell, M. Dümpelmann, M.P. Richardson, D.R. Freestone, B.H. Brinkmann, Ambulatory seizure forecasting with a wrist-worn device using long-short term memory deep learning, Sci. Rep. 11 (2021) 21935, https://doi.org/10.1038/s41598-021-01449-2.

[8] M. Aboubakar, M. Kellil, P. Roux, A review of IoT network management: current status and perspectives, J. King Saud Univ. - Comput. Inform. Sci. (2021), https://doi.org/10.1016/j.jksuci.2021.03.006.

[9] W. Brütting, C. Adachi, Physics of Organic Semiconductors, second ed., 2012, ISBN: 978-3-527-41053-8. Completely New Revised Edition.

[10] M.V. Jacob, Organic semiconductors: past, present and future, Electronics 3 (4) (2014) 594–597, https://doi.org/10.3390/electronics3040594.

[11] S. Chen, J. Yu, Introduction to organic light-emitting display technologies, Frontiers in Electrical Engineering, Volume 1: Active-Matrix Organic Light-Emitting Display Technologies, Bentham Science, 2015, pp. 3–39.

[12] N. Heydari, S.M.B. Ghorashi, W. Han, H.-H. Park, Quantum dot-based light emitting diodes (QDLEDs): new progress, IntechOpen (2017), https://doi.org/10.5772/intechopen.69014.

[13] Z.-K. Tan, R.S. Moghaddam, M.L. Lai, P. Docampo, R. Higler, F. Deschler, M. Price, A. Sadhanala, L.M. Pazos, D. Credgington, F. Hanusch, T. Bein, H.J. Snaith, R.H. Friend, Bright light-emitting diodes based on organometal halide perovskite, Nat. Nanotechnol. 9 (2014) 687–692, https://doi.org/10.1038/nnano.2014.149.

[14] Y. Cao, N. Wang, H. Tian, J. Guo, Y. Wei, H. Chen, Y. Miao, W. Zou, K. Pan, Y. He, H. Cao, Y. Ke, M. Xu, Y. Wang, M. Yang, K. Du, Z. Fu, D. Kong, D. Dai, Y. Jin, G. Li, H. Li, Q. Peng, J. Wang, W. Huang, Perovskite light-emitting diodes based on

spontaneously formed submicrometre-scale structures, Nature 562 (2018) 249–253,- https://doi.org/10.1038/s41586-018-0576-2.

[15] K. Lin, J. Xing, L.N. Quan, F.P.G. de Arquer, X. Gong, J. Lu, L. Xie, W. Zhao, D. Zhang, C. Yan, W. Li, X. Liu, Y. Lu, J. Kirman, E.H. Sargent, Q. Xiong, Z. Wei, Perovskite light-emitting diodes with external quantum efficiency exceeding 20 percent, Nature 562 (2018) 245–248, https://doi.org/10.1038/s41586-018-0575-3.

[16] Y. He, J. Yan, L. Xu, B. Zhang, Q. Cheng, Y. Cao, J. Zhang, C. Tao, Y. Wei, K. Wen, Z. Kuang, G.M. Chow, Z. Shen, Q. Peng, W. Huang, J. Wang, Perovskite light-emitting diodes with near unit internal quantum efficiency at low temperatures, Adv. Mater. 33 (2021) 2006302, https://doi.org/10.1002/adma.202006302.

[17] S. Wu, Z. Chen, H.-L. Yip, A.K.-Y. Jen, The evolution and future of metal halide perovskite-based optoelectronic devices, Matter 4 (2021) 3814–3834, https://doi.org/10.1016/j.matt.2021.10.026.

[18] Y. Zhou, M. Saliba, Zooming in on metal halide perovskites: new energy frontiers emerge, ACS Energy Lett. 6 (2021) 2750–2754, https://doi.org/10.1021/acsenergylett.1c01281.

[19] L.M. Herz, Charge-carrier mobilities in metal halide perovskites: fundamental mechanisms and limits, ACS Energy Lett. 2 (2017) 1539–1548, https://doi.org/10.1021/acsenergylett.7b00276.

[20] C. Motta, F. El-Mellouhi, S. Sanvito, Charge carrier mobility in hybrid halide perovskites, Sci. Rep. 5 (2015) 12746, https://doi.org/10.1038/srep12746.

[21] X. Zhao, Z.-K. Tan, Large-area near-infrared perovskite light-emitting diodes, Nat. Photon. 14 (2020) 215–218, https://doi.org/10.1038/s41566-019-0559-3.

[22] R. Liu, K. Xu, Blue perovskite light-emitting diodes (LEDs): a minireview, Instrum. Sci. Technol. 48 (2020) 616–636, https://doi.org/10.1080/10739149.2020.1762643.

[23] J. Lu, Z. Wei, The strategies for preparing blue perovskite light-emitting diodes, J. Semicond. 41 (2020) 051203, https://doi.org/10.1088/1674-4926/41/5/051203.

[24] Y. Yang, S. Xu, Z. Ni, C.H. Van Brackle, L. Zhao, X. Xiao, X. Dai, J. Huang, Highly efficient pure-blue light-emitting diodes based on rubidium and chlorine alloyed metal halide perovskite, Adv. Mater. 33 (2021) 2100783, https://doi.org/10.1002/adma.202100783.

[25] Z. Li, K. Cao, J. Li, Y. Tang, X. Ding, B. Yu, Review of blue perovskite light emitting diodes with optimization strategies for perovskite film and device structure, Opto-Electron. Adv. 4 (2021) 200019, https://doi.org/10.29026/oea.2021.200019.

[26] M. Worku, A. Ben-Akacha, T.B. Shonde, H. Liu, B. Ma, The past, present, and future of metal halide perovskite light-emitting diodes, Small Sci. 1 (2021) 2000072, https://doi.org/10.1002/smsc.202000072.

[27] T.G. Liashenko, A.P. Pushkarev, A. Naujokaitis, V. Pakštas, M. Franckevičius, A.A. Zakhidov, S.V. Makarov, Suppression of electric field-induced segregation in sky-blue perovskite light-emitting electrochemical cells, Nanomaterials 10 (2020) 1937, https://doi.org/10.3390/nano10101937.

[28] Q. Chen, N.D. Marco, Y.M. Yang, T.-B. Song, C.C. Chen, H. Zhao, Z. Hong, H. Zhou, Y. Yang, Under the spotlight: The organic—inorganic hybrid halide perovskite for optoelectronic applications, Nanotoday 10 (2015) 355–396, https://doi.org/10.1016/j.nantod.2015.04.009.

[29] S. Krishnamurthy, P. Pandey, J. Kaur, S. Chakraborty, P.K. Nayak, A. Sadhanala, S. Ogale, Organic–inorganic hybrid and inorganic halide perovskites: structural and chemical engineering, interfaces and optoelectronic properties, J. Phys. D Appl. Phys. 54 (2021) 133002, https://doi.org/10.1088/1361-6463/abd0ad.

[30] X. Li, X. Gao, X. Zhang, X. Shen, M. Lu, J. Wu, Z. Shi, V.L. Colvin, J. Hu, X. Bai, W.W. Yu, Y. Zhang, Lead-free halide perovskites for light emission: recent advances and perspectives, Adv. Sci. 8 (2021) 2003334, https://doi.org/10.1002/advs.202003334.
[31] J. Lu, C. Yan, W. Feng, X. Guan, K. Lin, Z. Wei, Lead-free metal halide perovskites for light-emitting diodes, Eco Mat. 3 (2021) e12082, https://doi.org/10.1002/eom2.12082.
[32] Y. Gao, Y. Pan, F. Zhou, G. Niu, C. Yan, Lead-free halide perovskites: a review of the structure–property relationship and applications in light emitting devices and radiation detectors, J. Mater. Chem. A 9 (2021) 11931–11943, https://doi.org/10.1039/D1TA01737C.
[33] R. Wang, J. Wang, S. Tan, Y. Duan, Z.-K. Wang, Y., Yang opportunities and challenges of lead-free perovskite optoelectronic devices, Trends Chem. 1 (2019) 368–379, https://doi.org/10.1016/j.trechm.2019.04.004.
[34] D. Yang, G. Zhang, R. Lai, Y. Cheng, Y. Lian, M. Rao, D. Huo, D. Lan, B. Zhao, D. Di, Germanium-lead perovskite light-emitting diodes, Nat. Commun. 12 (2021) 4295, https://doi.org/10.1038/s41467-021-24616-5.
[35] P. Xiao, Y. Yu, J. Cheng, Y. Chen, S. Yuan, J. Chen, J. Yuan, B. Liu, Advances in perovskite light-emitting diodes possessing improved lifetime, Nanomaterials 11 (1) (2021) 103, https://doi.org/10.3390/nano11010103.
[36] D. Luo, Q. Chen, Y. Qiu, M. Zhang, B. Liu, Device engineering for all-inorganic perovskite light-emitting diodes, Nanomaterials 9 (7) (2019) 1007, https://doi.org/10.3390/nano9071007.
[37] M. Li, J. Wang, C. Mai, Y. Cun, B. Zhang, G. Huang, D. Yu, J. Li, L. Mu, L. Cao, D. Li, J. Wang, J. Wang, J. Peng, Bifacial passivation towards efficient $FAPbBr_3$-based inverted perovskite light-emitting diodes, Nanoscale 12 (2020) 14724–14732, https://doi.org/10.1039/D0NR02323J.
[38] Z. Chen, Z. Li, T.R. Hopper, A.A. Bakulin, H.-L. Yip, Materials, photophysics and device engineering of perovskite light-emitting diodes, Rep. Prog. Phys. 84 (2021) 046401, https://doi.org/10.1088/1361-6633/abefba.
[39] K. Ji, M. Anaya, S.D. Stranks, Perovskite Light—Emitting Diode Technologies, Wiley, 2021, https://doi.org/10.1002/9783527826391.ch12. Online ISBN:9783527826391.
[40] K. Ji, M. Anaya, A. Abfalterer, S.D. Stranks, Halide perovskite light-emitting diode technologies, Adv. Opt. Mater. 9 (2021) 2002128, https://doi.org/10.1002/adom.202002128.
[41] Y.-H. Kim, J.S. Kim, T.-W. Lee, Strategies to improve luminescence efficiency of metal-halide perovskites and light-emitting diodes, Adv. Mater. 31 (2019) 1804595, https://doi.org/10.1002/adma.201804595.
[42] Y. Meng, M. Ahmadi, X. Wu, T. Xu, L. Xu, Z. Xiong, P. Chen, High performance and stable all-inorganic perovskite light emitting diodes by reducing luminescence quenching at PEDOT:PSS/perovskites interface, Org. Electron. 64 (2019) 47–53,-https://doi.org/10.1016/j.orgel.2018.10.014.
[43] J. Zhang, L. Jiang, J. Huang, X. Luo, Z. Luo, D. Zhou, H.-S. Kwok, P. Xu, G. Li, Unravelling the role of band-offset landscape on the recombination zone dynamics in perovskite light-emitting diodes, Nanoselect 2 (2021) 624–631, https://doi.org/10.1002/nano.202000154.
[44] V. Prakasam, F.D. Giacomo, R. Abbel, D. Tordera, M. Sessolo, G. Gelinck, H.J. Bolink, Efficient perovskite light emitting diodes: effect of composition, morphology and transport layers, ACS Appl. Mater. Interfaces 10 (2018) 41586–41591, https://doi.org/10.1021/acsami.8b15718.

[45] J.C. Yu, J.H. Park, S.Y. Lee, M.H. Song, Effect of perovskite film morphology on device performance of perovskite light-emitting diodes, Nanoscale 11 (2019) 1505–1514,- https://doi.org/10.1039/C8NR08683D.

[46] J.-E. Jeong, J.H. Park, C.H. Jang, M.H. Song, H.Y. Woo, Multifunctional charge transporting materials for perovskite light-emitting diodes, Adv. Mater. 32 (2020) 2002176, https://doi.org/10.1002/adma.202002176.

[47] S. Zhuang, X. Ma, D. Hu, X. Dong, B. Zhang, Air-stable all inorganic green perovskite light emitting diodes based on ZnO/CsPbBr3/NiO heterojunction structure, Ceram. Int. 44 (2018) 4685–4688, https://doi.org/10.1016/j.ceramint.2017.12.048.

[48] M. Righetto, F. Cacialli, Chapter 5 Charge Transport Layers in Halide Perovskite Photonic Devices, AIP Publishing, 2021, https://doi.org/10.1063/9780735423633_005. ISBN (Online): 978-0-7354-2363-3.

[49] X.-K. Liu, W. Xu, S. Bai, Y. Jin, J. Wang, R.H. Friend, F. Gao, Metal halide perovskites for light-emitting diodes, Nat. Mater. 20 (2021) 10–21, https://doi.org/10.1038/s41563-020-0784-7.

[50] L. Tang, J. Qiu, Q. Wei, H. Gu, B. Du, H. Du, W. Hui, Y. Xia, Y. Chen, W. Huang, Enhanced performance of perovskite light-emitting diodes via diamine interface modification, ACS Appl. Mater. Interfaces 11 (2019) 29132–29138, https://doi.org/10.1021/acsami.9b11866.

[51] L. Zhao, K. Roh, S. Kacmoli, K. Al Kurdi, S. Jhulki, S. Barlow, S.R. Marder, C. Gmachl, B.P. Rand, Thermal management enables bright and stable perovskite light-emitting diodes, Adv. Mater. 32 (2020) 2000752, https://doi.org/10.1002/adma.202000752.

[52] Y. Lu, Z. Wang, J. Chen, Y. Peng, X. Tang, Z. Liang, F. Qi, W. Chen, Tuning hole transport layers and optimizing perovskite films thickness for high efficiency CsPbBr3 nanocrystals electroluminescence light-emitting diodes, JOL 234 (2021) 117952, https://doi.org/10.1016/j.jlumin.2021.117952.

[53] Y. Liu, T. Wu, Y. Liu, T. Song, B. Sun, Suppression of non-radiative recombination toward high efficiency perovskite light emitting diodes, APL Mater. 7 (2019) 021102, https://doi.org/10.1063/1.5064370.

[54] M. Singh, C.W. Chu, A. Ng, Perspective on predominant metal oxide charge transporting materials for high-performance perovskite solar cells, Front. Mater. 8 (2021) 1–9, https://doi.org/10.3389/fmats.2021.655207.

[55] S.S. Shin, S.J. Lee, S. Il Seok, Metal oxide charge transport layers for efficient and stable perovskite solar cells, Adv. Funct. Mater. 29 (2019) 1900455, https://doi.org/10.1002/adfm.201900455.

[56] S. Rajendran, J. Qin, F. Gracia, E. Lichtfouse, Metal and Metal Oxides For Energy and Electronics, Springer, 2021, https://doi.org/10.1007/978-3-030-53065-5. ISBN: 978-3-030-53065-5.

[57] Y. Wang, A.B. Djurišić, W. Chen, F. Liu, R. Cheng, S.P. Feng, A.M.C. Ng, Z. He, Metal oxide charge transport layers in perovskite solar cells—optimising low temperature processing and improving the interfaces towards low temperature processed, efficient and stable devices, J. Phys. Energy 3 (2021) 012004, https://doi.org/10.1088/2515-7655/abc73f.

[58] Z. Wang, Z. Luo, C. Zhao, Q. Guo, Y. Wang, F. Wang, X. Bian, A. Alsaedi, T. Hayat, Z. Tan, Efficient and stable pure green all-inorganic perovskite CsPbBr3 light-emitting diodes with solution-processed NiOx interlayer, J. Phys. Chem. C 121 (2017) 28132–28138, https://doi.org/10.1021/acs.jpcc.7b11518.

[59] C.-Y. Huang, S.-P. Chang, A.G. Ansay, Z.-H. Wang, C.-C. Yang, Ambient-processed, additive-assisted CsPbBr perovskite light-emitting diodes with colloidal NiOx

nanoparticles for efficient hole transporting, Coatings 10 (4) (2020) 336, https://doi.org/10.3390/coatings10040336.
[60] T. Hu, D. Li, Q. Shan, Y. Dong, H. Xiang, W.C.H. Choy, H. Zeng, Defect behaviors in perovskite light-emitting diodes, ACS Mater. Lett. 3 (2021) 1702–1728, https://doi.org/10.1021/acsmaterialslett.1c00474.
[61] H. Jin, E. Debroye, M. Keshavarz, I.G. Scheblykin, M.B.J. Roeffaers, J. Hofkens, J.A. Steele, It's a trap! on the nature of localised states and charge trapping in lead halide perovskites, Mater. Horiz. 7 (2020) 397–410, https://doi.org/10.1039/C9MH00500E.
[62] L. Liu, Z. Wang, W. Sun, J. Zhang, S. Hu, T. Hayat, A. Alsaedi, Z. Tan, All-solution-processed perovskite light-emitting diodes with all metal oxide transport layers, Chem. Commun. 54 (2018) 13283–13286, https://doi.org/10.1039/C8CC07821A.
[63] A.H. Hammad, M.S.A. Wahab, S. Vattamkandathil, A.R. Ansari, Growth and correlation of the physical and structural properties of hexagonal nanocrystalline nickel oxide thin films with film thickness, Coatings 9 (10) (2019) 615, https://doi.org/10.3390/coatings9100615.
[64] Y. Gong, S. Zhang, H. Gao, Z. Ma, S. Hu, Z. Tan, Recent advances and comprehensive insights on nickel oxide in emerging optoelectronic devices, Sustain. Energy Fuels 4 (2020) 4415–4458, https://doi.org/10.1039/D0SE00621A.
[65] X. Qiu, Y. Liu, W. Li, Y. Hu, Traps in metal halide perovskites: characterization and passivation, Nanoscale 12 (2020) 22425–22451, https://doi.org/10.1039/D0NR05739H.
[66] M.H. Futscher, C. Deibel, Defect spectroscopy in halide perovskites is dominated by ionic rather than electronic defects, ACS Energy Lett. 7 (2022) 140–144, https://doi.org/10.1021/acsenergylett.1c02076.
[67] H. Wang, H. Yuan, J. Yu, C. Zhang, K. Li, M. You, W. Li, J. Shao, J. Wei, X. Zhang, R. Chen, X. Yang, W. Zhao, Boosting the efficiency of NiOx-based perovskite light-emitting diodes by interface engineering, ACS Appl. Mater. Interfaces 12 (2020) 53528–53536, https://doi.org/10.1021/acsami.0c16139.
[68] H.-B. Kim, Y.J. Yoon, J. Jeong, J. Heo, H. Jang, J.H. Seo, B. Walker, J.Y. Kim, Peroptronic devices: perovskite-based light-emitting solar cells, Energ. Environ. Sci. 10 (2017) 1950–1957, https://doi.org/10.1039/C7EE01666B.
[69] N. Koch Dr, Organic electronic devices and their functional interfaces, ChemPhysChem 8 (2007) 1438–1455, https://doi.org/10.1002/cphc.200700177.
[70] S. Zhuang, J. He, X. Ma, Y. Zhao, H. Wang, B. Zhang, Fabrication and optimization of hole transport layer NiO for all inorganic perovskite light emitting diodes, Mater. Sci. Semicond. Process. 109 (2020) 104924, https://doi.org/10.1016/j.mssp.2020.104924.
[71] S.S. Shin, S.J. Lee, S.I. Seok, Exploring wide bandgap metal oxides for perovskite solar cells, APL Mater. 7 (2019) 022401, https://doi.org/10.1063/1.5055607.
[72] M. Singh, R.-T. Yang, D.-W. Weng, H. Hu, A. Singh, A. Mohapatra, Y.-T. Chen, Y.-J. Lu, T.-F. Guo, G. Li, H.-C. Lin, C.W. Chu, Low-temperature processed bipolar metal oxide charge transporting layers for highly efficient perovskite solar cells, Sol. Energ. Mater. Sol. Cells 221 (2021) 110870, https://doi.org/10.1016/j.solmat.2020.110870.
[73] M. Thambidurai, F. Shini, P.C. Harikesh, N. Mathews, C. Dang, Highly stable and efficient planar perovskite solar cells using ternary metal oxide electron transport layers, J. Power Sources 448 (2020) 227362, https://doi.org/10.1016/j.jpowsour.2019.227362.
[74] H. Wang, H. Yu, W. Xu, Z. Yuan, Z. Yan, C. Wang, X. Liu, M. Fahlman, J.-M. Liu, X.-K. Liu, F. Gao, Efficient perovskite light-emitting diodes based on a solution-processed tin dioxide electron transport layer, J. Mater. Chem. C 6 (2018) 6996–7002, https://doi.org/10.1039/C8TC01871E.

[75] S. Rhee, K. An, K.-T. Kang, Recent advances and challenges in halide perovskite crystals in optoelectronic devices from solar cells to other applications, Crystals 11 (1) (2021) 39, https://doi.org/10.3390/cryst11010039.
[76] W. Li, Y.-X. Xu, D. Wang, F. Chen, Z.-K. Chen, Inorganic perovskite light emitting diodes with ZnO as the electron transport layer by direct atomic layer deposition, Org. Electron. 57 (2018) 60–67, https://doi.org/10.1016/j.orgel.2018.02.032.
[77] D.B. Kim, J.C. Yu, Y.S. Nam, D.W. Kim, E.D. Jung, S.Y. Lee, S. Lee, J.H. Park, A.-Y. Lee, B.R. Lee, D.D. Nuzzo, R.H. Friend, M. Hoon Song, Improved performance of perovskite light-emitting diodes using PEDOT:PSS and MoO3 composite layers, J. Mater. Chem. C 4 (2016) 8161–8165, https://doi.org/10.1039/C6TC02099B.
[78] J. Aarik, A. Aidla, V. Sammelselg, T. Uustare, M. Ritala, M. Leskelä, Characterization of titanium dioxide atomic layer growth from titanium ethoxide and water, Thin Solid Films 370 (1–2) (2000) 163–172, https://doi.org/10.1016/S0040-6090(00)00911-1.
[79] Q. Jiang, X. Zhang, J. You, SnO_2: a wonderful electron transport layer for perovskite solar cells, Small 14 (2018) 1801154, https://doi.org/10.1002/smll.201801154.
[80] S.W. Xue, X.T. Zu, W.L. Zhou, H.X. Deng, X. Xiang, L. Zhang, H. Deng, Effects of post-thermal annealing on the optical constants of ZnO thin film, J. Alloys Compd. 448 (2008) 21–26, https://doi.org/10.1016/j.jallcom.2006.10.076.
[81] D.C. Look, D.C. Reynolds, J.R. Sizelove, R.L. Jones, C.W. Litton, G. Cantwell, W.C. Harsch, Electrical properties of bulk ZnO, Solid State Commun. 105 (1998) 399–401,-https://doi.org/10.1016/S0038-1098(97)10145-4.
[82] N. Özer, D.-G. Chen, C.M. Lampert, Preparation and properties of spin-coated Nb_2O_5 films by the sol-gel process for electrochromic applications, Thin Solid Films 277 (1996) 162–168, https://doi.org/10.1016/0040-6090(95)08011-2.
[83] D.L. Young, H. Moutinho, Y. Yan, T.J. Coutts, Growth and characterization of radio frequency magnetron sputter-deposited zinc stannate, Zn_2SnO_4, thin films, J. Appl. Phys. 92 (2002) 310, https://doi.org/10.1063/1.1483104.
[84] N.P. Reddy, R. Santhosh, J.M. Fernandes, R. Muniramaiah, B. Murali, D.P. Joseph, Nanocrystalline Sb-doped-BaSnO3 perovskite electron transport layer for dye-sensitized solar cells, Mater. Lett. 311 (2022) 131629, https://doi.org/10.1016/j.matlet.2021.131629.
[85] D. Cherrad, M. Maouche, M. Maamache, L. Krache, Influence of valence electron concentration on elastic, electronic and optical properties of the alkaline-earth tin oxides ASnO3 (A= Ca, Sr and Ba): a comparative study with ASnO3 compounds, Phys. B 406 (2011) 2714–2722, https://doi.org/10.1016/j.physb.2011.04.014.
[86] S.-M. Lee, C.S. Choi, K.C. Choi, H.-C. Lee, Low resistive transparent and flexible ZnO/Ag/ZnO/Ag/WO electrode for organic light-emitting diodes, Org. Electron. 13 (2012) 1654–1659, https://doi.org/10.1016/j.orgel.2012.05.014.
[87] J.A. Crawford, R.W. Vest, Electrical conductivity of single-crystal Cr2O3, J. Appl. Phys. 35 (1964) 2413, https://doi.org/10.1063/1.1702871.
[88] M. Julkarnain, J. Hossain, K.S. Sharif, K.A. Khan, Optical properties of thermally evaporated Cr2O3 thin films, Can. J. Chem. Eng. Tech. 3 (2012) 81–85.
[89] C. Zhao, D. Zhang, C. Qin, Perovskite light-emitting diodes, CCS Chem. 2 (2020) 859–869, https://doi.org/10.31635/ccschem.020.202000216.
[90] H. Shi, Z. Wang, H. Ma, H. Jia, F. Wang, C. Zou, S. Hu, H. Li, Z. Tan, High-efficiency red perovskite light-emitting diodes based on collaborative optimization of emission layer and transport layers, J. Mater. Chem. C 9 (2021) 12367–12373, https://doi.org/10.1039/D1TC02504J.

[91] J. Zeng, Y. Qi, Y. Liu, D. Chen, Z. Ye, Y. Jin, ZnO-based electron-transporting layers for perovskite light-emitting diodes: controlling the interfacial reactions, J. Phys. Chem. Lett. 13 (2022) 694–703, https://doi.org/10.1021/acs.jpclett.1c04117.
[92] X. Wang, X. Chen, X. Wang, J. Hu, Y. Wu, W.-H. Zhang, Surface polarity engineering of ZnO layer for improved photoluminescence of $CsPbBr_3$ quantum dot films, Chem. Phys. Lett. 750 (2020) 137454, https://doi.org/10.1016/j.cplett.2020.137454.
[93] G.K. Dalapati, H. Sharma, A. Guchhait, N. Chakrabarty, P. Bamola, Q. Liu, G. Saianand, A.M.S. Krishna, S. Mukhopadhyay, A. Dey, T.K.S. Wong, S. Zhuk, S. Ghosh, S. Chakrabortty, C. Mahata, S. Biring, A. Kumar, C.S. Ribeiro, S. Ramakrishna, A.K. Chakraborty, S. Krishnamurthy, P. Sonar, M. Sharma, Tin oxide for optoelectronic, photovoltaic and energy storage devices: a review, J. Mater. Chem. A 9 (2021) 16621–16684, https://doi.org/10.1039/D1TA01291F.
[94] Y.W. Noh, I.S. Jin, K.S. Kim, S.H. Park, J.W. Jung, Reduced energy loss in SnO2/ZnO bilayer electron transport layer-based perovskite solar cells for achieving high efficiencies in outdoor/indoor environments, J. Mater. Chem. A 8 (2020) 17163–17173, https://doi.org/10.1039/D0TA04721J.
[95] U. Khan, T. Iqbal, M. Khan, R. Wu, SnO2/ZnO as double electron transport layer for halide perovskite solar cells, Solar Energy 223 (2021) 346–350, https://doi.org/10.1016/j.solener.2021.05.059.
[96] Y. Shynkarenko, M.I. Bodnarchuk, C. Bernasconi, Y. Berezovska, V. Verteletskyi, S.T. Ochsenbein, M.V. Kovalenko, Direct synthesis of quaternary alkylammonium-capped perovskite nanocrystals for efficient blue and green light-emitting diodes, ACS Energy Lett. 4 (2019) 2703–2711, https://doi.org/10.1021/acsenergylett.9b01915.
[97] Y. Xia, G. Yan, J. Lin, Review on tailoring PEDOT:PSS layer for improved device stability of perovskite solar cells, Nanomaterials 11 (2021) 3119, https://doi.org/10.3390/nano11113119.
[98] L. Zhang, X. Yang, Q. Jiang, P. Wang, Z. Yin, X. Zhang, H. Tan, Y.M. Yang, M. Wei, B. R. Sutherland, E.H. Sargent, J. You, Ultra-bright and highly efficient inorganic based perovskite light-emitting diodes, Nat. Commun. 8 (2017) 15640, https://doi.org/10.1038/ncomms15640.
[99] Y. Zhou, S. Mei, D. Sun, N. Liu, W. Shi, J. Feng, F. Mei, J. Xu, Y. Jiang, X. Cao, Improved efficiency of perovskite light-emitting diodes using a three-step spin-coated $CH_3NH_3PbBr_3$ emitter and a PEDOT:PSS/MoO3-ammonia composite hole transport layer, Micromachines 10 (2019) 459, https://doi.org/10.3390/mi10070459.
[100] Z. Shi, Y. Li, Y. Zhang, Y. Chen, X. Li, D. Wu, T. Xu, C. Shan, G. Du, High-efficiency and air-stable perovskite quantum dots light-emitting diodes with an all-inorganic heterostructure, Nano Lett. 17 (2017) 313–321, https://doi.org/10.1021/acs.nanolett.6b04116.
[101] X. Xiao, K. Wang, T. Ye, R. Cai, Z. Ren, D. Wu, X. Qu, J. Sun, S. Ding, X.W. Sun, W.C. H. Choy, Enhanced hole injection assisted by electric dipoles for efficient perovskite light-emitting diodes, Commun. Mater. 1 (2020) 81, https://doi.org/10.1038/s43246-020-00084-0.
[102] S.Y. Kim, H. Kang, K. Chang, H.J. Yoon, Case studies on structure-property relations in perovskite light-emitting diodes via interfacial engineering with self-assembled monolayers, ACS Appl. Mater. Interfaces 13 (2021) 31236–31247, https://doi.org/10.1021/acsami.1c03797.
[103] Y. Wang, Y. Liu, J. Tong, X. Shi, L. Huang, Z. Xiao, G. Wang, D. Pan, A general and facile solution approach for deposition of high-quality metal oxide charge transport layers, Sol. Energ. Mat. Sol. Cells 236 (2022) 111511, https://doi.org/10.1016/j.solmat.2021.111511.

[104] R.L.Z. Hoye, M.R. Chua, K.P. Musselman, G. Li, M.-L. Lai, Z.-K. Tan, N.C. Greenham, J.L. MacManus-Driscoll, R.H. Friend, D. Credgington, Enhanced performance in fluorene-free organometal halide perovskite light-emitting diodes using tunable, low electron affinity oxide electron injectors, Adv. Mater. 27 (2015) 1414–1419, https://doi.org/10.1002/adma.201405044.

[105] I. Bretos, R. Jiménez, J. Ricote, M.L. Calzada, Low-temperature crystallization of solution-derived metal oxide thin films assisted by chemical processes, Chem. Soc. Rev. 47 (2018) 291–308, https://doi.org/10.1039/C6CS00917D.

[106] K.C. Icli, M. Ozenbas, Fully metal oxide charge selective layers for n-i-p perovskite solar cells employing nickel oxide nanoparticles, Electrochim. Acta 263 (2018) 338–345,- https://doi.org/10.1016/j.electacta.2018.01.073.

[107] P. Du, J. Li, L. Wang, L. Sun, X. Wang, X. Xu, L. Yang, J. Pang, W. Liang, J. Luo, Y. Ma, J. Tang, Efficient and large-area all vacuum-deposited perovskite light-emitting diodes via spatial confinement, Nat. Commun. 12 (2021) 4751, https://doi.org/10.1038/s41467-021-25093-6.

[108] S. Xie, A. Osherov, V. Bulović, All-vacuum-deposited inorganic cesium lead halide perovskite light-emitting diodes, APL Mater. 8 (2020) 051113, https://doi.org/10.1063/1.5144103.

[109] A. Herklotz, K. Dörr, T.Z. Ward, G. Eres, H.M. Christen, M.D. Biegalski, Stoichiometry control of complex oxides by sequential pulsed-laser deposition from binary-oxide targets, Appl. Phys. Lett. 106 (2015) 131601, https://doi.org/10.1063/1.4916948.

[110] E. Alfonso, J. Olaya, G. Cubillos, Thin film growth through sputtering technique and its applications, IntechOpen (2012), https://doi.org/10.5772/35844. *https://www.intechopen.com/chapters/39143*.

[111] M.R. Leyden, L. Meng, Y. Jiang, L.K. Ono, L. Qiu, E.J. Juarez-Perez, C. Qin, C. Adachi, Y. Qi, Methylammonium lead bromide perovskite light emitting diodes by chemical vapor deposition, J. Phys. Chem. Lett. 8 (2017) 3193–3198, https://doi.org/10.1021/acs.jpclett.7b01093.

[112] P. Jia, M. Lu, S. Sun, Y. Gao, R. Wang, X. Zhao, G. Sun, V.L. Colvin, W.W. Yu, Recent advances in flexible perovskite light-emitting diodes, Adv. Mater. Interfaces 8 (2021) 2100441, https://doi.org/10.1002/admi.202100441.

[113] P. Du, L. Gao, J. Tang, Focus on performance of perovskite light-emitting diodes, Front. Optoelectron. 13 (2020) 235–245, https://doi.org/10.1007/s12200-020-1042-y.

[114] H.H. Park, Inorganic materials by atomic layer deposition for perovskite solar cells, Nanomaterials 11 (2021) 88, https://doi.org/10.3390/nano11010088.

[115] Z. Cui, G. Korotcenkov, Solution Processed Metal Oxide Thin Films for Electronic Applications, 2020. ISBN: 978-0-12-814930-0.

[116] Q. He, K. Yao, X. Wang, X. Xia, S. Leng, F. Li, Room-temperature and solution-processable Cu-doped nickel oxide nanoparticles for efficient hole-transport layers of flexible large-area perovskite solar cells, ACS Appl. Mater. Interfaces 9 (2017) 41887–41897, https://doi.org/10.1021/acsami.7b13621.

[117] D.D. Girolamo, F. Matteocci, M. Piccinni, A.D. Carlo, D. Dini, Anodically electrodeposited NiO nanoflakes as hole selective contact in efficient air processed p-i-n perovskite solar cells, Sol. Energ. Mater. Sol. Cells 205 (2020) 110288, https://doi.org/10.1016/j.solmat.2019.110288.

[118] Q. Shan, J. Li, J. Song, Y. Zou, L. Xu, J. Xue, Y. Dong, C. Huo, J. Chen, B. Han, H. Zeng, All-inorganic quantum-dot light-emitting diodes based on perovskite emitters with low turn-on voltage and high humidity stability, J. Mater. Chem. C 5 (2017) 4565–4570,- https://doi.org/10.1039/C6TC05578H.

[119] H. Wu, Y. Zhang, X. Zhang, M. Lu, C. Sun, X. Bai, T. Zhang, G. Sun, W.W. Yu, Fine-tuned multilayered transparent electrode for highly transparent perovskite light-emitting devices, Adv. Electron. Mater. 4 (2018) 1700285, https://doi.org/10.1002/aelm.201700285.
[120] A. Subramanian, Z. Pan, Z. Zhang, I. Ahmad, J. Chen, M. Liu, S. Cheng, Y. Xu, J. Wu, W. Lei, Q. Khan, Y. Zhang, Interfacial energy-level alignment for high-performance all-inorganic perovskite CsPbBr3 quantum dot-based inverted light-emitting diodes, ACS Appl. Mater. Interfaces 10 (2018) 13236–13243, https://doi.org/10.1021/acsami.8b01684.
[121] K. Sim, T. Jun, J. Bang, H. Kamioka, J. Kim, H. Hiramatsu, H. Hosono, Performance boosting strategy for perovskite light-emitting diodes, Appl. Phys. Rev. 6 (2019) 031402, https://doi.org/10.1063/1.5098871.
[122] S. Lee, D.B. Kim, I. Hamilton, M. Daboczi, Y.S. Nam, B.R. Lee, B. Zhao, C.H. Jang, R. H. Friend, J.-S. Kim, M.H. Song, Control of interface defects for efficient and stable quasi-2D perovskite light-emitting diodes using nickel oxide hole injection layer, Adv. Sci. 5 (2018) 1801350, https://doi.org/10.1002/advs.201801350.
[123] K. Qasim, B. Wang, Y. Zhang, P. Li, Y. Wang, S. Li, S.-T. Lee, L.-S. Liao, W. Lei, Q. Bao, Solution-processed extremely efficient multicolor perovskite light-emitting diodes utilizing doped electron transport layer, Adv. Funct. Mater. 27 (2017) 1606874, https://doi.org/10.1002/adfm.201606874.
[124] J.-L. Wu, Y.-J. Dou, J.-F. Zhang, H.-R. Wang, X.-Y. Yang, Perovskite light-emitting diodes based on solution-processed metal-doped nickel oxide hole injection layer, Acta Phys. Sin. 69 (1) (2020) 018101, https://doi.org/10.7498/aps.69.20191269.
[125] Y. Shen, L.-P. Cheng, Y.-Q. Li, W. Li, J.-D. Chen, S.-T. Lee, J.-X. Tang, High-efficiency perovskite light-emitting diodes with synergetic outcoupling enhancement, Adv. Mater. 31 (2019) 1901517, https://doi.org/10.1002/adma.201901517.
[126] C. Tang, X. Shen, X. Wu, Y. Zhong, J. Hu, M. Lu, Z. Wu, Y. Zhang, W.W. Yu, X. Bai, Optimizing the performance of perovskite nanocrystal LEDs utilizing cobalt doping on a ZnO electron transport layer, J. Phys. Chem. Lett. 12 (2021) 10112–10119, https://doi.org/10.1021/acs.jpclett.1c03060.
[127] S.M. Nejad, S. Ahadzadeh, M.N. Rezaie, Effect of ZnO nanorods and nanotubes on the electrical and optical characteristics of organic and perovskite light-emitting diodes, 2021, https://doi.org/10.1088/1361-6528/abe893.
[128] C.-Y. Huang, S.-H. Huang, C.-L. Wu, Z.-H. Wang, C.-C. Yang, Cs4PbBr6/CsPbBr3 nanocomposites for all-inorganic electroluminescent perovskite light-emitting diodes, ACS Appl. Nano Mater. 3 (2020) 11760–11768, https://doi.org/10.1021/acsanm.0c02274.
[129] Q. Dong, L. Lei, J. Mendes, F. So, Operational stability of perovskite light emitting diodes, J. Phys. Mater. 3 (2020) 012002, https://doi.org/10.1088/2515-7639/ab60c4.
[130] Z. Ren, X. Xiao, R. Ma, H. Lin, K. Wang, X.W. Sun, W.C.H. Choy, Hole transport bilayer structure for quasi-2D perovskite based blue light-emitting diodes with high brightness and good spectral stability, Adv. Funct. Mater. 29 (2019) 1905339, https://doi.org/10.1002/adfm.201905339.
[131] Z. Yuan, Y. Miao, Z. Hu, W. Xu, C. Kuang, K. Pan, P. Liu, J. Lai, B. Sun, J. Wang, S. Bai, F. Gao, Unveiling the synergistic effect of precursor stoichiometry and interfacial reactions for perovskite light-emitting diodes, Nat. Commun. 10 (2019) 2818, https://doi.org/10.1038/s41467-019-10612-3.
[132] Y.C. Kim, S.-D. Baek, J.-M. Myoung, Enhanced brightness of methylammonium lead tribromide perovskite microcrystal-based green light-emitting diodes by adding hydrophilic polyvinylpyrrolidone with oleic acid-modified ZnO quantum dot electron

transporting layer, J. Alloys Compd. 786 (2019) 11–17, https://doi.org/10.1016/j.jallcom.2019.01.317.

[133] R. Chakraborty, H. Bhunia, S. Chatterjee, A.J. Pal, Surface-modification of Cu2O nanoparticles towards band-optimized hole-injection layers in CsPbBr3 perovskite light-emitting diodes, J. Solid State Chem. 281 (2020) 121021, https://doi.org/10.1016/j.jssc.2019.121021.

[134] Z. Huang, D. Ouyang, C.-J. Shih, B. Yang, W.C.H. Choy, Solution-processed ternary oxides as carrier transport/injection layers in optoelectronics, Adv. Energy. Mater. 10 (2020) 1900903, https://doi.org/10.1002/aenm.201900903.

[135] J.H. Heo, M.S. You, M.H. Chang, W. Yin, T.K. Ahn, S.-J. Lee, S.-J. Sung, D.H. Kim, S.H. Im, Hysteresis-less mesoscopic CH3NH3PbI3 perovskite hybrid solar cells by introduction of Li-treated TiO2 electrode, Nano Energy 15 (2015) 530–539, https://doi.org/10.1016/j.nanoen.2015.05.014.

[136] R. Ramarajan, J.M. Fernandes, M. Kovendhan, G. Dasi, N.P. Reddy, K. Thangaraju, D.P. Joseph, Boltzmann conductivity approach for charge transport in spray-deposited transparent Ta-doped SnO2 thin films, J. Alloys Compd. 897 (2022) 163159, https://doi.org/10.1016/j.jallcom.2021.163159.

[137] R.H. Bube, Trap density determination by space-charge-limited currents, J. Appl. Phys. 33 (1962) 1733, https://doi.org/10.1063/1.1728818.

[138] E. Yoon, K.Y. Jang, J. Park, T.-W. Lee, Understanding the synergistic effect of device architecture design toward efficient perovskite light-emitting diodes using interfacial layer engineering, Adv. Mater. Interfaces 8 (2021) 2001712, https://doi.org/10.1002/admi.202001712.

[139] C.-H. Lin, L. Hu, X. Guan, J. Kim, C.-Y. Huang, J.-K. Huang, S. Singh, T. Wu, Electrode engineering in halide perovskite electronics: plenty of room at the interfaces, Adv. Mater. 2108616 (2022), https://doi.org/10.1002/adma.202108616.

[140] M. Guo, Y. Lu, X.-Y. Cai, Y. Shen, X.-Y. Qian, H. Ren, Y. Li, W. Wang, J. Tang, Interface engineering improves the performance of green perovskite light-emitting diodes, J. Mater. Chem. C (2022), https://doi.org/10.1039/D1TC05706E.

[141] U. Farooq, M. Ishaq, U.A. Shah, S. Chen, Z.-H. Zheng, M. Azam, Z.-H. Su, R. Tang, P. Fan, Y. Bai, G.-X. Liang, Bandgap engineering of lead-free ternary halide perovskites for photovoltaics and beyond: Recent progress and future prospects, Nano Energy 92 (2022) 106710, https://doi.org/10.1016/j.nanoen.2021.106710.

Antireflective coatings and optical filters

12

Animesh M. Ramachandran[a,b], *Manjit Singh*[a,b], *Adhithya S. Thampi*[a,b], *and Adersh Asok*[a,b]
[a]Photosciences and Photonics, Chemical Sciences and Technology Division, CSIR-National Institute for Interdisciplinary Science and Technology (NIIST), Thiruvananthapuram, Kerala, India, [b]Academy of Scientific and Innovative Research (AcSIR), Ghaziabad, Uttar Pradesh, India

Chapter outline

1 Introduction

Humans were always fascinated with the nature and theory of light and its travel throughout history. Understanding and controlling light were always a cup of tea for science and technology, leading to many inventions that changed the world in its own application levels, especially in the field of energy, healthcare, environment, and astronomy, to name a few. History says that the usage of mirrors has been dated long before the prehistoric period [1]. In the modern era, application-based light energy control has improvised many technological advancements. Optical coatings/materials utilize the fundamental nature of light to reflect, refract, polarize and absorb by materials, to introduce specific utility. Antireflective coatings (ARC) and optical filters (OF) are classes of optical elements that are distinct in their functionalities. Former is used for the passage of light of a specific wavelength through a material, which

Metal Oxides for Next-Generation Optoelectronic, Photonic, and Photovoltaic Applications
https://doi.org/10.1016/B978-0-323-99143-8.00004-3

will be either absorbed or transmitted, avoiding reflection. The latter is being used to block/transmit a certain wavelength band of light, either by reflection or absorption.

2 Metal oxides as an optical material

The scientific community has shown immense interest in metal oxide materials for a wide variety of applications, by exploiting their optical, electrical and magnetic behavior for the past centuries. Metal oxide materials usually range in their properties, such as optical properties (refractive index), electrical properties (from insulators to semiconductors and conductors), etc. Understanding and utilizing metal oxide materials in their structural, electronic, dielectric, and optical levels paved the way for innovative and precise technological applications, in the field of optics, sensors, energy, display, catalysis, and so on.

A broad range of ARC materials with varied refractive indexes (one of the design parameters) are currently utilized for different applications based on their transparency, durability, mechanical properties, and other functional advantages. The material library ranges from conventional materials like metals, metal oxides, metal fluorides, polymers, and silicon-based coatings to advanced materials like CNT, graphene, fullerenes, etc. Metal oxides (ZnO, TiO_2, Al_2O_3, MgO, Ta_2O_5, ZrO_2, HfO_2, Y_2O_3, and Indium Tin Oxide (ITO) are some of the common metal oxides) is one of the widely used materials for ARC because of its easy adaptability, material inertness of most of the metal oxides to the harsh environment, mechanical and optical properties, along with simple fabrication techniques for thin-film coating as well as texturing. However, material selection depends on environmental considerations; for example, metal oxide like MgO is not used as the atmospheric exposed layer due to its reactivity toward CO_2 and subsequent formation of hazy surfaces. Among the metal oxides, TiO_2 and its composites are the most widely adopted, especially in photovoltaic applications, due to the as-mentioned advantages [2]. TiO_2 ARC is also administered to induce additional application settings, like self-cleaning or superhydrophilic properties [3].

Metal oxide thin film is composed of deposits of the oxide, which have continuous and defined thicknesses over a specific surface. The thickness of the film can range from single atomic or molecular layers to nanometers and micrometers, depending upon the application [4]. Apart from those, creating morphological and structural distinctions in metal oxides also induces varied optical properties. Porous surfaces created with metal oxide materials like TiO_2 are also explored for tuning the effective refractive index [5]. Apart from TiO_2, ZnO AR materials are also explored globally, especially for structured ARC like nano-rods [6]. Research suggests that the anti-reflectivity and efficiency of solar cells have been greatly enhanced with both ZnO material and nano-structures [7,8]. Apart from TiO_2 and ZnO, studies are also conducted on other metal oxides like Indium tin oxide (ITO), Ta_2O_5, etc. [9,10], especially for specialty applications, like the requirement of a transparent conductive surface as ARC using ITO. ITO possesses superior conductivity and optical properties of transmittance and refractive index for ARC. ITO has also been utilized for both thin-film and nano-structured coatings [11,12].

Metal oxides are commonly used materials for manufacturing "colored glass" or "stained glass," which remained one of the elements of ancient architecture and artistry. Infusion of metal oxide powders like iron oxide (for green glass), copper oxide (for blue glass), arsenic oxide, and tin oxide with antimony (for white glass) in molten glass created the oldest form of optical filters, which utilizes light absorption properties of the material. However, modern OF ranges from absorptive to more efficient thin-film coatings and nano-structured surfaces for optical filtration and thus extended from aesthetics to more niche applications for energy, optics, and other sectors. Metal oxide materials in bulk and thin film form, and metal oxide nanostructures exhibit a wide variety of functional properties which make them utilizable for many applications. Metal oxide's functional properties are dependent on its crystal structure, composition, native defects, doping, etc., and govern its optical, electrical, chemical, and mechanical characteristics. With its multifunctional nature as OF, metal oxides can be exceptionally utilized in applications like solar cells, optoelectronic devices, transparent conductive oxides (TCOs), plasmonic photonic integrated circuits, chemical protection, corrosion protection, thermal protection, healthcare, energy conversion, and storage. For half a century, thin-film metal oxide coatings are widely used as electromagnetic filters from UV to infrared. These thin-film coatings provide better or varied optical performance when applied to different substrates like glass, plastics, metals, semiconductors, and ceramics [13], and are one of the important components used in precision optics applications like optics, avionics, sensors, fiber optics, and space applications [14]. Deposition onto a substrate was typically accomplished using various techniques and methods and is crucial for precision filter development. Various metal oxide thin films, including TiO_2, SnO_2, MoO_3, and NiO have been optimized for implementation as multilayered structures for niche applications like infrared filtering [15]. Semiconductor metal oxides like TiO_2 and ZnO always persist in the market as UV filters, especially in cosmetic applications [16]. Nano-structured or patterned optic filters are a very competing research area with its wide applicability of tuning optical properties with the same material of concern [17,18]. With their varied index of refraction and ease of fabrication through different techniques, metal oxides have been explored profoundly as they can cater to a wide range of applications [19].

3 Antireflective coatings

It's well known that even transparent mediums such as glass show reflection at the transmitting interfaces, and it increases upon the incident angle of light. This can be attributed to the Fresnel loss of reflection. The undesirable Fresnel reflection can be reduced by administering an engineered optical layer at the fore-end, broadly known as antireflective coatings. In 1886, Lord Rayleigh discovered the antireflective effect on a tarnished old piece of glass, when a layer (glass reacted with the air and moisture) becomes tarnished and has a refractive index between air and glass) is formed between the glass and the air medium. Later in 1935, Alexander Smakula discovered antireflective coatings based on light interference phenomena. If the reflected light from the two interfaces (air-coating and the coating-glass) interferes destructively, then the reflection can be reduced for the particular wavelength of light. However, ARC in modern times has become versatile with its architecture, operating wavelength band, application, the surface of application, and efficiency. Most of the antireflective layers follow the physics of interference of the light and the refractive index modulation at the interfaces, such as a single layer, multiple layers, gradient refractive index layer, textured layer, and so on. With technological and research evolution in the field and manufacturing processes, application-based antireflective layers are more sought-after. Today, ARC is continually serving many industries and applications, especially photovoltaic industries (to minimize efficiency loss due to Fresnel reflection), eyewear industries (to avoid glare and enhance visual comfort), architectural and automobile glass industry, optoelectronic devices, camera lenses, airborne imaging to collect data from the surrounding in the IR and visible spectra, etc. [20–25]. Further, multifunctional ARC, such as self-cleaning AR layers, thermochromic AR coatings, etc., is also alluring to the ARC market [26].

3.1 Defining a perfect antireflective coating

The conditions for a perfect ARC depend on the application requirements. However, in general, the ARC functions should have these attributes.

(1) Should show minimal reflection at the required wavelength/wavelength range
(2) Should be omnidirectional in the working range of the angle of incidence
(3) Should possess high polarization insensitivity
(4) Modern ARC is based on multifunctional aspects to serve a particular/number of functions
(5) Functional stability and durability of coatings, especially for conditions of harsh environment
(6) Ease of fabrication
(7) Low cost of production

3.2 Theory of antireflective coatings

When light travels from one optical medium to another, partially light gets reflected (both in s and p polarizations) at the interface, known as Fresnel reflection. The Fresnel law defines the coefficient of reflection in a nonabsorbing medium as

$$r_s = \frac{n_1 \cos\theta_i - n_2 \cos\theta_t}{n_1 \cos\theta_i + n_2 \cos\theta_t} \tag{1}$$

$$r_p = \frac{n_2 \cos\theta_i - n_1 \cos\theta_t}{n_2 \cos\theta_i + n_1 \cos\theta_t} \tag{2}$$

where r_s and r_p are the coefficient of reflection for s and p polarizations, n_1 and n_2 are the refractive indices creating the interface of reflection, θ_i and θ_t are the angle of incidence and transmission, as shown in Fig. 1.

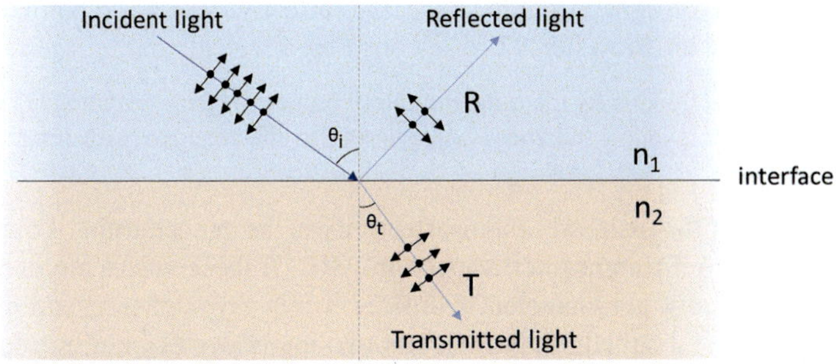

Fig. 1 Light transmission through an interface of two media with different refractive index.

The reflectance value can be calculated as,

$$R_s = |r_s|^2 \tag{3}$$

$$R_p = |r_p|^2 \tag{4}$$

$$R = \frac{1}{2}\left(R_s + R_p\right) \tag{5}$$

The net transmittance,

$$T = 1 - R \tag{6}$$

The fundamental physics involved in ARC are either to (1) minimize reflection at the interface of two media (or) (2) create destructive interference in the wavelength range when the light reflects back from the subsequent interfaces of mediums.

(1) To minimize the reflection at the interface of media: This can be achieved by reducing the intensity of Fresnel reflection caused at the different interfaces by refractive index modulation.

For normal incidence Eq. (5) gives

$$R = r^2 = \left|\frac{n_1 - n_2}{n_1 + n_2}\right|^2 \tag{7}$$

The above relation (7) suggests that if the difference between the refractive index of consecutive media is lesser, then the reflectance will be minimum. However, for the introduction of a third medium as a sandwiched ARC layer, the optimal refractive index for minimal effective reflectance is

$$n_{\text{ARC}} = \sqrt{n_1 n_2} \tag{8}$$

If we consider air ($n_1 = 1$) as medium and substrate with refractive index n, then the refractive index of ARC should be close to $\sqrt{n}$ to minimize the reflection. In the case of oblique incidence of light at the interface, the reflectivity of s and p polarized light individually along with Brewster's law is considered for calculating the reflectance.

(2) To minimize the reflection by the thin-film interference of reflected light: This is achieved by modulating the coating thickness with respect to the required wavelength range for antireflectivity.

The key principle of this method of antireflectivity is the cancellation of out-of-phase waves reflected from different interfaces of an ARC. If these waves are out of phase, they partially or totally get canceled. Consider a thin layer of transparent material coated on a transparent substrate. Now it has two interfaces (1) medium-layer interface and (2) layer-substrate interface. Each interface will show Fresnel reflection, according to Eqs. (1), (2), and (5). If the phase difference between the light reflecting from two consecutive interfaces is an odd multiple of 180 degrees (π) and with the same intensity, it will face destructive interference and give zero reflectance. 180 degrees phase difference can be achieved with a coating of quarter wavelength (wavelength for antireflection property). By adjusting the thickness of the layer, we can adjust the phase difference for different wavelength considerations.

Consider a coating of thickness x; the phase difference for the light of wavelength λ is

$$\varphi = \frac{2\pi m x \cos\theta}{\lambda} \tag{9}$$

where m is a natural number, and θ is the angle of refraction.

The destructive interference reduces the reflected intensity for a particular wavelength. By using the relation (8) we can calculate the thickness (d) of the layer for the destructive interference. For odd multiple of π phase difference, path difference will come as $\lambda/2$, $3\lambda/2$, $5\lambda/2$, …, $n\lambda/2$ and the thickness will be half of the path difference that is $\lambda/4$, $3\lambda/4$, $5\lambda/4$, …, $n\lambda/4$ (Fig. 2).

For multiple layers deposited on the substrate having refractive index n_1, n_2, n_3…, each has thickness $\lambda/4$, $3\lambda/4$, $5\lambda/4$…, the phase difference of each layer is

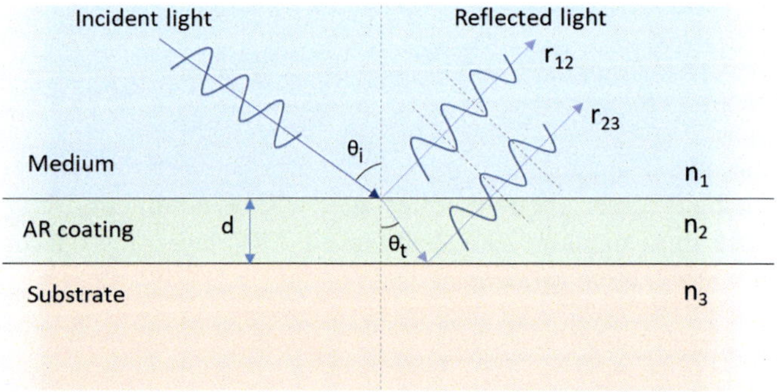

Fig. 2 Destructive interference of reflected rays (with π phase difference) from medium-ARC interface and ARC-substrate interface.

$$\phi_n = \frac{2\pi m_n x_n \cos\theta_n}{\lambda} \tag{10}$$

Consider x_n is the thickness of each layer with refractive index n, $r_{(n-1)(n)}$ is the coefficient of reflection at the different interfaces. Combining phase difference and the coefficient of reflection of interfaces to get net reflection at interfaces as

$$r = \left|\frac{n_1 - n_2}{n_1 + n_2}\right|$$

$$r'_{(n-1)(n)} = |r_{(n-1)(n)}|e^{-2(\Phi_1+\Phi_2+\Phi_3+\Phi_{n-1})} \tag{11}$$

where $\Phi_1, \Phi_2, \Phi_3, \ldots, \Phi_{n-1}$ give the phase difference for each layer. The net reflection is the sum of all reflection coefficient values, which should be kept minimum for better antireflection [27].

Apart from these layered ARC, AR structures based on different surface topography are also employed more recently, that use surface patterns/structures like porosity, surface texturing, nano piles, nanostructures like pyramids, cones, paraboloids [28–31], etc., that create broadband antireflection. The different types of ARC and AR surfaces are discussed below.

3.3 Types of antireflective coatings and surfaces

There are two major categories of ARC, based on their design and functional architecture: (1) layer composition (class of ARC that utilizes refractive index tuning and interference of waves from multiple layers to create antireflection) and (2) surface topography based (class of ARC that utilizes surface patterns to create broadband antireflection), as shown in Table 1.

Single layer ARC (SLARC): SLARC utilizes a single layer of ARC, sandwiched between the medium and substrate, for reducing the Fresnel reflection intensity for a particular wavelength. The coating thickness is designed (as $\lambda/4$) for occurring destructive interference of waves reflecting at the medium-coating and coating-

Table 1 Classification of antireflective coatings.

Antireflective coatings	
Layer composition	**Surface topography**
(1) Single layer ARC	(1) Porous ARC
(2) Double layer ARC	(2) Photonic nanostructures
(3) Multiple layer ARC	(3) Textured surface
(4) Gradient refractive index (GRIN) ARC	

substrate interfaces for the wavelength λ, and the refractive index of coating can be designed to geometrical mean of refractive indices of medium and substrate. However, SARC is not applicable for high antireflection in a broad range of wavelengths [32]. Also, for SLARC, acquiring the perfect material (refractive index as the geometric mean of medium and substrate) is challenging, especially for glass industries, which is one of the major sectors utilizing ARC. However, SLARC can be used for applications that require only a meager reduction in reflection for a broader range of wavelengths, especially for substrates with a high refractive index [27].

Double layer ARCs (DLARC): DLARC employs two layers of the same/different materials, with the same/different thickness to employ the antireflection. The pattern of loss in reflection will be maximum at a target wavelength and will gradually reduce with the change from the target wavelength, creating a V-curve, making DLARC also known as the V-type layer. For attaining zero reflectance for the same coating thicknesses in DLARC the condition of refractive indices is

$$\frac{n_1}{n_2} = \sqrt{\frac{n_m}{n_s}} \tag{12}$$

where n_1, and n_2 are the refractive index of ARC layers, n_m, and n_s are the refractive index of medium and substrate respectively. The thickness of layers is commonly taken as $\lambda/2$, $\lambda/4$ [33]. Compared to SLARC, DLARC offers more antireflectivity at specific wavelengths and is useful in laser applications. Hiroyuki Kanda et al. manufactured TiO_2/Al_2O_3 DLARCs by spray pyrolysis for crystalline silicon solar cells and this gives the 0.4% reflectance at one of the most utilized wavelengths of 600 nm by silicon cells [34]. Vicent et al. reported the fabrication of metal oxide double layer coating of SiO_2 and TiO_2 on solar glass cover by sol-gel process, and was able to achieve more than 98% light transmission in the range of 500–600 nm [35].

Multiple layers ARC (MARC): To achieve antireflection in a much broader range of wavelengths, the MARC is administered rather than single or double-layered ARC. Similar to DLARC, here also transmittance is optimized by thickness and the refractive index of the layers [32]. Chattopadhyay et al. conducted a study of the single-layer (MgF_2), double-layer (MgF_2/Al_2O_3), and triple-layer ($MgF_2/ZrO_2/CeF_3$) ARCs on a glass substrate [36]. The study was conducted within the range of 400–800 nm, as shown in Fig. 3. The SLARC showed a reduction in reflectance from 4.26% to 1.5%, compared to bare glass. DLARC showed a much reduced effective reflectance in the range of 500–600 nm. However, the V-curve of DLARC shows enhanced reflection at the two extremes of the graph (below 475 nm and above 700 nm). With the increase of the layer number, triple-layer ARC showed an improved range of antireflectivity.

Gradient refractive index (GRIN) ARC: In this, the different layers of ARC are created as graded refractive index as we move along the thickness, to improve the range of antireflectivity. Various gradation profiles have been investigated and reviewed by researchers in this regard, such as linear, exponential, cubic, parabolic, quintic, etc. [32,36]. Zhongyang Ge et al. coated aluminum-doped zinc oxide (AZO) coated glass

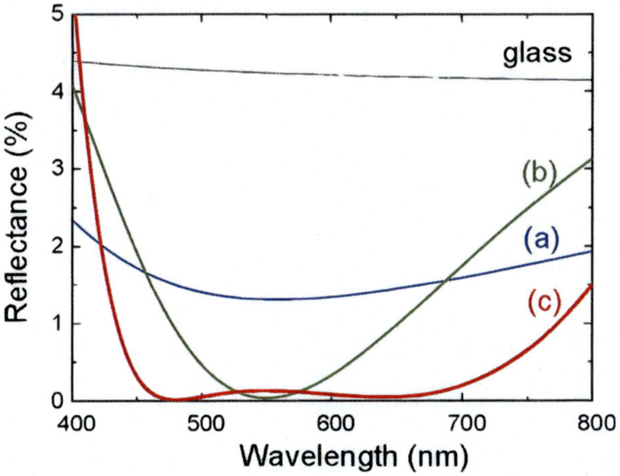

Fig. 3 Reflectance spectra of AR coatings with air as medium and glass as substrate (A) SLARC of $0.25\lambda_t$ thickness MgF_2; (B) DLARC of $0.25\lambda_t$ thickness MgF_2 and $0.25\lambda_t$ thickness Al_2O_3; (C) MLARC (3-layer) of $0.25\lambda_t$ thickness MgF_2, $0.5\lambda_t$ thickness ZrO_2, and $0.25\lambda_t$ thickness ZrO_2. λ_t is the target wavelength of minimal reflection.
Source: Reused with permission from S. Chattopadhyay, Y.F. Huang, Y.J. Jen, A. Ganguly, K. H. Chen, L.C. Chen, Anti-reflecting and photonic nanostructures, Mater. Sci. Eng. R Rep. 69 (2010) 1–35. https://doi.org/10.1016/J.MSER.2010.04.001. Copyright © 2010, Elsevier.

surface with a GRIN ARC coating of AZO and SiO_2 with a linear profile of refractive index ranging from 1.9 to 1.5 using simultaneous sputtering technique [37]. Fig. 4A shows the refractive index profile of the whole system, Fig. 4B shows the cross-sectional SEM image of the coating revealing the GRIN ARC and AZO coating thickness. Fig. 4C shows the reflectance spectra of GRIN-coated and uncoated samples at normal incidence, indicating the reduction in reflectance in the coated sample.

Porous layer: Creating nano-pores of size lesser than the wavelength of consideration can enhance the antireflectivity of AR coatings/materials [38].

The refractive index of the material can be related to the porosity as,

$$n_p = \left[\left(1 - \frac{p}{100}\right)\left(n_d{}^2 - 1\right) + 1\right]^{\frac{1}{2}} \tag{13}$$

where n_p, n_d, are the refractive index of porous material and dense material, and p is the percentage of porosity [27]. By creating varied porous structures, the refractive index gradation can also be increased, enhancing the range of antireflectivity. Film thickness and the density of porous material are important factors for gradient refractive index, and it works for wide incident angle [39–41]. A high transmittance broadband layer is fabricated by Walheim et al. in their study, where they also varied the refractive indices from 1.2 to 1.05. The fabricated film showed high light transmittance of 99.7% [42].

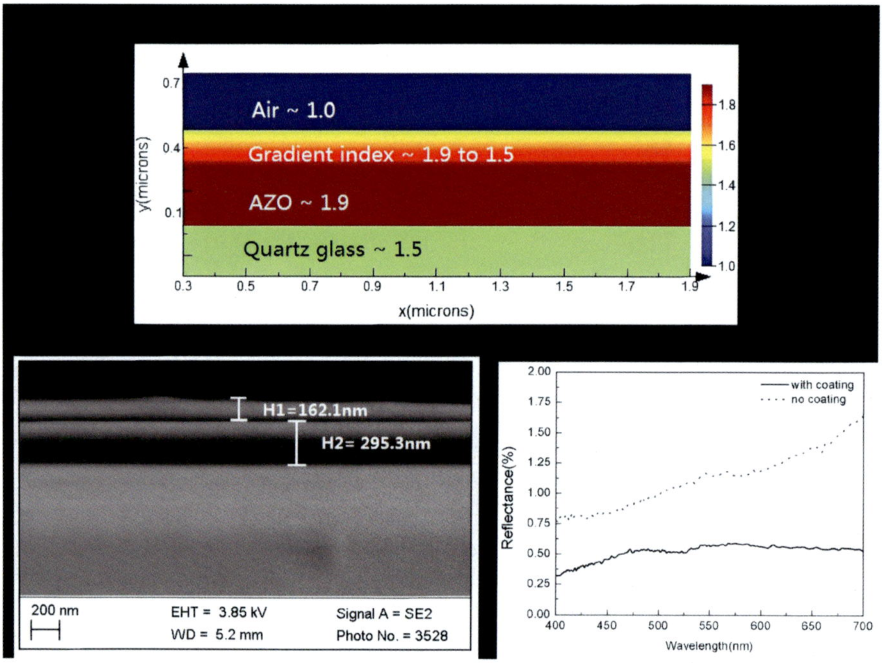

Fig. 4 (A) Refractive index profile of GRIN ARC on AZO-coated quartz glass, with air as a medium; (B) SEM image of GRIN ARC on AZO-coated quartz glass, where H1 and H2 represent the thickness of GRIN ARC and AZO respectively; (C) Reflectance spectra of AZO coated sample with and without GRIN coating.
Source: Reused with permission from Z. Ge, P. Rajbhandari, J. Hu, A. Emrani, T.P. Dhakal, C. Westgate, et al., Enhanced omni-directional performance of copper zinc tin sulfide thin film solar cell by gradient index coating, Appl. Phys. Lett. 104 (2014) 101104. https://doi.org/10.1063/1.4868104. Copyright © 2014, AIP Publishing LLC.

The creation of nano-pores in the AR layers is commonly done by etching [27]. Sobahan et al. used electron beam evaporation, to produce SiO_2 porous coating by the Glancing angle deposition technique (GLAD) [43]. The morphology of isolated nanocolumns of SiO_2 and gaps were created yielding a very low refractive index of 1.08 (measured at 633 nm). The four-layered structure observed a very low reflection of 0.7% in the visible range, at angles ranging from 0 degrees to 45 degrees.

Photonic nanostructures: Motivated by the discovery of moth's eye nanostructures by Bernhard, which is creating antireflectivity, scientists started to mimic and explore more such subwavelength structures for the application [44]. The moth's eye has submicron height distant pillar-like structures arranged in an array. Such a structure creates refractive index variation along the depth between air and substrate, which creates a GRIN-like effect. Fig. 5 shows the principle of formation of varied effective refractive index in moth-eye structure with n_m and n_s as the refractive index of medium and substrate respectively.

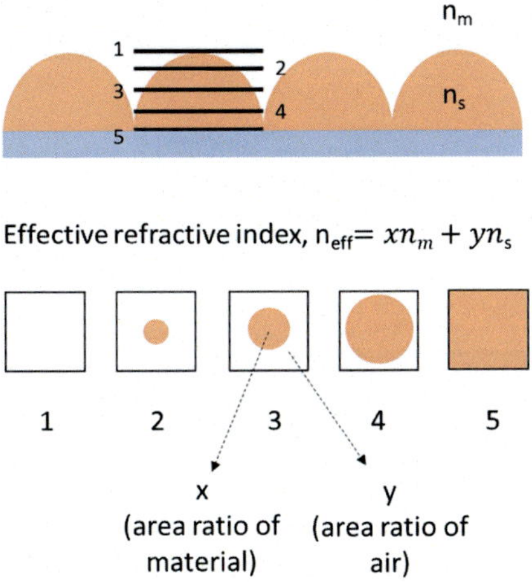

Fig. 5 Effective refractive index at the different cross sections of a moth-eye ripple structure. n_m and n_s are the refractive indexes of medium and substrate respectively.

However, different models such as the Maxwell-Garnett model and Bruggeman approximation, have been proposed for calculating the effective refractive index of inhomogeneous structures for predicting optical behavior [27]. For the case of broadband antireflection, the coating can be optimized by controlling the space as finer as it can be and by increasing the height of nanostructures. These nanostructures could be of different structural shapes such as conical, paraboloid, gaussian, etc. [45–47]. The height of nano-structures is a major factor of reduction in reflection, rather than the width, to create a perfect gradation of refractive indices for broadband antireflection applications [48]. The surface structural importance of AR materials is clearly exhibited by Seung-Yeol Han et al., by fabricating the ZnO nanorod structures over pyramidal silicon structures. The pyramidal structure reduced the reflectance from 30.8% to 10.6% between 400 and 900 nm and with the addition of ZnO nanostructures, the reflectance was further reduced to 3.4% [49]. Qu et al. also conducted a study on ZnO nano-structures for ARC (ZnO nano-rods) [8]. They created ZnO nano-rods of diameter between 40 and 50 nm and compared the antireflectivity with the bare substrate (Si solar cell) and ZnO thin film-coated Si solar cells (sample schematics shown in Fig. 6A). The nano-rods tend to reflect the incoming radiation within themselves like a light trap, and increase the light transmission to the substrate, as shown in the schematic of Fig. 6B. The results showed better performance of antireflectivity (15.9%) for samples with ZnO nano-rods, compared with bare solar cells (41.6%) and ZnO-coated solar cells (18.1%), as shown in Fig. 6C.

Textured surface ARCs: Texturization simply means modifying the smooth plane surface by making grooves, structures, or any other pattern on the surface. For the antireflection to happen, one case is that the period of texturization is lesser than the wavelength of consideration [50,51]. In such cases, textured surfaces work similarly to moth-eye AR materials, by the creation of gradual refractive index change to the depth. However, if the period's size is bigger than the wavelength (macro-structures),

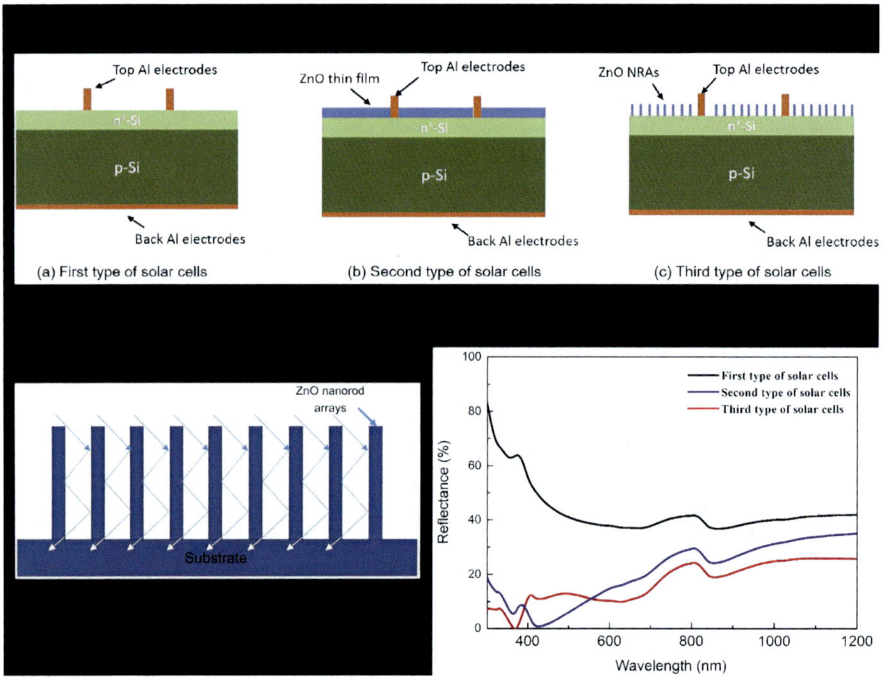

Fig. 6 (A) Structure of samples: bare solar cells (1st type), solar cells with ZnO thin film (2nd type), and solar cell with ZnO nano-rod array (3rd type) (from left to right); (B) principle of light trapping in ZnO nano-rods; (C) reflectance spectra of the three samples in 300–1200 nm wavelength range.

Source: Reused with permission from Y. Qu, X. Huang, Y. Li, G. Lin, B. Guo, D. Song, et al., Chemical bath deposition produced ZnO nanorod arrays as an antireflective layer in the polycrystalline Si solar cells, J. Alloys Compd. 698 (2017) 719–724. https://doi.org/10.1016/J.JALLCOM.2016.12.265. Copyright © 2016, Elsevier.

light can be treated as rays and the antireflection phenomenon will be governed by the ray optics, as it will suffer multiple reflections and get trapped inside crevices. The possibilities of macrostructures can be extended to different structural architectures like honeycomb, spherical, etc. [52–54].

4 Optical filters

Optical filters (OF) are devices that allow the transmission of a specific wavelength or a band of wavelengths by filtering out undesired wavelengths. Therefore, this can be utilized to select a portion of the incoming light to use as an output from the system, or, an input for any other optical system. They can also be used to filter out background noise from the rest of the optical spectrum. The process of filtering may be accomplished by absorption, or reflection of undesired radiations and transmitting the required ones. OF are available in different physical forms. Filtering elements can be deposited/ integrated either on exterior windows of optical systems or on geometrical optics within the optical system. Complex optical filtering systems can be deposited/integrated on a single substrate or deposited on several separate substrates and laminated together to form a single element. OF have applications in a large number of areas like healthcare application (UV absorbing coatings), night vision (IR coatings and night driving filter), coatings for windscreens, analytic instrumentation, laser applications, ophthalmic applications, lens coatings, military applications, surveillance and targeting systems, binocular coatings, astronomy, etc. [55–62].

4.1 Classification of optical filters (mechanism of operation)

Optical filters are classified according to the mechanism of operation or with their functional distinction as represented in Table 2.

Absorptive filters: Absorptive filters utilize different organic and inorganic materials that absorb certain wavelengths of light and allow the desired wavelengths to pass through [63]. These filters mostly provide attenuated filtration, and the spectral performance is a function of the physical thickness of the filter and the amount of absorptive material present in the glass. The absorptive material can be infused in optical materials like glass, PMMA, etc., or can be coated onto the surface. The technique is simpler, less costly, and has high angular acceptance compared to thin film techniques. However, the efficiency of absorptive filters may not lead to precision applications. Since it is an energy absorption technique, the heating of material will be higher, especially for IR filters and high-power applications.

The optical bandgap is the energy threshold for light absorption of the material and it can be calculated through the analysis of the absorption edge. The threshold energy where optical absorption can take place in the absence of scattering can be calculated from the absorption coefficient through the relation

$$\alpha h\vartheta = A\left(h\vartheta - E_g\right)^n \tag{14}$$

where α is the absorption coefficient of the material, A is a constant, E_g is the effective optical bandgap, $h\vartheta$ is the photon energy and the exponent n value is related to the type

Table 2 Classification of optical filters.

Optical filters	
Mechanism of operation	**Functional**
(1) Absorptive filters (2) Nonabsorptive filters (a) Dichroic filters (b) Febry-Perot filters (c) Fiber-Bragg grating filter	(1) Monochromatic filters (2) Infrared filters (3) Ultraviolet filters (4) Neutral density filters (5) Bandpass filters (6) Photonic based filters

of transition. The absorption coefficient determines the depth at which a given wavelength penetrates the material.

The efficiency of the adsorption filters is expressed in terms of optical density (OD), which is correlated to the amount of energy blocked by the filter, expressed in terms of light transmittance of the system.

$$D = -\log\left(\frac{T}{100}\right) \tag{15}$$

Dichroic filters: Also called "reflective" or "thin film" or "interference" filters. In its principle, dichroic filters differ from absorptive filters by the mode of rejection of the light band. Rather than absorbing the light as in absorptive filters, dichroic filters produce sharp reflections, making it utilizable for high-power applications too, like in Raman spectroscopy. A series of optical coatings are placed with a specific thickness and refractive indices, resulting in the reflection of a certain wavelength range. The thickness of each layer is chosen so that the reflections of the selected wavelength either interfere constructively or destructively with themselves. The change of the light's phase is important for determining whether it will interfere constructively or destructively with itself. Light from a single source can interfere with itself when it is split by reflection from multiple optical layers. A quarter-wave layer (QWL (λ/4)) is designed to reflect a range of wavelengths of light and is commonly used in OF. A QWL's thickness and index of refraction are chosen so that the phase of a specified wavelength changes by one-fourth of a full rotation (90 degrees) as it passes through the material. The faces of QWL provide two reflections: one from the front side and one from the backside. The light that reflects off the front side of the QWL has a phase rotation of 180 degrees due to reflection from a material with a higher index of refraction. The light that reflects from the backside of a QWL and returns through the front layer has a phase difference of half a rotation (180 degrees) since it has traveled a distance equal to twice the layer thickness. If the change of phase ($\Delta\theta$) is even multiple 180 degrees, then the light interferes constructively, resulting in a high intensity. If $\Delta\theta$ is an odd multiple 180 degrees, it interferes destructively, resulting in a very low intensity. With the advantage of wavelength selection and better filtering efficiency, however, dichroic filters are high angle depended to screen the wavelengths. Shifting in direction can lead to spectral shifts for the reflected radiation. Hence, they are preferred with single-angle applications.

Fabry-Perot filter: A Fabry-Perot filter is one of the simplest filters in principle, which is based on the Fabry-Perot interferometer, and has a resonant cavity bounded on each end by parallelly placed partially-silvered mirror as shown in Fig. 7. Light input is from one of the mirrors, normal to the surface, and the output of filtered radiation is collected from the second mirror. The required transmission wavelength is to be made resonant at the output end. When incident light of a nonresonant wavelength reaches the mirror portion, part of it (will be reflected and part will be transmitted, and the amount that gets through into the cavity will bounce around for a span and then it will ultimately exit through one of the end mirrors. When the light of the resonant wavelength enters the mirror, it passes through into the cavity without loss. The

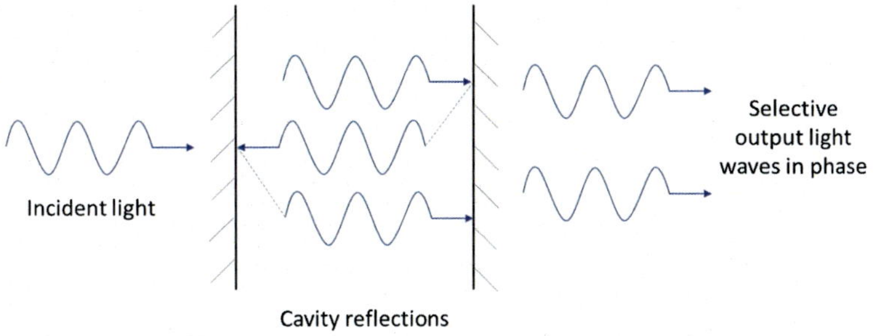

Fig. 7 Working principle of Fabry-Perot filter.

incident light suffers multiple reflections between the silver-coated mirrors and if the cavity width is equal to an odd multiple of a half wavelength of the incident light, the reflected beams interfere constructively. For other cavity widths, the reflected beams interfere destructively and the optical transmission can be reduced toward zero. Dielectric mirrors are also used instead of metallic mirrors for high reflectivity and high transmittance [64].

Fiber Bragg grating (FBG) filter: A FBG filter is a distributed Bragg reflector made in a short segment of optical fiber which reflects particular wavelengths of light and transmits all others. This is attained by creating a periodic variation in the refractive index of the fiber core, and it generates a wavelength-specific dielectric mirror (as shown in Fig. 8). Therefore, a fiber Bragg grating can be used in line with optical fiber which can block certain wavelengths. That is, it can be used as a wavelength-specific reflector. Even though it resembles a thin-film coating, there are significant differences. In a thin-film filter, the changes in refractive index are in the order of 10^{-1} and also discontinuous. But in the grating, the refractive index variation is smaller and continuous [65]. The wavelength of maximum reflectance depends not only on the Bragg grating period but also on temperature and mechanical strain. Hence, it can be used in temperature and strain sensors also [66]. The reflected wavelength (λ_{ref}) is dependent on the grating period (Λ) and effective refractive index (n_{eff}) as follows,

$$\lambda_{\text{ref}} = 2n_{\text{eff}}\Lambda \tag{16}$$

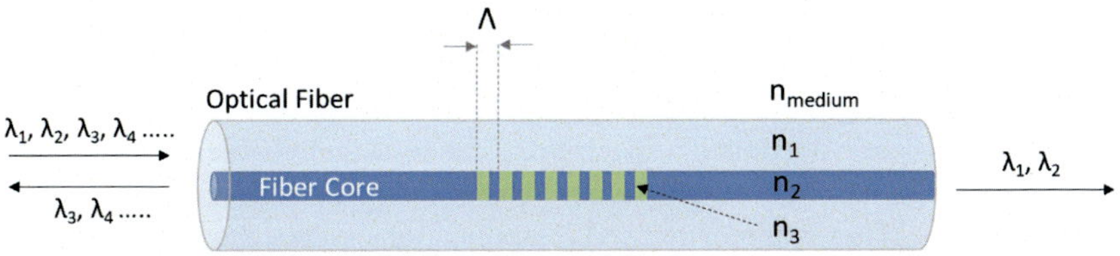

Fig. 8 Structure of a typical Fiber Bragg grating filter.

The grating period can be functioned with different patterns and the refractive index can be made uniform or graded, accordingly different light reflectance spectra can be obtained.

4.2 Classification of optical filters (functional)

Monochromatic filters: Monochromatic filters transmit only a single wavelength or a very narrow wavelength range. Absorptive type filters are not utilized for this, as the transmitted wavelength selection is very broad. Nonabsorptive filters such as interference filters with multiple layers can be used for monochromatic filters. Shakya et al. demonstrated monochromatic filter tuning using a plasmonic color filter array based on a resonant grating structure [67]. The grating is made of Al metal with SiN as waveguide on SiO_2 substrate as shown in Fig. 9A (where *P* is the period of grating which is varied from 260 to 450nm; *G* is the gap between fingers, 60nm; Tm is the thickness of the metal, 40nm; Tw is the thickness of the waveguide, 90nm; and Tb is the distance between grating and waveguide, 60nm). Fig. 9C shows the SEM image of one of the grating systems. The filter wavelength is tuned in the system with varying the period of grating from 260nm to 450nm with a 10nm step. Each step created a spectral shift as shown in the reflectance spectra (Fig. 9D). Fig. 9B shows the optical microscopy image indicating the RGB exposure.

Infrared filters: IR filters can be used for transmitting or blocking IR radiations by effectively cutting the undesired wavelength range. IR-blocking filters are mostly used in applications where the heating is undesirable. Absorptive IR-blocking filters are preferred less due to the heating effect it is produced in the filter material. Thad Druffel et al. fabricated a 15-layer stacking of metal oxide polymer nanocomposite films (consisting of spherical TiO_2/SiO_2 nanoparticles dispersed in UV-cured acrylate polymer matrix) to create an IR mirror with peak reflectance maxima of greater than 90% at 1060nm [13]. The theoretical and predicted response tend to follow the same pattern, as shown in Fig. 10. IR filters have a wide variety of applications including night vision technology, meteorology, thermography, heating, and spectroscopy.

Ultraviolet filters: UV filters can be classified into two groups according to their nature. The inorganic UV filters also called physical UV filters, principally work by reflecting and scattering the UV radiation, while the organic UV filters, also called chemical UV filters, absorb the light [16]. Pei Guo et al. fabricated a UV protective hetero-coatings (long pass UV filter (refer bandpass filter)) with 3% Ga doped ZnO as a UV absorber fabricated by metal-organic deposition and Ge films as visible light absorber fabricated by electron-beam evaporation on a quartz substrate [68]. Nonabsorptive filters are also utilized for UV screening. Dai et al. designed and fabricated a multilayer bandpass filter in the UV region at different positions using a SiO_2/Si_3N_4 dielectric distributed Bragg reflector structure [69].

Neutral density filters: Neutral density (ND) filters are designed for lowering the transmission of a certain portion of a spectrum, ultraviolet and visible, visible, or infrared. ND filters can also be of two types: absorptive and reflective (by partially reflecting mirrors). They are mostly used in applications where the intensity of radiation is necessary for a lower value, rather than blocking out, such as in cameras to

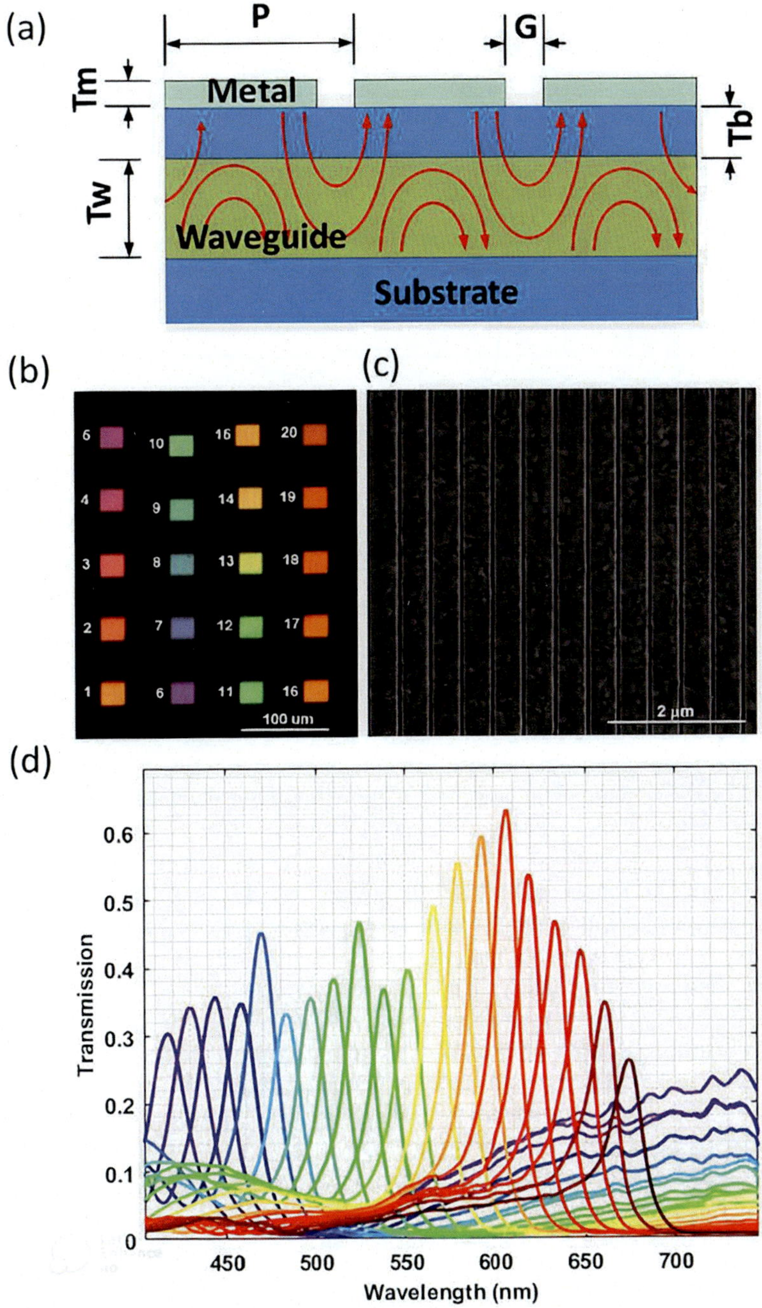

Fig. 9 (A) Schematic of plasmonic grating filter (Al metal with SiN as waveguide on SiO_2 substrate). *P* is the period of grating, *G* is the gap between fingers, Tm is the thickness of the metal, Tw is the thickness of the waveguide, and Tb is the distance between grating and waveguide; (B) RGB images of the filters, taken using optical microscopy; (C) SEM image of one of the Plasmonic Grating Filter; (D) Optical transmission spectra of the filters.
Source: Reused with permission from J.R. Shakya, F.H. Shashi, A.X. Wang, Plasmonic color filter array based visible light spectroscopy. Sci. Rep. 11 (2021). https://doi.org/10.1038/S41598-021-03092-3. Copyright © 2021, Springer Nature Limited.

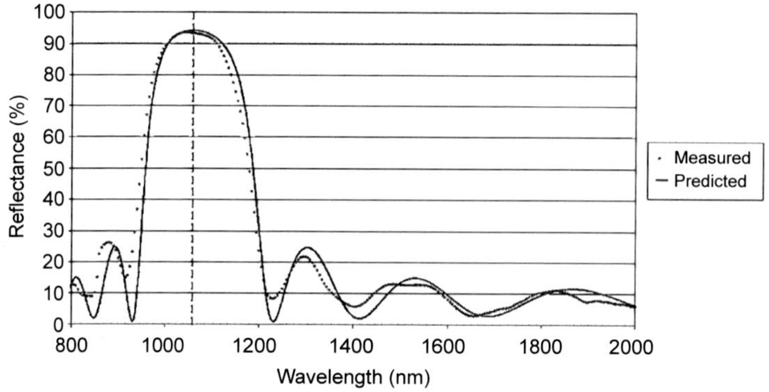

Fig. 10 Reflectance spectra (measured and predicted) of 15-layer metal oxide polymer nanocomposite films (TiO_2/SiO_2 nanoparticles dispersed in UV-cured acrylate polymer). Reused with permission from T. Druffel, N. Mandzy, M. Sunkara, E. Grulke, Polymer nanocomposite thin film mirror for the infrared region, Small 4 (2008) 459–461. https://doi.org/10.1002/SMLL.200700680. Source: Copyright © 2008, Wiley-VCH Verlag GmbH & Co. KGaA, Weinheim.

reduce the light intensity. ND absorptive filters are mostly used, with a rating of optical density. Common metallic glasses with Vi, Cr, Co, Fe, etc., and metal oxides are the common ND filters in the visible range. Broadband ND filters covering both visible and IR regions have great commercial potential in products like sunglasses, goggles, and other solar filtering optical elements. Zhang et al. utilized the electromagnetic filtering capability of graphene material to develop broadband ND filter, covering the range of 400–2500 nm. Fig. 11A shows the photographs of the graphene-based ND filter prepared using atmospheric pressure chemical vapor deposition process in quartz glass substrate. The sample coding 9C, 11C, 12C, and 13C represent samples with an average graphene layer thickness of 2.5, 5, 20, and 28 nm respectively. The transmission spectra of corresponding samples are shown in Fig. 11B showing varied amount of filtration in the complete solar spectrum [70].

Bandpass filters: Bandpass filters are used as application-specific, where the required band is transmitted across the filter. They are many types of band filters that use the absorptive or reflection principle. Long pass OF transmit all wavelengths longer than the specified cut-off wavelength and it includes cold mirrors, colored glass filters, Thermoset ADC (optical cast plastic) filters, etc. [71]. Short-pass filters transmit all wavelengths shorter than the specified cut-off wavelength and they include IR cut-off filters, hot mirrors, and heat-absorbing glass [72]. Apart from that broad and narrow bandpass filters are also there, based on the width of the transmitted wavelength window. Multiple bandpass filters can also be utilized to create the required wavelength range. For example, a long and short pass band filter, when properly selected can produce a narrow bandpass filter. Guo et al. conducted a numerical study on metal-dielectric subwavelength grating structure for narrow bandpass filters, in NIR region [73]. The metal-dielectric subwavelength grating is a 1D periodic array of air, gold, and Al_2O_3 elements as shown in Fig. 12A (where w_1 is the width of

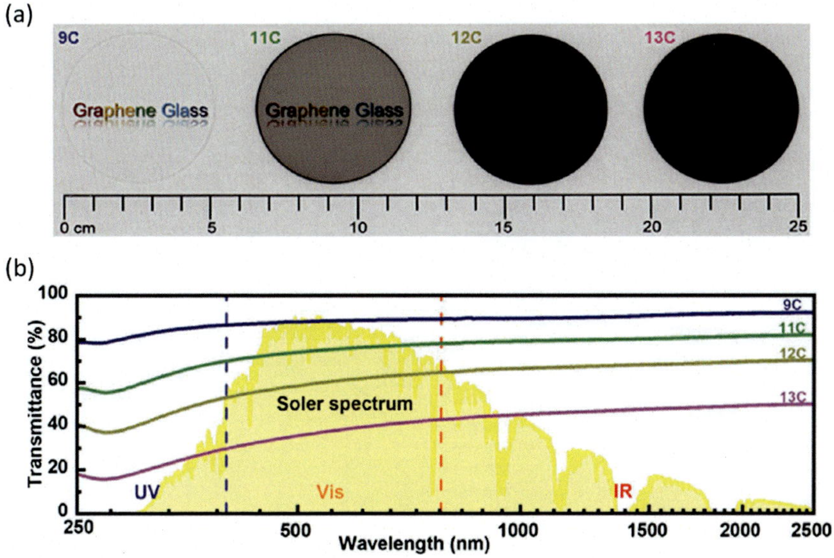

Fig. 11 (A) Photographs and (B) UV-Vis-NIR transmittance spectra of graphene-based broadband ND filter, prepared on quartz glass substrate. The samples 9C, 11C, 12C, and 13C have an average graphene layer thickness of 2.5, 5, 20, and 28 nm respectively.
Source: Reused with permission from Z. Zhang, F. Zhou, P. Yang, B. Jiang, J. Hu, X. Gao, et al., Direct growth of multi-layer graphene on quartz glass for high-performance broadband neutral density filter applications, Adv. Opt. Mater. 8 (2020) 2000166. https://doi.org/10.1002/ADOM.202000166. Copyright © 2020, Wiley-VCH Verlag GmbH & Co. KGaA, Weinheim.

Al_2O_3, w_2 is the width of air, w_{Au} is the width of gold metal, and h is the grating height; period is equal to $(w_1+w_2+2w_{Au})$). The simulation results yielded band pass filtering with ~10 nm line width in the NIR region (Fig. 12B). Further, the spectral band has been tuned in the NIR region with a varying height of grating, dielectric grating period, and dielectric width ratio and analyzed with optical spectroscopy (reflectance) as shown in Fig. 12C, D, and E, respectively.

Photonics-based OF: Photonics is the smallest unit of light and in its application level a single or banded wavelength of light is a necessity associated with the generation, transmission, manipulation, and detection of the radiation. Many photonic-based filters are used such as microwave filters, RF filters, etc., in the field of communication, optical interconnection network, and ultra-high-speed information processing.

5 Fabrication techniques for optical materials

Optical materials can be broadly classified as material-infused and coatings. In material infusion, the optical particle/material is infused into the matrix of concern at the time of solidification (for example, molten glass or curing of polymer). Optical

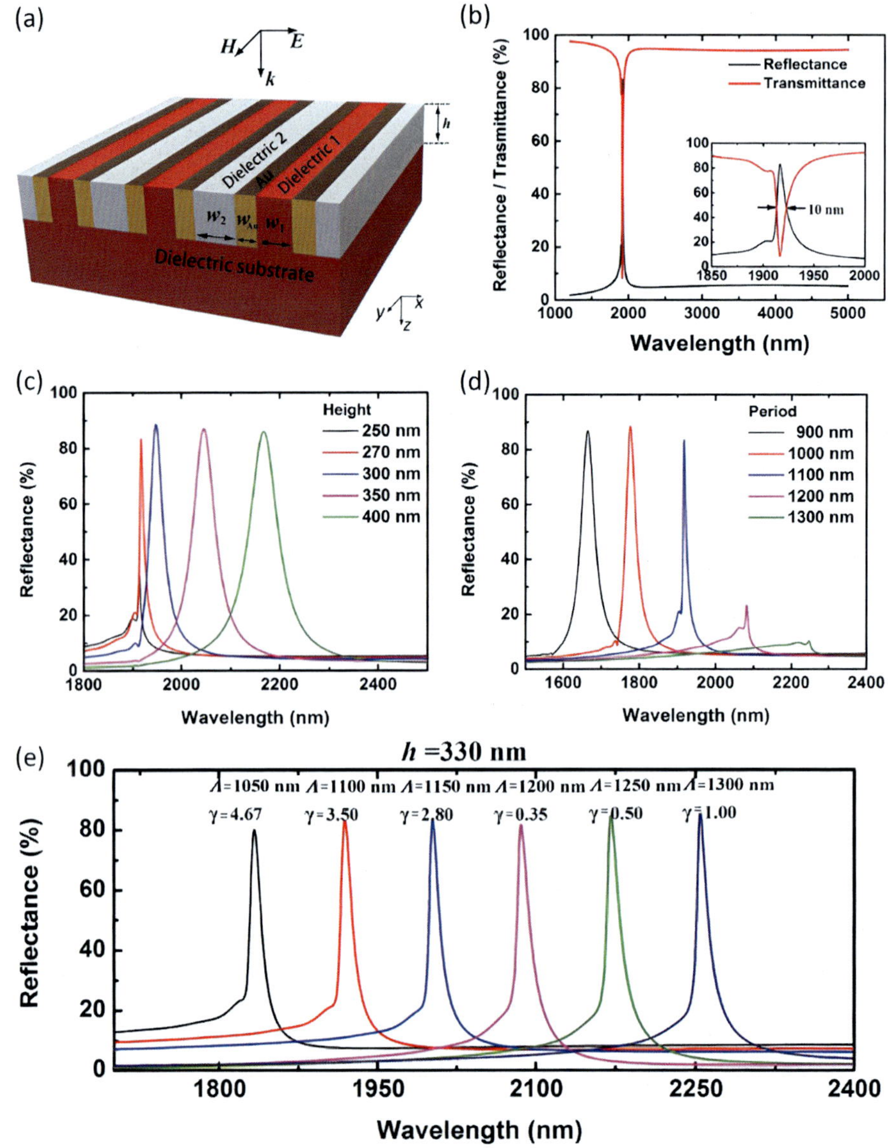

Fig. 12 (A) Schematic of metal-dielectric subwavelength grating structure for narrow band pass filter with dimensional notations; (B) reflection and transmission spectra of a grating structure with period 1100 nm, w_1 450 nm, w_2 450 nm, w_{Au} 100 nm, and h 270 nm. The inset shows the FWHM linewidth of the filter; (C) reflectance spectrum of filters with a varying height of the grating (the period is 1100 nm); (D) reflectance spectrum of filters with a varying period of the grating (height is 270 nm); (E) reflectance spectrum of filters with varying dielectric width ratio, γ (w_1/w_2) (height is 330 nm).

Source: Reused with permission from J. Guo, R. Zhang, Z. Wang, Quadrupole mode plasmon resonance enabled subwavelength metal-dielectric grating optical reflection filters, Opt. Exp. 26 (2018) 496–504. https://doi.org/10.1364/OE.26.000496. Copyright © 2018, Optical Society of America.

Table 3 Classification of fabrication techniques for optical coatings.

Fabrication techniques for optical coatings		
Bottom-up approach	**Top-down approach**	**Other unconventional approaches**
(1) Sol-gel method (2) Thermal evaporation (PVD) (3) Sputtering (PVD) (4) GLAD (PVD) (5) CVD	Etching (wet, dry)	(1) Lithography (2) Soft lithography (3) Miscellaneous (photo-aligning, photo-patterning, etc.)

coating fabrication can be conducted by the methods categorized as conventional (bottom-up, top-down) and unconventional approaches. The different techniques are classified and enlisted in Table 3. The bottom-up approach is generally used for thin film optical coatings, except for GLAD, while top-down and unconventional techniques like lithography can be used for creating patterned and porous optical coatings at the nano-level. Bottom-up approaches are mostly used for metal oxide optical coatings.

Sol-gel method: Sol-gel is one of the most adopted techniques for metal oxide synthesis, and the process can be considered cheaper, but the uniformity of coating is comparatively poor [74]. Sol-gel process for metal oxide involves the use of inorganic salts as a precursor. When the precursor is mixed with water or organic solvent then hydrolysis takes place to form metal hydroxides and condensation reactions take place which form the gel for coating. Once the coating is applied through different techniques like dip coating (through dipping of the substrate at a controlled rate), spin coating (deposition by spinning the substrate with the gel at high angular velocity), meniscus coating (coating by flowing the gel through a porous applicator), etc., then the coating needs heating to solidify and make oxide compound on the substrate. The sol-gel method is one of the most explored techniques for metal oxide optical materials. Jung et al. reported a double layer Al_2O_3/TiO_2 coating fabricated by the sol-gel technique with spin coating, which showed a reflectance of 4.75% in the range of 400–1000 nm wavelength [75]. Additional features of porosity are also explored with the sol-gel coating technique [76].

Thermal evaporation: In the thermal evaporation technique, the vapor of material is produced by heating the material till it evaporates and made to condense in the form of a thin film onto the substrate surface, making it a physical vapor deposition (PVD) process. The evaporation of material is carried out by resistive heating, electron beam heating, etc. Zaier et al. prepared ZnO thin films using the technique for obtaining coating with more than 90% visible transmittance [77]. Dwivedi et al. deposited 30 nm thin film coatings of Ag and ZnO and compared the optical transmission for its usage as OF in the 300 to 900 nm range.

Sputtering: Sputtering is also a PVD process, in which atoms are ejected from a solid target material and get deposited on the substrate by bombarding the target with high-energy particles like atoms or ions. Momentum transfer from the bombarding atoms or ions is employed to eject target material to be deposited on the substrate. Commonly, the target is bombarded with inert elements like argon ions to cause the target materials to come into the vapor phase. Different sputtering systems are available such as magnetron sputtering, ion-beam sputtering, reactive sputtering, etc., for the fabrication of thin film coatings. Wang et al. fabricated TiO_2 two-layer ARC coating with different methods of sputtering: direct reactive magnetron sputtering for creating a low refractive index layer and used energy filtrating magnetron sputtering for a higher refractive index layer [78]. The system showed average transmittance of 88.4% in the visible region and is used for photocatalytic application.

Glancing angle deposition (GLAD): GLAD is a PVD technique, with the prospectus of creating substrate inclination during the time of deposition, making a variation in deposition density, and also helping in creating morphological structures, unlike the other PVD techniques discussed [79]. Steele et al. used electron beam PVD with GLAD setup for fabricating bandpass interference filter with TiO_2 material by oscillating angle variations creating a varied porous, narrow-bandpass interference filter for humidity sensing application [80].

Apart from thermal evaporation, sputtering, and GLAD, other PVD methods used for optical coating fabrication are molecular beam epitaxy, pulsed laser deposition, cathodic arc deposition, electrohydrodynamic deposition (Electrospray deposition), etc.

Chemical vapor deposition (CVD): CVD generally uses a gas-phase precursor, often a halide, hydride or oxide of the materials to be deposited in which the chemical precursors are transported in the vapor phase to decompose, combine or react with other precursors on a heated substrate to form a film. The films can be epitaxial, polycrystalline, or amorphous depending on the reactor conditions and materials [81]. Depending on the vacuum conditions maintained inside the CVD reactor, the activation process of the precursors, as well as the type of thin films formed on the substrate, they are differentiated as low-pressure CVD, atmospheric pressure CVD, ultra-high vacuum CVD, plasma enhanced CVD, metal-organic CVD, etc. CVD is one of the widely used methods for film deposition in the semiconductor industry due to its high throughput, high purity, and low cost of operation. Neuman prepared SiO_2/TiO_2 GRIN ARC via atmospheric pressure CVD, and the results show a low reflectance of 0.5% [82]. Improved efficiency in multicrystalline silicon solar cells has been achieved by Hocine et al. by preparing TiO_2 ARC with existing atmospheric pressure CVD process, emphasizing on industrial compatibility and cost-effectiveness. The cell efficiency was improved from 11.24% to 14.26% [83].

Other general fabrication techniques for optical materials include:

Etching: The etching process basically involves material removal from the surface, or subsurface region, to create the desired morphology in the optical material. The etching process can be controlled as a uniform or nonuniform material removal. Two types of etching processes are:

(1) Wet etching: It involves the substrate being immersed inside the etching solution. The substrate should be inert or should be masked before the etching process. Wet etching has characteristics of a fast etching rate and precise and selective material removal.
(2) Dry etching: It involves ablation or volatilization of material through plasma energy or ion bombardment.

Lithography: Lithography is a method of printing patterns on surfaces, e.g., grooves, an array of nanostructures, etc., which will generate the optical conditions of antireflection or filtration. There are several types of lithographic techniques used to generate the patterns such as photolithography, ion beam lithography (IBL), electron lithography, mold lithography (nanoimprint lithography (NIL)), scanning probe lithography, etc. [84]. Photolithography, IBL, and NIL are the most commonly used lithographic technique for optical coating fabrication [19]. Photolithography involves the coating of photoresist material on the previously coated material (the optical layer). A beam of electromagnetic radiation is then made to fall on the photoresist, through a mask of the required pattern. When light falls on uncovered photoresist material, the solubility of photoresist in a developer solution changes, and the exposed area will be subsequently removed. The optical layer not covered with the photoresist is then etched out and finally, the extra photoresist material is also removed to obtain the required pattern [85]. IBL uses a similar technique with the replacement of electromagnetic radiation with an electron beam. More fine patterning and precision can be attained in IBL as there will be less backscattering. NIL is a molding technique, where precise hard mold is used to develop their imprint in optical materials. J.Y. Chen et al. fabricated nanostructured PMMA optical layers using NIL with spin-coating replication/hot-embossing technique. The initial Si mold for NIL is fabricated using electron beam lithography and subsequent chemical wet etching. The nanostructured PMMA ARC produced an improved cell efficiency from 10.4% to 13.5% [86].

Soft lithography: It is an extension of lithography techniques, except the method utilizes stamp (elastomeric) or mold to deposit optical material as a single ink layer or as bulk on the substrate. It has additional possibilities of using soft materials such as polymers, gels, etc., and unlike lithographic processes, the cost factor is also favorable for soft lithography. Different modes of soft lithography processes are there like replica molding, capillary molding, microcontact printing, microtransfer molding, etc. Lim et al. fabricated an inversely tapered nanohole as an optical layer on PDMS by soft lithography with a reverse pattern of conical Si nanopillar mold. The solar-weighted reflectance was obtained to be a low value of 7.1% at 0 degrees incident angle and an average of 8.5% at 20 degrees to 70 degrees angle range [87].

6 Summary and outlook

In summary, this chapter puts a broader view of different antireflective coatings (ARC) and optical filters (OF) with their working principle, material selection, and fabrication techniques, focused mainly on metal oxides as the optical material. With the wide range of applicability of both ARC and OF, the technology selection seems to be crucial and challenging. The required efficiency for the functional specificity and additional functionalities is becoming the key to extending the application of ARC and OF. However, designing and manufacturing a perfect optical material/surface for antireflection and wavelength filtering becomes far-fetched. Low-cost manufacturing feasibility of the optical elements seems to be the most challenging for their widespread applicability. Apart from that, the optics/material stability, mechanical stability, ease of fabrication are extremely critical in defining the specific ARC/OF for an application. In conclusion, the book chapter is expected to serve as a quick guide for researchers and associated industries to review different aspects of ARC and OF, for their rational selection based on the functional output, working environment, fabrication process, and material of selection.

References

[1] J.M. Enoch, History of mirrors dating back 8000 years, Optom. Vis. Sci. 83 (2006) 775–781, https://doi.org/10.1097/01.OPX.0000237925.65901.C0.

[2] A.J. Haider, A.A. Najim, M.A.H. Muhi, TiO_2/Ni composite as antireflection coating for solar cell application, Opt. Commun. 370 (2016) 263–266, https://doi.org/10.1016/J.OPTCOM.2016.03.034.

[3] S. Prabhu, L. Cindrella, O. Joong Kwon, K. Mohanraju, Superhydrophilic and self-cleaning rGO-TiO2 composite coatings for indoor and outdoor photovoltaic applications, Sol. Energy Mater. Sol. Cells 169 (2017) 304–312, https://doi.org/10.1016/J.SOLMAT.2017.05.023.

[4] C. Glynn, C. O'Dwyer, Solution processable metal oxide thin film deposition and material growth for electronic and photonic devices, Adv. Mater. Interfaces 4 (2017) 1600610, https://doi.org/10.1002/ADMI.201600610.

[5] D. Adak, S. Ghosh, P. Chakraborty, K.M.K. Srivatsa, A. Mondal, H. Saha, et al., Non lithographic block copolymer directed self-assembled and plasma treated self-cleaning transparent coating for photovoltaic modules and other solar energy devices, Sol. Energy Mater. Sol. Cells 188 (2018) 127–139, https://doi.org/10.1016/J.SOLMAT.2018.08.011.

[6] R.E. Nowak, M. Vehse, O. Sergeev, T. Voss, M. Seyfried, K. von Maydell, et al., ZnO nanorods with broadband antireflective properties for improved light management in silicon thin-film solar cells, Adv. Opt. Mater. 2 (2014) 94–99, https://doi.org/10.1002/ADOM.201300455.

[7] Y.F. Makableh, R. Vasan, J.C. Sarker, A.I. Nusir, S. Seal, M.O. Manasreh, Enhancement of GaAs solar cell performance by using a ZnO sol–gel anti-reflection coating, Sol. Energy Mater. Sol. Cells 123 (2014) 178–182, https://doi.org/10.1016/J.SOLMAT.2014.01.007.

[8] Y. Qu, X. Huang, Y. Li, G. Lin, B. Guo, D. Song, et al., Chemical bath deposition produced ZnO nanorod arrays as an antireflective layer in the polycrystalline Si solar cells, J. Alloys Compd. 698 (2017) 719–724, https://doi.org/10.1016/J.JALLCOM.2016.12.265.

[9] J.H. Yun, E. Lee, H.H. Park, D.W. Kim, W.A. Anderson, J. Kim, et al., Incident light adjustable solar cell by periodic nanolens architecture, Sci. Rep. 4 (1) (2014) 1–8, https://doi.org/10.1038/srep06879.

[10] F. Rubio, J. Denis, J.M. Albella, J.M. Martinez-Duart, Sputtered Ta2O5 antireflection coatings for silicon solar cells, Thin Solid Films 90 (1982) 405–408, https://doi.org/10.1016/0040-6090(82)90545-4.

[11] J. Ham, J.Y. Park, W.J. Dong, G.H. Jung, H.K. Yu, J.L. Lee, Antireflective indium-tin-oxide nanobranches for efficient organic solar cells, Appl. Phys. Lett. 108 (2016) 073903, https://doi.org/10.1063/1.4942399.
[12] W.C. Tien, A.K. Chu, Double-layer ITO antireflection electrodes fabricated at low temperature, Sol. Energy Mater. Sol. Cells 100 (2012) 258–262, https://doi.org/10.1016/J.SOLMAT.2012.01.029.
[13] T. Druffel, N. Mandzy, M. Sunkara, E. Grulke, Polymer nanocomposite thin film mirror for the infrared region, Small 4 (2008) 459–461, https://doi.org/10.1002/SMLL.200700680.
[14] W. Xie, M. Yang, Y. Cheng, D. Li, Y. Zhang, Z. Zhuang, et al., Optical fiber relative-humidity sensor with evaporated dielectric coatings on fiber end-face, Opt. Fiber Technol. 20 (2014) 314–319, https://doi.org/10.1016/J.YOFTE.2014.03.008.
[15] M.I. Hossain, A. Khandakar, M.E.H. Chowdhury, S. Ahmed, M.M. Nauman, B. Aïssa, Numerical and experimental investigation of infrared optical filter based on metal oxide thin films for temperature mitigation in photovoltaics, J. Electron. Mater. 51 (2022) 179–189, https://doi.org/10.1007/S11664-021-09269-W/FIGURES/10.
[16] A. Chisvert, A. Salvador, UV filters in sunscreens and other cosmetics. Regulatory aspects and analytical methods, Anal. Cosmet. Prod. (2007) 83–120, https://doi.org/10.1016/B978-044452260-3/50028-0.
[17] F. Gildas, Y. Dan, Review of nanostructure color filters, J. Nanophoton. 13 (2019) 020901, https://doi.org/10.1117/1.JNP.13.020901.
[18] F. Flory, L. Escoubas, G. Berginc, Optical properties of nanostructured materials: a review, 5 (2011) 052502, https://doi.org/10.1117/1.3609266.
[19] N. Shanmugam, R. Pugazhendhi, R.M. Elavarasan, P. Kasiviswanathan, N. Das, Anti-reflective coating materials: a holistic review from PV perspective, Energies 13 (2020) 2631, https://doi.org/10.3390/EN13102631.
[20] T.P. Mollart, K.L. Lewis, Transition metal oxide anti-reflection coatings for airborne diamond optics, Diam. Relat. Mater. 10 (2001) 536–541, https://doi.org/10.1016/S0925-9635(00)00516-1.
[21] I. Arabatzis, N. Todorova, I. Fasaki, C. Tsesmeli, A. Peppas, W.X. Li, et al., Photocatalytic, self-cleaning, antireflective coating for photovoltaic panels: characterization and monitoring in real conditions, Sol. Energy 159 (2018) 251–259, https://doi.org/10.1016/J.SOLENER.2017.10.088.
[22] K.C. Kim, Effective graded refractive-index anti-reflection coating for high refractive-index polymer ophthalmic lenses, Mater. Lett. 160 (2015) 158–161, https://doi.org/10.1016/J.MATLET.2015.07.108.
[23] R. Prado, G. Beobide, A. Marcaide, J. Goikoetxea, A. Aranzabe, Development of multifunctional sol–gel coatings: anti-reflection coatings with enhanced self-cleaning capacity, Sol. Energy Mater. Sol. Cells 94 (2010) 1081–1088, https://doi.org/10.1016/J.SOLMAT.2010.02.031.
[24] D. Przybylski, S. Patela, Modelling of a two-dimensional photonic crystal as an anti-reflection coating for optoelectronic applications, Opto-Electron. Rev. 27 (2019) 79–89, https://doi.org/10.1016/J.OPELRE.2019.02.004.
[25] T. Murata, H. Ishizawa, A. Tanaka, High-performance antireflective coatings with a porous nanoparticle layer for visible wavelengths, Appl. Opt. 50 (9) (2011) C403–C407, https://doi.org/10.1364/AO.50.00C403.
[26] S.B. Khan, H. Wu, C. Pan, Z. Zhang, A mini review: antireflective coatings processing techniques, applications and future perspective, Res. Rev. J. Mater. Sci. 5 (2017) 1–19, https://doi.org/10.4172/2321-6212.1000192.

[27] H.K. Raut, V.A. Ganesh, A.S. Nair, S. Ramakrishna, Anti-reflective coatings: a critical, in-depth review, Energy Environ. Sci. 4 (2011) 3779–3804, https://doi.org/10.1039/C1EE01297E.
[28] P.B. Clapham, M.C. Hutley, Reduction of lens reflexion by the "moth eye" principle, Nature 244 (1973) 281–282, https://doi.org/10.1038/244281a0.
[29] S.I. Bae, Y. Lee, Y.H. Seo, K.H. Jeong, Antireflective structures on highly flexible and large area elastomer membrane for tunable liquid-filled endoscopic lens, Nanoscale 11 (2019) 856–861, https://doi.org/10.1039/C8NR06553E.
[30] K. Anissa Tabet Aoul, D. Efurosibina Attoye, L. Al Ghatrif, Performance of electrochromic glazing: state of the art review, IOP Conf. Ser. Mater. Sci. Eng. 603 (2019) 022085, https://doi.org/10.1088/1757-899X/603/2/022085.
[31] M.K. Hedayati, M. Elbahri, Antireflective coatings: conventional stacking layers and ultrathin plasmonic metasurfaces, a mini-review, Materials 9 (2016) 497, https://doi.org/10.3390/MA9060497.
[32] M. Moayedfar, M.K. Assadi, Various types of anti-reflective coatings (ARCS) based on the layer composition and surface topography: a review, Rev. Adv. Mater. Sci. 53 (2018) 187–205, https://doi.org/10.1515/RAMS-2018-0013.
[33] F. Zhan, Z. Li, X. Shen, H. He, J. Zeng, Design multilayer antireflection coatings for terrestrial solar cells, Sci. World J. (2014) 2014, https://doi.org/10.1155/2014/265351.
[34] H. Kanda, A. Uzum, N. Harano, S. Yoshinaga, Y. Ishikawa, Y. Uraoka, et al., Al2O3/TiO2 double layer anti-reflection coating film for crystalline silicon solar cells formed by spray pyrolysis, Energy Sci. Eng. 4 (2016) 269–276, https://doi.org/10.1002/ESE3.123.
[35] W. Lin, J. Zheng, L. Yan, X. Zhang, Sol-gel preparation of self-cleaning SiO2-TiO2/SiO2-TiO2 double-layer antireflective coating for solar glass, Results Phys. 8 (2018) 532–536, https://doi.org/10.1016/J.RINP.2017.12.058.
[36] S. Chattopadhyay, Y.F. Huang, Y.J. Jen, A. Ganguly, K.H. Chen, L.C. Chen, Anti-reflecting and photonic nanostructures, Mater. Sci. Eng. R Rep. 69 (2010) 1–35, https://doi.org/10.1016/J.MSER.2010.04.001.
[37] Z. Ge, P. Rajbhandari, J. Hu, A. Emrani, T.P. Dhakal, C. Westgate, et al., Enhanced omni-directional performance of copper zinc tin sulfide thin film solar cell by gradient index coating, Appl. Phys. Lett. 104 (2014) 101104, https://doi.org/10.1063/1.4868104.
[38] F.E. Williams, F.H. Nicoll, Properties of low reflection films produced by the action of hydrofluoric acid vapor, JOSA 33 (1943) 434–435, https://doi.org/10.1364/JOSA.33.000434.
[39] S.F. Monaco, Reflectance of an inhomogeneous thin film, JOSA 51 (3) (1961) 280–282, https://doi.org/10.1364/JOSA.51.000280.
[40] C.-H. Kuo, T.-C. Teng, Y.-J. Li, Planar solar concentrator composed of stacked waveguides with arc-segment structures and movable receiving assemblies, Opt Exp. 28 (23) (2020) 34362–34377, https://doi.org/10.1364/OE.405909.
[41] M.J. Minot, The angluar reflectance of single-layer gradient refractive-index films, JOSA 67 (8) (1977) 1046–1050, https://doi.org/10.1364/JOSA.67.001046.
[42] S. Walheim, E. Schäffer, J. Mlynek, U. Steiner, Nanophase-separated polymer films as high-performance antireflection coatings, Science 283 (1999) 520–522, https://doi.org/10.1126/SCIENCE.283.5401.520.
[43] K.M.A. Sobahan, Y.J. Park, J.J. Kim, C.K. Hwangbo, Nanostructured porous SiO2 films for antireflection coatings, Opt. Commun. 284 (2011) 873–876, https://doi.org/10.1016/J.OPTCOM.2010.09.075.
[44] Z.W. Han, Z. Wang, X.M. Feng, B. Li, Z.Z. Mu, J.Q. Zhang, et al., Antireflective surface inspired from biology: a review, Biosurf. Biotribol. 2 (2016) 137–150, https://doi.org/10.1016/J.BSBT.2016.11.002.

[45] S.A. Boden, D.M. Bagnall, Tunable reflection minima of nanostructured antireflective surfaces, Appl. Phys. Lett. 93 (2008) 133108, https://doi.org/10.1063/1.2993231.
[46] D.G. Stavenga, S. Foletti, G. Palasantzas, K. Arikawa, Light on the moth-eye corneal nipple array of butterflies, Proc. R. Soc. B Biol. Sci. 273 (2006) 661, https://doi.org/10.1098/RSPB.2005.3369.
[47] A.R. Parker, Z. Hegedus, R.A. Watts, Solar-absorber antireflector on the eye of an Eocene fly (45 Ma), Proc. R. Soc. B Biol. Sci. 265 (1998) 811, https://doi.org/10.1098/RSPB.1998.0364.
[48] A. Yoshida, M. Motoyama, A. Kosaku, K. Miyamoto, Antireflective nanoprotuberance array in the transparent wing of a hawkmoth, *Cephonodes hylas*, Zool. Sci. 14 (1997) 737–741, https://doi.org/10.2108/ZSJ.14.737.
[49] S.Y. Han, B.K. Paul, C.H. Chang, Nanostructured ZnO as biomimetic anti-reflective coatings on textured silicon using a continuous solution process, J. Mater. Chem. 22 (2012) 22906–22912, https://doi.org/10.1039/C2JM33462C.
[50] H.G. Craighead, R.E. Howard, J.E. Sweeney, D.M. Tennant, Textured surfaces: optical storage and other applications, J. Vac. Sci. Technol. 20 (1998) 316, https://doi.org/10.1116/1.571291.
[51] H.G. Craighead, R.E. Howard, D.M. Tennant, Selectively emissive refractory metal surfaces, Appl. Phys. Lett. 38 (1981) 74–76, https://doi.org/10.1063/1.92253.
[52] J.I. Gittleman, E.K. Sichel, H.W. Lehmann, R. Widmer, Textured silicon: a selective absorber for solar thermal conversion, Appl. Phys. Lett. 35 (2008) 742, https://doi.org/10.1063/1.90953.
[53] Y. Saito, T. Kosuge, Honeycomb-textured structures on crystalline silicon surfaces for solar cells by spontaneous dry etching with chlorine trifluoride gas, Sol. Energy Mater. Sol. Cells 91 (2007) 1800–1804, https://doi.org/10.1016/J.SOLMAT.2007.06.009.
[54] K.V. Baryshnikova, M.I. Petrov, V.E. Babicheva, P.A. Belov, Plasmonic and silicon spherical nanoparticle antireflective coatings, Sci. Rep. 6 (2016) 1–11, https://doi.org/10.1038/srep22136.
[55] A. Sarangan, K. Hirakawa, R. Heenkenda, Tunable optical filter using phase change materials for smart IR night vision applications, Opt. Exp. 29 (21) (2021) 33795–33803, https://doi.org/10.1364/OE.440299.
[56] R.I. Álvarez-Tamayo, M. Durán-Sánchez, O. Pottiez, E.A. Kuzin, B. Ibarra-Escamilla, A. Flores-Rosas, Theoretical and experimental analysis of tunable Sagnac high-birefringence loop filter for dual-wavelength laser application, Appl. Opt. 50 (3) (2011) 253–260, https://doi.org/10.1364/AO.50.000253.
[57] T. Erdogan, Optical filters for wavelength selection in fluorescence instrumentation, Curr. Protoc. Cytom. 56 (2011) 2.4.1–2.4.25, https://doi.org/10.1002/0471142956.CY0204S56.
[58] T. Karlessi, M. Santamouris, Improving the performance of thermochromic coatings with the use of UV and optical filters tested under accelerated aging conditions, Int. J. Low-Carbon Technol. 10 (2015) 45–61, https://doi.org/10.1093/IJLCT/CTT027.
[59] B. Tupper, D. Miller, R. Miller, The effect of a 550 nm cutoff filter on the vision of cataract patients, Ann. Ophthalmol. 17 (1985) 67–72.
[60] F. Shen, Q. Kang, J. Wang, K. Guo, Q. Zhou, Z. Guo, Dielectric metasurface-based high-efficiency mid-infrared optical filter, Nanomater 8 (2018) 938, https://doi.org/10.3390/NANO8110938.
[61] S. Mariani, V. Robbiano, R. Iglio, A.A. La Mattina, P. Nadimi, J. Wang, et al., Moldless printing of silicone lenses with embedded nanostructured optical filters, Adv. Funct. Mater. 30 (2020) 1906836, https://doi.org/10.1002/ADFM.201906836.
[62] J. Pool, S.N. Osterman, H. Lovelady, K.J. Ryan, Design, fabrication, and test of a patterned optical filter array for the Europa imaging system (EIS), in: SPIE Astronomical

Telescopes + Instrumentation, vol. 10706, 2018, pp. 1477–1486, https://doi.org/10.1117/12.2312754.

[63] R.A. Anjali, A review on different optical filters, J. Manag. Eng. Inf. Technol. 7 (2020) 2394–8124.

[64] J.T. Trauger, Broadband dielectric mirror coatings for Fabry-Perot spectroscopy, Appl. Opt. 15 (12) (1976) 2998–3005, https://doi.org/10.1364/AO.15.002998.

[65] Fiber Bragg Gratings Filter WDM Signals | Features, Photonics Spectra, 2003, March. https://www.photonics.com/Articles/Fiber_Bragg_Gratings_Filter_WDM_Signals/a15318. (Accessed 21 March 2022).

[66] J.K. Sahota, N. Gupta, D. Dhawan, Fiber Bragg grating sensors for monitoring of physical parameters: a comprehensive review, Opt. Eng. 59 (2020) 060901, https://doi.org/10.1117/1.OE.59.6.060901.

[67] J.R. Shakya, F.H. Shashi, A.X. Wang, Plasmonic color filter array based visible light spectroscopy, Sci. Rep. (2021) 11, https://doi.org/10.1038/S41598-021-03092-3.

[68] P. Guo, J. Xiong, X. Zhao, B. Tao, X. Liu, Nano-structured optical hetero-coatings for ultraviolet protection, Mater. Lett. 152 (2015) 290–292, https://doi.org/10.1016/J.MATLET.2015.03.129.

[69] J. Dai, W. Gao, B. Liu, X. Cao, T. Tao, Z. Xie, et al., Design and fabrication of UV band-pass filters based on SiO2/Si3N4 dielectric distributed bragg reflectors, Appl. Surf. Sci. 364 (2016) 886–891, https://doi.org/10.1016/J.APSUSC.2015.12.222.

[70] Z. Zhang, F. Zhou, P. Yang, B. Jiang, J. Hu, X. Gao, et al., Direct growth of multi-layer graphene on quartz glass for high-performance broadband neutral density filter applications, Adv. Opt. Mater. 8 (2020) 2000166, https://doi.org/10.1002/ADOM.202000166.

[71] S.M. El-Bashir, I.S. Yahia, M.A. Binhussain, M.S. AlSalhi, Designing of PVA/Rose Bengal long-pass optical window applications, Results Phys. 7 (2017) 1238–1244, https://doi.org/10.1016/J.RINP.2017.03.033.

[72] V. Lehmann, R. Stengl, H. Reisinger, R. Detemple, W. Theiss, Optical shortpass filters based on macroporous silicon, Appl. Phys. Lett. 78 (2001) 589, https://doi.org/10.1063/1.1334943.

[73] J. Guo, R. Zhang, Z. Wang, Quadrupole mode plasmon resonance enabled subwavelength metal-dielectric grating optical reflection filters, Opt. Exp. 26 (1) (2018) 496–504, https://doi.org/10.1364/OE.26.000496.

[74] X. Sun, X. Xu, J. Tu, P. Yan, G. Song, L. Zhang, et al., Research status of antireflection film based on TiO2, IOP Conf. Ser. Mater. Sci. Eng. 490 (2019) 022074, https://doi.org/10.1088/1757-899X/490/2/022074.

[75] J. Jung, A. Jannat, M.S. Akhtar, O.-B. Yang, Sol–gel deposited double layer TiO_2 and Al_2O_3 anti-reflection coating for silicon solar cell, J. Nanosci. Nanotechnol. 18 (2017) 1274–1278, https://doi.org/10.1166/JNN.2018.14928.

[76] M. Ramirez-Del-Solar, E. Blanco, Porous thin films from sol-gel, Submicron Porous Mater. (2017) 157–188, https://doi.org/10.1007/978-3-319-53035-2_6.

[77] A. Zaier, A. Meftah, A.Y. Jaber, A.A. Abdelaziz, M.S. Aida, Annealing effects on the structural, electrical and optical properties of ZnO thin films prepared by thermal evaporation technique, J. King Saud Univ. – Sci. 27 (2015) 356–360, https://doi.org/10.1016/J.JKSUS.2015.04.007.

[78] Z. Wang, N. Yao, X. Hu, Single material TiO2 double layers antireflection coating with photocatalytic property prepared by magnetron sputtering technique, Vacuum 108 (2014) 20–26, https://doi.org/10.1016/J.VACUUM.2014.05.009.

[79] A. Barranco, A. Borras, A.R. Gonzalez-Elipe, A. Palmero, Perspectives on oblique angle deposition of thin films: from fundamentals to devices, Prog. Mater. Sci. 76 (2016) 59–153, https://doi.org/10.1016/J.PMATSCI.2015.06.003.

[80] J.J. Steele, A.C. van Popta, M.M. Hawkeye, J.C. Sit, M.J. Brett, Nanostructured gradient index optical filter for high-speed humidity sensing, Sensors Actuators B Chem. 120 (2006) 213–219, https://doi.org/10.1016/J.SNB.2006.02.003.

[81] S.A. Campbell, R.C. Smith, Chemical vapor deposition, in: High-k Gate Dielectrics, 2002, pp. 277–355, https://doi.org/10.1016/B978-012524975-1/50009-4.

[82] G.A. Neuman, Anti-reflective coatings by APCVD using graded index layers, J. Non-Cryst. Solids 218 (1997) 92–99, https://doi.org/10.1016/S0022-3093(97)00160-9.

[83] D. Hocine, M.S. Belkaid, M. Pasquinelli, L. Escoubas, J.J. Simon, G.A. Rivière, et al., Improved efficiency of multicrystalline silicon solar cells by TiO2 antireflection coatings derived by APCVD process, Mater. Sci. Semicond. Process. 16 (2013) 113–117, https://doi.org/10.1016/J.MSSP.2012.06.004.

[84] N. Kooy, K. Mohamed, L.T. Pin, O.S. Guan, A review of roll-to-roll nanoimprint lithography, Nanoscale Res. Lett. 9 (2014) 1–13, https://doi.org/10.1186/1556-276X-9-320/FIGURES/20.

[85] S. Zhou, M. Hu, Q. Guo, X. Cai, X. Xu, J. Yang, Solvent-transfer assisted photolithography of high-density and high-aspect-ratio superhydrophobic micropillar arrays, J. Micromech. Microeng. 25 (2015) 025005, https://doi.org/10.1088/0960-1317/25/2/025005.

[86] J.Y. Chen, K.W. Sun, Enhancement of the light conversion efficiency of silicon solar cells by using nanoimprint anti-reflection layer, Sol. Energy Mater. Sol. Cells 94 (2010) 629–633, https://doi.org/10.1016/J.SOLMAT.2009.11.028.

[87] J.H. Lim, J.W. Leem, J.S. Yu, Solar power generation enhancement of dye-sensitized solar cells using hydrophobic and antireflective polymers with nanoholes, RSC Adv. 5 (2015) 61284–61289, https://doi.org/10.1039/C5RA10269C.

Colloidal metal oxides and their optoelectronic and photonic applications

13

Sangeetha M.S.[a,b], *Sayoni Sarkar*[c], *Ajit R. Kulkarni*[c,d], *and Adersh Asok*[a,b]
[a]Photosciences and Photonics, Chemical Sciences and Technology Division, CSIR-National Institute for Interdisciplinary Science and Technology (NIIST), Thiruvananthapuram, Kerala, India, [b]Academy of Scientific and Innovative Research (AcSIR), Ghaziabad, Uttar Pradesh, India, [c]Centre for Research in Nano Technology and Science, Indian Institute of Technology Bombay, Mumbai, Maharashtra, India, [d]Metallurgical Engineering and Materials Science, Indian Institute of Technology Bombay, Mumbai, Maharashtra, India

Chapter outline

1 Introduction

Metal oxide forms an integral part of the optoelectronic and photonic devices that are extensively used in various fields such as communication, entertainment, health, and security, to name a few [1]. Currently, these intelligent devices are the major drivers for improving the human living standard and its advancement in many ways. Due to its increasing demand, the global photonics and optoelectronics market is expected to generate a revenue of ~USD 1.02 trillion by 2026 [2]. To briefly introduce optoelectronics, it is a subdiscipline of photonics that deals with the study and application of light-emitting or light-detecting devices typically in the visible and NIR regions of the

Metal Oxides for Next-Generation Optoelectronic, Photonic, and Photovoltaic Applications
https://doi.org/10.1016/B978-0-323-99143-8.00011-0

electromagnetic spectrum [3]. In principle, an optoelectronic device functions based on the photoelectric effect due to the light-matter interaction; thus, these devices will act as an electrical-to-optical or optical-to-electrical transducer [4].

Metal oxide semiconductors (MOS) have received significant attention from researchers working on optoelectronic and photonic devices due to their superior structural, electrical, and optical properties. Hence, this chapter attempts to introduce the theory of metal oxide semiconductors and their electrical and optical features that are essential to explore their feasibility and applicability as a working material in various optoelectronics and photonics applications. Taking account of its excellent processability for device fabrication and flexibility of synthesis approaches for tuning the electrical and optical properties, special attention has been given to colloidal metal oxides (CMOs) in the subsequent sections of this chapter. Various synthesis techniques and application-specific characterization of CMOs will be elaborated along with their massive potential in the optoelectronic and photonic sector, for example, in light-emitting devices (LED), photodetectors (PD), solar cells, and thin-film transistors (TFT). Finally, we also present some existing challenges and expectations of CMOs to meet the requirements of next-generation materials and devices for the photonics and optoelectronics industry.

1.1 Metal oxide semiconductors (MOS)

Metal oxides (MO) are the most plentiful materials in the Earth's crust as bulk, thin-film forms, and their nanostructures evince a great variety of functional properties, making them classic for optoelectronic applications and photonic devices. The first MOS-based thin-film transistor (TFT) was reported in 1964 using SnO [5]. Metal oxide semiconductors are remarkably different from conventional inorganic semiconductors (silicon and III–V compounds), in which the interaction between the metal and oxide orbitals (Fig. 1) resulted in a significant discrepancy in the charge carrier transport. According to the traditional bandgap theory, larger carrier mobility occurs when the effective masses of electron and hole are smaller to cause greater hybridization between valence band maximum (VBM) and conduction band minimum (CBM)

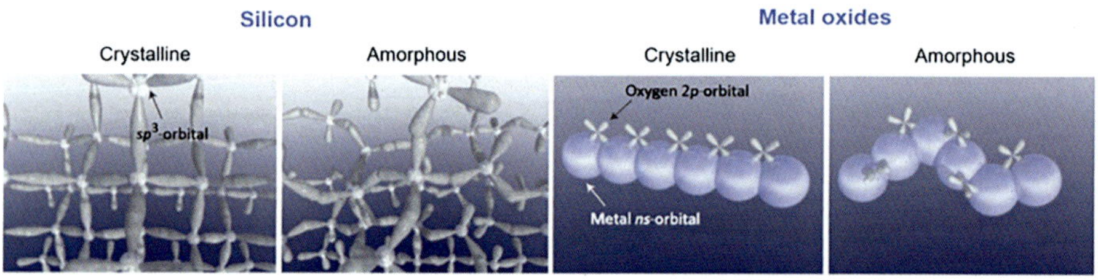

Fig. 1 Orbital schematic for the interaction of valence and conduction bands in crystalline and amorphous Si and metal oxides.

Source: Reused with permission from T. Kamiya, K. Nomura, H. Hosono, Origins of high mobility and low operation voltage of amorphous oxide TFTs: electronic structure, electron transport, defects and doping, J. Display Technol. 5 (7) (2009) 273–288. Copyright © 2016, Nature Publishing Group.

[6]. In the case of MOS, CBM is the spatially extended, spherically symmetrical metal *ns* bonding state with minimum lattice distortions, while VBM is localized. Hence, CBM carries small electron effective masses resulting in efficient electron transport compared to hole transport which is prevented by localized oxygen 2*p* orbitals and deep VBM levels where oxygen ions trap holes. Hence, this is reported as one of the primary reasons for the predominant n-type conductivity observed in most metal oxide semiconductors reported to date. [7,8]

In contrast, conventional Si semiconductors with spatially directional sp^3 σ states experience reduced carrier mobilities (~0.5–1.0 $cm^2\ V^{-1}\ s^{-1}$) in the amorphous state. Besides, they suffer optical opacity, poor current-carrying capacity, and modest mechanical flexibility. However, MO exhibits unique properties, including excellent carrier mobilities even in the amorphous state, compatibility with organic dielectric and photoactive materials, and high optical transparency. These potential characteristics are responsible for overcoming the challenges presented by conventional semiconductor materials, thereby enabling both conventional and entirely new functions. Furthermore, concerning electronic structure, charge transport mechanisms, defect states, thin-film processing strategies, and optoelectronic properties, MOS shows uniqueness and universality compared with other unconventional organic electronic materials, CNTs, and 2D materials. Therefore, they offer vast application potential.

It is observed that the crystal structure, composition, native defects and doping, and the synthetic routes and growth parameters that control MO physicochemical and morphostructural properties significantly affect their functional properties [5].

1.2 Colloidal metal oxide semiconductor (CMOS)

Recently, solution-processed optoelectronic and photonic devices have attracted much interest as they offer notable advantages such as easy manufacture, considerable device area, physical flexibility, and affordability. Among MO semiconductors, colloidal metal oxide (CMO) nanocrystals come up with a special package of exceptional solution processability at low temperature, versatile optoelectronic properties, and intrinsic stability [9]. Colloidal nanocrystals (NCs) are fragments of the corresponding crystalline lattice of bulk inorganic solids, generally in the range of ~2–20 nm in length. As the size of NCs approaches the size of semiconductor exciton, it starts exhibiting a quantum confinement effect; such colloids are often known as quantum dots [10]. The important fabrication techniques for solution-processed metal oxide devices are the precursor and nanocrystal approaches. The nanocrystal approach shows an edge-over precursor approach concerning freedom of development of chemistry for solution processing and device design architecture. Crystallization of oxide materials from the film deposition process can be decoupled in the nanocrystal approach. The nanocrystal approach has no restriction on applying harsh reaction conditions, purity methods of products, and modulation of solubility with low-temperature processability. Furthermore, the requirement of removal of unwanted byproducts limits the applicability of the precursor approach while drawing much significance to the chemistry of the colloidal nanocrystal approach [9]. Various solution process methods are known, such as metering rods, spin coating, dip coating, inkjet printing, spray coating, and roll-to-roll techniques, which will be detailed later [11]. These processes are cost-effective and large-area fabrication is easier in comparison with vapor-based processes employed in most traditional semiconductors. Moreover, solution processability provides high-quality CMO thin films that can be completed at ambient conditions (25°C under air), enabling compatibility with flexible and lightweight plastic substrates [12–14].

Coming to the unique optoelectronic properties, colloidal metal oxides (CMOs) exhibit tunable electronic properties and optical properties (e.g., absorption and photoluminescence) such as work function (WF), resistivity, dielectric constant (K), and bandgap [15]. In addition to this, CMOs present attractive optical phenomena such as superior and tunable localized surface plasmon resonance (LSPR) effect. All these features will be explored in the next section [16].

Engineering of the optoelectronic properties in CMOs can be achieved by the development of new material chemistry routes; synthetic chemistry, ligand chemistry, and postdeposition chemistry. Synthetic chemistry enables the custom-tailoring of optoelectronic properties by changing the size, shape, composition, and structure of individual nanocrystals, including different reaction approaches such as hydrolysis and alcoholysis (Fig. 2), aminolysis, solvothermal, hydrothermal, and halide elimination. Ligand chemistry offers essential colloidal stability as well as processability, and influences its optoelectronic properties in different ways. The bonding, sizes, and ligand-induced surface dipoles of surface ligands modulate the intragap states, charge transport processes (Fig. 3), and absolute energy levels of oxide nanocrystals, respectively. Postdeposition treatment is another technique that can be used to modify the properties of oxide-nanocrystal films, such as work function, conductivity, intragap states, and surface-wetting properties [9].

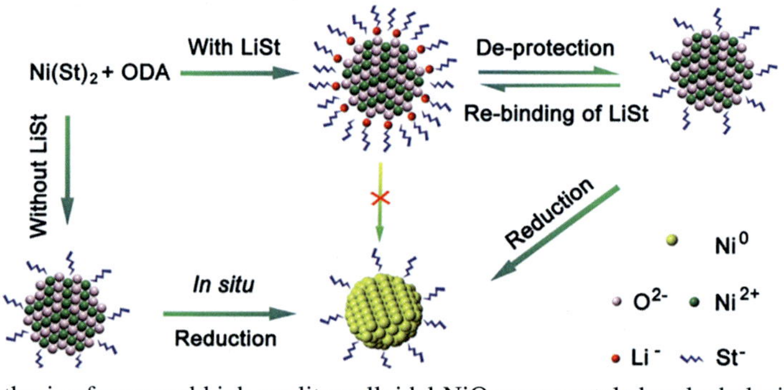

Fig. 2 Synthesis of pure and high-quality colloidal NiO nanocrystals by alcoholysis using ligand protection strategy, with LiSt as the protecting ligand. Abbreviations: ODA, 1-octadecanol; LiSt, lithium stearate; $Ni(St)_2$, nickel stearate.

Source: Reused with permission from X. Liang, S. Bai, X. Wang, X. Dai, F. Gao, B. Sun, Z. Ning, Z. Ye, Y. Jin, Colloidal metal oxide nanocrystals as charge transporting layers for solution-processed light-emitting diodes and solar cells. Chem. Soc. Rev. 46 (6) (2017) 1730–1759. Copyright © 2017, The Royal Society of Chemistry.

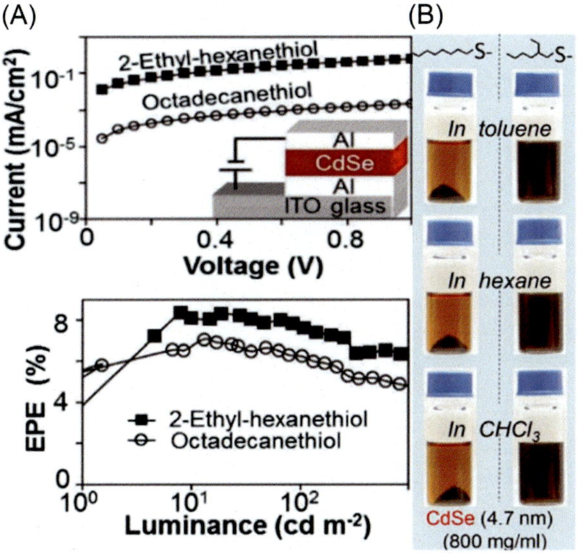

Fig. 3 (A) Enhancement of charge transport of the CdSe QD films by the use of entropic ligand 2-Ethyl-hexanethiol in comparison with Octadecanethiol. (B) Digital photos of efficiently processable CdSe nanocrystals with stearate (left column) or 2-hexyl-ecanoate (right column) as ligands in different organic solvents at 303 K.
Source: Reused with permission from X. Liang, S. Bai, X. Wang, X. Dai, F. Gao, B. Sun, Z. Ning, Z. Ye, Y. Jin, Colloidal metal oxide nanocrystals as charge transporting layers for solution-processed light-emitting diodes and solar cells. Chem. Soc. Rev. 46 (6) (2017) 1730–1759. Copyright © 2017, The Royal Society of Chemistry.

On the whole, substantial developments in engineering and science of the colloidal oxide nanocrystals in conjunction with their intrinsic stability, which led to the enhancement of device lifetime, make CMOs a unique class of materials. They drive the prodigious generation of economic, large-area, and flexible devices fabricated by solution-based techniques.

A comprehensive understanding of the CMOs for device engineering is given below (Fig. 4).

1.2.1 Electrical properties

Control of electrical properties is crucial in engineering optoelectronic properties and promoting the fast development of optoelectronic devices. One of the most essential qualities of CMOs is their tunable electrical property through the alterations in doping and inducing defect states. Resistivity control (p-type or n-type) of CMOs can be achieved by implementing doping [17]. With a high valence ion implementation of doping, conductivity is enhanced by providing more free electrons. This phenomenon can be observed for ZnO nanocrystals doped with group III elements of Al^{3+}, Ga^{3+}, or In^{3+} and TiO_2 nanocrystals doped with Nb^{5+} [18]. The presence of O vacancy and Zn defects in the wurtzite structure of ZnO leads to n-type conductivity [19]. This can be

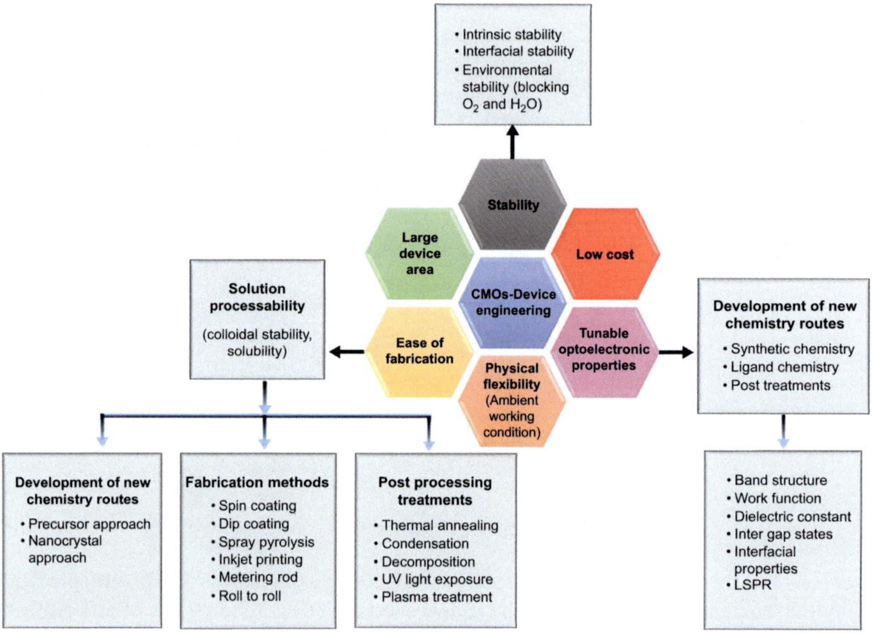

Fig. 4 A comprehensive understanding of the CMOs for device engineering describing the interrelationship of synthetic strategies, optical, and electrical properties.

Fig. 5 Photographs showing the stabilities of Al or In or Ga-doped ZnO along with other doping transparent conductive oxides (TCO) NC inks dispersed in toluene.
Source: Reused with permission from J. Song, Colloidal metal oxides in electronics and optoelectronics, in Colloidal Metal Oxide Nanoparticles, Elsevier, 2020, pp. 203–246. Copyright © 2020 Elsevier.

altered into the p-type via doping elements in group IA(Li, Na, K), IB(Cu, Ag), and VA(N, P, As, Sb). In another study, stable p-type conductivity for a long time was experimentally verified for N-doped ZnO nanoparticles (NPs) having 20% of Zn vacancies synthesized through ammonolysis at a low temperature of ZnO_2 [20].

For instance, Fig. 5 shows the photographs of Al/In/Ga-doped ZnO NC-based transparent conductive oxides (TCO) inks synthesized by heating a mixture of metal-organic precursors dispersed in toluene, which shows good stability of more than one year. Fig. 6 shows the relationship between the sheet resistance and the concentration of dopants, showing that sheet resistance decreases with an increase in dopant concentration [21]. The n-type doping will also influence the work function of the CMO. For example, the work function of the MoO_3-Al composite film can be tailored

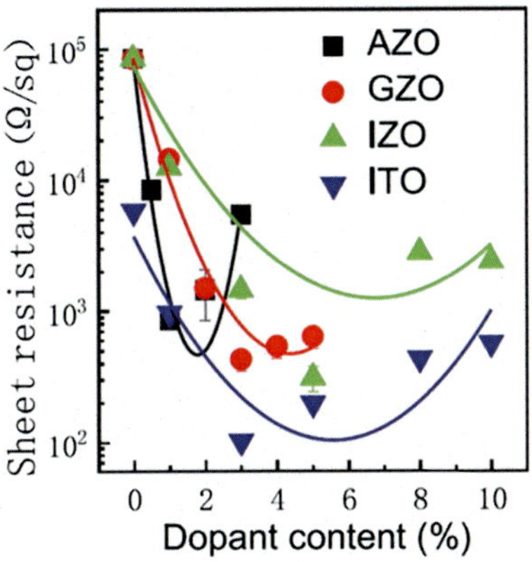

Fig. 6 Variation of the sheet resistance with dopant concentration.
Source: Reused with permission from J. Song, Colloidal metal oxides in electronics and optoelectronics, in Colloidal Metal Oxide Nanoparticles, Elsevier, 2020, pp. 203–246. Copyright © 2020 Elsevier.

(5.49–4.04 eV) by changing the concentration of Al in the films, which has a significant role in optoelectronic devices [22]. Other than ZnO, the electrical properties of many other CMOs, such as TiO_2, NiO, cadmium oxide (CdO), In_2O_3, SnO_2, and HfO_2, can also be tuned by doping. Extra free carriers generated in films processed from the oxide nanocrystals with aliovalent doping have improved the conductivity and work function. As mentioned earlier (in Section 1.2), different posttreatments will also provide new routes to modify the electrical properties of CMOs. Chemical treatment is one such technique to attain better carrier mobility of CMOs with the aid of doping and postprocessing technologies making use of annealing temperature (Ta). It was shown that via controlled annealing conditions, significant variations in the electrical properties (electron mobility and sheet resistance) of ZnO-nanocrystal films were observed [23].

CdO thin films fabricated via the sol-gel method showed a decrease in the resistivity to $6 \times 10^{-4}\,\Omega$cm for thermal treatment (350°C) and could be used as a potential tool to develop transparent conducting electrodes [24]. There is a decrease in resistivity (from 4.2×10^{-2} to $3.2 \times 10^{-2}\,\Omega$cm) for dip-coated Al-doped ZnO (AZO) polycrystalline thin films, which shows a treatment temperature enhancement from 300°C to 500°C due to the grain boundaries and the deficiencies in the crystal lattice of the films [25]. The electrical properties of HfO_2 carried out at different treatment temperatures showed an enhancement in dielectric properties, turning HfO_2 as a potential alternative for gate dielectric [26]. Similarly, an enhancement in the work function can be observed by O_2-plasma treatment on NiO-nanocrystal thin films [27]. Additionally, the work function of ZnO films can be controlled by using covalently bonded self-assembled monolayers (SAMs) of molecules derived from benzoic acid derivatives with different dipole orientations [28]. Surface modification of oxide nanocrystals by covalent bonding of additional molecules can effectively alter the

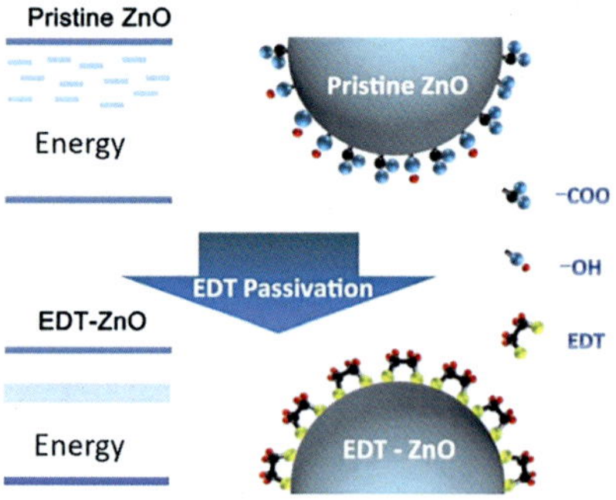

Fig. 7 Schematics of generation of the intragap states on ZnO-nanocrystal films by passivation method (EDT treatment).
Source: Reused with permission from K. Manthiram, A.P. Alivisatos, Tunable localized surface plasmon resonances in tungsten oxide nanocrystals, J. Am. Chem. Soc. 134 (9) (2012) 3995–3998. Copyright © 2017, The Royal Society of Chemistry.

intragap states of the films. A new intragap band can be developed by the passivation method based on ethanedithiol (EDT) treatment for ZnO-nanocrystal films (Fig. 7) [29].

1.2.2 Optical properties

CMOs have unique optical properties that are significant for photonic and optoelectronic devices. Optical properties are displayed by the CMOs mainly in the form of absorption, luminescence, and LSPR. Doping, surface modification, and nonequilibrium processing bringing in the defect level will act as tools for tuning the optical bandgap of MOs which has a direct influence on the absorption and luminescence properties. Due to the long-term stability of the optical (luminescence and absorption) properties of the NPs at various conditions, they can be used as a potential candidate in display devices [30]. ZnO NP doped with Mg^{2+} showed controlled emission wavelength from yellow to blue with varying Mg/Zn molar ratios [31]. Bandgap tuning can also be done via doping with various other dopants like Cd, Mg, Mn, and Fe ions in the range of 2.9–3.8 eV [32]. Doping in other oxides such as TiO_2, NiO, and CdO will alter the luminescence and absorption properties. Surface modification is another method to modify the wide bandgap. Controlling the size of polymer (PEGME)-grafted ZnO nanoparticle from 1 to 4 nm by modified sol-gel technique, photoluminescence from the blue (450 nm) to the yellow (570 nm) was observed (Fig. 8A and B) [33]. Moreover, in terms of quantum yield and stability of emission, polyether-grafted ZnO sols showed higher values than conventional acetate-modified ZnO colloids. Ligand modification will also influence the luminescence of ZnO NCs, (blue, green, and yellow) and other oxides, including TiO_2, Fe_2O_3, and ZrO_2 [34–38]. A review of the defect levels of ZnO, introduced by the nonequilibrium process, reported controllable selection and c-emission of visible emissions under different excitations [39]. The approach of combined annealing and excitation route adopted

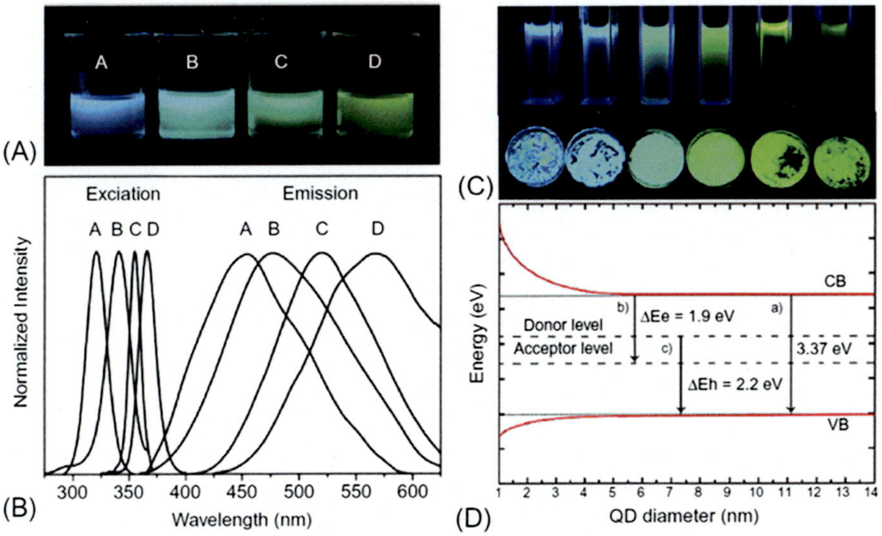

Fig. 8 (A) UV (254 nm) exposed photograph, (B) PL spectra, (C) UV (254 nm) exposed photographs of ethanolic solutions (upper) and dried powder (lower) obtained at different intervals, (D) Schematic illustration of band-gap enlargement with QD size and possible transitions corresponding to different emissions of polymer grafted ZnO (PEGME350) ethanol colloid samples.
Source: Reused with permission from J. Song, Colloidal metal oxides in electronics and optoelectronics, in Colloidal Metal Oxide Nanoparticles, Elsevier, 2020, pp. 203–246. Copyright © 2020 Elsevier.

in another work is found to be one of the facile methods to control these polychromic visible emissions of nanostructured ZnO. This leads to emissions from single-band, multiband, and broad regions resulting in divergent monochromes, mixed colors, and white accordingly [40]. Research on the properties of visible emissions of ZnO QDs synthesized via a sol-gel route showed a blue shift with the decreasing ZnO QD sizes, as shown in Fig. 8C. Fig. 8D describes the possible transitions along with variation in band-gap with different surface defects of ZnO QDs [21]. Nanostructures produce the quantum confinement effect which is described as the spatial confinement of electron-hole pairs (excitons) that result in the formation of discrete electronic energy levels. The effect of quantum confinement that can be experimentally probed on the MOs is related to the energy shift of exciton levels and optical bandgap. An investigation on the nonlinear optical properties of ZnO nano colloids showed an enhanced optical limiting response in the diameter range of 6–18 nm, with the increase of particle size [41].

LSPR is an attractive optical feature of NPs arising from surface plasmon confinement. On the incidence of electromagnetic radiation, the oscillating electric field causes the conduction electrons to oscillate coherently due to the coulombic attraction between conduction electrons and nuclei. The phenomenon of LSPR, when an electric field of radiation interacts with NP, is schematically shown in (Fig. 9A). LSPR leads to

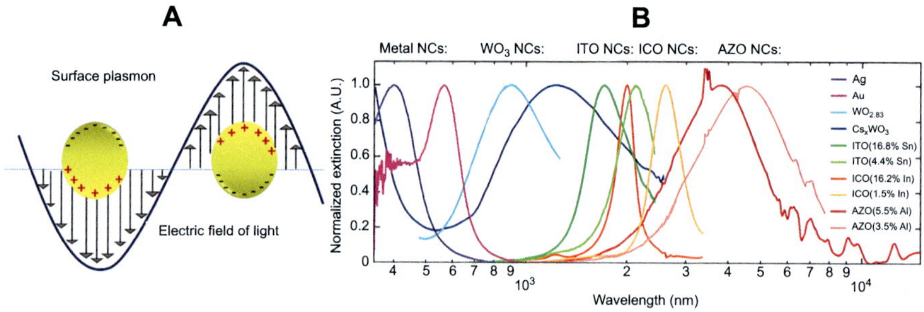

Fig. 9 (A) Schematics of LSPR phenomenon for a nanocrystal. (B) Normalized optical extinction caused by LSPR in solutions and films of metal and MO nanocrystals.
Source: Reused with permission from S.D. Lounis, E.L. Runnerstrom, A. Llordes, D.J. Milliron, Defect chemistry and plasmon physics of colloidal metal oxide nanocrystals, J. Phys. Chem. Lett. 5 (9) (2014) 1564–1574. Copyright © 2014, American Chemical Society.

resonant absorption, scattering, and near-field improvement around NP [42]. In CMOs, LSPRs can be tuned by varying the doping amount and postsynthetic routes like electrochemical and photochemical control. The tunability of LSPR provides new functional devices across visible, near-infrared (NIR), and mid-IR spectral regions. The LSPR absorption of ITO film synthesized by high-temperature spin coating was tuned to a high wavelength region by changing the doping level of Sn (3–30 mol%) [43]. A recent study on the defect chemistry relation with plasmon physics of colloidal MO NCs demonstrated plasmonic effects from visible- to mid-IR wavelengths that will lead to new applications (Fig. 9B) [44].

2 Synthesis of colloidal metal oxides (MOs) for optoelectronic and photonic applications

In recent years, the ingenious synthesis routes for MOs and allied materials fabrication have attracted immense attention. These methods are of particular interest to massively modulate the material properties, which is crucial for widening their range of applications in exigent fields of optoelectronics and photonics. It is noteworthy to mention here, among the wet chemical synthesis processes for MOs, that the broadly recognized sol-gel synthesis involves the evolution from a sol consisting of a colloidal suspension of particles to a solid-liquid phase entity. The synthesis protocol often involves the hydrolysis of the precursor under ambient conditions of temperature, pH, and other process parameters. The kinetics of the reaction is often dependent on the solvent system chosen, wherein long-chain organic compounds slow down the reaction rate. This reaction environment is conducive to fabricating different morphologies of gel-based material systems, powdered MOs, films, and even fibers. In this cost-effective, low-temperature process, additional steps such as densification of the end product and thermal treatments that ensure fine-tuning of the gel's desirable complex geometries are often involved. In recent years, this synthesis process has been a game-changer in the fabrication of MOs films, leading to enormous advancements in printed electronics and other optoelectronic devices [45,46]. Various groups have outlined the sol-gel process fabrication of bulk MOs or films of MOs. In a study, the preparation of ZnO films by this method involved hydrolysis of Zn precursor salts accompanied by a heat treatment that allowed the formation of film in a simple yet fast and efficient manner. This was utilized for cathode interlayers in polymer solar cells without hampering the performance of the cell [47]. Some fascinating observations have been made on WO_3-buffered layers for small-molecule organic LEDs synthesized by sol-gel route in an inert niche from tungsten ethoxide precursors. Here, the processing environment was vital to prevent any form of contagion with oxygen and thus rationalized a facile synthesis methodology for OLED fabrication in an oxygen-free milieu [48]. In another distinct work by Kumar et al., SnO_2 leaf-shaped nanostructures were obtained for optoelectronic applications in humidity sensing wherein the precursor of stannous chloride dehydrate was hydrolyzed in the presence of ammonia. The process was performed under optimized pH and was vigorously stirred before being subjected to postsynthesis treatments at high temperatures. The synthesized nanostructures exhibited superior sensitivity toward a particular humidity range at room temperature [49]. Notable inventions have led to the devising of unique sol-gel routes that subdue the hurdles of low dispersibility and low endurance of MOs to preserve their favorable properties without affecting them. Such a novel synthesis methodology has been adopted in an investigation involving a hot-injection process via a sol-gel mechanism because of its added benefits of ease of governing the stringent synthesis conditions and improved yield, among others. This single-step hot-injection facilitates the dispersion of MOs nanoparticles in the matrix of a host material where it gets enclosed [50]. This suggests that the process intensification of the sol-gel method holds great promise in advanced fields of science and technology.

Among other new MOs synthetic strategies, the microwave-assisted approach has gained popularity due to its kinetically controlled nonequilibrium environment and its ability to efficiently maintain volumetric heating that accelerates reaction kinetics and makes it a highly energy-efficient process, thereby achieving the principles of sustainable and green chemistry. In one of the studies, it is reported that SnO_2 thin-film transistors synthesized by microwave-annealing protocol postdeposition had enhanced performance concerning its electrical properties, optical characteristics including transmittance, and surprisingly elevated mobility compared with its thermally annealed equivalent. This technique aids in revamping the features of the pristine device, such as extremely minimal leakage current, by initiating a drastic reduction in the number of defects that get incorporated. Thus, it signifies the importance of microwave-irradiation-mediated fabrication processes for designing affordable devices with impressive performance in flexible transparent OLED display applications [51]. In a similar investigation, the importance of microwave irradiation on accomplishing eminent device characteristics by assessing the potential of ZnO thin-film transistors fabricated by this route was analyzed. Several structural, optical, and chemical characterization and analysis methods revealed that the impact of microwave irradiation boosts the thermodynamics governing crystallization and dehydroxylation. Therefore, this mode of fabrication resolves all the challenges associated with conventional thermal treatments postprocessing, which is responsible for the poor efficiency of the device [52]. The versatility of this synthesis route has been exploited further for obtaining stable colloidal nanocrystals of MoO_x at room temperature. These nanosuspensions of MoO_x were employed in organic solar cells. Surprisingly, the efficiency of the device increased upon certain facile chemical treatments on these MoO_x spin-coated hole transport layers [53].

In order to fine-tune the material properties of the MOs crystals, it is an absolute necessity to have an in-depth understanding of the synthesis-structure-property correlation. Detailed investigations on crystallization methods of colloidal MOs are underway. This has opened up newer possibilities for improving the functionality of desired applications by designing synthesis processes that yield good-quality MOs. A state-of-the-art synthesis by coupling pulsed laser deposition (PLD) and colloidal templates fabricated two-dimensional ZnO-NiO ordered arrays by Moon and his coworkers. The processing conditions were optimized via structural and compositional characterizations. The PLD target used was quite distinct and was specifically built for this work. This ensured that the grains of this mixed oxide resulted in outstanding electrical properties for sensor applications. The results signify the notable changes in the structure of the ZnO-NiO mixed oxide due to the simplicity of the fabrication technique that largely alters the electrical properties dominated by the oxide grain junctions [54]. Currently, for the high precision preparation of two-dimensional MOs, the vapor phase deposition route has gained immense attention. For reliability and better control, chemical vapor deposition (CVD) attunes some important experimental conditions, including temperature, ambience under which the reaction is taking place, precursor, and substrate material system. 2D single crystals of MoO_2 were synthesized of micrometer dimensions by CVD. In this process, the reduction of the metal oxide by hydrogen under atmospheric pressure conditions formed single crystals of three different morphologies. Further analysis of the electrical properties of the as-fabricated MoO_2 crystals revealed their superior conductivity which has huge scope in conducting films. The synthesis process demonstrated ease in operation and high reproducibility, facilitating large-scale production. Delving deeper into the nano-regime, it is observed that the surface chemistry, electrical, optical, and thermal characteristics become notably altered compared to their bulk equivalent. Thus, large-scale fabrication of nanomaterials with imminent opportunities in optoelectronics and photonics will help in overcoming the challenges of poor yield, operational hazards, and disrupted reaction dynamics. Besides, for the smooth functioning of photonic and electronic appliances, it is essential to have good control over the defects and nanoscale dimensions of the materials being synthesized. On these grounds, sophisticated synthesis procedures involving metalorganic vapor phase epitaxy have garnered immense attention. This method has its unique ability to grow in large areas and accurately control the dimensions of nanoarchitecture. For example, Park and others exhibited this synthesis route of metal-organic vapor phase epitaxy where ZnO nanorods were grown vertically. Moreover, the processing conditions concerned low temperature and no additional catalyst contributing to homogeneity in size, excellent crystallinity, and superior optical characteristics. At the same time, the PL spectroscopic studies demonstrated their poor deep-level emission intensity, suggesting the presence of a minimum number of defects. This renders the nanorods of ZnO with favorable optical profiles [55]. In a new-fangled synthesis of colloidal Ga-doped ZnO nanocrystals, hot injection of the precursors followed by temperature-dependent decomposition resulted in unique optical and electrical properties. The synthesized nanocrystals were employed to form thin films by various routes: spray coating and spinning. Refinement of the optoelectronic characteristics of the fabricated films required certain postprocessing treatments. Under ambient conditions, the thermally annealed films exhibit increased performance almost comparable to the commercially available coatings. The films of Ga-doped ZnO nanocrystals evince 90% transmittance in the visible region and strong absorbance in the near-IR range, which is attributed to the varying Ga doping that results in free charge carries responsible for the plasmon resonance [56]. Many of these synthesis approaches have bridged the gap between lab-scale processing and industrial-scale production by excellent process intensification, ease in down streaming, and demonstrating a good control over the reaction conditions that significantly influence the material properties.

3 Applications and specific characterization techniques of colloidal metal oxide in optoelectronic and photonic fields

Colloidal MO has garnered immense attention for their exceptional optical and electrical properties, making this material system the forerunner in cutting-edge optoelectronic and photonics applications in numerous fields. Designing neoteric smart MOs focuses majorly on enriching the novel optical and electrical traits through various analysis and characterization methods. Emphasis is laid particularly on the molar absorptivity, transmittance, luminescence, polarizability, and localized surface plasmon resonance exhibited by MOs. Doping, size variation, and thermodynamics of synthesis processes that lead to defect incorporation and modification of the surface attune the band structure-related UV-visible absorption and luminescence characteristics immensely. For upgraded device performance, certain important parameters such as electrical conductivity, trap density, the density of dopants, and diffusion length of the carriers can be easily tailored by surface refinement methodologies. Thus, a meticulous investigation of these key attributes concerning the efficiency of materials for optoelectronic and photonic applications is indispensable. For instance, comprehensive examination of the microstructure, applying tools like X-ray diffraction (XRD), energy-dispersive X-ray spectroscopy (EDX), atomic force microscopy (AFM), high-resolution transmission electron microscopy (HR-TEM), selected-area electron diffraction (SAED), and field emission-scanning electron microscopy (FE-SEM) aids in establishing a relationship with the altered and exceptional properties of novel MOs.

Here, we explore the three important applications of CMOs, namely light-emitting devices, photodetectors, and thin-film transistors, which are the building blocks of modern optoelectronics and photonics.

3.1 Light-emitting devices (LED)

For the state-of-the-art solution-processed LEDs, the combination of multiple extraordinary functional components is needed.

A typical LED consists of a cathode, anode, n-metal oxides (act as electron transport layer (ETL)), an active layer, and p-metal oxides (act as hole transport layer (HCL)). The electrons are injected from the cathode to ETL and holes are injected from the anode to HTL. Electrons and holes are transported to the active layer by ECL and HCL; excitons recombination takes place within the active layer and forms photons with the radiative decay, that are emitted out of the devices (Fig. 10). Solution-processed CMOs offer affordable and high-performance LEDs owing to their excellent optoelectronic traits such as conductivity, charge carrier mobility, work function, transparency, and photoluminescence. CMOs can be used in different operational parts of LEDs such as transparent conductive electrodes, charge transporting layers (ETL & HTL), emission layer, and other parts.

CMO thin films become attractive as transparent electrode materials in modern LEDs because of their extremely high conductivities (10^4 S/cm), excellent transparencies (>85%) in the visible range, and work functions (4.3–6.1 eV) [57]. The best performing metal oxides such as indium tin oxide (ITO) and fluorine tin oxide (FTO) contain toxic and rare elements (In). The nontoxic and solution processable ZnO, Ga-doped ZnO (GZO), and Al-doped ZnO (AZO) were developed to replace ITO. It is found that GZO and AZO achieved comparable conductivity ($\sim 10^{-5}\,\Omega$cm) and transparency (>90%) of those of the efficient ITO and also much better than those of FTO. In-doped ZnO (IZO), Ga-doped In_2O_3 (GIO), Mo-doped In_2O_3 (IMO), and Ti-doped In_2O_3 (ITiO) have also been studied but In-based ones are certainly more expensive. Sb-doped SnO_2 (ATO) and single-wall carbon nanotubes (SWCNT) as TCOs need further development to compete with GZO/AZO. GZO and AZO are the promising TCOs to replace ITO and FTO due to not only their high conductivity

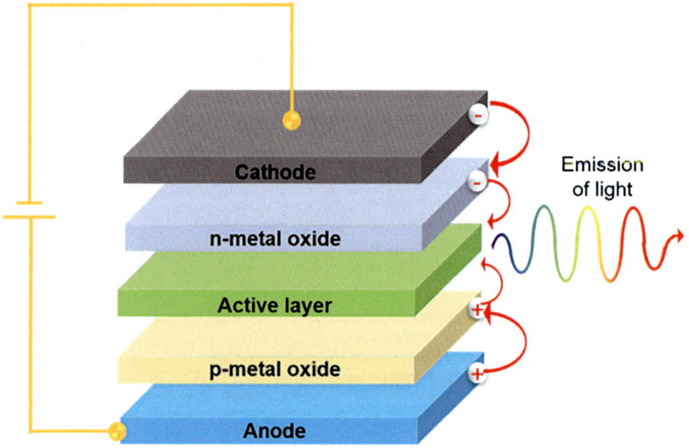

Fig. 10 Schematic diagram of a typical LED device showing the emission with different operational parts.

and transparency but also high resistance to hydrogen-rich plasmas. Moreover, improving light extraction in LEDs is possible via nanostructured ZnO surface, synthesized by controlling growth conditions. A plenteous amount of the raw materials needed for AZO, and to a lesser extent for GZO, together with the abovementioned qualities, makes AZO and GZO potential tools for the next-generation TCOs [58]. The optoelectronic properties of many thin films significantly enhance upon the addition of dopants with favorable intrinsic characteristics. Uniform dopant concentration is essential for the better performance of transparent conducting oxide thin films. It is reported that the analysis of absolute concentrations of Sn dopant added to indium oxide can be done by inductively coupled plasma optical emission spectroscopy (ICP-OES). This was further validated through XPS analysis, wherein the atomic percentage of surface Sn was relatively similar to that obtained for bulk Sn. These results imply the homogeneity in the doping concentration of the nanoparticles fabricated [59].

Transparent films of MOs have attracted attention as charge transport layers for LEDs. An interesting study on nanoparticulate SnO_2 use as an electron transport layer in LEDs for optoelectronics demonstrated its excellent properties that can easily substitute ZnO. Optical bandgaps of as-synthesized SnO_2 NPs were determined and compared with ZnO NPs by UV-visible absorption spectroscopy. The spectra showed an increment in the bandgap value when compared to their bulk equivalents. This signified that in the case of SnO_2 NPs, there was an upward shift in their conduction band, making it larger than that of ZnO NPs. This upward shift directly improved the resistivity and mobility of quantum-confined SnO_2 NPs. To supplement the potential of SnO_2 NPs in LEDs as electron transport layers, photoluminescence spectra were recorded on the emitting thin films of SnO_2 and ZnO. Quenching in the PL intensity was detected for the QDs on the thin films of SnO_2 and ZnO compared to the QDs deposited on a glass substrate. This drop was similar for both SnO_2 and ZnO NPs, implying that both exhibit similar performance. Thus, these SnO_2 NPs serve as an excellent alternative to ZnO NPs for optoelectronic devices [60]. Oftentimes, doping MOs drastically improves the performance of the device. For instance, ZnO nanocrystals doped with lithium and magnesium result in much better device efficiency owing to the concentration-controlled tuning of the bandgap. In such cases, an excellent balance between the charge carriers-electrons and holes is achieved, showing a remarkable increase in the LED's performance. The summarized results from various characterizations for optical property analysis signify the change in the bandgap in Li-doped, Mg-doped, and co-doped ZnO nanocrystals. One of the leading causes behind the poor efficiency of the device is more often than not due to the presence of hydroxyl bonds at the surface of MOs, which disturbs the time taken for exciton decay. Thus, time-resolved photoluminescence spectroscopy that determines the decay time helped ascertain that of the doped ZnO nanocrystals containing the minimal hydroxyl bond. Hence, the longest exciton decay time can be correlated to superior LED performance [61]. Besides, the notable optical transmittance necessary for good display devices was observed in SnO_2 NPs, which was enhanced by adding dopants like antimony and zinc. The optical characteristics were evaluated by diffuse reflectance UV-vis spectroscopy and photoluminescence spectroscopy for the SnO_2 NPs. The studies revealed that for samples with antimony as the dopant, contraction of the bandgap occurred while absorption in the visible region improved. At the same time, shifting of the band-edge to a higher wavelength was observed for NPs doped with Zn. Nevertheless, PL spectra recorded were used to assess the phenomenon of recombination of electrons and holes. A bathochromic shift in the spectra of antimony-doped SnO_2 NPs matches well with the noticeable change in their absorbance spectra, indicating shifted band-edge. Moreover, a drop in the PL intensity upon doping depicts that the electron-hole recombination is influenced due to the presence of oxygen vacancy defects (most probable cause in MOs). These altered properties make doped SnO_2 NPs promising materials for a wide range of photonics and optoelectronic devices [62]. Interestingly, research on ZnO-doped thin films for LEDs and their intriguing optical properties were probed by measuring the scattering energy by the process of transmission and reflection within a particular range of spatial angles as shown in Fig. 11. The addition of ZnO as a dopant notably improved the scattering effect manifested in the orderly distribution of scattering energy, and the dopant concentration influences the transmittance to a large extent. Besides, from the CCT measurements for varying spatial angles, the uniformity observed for the LEDs in which ZnO was incorporated was enhanced, implying a great scattering effect [63].

More incredible advancements in localized surface plasmon-based LED applications have played a pivotal role in deep-tech innovations. One such study explained how the device performance could be improved when the close proximity of the localized surface plasmon layer to the LED active layer initiates the matching of the resonance energy of the former to the emission wavelength of the latter. This coupling of energy increased the quantum efficiency. The UV-vis and photoluminescence spectra of the Ag/SiO_2 nanoparticles that exhibited localized surface plasmon resonance and the LEDs containing Ag/SiO_2 nanoparticles layer illustrated that the wavelength of LSPR absorption coordinates with the PL peak wavelength of the device. Moreover, the NP containing LEDs proclaimed an intense PL peak with much higher intensity than the bare LEDs. The time-resolved photoluminescence spectroscopy measurements showed the decrease in the decay lifetime for Ag/SiO_2 nanoparticles-coupled LEDs, and this was ascribed to the localized surface plasmon resonance [64]. Enhancement of LEDs efficiency by incorporating MOs layer showing LSPR has strengthened the devices' potential in cutting-edge innovations in optoelectronic chip technology, lighting, and display technology, over the years. Fig. 12 depicts the crux of the viable design of newer material systems for LSPR combined with LEDs. The utility of LEDs in flat panel displays for television and other applications has necessitated the need to explore innovative materials and synthesis routes that will yield superior performance of LEDs. Researchers have leaned upon the phenomenon of LSPR for higher quantum efficiency by embedding a layer of MOs that supports as the top emitter [65].

Thin films of indium-doped zinc oxide NPs as electron transport layers for optoelectronic devices have gained popularity owing to their flexible nature and stable performance. In a report on this electron transport layer by Liang et al., the proposed synthesis process fabricated indium-doped ZnO NPs whose work function depended on the doping concentration. Analysis of the ultraviolet photoelectron spectroscopy

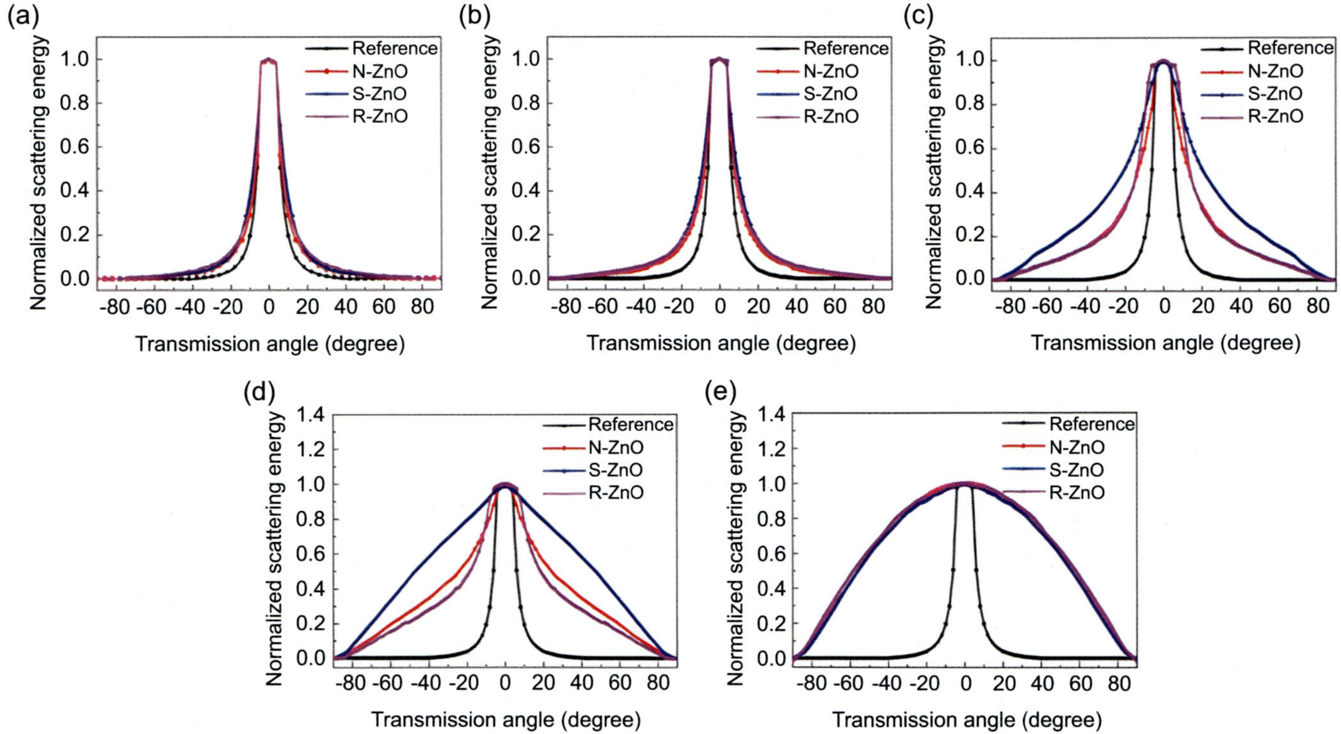

Fig. 11 ZnO nanostructured-doped films exhibiting normalized distribution of scattering energy determined by the transmission method for varying doping concentrations as well as morphologies.
Source: Reused with permission from L. Rao, Y. Tang, Z. Li, X. Ding, J. Li, S. Yu, C. Yan, H. Lu, Effect of ZnO nanostructures on the optical properties of white light-emitting diodes, Opt. Express 25 (8) (2017) A432–A443. Copyright © 2017, Optical Society of America.

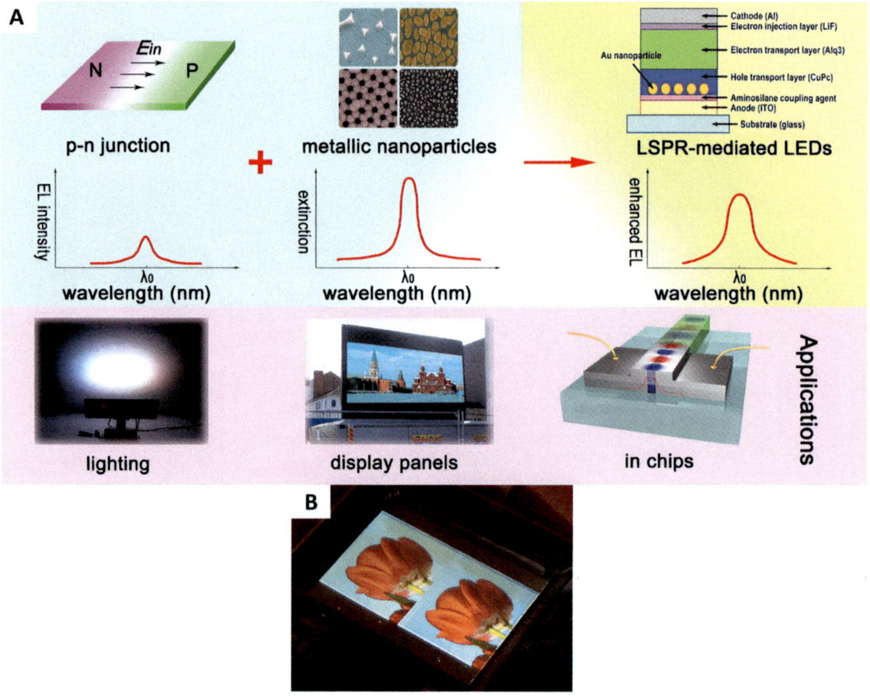

Fig. 12 (A) Metallic NPs exhibiting localized surface plasmon resonance, embedded on LEDs to achieve superior quantum efficiency. (B) Image showing an ink-jet printed display which was designed by Philips Electronics. It is a culmination of the presence of pixels emitting fluorescent light upon excitation.
Source: Reused with permission from X. Gu, T. Qiu, W. Zhang, P.K. Chu, Light-emitting diodes enhanced by localized surface plasmon resonance, Nanoscale Res. Lett. 6 (1) (2011) 1–12. Copyright © 2011, Gu et al; licensee Springer.

data demonstrated that the differing work function of samples is because of the varying concentrations of the free electrons which tend to occupy the conduction band, thus shifting the Fermi level. Application of this MOs thin film in LEDs and comparison of the device characteristics reveal lower turn-on voltage of the device consisting of doped ZnO NPs thin film than the pristine ZnO thin film. In addition to this, the efficiency of the device and the luminance are far superior for the indium-doped ZnO NPs thin film as the electron transport layer. Further, the scanning Kelvin probe microscopy substantiates the enhanced performance of the indium-doped ZnO NPs thin film because of its reduced work function that allows no barrier for electron injection into the emitting layer of the LEDs [66]. Unique and wide bandgap along with defect states such as oxygen vacancy and metal vacancy facilitate CMO nanocrystals making as emission layers. Hybrid device structure of ZnO core and quantum dots (QDs) prepared by simple solution method is investigated as emissive material in LED [67].

Upgrading the electrical conductivity of MOs by tuning their energy band structure for their implementation in optoelectronics and photonics devices is often riddled with several challenges. To overcome these limitations, a novel approach of fabricating heterostructured MOs was adopted by Kim et al. By this method, the unique nanoregimes formed and exhibited specific energy band structure, which was verified by choosing an appropriate material system. Kim et al. preferred MOs with increased work function like MoO_3 NPs as the dopant and a p-type MOs matrix like NiO. Modulation of the energy band structure by this MOs heterostructure charge transfer complex improves the charge injection. The addition of nanoparticulate dopant alters the band's position by interaction with the NiO matrix throughout the film. The optimized device efficiency was accomplished by accurately matching the energy structure by playing with the ratio of MoO_3 NPs in the matrix. Furthermore, a comprehensive understanding of the charge transfer phenomenon resulted in a significantly different work function in the heterostructured NiO:MoO_3 is examined by ultraviolet photoelectron spectroscopy. This change in the work function is a consequence of the charge transfer because of the interaction between the thermodynamically favorable and metastable heterostructure of NiO and MoO_3 NPs. Computation of the electrical conductivity showed an admirable increment in the current density, which increased with the increasing concentration of MoO_3 NPs. The current density-voltage relationship and the Hall effect determination suggested the increasing conductivity because of the improved carrier concentration arising from enhancement in the concentration of MoO_3 NPs. Additionally, charge carriers were formed due to the charge transfer process and vacancies generated during the fabrication. Employing this heterostructured NiO, MoO_3 in LEDs as hole injection layer demonstrated its potential by enhanced current efficiency and green and blue electroluminescence spectra of the LEDs as shown in Fig. 13. These encouraging results are ascribed to the modulation of the band energy structure that optimized the electron-hole charge balance and brought about the stupendous device efficiency. This gave the proof-of-concept that multilayered optoelectronics from heterostructured MOs is a great alternative strategy for high-performing devices [68]. p-type CMOs are limited in number, yet they play significant role as HTL. Solution-processed CuO and WO_3 are rare p-type semiconductors, both of which can be used as HTL in QLEDs with comparable external quantum efficiency (EQE) and brightness as that of commonly used PEDOT:PSS, which paves a new step toward the practical application of QLED technology [69,70]. Cu-doped NiO films from solution processing used as p-metal oxide show improved hole injection capability resulting in enhanced current efficiency (45.7 cd/ A) and EQE (10.5%) which altogether makes it as a best alternative for PEDOT:PSS-based LEDs [71].

Although much efforts have been done by the researchers to investigate the ability of CMOs as CTLs and emissive layers, still more understanding and optimization of the chemical and physical phenomena at the emitter-CTL interfaces, energy transfer, and carrier injection from CTLs into the emitters is necessary [9].

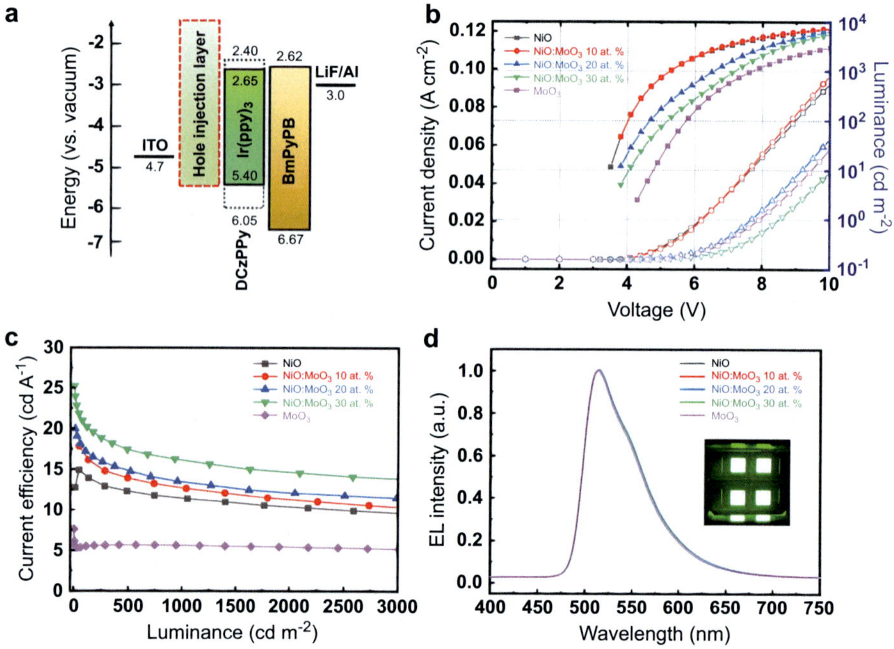

Fig. 13 (A) OLED exhibiting blue phosphorescent and their corresponding energy level diagram. Plots demonstrating the (B) current density-voltage-luminance characteristics and (C) current efficiency and luminance behavior of NiO:MoO_3 complex and HATCN. (C) The electroluminescence spectra (normalized) of NiO:MoO_3 complex and HATCN. The inset of (D) shows the blue OLED consisting of heterostructured NiO:MoO_3.
Source: Reused with permission from M. Kim, B.H. Kwon, C.W. Joo, M.S. Cho, H. Jang, H. Cho, D.Y. Jeon, E.N. Cho, Y.S. Jung, Metal oxide charge transfer complex for effective energy band tailoring in multilayer optoelectronics, Nat. Commun. 13 (1) (2022) 1–9. Copyright © 2022, Springer Nature Limited.

3.2 *Photodetectors (PD)*

A PD is an optical-to-electrical transducer that converts incident electromagnetic radiations such as UV, visible, and IR to electrical signals. On illumination, there occurs excitation of electron-hole pair that results in corresponding photocurrent. Owing to the wide bandgap and large exciton binding energy, CMOs find potential role in PDs by providing easy fabrication at large-scale, economic, and unique physical and chemical properties [72,73].

In case of organic solar cells, CTLs are the essential parts. Desirable values of bandgap and work function of n-type TiOx and ZnO films find applications in organic solar cells as ETLs [74,75]. The incorporation of metal (Au or Ag) nanoparticles and doping with Cs_2CO_3 resulted in enhancement of the electronic properties of TiOx-nanocrystal films [76,77]. Alcoholic dispersion of ZnO nanocrystals finds applications as top ETLs and bottom ETLs in conventional and inverted structure,

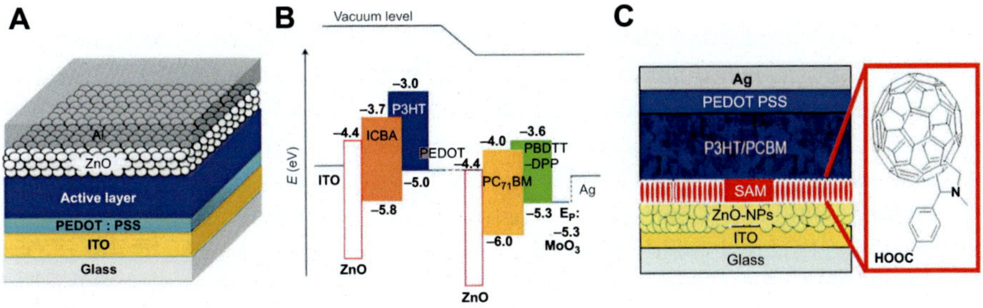

Fig. 14 Schematic representation of an organic solar cell consists of ZnO nanocrystals as ETLs. (A) Top ETLs of ZnO NC films in conventional structure (B) Energy level diagram of organic SC with PEDOT:PSS and ZnO nanocrystals as ICL. (C) Bottom ETLs of fullerene-SAM modified ZnO NC films in inverted structure.
Source: Reused with permission from X. Liang, S. Bai, X. Wang, X. Dai, F. Gao, B. Sun, Z. Ning, Z. Ye, Y. Jin, Colloidal metal oxide nanocrystals as charge transporting layers for solution-processed light-emitting diodes and solar cells. Chem. Soc. Rev. 46 (6) (2017) 1730–1759. Copyright © 2017, The Royal Society of Chemistry.

respectively [78,79]. It is because of the potential of hole blocking and small contact resistance shown by ZnO nanocrystals that allow better recombination property of photo-generated excitons at the interface, advantageous for efficient device performance (shown in Fig. 14) [80]. ETLs with low surface irregularities are also available with beneficial conductivity and compatibility using colloidal AZO nanocrystals [81]. The n-type oxide layer plays a significant role in heterojunction quantum dot solar cells (QDSCs) such as to predict the depletion region thickness in the QD layer, the built-in voltage, electron extraction, and transport. NiO_x serves as HTL and thus widely explored and proved as a promising material in inverted flexible perovskite solar cells (PCE-14.5%).

MOs as hole extraction layers in solar cells has gained impetus over the last few years. Recently published work on transition metal oxide layers revealed the high work function; however, specific postprocessing treatments hampered the device efficiency. Rising through such challenges, novel V_2O_5 hole extraction layers were fabricated for solar cell applications that disclosed high work functions without any requirement of annealing or thermal treatments postsynthesis. The vital electrical properties were evaluated by the Kelvin probe measurements. The studies show that the work function of the pristine V_2O_5 hole extraction layers is higher than the annealed ones. Besides, the solar cell designed with varying thickness of V_2O_5 as the hole extraction layer was analyzed to optimize their power conversion efficiency, open-circuit voltage, filling factor, and current-voltage characteristics. The observations evince that the power conversion efficiency of the solar cell developed is dependent on the thickness of the V_2O_5 layers, while the electrical conductivity shows no relation with the hole extraction layer thickness. Here, what is of utmost significance is that no postprocessing is needed to enhance the device characteristics; instead, the work function and the device stability are sufficiently high for pristine V_2O_5 layers. Therefore, this concept proposes a methodology to cut processing costs without compromising device performance [82]. In a study, FT-IR and Raman spectroscopy were used to reveal the chemical composition of the surface of synthesized NiO_x structures. The Raman spectra recorded at varying temperatures showed the presence of peaks corresponding to the nitrate ions, O-H stretching band of $Ni(OH)_2$ at higher wavenumbers, and Ni-O stretching in the fingerprint region of $500\,cm^{-1}$. Besides, additional evaluation by FT-IR spectroscopy determined the chemisorbed sodium and nitrate ions that contributed to the amorphous nature. At the same time, TGA helped identify the unreacted compounds responsible for the stability of the NiO_x dispersion. The stabilized dispersion of NiO_x was used as the hole transporting layer for perovskite solar cells with increased optoelectronic efficiency and low cost [83].

Novel zinc stannate NPs were employed in dye-sensitized solar cells because of the wide bandgap that makes them more photostable. The performance of the cells with varying thickness of the films was dealt with in detail through several characterizations for analyzing the current density-voltage behavior and overall efficiency. Comparison with TiO_2 cells denotes a significant enhancement of the device characteristics, like higher photocurrent density and improved overall efficiency for zinc stannate NPs. The efficiency achieved for Zn_2SnO_4 NPs is much higher than the other binary MOs as well as their stability is strikingly boosted [84]. Similarly, when Mg was added as a dopant to TiO_2 NPs, the bandgap broadened therefore stimulating its potential as charge transport layers in perovskite solar cells with a remarkable increase in open-circuit voltage and fill factor. On top of that, the efficiency of converting the energy also became better. Thus, tailoring the energy band structure modulates and optimizes the device performance to a large extent [85].

MOs-based photoelectronic devices, including transparent displays, solar cells, photodetectors, and sensors, oftentimes outweigh their rivals for their exclusive electronic characteristics, affordability, consistent performance, and sustainability. The elevated carrier mobility, surface functionalization yielding minimal leakage current, high current on-off ratio, and low power consumption exhibited by MOs have set the stone rolling for contemporary avenues of research. To investigate the viable causes for such phenomenal properties of MOs unique to an application, specialized and sophisticated measurements and meticulous analysis of the results are vital. Additionally, the electronic properties are tuned by adding dopants or annealing and other thermal treatments postprocessing. In another example, the device characteristics of an unconventional ultraviolet photodetector consisting of a layer of NiO (p-type), ZnO layer (n-type), indium tin oxide as anode, and aluminum as the cathode were studied. Measurement of the current-voltage behavior of the device was performed in the dark under UV illumination. Certain annealing treatments on the spin-coated NiO and ZnO thin films suffice for the optimum rectification ratio of the p-n junction were formed. The lock-in amplifier determined the quantum efficiency to ensure excellent device performance, and the salient photocurrent response within the UV range highlights its potential as a UV photodetector. The distinct elevated gain in the detector is attributed to the presence of defects/photo-generated holes in NiO thin film, which is correlated to the annealing temperature. Hence, extensive characterization and evaluation are fundamental in designing state-of-the-art devices [86].

CMONWs and nanobelts (NBs) are widely used in UV PDs. UV photodetector fabricated from thin SnO_2 NWs synthesized via vapor-liquid-solid (VLS) process displayed excellent device performance such as photostability and selectiveness. Moreover, it shows an extreme EQE value of 1.32×10^7 (1.32×10^9%) which is several orders of magnitude higher than that of other potential materials such as 1D inorganic semiconductor nanostructure-based PDs [87]. Potential values of light-to-dark current ratio, photoconductive gain, responsivity, specific detectivity, and stability, Zn_2SnO_4 (ZTO) NWs modified with ZnO QDs (ZnO-ZTO) show immense UV photodetection functionality (Fig. 15D). The corresponding schematics of the carrier

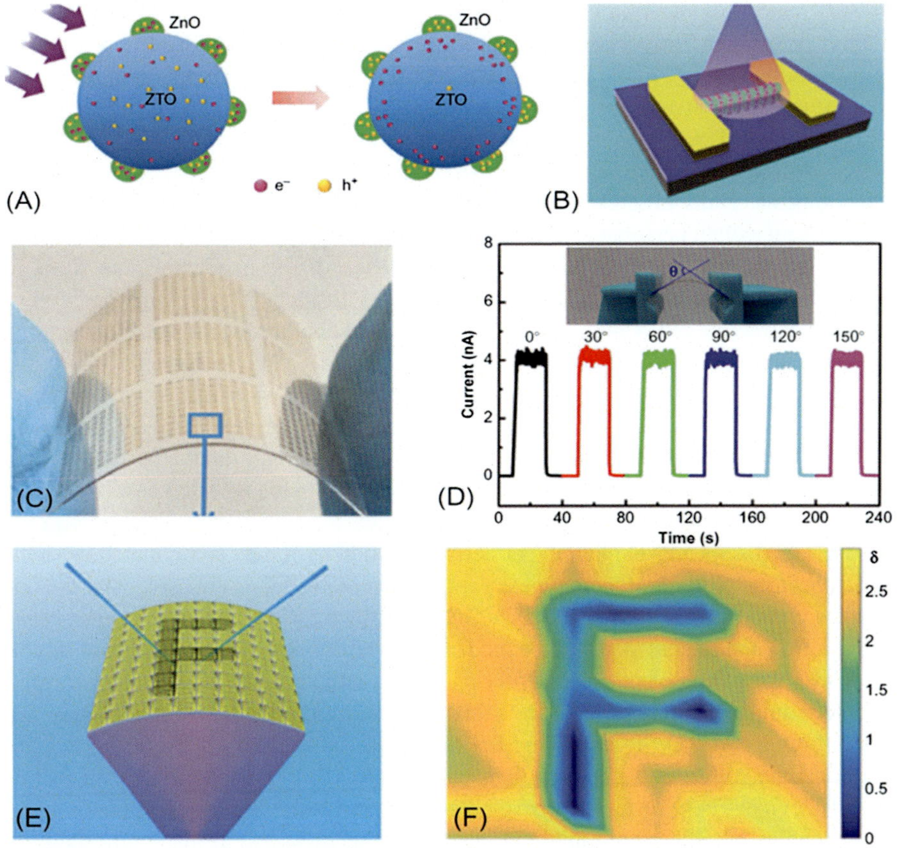

Fig. 15 (A) UV illuminated schematics of the carrier separation mechanism in the PD. (B) Schematic representation of PD fabricated using ZnO QD modified ZTONW. (C) Optical images of flexible PD arrays on a PET substrate. (D) Plot of I vs T of the flexible PD bent with different bending angles providing a bias voltage of 1 V. Schematic of the bending angle is given as inset. (E) Representation of flexible PD array for sensing the letter "F" under tensile bending directions. (F) Resulting images of letters "F" by the flexible PD array.
Source: Reused with permission from J. Song, Colloidal metal oxides in electronics and optoelectronics, in Colloidal Metal Oxide Nanoparticles, Elsevier, 2020, pp. 203–246. Copyright © 2020, Elsevier.

separation mechanism in the PD upon UV illumination and optical images of PD along with the resulted images of a sensed letter "F" by flexible PD array are shown in Fig. 15A, B, C, E and F [21]. It can be investigated that ZnO-ZTO displayed 10 times higher photocurrent and responsivity compared with pristine ZTO NWs [88]. High-quality single crystalline $In_2Ge_2O_7$ is another promising CMO that finds application as NBs for deep UV solar-blind PDs, optical sensing, switches, and communication systems owing to the enhanced photoconductive performance in terms of reproducibility, sensitivity, selectivity, stability, fast decay times ($\sim$3 ms), large responsivity ($3.9 \times 10^5 AW^{-1}$), and quantum efficiency ($2.0 \times 10^8\%$) [89]. Besides, strong absorption, scattering, and local field enhancement that result from LSPR of spin-coated Au-ZnO nanofilms show an improved photo response by 80 times at 335 nm [90].

Transparent thin films of MOs have become popular for their use in nonlinear optics, antireflection coatings, and varied optical domains where transparency and refractive index of the films are essential. One such example of MOs films with high efficacy has been reported by Xia et al., wherein surface-modified ZrO_2 films were fabricated. PVA was used to optimize and enhance the transparency and refractive index of the nanocomposite films of ZrO_2. The measured optical transmittance showed that with an increasing weight percentage of ZrO_2 in the dispersion, there was a negligible change in the transparency. A study of the refractive index of the nanocomposite films of PVA-ZrO_2 using the ellipsometer elucidates that with increasing volume fraction of ZrO_2, the refractive index improved to 1.754 [91]. In another recent study, the carbon nanodot-functionalized surface of nanowire-shaped ZnO exhibited admirable photoresponse in the visible and UV range of the spectra. UV-visible absorption spectroscopy and PL spectroscopy were employed to investigate the optical properties. It was seen that the presence of a prominent peak at 5.6 eV was attributed to the electronic transitions from HOMO to the LUMO. Further, the shift in the PL spectra was ascribed to the charge carriers at the surface. Delving deeper, the PL spectra of the carbon dot functionalized ZnO NPs explain the effective attachment of the carbon nanodots to the surface because of the presence of peaks at the band edge, defect emission, and the one from the carbon dots. Besides, the experimental investigations show the longer decay time for ZnO nanowires upon surface modification which opens up a plethora of opportunities for green optoelectronics involving carbon nanodots [92].

Hetero junction photodiodes are fabricated with Ag NPs-decorated Al-Ga co-doped ZnO/Si (AGZO/Si) and found that there is a substantial improvement in photoresponsivity (58%) and photo-to-dark current ratio (147%) compared with that of the bare device. These results offer Ag NPs-decorated AGZO/Si as a promising material for future optoelectronic applications [93]. The InGaZnO films synthesized by dynamic spin coating method with varying Ga concentration as an interlayer in the Al/ InGaZnO/p-Si/Al photodiode exhibited high transparency in the visible region (84%–92%). With the varying Ga content in InGaZnO resulted coincidence of energy bandgap of InGaZnO with that of semiconductor and the electrolyte leads to high sensitivity on illumination [94]. Comparative analysis of photo-sensing properties of CuO, CuO@Ag core-shell and CuO@Ag:La NPs fabricated for thin film device showed a bias voltage similar to photoresistor where CuO@Ag:La NPs are found to be efficient due to better sensitivity toward light [95].

3.3 Thin-film field effect transistors

Thin-film transistors (TFT) and metal oxide-semiconductor field effect transistors (MOSFET) are similar in operation and differ by device structure such that in MOSFET silicon serves as a device substrate where as in TFT nonactive materials, like glass or plastic, are used. The main components in TFT are the dielectric, semiconductor layer, and the source, drain (electrodes), and gate electrodes as thin films on an inert supporting substrate [8]. In an ideal condition, the current through the gate electrode must be negligible in comparison with the current between the drain and source electrodes [96]. Nowadays, TFTs based on various MOS, such as In, Ga and ZnO, have gained immense attraction owing to superior electrical performance such as high K, mobility, and low carrier concentration, to be used in switching devices such as smart window, transparent electronic wall, mobile display, and smart sign [97].

From the perspective of high-quality MOs for thin-film-transistors, rationally designed facile synthesis approaches substantially fine-tune the nucleation-growth mechanisms, and therefore, the structure-related characteristics, markedly influenced by the processing temperature. To support the effect of growth temperature on the solution stability and efficiency of nanoparticulate indium oxide for thin-film transistor applications, TEM images examined for different batches of the samples revealed that for samples which were allowed to grow at low temperatures manifested polydispersity, as well as the solubility was affected. Morphology, thickness, and surface roughness of the prepared films were further analyzed by atomic force microscope, and the crystalline structure of the samples were understood from the XRD patterns [98]. Remarkable advancements have emerged in the fabrication of solution processed thin-film transistors for the active layers of MO semiconductor (n-type and p-type) and dielectrics to achieve better device performance. Depending upon the low temperature and large area manufacturing of solution processed thin films, including post-deposition treatments, device performance will be varied. The solution processed In_2O_3 TFTs annealed at 500°C in air after spin coating, exhibited a high field-effect mobility of 55.3 cm^2/Vs. The ZnO-TFTs for the same conditions of temperature and processing time (3 min at 140°C) showed a field effect mobility of 0.72 cm^2/Vs for microwave-assisted annealing, whereas hot plate annealing showed a reduced mobility of 1.5×10^{-5} cm^2/Vs. [99] Film densification of deep UV radiation annealing (DUV) method after spin coating has been able to show oxide TFTs with high electron mobility (7 cm^2/Vs). Recently, Cu_2O and SnO are the two potential materials for the use in p-type TFTs. High-quality epitaxial Cu_2O films and high-k dielectrics (HfON and HfO_2) developed by pulser laser deposition (PLD) with substrate temperatures of 500–700°C exhibited TFTs with mobilities of up to 4.3 cm^2/Vs and I_D (on)/I_D (off) ratio in the order of 10^6 (Fig. 16) [8]. A comparative study of the IGZO TFTs synthesized via bar-coating technique and spin-coating technique showed that bar-coated one have a higher average mobility value due to the increased thin film quality that arises from longer evaporation time [100]. Solution-processed amorphous zinc tin oxide (ZTO) semiconductor TFTs with alkali metals (Li,Na) as a dopant showed an enhancement of the electrical performance such as high field-effect mobility than

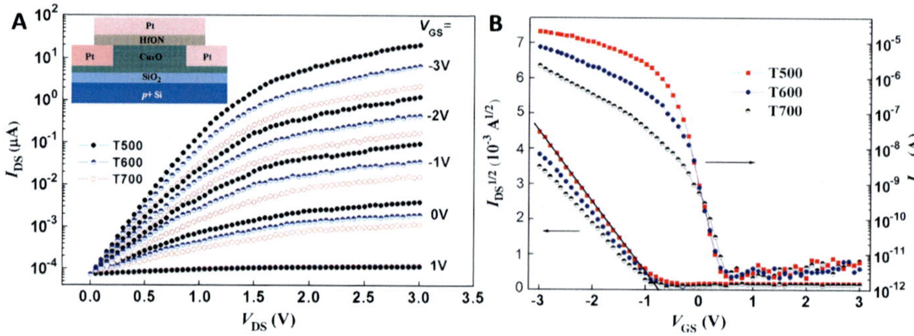

Fig. 16 (A) I-V curves for field-effect and drain current saturation (inset TFT structure) and (B) for transfer characteristics of the device.
Source: Reused with permission from S.R. Thomas, P. Pattanasattayavong, T.D. Anthopoulos, Solution-processable metal oxide semiconductors for thin-film transistor applications, Chem. Soc. Rev. 42 (16) (2013) 6910–6923. Copyright © 2013, The Royal Society of Chemistry.

that of intrinsic amorphous ZTO semiconductor [97]. In another study, replacement of the solution deposited In_2O_3 layer with a photonically processed In_2O_3/ZnO heterojunction enabled to increase the electron mobility from 6 cm^2/Vs to 36 cm^2/Vs, maintaining the low-voltage operation. The photonic treatment approach helped to decrease the processing time of less than 18 s per layer [101]. A large number of CMOs is known with high-*k* dielectrics including Al_2O_3, ZrO_2, HfO_2, Y_2O_3, La_2O_3, and others.

Schottky emission acts as a barrier for device performance via large leakage current by influencing the bandgap and dielectric constant that should be as high as possible. Dielectrics can be of two types such as binary (Al_2O_3, ZrO_2, HfO_2, Y_2O_3, La_2O_3, Ga_2O_3, Sc_2O_3, MgO, Li_2O, Nd_2O_3, Gd_2O_3, Yb_2O_3) and ternary (Zr-Al-O, La-Al-O, Ti-Al-O, Hf-Si-O). Ternary oxides with high-*k* values are more potent to maximize the dielectric constant-bandgap product [102].

Fabrication of MOs by high-pressure annealing methods too follows a cost-saving route with commercial potential and ease in processing as it does not require very high temperatures (posttreatment temperature). Rim et al., in their work, revealed exceptional electrical properties of thin-film transistors made of high-quality MOs synthesized by the solution processing route. High-pressure annealing annihilated the organic impurities, thereby preventing interaction among the groups containing carbon and hydrogen, which ultimately deteriorates the film quality. High-pressure annealing led to a commendable improvement in the transfer characteristics of indium zinc oxide-based thin film transistors, with a notable increase in field-effect mobility and on-off ratio. An important point that needs to be noted here is that the performance improved only for an optimized pressure condition, but when the pressure was increased further, the device performance degraded. Under high-pressure annealing conditions, interestingly, the threshold voltage was lowered, which is probably due to the improved densification and decreased thickness of the film. These indium zinc oxide films were employed in a flexible electronics setup to prove their economic impact (Fig. 17) [103].

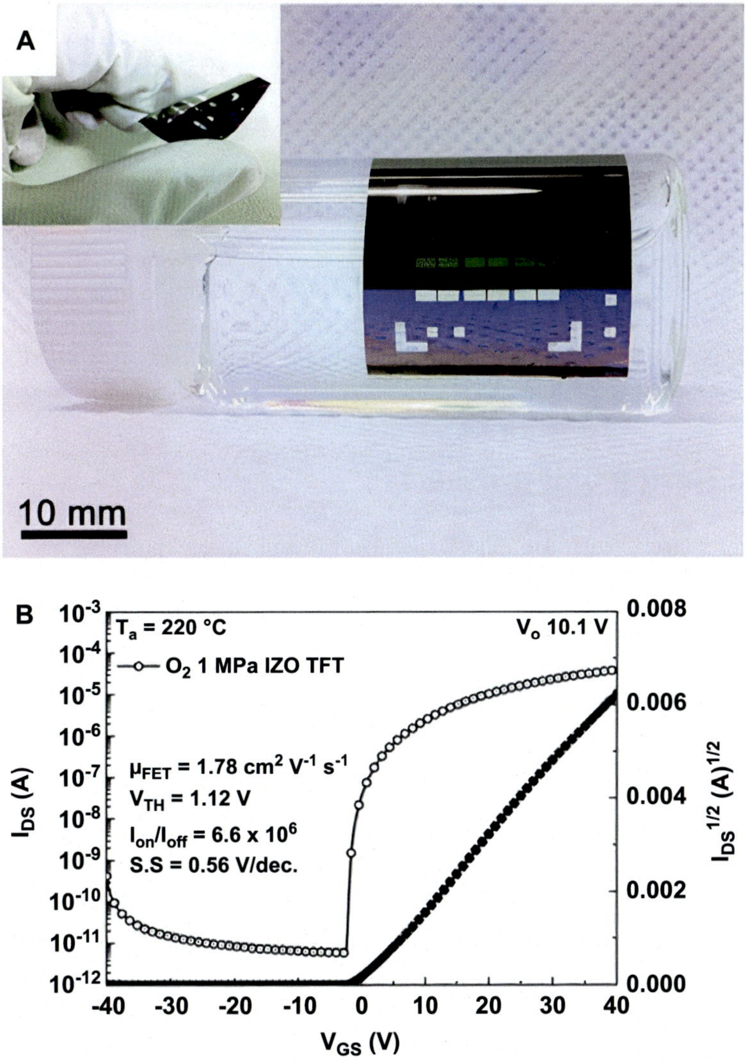

Fig. 17 (A) Digital images showing high-quality indium zinc oxide thin film transistors fabricated by solution processing routes followed by high-pressure annealing. (B) Transfer characteristics of the high-pressure annealed indium zinc oxide flexible thin film transistors. *Source*: Reused with permission from Y.S. Rim, W.H. Jeong, D.L. Kim, H.S. Lim, K.M. Kim, H.J. Kim, Simultaneous modification of pyrolysis and densification for low-temperature solution-processed flexible oxide thin-film transistors, J. Mater. Chem. 22 (25) (2012) 12491–12497. Copyright © 2020, Royal society of chemistry.

Low-temperature processing of MOs for optoelectronics and photonics has attracted researchers' attention for its green chemistry protocol that stands out in terms of sustainability and device efficiency. Cu:NiO thin films fabricated by a unique low-temperature approach manifested enhanced transfer characteristics involving increased channel currents ascribed to the high carrier concentration. What's more, with the increasing doping concentration of Cu, the on-off ratio was enlarged along with hole mobility. On top of that, the bias stress measurements signify good device compatibility. The electrical conductivity also showed an increasing behavior with increment in Cu concentration, determined by the four-probe measurement. Conjointly, investigation of the Cu:NiO thin-film transistor-based device performance in logic circuits was executed by cycles of gate voltage. From the application perspective, p-type MOs thin-film transistors serve as an excellent option for hole-injection layers in LEDs. This was also validated by incorporation in green LEDs. Another aspect that explored the viability of these MOs thin film transistors in transparent displays proclaimed that Cu:NiO thin film can be integrated with indium tin oxide/glass substrate successfully, however it is conditioned to the thickness of ZrO_2 acting as the dielectric layer. Thus, it emphasized the rapid progress in domains of MOs-based devices for varied developing technologies [104].

4 Conclusion and prospects

This chapter discussed and emphasized the advancements and potential use of colloidal metal oxide (CMO) nanocrystals (NCs) for significant optoelectronic and photonic applications, mainly focusing on LEDs, PDs, and TFTs. Low-temperature solution processability, tunable optoelectronic properties, large area flexibility with polymer substrates, ease of synthesis, and intrinsic stability make CMOs the building blocks of modern optoelectronic technology. These superior qualities of CMOs over conventional Si-based semiconductors attracted enormous attention toward low-cost, high-performance device fabrication. Various solution-processed techniques have been elaborated and demonstrated advantages over vapor-based processes, such as cost-effectiveness and compatibility with flexible and plastic substrates by providing ambient working conditions. Although CMOs display proficiency in device fabrication and operation, a thorough understanding of the interrelationship between the properties and device performance is essential to fully exploiting specific applications. Apart from the developed chemical approaches, new on-demand material chemistry for colloidal oxide nanocrystals is needed to facilitate guidance for the material design and device operations, some of which are currently unavailable. Pointing to the chemistry approaches, it is relevant to link the impact of surface modifications on the CMO and the resulting electrical and optical properties. Investigations by the ligand chemistry approach provide more insight and can be confirmed by the various characterization techniques. Another area that needs attention is p-type CMOs, as they are scarce, and the conversion of n-type MOs to p-type needs new synthetic strategies. For example, more delicate synthesis strategies that enable control over the band structure, carrier density, intragap states, stoichiometry, and optical properties of the p-type NCs-based hole transport layers shall be pursued while taking care of in-depth mechanistic understanding. Despite taking advantage of the solution processability of CMOs, it necessitates further research for precision control of film thickness, surface irregularities, and coverage uniformity for the solution-processed thin films. For the industrial application, the challenges exhibited by CMOs should be mitigated by achieving precision control of doping concentration, reproducible surface chemistry, and viable post-treatment techniques. Exploratory advancements such as colloidal syntheses of oxide nanomaterials and their integration into devices may result in newer challenges for predicting device characteristics. Resolving these will further improve the fundamental understanding, leading to novel CMOs deposition strategies and applications. In conclusion, this chapter attempted to provide insights into commonly adopted synthesis approaches, application-oriented process optimization, characterization, and to provide the readers with the prospectus of CMOs in the photonics industry.

Acknowledgments

Authors acknowledge the Department of Science and Technology, Govt. of India for the financial support under DST-INSPIRE faculty award scheme.

References

[1] I. Konidakis, A. Karagiannaki, E. Stratakis, Advanced composite glasses with metallic, perovskite, and two-dimensional nanocrystals for optoelectronic and photonic applications, Nanoscale 14 (8) (2022) 2966–2989.

[2] Photonics Market—Growth, Trends, COVID-19 Impact, and Forecasts (2021–2026). https://www.researchandmarkets.com/reports/4771762/photonics-market-growth-trends-covid-19. (Accessed 4 September 2022).

[3] Optoelectronics, Optoelectronics: Emerging Technology Focused on Light-detecting Devices. *https://blog.ttelectronics.com/optoelectronics*. (Accessed 4 September 2022).

[4] Photonics, Electro Optics or Optoelectronics? | Quality Magazine. https://www.qualitymag.com/articles/90749-photonics-electro-optics-or-optoelectronics. (Accessed 4 September 2022).

[5] X. Yu, T.J. Marks, A. Facchetti, Metal oxides for optoelectronic applications, Nat. Mater. 15 (4) (2016) 383–396.

[6] T. Kamiya, K. Nomura, H. Hosono, Origins of high mobility and low operation voltage of amorphous oxide TFTs: electronic structure, electron transport, defects and doping, J. Display Technol. 5 (7) (2009) 273–288.

[7] K. Nomura, H. Ohta, A. Takagi, T. Kamiya, M. Hirano, H. Hosono, Room-temperature fabrication of transparent flexible thin-film transistors using amorphous oxide semiconductors, Nature 432 (7016) (2004) 488–492.

[8] S.R. Thomas, P. Pattanasattayavong, T.D. Anthopoulos, Solution-processable metal oxide semiconductors for thin-film transistor applications, Chem. Soc. Rev. 42 (16) (2013) 6910–6923.

[9] X. Liang, S. Bai, X. Wang, X. Dai, F. Gao, B. Sun, Z. Ning, Z. Ye, Y. Jin, Colloidal metal oxide nanocrystals as charge transporting layers for solution-processed light-emitting diodes and solar cells, Chem. Soc. Rev. 46 (6) (2017) 1730–1759.

[10] C.R. Kagan, Flexible colloidal nanocrystal electronics, Chem. Soc. Rev. 48 (6) (2019) 1626–1641.

[11] A.P. Caricato, S. Capone, M. Epifani, M. Lomascolo, A. Luches, M. Martino, F. Romano, R. Rella, P. Siciliano, J. Spadavecchia, A. Taurino, Nanoparticle thin films deposited by MAPLE for sensor applications, Fund. Laser Assist. Micro Nanotechnol., SPIE 6958 (2008) 123–135.

[12] H. Sirringhaus, Device physics of solution-processed organic field-effect transistors, Adv. Mater. 17 (20) (2005) 2411–2425.

[13] G. Konstantatos, I. Howard, A. Fischer, S. Hoogland, J. Clifford, E. Klem, L. Levina, E. H. Sargent, Ultrasensitive solution-cast quantum dot photodetectors, Nature 442 (7099) (2006) 180–183.

[14] T. Karlessi, M. Santamouris, Improving the performance of thermochromic coatings with the use of UV and optical filters tested under accelerated aging conditions, Int. J. Low-Carbon Technol. 10 (1) (2015) 45–61.

[15] R. Buonsanti, A. Llordes, S. Aloni, B.A. Helms, D.J. Milliron, Tunable infrared absorption and visible transparency of colloidal aluminum-doped zinc oxide nanocrystals, Nano Lett. 11 (11) (2011) 4706–4710.

[16] X. Liu, M.T. Swihart, Heavily-doped colloidal semiconductor and metal oxide nanocrystals: an emerging new class of plasmonic nanomaterials, Chem. Soc. Rev. 43 (11) (2014) 3908–3920.

[17] M.J. Alam, D.C. Cameron, Preparation and properties of transparent conductive aluminum-doped zinc oxide thin films by sol-gel process, J. Vac. Sci. Technol. A 19 (4) (2001) 1642–1646.

[18] J. Song, S.A. Kulinich, J. Li, Y. Liu, H. Zeng, A general one-pot strategy for the synthesis of high-performance transparent-conducting-oxide nanocrystal inks for all-solution-processed devices, Angew. Chem. 127 (2) (2015) 472–476.

[19] Z.L. Wang, ZnO nanowire and nanobelt platform for nanotechnology, Mater. Sci. Eng.: R: Rep. 64 (3–4) (2009) 33–71.

[20] B. Chavillon, L. Cario, A. Renaud, F. Tessier, F. Cheviré, M. Boujtita, Y. Pellegrin, E. Blart, A. Smeigh, L. Hammarstrom, F. Odobel, P-type nitrogen-doped ZnO nanoparticles stable under ambient conditions, J. Am. Chem. Soc. 134 (1) (2012) 464–470.

[21] J. Song, Colloidal metal oxides in electronics and optoelectronics, in: Colloidal Metal Oxide Nanoparticles, Elsevier, 2020, pp. 203–246.

[22] J. Liu, S. Shao, G. Fang, B. Meng, Z. Xie, L. Wang, High-efficiency inverted polymer solar cells with transparent and work-function tunable MoO3-Al composite film as cathode buffer layer, Adv. Mater. 24 (20) (2012) 2774–2779.

[23] M. Hilgendorff, L. Spanhel, C. Rothenhäusler, G. Müller, From ZnO colloids to nanocrystalline highly conductive films, J. Electrochem. Soc. 145 (10) (1998) 3632.

[24] J. Santos-Cruz, G. Torres-Delgado, R. Castanedo-Perez, S. Jiménez-Sandoval, O. Jiménez-Sandoval, C.I. Zuniga-Romero, J.M. Marín, O. Zelaya-Angel, Dependence of electrical and optical properties of sol–gel prepared undoped cadmium oxide thin films on annealing temperature, Thin Solid Films 493 (1–2) (2005) 83–87.

[25] H.M. Zhou, D.Q. Yi, Z.M. Yu, L.R. Xiao, J. Li, Preparation of aluminum doped zinc oxide films and the study of their microstructure, electrical and optical properties, Thin Solid Films 515 (17) (2007) 6909–6914.

[26] B.H. Lee, L. Kang, R. Nieh, W.J. Qi, J.C. Lee, Thermal stability and electrical characteristics of ultrathin hafnium oxide gate dielectric reoxidized with rapid thermal annealing, Appl. Phys. Lett. 76 (14) (2000) 1926–1928.

[27] J. Zhang, J. Wang, Y. Fu, B. Zhang, Z. Xie, Efficient and stable polymer solar cells with annealing-free solution-processible NiO nanoparticles as anode buffer layers, J. Mater. Chem. C 2 (39) (2014) 8295–8302.

[28] Y.E. Ha, M.Y. Jo, J. Park, Y.C. Kang, S.I. Yoo, J.H. Kim, Inverted type polymer solar cells with self-assembled monolayer treated ZnO, J. Phys. Chem. C 117 (6) (2013) 2646–2652.

[29] K. Manthiram, A.P. Alivisatos, Tunable localized surface plasmon resonances in tungsten oxide nanocrystals, J. Am. Chem. Soc. 134 (9) (2012) 3995–3998.

[30] Y.S. Wang, P.J. Thomas, P. O'Brien, Optical properties of ZnO nanocrystals doped with Cd, Mg, Mn, and Fe ions, J. Phys. Chem. B. 110 (43) (2006) 21412–21415.

[31] H.M. Xiong, D.G. Shchukin, H. Möhwald, Y. Xu, Y.Y. Xia, Sonochemical synthesis of highly luminescent zinc oxide nanoparticles doped with magnesium (II), Angew. Chem. Int. Ed. 48 (15) (2009) 2727–2731.

[32] M.C. Wang, H.J. Lin, T.S. Yang, Characteristics and optical properties of iron ion (Fe3 +)-doped titanium oxide thin films prepared by a sol–gel spin coating, J. Alloys Compd. 473 (1–2) (2009) 394–400.

[33] H.M. Xiong, D.P. Liu, Y.Y. Xia, J.S. Chen, Polyether-grafted ZnO nanoparticles with tunable and stable photoluminescence at room temperature, Chem. Mater. 17 (12) (2005) 3062–3064.

[34] X. Liu, M.T. Swihart, A general single-pot heating method for morphology, size and luminescence-controllable synthesis of colloidal ZnO nanocrystals, Nanoscale 5 (17) (2013) 8029–8036.

[35] L. Yu, X. Ren, Z. Yang, Y. Han, Z. Li, The preparation and assembly of CdS_xSe_{1-x} alloyed quantum dots on TiO2 nanowire arrays for quantum dot-sensitized solar cells, J. Mater. Sci. Mater. Electron. 27 (7) (2016) 7150–7160.

[36] S. Mitra, S. Das, K. Mandal, S. Chaudhuri, Synthesis of a α-Fe2O3 nanocrystal in its different morphological attributes: growth mechanism, optical and magnetic properties, Nanotechnology 18 (27) (2007), 275608.

[37] H.M. Fan, G.J. You, Y. Li, Z. Zheng, H.R. Tan, Z.X. Shen, S.H. Tang, Y.P. Feng, Shape-controlled synthesis of single-crystalline Fe_2O_3 hollow nanocrystals and their tunable optical properties, J. Phys. Chem. C 113 (22) (2009) 9928–9935.

[38] T. Rajh, L.X. Chen, K. Lukas, T. Liu, M.C. Thurnauer, D.M. Tiede, Surface restructuring of nanoparticles: an efficient route for ligand– metal oxide crosstalk, J. Phys. Chem. B. 106 (41) (2002) 10543–10552.

[39] H. Zeng, G. Duan, Y. Li, S. Yang, X. Xu, W. Cai, Blue luminescence of ZnO nanoparticles based on non-equilibrium processes: defect origins and emission controls, Adv. Funct. Mater. 20 (4) (2010) 561–572.

[40] A.B. Djurišić, Y.H. Leung, K.H. Tam, L. Ding, W.K. Ge, H.Y. Chen, S. Gwo, Green, yellow, and orange defect emission from ZnO nanostructures: Influence of excitation wavelength, Appl. Phys. Lett. 88 (2006), 103107.

[41] L. Irimpan, V.P.N. Nampoori, P. Radhakrishnan, B. Krishnan, A. Deepthy, Size-dependent enhancement of nonlinear optical properties in nanocolloids of ZnO, J. Appl. Phys. 103 (3) (2008), 033105.

[42] N. Pradhan, S. Das Adhikari, A. Nag, D.D. Sarma, Luminescence, plasmonic, and magnetic properties of doped semiconductor nanocrystals, Angew. Chem. Int. Ed. 56 (25) (2017) 7038–7054.

[43] M. Kanehara, H. Koike, T. Yoshinaga, T. Teranishi, Indium tin oxide nanoparticles with compositionally tunable surface plasmon resonance frequencies in the near-IR region, J. Am. Chem. Soc. 131 (49) (2009) 17736–17737.

[44] S.D. Lounis, E.L. Runnerstrom, A. Llordes, D.J. Milliron, Defect chemistry and plasmon physics of colloidal metal oxide nanocrystals, J. Phys. Chem. Lett. 5 (9) (2014) 1564–1574.

[45] J. Perelaer, P.J. Smith, D. Mager, D. Soltman, S.K. Volkman, V. Subramanian, J.G. Korvink, U.S. Schubert, Printed electronics: the challenges involved in printing devices,

interconnects, and contacts based on inorganic materials, J. Mater. Chem. 20 (39) (2010) 8446–8453.

[46] R. Steim, F.R. Kogler, C.J. Brabec, Interface materials for organic solar cells, J. Mater. Chem. 20 (13) (2010) 2499–2512.

[47] P. De Bruyn, D.J.D. Moet, P.W.M. Blom, A facile route to inverted polymer solar cells using a precursor based zinc oxide electron transport layer, Org. Electron. 11 (8) (2010) 1419–1422.

[48] S. Höfle, M. Bruns, S. Strässle, C. Feldmann, U. Lemmer, A. Colsmann, Tungsten oxide buffer layers fabricated in an inert sol-gel process at room-temperature for blue organic light-emitting diodes, Adv. Mater. 25 (30) (2013) 4113–4116.

[49] P. Kumar, S. Khadtare, J. Park, B.C. Yadav, Fabrication of leaf shaped SnO2 nanoparticles via sol–gel route and its application for the optoelectronic humidity sensor, Mater. Lett. 278 (2020), 128451.

[50] A. Barhoum, G. Van Assche, H. Rahier, M. Fleisch, S. Bals, M.P. Delplancked, F. Leroux, D. Bahnemann, Sol-gel hot injection synthesis of ZnO nanoparticles into a porous silica matrix and reaction mechanism, Mater. Des. 119 (2017) 270–276.

[51] K.W. Jo, S.W. Moon, W.J. Cho, Fabrication of high-performance ultra-thin-body SnO2 thin-film transistors using microwave-irradiation post-deposition annealing, Appl. Phys. Lett. 106 (4) (2015), 043501.

[52] T. Jun, K. Song, Y. Jeong, K. Woo, D. Kim, C. Bae, J. Moon, High-performance low-temperature solution-processable ZnO thin film transistors by microwave-assisted annealing, J. Mater. Chem. 21 (4) (2011) 1102–1108.

[53] Y.J. Lee, J. Yi, G.F. Gao, H. Koerner, K. Park, J. Wang, K. Luo, R.A. Vaia, J.W. Hsu, Low-temperature solution-processed molybdenum oxide nanoparticle hole transport layers for organic photovoltaic devices, Adv. Energy Mater. 2 (10) (2012) 1193–1197.

[54] J. Moon, J.A. Park, S.J. Lee, H.Y. Chu, T. Zyung, Fabrication of ordered hollow ZnO–NiO oxide arrays, J. Nanosci. Nanotechnol. 11 (5) (2011) 4394–4399.

[55] W.I. Park, D.H. Kim, S.W. Jung, G.C. Yi, Metalorganic vapor-phase epitaxial growth of vertically well-aligned ZnO nanorods, Appl. Phys. Lett. 80 (22) (2002) 4232–4234.

[56] E. Della Gaspera, M. Bersani, M. Cittadini, M. Guglielmi, D. Pagani, R. Noriega, S. Mehra, A. Salleo, A. Martucci, Low-temperature processed Ga-doped ZnO coatings from colloidal inks, J. Am. Chem. Soc. 135 (9) (2013) 3439–3448.

[57] A.I. Hofmann, E. Cloutet, G. Hadziioannou, Materials for transparent electrodes: from metal oxides to organic alternatives, Adv. Electron. Mater. 4 (10) (2018) 1700412.

[58] H. Liu, V. Avrutin, N. Izyumskaya, Ü. Özgür, H. Morkoç, Transparent conducting oxides for electrode applications in light emitting and absorbing devices, Superlatt. Microstruct. 48 (5) (2010) 458–484.

[59] B.M. Crockett, A.W. Jansons, K.M. Koskela, M.C. Sharps, D.W. Johnson, J.E. Hutchison, Influence of nanocrystal size on the optoelectronic properties of thin, solution-cast Sn-doped In_2O_3 films, Chem. Mater. 31 (9) (2019) 3370–3380.

[60] Y. Liu, S. Wei, G. Wang, J. Tong, J. Li, D. Pan, Quantum-sized SnO2 nanoparticles with upshifted conduction band: a promising electron transportation material for quantum dot light-emitting diodes, Langmuir 36 (23) (2020) 6605–6609.

[61] H.M. Kim, S. Cho, J. Kim, H. Shin, J. Jang, Li and Mg Co-doped zinc oxide electron transporting layer for highly efficient quantum dot light-emitting diodes, ACS Appl. Mater. Interfaces 10 (28) (2018) 24028–24036.

[62] R. Medhi, C.H. Li, S.H. Lee, M.D. Marquez, A.J. Jacobson, T.C. Lee, T.R. Lee, Uniformly spherical and monodisperse antimony-and zinc-doped tin oxide nanoparticles for optical and electronic applications, ACS Appl. Nano Mater. 2 (10) (2019) 6554–6564.

[63] L. Rao, Y. Tang, Z. Li, X. Ding, J. Li, S. Yu, C. Yan, H. Lu, Effect of ZnO nanostructures on the optical properties of white light-emitting diodes, Opt. Express 25 (8) (2017) A432–A443.
[64] J.H. Yun, H.S. Cho, K.B. Bae, S. Sudhakar, Y.S. Kang, J.S. Lee, A.Y. Polyakov, I.H. Lee, Enhanced optical properties of nanopillar light-emitting diodes by coupling localized surface plasmon of Ag/SiO2 nanoparticles, Appl. Phys. Expr. 8 (9) (2015), 092002.
[65] X. Gu, T. Qiu, W. Zhang, P.K. Chu, Light-emitting diodes enhanced by localized surface plasmon resonance, Nanoscale Res. Lett. 6 (1) (2011) 1–12.
[66] X. Liang, Y. Ren, S. Bai, N. Zhang, X. Dai, X. Wang, H. He, C. Jin, Z. Ye, Q. Chen, L. Chen, Colloidal indium-doped zinc oxide nanocrystals with tunable work function: rational synthesis and optoelectronic applications, Chem. Mater. 26 (17) (2014) 5169–5178.
[67] D.I. Son, B.W. Kwon, D.H. Park, W.S. Seo, Y. Yi, B. Angadi, C.L. Lee, W.K. Choi, Emissive ZnO–graphene quantum dots for white-light-emitting diodes, Nat. Nanotechnol. 7 (7) (2012) 465–471.
[68] M. Kim, B.H. Kwon, C.W. Joo, M.S. Cho, H. Jang, H. Cho, D.Y. Jeon, E.N. Cho, Y.S. Jung, Metal oxide charge transfer complex for effective energy band tailoring in multilayer optoelectronics, Nat. Commun. 13 (1) (2022) 1–9.
[69] T. Ding, X. Yang, L. Bai, Y. Zhao, K.E. Fong, N. Wang, H.V. Demir, X.W. Sun, Colloidal quantum-dot LEDs with a solution-processed copper oxide (CuO) hole injection layer, Org. Electron. 26 (2015) 245–250.
[70] X. Yang, E. Mutlugun, Y. Zhao, Y. Gao, K.S. Leck, Y. Ma, L. Ke, S.T. Tan, H.V. Demir, X.W. Sun, Solution processed tungsten oxide interfacial layer for efficient hole-injection in quantum dot light-emitting diodes, Small 10 (2) (2014) 247–252.
[71] F. Cao, H. Wang, P. Shen, X. Li, Y. Zheng, Y. Shang, J. Zhang, Z. Ning, X. Yang, High-efficiency and stable quantum dot light-emitting diodes enabled by a solution-processed metal-doped nickel oxide hole injection interfacial layer, Adv. Funct. Mater. 27 (42) (2017) 1704278.
[72] S. Lee, Y. Jeong, S. Jeong, J. Lee, M. Jeon, J. Moon, Solution-processed ZnO nanoparticle-based semiconductor oxide thin-film transistors, Superlatt. Microstruct. 44 (6) (2008) 761–769.
[73] M. Wang, Y. Lian, X. Wang, PPV/PVA/ZnO nanocomposite prepared by complex precursor method and its photovoltaic application, Curr. Appl. Phys. 9 (1) (2009) 189–194.
[74] Y. Sun, J.H. Seo, C.J. Takacs, J. Seifter, A.J. Heeger, Inverted polymer solar cells integrated with a low-temperature-annealed sol-gel-derived ZnO film as an electron transport layer, Adv. Mater. 23 (14) (2011) 1679–1683.
[75] J.Y. Kim, K. Lee, N.E. Coates, D. Moses, T.Q. Nguyen, M. Dante, A.J. Heeger, Efficient tandem polymer solar cells fabricated by all-solution processing, Science 317 (5835) (2007) 222–225.
[76] D. Zhang, W.C. Choy, F. Xie, W.E. Sha, X. Li, B. Ding, K. Zhang, F. Huang, Y. Cao, Plasmonic electrically functionalized TiO2 for high-performance organic solar cells, Adv. Funct. Mater. 23 (34) (2013) 4255–4261.
[77] M.H. Park, J.H. Li, A. Kumar, G. Li, Y. Yang, Doping of the metal oxide nanostructure and its influence in organic electronics, Adv. Funct. Mater. 19 (8) (2009) 1241–1246.
[78] Z. Liang, Q. Zhang, L. Jiang, G. Cao, ZnO cathode buffer layers for inverted polymer solar cells, Energ. Environ. Sci. 8 (12) (2015) 3442–3476.
[79] R.M. Hewlett, M.A. McLachlan, Surface structure modification of ZnO and the impact on electronic properties, Adv. Mater. 28 (20) (2016) 3893–3921.

[80] S.B. Dkhil, D. Duché, M. Gaceur, A.K. Thakur, F.B. Aboura, L. Escoubas, J.J. Simon, A. Guerrero, J. Bisquert, G. Garcia-Belmonte, Q. Bao, Interplay of optical, morphological, and electronic effects of ZnO optical spacers in highly efficient polymer solar cells, Adv. Energy Mater. 4 (18) (2014) 1400805.

[81] M. Gaceur, S.B. Dkhil, D. Duché, F. Bencheikh, J.J. Simon, L. Escoubas, M. Mansour, A. Guerrero, G. Garcia-Belmonte, X. Liu, M. Fahlman, Ligand-free synthesis of aluminum-doped zinc oxide nanocrystals and their use as optical spacers in color-tuned highly efficient organic solar cells, Adv. Funct. Mater. 26 (2) (2016) 243–253.

[82] K. Zilberberg, S. Trost, H. Schmidt, T. Riedl, Solution processed vanadium pentoxide as charge extraction layer for organic solar cells, Adv. Energy Mater. 1 (3) (2011) 377–381.

[83] J. Ciro, D. Ramírez, M.A. Mejía Escobar, J.F. Montoya, S. Mesa, R. Betancur, F. Jaramillo, Self-functionalization behind a solution-processed NiO_x film used as hole transporting layer for efficient perovskite solar cells, ACS Appl. Mater. Interfaces 9 (14) (2017) 12348–12354.

[84] B. Tan, E. Toman, Y. Li, Y. Wu, Zinc stannate (Zn_2SnO_4) dye-sensitized solar cells, J. Am. Chem. Soc. 129 (14) (2007) 4162–4163.

[85] K. Manseki, T. Ikeya, A. Tamura, T. Ban, T. Sugiura, T. Yoshida, Mg-doped TiO_2 nanorods improving open-circuit voltages of ammonium lead halide perovskite solar cells, RSC Adv. 4 (19) (2014) 9652–9655.

[86] D.Y. Kim, J. Ryu, J. Manders, J. Lee, F. So, Air-stable, solution-processed oxide p–n heterojunction ultraviolet photodetector, ACS Appl. Mater. Interfaces 6 (3) (2014) 1370–1374.

[87] L. Hu, J. Yan, M. Liao, L. Wu, X. Fang, Ultrahigh external quantum efficiency from thin SnO2 nanowire ultraviolet photodetectors, Small 7 (8) (2011) 1012–1017.

[88] L. Li, L. Gu, Z. Lou, Z. Fan, G. Shen, ZnO quantum dot decorated Zn2SnO4 nanowire heterojunction photodetectors with drastic performance enhancement and flexible ultraviolet image sensors, ACS Nano 11 (4) (2017) 4067–4076.

[89] L. Li, P.S. Lee, C. Yan, T. Zhai, X. Fang, M. Liao, Y. Koide, Y. Bando, D. Golberg, Ultrahigh-performance solar-blind photodetectors based on individual single-crystalline $In_2Ge_2O_7$ nanobelts, Adv. Mater. 22 (45) (2010) 5145–5149.

[90] N. Gogurla, A.K. Sinha, S. Santra, S. Manna, S.K. Ray, Multifunctional Au-ZnO plasmonic nanostructures for enhanced UV photodetector and room temperature NO sensing devices, Sci. Rep. 4 (1) (2014) 1–9.

[91] Y. Xia, C. Zhang, J.X. Wang, D. Wang, X.F. Zeng, J.F. Chen, Synthesis of transparent aqueous ZrO2 nanodispersion with a controllable crystalline phase without modification for a high-refractive-index nanocomposite film, Langmuir 34 (23) (2018) 6806–6813.

[92] D. Cammi, K. Zimmermann, R. Gorny, A. Vogt, F. Dissinger, A. Gad, N. Markiewcz, A. Waag, J.D. Prades, C. Ronning, S.R. Waldvogel, Enhancement of the sub-bandgap photoconductivity in ZnO nanowires through surface functionalization with carbon nanodots, J. Phys. Chem. C 122 (3) (2018) 1852–1859.

[93] N.E. Koksal, M. Sbeta, A.B.D.U.L.L.A.H. Yildiz, Ag nanoparticles-decorated Al-Ga codoped ZnO based photodiodes, Optik 224 (2020), 165523.

[94] A. Karabulut, A. Dere, A.G. Al-Sehemi, A.A. Al-Ghamdi, F. Yakuphanoglu, Zinc oxide based 3-components semiconductor oxide photodiodes by dynamic spin coating method, Mater. Sci. Semicond. Process. 134 (2021), 106034.

[95] S. Arya, A. Singh, R. Kour, Comparative study of CuO, CuO@ Ag and CuO@ Ag: La nanoparticles for their photosensing properties, Mater. Res. Expr. 6 (11) (2019), 116313.

[96] J.P. Braga, G.R. De Lima, G. Gozzi, L.F. Santos, Electrical characterization of thin-film transistors based on solution-processed metal oxides, in: Design, Simulation and Construction of Field Effect Transistors, 2018, p. 8.

[97] K.H. Lim, K. Kim, S. Kim, S.Y. Park, H. Kim, Y.S. Kim, UV–visible spectroscopic analysis of electrical properties in alkali metal-doped amorphous zinc tin oxide thin-film transistors, Adv. Mater. 25 (21) (2013) 2994–3000.
[98] S.L. Swisher, S.K. Volkman, V. Subramanian, Tailoring indium oxide nanocrystal synthesis conditions for air-stable high-performance solution-processed thin-film transistors, ACS Appl. Mater. Interfaces 7 (19) (2015) 10069–10075.
[99] S. Jeong, J. Moon, Low-temperature, solution-processed metal oxide thin film transistors, J. Mater. Chem. 22 (4) (2012) 1243–1250.
[100] C. Glynn, C. O'Dwyer, Solution processable metal oxide thin film deposition and material growth for electronic and photonic devices, Adv. Mater. Interfaces 4 (2) (2017) 1600610.
[101] K. Tetzner, Y.H. Lin, A. Regoutz, A. Seitkhan, D.J. Payne, T.D. Anthopoulos, Sub-second photonic processing of solution-deposited single layer and heterojunction metal oxide thin-film transistors using a high-power xenon flash lamp, J. Mater. Chem. C 5 (45) (2017) 11724–11732.
[102] W. Xu, H. Li, J.B. Xu, L. Wang, Recent advances of solution-processed metal oxide thin-film transistors, ACS Appl. Mater. Interfaces 10 (31) (2018) 25878–25901.
[103] Y.S. Rim, W.H. Jeong, D.L. Kim, H.S. Lim, K.M. Kim, H.J. Kim, Simultaneous modification of pyrolysis and densification for low-temperature solution-processed flexible oxide thin-film transistors, J. Mater. Chem. 22 (25) (2012) 12491–12497.
[104] A. Liu, H. Zhu, Z. Guo, Y. Meng, G. Liu, E. Fortunato, R. Martins, F. Shan, Solution combustion synthesis: low-temperature processing for p-type Cu: NiO thin films for transparent electronics, Adv. Mater. 29 (34) (2017) 1701599.

Metal oxides in quantum-dot-based LEDs and their applications 14

Irfan Ayoub[a], *Umer Mushtaq*[a], *Hendrik C. Swart*[b], *and Vijay Kumar*[a,b]
[a]Department of Physics, National Institute of Technology Srinagar, Hazratbal, Srinagar, Jammu and Kashmir, India, [b]Department of Physics, University of the Free State, Bloemfontein, Free State, South Africa

Chapter outline

1 Introduction

The proliferation of different technologies that are connected to the Internet of Things (IoT) has led to the development of a broad variety of electronic gadgets incorporated into a wide variety of platforms for interaction with the surrounding environment [1–4]. In particular, optoelectronic devices and photonic systems have been the subject of extensive research in various application sectors. These applications include light-emitting diodes (LEDs), sensors, displays, health-care systems relying on biomedicine, and similar technologies [5–14]. Silicon-based complementary metal oxide semiconductor (CMOS) technology has enabled the production of small, fast, and low-cost photonic integrated circuits and systems [15–19]. However, the drawbacks of traditional silicon-based optoelectronic technology have prevented it from completely satisfying

Metal Oxides for Next-Generation Optoelectronic, Photonic, and Photovoltaic Applications
https://doi.org/10.1016/B978-0-323-99143-8.00010-9

the growing demand for further advanced and multifunctional state-of-the-art photonic devices. The primary drawbacks of this technology are the tactility and limited sensing characteristics of silicon-based optoelectronic devices. Furthermore, the low rigidity and fragility of silicon devices prevent their use in applications that require flexibility and wearability. As a result, a wide range of extremely flexible and sensitive photonic materials have remained the subject of intensive research to alleviate technological constraints. Researchers have claimed that there are a variety of nanoscale materials that possess exceptional capabilities such as being extremely sensitive and having multifunctional features, thereby fulfilling the demands of the next generation of photonic technology [20–25]. Over the past several years, new photonic devices employing a wide variety of unique materials have been developed. These include amorphous oxides, organic semiconductors, and low-dimensional materials [5,6,9,12,26–32]. Although semiconductors based on amorphous oxides possess efficient characteristic features such as high carrier mobility, low off-current level, and good reliability, their large bandgap means that they possess a small absorption coefficient in the visible region, which prevents their use in different photonic and optoelectronic applications [27,29–33]. In addition, organic semiconductors have proven to be attractive candidates for being used in different applications owing to their ease of fabrication and a wide variety of sensing capabilities [34,35]. However, their use in various applications is limited due to their poor stability and low carrier mobility. In this regard, researchers have found that quantum dots (QDs) are a potential alternative owing to their exceptional characteristics for different multifunctional applications [36–42]. These QDs are nanocrystalline particles with a quasi-zero-dimensional structure of nanometer size (1–20nm) [43]. Owing to their significant optical and electrical characteristics, such as efficient photoluminescence quantum yield (PLQY), broad absorption spectrum, narrow emission bands, better stability, and ease of synthesis, they have garnered significant attention in the scientific community [44]. These materials are currently being used for the fabrication of quantum-dot light-emitting diodes (QLEDs), photodiodes/phototransistors, solar cells, and memory devices, and find applications in flexible image sensors, biosensors, upconversion photodetectors, and lightning applications [45–48]. Although these materials have been used in every field, significant progress has been made in the design of efficient LEDs. Because of their efficient characteristics, QDs proved to be a potentially useful alternative emissive material for the development of different LEDs [42,49,50]. It has been observed that the development of the LEDs based on QDs has shown significant progress in the functionality of the device, that is, brightness, efficiency, and operating lifespan. This progress has been accomplished through the design of new materials and the creation of new device structures. The most advanced QLEDs, in particular red and green, have shown an external quantum efficiency of the order of 20% along with an operating lifespan of 1,000,000h [51].

In general, QLEDs are made up of numerous organic and inorganic semiconducting layers sandwiched between two electrodes with a thickness of less than 100nm. Different functional layers as well as the interface between them have a significant impact on the functionality of QLEDs, especially on their operational lifespan. The presence of multiple interfaces makes it more difficult to understand how the charge carrier dynamics and device functioning mechanisms are related to one another [49,51]. QDs whose radii are smaller than the Bohr radius reveal a significant quantum confinement effect

[52–56]. The capability of controlling the size of QDs ensures the feasibility of tailoring the emission range from the ultraviolet to the infrared regions of the spectrum in the QDs [55,56]. For tuning of the optical characteristics and passivation of the defective cores, the bare QDs are coated with inorganic shells [57–59]. To date, there exist three different categories of coatings: single-shell, multishell, and graded alloy structures. It is the bandgap and the relative position of the electronic energy levels of the structure (core/shell semiconductor) that determine the type of coating to be applied [57]. For the single-shelled QDs, generally, type I and type II core and shell architectures are used. Type I structures are characterized by the confinement of the carrier wave within the core. Among the type I structures are QDs belonging to the II-IV group such as CdSe/ZnS and CdSe/ZnSe, which have been extensively examined as core or shell structures [60–63]. Discordant core/shell alignment was discovered in QDs with type II structures. In these structures, the electron and hole wave functions were found to remain confined in separate regimes. Bawendi and his team [64] were the first to investigate QDs with type II structures. In their earlier attempts, they studied CdTe/CdSe and CdSe/ZnTe QDs, and then they expanded their circle of investigations and tried to study other QDs such as CdS/ZnTe and ZnTe/CdTe [65,66]. While working the graded alloy QDs, the bandgap engineering is accomplished by either annealing the QD core/shell structure at higher temperatures or by employing the single-pot synthesis approach for the development of the core/shell structure. A wide variety of alloyed QDs have been explored by different researchers, such as CdSeTe, CdSeS, ZnCdSe, and ZnCdS [67–72]. In addition to these, many researchers have also studied interfused alloy structures like CdSe/ZnS [73]. While studying the multishell QDs, Lim et al. [74] have observed that the interfacial alloyed layer in the CdSe0.5S0.5 has the ability to control the lattice misfit, thereby reducing the lattice stress in the QDs. Among the different varieties of available shell coatings, QDs coated with ZnS are the most popular ones and are available in a variety of configurations, such as CdSe/ZnSe/ZnS [75,76], CdSe/CdS/ZnS [76–78], and CdTe/CdSe/CdS/ZnS [79]. The quantum yield (QY) of QD-based LEDs, that is, QLEDs, was discovered to be approximately 100%, which is fantastic and can be attributed to established synthetic techniques as well as shell engineering [80–82]. With the remarkable results of these QDs in mind, this chapter is intended to provide various fundamental insights into these QD structures, in particular in the fabrication of LEDs. A brief discussion about the design of the QDs will be provided. Furthermore, discussions about the need for and importance of QD-based LEDs and the related aspects of electroluminescence and its properties will be discussed. In addition to having a brief discussion on the different applications of the QLEDs in different fields, an overview of the different challenges faced by the QLEDs in different applications will also be presented.

2 Design of quantum-dots light-emitting diode (QLEDs)

QLEDs are generally categorized on the basis of the charge-transporting materials incorporated into the device structure. Over the course of time, there existed three distinct classes of structures for the QLEDs, namely, all-organic, all-inorganic, and hybrid, as depicted in Fig. 1 [51].

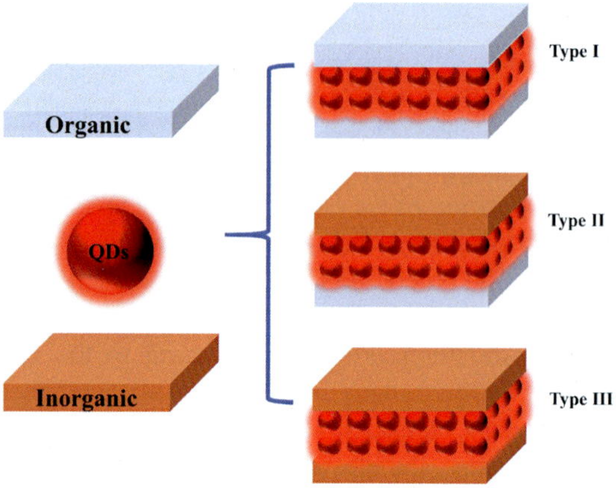

Fig. 1 Pictorial representation of different QLEDs structure: Type I (all-organic), Type II (all-inorganic), Type III (Hybrid).
Reproduced with permission from Q. Yuan, T. Wang, P. Yu, H. Zhang, H. Zhang, W. Ji, A review on the electroluminescence properties of quantum-dot light-emitting diodes, Org. Electron. 90 (2021) 106086. https://doi.org/10.1016/J.ORGEL.2021.106086, Copyright 2021, Elsevier.

In most of the cases, the term "all-inorganic" QLEDs becomes misleading due to the fact that none of the transporting materials, that is, electrons and holes, are made up of inorganic material, not even the ZnO nanoparticles (NPs) that are being widely utilized. In most cases, these layers are accompanied by organic ligands. In spite of this, the phrase "all-inorganic QLED" is widely used and recognized. The very first QLED was reported by Colvin in 1994 and was found to possess a very low EQE of the order of 0.01 percent [49]. This QLED was designed in such a way that the QD-emissive layer was placed in between the indium tin oxide anode and metal cathode. The presence of the metal electrode was observed to induce a significant quenching effect for the QD emission and was the primary cause of the performance bottleneck experienced by the device. To eradicate this negative quenching effect, Coe et al. [49] developed a three-layer QLED with a monolayer sandwiched between the organic charge transport layers, as depicted in Fig. 1 (Type I). The influence of the metal electrode's quenching on the emission of QLEDs is mitigated by the incorporation of the organic electron transport layer (ETL). In this manner, an unprecedented EQE of 0.5 percent was accomplished through the emission of QD. Taking the strong physical and chemical durability of inorganic materials into account, researchers have continuously attempted to design different QLEDs. In this regard, Muller et al. [83] in 2005 developed a QLED with a structure of Au/p-GaN/CdSe/n-GaN/In as depicted in Fig. 1 (Type II) and found it to possess a higher EQE than the previous ones. The third type, the hybrid design (Fig. 1, Type III), of QLEDs, which is now considered to be the most effective, has received substantial attention since 2011 owing to the remarkable brightness and efficiency shown by them for the different QLED devices [84–87].

The inclusion of n-type ZnO NPs as ETL in the hybrid QLEDs has marked a significant step forward in the further development of these devices, as it has resulted in record brightness of the order of 68,000 cd/m2 along with good stability [86]. Due to advancements in technology, significant developments have been made in the field of QLEDs, where the EQE of these devices has crossed the theoretically estimated value of 20%. In addition to this, enhancements in the luminance of the order of 356,000, 1,680,000, and 62,600, corresponding to red, green, and blue devices, respectively, have been achieved [85,88–91]. Furthermore, to improve the functionality of these devices, tandem-based QLEDs were designed, which include multiple light-emitting units connected to one another through an intervening layer [92–95]. In addition to this, significantly efficient white QLEDs were also accomplished through the use of tandem architecture [94,95]. Nowadays top-emitting structures are commonly utilized because of their large aperture ration and effective light outcoupling [96,97]. In this design, the light is emitted from the electrodes residing at the top of the device. In addition, devices with color tunability and transparency are being successfully manufactured for use in display and lighting applications. As the efficiency of the fabricated devices continues to improve, the technology for the development of QLEDs has become an increasingly important factor. Significant developments in technology need to be made so that it becomes compatible with the framework of the industry and that mass production can become reliable [97,98].

3 QD based light-emitting devices (LEDs)

The emission spectra of the QDs can either be photoluminescence (PL) or electroluminescence (EL), depending on whether the source of excitation is optical or electrical in nature. Among these, optically stimulated QD-based LEDs act as nanophosphors, in which the light is created by using multiple QDs with distinct emission colors. These QDs have proven to be significant candidates for being used in LED-backlit LCDs [57]. In the case of the electrically simulated QDs, different carriers are being directly inoculated into them. In these QDs, light is generated by the radiative recombination of the injected carriers present within the QDs with different sizes and colors of emission. Optically stimulated QD-based LEDs are being employed to back-light LCDs in three different ways. The first arrangement is known as the "film configuration" in which the QD is placed on top of the light guide plate. However, due to the distance between the film and the chip, significant intensity flux and heat are produced, both of which are harmful for the QDs. The second kind of design is known as "edge configuration," but due to the fragility of the QD tube in this design, it is hardly used for the fabrication of the backlight displays. The last arrangement is known as an "on-chip" configuration, and it involves spraying the QD directly onto a blue GaN chip. This design is preferred over the others because it only requires a few QDs to design the configuration; however, these QDs must be able to withstand the intense heat and flux. More importantly for the applicability of QD-based LEDs for backlighting, the QD color gamut is regulated by the National Television Standards Committee

(NTSC). Perovskite QDs, which possess narrowband emission, have also proved to be an efficient downconverter option for LCD backlighting [99]. Recently, Wang et al. [100] reported the use of wLEDs based on CsPbBr3 perovskite QD for backlight purposes and observed interference of the color gamut with the NTSC space of the order of 113%. This result of interference significantly surpassed the obtained results with wLEDs relying on phosphor (86%) and Cd-based QDs (104%), respectively. But due to the fact that QD-based wLEDs employed in liquid crystal displays possess high energy consumption, it is challenging to utilize them for thin and flexible displays [101]. Likewise, for optically stimulated QDs, significant advancements have been made in QD materials, charge transport materials, and the synthesis techniques of the electrically stimulated QD-based QLED technology. Because of the high external quantum efficiency (EQE) of white QLEDs as compared to the usual RGB (which exceeds by 10%), QLEDs are considered a viable alternative for the replacement of OLEDs in future applications for thin and flexible displays. The achievement of effective active-matrix QLED (AM-QLED) displays, which confronts numerous fundamental hurdles, is one of the most crucial attributes for the real-world use of QLEDs. AM-QLED displays have proven to be an effective self-emitting device working on the principle of current-driving mode [101]. The excellent performance of different QD-based red, blue, green, and wLEDs with external quantum efficiency >10% is presented in Table 1. Furthermore, researchers have reported that QLEDs with an external quantum efficiency greater than 20% can be created by using an ultra-thin layer of ploy (methyl methacrylate) for charge balance enhancement and tandem structures [93,94].

Table 1 Different characteristics parameters of red, green, blue and white quantum-dot based LEDs.

Emitting Layer	Colour of QLED	EL λ_{max} (nm)	PL QY (%)	Peak EQE (%)	Peak CE (cd A^{-1})	Max. L (cd m^{-2})	Ref.
CdSe/CdS/ZnS	Red	633	83	13.57	18.69	23,590	[102]
CdZnSe/ZnS/OT	Red	622	85	23.1	41.5	65,900	[94]
$Cd_{1-x}Zn_xSe_{1-x}S_y$	Red	625	-	12	15	21,000	[103]
CdZnSe/ZnS/ oleic acid	Green	534	85	27.6	121.5	115,500	[94]
CdSe@ZnS/ZnS	Green	–	90	15.6	65.3	216,500	[104]
CdZnSeS/ZnS	Green	538	–	10.5	45.7	61,030	[105]
CdZnSeS/ZnS/ OT	Blue	474	85	21.4	17.9	26,800	[94]
Cd1-xZnxS	Blue	455	–	10.7	4.4	4000	[103]
CdZnS/ZnS CdZnSeS/ ZnSCdSeS/ZnS	White	–	82 78 75	10.9	21.8	23,352	[106]

4 Electroluminescence mechanism in QLEDs

With continuous advancements in engineering and optimization, significant enhancements in the efficiency of QLEDs have been achieved. Currently, the most commonly used prototype QLED is the hybrid architecture. The QD is located between the electron transport layer (ETL) and the hole transporting layer (HTL) in this QLED design. Although an enormous number of achievements have been made, the mechanism responsible for the EL process in QLEDs has not yet been fully elucidated. For enhancing the characteristics and features of QLEDs and their usage in different fields, it is vital to have a comprehension of the electroluminescence (EL) mechanism. Efforts are currently being made to develop QLED-based electrically pumped lasers using solution-processed QD layers. Development of these types of lasers will not only ease manufacturing but also broaden the application spectrum of laser technologies in various fields like on-chip integrated photonic circuits, on-chip based diagnostics, imaging techniques, and sensing. [107–110]. The process of electrically pumping QDs relies on multiple phenomena, including charge injection and transport, generation and annihilation of excitons, and the switching (on/off) properties of the device. It is critical to understand different exciton processes in order to maximize the device's efficiency by modifying the design and ensuring the blockage of dominating route losses. The excitation process of the QDs in QLEDs mainly occurs through three different pathways: energy transfer (ET) [111–115], electric field ionization [116,117], and charge injection in QDs [84–87,118]. A brief illustration of all these pathways is provided below:

4.1 Energy transfer

When a voltage is applied to QLED, electrons and holes begin to generate from the respective electrodes. Among the generated charge carriers, only one type, either an electron or a hole, travels across the QD layer to the nearest neighboring donor layer. Soon after the charge carriers reach the adjacent donor layer, it results in the creation of the excitons by combining with the opposite charge carrier [39,86,119]. Rather than going through the radiative recombination process for the generation of photons, the generated excitons transfer their energy to QDs through the Froster resonant ET process [120,121]. Because of the presence of a thin emissive layer in organic QLEDs, the mechanism of ET is responsible for EL in these QLEDs [122]. Significantly efficient energy transfer from HTL to QDs results in enhancing both the functionality and reliability of the fabricated device. Because the phenomenon of ET is also dependent on the distance between the acceptor and donor, it is critical to regulate the distance between them in order to achieve a significant ET. The phenomenon of ET from the QD films to the QDs surrounded by organic compounds was first observed and explained by Madigan and his co-workers [121]. As the QD monolayers are embedded into the HTL with a separation of 10 nm from the HTL/ETL interface, observations have shown that the equivalent efficiency (EQE) of the device increases by more than 50%, which is in accordance with the Froster distance (DF) of 11 nm. There is always a tendency that the excitons produced in the HTL may leak into the device from the HTL to the anode. Thus, to make use of the leakage charge, G. H. Liu et al. [113] have established an ET mechanism through which the leaked excitons are used to increase the efficiency of the device, as depicted in Fig. 2A. For attaining a better ET and an efficient number, materials that possess a long phosphorescence lifespan are used. By controlling the donor concentration and optimizing the distance between the donor and acceptor, significant ET from FIrPic to QDs is attained, resulting in a significant enhancement of EQE from 4.9 cd/A (3.17%) to 6.6 cd/A (4.23%). Furthermore, by retrieving the leakage charges with phosphorescent organic molecules, heat dissipation caused by leakage charges gets slowed down, which in turn enhances the lifespan of HTL operation. Similarly, effective ET between the polymers and QDs has proven efficient in developing infrared-based QLEDs. Effective ET in these types of QLEDs occurs as a consequence of the slightly different relationship between the diameter and constituents of various QDs [123]. It has been observed that excitons produced in higher bandgap semiconductors have a tendency to transmit their lower bandgap counterparts, which significantly reduces the functionality of the device. On comparing the emission from QD film and QD-disseminated solvents, it has been observed that the ET process offers a significant red shift [124,125]. It has been observed that ET processes occurring in white QLEDs significantly rely on diversely colored QDs along with the single emissive layer attained by the intermixing of multiple colored QDs and single-layer stacked QDs possessing varied emission wavelengths [67]. Recently, success has been achieved by Lee et al. [106] in the fabrication of trichromatic QLEDs through progressive layering of HTL, three different kinds of QD mixed layers, and ZNO as ETL. Despite the fact that white QLEDs have displayed high efficiency (Fig. 2B) as compared to others, it has been observed that the ET processes occurring between variously colored QDs still exist there. In order to avoid these, Zhang et al. [126] constructed a tandem white QLED by utilizing different interconnecting layers for linking multiple light-emitting units. Through this arrangement, unavoidable ET processes were entirely sidestepped due to the fact that in this arrangement it is possible to maintain a significant distance among various emissive components.

4.2 Field ionization

Field-driven QLED is another prominent device with which QD emission is achieved. In these types of QLEDs, an emissive layer is placed between insulating layers. These insulating layers are usually made up of large-bandgap oxide materials. These devices are comparable to alternating-current thin film-based electroluminescence devices in which phosphor materials such as ZnS:Mn and SrS:Ce are used as emissive layers [127]. In these types of QLEDs, it has been observed that upon applying appropriate voltage to the device, charge carriers are not inserted, unlike in conventional QLEDs. Unlike conventional QLEDs, instead of current excitation, these are driven by an intense electric field, which results in the ionization of QDs. In these QLEDs, as the voltage applied to the QD reached a value greater than its bandgap energy, the ejection and transfer of an electron from the valence band of one QD to the conduction

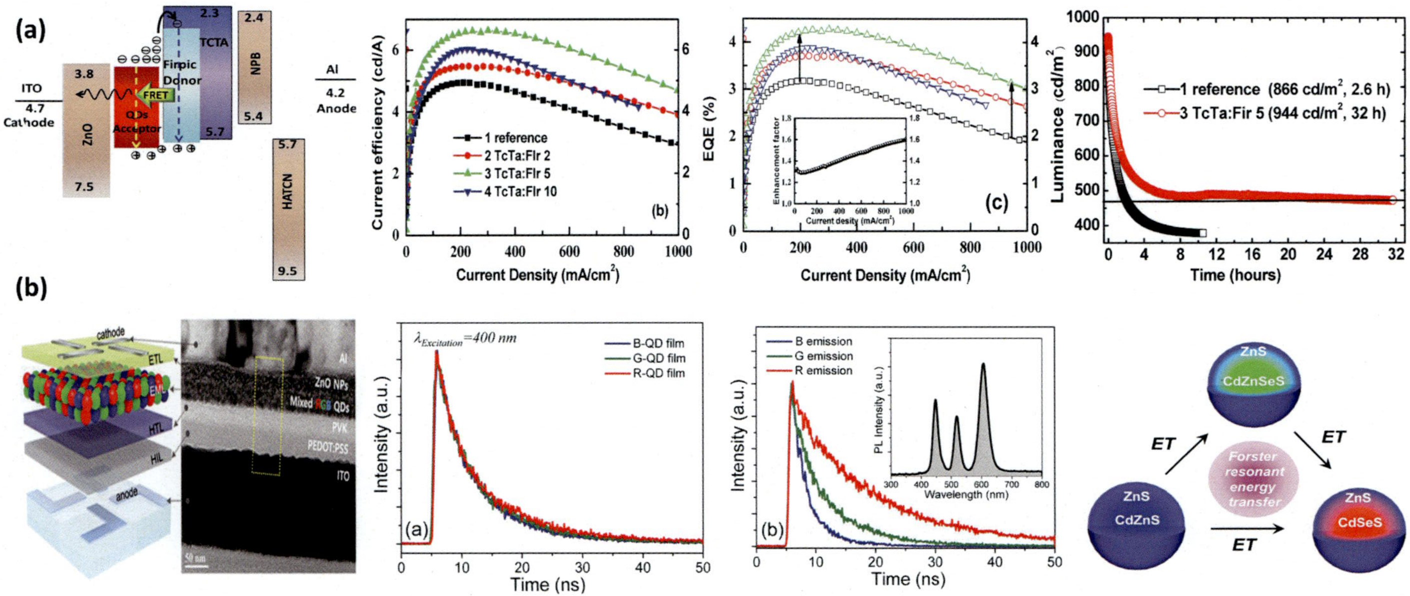

Fig. 2 (A) Design of QLED consists of phosphorescent material and mechanism of ET along with its luminous lifetime (B) cross-sectional view of the white QLEDs structure developed by using mixed QDs along with TEM micrograph. Depiction of FRET and PL variation in three different QDs. Reproduced with permission from K.H. Lee, C.Y. Han, H.D. Kang, H. Ko, C. Lee, J. Lee, N.S. Myoung, S.Y. Yim, H. Yang, Highly efficient, color-reproducible full-color electroluminescent devices based on red/green/blue quantum dot-mixed multilayer, ACS Nano 9 (2015) 10941–10949. https://doi.org/10.1021/ACSNANO.5B05513/SUPPL_FILE/NN5B05513_SI_001.PDF and G. Liu, X. Zhou, X. Sun, S. Chen, Performance of inverted quantum dot light-emitting diodes enhanced by using phosphorescent molecules as exciton harvesters, J. Phys. Chem. C 120 (2016) 4667–4672. https://doi.org/10.1021/acs.jpcc.5b12692, Copyright 2016 & 2015, American Chemical Society, respectively.

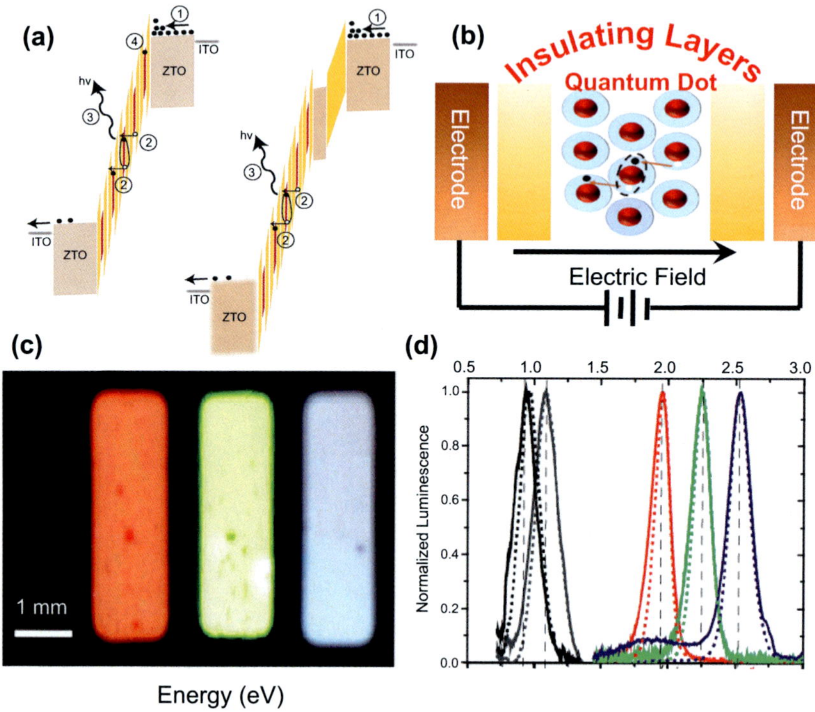

Fig. 3 (A) Energy band structure of QLED device with n-i-n design. (B) Working mechanism for the field-driven QLED device with a structure in which QD is sandwiched in between insulating layers. (C–D) Different emitted colors and an EL spectrum of device varying form the infrared to visible ranges.

(A, C, D) Reproduced with permission from V. Wood, M.J. Panzer, D. Bozyigit, Y. Shirasaki, I. Rousseau, S. Geyer, M.G. Bawendi, V. Bulović, Electroluminescence from nanoscale materials via field-driven ionization, Nano Lett. 11 (2011) 2927–2932. https://doi.org/10.1021/nl2013983 and V. Wood, M.J. Panzer, J.M. Caruge, J.E. Halpert, M.G. Bawendi, V. Bulović, Air-stable operation of transparent, colloidal quantum dot based LEDs with a unipolar device architecture, Nano Lett. 10 (2010) 24–29. https://doi.org/10.1021/nl902425g, Copyright 2010 & 2011, American Chemical Society, respectively.

band of its nearest neighboring QD took place, as depicted in Fig. 3A [128]. This process results in the production of an electron-hole pair in between the neighboring QDs. In the overall process, excitons are produced through multiple ionization processes. Furthermore, Wood et al. [116] developed field-driven n-i-n QLED, which consists of a layer of QD emissive materials sandwiched between ZnO and SnO2 alloy layers. While studying the functioning mechanism of the device, they have reported that only electrons can be injected into it. In addition, the insertion of ZnS blocking layers into the device results in boosting the efficiency as these layers cause the electrons to accumulate at the ZTO/ZnS interfaces. The bocking layers also tend to significantly decrease the voltage applied to QDs, which thereby proves to be beneficial for the ionization of QDs. In addition to this, they have also studied the charging behavior through the PL efficiency measurements of QD films in multiple structures while they are being subjected to an electric field [51]. Fig. 3B–D shows the emission profiles of QLEDs with QD configurations placed between the insulating layers, which span the infrared to visible regions [116]. Without the insertion of charge carriers, in the field-driven QLEDs, the emission was accomplished solely through the recombination of the carriers produced by field ionization. QLEDs with the mentioned design do not require materials that possess strong charge transfer capacity or match energy levels in structures. This design also ensures the eradication of other demerits related to QD charging and carrier imbalance, which decrease the efficiency of the device. By employing this design, it also became possible to easily utilize the different stable materials as insulating layers, with the end goal of enhancing the stability and functionality of the device. However, it has been observed that in many cases, it significantly reduces the device's effective functionality when compared to conventional ones. High operating voltage and the electric field in the device have a significant negative impact on the photoelectric properties of QDs, efficiency, and operating stability. Thus, it is clearly evident from the above discussion that enhancing the electrical stability of QDs is the prime concern in designing QLED devices. In addition, it is important to note that field-driven QLEDs possess potential and can prove to be an excellent platform for research into the effect of an electric field on the exciton dynamics of QDs [51].

4.3 Insertion of charges

Insertion of charges directly into QDs is recognized as one of the preeminent modes of excitation for hybrid QLEDs [39,86]. In this design, the electrons and holes inserted into the QD are generated from the respective ETL and HTL layers. This process is followed by the development of excitons in QDs, which are then recombined either through a radiative (emitting photons) or non-radiative (heating) approach. In addition to the injection barrier that exists between the CTL and QD, the carrier mobility of the HTL and ETL are the primary factors that determine the effectiveness of charges inserted into QDs. To avoid the deterioration of the HTL, which arises as a consequence of the processing and intermixing of QDs and HTL, an inverted design that utilizes metal oxide NPs as ETL was suggested for device fabrication [129,130]. The development of HTL layers for an inverted structural device by employing vacuum evaporation deposition technology increases the options for selecting different hole transporting materials for controlling the hole transfer and other characteristics of the device. Inverted structural design also helps in understanding the charge-insertion mechanism of QLEDs [131]. During the charge insertion mechanism, there is the emergence of sub-bandgap EL in QLEDs, which is one of the most interesting and fascinating phenomena. Because of the presence of multiple barriers and multiple ways for energy dissipation, it is necessary that the turn-on voltage, that is, the voltage under luminance of 0.1 cd/cm^2, for the device be slightly higher than the bandgap voltage. However, in most QLEDs, the converse of this is also observed, which is a phenomenon referred to as sub-bandgap turn-on voltage [39,86,119,131]. Qian et al. [119] were the first to report this phenomenon while fabricating an ZnO-based EL device.

They have attributed this phenomenon to the insertion of a hole through Auger-based energy up-conversion (UC). As these devices make use of the ZnO ETL, there is an enormous increase in the number of electrons at the boundary between the emissive layer and HTL. Due to the increase in carrier concentration at the boundary, there is a significant increase in the occurrence of tri-carrier Auger events. Mashford et al. [39] have reported the fabrication of an effective inverted hybrid QLED in which they have utilized ZnO NPs as ETL. In their study, they have observed strong interaction among the QDs and nearby ZnO ETL, which significantly enhances the insertion of electrons, which in turn triggers the Auger assisted hole insertion process. This process results in the creation of sub-bandgap EL, followed by high efficiency values of 18% and 90%, respectively. As a result, the discussed method of UC insertion of charges necessitates further investigation and comprehension for proper comprehension. Researchers have also focused on studying electro-absorption spectroscopy for evaluating the flat-band voltage of red QLEDs. In this regard, Luo et al. [132] have demonstrated that sub-bandgap turn-on voltage is equivalent to flat-band voltage for different QD emissive layers. However, they remained unable to prove it for green and blue devices. Assessing the lag, it was discovered that it is caused by electro-absorption limitations, which impede the achievement of accurate flat-band voltage in QDs. As a result, further efforts must be made in the future to bring light to the riddle.

5 Luminescence properties of QDs

The luminescence properties of any device depend on the process through which the process occurs. Similarly, in the case of QDs, there exist two primary types of luminous mechanisms responsible for the depiction of the luminous property in them. In the first case, the electron residing in the ground state tends to absorb a photon that has an energy greater than the bandgap and gains the energy to jump into the excited state, where it forms the electron-hole pair. During the de-excitation process (returning to the ground state), the electrons release the excess energy in the form of light, and the process is called photoluminescence, as depicted in Fig. 4A [43]. The second process is called electroluminescence, which occurs in QLEDs. In this process, electrons and holes are inserted into the materials that act as the anode and cathode in the device. The available electrons and holes are being driven by the electric field toward the emitting layer of QD, where they recombine radiatively and result in the production of photons (Fig. 4A) [43]. The electrons residing in the excited state, in particular those present at higher energy levels, are considered to be very unstable as their potential energy remains far greater than that possessed by the band edge excitons in QDs. Such excitons return to the lower energy of the excited state through the phonon relaxation process, also called hot exciton relaxations, as depicted in Fig. 4B. The relaxation process is so fast that its time period is usually measured in picoseconds. As long as the rules are not violated, there is a chance that most energetic excitons can jump directly to the valence band (VB), thereby resulting in radiative emission. This suggests that there is a possibility of a widening of the fluorescence spectrum in QDs even though they possess regular particle sizes and morphologies [133]. Soon

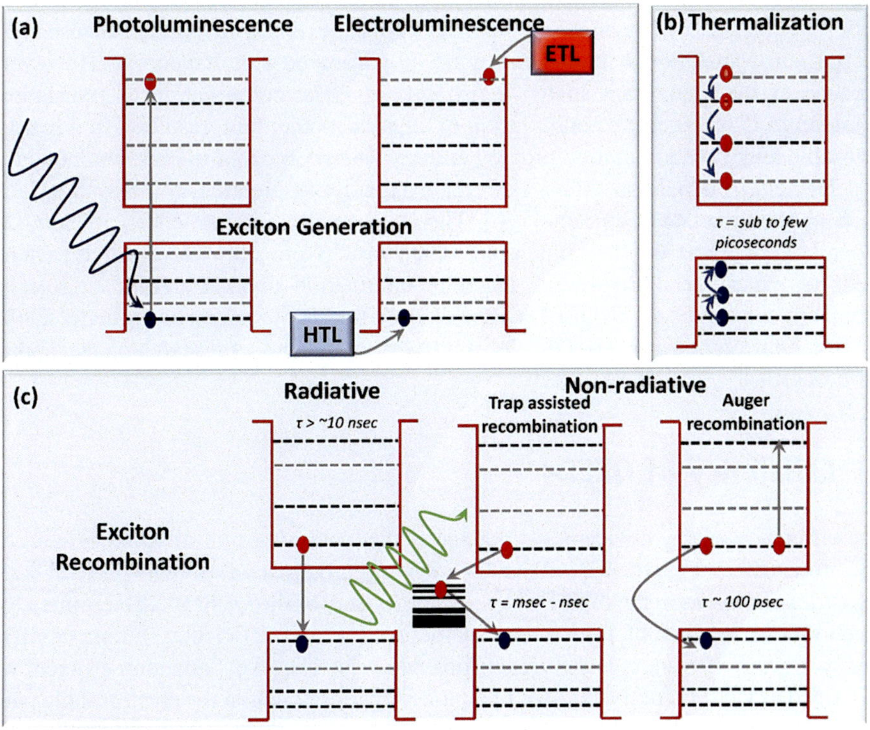

Fig. 4 Pictorial representation of excitons dynamics in QDs: (A) Mechanism through which photo genetic and field induced excitons are created. (B) Pathways for the relaxation of excitons. (C) Possible pathways for the recombination of excitons: radiative, trap-assisted and auger recombination's respectively.

after the electrons and holes get relaxed to their respective places, that is, the conduction band minima (CBM) or valence band maxima (VBM), they can interact radiatively or non-radiatively. Non-radiative recombinations, or auger recombinations, on the other hand, are preferred for display applications [134,135]. In general, there are three primary recombination mechanisms: radiative, trap-assisted, and auger recombinations, through which the excitons tend to recombine with one another, as depicted in Fig. 4C.

Both the impurities and defects result in the creation of defect energy levels in semiconductors, which assist the defect recombination process. In this process, the excited electron residing in the CB bands relaxes to the VB state via the defect state. Thus, in this deexcitation process, the defect state works as the intermediate state, and the electron, after reaching the VB, gets neutralized by combining with a hole. Another recombination process called "Auger" typically requires the participation of three carriers. Firstly, the electron and hole get recombined, after which they transfer the energy to a third carrier, which can be either another electron or hole residing in the CB. As a consequence of the supplied energy, this electron gets shifted toward the edge of the CB, where it dissipates the energy in the form of heat [43]. In the case of bulk semiconductors, the auger recombination process is less favorable because both energy and translational momentum must be conserved simultaneously. However, because of the significant spatial constraints in QDs, conservation of translation momentum is replaced by conservation of angular momentum, resulting in a highly favorable auger recombination process in them when compared to bulk semiconductors. However, spatial constraints exacerbate auger recombination by increasing Coulomb coupling between carriers [136]. The auger recombination tendencies in QDs depend on their size and their time span, which varies from a few picoseconds to hundreds of nanometers. In general, auger recombination plays a significant role in enhancing the efficiency of QLEDs [84,137,138]. In order to have a better understanding of the different aspects of QLEDs, it is essential to conduct research on their optical properties.

6 Efficiency of QLEDs

The efficiency of any device is the parameter that determines its survival and usage. The more efficient the device will be, the more effectively it will find its place in both industries as well as in the market. The total efficiency of the QLEDs is determined by an enormous number of parameters, including quantum efficiency, luminous efficiency, power efficiency, and current efficiency [43]. The term "quantum efficiency" is defined as the ratio of the number of emitted photons to that of the electron-hole pair injected into a system under the influence of the external electric field. Depending on the mechanism of photon emission, the quantum efficiency has been further categorized into two well-known types: internal and external quantum efficiency, designated as η_{int} and η_{ext}. η_{int} is defined as the ration of photons generated from the emitting layer of QLED to the number of externally introduced number of electron-hole pairs and is mathematically written as: $\eta_{int}=\gamma\chi\eta_r$. In the equation the variable γ, χ, η_r, represents the charge balance factor, the proportion of the excitons that enables the optically permitted transitions, and the efficiency of the radiative recombinations in the QDs, respectively. For designing efficient QLEDs, the maximum number of electrons and holes should reside in the QD emitting layer. The charge balancing factor configuration of the device and the characteristics of the charge transport layer. It has the potential to attain 100 percent accuracy, depending on the balance of the electron-hole pair externally injected into the device. The theoretical value of χ for II-VI group-based semiconductor QDs is approximately equal to 100% due to the spin-orbit coupling phenomenon [139,140]. It is possible that η_{int} will be close to 100% in ideal circumstances. But due to the layered structure of the QLEDs, different types of optical losses occur during the processes of generation and propagation of the photons from the emitting layer to the surface of the device via absorption, reflection, and refraction. Therefore, only a limited portion of the emitted light remains available for impactful use [43].

Therefore, for attaining the effective total efficiency of QLEDs, the external quantum efficiency (η_{ext}) is considered as an essential statistic parameter for determining the efficiency of QLEDs. Mathematically, η_{ext} is defined as: $\eta_{ext}=\eta_{coupling}\eta_{int}$. In the mentioned equation, $\eta_{coupling}$ represents the optical outcoupling of QLED. Till date, the different components involved in the planar QLEDs are metal electrodes, glass substrate, and thin film material. for reducing the different optical losses, refractive index of the materials used within QLEDs is usually larger than 1.6; however, for the glass substrate in is greater than 1.5. Due to the presence of different indexed materials in QLED devices, a significant portion of light reaches the surface as losses get reduced through total internal reflection [43]. Furthermore, the different processes that occur within the QLEDs result in energy dissipation. The loss in energy occurring within the QDs can be broadly classified into four different modes, as depicted in Fig. 5 [139,141,142]. The first one is direct emission, defined as the amount of light released into the air via the substrate as a consequence of $\eta_{coupling}$. The phenomenon of direct emission depends on the width of the charge transporting layer, the refractive index, and the stance of the emitting layer with respect to the electrode. On average, it accounts for about 20% of the total radiation. The second mode through which the energy loss occurs is called the substrate mode. This mode described the phenomenon of light flowing through the interfaces of the ITO/glass substate and its reflection back into the device because of the occurrence of complete reflection at the interface between the substate and air. Waveguide mode refers to the third process that results in energy dissipation. In this mode, light is not allowed to reach the glass substrate as

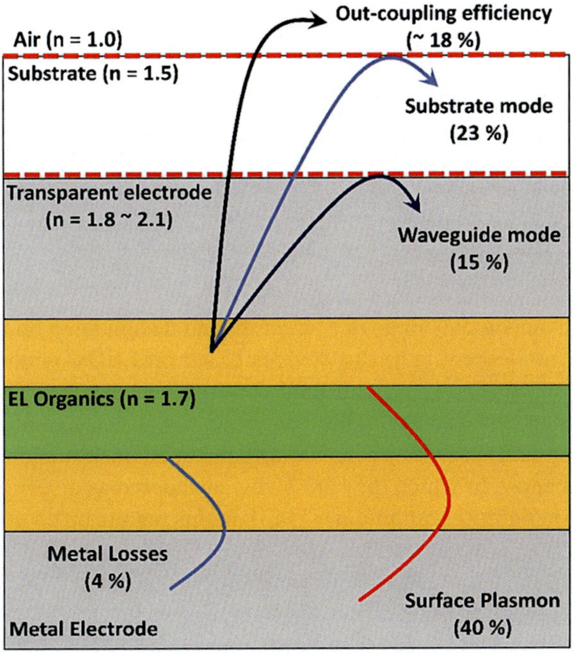

Fig. 5 Different types of energy dissipations experienced by the regular structured quantum-dot based light-emitting diodes.

the waveguide effect prevents it from passing in between the ITO and cathode. As the whole process of light blockage took place within the functional layers, all the lost energy eventually got converted into heat energy, which has a significant impact on the functionality and longevity of the device. The last mode through which the energy dissipation occurs is the phenomenon of "surface plasmons." This is a collective resonance phenomenon and is observed at the interface of metal-medium interaction. This phenomenon occurs as a consequence of interaction among the light-wave electromagnetic fields and free electrons. During this phenomenon, there is a coupling effect between the electromagnetic radiation and the free charges available on the surface. This interaction results in the generation of an acoustic wave with an exciting nature, which in turn causes surface plasmas to attain high momentum as compared to the free photons in organic layers. Because of the acquired enhanced momentum, there is a significant mismatch between the wave vectors of surface plasmon polaritons (SSP) and light waves. Due to the mismatch, the formed acoustic waves do not propagate along the interface, thereby resulting in the absence of planar SSPs and non-radiative modes [141].

7 Applications

In the current scenario of technological world, every reach work remains devoted toward some applicative output. Researchers working in the different field work with the aim of designing the materials that needs be efficient, eco-friendly, cost-effective, and easy to operate, as compared to the previous technology [143]. Similarly, the marvelous characteristics of the QDs and QD-based LEDs has led their usage in an enormous number application like in lighting, photodetectors, flexible displays, sensors, and photoemission. Among the large number of applicatives, few important ones will be briefly discussed below:

7.1 Lighting purposes

Lighting has always remained an essential part in human life. With the advancement of technology, advancements in device structure and design keep on modifying from the lanterns to incandescent lamps to devices based on LEDs. Among the different types of LEDs, recent years have witnessed the excessive usage of QLEDs semiconductors for designing devices for lighting purposes. QLEDs offer substantial benefits over other types of LEDs in terms of photoluminescence quantum yield, color purity, and efficiency, because of which they are being considered as an efficient alternative for next-generation lighting technologies [143]. In this regard, an enormous number of efforts have been made by different researchers in this domain. Lin et al. [144] have developed hybrid-type white LEDs from inorganic halide-based QDs. In designing the LEDs, they have opted for three different geometries: solid, liquid, and hybrid. They have developed the white LED using the inorganic green ($CsPbBr_3$) and red ($CsPbBr_{1.2}I_{1.8}$) halides as constituents by employing the hot-insertion technique. The fabricated LEDs were stimulated by using blue LEDs as an excitation source.

In spite of the fact that the solid-state geometry-based QLEDs had shown excellent area coverage in the color gamut of the order of 120% as per NTSC and 90% as per REC 2020 standards, they remained unable to compete with the hybrid type, which has sown much more excellent results (value for the area coverage in the color gamut of 122% as per NTSC and 91% as per REC 2020 standards). Therefore, the reduction or increase in productivity caused by the design employed for the fabrication [145] as the efficiency of the QDs relies on the structure and their arrangement within it. It has been observed that the QDs in the LEDs that hold the solid-type structure remain enclosed in the silicone resin, which causes dispersion and re-absorption processes that, in turn, cause a decrease in the LED's efficiency [146,147]. However, the hybrid type structures have demonstrated enhanced performances along with a broad color spectrum as compared to both solid and liquid types. The emission spectra of the hybrid structure appear to be unaltered even after 200 h of age, resulting in a practically constant region of color gamut as a result of the constant spectra shape. In addition, during the same working period, it has been observed that there is a gradual decrease in the luminescence efficiency of devices, with an estimated value of the order of 28% and 12.2% for the solid and hybrid type structures, respectively. While analyzing the current efficiency of both the solid and hybrid structures, it has been found that wLEDs corresponding to both types possess the current efficiencies of 41 and 51 lm/W, respectively, thereby illustrating the dominance of hybrid-type wLEDs for lighting purposes. While investigating the causes of lower current efficiency in solid state type structures, it was discovered that self-aggregation of QDs is to blame [148]. The above assessments clearly demonstrate the hybrid type's stability as well as its frequent use for lighting purposes when compared to solid type structures. Similarly, Kang et al. [149] have reported the synthesis of wLEDs by utilizing perovskite-based QDs (PQDs), which were found to exhibit extremely high luminous intensities, broad color ranges, and extended lifetimes, and thus favor their usage for lighting purposes. Although QLEDs are now frequently employed for lighting purposes because of their efficient characteristics, there are still some hurdles, such as anion-exchange reactions, surface oxidation of QDs, photo-bleaching, and aggregation of the QDs, that restrict their usage [150,151]. Thus, for their consistent progress and development efforts must be devoted to their further development in the future.

7.2 QLED based display

wLEDs discussed above were found to possess a highly brittle nature due to the insufficient bonding strength among the QD powders and substrate materials. The brittle nature of the substrate tends to reduce its extensibility, which is necessary for the fabrication of flexible display devices. Thus, the search and development of white light-emitting materials with strong and elastic characteristic features for the fabrication of flexible displays is of utmost importance [152]. Kim et al. [38] fabricated a full-color QD display (transfer printing) without using a solvent for patterning the red, green, and blue emitting QDs on the pixelated display screen. The transfer orientation technology proved to be much more efficient as compared to previous ones such as spin-coating and others. The spin-methods faces the limitation of cross contamination among multiple sizes. Similarly, other methods were found to yield inconsistent films with respect to their uniformity, which results in surface roughness. As a consequence of the contamination, non-uniformity, and surface roughness, there is a significant reduction in the amount of charge transfer as well as a decrease in quantum efficiency. As compared to other methods in the transfer printing process, the QDs were first spin-coated onto the substrate. In the next step, polydimethylsiloxane (PDMS) stamps are applied to QD films.

On applying a substantial amount of pressure, the QDs are picked up because of the lower surface energy of the stamp as compared to the donor substrate. The QDs are quickly moved to the device track, where they are organized in an array of thin strips. It has been observed that the density of the QD films developed by printing technology is 20% higher than that of the QD films developed via spin-coating [38]. When printed red, green, and blue QDs are incorporated into organic or inorganic hybrid LED structures, their brightness increases significantly, reaching up to 16,380, 6425, and 423 $cd\,m^{-1}$, respectively, with current efficiencies of the order of 3.00, 0.55, and 0.05 $cd\,A^{-1}$. The observed results were approximately 25%–52% higher than the values obtained from spin-coating devices. With the help of the QD printing technique, the researchers were able to construct a 4-inch full-color active matrix QD display, which displayed almost flawless pictures. Due to the use of PDMS stamps, the printing technique enables the transformation of QD into flexible substrates [152]. Visualizing the importance of printing technology, Kim et al. [153] have conducted extensive research on flexible devices by employing this technology for flexible luminaires. However, it was later discovered that using the referred technology made it difficult to construct heavy devices. Later, it was found that researchers opted for the combination of solid-state lighting and curved light-guiding technology for the fabrication of flexible devices. However, due to the rigid nature of curved light guides, it was found that it is not suitable for bending the light multiple times within the guide, which is very necessary for flexible devices [154]. Similarly, Sher et al. [155] have successfully fabricated the flexible lighting device by using flip-flops, anisotropic conductive glue, and phosphor film. By using the mentioned components, it was found that the fabricated flexible LED possesses enhanced lighting efficiency along with the broad bending range without causing any reduction in overall amount of light emitted. The working efficiency of the device under different curvature modes is illustrated in Fig. 6A. The fabricated wLED had a cross-sectional area of approximately 5 cm^2 and a dimension of approximately 5×5 cm. In addition, the top and bottom portions of the fabricated devices before and after attaching the phosphor film and blue LED array are depicted in Fig. 6A. While assessing the change in voltage or luminous efficiency versus current under different curvatures, it was found that up to a diameter of 3 cm there is no evident change, as depicted in Fig. 6B. Based on the observed characteristic, it is revealed that a large array of flexible LEDs can prove to be an excellent replacement for use in flexible displays and other related gadgets. From the above discussion, it is clearly evident that the development and incorporation of QDs in the domain of display technology have led to efficient results and also appear to be quite promising in the near future. In the coming years, it is anticipated that many new unexpected and

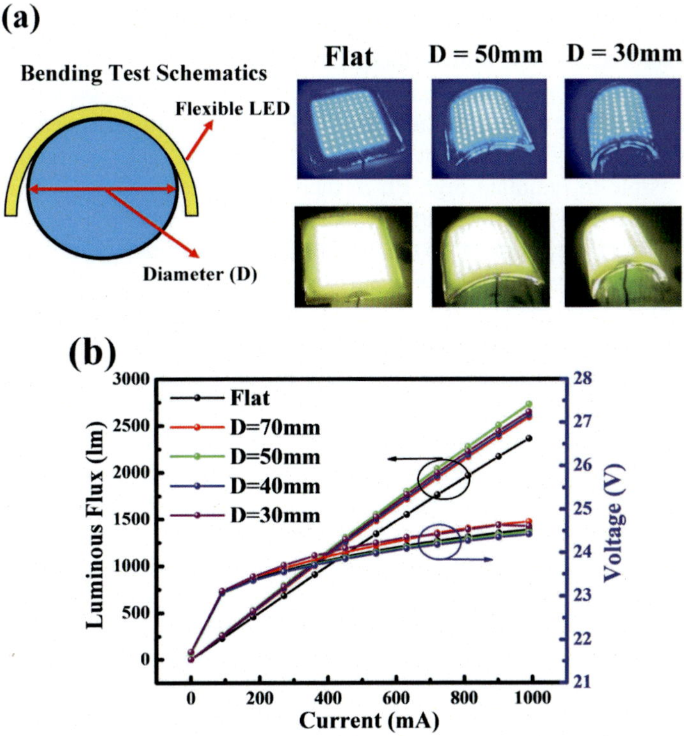

Fig. 6 (A) Depiction of flexibility of the device and images on top and bottom on top right side shows nature of adhesive phosphor film and blue LED arrays before and after bonding.
(B) Depiction of relationship between voltage and light output for a bendable white LED at varying diameters.
Reproduced with permission from C.W. Sher, K.J. Chen, C.C. Lin, H.-V. Han, H.Y. Lin, Z.-Y. Tu, H.H. Tu, K. Honjo, H.Y. Jiang, S.L. Ou, R.H. Horng, X. Li, C.C. Fu, H.C. Kuo, Large-area, uniform white light LED source on a flexible substrate, Opt. Express 23 (2015) A1167. https://doi.org/10.1364/oe.23.0a1167, @ The Optical Society.

fascinating achievements will be made, and it is also believed that the time is not far off when everyone will have access to display technology that is highly efficient and reliable.

7.3 Photoemission

Due to the exceptional electrical and optical qualities possessed by the different forms of pure QDs, they are being widely employed in the field of photoemission. In addition, because of their excellent PL properties and EQE, they are being described as a viable choice for different photoemission devices such as LEDs and lasers [156–158]. It has been observed that in the case of the halide-based QDs, the use of templates during the synthesis process assists in maintaining the size of the QDs. In addition, utilization of the protection layer of silicon or polymer for the composite structures embedded in a matrix helps in improving the stability of the QDs against different environmental factors and allows them to function in water, and solar solvents. Because of the mentioned benefits, QD-based composites act as promising constituents for the development of luminous inks [159]. Researchers have developed different QLEDs by employing different oxide and polymer matrixes, and have observed that it is possible to maintain high PL in them even after dealing with water, heat, or UV light [159–161]. Additionally, high color purity has also been attained via a narrow FWHM of the order of 25 nm and a luminous efficiency of 80 lm/W [162]. The QD-based composites of SiO_2 and glass have shown powerful scattering qualities and increased optical gain [163,164]. Because of the mentioned characteristics, these QD-based composites were used in the random lasing emissions, where it has been observed that they reduce the threshold by 50% and improve the efficiency up to 388% [165]. The heterojunction structures of QD composites and QDs encapsulated on the surface of oxides were also employed for the photoemission, which depicted efficient results. Among the oxide-based QD materials, the larger bandgap and ionization potential of Al_2O_3 and ZrO_2 assisted in the rapid insertion of the electrons, thereby leading to strong emission [166,167]. Among the dual-face perovskite-based QDs, it was found that the incorporation of fabricated lead-rich $CsPb_2Br_5$ QDs in the pristine $CsPbBr_3$ tends to reduce its exciton emission and ionic conductivity, which helps in enhancing its efficiency in the LED devices via enhancing the lifetime, an EQE of the order of 2.21%, and a narrowing of the FWHM up to 19 nm [168,169]. Furthermore, for enhancing the thermal stability of different QDs, ion doping via composites has proven to be very effective. Ion-doping enhances the stability by increasing the formation energy, efficiency, and strength of PL emission by changing the kinetics, which resulted in enhancing the EQE by 4.4% [170–172]. In addition, the introduction of ions was capable of reducing the toxic effects of Pb^{2+} by replacing its content with the doping ions. The replacement of ions also tends to shift the peak location of emission peaks in QDs [173]. In the study that was done by Yuan et al. [174] on Eu^{3+}:$CsPbBr_3$ QDs, there was a partial replacement of Pb^{2+} with Eu^{3+}, which caused the peak to shift to blue while other QDs continued to show red light emission. As the concentration of Eu^{3+} ions in the solution increased, more ions penetrated the cell structure of QDs. As a result, the emission peaks widens, resulting in a color change of light from green to blue and finally to red.

8 Challenges of QDs for LED applications

By the advent of shell engineering, multiple advancements has been made in different synthetic approaches that led to an increase in the quantum yield (QY) of QDs in solution up to 100%. However, it has been observed that the efficiency of the QDs employed in different devices is not still up to the mark. The external quantum efficiency (EQE) of the devices relies on many factors that is QY of QD employed, number of charges inserted into the QDs from the excitons, and portion of emitted photons that gets paired outside the device [175]. While comparing with the EQE of these

QLEDs with the different organic LEDs, it is there in need of further advancements so that the EQE of the QLEDs will get increased up to the mark [176–179]. Due to high surface to volume ratio of QDs, they are very susceptible to the residing environmental conditions. Because of this, QDs are prone to damage on being embedded into layered structures. Similarly, most of the case solid-state QD films goes through different non-radiative processes such as surface trapping, auger recombination, and FRET. It has been observed that most of the excitonic nonradiative recombination rates arises from the mentioned non-radiative process, which in turn results in decreasing the QY of the QDs that decreases the EQE of the device [44,180]. Another significant aspect to take into account is the fact that it is the electronic structure of semiconductor materials that ascertains their optical characteristics. This fact tends to remind that electroluminescent LEDs are governed by an intense electric field (greater than $1\,MV/cm^2$). This demonstrates that the effectiveness of QLEDs is affected not only by their physical surroundings but also by their electronic structures.

9 Conclusion

Recently, there has been enormous growth in the commoditization of QDs, which has been attributed to improvements in their characteristics such as stability, brightness, solubility, and enhanced efficiency. Since its establishment in 2004, the QD vision has been dedicated to the manufacturing of nanomaterials with applications in different fields, especially lighting and display technologies. The photoluminescence QY of QDs approaches 100 percent, particularly in the case of alloyed QDs, and they are thus considered excellent candidates for the development of efficient QLEDs. Extensive research has been conducted in this field, which has led to a proper understanding of electroluminescence and its degradation mechanisms. This chapter provides an overview of the establishment of QDs and their implications for solid-state lighting technology. In general, QLEDs are considered a highly competitive EL technology because they offer numerous advantages over other technologies. Although preliminary research has been fruitful, scaling up the manufacturing process while maintaining QD quality requires enormous effort. Despite this progress, the conditions for commoditization are only partially satisfied in terms of device capacity, efficiency, and stability. Advancements in quantum efficiency and improved manufacturing and use of novel materials can help in the development of QLEDs with extended emission in the IR region of the spectrum. Thus, it is predicted that this technology will broadly penetrate the market in the near future.

References

[1] Y. Jung, H. Park, J.A. Park, J. Noh, Y. Choi, M. Jung, K. Jung, M. Pyo, K. Chen, A. Javey, G. Cho, Fully printed flexible and disposable wireless cyclic voltammetry tag, Sci. Rep. 5 (2015) 8105, https://doi.org/10.1038/srep08105.

[2] L. Atzori, A. Iera, G. Morabito, The internet of things: a survey, Comput. Netw. 54 (2010) 2787–2805, https://doi.org/10.1016/j.comnet.2010.05.010.

[3] C. Wang, D. Hwang, Z. Yu, K. Takei, J. Park, T. Chen, B. Ma, A. Javey, User-interactive electronic skin for instantaneous pressure visualization, Nat. Mater. 12 (2013) 899–904, https://doi.org/10.1038/nmat3711.

[4] M.L. Hammock, A. Chortos, B.C.K. Tee, J.B.H. Tok, Z. Bao, 25th anniversary article: the evolution of electronic skin (E-Skin): a brief history, design considerations, and recent progress, Adv. Mater. 25 (2013) 5997–6038, https://doi.org/10.1002/adma.201302240.

[5] T. Yokota, P. Zalar, M. Kaltenbrunner, H. Jinno, N. Matsuhisa, H. Kitanosako, Y. Tachibana, W. Yukita, M. Koizumi, T. Someya, Ultraflexible organic photonic skin, Sci. Adv. 2 (2016), https://doi.org/10.1126/sciadv.1501856.

[6] S. Park, K. Fukuda, M. Wang, C. Lee, T. Yokota, H. Jin, H. Jinno, H. Kimura, P. Zalar, N. Matsuhisa, S. Umezu, G.C. Bazan, T. Someya, Ultraflexible near-infrared organic photodetectors for conformal photoplethysmogram sensors, Adv. Mater. 30 (2018) 1802359, https://doi.org/10.1002/adma.201802359.

[7] S. Chen, Z. Lou, D. Chen, G. Shen, An artificial flexible visual memory system based on an UV-motivated memristor, Adv. Mater. 30 (2018) 1705400, https://doi.org/10.1002/adma.201705400.

[8] S. Jeon, S.E. Ahn, I. Song, C.J. Kim, U.I. Chung, E. Lee, I. Yoo, A. Nathan, S. Lee, J. Robertson, K. Kim, Gated three-terminal device architecture to eliminate persistent photoconductivity in oxide semiconductor photosensor arrays, Nat. Mater. 11 (2012) 301–305, https://doi.org/10.1038/nmat3256.

[9] S.E. Ahn, I. Song, S. Jeon, Y.W. Jeon, Y. Kim, C. Kim, B. Ryu, J.H. Lee, A. Nathan, S. Lee, G.T. Kim, U.I. Chung, Metal oxide thin film phototransistor for remote touch interactive displays, Adv. Mater. 24 (2012) 2631–2636, https://doi.org/10.1002/adma.201200293.

[10] S. Seo, S.H. Jo, S. Kim, J. Shim, S. Oh, J.H. Kim, K. Heo, J.W. Choi, C. Choi, S. Oh, D. Kuzum, H.S.P. Wong, J.H. Park, Artificial optic-neural synapse for colored and color-mixed pattern recognition, Nat. Commun. 9 (2018) 1–8, https://doi.org/10.1038/s41467-018-07572-5.

[11] H. Han, H. Yu, H. Wei, J. Gong, W. Xu, Recent progress in three-terminal artificial synapses: from device to system, Small 15 (2019) 1900695, https://doi.org/10.1002/smll.201900695.

[12] Y. Lee, J.Y. Oh, W. Xu, O. Kim, T.R. Kim, J. Kang, Y. Kim, D. Son, J.B.H. Tok, M.J. Park, Z. Bao, T.W. Lee, Stretchable organic optoelectronic sensorimotor synapse, Sci. Adv. 4 (2018) 7387–7410, https://doi.org/10.1126/sciadv.aat7387.

[13] M. Kimura, T. Shima, T. Okuyama, S. Utsunomiya, W. Miyazawa, S. Inoue, T. Shimoda, Artificial retina using thin-film photodiodes and thin-film transistors, Japanese J. Appl. Physics, Part 1 Regul. Pap. Short Notes Rev. Pap. 45 (2006) 4419–4422, https://doi.org/10.1143/JJAP.45.4419.

[14] Y. Miura, T. Hachida, M. Kimura, Artificial retina using thin-film transistors driven by wireless power supply, IEEE Sensors J. 11 (2011) 1564–1567, https://doi.org/10.1109/JSEN.2010.2096807.

[15] M.T. Bohr, I.A. Young, CMOS scaling trends and beyond, IEEE Micro 37 (2017) 20–29, https://doi.org/10.1109/MM.2017.4241347.

[16] R. Maurand, X. Jehl, D. Kotekar-Patil, A. Corna, H. Bohuslavskyi, R. Laviéville, L. Hutin, S. Barraud, M. Vinet, M. Sanquer, S. De Franceschi, A CMOS silicon spin qubit, Nat. Commun. 7 (2016) 1–6, https://doi.org/10.1038/ncomms13575.

[17] M. Urdampilleta, D.J. Niegemann, E. Chanrion, B. Jadot, C. Spence, P.A. Mortemousque, C. Bäuerle, L. Hutin, B. Bertrand, S. Barraud, R. Maurand, M. Sanquer,

X. Jehl, S. De Franceschi, M. Vinet, T. Meunier, Gate-based high fidelity spin readout in a CMOS device, Nat. Nanotechnol. 14 (2019) 737–741, https://doi.org/10.1038/s41565-019-0443-9.

[18] M. Veldhorst, H.G.J. Eenink, C.H. Yang, A.S. Dzurak, Silicon CMOS architecture for a spin-based quantum computer, Nat. Commun. 8 (2017) 1–8, https://doi.org/10.1038/s41467-017-01905-6.

[19] W. Haensch, E.J. Nowak, R.H. Dennard, P.M. Solomon, A. Bryant, O.H. Dokumaci, A. Kumar, X. Wang, J.B. Johnson, M.V. Fischetti, Silicon CMOS devices beyond scaling, IBM J. Res. Dev. 50 (2006) 339–361, https://doi.org/10.1147/rd.504.0339.

[20] H. Chang, H. Wu, Graphene-based nanomaterials: synthesis, properties, and optical and optoelectronic applications, Adv. Funct. Mater. 23 (2013) 1984–1997, https://doi.org/10.1002/adfm.201202460.

[21] X. Li, M. Rui, J. Song, Z. Shen, H. Zeng, Carbon and graphene quantum dots for optoelectronic and energy devices: a review, Adv. Funct. Mater. 25 (2015) 4929–4947, https://doi.org/10.1002/adfm.201501250.

[22] S. Jang, E. Hwang, Y. Lee, S. Lee, J.H. Cho, Multifunctional graphene optoelectronic devices capable of detecting and storing photonic signals, Nano Lett. 15 (2015) 2542–2547, https://doi.org/10.1021/acs.nanolett.5b00105.

[23] J.S. Ponraj, Z.Q. Xu, S.C. Dhanabalan, H. Mu, Y. Wang, J. Yuan, P. Li, S. Thakur, M. Ashrafi, K. McCoubrey, Y. Zhang, S. Li, H. Zhang, Q. Bao, Photonics and optoelectronics of two-dimensional materials beyond graphene, Nanotechnology 27 (2016) 462001, https://doi.org/10.1088/0957-4484/27/46/462001.

[24] D. Kufer, G. Konstantatos, Photo-FETs: phototransistors enabled by 2D and 0D nanomaterials, ACS Photon. 3 (2016) 2197–2210, https://doi.org/10.1021/acsphotonics.6b00391.

[25] Y. Li, F. Qian, J. Xiang, C.M. Lieber, Nanowire electronic and optoelectronic devices, Mater. Today 9 (2006) 18–27, https://doi.org/10.1016/S1369-7021(06)71650-9.

[26] J. Kim, H.C. Lee, K.H. Kim, M.S. Hwang, J.S. Park, J.M. Lee, J.P. So, J.H. Choi, S.H. Kwon, C.J. Barrelet, H.G. Park, Photon-triggered nanowire transistors, Nat. Nanotechnol. 12 (2017) 963–968, https://doi.org/10.1038/nnano.2017.153.

[27] Y.H. Kim, J.S. Heo, T.H. Kim, S. Park, M.H. Yoon, J. Kim, M.S. Oh, G.R. Yi, Y.Y. Noh, S.K. Park, Flexible metal-oxide devices made by room-temperature photochemical activation of sol-gel films, Nature 489 (2012) 128–132, https://doi.org/10.1038/nature11434.

[28] W.T. Chen, H.W. Zan, High-performance light-erasable memory and real-time ultraviolet detector based on unannealed indium-gallium-zinc-oxide thin-film transistor, IEEE Electron Device Lett. 33 (2012) 77–79, https://doi.org/10.1109/LED.2011.2171316.

[29] A.D. Mottram, Y.H. Lin, P. Pattanasattayavong, K. Zhao, A. Amassian, T.D. Anthopoulos, Quasi two-dimensional dye-sensitized In_2O_3 phototransistors for ultrahigh responsivity and photosensitivity photodetector applications, ACS Appl. Mater. Interfaces 8 (2016) 4894–4902, https://doi.org/10.1021/acsami.5b11210.

[30] I. Hwang, J. Kim, M. Lee, M.W. Lee, H.J. Kim, H.I. Kwon, D.K. Hwang, M. Kim, H. Yoon, Y.H. Kim, S.K. Park, Wide-spectral/dynamic-range skin-compatible phototransistors enabled by floated heterojunction structures with surface functionalized SWCNTs and amorphous oxide semiconductors, Nanoscale 9 (2017) 16711–16721, https://doi.org/10.1039/c7nr05729f.

[31] S. Park, S.W. Heo, W. Lee, D. Inoue, Z. Jiang, K. Yu, H. Jinno, D. Hashizume, M. Sekino, T. Yokota, K. Fukuda, K. Tajima, T. Someya, Self-powered ultra-flexible electronics via nano-grating-patterned organic photovoltaics, Nature 561 (2018) 516–521, https://doi.org/10.1038/s41586-018-0536-x.

[32] J. Lee, B. Yoo, H. Lee, G.D. Cha, H.S. Lee, Y. Cho, S.Y. Kim, H. Seo, W. Lee, D. Son, M. Kang, H.M. Kim, Y. Il Park, T. Hyeon, D.H. Kim, Ultra-wideband multi-dye-sensitized upconverting nanoparticles for information security application, Adv. Mater. 29 (2017) 1603169, https://doi.org/10.1002/adma.201603169.

[33] M.G. Kim, M.G. Kanatzidis, A. Facchetti, T.J. Marks, Low-temperature fabrication of high-performance metal oxide thin-film electronics via combustion processing, Nat. Mater. 10 (2011) 382–388, https://doi.org/10.1038/nmat3011.

[34] J. Kim, J. Kim, K.T. Kim, Y.H. Kim, S.K. Park, Monolithic integration and design of solution-processed metal-oxide circuitry in organic photosensor arrays, IEEE Electron Device Lett. 37 (2016) 671–673, https://doi.org/10.1109/LED.2016.2546960.

[35] J. Kim, J. Kim, S. Jo, J. Kang, J.W. Jo, M. Lee, J. Moon, L. Yang, M.G. Kim, Y.H. Kim, S.K. Park, Ultrahigh detective heterogeneous photosensor arrays with in-pixel signal boosting capability for large-area and skin-compatible electronics, Adv. Mater. 28 (2016) 3078–3086, https://doi.org/10.1002/adma.201505149.

[36] A. Pierre, A. Gaikwad, A.C. Arias, Charge-integrating organic heterojunction phototransistors for wide-dynamic-range image sensors, Nat. Photonics 11 (2017) 193–199, https://doi.org/10.1038/nphoton.2017.15.

[37] Q. Sun, Y.A. Wang, L.S. Li, D. Wang, T. Zhu, J. Xu, C. Yang, Y. Li, Bright, multicoloured light-emitting diodes based on quantum dots, Nat. Photonics 1 (2007) 717–722, https://doi.org/10.1038/nphoton.2007.226.

[38] T.H. Kim, K.S. Cho, E.K. Lee, S.J. Lee, J. Chae, J.W. Kim, D.H. Kim, J.Y. Kwon, G. Amaratunga, S.Y. Lee, B.L. Choi, Y. Kuk, J.M. Kim, K. Kim, Full-colour quantum dot displays fabricated by transfer printing, Nat. Photonics 5 (2011) 176–182, https://doi.org/10.1038/nphoton.2011.12.

[39] B.S. Mashford, M. Stevenson, Z. Popovic, C. Hamilton, Z. Zhou, C. Breen, J. Steckel, V. Bulovic, M. Bawendi, S. Coe-Sullivan, P.T. Kazlas, High-efficiency quantum-dot light-emitting devices with enhanced charge injection, Nat. Photonics 7 (2013) 407–412, https://doi.org/10.1038/nphoton.2013.70.

[40] Y. Shirasaki, G.J. Supran, M.G. Bawendi, V. Bulović, Emergence of colloidal quantum-dot light-emitting technologies, Nat. Photonics 7 (2013) 13–23, https://doi.org/10.1038/nphoton.2012.328.

[41] J. Bao, M.G. Bawendi, A colloidal quantum dot spectrometer, Nature 523 (2015) 67–70, https://doi.org/10.1038/NATURE14576.

[42] C.R. Kagan, E. Lifshitz, E.H. Sargent, D.V. Talapin, Building devices from colloidal quantum dots, Science 80 (2016) 353, https://doi.org/10.1126/science.aac5523.

[43] H. Jia, F. Wang, Z. Tan, Material and device engineering for high-performance blue quantum dot light-emitting diodes, Nanoscale 12 (2020) 13186–13224, https://doi.org/10.1039/d0nr02074e.

[44] J.M. Pietryga, Y.S. Park, J. Lim, A.F. Fidler, W.K. Bae, S. Brovelli, V.I. Klimov, Spectroscopic and device aspects of nanocrystal quantum dots, Chem. Rev. 116 (2016) 10513–10622, https://doi.org/10.1021/acs.chemrev.6b00169.

[45] J. Kim, S.M. Kwon, C. Jo, J.S. Heo, W. Bin Kim, H.S. Jung, Y.H. Kim, M.G. Kim, S.K. Park, Highly efficient photo-induced charge separation enabled by metal-chalcogenide interfaces in quantum-dot/metal-oxide hybrid phototransistors, ACS Appl. Mater. Interfaces 12 (2020) 16620–16629, https://doi.org/10.1021/acsami.0c01176.

[46] J. Yang, D. Hahm, K. Kim, S. Rhee, M. Lee, S. Kim, J.H. Chang, H.W. Park, J. Lim, M. Lee, H. Kim, J. Bang, H. Ahn, J.H. Cho, J. Kwak, B.S. Kim, C. Lee, W.K. Bae, M.S. Kang, High-resolution patterning of colloidal quantum dots via non-destructive, light-driven ligand crosslinking, Nat. Commun. 11 (2020) 1–9, https://doi.org/10.1038/s41467-020-16652-4.

[47] D.K. Hwang, Y.T. Lee, H.S. Lee, Y.J. Lee, S.H. Shokouh, J.H. Kyhm, J. Lee, H.H. Kim, T.H. Yoo, S.H. Nam, D.I. Son, B.K. Ju, M.C. Park, J.D. Song, W.K. Choi, S. Im, Ultrasensitive PbS quantum-dot-sensitized InGaZnO hybrid photoinverter for near-infrared detection and imaging with high photogain, NPG Asia Mater. 8 (2016) e233, https://doi.org/10.1038/am.2015.137.

[48] D. Jia, J. Chen, M. Yu, J. Liu, E.M.J. Johansson, A. Hagfeldt, X. Zhang, Dual passivation of CsPbI3 perovskite nanocrystals with amino acid ligands for efficient quantum dot solar cells, Small 16 (2020) 2001772, https://doi.org/10.1002/smll.202001772.

[49] V.L. Colvin, M.C. Schlamp, A.P. Alivisatos, Light-emitting diodes made from cadmium selenide nanocrystals and a semiconducting polymer, Nature 370 (1994) 354–357, https://doi.org/10.1038/370354a0.

[50] W.K. Bae, S. Brovelli, V.I. Klimov, Spectroscopic insights into the performance of quantum dot light-emitting diodes, MRS Bull. 38 (2013) 721–730, https://doi.org/10.1557/mrs.2013.182.

[51] Q. Yuan, T. Wang, P. Yu, H. Zhang, H. Zhang, W. Ji, A review on the electroluminescence properties of quantum-dot light-emitting diodes, Org. Electron. 90 (2021) 106086, https://doi.org/10.1016/J.ORGEL.2021.106086.

[52] B. Ai, C. Liu, J. Wang, J. Xie, J. Han, X. Zhao, J. Heo, Precipitation and optical properties of $CsPbBr_3$ quantum dots in phosphate glasses, J. Am. Ceram. Soc. 99 (2016) 2875–2877, https://doi.org/10.1111/jace.14400.

[53] M. Chamarro, C. Gourdon, P. Lavallard, O. Lublinskaya, A. Ekimov, Enhancement of electron-hole exchange interaction in CdSe nanocrystals: a quantum confinement effect, Phys. Rev. B Condens. Matter Mater. Phys. 53 (1996) 1336–1342, https://doi.org/10.1103/PhysRevB.53.1336.

[54] Y.W. Jun, C.S. Choi, J. Cheon, Size and shape controlled ZnTe nanocrystals with quantum confinement effect, Chem. Commun. 0 (2001) 101–102, https://doi.org/10.1039/b008376n.

[55] A.L. Rogach, D.V. Talapin, E.V. Shevchenko, A. Kornowski, M. Haase, H. Weller, Organization of matter on different size scales: monodisperse nanocrystals and their superstructures, Adv. Funct. Mater. 12 (2002) 653–664, https://doi.org/10.1002/1616-3028(20021016)12:10<653::AID-ADFM653>3.0.CO;2-V.

[56] P.O. Anikeeva, J.E. Halpert, M.G. Bawendi, V. Bulović, Quantum dot light-emitting devices with electroluminescence tunable over the entire visible spectrum, Nano Lett. 9 (2009) 2532–2536, https://doi.org/10.1021/nl9002969.

[57] Z. Yang, M. Gao, W. Wu, X. Yang, X.W. Sun, J. Zhang, H.C. Wang, R.S. Liu, C.Y. Han, H. Yang, W. Li, Recent advances in quantum dot-based light-emitting devices: challenges and possible solutions, Mater. Today 24 (2019) 69–93, https://doi.org/10.1016/j.mattod.2018.09.002.

[58] B.O. Dabbousi, J. Rodriguez-Viejo, F.V. Mikulec, J.R. Heine, H. Mattoussi, R. Ober, K. F. Jensen, M.G. Bawendi, (CdSe)ZnS core-shell quantum dots: synthesis and characterization of a size series of highly luminescent nanocrystallites, J. Phys. Chem. B 101 (1997) 9463–9475, https://doi.org/10.1021/jp971091y.

[59] J.J. Li, Y.A. Wang, W. Guo, J.C. Keay, T.D. Mishima, M.B. Johnson, X. Peng, Large-scale synthesis of nearly monodisperse CdSe/CdS core/shell nanocrystals using air-stable reagents via successive ion layer adsorption and reaction, J. Am. Chem. Soc. 125 (2003) 12567–12575, https://doi.org/10.1021/ja0363563.

[60] D.V. Talapin, A.L. Rogach, A. Kornowski, M. Haase, H. Weller, Highly luminescent monodisperse CdSe and CdSe/ZnS nanocrystals synthesized in a hexadecylamine-trioctylphosphine oxide-trioctylphospine mixture, Nano Lett. 1 (2001) 207–211, https://doi.org/10.1021/nl0155126.

[61] S. Kudera, M. Zanella, C. Giannini, A. Rizzo, Y. Li, G. Gigli, R. Cingolani, G. Ciccarella, W. Spahl, W.J. Parak, L. Manna, Sequential growth of magic-size CdSe nanocrystals, Adv. Mater. 19 (2007) 548–552, https://doi.org/10.1002/adma.200601015.
[62] M. Danek, K.F. Jensen, C.B. Murray, M.G. Bawendi, Synthesis of luminescent thin-film CdSe/ZnSe quantum dot composites using CdSe quantum dots passivated with an overlayer of ZnSe, Chem. Mater. 8 (1996) 173–180, https://doi.org/10.1021/cm9503137.
[63] P. Reiss, J. Bleuse, A. Pron, Highly luminescent CdSe/ZnSe core/shell nanocrystals of low size dispersion, Nano Lett. 2 (2002) 781–784, https://doi.org/10.1021/nl025596y.
[64] S. Kim, B. Fisher, H.J. Eisler, M. Bawendi, Type-II quantum dots: CdTe/CdSe(core/shell) and CdSe/ZnTe(core/shell) heterostructures, J. Am. Chem. Soc. 125 (2003) 11466–11467, https://doi.org/10.1021/ja0361749.
[65] R. Xie, X. Zhong, T. Basche, Synthesis, characterization, and spectroscopy of type-II core/shell semiconductor nanocrystals with ZnTe cores, Adv. Mater. 17 (2005) 2741–2745, https://doi.org/10.1002/adma.200501029.
[66] S.A. Ivanov, A. Piryatinski, J. Nanda, S. Tretiak, K.R. Zavadil, W.O. Wallace, D. Werder, V.I. Klimov, Type-II core/shell CdS/ZnSe nanocrystals: synthesis, electronic structures, and spectroscopic properties, J. Am. Chem. Soc. 129 (2007) 11708–11719, https://doi.org/10.1021/JA068351M/SUPPL_FILE/JA068351MSI20061121_025926.PDF.
[67] X. Zhong, Y. Feng, W. Knoll, M. Han, Alloyed $Zn_xCd_{1-x}S$ nanocrystals with highly narrow luminescence spectral width, J. Am. Chem. Soc. 125 (2003) 13559–13563, https://doi.org/10.1021/ja036683a.
[68] W.K. Bae, K. Char, H. Hur, S. Lee, Single-step synthesis of quantum dots with chemical composition gradients, Chem. Mater. 20 (2008) 531–539, https://doi.org/10.1021/cm070754d.
[69] L.E. Shea Rohwer, J.E. Martin, X. Cai, D.F. Kelley, Red-emitting quantum dots for solid-state lighting, ECS J. Solid State Sci. Technol. 2 (2013) R3112–R3118, https://doi.org/10.1149/2.015302JSS/XML.
[70] R.E. Bailey, S. Nie, Alloyed semiconductor quantum dots: tuning the optical properties without changing the particle size, J. Am. Chem. Soc. 125 (2003) 7100–7106, https://doi.org/10.1021/ja035000o.
[71] E. Jang, S. Jun, L. Pu, High quality CdSeS nanocrystals synthesized by facile single injection process and their electroluminescence, Chem. Commun. 3 (2003) 2964–2965, https://doi.org/10.1039/b310853h.
[72] Y. Zheng, Z. Yang, J.Y. Ying, Aqueous synthesis of glutathione-capped ZnSe and $Zn_{1-x}Cd_xSe$ alloyed quantum dots, Adv. Mater. 19 (2007) 1475–1479, https://doi.org/10.1002/ADMA.200601939.
[73] S. Jun, E. Jang, Interfused semiconductor nanocrystals: brilliant blue photoluminescence and electroluminescence, Chem. Commun. (2005) 4616–4618, https://doi.org/10.1039/b509196a.
[74] J. Lim, M. Park, W.K. Bae, D. Lee, S. Lee, C. Lee, K. Char, Highly efficient cadmium-free quantum dot light-emitting diodes enabled by the direct formation of excitons within InP@ZnSeS quantum dots, ACS Nano 7 (2013) 9019–9026, https://doi.org/10.1021/nn403594j.
[75] P. Reiss, S. Carayon, J. Bleuse, A. Pron, Low polydispersity core/shell nanocrystals of CdSe/ZnSe and CdSe/ZnSe/ZnS type: preparation and optical studies, Synth. Met. 139 (2003) 649–652, https://doi.org/10.1016/S0379-6779(03)00335-7.
[76] D.V. Talapin, I. Mekis, S. Götzinger, A. Kornowski, O. Benson, H. Weller, CdSe/CdS/ZnS and CdSe/ZnSe/ZnS core-shell-shell nanocrystals, J. Phys. Chem. B 108

(2004) 18826–18831, https://doi.org/10.1021/JP046481G/SUPPL_FILE/JP046481GSI20040917_094513.PDF.
[77] X. Wang, W. Li, K. Sun, Stable efficient CdSe/CdS/ZnS core/multi-shell nanophosphors fabricated through a phosphine-free route for white light-emitting-diodes with high color rendering properties, J. Mater. Chem. 21 (2011) 8558–8565, https://doi.org/10.1039/c1jm00061f.
[78] J. Lim, S. Jun, E. Jang, H. Baik, H. Kim, J. Cho, Preparation of highly luminescent nanocrystals and their application to light-emitting diodes, Adv. Mater. 19 (2007) 1927–1932, https://doi.org/10.1002/adma.200602642.
[79] C. Wang, Y. Zhang, A. Wang, Q. Wang, H. Tang, W. Shen, Z. Li, Z. Deng, Controlled synthesis of composition tunable formamidinium cesium double cation lead halide perovskite nanowires and nanosheets with improved stability, Chem. Mater. 29 (2017) 2157–2166, https://doi.org/10.1021/acs.chemmater.6b04848.
[80] H. Qin, Y. Niu, R. Meng, X. Lin, R. Lai, W. Fang, X. Peng, Single-dot spectroscopy of zinc-blende CdSe/CdS core/shell nanocrystals: nonblinking and correlation with ensemble measurements, J. Am. Chem. Soc. 136 (2014) 179–187, https://doi.org/10.1021/ja4078528.
[81] S. Jun, E. Jang, Bright and stable alloy core/multishell quantum dots, Angew. Chem. 125 (2013) 707–710, https://doi.org/10.1002/ange.201206333.
[82] O. Chen, J. Zhao, V.P. Chauhan, J. Cui, C. Wong, D.K. Harris, H. Wei, H.S. Han, D. Fukumura, R.K. Jain, M.G. Bawendi, Compact high-quality CdSe-CdS core-shell nanocrystals with narrow emission linewidths and suppressed blinking, Nat. Mater. 12 (2013) 445–451, https://doi.org/10.1038/nmat3539.
[83] A.H. Mueller, M.A. Petruska, M. Achermann, D.J. Werder, E.A. Akhadov, D.D. Koleske, M.A. Hoffbauer, V.I. Klimov, Multicolor light-emitting diodes based on semiconductor nanocrystals encapsulated in GaN charge injection layers, Nano Lett. 5 (2005) 1039–1044, https://doi.org/10.1021/NL050384X/SUPPL_FILE/NL050384XSI20050301_113926.PDF.
[84] W.K. Bae, Y.S. Park, J. Lim, D. Lee, L.A. Padilha, H. Mc Daniel, I. Robel, C. Lee, J.M. Pietryga, V.I. Klimov, Controlling the influence of Auger recombination on the performance of quantum-dot light-emitting diodes, Nat. Commun. 4 (2013), https://doi.org/10.1038/NCOMMS3661.
[85] X. Dai, Z. Zhang, Y. Jin, Y. Niu, H. Cao, X. Liang, L. Chen, J. Wang, X. Peng, Solution-processed, high-performance light-emitting diodes based on quantum dots, Nature 515 (2014) 96–99, https://doi.org/10.1038/nature13829.
[86] L. Qian, Y. Zheng, J. Xue, P.H. Holloway, Stable and efficient quantum-dot light-emitting diodes based on solution-processed multilayer structures, Nat. Photonics 5 (2011) 543–548, https://doi.org/10.1038/nphoton.2011.171.
[87] Z. Zhang, Y. Ye, C. Pu, Y. Deng, X. Dai, X. Chen, D. Chen, X. Zheng, Y. Gao, W. Fang, X. Peng, Y. Jin, High-performance, solution-processed, and insulating-layer-free light-emitting diodes based on colloidal quantum dots, Adv. Mater. 30 (2018), https://doi.org/10.1002/ADMA.201801387.
[88] Y.H. Won, O. Cho, T. Kim, D.Y. Chung, T. Kim, H. Chung, H. Jang, J. Lee, D. Kim, E. Jang, Highly efficient and stable InP/ZnSe/ZnS quantum dot light-emitting diodes, Nature 575 (2019) 634–638, https://doi.org/10.1038/s41586-019-1771-5.
[89] H. Shen, Q. Gao, Y. Zhang, Y. Lin, Q. Lin, Z. Li, L. Chen, Z. Zeng, X. Li, Y. Jia, S. Wang, Z. Du, L.S. Li, Z. Zhang, Visible quantum dot light-emitting diodes with simultaneous high brightness and efficiency, Nat. Photonics 13 (2019) 192–197, https://doi.org/10.1038/s41566-019-0364-z.

[90] Y. Sun, Q. Su, H. Zhang, F. Wang, S. Zhang, S. Chen, Investigation on thermally induced efficiency roll-off: toward efficient and ultrabright quantum-dot light-emitting diodes, ACS Nano 13 (2019) 11433–11442, https://doi.org/10.1021/ACSNANO.9B04879/SUPPL_FILE/NN9B04879_SI_001.PDF.

[91] Y. Fu, W. Jiang, D. Kim, W. Lee, H. Chae, Highly efficient and fully solution-processed inverted light-emitting diodes with charge control interlayers, ACS Appl. Mater. Interfaces 10 (2018) 17295–17300, https://doi.org/10.1021/ACSAMI.8B05092/SUPPL_FILE/AM8B05092_SI_001.PDF.

[92] P. Shen, F. Cao, H. Wang, B. Wei, F. Wang, X.W. Sun, X. Yang, Solution-processed double-junction quantum-dot light-emitting diodes with an EQE of over 40%, ACS Appl. Mater. Interfaces 11 (2019) 1065–1070, https://doi.org/10.1021/ACSAMI.8B18940.

[93] H. Zhang, X. Sun, S. Chen, Over 100 cd A^{-1} efficient quantum dot light-emitting diodes with inverted tandem structure, Adv. Funct. Mater. 27 (2017), https://doi.org/10.1002/ADFM.201700610.

[94] H. Zhang, S. Chen, X.W. Sun, Efficient red/green/blue tandem quantum-dot light-emitting diodes with external quantum efficiency exceeding 21%, ACS Nano 12 (2018) 697–704, https://doi.org/10.1021/acsnano.7b07867.

[95] C. Jiang, J. Zou, Y. Liu, C. Song, Z. He, Z. Zhong, J. Wang, H.L. Yip, J. Peng, Y. Cao, Fully solution-processed tandem white quantum-dot light-emitting diode with an external quantum efficiency exceeding 25%, ACS Nano 12 (2018) 6040–6049, https://doi.org/10.1021/ACSNANO.8B02289/SUPPL_FILE/NN8B02289_SI_001.PDF.

[96] X. Yang, E. Mutlugun, C. Dang, K. Dev, Y. Gao, S.T. Tan, X.W. Sun, H.V. Demir, Highly flexible, electrically driven, top-emitting, quantum dot light-emitting stickers, ACS Nano 8 (2014) 8224–8231, https://doi.org/10.1021/NN502588K/SUPPL_FILE/NN502588K_SI_001.PDF.

[97] T. Lee, D. Hahm, K. Kim, W.K. Bae, C. Lee, J. Kwak, Highly efficient and bright inverted top-emitting InP quantum dot light-emitting diodes introducing a hole-suppressing interlayer, Small 15 (2019), https://doi.org/10.1002/SMLL.201905162.

[98] H. Zhang, Q. Su, S. Chen, Quantum-dot and organic hybrid tandem light-emitting diodes with multi-functionality of full-color-tunability and white-light-emission, Nat. Commun. 11 (2020), https://doi.org/10.1038/S41467-020-16659-X.

[99] J. He, H. Chen, H. Chen, Y. Wang, S.-T. Wu, Y. Dong, Hybrid downconverters with green perovskite-polymer composite films for wide color gamut displays, Opt. Express 25 (2017) 12915, https://doi.org/10.1364/oe.25.012915.

[100] H.C. Wang, S.Y. Lin, A.C. Tang, B.P. Singh, H.C. Tong, C.Y. Chen, Y.C. Lee, T.L. Tsai, R.S. Liu, Mesoporous silica particles integrated with all-inorganic $CsPbBr_3$ perovskite quantum-dot nanocomposites (MP-PQDs) with high stability and wide color gamut used for backlight display, Angew. Chem. Int. Ed. 55 (2016) 7924–7929, https://doi.org/10.1002/anie.201603698.

[101] X. Dai, Y. Deng, X. Peng, Y. Jin, Quantum-dot light-emitting diodes for large-area displays: towards the dawn of commercialization, Adv. Mater. 29 (2017) 1607022, https://doi.org/10.1002/adma.201607022.

[102] Y. Sun, Y. Jiang, H. Peng, J. Wei, S. Zhang, S. Chen, Efficient quantum dot light-emitting diodes with a $Zn_{0.85}Mg_{0.15}O$ interfacial modification layer, Nanoscale 9 (2017) 8962–8969, https://doi.org/10.1039/c7nr02099f.

[103] Y. Yang, Y. Zheng, W. Cao, A. Titov, J. Hyvonen, J.R. Manders, J. Xue, P.H. Holloway, L. Qian, High-efficiency light-emitting devices based on quantum dots with tailored nanostructures, Nat. Photonics 9 (2015) 259–265, https://doi.org/10.1038/nphoton.2015.36.

[104] D. Kim, Y. Fu, S. Kim, W. Lee, K.H. Lee, H.K. Chung, H.J. Lee, H. Yang, H. Chae, Polyethylenimine ethoxylated-mediated all-solution-processed high-performance flexible inverted quantum dot-light-emitting device, ACS Nano 11 (2017) 1982–1990, https://doi.org/10.1021/acsnano.6b08142.

[105] F. Cao, H. Wang, P. Shen, X. Li, Y. Zheng, Y. Shang, J. Zhang, Z. Ning, X. Yang, High-efficiency and stable quantum dot light-emitting diodes enabled by a solution-processed metal-doped nickel oxide hole injection interfacial layer, Adv. Funct. Mater. 27 (2017) 1704278, https://doi.org/10.1002/adfm.201704278.

[106] K.H. Lee, C.Y. Han, H.D. Kang, H. Ko, C. Lee, J. Lee, N.S. Myoung, S.Y. Yim, H. Yang, Highly efficient, color-reproducible full-color electroluminescent devices based on red/green/blue quantum dot-mixed multilayer, ACS Nano 9 (2015) 10941–10949, https://doi.org/10.1021/ACSNANO.5B05513/SUPPL_FILE/NN5B05513_SI_001.PDF.

[107] H. Jung, M. Lee, C. Han, Y. Park, K.-S. Cho, H. Jeon, Efficient on-chip integration of a colloidal quantum dot photonic crystal band-edge laser with a coplanar waveguide, Opt. Express 25 (2017) 32919, https://doi.org/10.1364/oe.25.032919.

[108] W. Xie, T. Stöferle, G. Rainò, T. Aubert, S. Bisschop, Y. Zhu, R.F. Mahrt, P. Geiregat, E. Brainis, Z. Hens, D. Van Thourhout, On-chip integrated quantum-dot–silicon-nitride microdisk lasers, Adv. Mater. 29 (2017) 1604866, https://doi.org/10.1002/adma.201604866.

[109] C. Grivas, M. Pollnau, Organic solid-state integrated amplifiers and lasers, Laser Photonics Rev. 6 (2012) 419–462, https://doi.org/10.1002/lpor.201100034.

[110] A. Rose, Z. Zhu, C.F. Madigan, T.M. Swager, V. Bulović, Sensitivity gains in chemosensing by lasing action in organic polymers, Nature 434 (2005) 876–879, https://doi.org/10.1038/nature03438.

[111] M. Achermann, M.A. Petruska, D.D. Koleske, M.H. Crawford, V.I. Klimov, Nanocrystal-based light-emitting diodes utilizing high-efficiency nonradiative energy transfer for color conversion, Nano Lett. 6 (2006) 1396–1400, https://doi.org/10.1021/nl060392t.

[112] L. Lan, B. Liu, H. Tao, J. Zou, C. Jiang, M. Xu, L. Wang, J. Peng, Y. Cao, Preparation of efficient quantum dot light-emitting diodes by balancing charge injection and sensitizing emitting layer with phosphorescent dye, J. Mater. Chem. C 7 (2019) 5755–5763, https://doi.org/10.1039/c8tc04991b.

[113] G. Liu, X. Zhou, X. Sun, S. Chen, Performance of inverted quantum dot light-emitting diodes enhanced by using phosphorescent molecules as exciton harvesters, J. Phys. Chem. C 120 (2016) 4667–4672, https://doi.org/10.1021/acs.jpcc.5b12692.

[114] W. Zheng, D. Song, S. Zhao, B. Qiao, Z. Xu, J. Chen, P. Wang, Y. Liang, All-solution processed inverted QLEDs with double hole transport layers and thermal activated delay fluorescent dopant as energy transfer medium, Org. Electron. 77 (2020) 105544, https://doi.org/10.1016/j.orgel.2019.105544.

[115] W. Zheng, Z. Xu, D. Song, S. Zhao, B. Qiao, J. Chen, P. Wang, X. Zheng, Enhancing the efficiency and the luminance of quantum dot light-emitting diodes by inserting a leaked electron harvesting layer with thermal-activated delayed fluorescence material, Org. Electron. 65 (2019) 357–362, https://doi.org/10.1016/j.orgel.2018.11.031.

[116] V. Wood, M.J. Panzer, D. Bozyigit, Y. Shirasaki, I. Rousseau, S. Geyer, M.G. Bawendi, V. Bulović, Electroluminescence from nanoscale materials via field-driven ionization, Nano Lett. 11 (2011) 2927–2932, https://doi.org/10.1021/nl2013983.

[117] D. Bozyigit, V. Wood, Y. Shirasaki, V. Bulovic, Study of field driven electroluminescence in colloidal quantum dot solids, J. Appl. Phys. 111 (2012) 113701, https://doi.org/10.1063/1.4720377.

[118] Y. Deng, X. Lin, W. Fang, D. Di, L. Wang, R.H. Friend, X. Peng, Y. Jin, Deciphering exciton-generation processes in quantum-dot electroluminescence, Nat. Commun. 11 (2020) 1–8, https://doi.org/10.1038/s41467-020-15944-z.
[119] L. Qian, Y. Zheng, K.R. Choudhury, D. Bera, F. So, J. Xue, P.H. Holloway, Electroluminescence from light-emitting polymer/ZnO nanoparticle heterojunctions at sub-bandgap voltages, Nano Today 5 (2010) 384–389, https://doi.org/10.1016/j.nantod.2010.08.010.
[120] Y.Q. Zhang, X.A. Cao, Electroluminescence of green CdSe/ZnS quantum dots enhanced by harvesting excitons from phosphorescent molecules, Appl. Phys. Lett. 97 (2010) 253115, https://doi.org/10.1063/1.3530450.
[121] P.O. Anikeeva, C.F. Madigan, J.E. Halpert, M.G. Bawendi, V. Bulović, Electronic and excitonic processes in light-emitting devices based on organic materials and colloidal quantum dots, Phys. Rev. B Condens. Matter Mater. Phys. 78 (2008) 085434, https://doi.org/10.1103/PhysRevB.78.085434.
[122] W. Ki Bae, J. Kwak, J. Lim, D. Lee, M. Ki Nam, K. Char, C. Lee, S. Lee, Multicolored light-emitting diodes based on all-quantum-dot multilayer films using layer-by-layer assembly method, Nano Lett. 10 (2010) 2368–2373, https://doi.org/10.1021/nl100168s.
[123] H. Zhang, Q. Su, S. Chen, Suppressing Förster resonance energy transfer in close-packed quantum-dot thin film: toward efficient quantum-dot light-emitting diodes with external quantum efficiency over 21.6%, Adv. Opt. Mater. 8 (2020), https://doi.org/10.1002/ADOM.201902092.
[124] O. Wang, L. Wang, Z. Li, Q. Xu, Q. Lin, H. Wang, Z. Du, H. Shen, L.S. Li, High-efficiency, deep blue ZnCdS/Cd: XZn1-xS/ZnS quantum-dot-light-emitting devices with an EQE exceeding 18%, Nanoscale 10 (2018) 5650–5657, https://doi.org/10.1039/c7nr09175c.
[125] J. Zhang, B. Ren, S. Deng, J. Huang, L. Jiang, D. Zhou, X. Zhang, M. Zhang, R. Chen, F. Yeung, H.S. Kwok, P. Xu, G. Li, Voltage-dependent multicolor electroluminescent device based on halide perovskite and chalcogenide quantum-dots emitters, Adv. Funct. Mater. 30 (2020) 1907074, https://doi.org/10.1002/adfm.201907074.
[126] H. Zhang, Q. Su, Y. Sun, S. Chen, Efficient and color stable white quantum-dot light-emitting diodes with external quantum efficiency over 23%, Adv. Opt. Mater. 6 (2018) 1800354, https://doi.org/10.1002/adom.201800354.
[127] J.F. Wager, P.D. Keir, Electrical characterization of thin-film electroluminescent devices, Annu. Rev. Mater. Sci. 27 (1997) 223–248, https://doi.org/10.1146/annurev.matsci.27.1.223.
[128] V. Wood, M.J. Panzer, J.M. Caruge, J.E. Halpert, M.G. Bawendi, V. Bulović, Air-stable operation of transparent, colloidal quantum dot based LEDs with a unipolar device architecture, Nano Lett. 10 (2010) 24–29, https://doi.org/10.1021/nl902425g.
[129] J. Kwak, W.K. Bae, D. Lee, I. Park, J. Lim, M. Park, H. Cho, H. Woo, D.Y. Yoon, K. Char, S. Lee, C. Lee, Bright and efficient full-color colloidal quantum dot light-emitting diodes using an inverted device structure, Nano Lett. 12 (2012) 2362–2366, https://doi.org/10.1021/NL3003254/SUPPL_FILE/NL3003254_SI_001.PDF.
[130] W. Ji, P. Jing, J. Zhao, X. Liu, A. Wang, H. Li, Inverted CdSe/CdS/ZnS quantum dot light emitting devices with titanium dioxide as an electron-injection contact, Nanoscale 5 (2013) 3474–3480, https://doi.org/10.1039/c3nr34168b.
[131] W. Ji, P. Jing, L. Zhang, D. Li, Q. Zeng, S. Qu, J. Zhao, The work mechanism and sub-bandgap-voltage electroluminescence in inverted quantum dot light-emitting diodes, Sci. Rep. 4 (2014), https://doi.org/10.1038/srep06974.

[132] H. Luo, W. Zhang, M. Li, Y. Yang, M. Guo, S.W. Tsang, S. Chen, Origin of subthreshold turn-on in quantum-dot light-emitting diodes, ACS Nano 13 (2019) 8229–8236, https://doi.org/10.1021/acsnano.9b03507.

[133] J. Cui, A.P. Beyler, I. Coropceanu, L. Cleary, T.R. Avila, Y. Chen, J.M. Cordero, S.L. Heathcote, D.K. Harris, O. Chen, J. Cao, M.G. Bawendi, Evolution of the single-nanocrystal photoluminescence linewidth with size and shell: implications for exciton-phonon coupling and the optimization of spectral linewidths, Nano Lett. 16 (2016) 289–296, https://doi.org/10.1021/acs.nanolett.5b03790.

[134] M.A. Boles, D. Ling, T. Hyeon, D.V. Talapin, The surface science of nanocrystals, Nat. Mater. 15 (2016) 141–153, https://doi.org/10.1038/nmat4526.

[135] A.L. Efros, Nanocrystals: almost always bright, Nat. Mater. 7 (2008) 612–613, https://doi.org/10.1038/nmat2239.

[136] A.W. Cohn, J.D. Rinehart, A.M. Schimpf, A.L. Weaver, D.R. Gamelin, Size dependence of negative trion auger recombination in photodoped cdse nanocrystals, Nano Lett. 14 (2014) 353–358, https://doi.org/10.1021/nl4041675.

[137] H. Shen, Q. Lin, W. Cao, C. Yang, N.T. Shewmon, H. Wang, J. Niu, L.S. Li, J. Xue, Efficient and long-lifetime full-color light-emitting diodes using high luminescence quantum yield thick-shell quantum dots, Nanoscale 9 (2017) 13583–13591, https://doi.org/10.1039/c7nr04953f.

[138] J.H. Chang, P. Park, H. Jung, B.G. Jeong, D. Hahm, G. Nagamine, J. Ko, J. Cho, L.A. Padilha, D.C. Lee, C. Lee, K. Char, W.K. Bae, Unraveling the origin of operational instability of quantum dot based light-emitting diodes, ACS Nano 12 (2018) 10231–10239, https://doi.org/10.1021/acsnano.8b03386.

[139] H. Liang, R. Zhu, Y. Dong, S.-T. Wu, J. Li, J. Wang, J. Zhou, Enhancing the outcoupling efficiency of quantum dot LEDs with internal nano-scattering pattern, Opt. Express 23 (2015) 12910, https://doi.org/10.1364/oe.23.012910.

[140] W.K. Bae, J. Lim, Nanostructured colloidal quantum dots for efficient electroluminescence devices, Korean J. Chem. Eng. 36 (2019) 173–185, https://doi.org/10.1007/s11814-018-0193-7.

[141] K. Hong, J.L. Lee, Review paper: recent developments in light extraction technologies of organic light emitting diodes, Electron. Mater. Lett. 7 (2011) 77–91, https://doi.org/10.1007/s13391-011-0601-1.

[142] R. Meerheim, M. Furno, S. Hofmann, B. Lüssem, K. Leo, Quantification of energy loss mechanisms in organic light-emitting diodes, Appl. Phys. Lett. 97 (2010) 253305, https://doi.org/10.1063/1.3527936.

[143] Y.M. Huang, K.J. Singh, A.C. Liu, C.C. Lin, Z. Chen, K. Wang, Y. Lin, Z. Liu, T. Wu, H. C. Kuo, Advances in quantum-dot-based displays, Nanomaterials 10 (2020) 1–29, https://doi.org/10.3390/nano10071327.

[144] C.H. Lin, A. Verma, C.Y. Kang, Y.M. Pai, T.Y. Chen, J.J. Yang, C.W. Sher, Y.Z. Yang, P.T. Lee, C.C. Lin, Y.C. Wu, S.K. Sharma, T. Wu, S.R. Chung, H.C. Kuo, Hybrid-type white LEDs based on inorganic halide perovskite QDs: candidates for wide color gamut display backlights, Photon. Res. 7 (2019) 579, https://doi.org/10.1364/prj.7.000579.

[145] B. Xie, R. Hu, X. Yu, B. Shang, Y. Ma, X. Luo, Effect of packaging method on performance of light-emitting diodes with quantum dot phosphor, IEEE Photon. Technol. Lett. 28 (2016) 1115–1118, https://doi.org/10.1109/LPT.2016.2531794.

[146] F. Fang, W. Chen, Y. Li, H. Liu, M. Mei, R. Zhang, J. Hao, M. Mikita, W. Cao, R. Pan, K. Wang, X.W. Sun, Employing polar solvent controlled ionization in precursors for synthesis of high-quality inorganic perovskite nanocrystals at room temperature, Adv. Funct. Mater. 28 (2018) 1706000, https://doi.org/10.1002/adfm.201706000.

[147] Y. Yuan, J. Zhang, G. Liang, X. Yang, Rapid fluorescent detection of neurogenin 3 by CdTe quantum dot aggregation, Analyst 137 (2012) 1775–1778, https://doi.org/10.1039/c2an16166d.

[148] K.S. Cho, E.K. Lee, W.J. Joo, E. Jang, T.H. Kim, S.J. Lee, S.J. Kwon, J.Y. Han, B.K. Kim, B.L. Choi, J.M. Kim, High-performance crosslinked colloidal quantum-dot light-emitting diodes, Nat. Photonics 3 (2009) 341–345, https://doi.org/10.1038/nphoton.2009.92.

[149] C.Y. Kang, C.H. Lin, C.H. Lin, T.Y. Li, S.W. Huang Chen, C.L. Tsai, C.W. Sher, T.Z. Wu, P.T. Lee, X. Xu, M. Zhang, C.H. Ho, J.H. He, H.C. Kuo, Highly efficient and stable white light-emitting diodes using perovskite quantum dot paper, Adv. Sci. 6 (2019) 1902230, https://doi.org/10.1002/advs.201902230.

[150] G. Nedelcu, L. Protesescu, S. Yakunin, M.I. Bodnarchuk, M.J. Grotevent, M.V. Kovalenko, Fast anion-exchange in highly luminescent nanocrystals of cesium lead halide perovskites ($CsPbX_3$, X = Cl, Br, I), Nano Lett. 15 (2015) 5635–5640, https://doi.org/10.1021/acs.nanolett.5b02404.

[151] Q.A. Akkerman, V. D'Innocenzo, S. Accornero, A. Scarpellini, A. Petrozza, M. Prato, L. Manna, Tuning the optical properties of cesium lead halide perovskite nanocrystals by anion exchange reactions, J. Am. Chem. Soc. 137 (2015) 10276–10281, https://doi.org/10.1021/jacs.5b05602.

[152] T. Frecker, D. Bailey, X. Arzeta-Ferrer, J. McBride, S.J. Rosenthal, Review—quantum dots and their application in lighting, displays, and biology, ECS J. Solid State Sci. Technol. 5 (2016) R3019–R3031, https://doi.org/10.1149/2.0031601jss.

[153] R.H. Kim, S. Kim, Y.M. Song, H. Jeong, T. Il Kim, J. Lee, X. Li, K.D. Choquette, J.A. Rogers, Flexible vertical light emitting diodes, Small 8 (2012) 3123–3128, https://doi.org/10.1002/smll.201201195.

[154] P.H. Huang, T.C. Huang, Y.-T. Sun, S.-Y. Yang, Large-area and thin light guide plates fabricated using UV-based imprinting, Opt. Express 16 (2008) 15033, https://doi.org/10.1364/oe.16.015033.

[155] C.W. Sher, K.J. Chen, C.-C. Lin, H.-V. Han, H.Y. Lin, Z.-Y. Tu, H.H. Tu, K. Honjo, H.Y. Jiang, S.L. Ou, R.H. Horng, X. Li, C.C. Fu, H.C. Kuo, Large-area, uniform white light LED source on a flexible substrate, Opt. Express 23 (2015) A1167, https://doi.org/10.1364/oe.23.0a1167.

[156] Q. Van Le, K. Hong, H.W. Jang, S.Y. Kim, Halide perovskite quantum dots for light-emitting diodes: properties, synthesis, applications, and outlooks, Adv. Electron. Mater. 4 (2018) 1800335, https://doi.org/10.1002/aelm.201800335.

[157] L. Zhang, X. Yang, Q. Jiang, P. Wang, Z. Yin, X. Zhang, H. Tan, Y.M. Yang, M. Wei, B. R. Sutherland, E.H. Sargent, J. You, Ultra-bright and highly efficient inorganic based perovskite light-emitting diodes, Nat. Commun. 8 (2017) 1–8, https://doi.org/10.1038/ncomms15640.

[158] X. Yang, X. Zhang, J. Deng, Z. Chu, Q. Jiang, J. Meng, P. Wang, L. Zhang, Z. Yin, J. You, Efficient green light-emitting diodes based on quasi-two-dimensional composition and phase engineered perovskite with surface passivation, Nat. Commun. 9 (2018) 1–8, https://doi.org/10.1038/s41467-018-02978-7.

[159] H. Liao, S. Guo, S. Cao, L. Wang, F. Gao, Z. Yang, J. Zheng, W. Yang, A general strategy for in situ growth of all-inorganic $CsPbX_3$ (X = Br, I, and Cl) perovskite nanocrystals in polymer fibers toward significantly enhanced water/thermal stabilities, Adv. Opt. Mater. 6 (2018) 1800346, https://doi.org/10.1002/adom.201800346.

[160] T. Song, X. Feng, H. Ju, T. Fang, F. Zhu, W. Liu, W. Huang, Enhancing acid, base and UV light resistance of halide perovskite CH3NH3PbBr3 quantum dots by encapsulation with ZrO_2 sol, J. Alloys Compd. 816 (2020) 152558, https://doi.org/10.1016/j.jallcom.2019.152558.

[161] S. Huang, Z. Li, L. Kong, N. Zhu, A. Shan, L. Li, Enhancing the stability of CH3NH3PbBr3 quantum dots by embedding in silica spheres derived from tetramethyl orthosilicate in "waterless" toluene, J. Am. Chem. Soc. 138 (2016) 5749–5752, https://doi.org/10.1021/jacs.5b13101.

[162] Z. Li, L. Kong, S. Huang, L. Li, Highly luminescent and ultrastable $CsPbBr_3$ perovskite quantum dots incorporated into a silica/alumina monolith, Angew. Chem. 129 (2017) 8246–8250, https://doi.org/10.1002/ange.201703264.

[163] S. Yuan, D. Chen, X. Li, J. Zhong, X. Xu, In situ crystallization synthesis of $CsPbBr_3$ perovskite quantum dot-embedded glasses with improved stability for solid-state lighting and random upconverted lasing, ACS Appl. Mater. Interfaces 10 (2018) 18918–18926, https://doi.org/10.1021/acsami.8b05155.

[164] X. Li, Y. Wang, H. Sun, H. Zeng, Amino-mediated anchoring perovskite quantum dots for stable and low-threshold random lasing, Adv. Mater. 29 (2017) 1701185, https://doi.org/10.1002/adma.201701185.

[165] Z. Hu, Z. Liu, Y. Bian, S. Li, X. Tang, J. Du, Z. Zang, M. Zhou, W. Hu, Y. Tian, Y. Leng, Enhanced two-photon-pumped emission from in situ synthesized nonblinking CsPbBr3/SiO2 nanocrystals with excellent stability, Adv. Opt. Mater. 6 (2018) 1700997, https://doi.org/10.1002/adom.201700997.

[166] A. Kojima, M. Ikegami, K. Teshima, T. Miyasaka, Highly luminescent lead bromide perovskite nanoparticles synthesized with porous alumina media, Chem. Lett. 41 (2012) 397–399, https://doi.org/10.1246/cl.2012.397.

[167] G. Longo, A. Pertegás, L. Martínez-Sarti, M. Sessolo, H.J. Bolink, Highly luminescent perovskite-aluminum oxide composites, J. Mater. Chem. C 3 (2015) 11286–11289, https://doi.org/10.1039/c5tc02447a.

[168] X. Zhang, B. Xu, J. Zhang, Y. Gao, Y. Zheng, K. Wang, X.W. Sun, All-inorganic perovskite nanocrystals for high-efficiency light emitting diodes: dual-phase CsPbBr3-CsPb2Br5 composites, Adv. Funct. Mater. 26 (2016) 4595–4600, https://doi.org/10.1002/adfm.201600958.

[169] W. Li, W. Deng, X. Fan, F. Chun, M. Xie, C. Luo, S. Yang, H. Osman, C. Liu, X. Zheng, W. Yang, Low toxicity antisolvent synthesis of composition-tunable luminescent all-inorganic perovskite nanocrystals, Ceram. Int. 44 (2018) 18123–18128, https://doi.org/10.1016/j.ceramint.2018.07.018.

[170] S. Zou, Y. Liu, J. Li, C. Liu, R. Feng, F. Jiang, Y. Li, J. Song, H. Zeng, M. Hong, X. Chen, Stabilizing cesium lead halide perovskite lattice through Mn(II) substitution for air-stable light-emitting diodes, J. Am. Chem. Soc. 139 (2017) 11443–11450, https://doi.org/10.1021/jacs.7b04000.

[171] J.S. Yao, J. Ge, B.N. Han, K.H. Wang, H. Bin Yao, H.L. Yu, J.H. Li, B.S. Zhu, J.Z. Song, C. Chen, Q. Zhang, H.B. Zeng, Y. Luo, S.H. Yu, Ce^{3+}-doping to modulate photoluminescence kinetics for efficient CsPbBr3 nanocrystals based light-emitting diodes, J. Am. Chem. Soc. 140 (2018) 3626–3634, https://doi.org/10.1021/jacs.7b11955.

[172] J. Wu, H. Bai, W. Qu, G. Cao, Y. Zhang, J. Yu, Preparation of Eu3+-doped CsPbBr3 quantum-dot microcrystals and their luminescence properties, Opt. Mater. (Amst.) 97 (2019) 109454, https://doi.org/10.1016/j.optmat.2019.109454.

[173] L. Liu, J. Li, J.A. Mcleod, Influence of Eu-substitution on luminescent CH3NH3PbBr3 quantum dots, Nanoscale 10 (2018) 11452–11459, https://doi.org/10.1039/c8nr01656a.
[174] R. Yuan, L. Shen, C. Shen, J. Liu, L. Zhou, W. Xiang, X. Liang, CsPbBr3:: X Eu3+ perovskite QD borosilicate glass: a new member of the luminescent material family, Chem. Commun. 54 (2018) 3395–3398, https://doi.org/10.1039/c8cc00243f.
[175] M.A. Baldo, D.F. O'Brien, Y. You, A. Shoustikov, S. Sibley, M.E. Thompson, S.R. Forrest, Highly efficient phosphorescent emission from organic electroluminescent devices, Nature 395 (1998) 151–154, https://doi.org/10.1038/25954.
[176] Z.B. Wang, M.G. Helander, J. Qiu, D.P. Puzzo, M.T. Greiner, Z.M. Hudson, S. Wang, Z. W. Liu, Z.H. Lu, Unlocking the full potential of organic light-emitting diodes on flexible plastic, Nat. Photonics 5 (2011) 753–757, https://doi.org/10.1038/nphoton.2011.259.
[177] S.Y. Kim, W.I. Jeong, C. Mayr, Y.S. Park, K.H. Kim, J.H. Lee, C.K. Moon, W. Brütting, J.J. Kim, Organic light-emitting diodes with 30% external quantum efficiency based on a horizontally oriented emitter, Adv. Funct. Mater. 23 (2013) 3896–3900, https://doi.org/10.1002/adfm.201300104.
[178] C.W. Lee, J.Y. Lee, Above 30% external quantum efficiency in blue phosphorescent organic light-emitting diodes using pyrido[2,3-b]indole derivatives as host materials, Adv. Mater. 25 (2013) 5450–5454, https://doi.org/10.1002/adma.201301091.
[179] C.Y. Kuei, W.L. Tsai, B. Tong, M. Jiao, W.K. Lee, Y. Chi, C.C. Wu, S.H. Liu, G.H. Lee, P.T. Chou, Bis-tridentate Ir(III) complexes with nearly unitary RGB phosphorescence and organic light-emitting diodes with external quantum efficiency exceeding 31%, Adv. Mater. 28 (2016) 2795–2800, https://doi.org/10.1002/adma.201505790.
[180] D. Bozyigit, V. Wood, Challenges and solutions for high-efficiency quantum dot-based LEDs, MRS Bull. 38 (2013) 731–736, https://doi.org/10.1557/mrs.2013.180.

Metal oxides for biophotonics

15

Umer Mushtaq[a], *Vijay Kumar*[a,b], *Vishal Sharma*[c], *and Hendrik C. Swart*[b]
[a]Department of Physics, National Institute of Technology Srinagar, Hazratbal, Srinagar, Jammu and Kashmir, India, [b]Department of Physics, University of the Free State, Bloemfontein, Free State, South Africa, [c]Institute of Forensic Science & Criminology, Panjab University, Chandigarh, India

Chapter outline

1 Introduction

Metal oxides, which can be understood as materials having a crystalline structure, are composed of a metal cation and an oxide anion. These types of compounds are currently one of the most well-studied and significant crystalline structures that can be understood as heterogeneous catalysts and can be used to catalyze acid–base and reduction–oxidation processes. Due to their unique outermost electronic arrangement, several classes of metals, especially metals involved in the d-block, have drawn a lot of interest. They are used for a lot of different chemical processes, such as oxidation, dehydration, and dehydrogenation, as well as isomerization. Metal oxides such as Nb_2O_5 [1], WO_3 [1], and TiO_2 [2] have been frequently utilized for heterogeneous acid nanocatalysts. Such d-block metal compositions possessing mesopore structures tend to allow substrates to enter the porous structure and catalyze the process. Such metal oxides with mesopores are distinguished by their increased surface-to-volume ratio, fluctuating size of pores, and mechanical stability. Biophotonics is a field related

Metal Oxides for Next-Generation Optoelectronic, Photonic, and Photovoltaic Applications
https://doi.org/10.1016/B978-0-323-99143-8.00001-8

to the investigation of optical processes in living organisms and can be understood as an interdisciplinary area of research wherein we study all elements of interaction between photons and biomaterials. The sensing, reflection, emission, modulation, absorption, production, and manipulation of light while they engage with biological components, species, molecules, tissues, and other materials are referred to as biophotonics [3–7]. It involves images and the detection or sensing of cells, tissues, or any particular organism. One of the most common examples of this type of research is the incorporation of fluorescent indicators into a biological substance for monitoring cellular kinetics and drug delivery to a specific location [8]. The use of biophotonics for the detection of different biological processes has potential applications in the fields of bioimaging, biosensing, and targeting drug delivery. Nanoparticles (NPs) exhibit distinctive physical as well as chemical properties owing to which such materials have proved to be viable contenders for biological applications, thanks to advances in nanomaterials and transdisciplinary studies. Nanoscience, in principle, entails the creation and management of materials with a size of just a few hundred nanometers or less that permit certain size-dependent features [9]. NPs used for biomedical purposes must possess a size, preferably below 200 nm [10]. NPs have improved colloidal stability and, as a result, improved bioaccessibility, displaying the potential to overcome the blood–brain interface, penetrate the pulmonary system, and adsorb via endothelial cells owing to their compact design and huge surface area [11]. Metal oxides (MOs) have several attributes because of which they have proved to be promising materials for biological applications, including improved chemical stability, ease of synthesis, ease of tenability to a particular size, structure, and pore size distribution, absence of variations in swelling patterns, effortless integration into hydrophobic and hydrophilic structures, and simple functionalization with the assistance of various molecules because of the presence of charge on the exterior of such compounds [12]. MOs respond in a variety of ways to in vivo systems based on their dimensions, structure, purity, stability, and surface qualities; hence, it is very important that their morphology be characterized. MOs are categorized as 0-D, 1-D, 2-D, or 3-D on the basis of the number of dimensions that are not limited to the nanoscale [13]. Some examples of nanostructures with zero dimensions include nanoclusters, quantum dots, and other nanostructures, as they possess all dimensions at the nanoscale level. The nanostructures that possess a single dimension outside the nanosize range, such as nanorods, nanotubes, nanowires, and nanofibers, are included in 1-D metal oxides. 2-D or planar nanostructures such as nanosheets, nanocoatings, nanofilms, and nanoplatelets are all included in 2-D metal oxide structures. 3-D metal oxide nanostructure NPs comprise dendrimers [14], bundles of nanowires or nanotubes, nanopillars, nanoflowers, multinanolayers, and tetrapods [15,16]. These emerge whenever nanostructures combine in all three orthogonal directions to a size >100 nm. Different alternatives for the manufacturing of such MO NPs have been proposed and explained in various research articles [17]. Overall, MO nanostructures are intensely ionic and could be structured into crystalline geometries having a variety of binding groups and vertices. Depositing MO nanostructures at the designated place on a base material with resolution on the nanoscale possesses the ability to realize nanotechnologies for a variety of uses in chemistry, optics, and biophotonics [18]. The medicinal utilization of nanostructures for diagnosis and therapies involving the delivery of drugs and the improved efficiency of devices used for clinical purposes has recently advanced quickly. The use of MO nanostructures in imaging and therapeutics has a number of benefits as they allow for better contrast imaging as well as targeted drug delivery. Nanostructures are currently employed for the diagnosis of a wide range of biomarkers, hereditary and immunological disorders, cancerous tumors, photodynamic therapy, and targeted delivery of medications [19].

MOs are used in bioimaging as functional luminescent labels, in magnetic resonance imaging (MRI), and in many other applications. Biophotonic mechanisms have found their application in the fields of bioimaging, noninvasive assessments of the veracity of biological characteristics such as glucose and oxygen levels present in blood, therapeutic photonic treatment of harmed and unhealthy cells or tissues, identification of harmed or unhealthy cells as well as tissues, tissue repair, and improvement in treatment. Finally, such mechanisms are also employed during surgery, involving laser cutting, destroying healthy tissues, and many other surgical procedures. The clinical utilization of nanomaterials for detection and therapies has advanced quickly in recent times. The use of MO nanomaterials in diagnostic and therapeutic applications has a number of benefits in modern medical science. Researchers have synthesized nanomaterials that are easily dispersed in water; thus, water-dispersible NPs have been engineered, allowing them to be used for a variety of biomedical studies. NPs are now also being used in imaging techniques for a variety of molecular biomarkers, congenital and autoimmune disorders, and cancerous tumors [19]. Apart from this, they are also being used as photosensitizers in photodynamic treatment [20,21] and targeted drug delivery [22]. Currently, nanomaterials are used in diagnostic applications as luminescence-based particles such as semiconductor quantum dots, MRI, and many other applications. In this chapter, we discuss the use of metal oxides in biophotonics.

2 Properties of metal oxides

Once nanoparticles are injected into a living subject, they interface with biological fluids as well as cellular molecules, enabling them to physically relocate into the internal cellular regions [23]. Since the biologic responsiveness to nanoparticles is influenced by a range of characteristics such as size, shape, and aggregation, controlled synthesis strategies aim to produce nanoparticles with specific morphology, dimensions, dispersion, and persistence. Some of the most important characteristics of metal oxides that affect their performance are discussed as follows:

2.1 *Dimensions of metal oxides*

Size is one of the main factors that determine the biodistribution of a particular drug as well as its absorption by determining the surface-to-volume proportion. Elsabahy and Wooley argue that intermediate-sized nanoparticles ranging from 20 to 100 nm are the best possible size for applications involving in vivo studies because they have an extended circulation time [24]. The underlying reason for this lies in the fact that generally cancerous cells have a vascular structure that is <200 nm, and the blood vessel structure in mammals possesses a porosity of almost 5 nm or less than that. Another very crucial element that influences biodistribution as well as the removal of nanoparticles throughout the body is their quasistatic size. According to scientific studies, iron oxide nanomaterials having a size >100 nm were observed to be easily captured by the liver and spleen via macrophage phagocytosis, whereas nanomaterials with a size <10 nm were removed via the renal pathway [25]. Multiple studies have revealed that nanoparticles with a size of 5 nm or less can easily pass through the cellular barrier via translocation or some other phenomenon, whereas large-sized particles enter the cell via pinocytosis, phagocytosis, or some other specific mechanism [19]. Oxide-based nanostructures have been synthesized to possess a variety of shapes, such as nanorods, nanospheres, nanocubes, nanowires, nanotubes, and much more. Hematite-based nanorods have been observed to be absorbed relatively fast as well as to a greater extent as compared to nanospheres, as observed during an in vitro comparison of iron oxide nanorods and nanospheres used for the treatment of tumor cells in the case of human cancer cells [26]. Andrade et al. [27] postulated that the increased contact area within the cellular membrane and rod-shaped nanostructures provide an explanation for the increased cellular intake of rod-shaped nanoparticles in comparison with nanoparticles having a spherical shape. While tripod-structured iron oxide nanoparticles are analyzed as opposed to sphere-shaped iron oxide nanostructures, it was discovered that the tripod nanostructures were less hazardous and possessed lower cytotoxicity in the case of HeLa as well as Hepa 1–6 cells within concentration levels of 0.022–0.35 mg Fe/mL [28]. Other researchers argued that nanoflower-shaped nanoparticles had an improved level of internalization as compared to single-core nanostructures [29]. As a result, an increased value of the ratio of length to width (aspect ratio) is crucial to how the morphology affects nanoparticle cytotoxicity. It is also necessary to consider the various arrangements of shapes and sizes that have specific characteristics.

2.2 Surface area, surface energy, and crystal structure

The extent of homogeneity, morphology, and size of a nanoparticle have been found to be significantly hampered because of a severe drop in thermodynamic stability, which is caused by the higher value of surface area and energy of a nanostructure [30]. Owing to distortion of crystalline symmetry, which results in violation of electroneutrality among anions and cations and, consequently, surface modification, the surface of crystalline nanostructured materials contains oxygen atoms with reduced coordination numbers [31]. The proportion of atoms at the surface to the total number of atoms increases as the size of nanoparticles decreases. These numerous corners and edges have been identified as potential areas for functional surface reactions. Because of their diminutive size, countless different active sites, and distinct shapes, nanoparticles have an increased surface-to-volume ratio, which can be used to modulate their reactivity. When nanoparticles interact with cells, those with a higher surface-to-volume ratio interact with the cell more frequently than those with a lower surface-to-volume ratio [32]. One of the most common reasons why nanoparticles interact toxically with living tissues is the generation of metallic ion species [19]. Crystalline structure, phase, strain, crystal size, crystal defects, composition of the nanomaterial, and some other factors can affect how the nanostructure dissolves inside the living body. As nanoparticles have a higher proportion of atoms at their edges and vertices, the elevated free energy of these regions tends to make it simpler to isolate ions from their periphery [33]. As in the case of zinc oxide nanostructures, the disintegration kinetics for polar-ended nanoparticles in pure water were observed to have an increased rate than those of those possessing nonpolar surface morphology because of the variance in Gibbs free energy associated with various faces [34]. According to He et al. [35], polar surfaces have three times as many bulk oxygen vacancies as nonpolar surfaces.

2.3 Physical and surface properties

Due to van der Waals interactions, increased surface energy, and magnetic forces, nanoparticles exhibit a pronounced ability toward agglomeration, while MO nanostructures exhibit a close correlation at concentrations of 1000 ppm [36]. The biodistribution, metabolic, and pharmacological actions of nanooxides are greatly influenced by the degree of aggregation. For instance, Lousinian et al. [37] discovered an inverse relationship between aggregate size and surface area that led to increased fibrinogen adsorption due to the low aggregate size of ZnO nanostructures. Additionally, it was found that groups of really small MgO nanoparticles with a size <5 nm may lessen the effectiveness of their interactions with microorganisms [38]. As a result, the charge on the surface of the entire system at various pH levels is determined by the organic or inorganic molecules that stabilize the designed nanostructures in colloidal solution, such as carbonate, cysteine, or surfactants. MO nanoparticles not being properly stabilized lead to colloidal instability, which in turn causes the particles to aggregate, which leads to the living body taking more time to clear such toxic compounds, ultimately leading to more cytotoxicity [39]. Due to the electrostatic force of attraction between the nanoparticles, which are deficient in electrons, and the cell membrane, which has an excess of electrons, such nanomaterials are considered to pose a higher chance of cytotoxicity [40]. On the other hand, the nanoparticles possessing a higher positive charge have been observed to demonstrate an improved potential for opsonins, which involves the adsorption of plasma proteins along with their interactions with immune cells, serum proteins, and other substances and thus tends to alter the shape and structure of molecules that are adsorbed, resulting in the alteration of their behavior. Research has pointed toward the fact that in the case of breast cancer in humans, positively charged nanomaterials have shown higher rates of mineralization than the nanoparticles possessing a negative charge, although both were equally swallowed by umbilical vein endothelial cells, as per reports on the relationship involving surface charge and cellular absorption [41]. According to research by Xiao et al. [42], micellar nanoparticles possessing larger charges are substantially integrated by macrophages, but nanoparticles with a negative charge of lower magnitude demonstrate improved tumor absorption as well as limited macrophage removal. The value of pH level at which surface charge is equal to zero and the acidity constant values of MO nanostructures, which are strongly associated with particle aggregation, are some more characteristics that describe their unique surface features [31,32,34,35,38,39,43].

2.4 Photocatalytic activity, chemical composition, and type of target cell

One of the widely accepted theories for photocatalytic activity involves the production of positively charged holes entrapped in surface defects and negatively charged electrons confined in tiny hydrogen-rich regions. The difference between the charges reduces the possibility of recombination of an electron and hole pair and thus boosts photocatalysis, electrochemical reactions, as well as antimicrobial, activities [44]. Molecules of water being absorbed from the atmosphere or the solution that are being oxidized capture the holes and thus create strong reactive oxygen species (ROSs) known as $^{\bullet}OH$ radicals. Oxygen molecules that are reduced to create superoxide ($O_2^{-\bullet}$) radicals, which then react with H^+ to generate peroxide radicals ($^{\bullet}OOH$) or H_2O_2, could also be trapped by the electrons present in the conduction band [45]. Certain MO nanostructures, such as zinc oxide, have the ability to form superoxide, hydroxide radicals, and singlet oxygen, whereas others simply generate singlet or couple, either generate no ROS or may even prevent the generation of ROS when H_2O_2 is present, indicating that they have antioxidant ability [46,47]. The kind of MO nanostructure observed to affect the cytotoxicity of nanoparticles may be related to the nanoparticles' capacity to produce ROS and hence generate metallic ions, as was previously mentioned. For instance, nanostructures of approximately 20 nm size such as ZnO, SiO_2, TiO_2, and Al_2O_3 have been observed to show varying degrees of cytotoxicity to human fetal lung fibroblasts (HFL1) [48]. MTT assays revealed that zinc oxide nanoparticles seemed to be more hazardous to HFL1 after 48 h of treatment at doses of 0.25, 0.50, 0.75, 1.00, and 1.50 mg/mL, preceded by TiO_2, SiO_2, and Al_2O_3 in decreasing order. Titania, alumina, ceria, and zirconia exhibit low to mild cytotoxicity in the absence of a significant relationship between toxicity and either specific surface area or equivalent spherical diameter, whereas copper and zinc oxide nanoparticles emerged to be more hazardous to two different categories of human pulmonary cells [49]. Diverse metabolic activities and, thus, various cellular death processes and responsiveness to MO nanoparticle exposure are frequently displayed by various cellular target specificities. SiO_2 nanostructures were discovered to have significantly higher cytotoxicity on human monocytes (THP-1) as compared to epithelial cells present in the lungs ($L-132$) [50]. It is considered that the ability of monocytes, to engage in phagocytosis, which is not found in lung epithelial cells, may be responsible for the specificity in choosing cellular targets. A549 alveolar cells have been observed to show less responsiveness as compared to macrophage THP-1 cells while taking the dose- and time-dependent toxicity of various MO nanostructures on these two cell types, i.e., A549 and monocytes to macrophage (THP-1) cells, into consideration [49]. It was predicted that macrophage sensitivity to nanoparticles would increase due to their increased sensitivity to responses and improved ability to participate in particle aggregation via phagocytic processes.

3 Application of metal oxides for biophotonics

MO nanostructures must fulfill specific conditions so that they can be employed for a specific purpose. For instance, MO nanoparticles used for drug delivery must be biodegradable and easy to remove from the body so that there is no need for any further medical treatment. Such nanomaterials must also possess pharmacokinetics that meet the standards for combating a specific disease. Particularly TiO_2, ZnO, CuO, Fe_2O_3, and Fe_3O_4 have proved to be quite harmless for humans, despite the wide range of MO nanostructures that are accessible [51]. Metal oxide nanomaterials have been used for a variety of applications over the years (Table 1).

3.1 Internal radiation therapy

The ability to influence several intracellular signal transduction pathways that control cell division, differentiation, movement, and death is often inextricably connected to the deployment of therapies [46]. MO nanomaterials can enter the human body through breathing, eating, ingesting, epidermal perforations, transfusion, direct injection, or transfer through nanotubes to any part of the human body. The ability to develop a nanomaterial with improved delivery capability because of its ability to transport through the cellular membranes is the key benefit of using nanomaterials for therapeutic strategies [12]. Improving pharmacological properties, extended and targeted distribution of several therapeutic drugs, stability, and bioavailability are some advantages of creating nanostructured carriers for drug delivery [66]. Material engineering has allowed researchers to improve the selectivity of nanostructures, reducing adverse reactions. A specific nanomaterial's toxicity level doesn't quite result in a familiar sequence of alterations once it enters a living cell. The circulatory interconnections and typical pathways of MO nanostructure-induced toxicity are shown in Fig. 1 [67]. Targeted nanomaterials might lessen the overall amount of medication being used for curing purposes or detection, thus minimizing the negative side effects by concentrating on particular areas. Reducing unintended interactions with certain other substances, damaging impacts on healthy tissues, and increasing specificity for tumor tissue or some other target cells are some of the problems being faced while using nanoparticles for therapeutic applications. Both delivering and targeting nanostructures used in this sort of targeted therapy can be synthetic polymer-type substances such as silica, MOs, hydroxyapatite (HAP), and polyester blends, or natural polymers, including polysaccharides, deoxyribonucleic acid (DNA), ribonucleic acid (RNA), polypeptides, enzymes, and proteins. Such functionalized nanomaterials are biocompatible because of the presence of a coated shell that is typically hydrophilic [68]. The exterior surface property is crucial for biokinetics as well as the distribution of the nanostructure inside a living system because it influences the total size of the colloidal particles [69]. Its shell may include biologically active molecules such as

Table 1 Various metal oxide nanostructures along with the synthesis method employed, the average size of the synthesized nanoparticle, and their corresponding applications.

S. No.	Metal oxide	Synthesis method	Size (nm)	Applications	References
1.	CNT/TiO_2	Modified sol–gel	Inner diameter = 16–29 nm Outer diameter = 20–45 nm	Photocatalytic degradation of organic contaminants in water	[52]
2.	Epoxidized natural rubber/TiO_2 hybrid	Sol–gel	10 wt% TiO_2 = 10–20 nm 30 wt% TiO_2 = 25–40 nm	Used in microoptical and optoelectronic devices	[53]
3.	TiO_2 nanotube–reduced graphene oxide	UV-assisted reduction of GO and subsequent TiO_2 nanotubes attachment to rGO sheets	Average inner diameter = 63 ± 13 nm. Average outer diameter = 134 ± 22 nm	Detection and treatment of breast cancer	[54]
4.	Zinc oxide (ZnO)–polyphenol nanohybrids (NHs)	Hydrothermal	Average diameter of 5 nm	Enhanced antioxidant activity with less cytotoxicity	[55]
5.	Polyimide/titania hybrid nanocomposite films	Sol–gel	10–30 nm		[56]
6.	Fe_3O_4/TiO_2/CuO nanohybrid	Low-temperature sol–gel technique		Removal of chromium from aqueous solution	[57]
7.	ZnO/Ag nanohybrid	Synthesized by using chitosan as mediator during electrostatic interaction	Grain size = 13 nm Particle size = 150–250 nm	Antibacterial activity	[58]

8.	Graphene/titanium dioxide nanocomposites	Sono-chemical-assisted solvent graphene exfoliation and hydrothermal synthesis technique	6.8–13.1 nm	Detection of disease biomarkers at very low levels	[59]
9.	Folic acid templated Fe_2O_3	Hydrothermal Synthesis technique		Diagnosis and therapy of colorectal cancer	[60]
10.	Graphene oxide-Folic acid-ZnO	Chemical precipitation method	~20 nm	Targeted photodynamic therapy under visible light irradiation	[61]
11.	MnO_2–Al_2O_3 nanocomposite	Chemical coprecipitation synthesis technique	MnO_2–Al_2O_3 pore size = 3.1 nm	White light photocatalyst and bactericidal agent	[62]
12.	ZnO–Au nanohybrids	Rapid microwave-assisted synthesis technique	2–3 nm	Catalytic activity for CO oxidation	[63]
13.	ZnO-encapsulated Eudragit FS30D nanohybrids	Solvent evaporation synthesis technique	Average particle size = 291.63–305.89 nm	Biodegradable drug delivery systems	[64]
14.	Chitosan (CS)-encapsulated titanium dioxide (TiO_2) nanohybrid	Chemical precipitation synthesis technique		Photocatalytic property	[65]

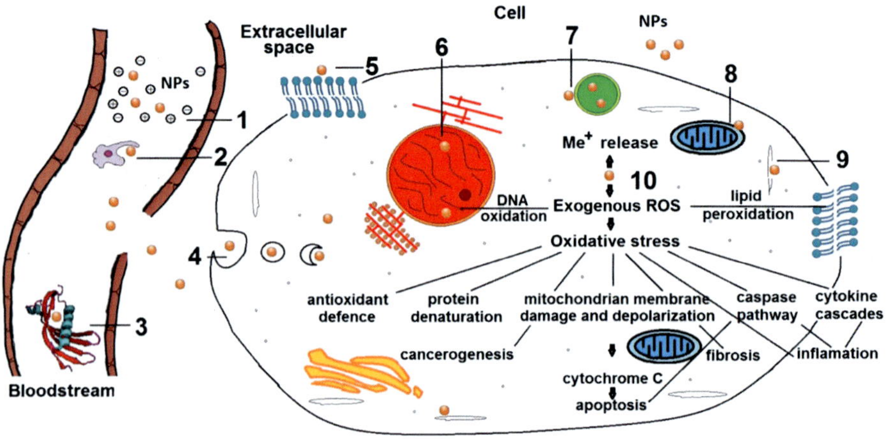

Fig. 1 The diagram represents the pathways by which MO nanostructures enter eukaryotic cells and hence cause cellular damage: (**1**) interaction of MO nanomaterials with ions present in the blood system; (**2**) phagocytic cell ingestion; (**3**) opsonization or enzymatic degradation; (**4**) internalization through the use of endocytosis after leaky vasculature and hence passage into the extravascular environment; or (**5**) membrane perforation and damage to its constituents and their purpose; (**6**) chromosomal rearrangements and modifications in cell; (**7**) lysosome rupture; (**8**) mitochondria damage; (**9**) lower growth rate, structural changes, and shorten lifetimes of microtubules of the cytoskeleton; (**10**) generation of ROS, oxidative stress, and subsequent processes [67].

organic acids, chitosan, gelatin, or liposome coatings, in addition to polymer-based substances such as polyethylene glycol (PRG), polyvinyl alcohol, and others, along with surfactants such as sodium dodecyl sulfate (SDS) and sodium oleate, which may guide the nanostructures or improve their action within a living thing. Theranostic treatment refers to an entire integrated platform, including both living and nonliving origins that diagnose, inhabit, or treat a particular ailment. Nanomaterials ought to possess a large load-bearing capacity in such a situation. To functionalize MO nanostructures, such as HAP, or to initiate an electrostatic force of attraction or repulsion with nanostructures possessing charge, drugs could well be bound using covalent bonds [70]. Antitumor, immunosuppressive, anticonvulsant, antiinflammatory, antimicrobial, antibacterial, antiviral, and other medications may also be associated with nanomaterials. By lowering the required dose and preventing the typical cellular cytotoxic activity, the MO nanostructure's capacity to localize the medication toward the target tissue could significantly improve the effectiveness of such medications. In addition to helping to keep MO nanostructures stable, using a nanohybrid material layer on the outer side of a nanomaterial may alter how long they stay in the body, adjust their inherent cytotoxicity due to the generation of ROS, and also modify them to be tissue specific [71]. When the drug delivery systems have been absorbed into the target tissue, the complete nanomaterial must exhibit strong penetration as well as bioaccumulation, and at the same time, it must be able to evade endosomes.

Fe_2O_3 nanomaterials possess magnetic properties that can be used for targeted drug administration, diagnosis, and separation of biological material and cellular components via the use of their magnetic properties [71]. Biomaterials designated as Fe_2O_3 nanostructures with magnetic properties can be focused and guided by an externally applied magnetic field, and they can be eliminated after the treatment is finished with the assistance of this magnetic property [72]. These iron oxide nanoparticles are often vulnerable to opsonins inside the blood, allowing subsequent detection and removal [73]. They are recognized by macrophages because of their significant vascular properties and porosity, which allow the reticular endothelium to absorb them. Such nanoparticles, on the other hand, are well known for their ability to eradicate tumor tissues while sparing normal tissue [74]. Magnetic nanoparticles mediate the focusing of thermal energy on tumors, particularly through changing magnetic fields, mechanically based fractional heating, or hysteresis heating, because malignant tissues are more susceptible to degradation through thermal energy in vivo via increased body temperature, a treatment known as hyperthermia [75]. Hilger et al. [76] implanted mammary gland tumor cells into an immune-deficient mouse subject and recorded that the temperature inside the tumor site elevated to 73°C when a magnetic field with a strength of 400 kHz was applied to it. Hurbankova et al. [77] examined the effects of Fe_2O_3 nanostructures on the vasculature structure of the respiratory system as well as the acute inflammation and cytotoxicity characteristics of bronchoalveolar lavage based on the broad hypothesis that nanoparticles can penetrate via numerous membranes. According to research, 4 weeks after implantation, Fe_3O_4 nanoparticles can be removed from the respiratory system by the body's immune system. Every cell in the human body requires iron to maintain equilibrium. Because raw Fe_2O_3 nanoparticles are colloidally unstable, various modifications improve their stability while also preventing the nanoparticles from opsonizing when they circulate in the bloodstream. Chitosan-coated Fe_2O_3 nanostructures were created with the help of biopolymers, for instance, polysaccharide chitosan, which exhibits hemostasis, antimicrobial activity, and hypoallergenic characteristics. These materials have been used to create pharmaceutical delivery systems to treat medical conditions characterized by low bone density, for instance, osteoporosis, or to prevent prosthetic devices from sagging [78]. Overall, the bulk of altered nanomaterials, according to the investigators, were detected in extracellular environments, whereas the untreated iron oxide nanostructures have mostly been found in intercellular environments. Although the iron oxide nanomaterials that were coupled to chitosan boosted osteoblast provability, differentiation, and rapid growth when the dose was over 300 g mL^{-1}, in addition to this, it also caused death in the cells. However, because chitosan-coated nanoparticles did not internalize in a concentration-dependent fashion, there could be a lower dosage constraint in practical use. According to additional hemolytic as well as cellular survival investigations, whenever iron oxide nanomaterials were coupled with BSA, enhanced specific absorption was seen because of the increased colloidal stability and decreased nanoparticle aggregation [79]. A majority of Fe_2O_3 nanoparticles targeting hyaluronic acid (HA) medication moieties were created primarily for macromolecular compounds for the application of tumor chemotherapy [80]. Besides having magnetic characteristics, magnetic nanotube structures also enable differentiating functionalization of the exterior and interior interfaces, which could be advantageous for targeted drug delivery with the help of an applied magnetic field. Folic acid was added to iron oxide nanorods having porosity by Yu et al. [81] in order to specifically administer a low-water-soluble chemotherapeutic medication called doxorubicin (DOX). They discovered that this increased both the drug's cytotoxicity as well as the HeLa cells' ability to absorb it. Because of the precise interaction of folic acid and folate receptors, such nanocomposites are suitable for targeted medication administration.

ZnO nanomaterials have also been widely used for radiation therapy and targeted drug delivery due to their enhanced drug loading capacity, variation in structure, and functionality. ZnO NPs have been observed to demonstrate preferential toxicity against tumor cells in vitro as well as in vivo when healthy tissue is taken as a reference [82, 83]. ZnO has thus been employed in treating cancer because of its preferential cytotoxicity toward tumor tissue, which results from a greater amount of released ions in the presence of media with a pH <7 and enhanced ROS generation. Moghaddam et al. [84] reported the death of cancer cells caused by internal and external apoptotic mechanisms in the case of the MCF-7 tumor cell line when investigating the antitumor efficiency of ZnO nanostructures. According to the investigators, ZnO nanoparticle-induced apoptosis appeared to be dependent on the mitochondria, as evidenced by the arrest of the cell cycle, a fivefold elevation in stress-responsive kinase activity, as well as depolarization of the mitochondrial potential of its membrane. By the use of commercially available ZnO nanoparticles and specially constructed ZnO nanorods, intense cytotoxic effects on human keratinocytes (HeCat) and epithelial cells (HeLa) were caused, thus triggering actin filament aggregation as well as structural reforms in microtubules that transform them into stiff tubulin microtubules [85]. Since ZnO nanostructures generate an initial cytoskeletal breakdown that results in necrosis and subsequent ROS-dependent apoptotic activities, they have a significant negative impact on cell proliferation and survival. Due to their inherent biodegradable properties, ZnO nanostructures have proved to be strong contenders for delivering drugs for the treatment of cancer [85]. It has been discovered that DOX-ZnO nanostructures function as effective vehicles for drug delivery against hepatocellular carcinoma tumors, thus improving treatment effectiveness by raising the quantity of DOX inside the intercellular environment [83]. This nanocomposite was also observed to show outstanding performance as a photodynamic therapeutic agent. By decreasing xenograft tumors within naked mice, hollowed ZnO spheres functionalized using folic acid while at the same time loaded with paclitaxel were observed to efficiently cause cytotoxic action against mammary gland cancer cells both in vitro and in vivo [86,87]. Like iron oxide nanoparticles, Dhivya et al. [88] created a copolymer by incorporating ZnO nanoparticles inside hydrophobic PMMA-AA that was heavily populated by the hydrophobic medication curcumin in order to increase the medication's solubility, bioavailability, and overall effectiveness. According to the findings, pH 5.4 released curcumin in a greater proportion than pH 6 or pH 7.2. Given that cancerous cells are capable of sustaining a pH of roughly 6, this finding was considered of critical importance. In addition, under similar laboratory conditions, the Cur/PMMA-AA/ZnO nanostructures demonstrated excellent cytotoxicity against AGS malignant tissue in comparison with certain other bionanocomposite substances. Zinc plays an important role in the production, preservation, and secretion of insulin [89]. Hence, when stabilized or bonded with specific chemicals, ZnO nanoparticles have been seen to display the capability of being an effective antidiabetic agent or may render diabetes complications less severe. In the year 2018, Hussein et al. [90] developed conjugated ZnO nanomaterials with sizes ranging from 20 to 27 nm using hydroxyethyl cellulose (HES), which is a polymer that is inherently biologically degradable and biocompatible. Male albino mice were treated with conjugated ZnO nanoparticles to minimize their diabetes effects. These findings showed that in the case of rat subjects suffering from diabetes, levels of CPR, proinflammatory IL-1, and ADMA rise significantly concurrent with a decrease in NO (an indicator molecule for regulation of angiogenesis, protein kinase G operation, protein phosphorylation, and other procedures linked with cellular proliferation, mobility, and vasodilatation), while the addition of ZnO nanoparticles suggests attenuation of such aspects, making ZnO nanocrystals extremely promising for improved specificity.

Given its capability to stimulate cellular attachment, osseointegration [91], and cellular migration, including tissue regeneration, TiO_2 is a commonly utilized substance for biological uses employed mostly in the skeleton and biomedical engineering [92]. In vivo tests using wavelength 365 nm illumination on unaltered TiO_2 revealed tumor development reduction in glioma-bearing rats as well as increased mouse survival [93]. At a dosage of 0.5 mg/mL, nitrogen-doped anatase nanoparticles caused about 93% apoptosis of tumor tissues during UV irradiation and also showed better absorption of visible light in comparison with unaltered TiO_2 [94]. The researchers additionally discovered that the susceptibility of tumor tissue for modifying nanoparticles may differ substantially depending on the specific kind of malignancy. Nevertheless, since UV radiation does not penetrate tissues well enough and might be hazardous, there are only a handful of applications for photodynamic therapy in the field of tissue overhealing [95]. TiO_2 nanostructures have been widely used to improve the overall specific surface area and, as a result, the loading capacity of some antiinflammatory medications such as ibuprofen on titania nanowires with a 100 ± 10 nm diameter [96]. Chitosan and poly (lactic-*co*-glycolic acid) (PLGA) coats having a thickness of 2.5 nm and synthesized using nanowires have been used to examine the overall releasing ability of a gentamicin-loaded oxide layer made of TiO_2 [97]. This engineered nano-oxide exhibited prolonged release up to 22 and 26 days for chitosan and PLGA, respectively, as compared to noncoated nanoparticles that deliver the medication after a duration of 2 weeks. Masoudi et al. [98] created TiO_2 NPs having significant DOX loading efficiency along with inherent luminous features in 2018. Osteosarcoma (SaOs-2) and breast cancer (MCF-7) cellular lines during in vitro cytotoxicity tests in humans showed increased antitumor activity and effectively improved imaging for intercellular monitoring of DOX-loaded nanostructures as compared to free DOX.

3.2 Metal oxide-based immunotherapy

Nanomaterials could boost or restrict the activity of the defense system of the body by interacting with various parts of this system [99]. Nanoparticles may be specially created to either enhance or reduce the resistance of the human body, as is the case when using a vaccination. The type of blood protein absorption and its volume determine the overall outcome of a nanostructure by interacting with some other molecule; this step initiates the detection of any foreign object with the assistance of nanomaterials. Nanoparticles possess the ability to start sending signals that result in provoking an immune response or even cytotoxicity when they attach to the cell membrane. Inflammatory responses of the body's defense system could generate cytokines following activation, which serve as facilitators of both regional and general inflammatory and hypersensitive responses [24]. For example, ROS generated by TiO_2 nanoparticles could activate downstream proinflammatory responses and thus suppress the innate immune system of macrophages because that can operate as a secondary messenger as well as an immune modulator [47]. Such findings imply that ROS-activated mechanisms can be utilized by MO nanomaterials possessing surface sensitivities to modify the body's immune system. Enzymes, amino acids, and some other charged compounds could be attached to or encapsulated within nanostructures to inhibit opsonin adsorption, the immune response, and the subsequent detection of foreign objects. Additionally, with the help of such compositions, cytotoxicity can be decreased by carefully designing their multiple aspects [100].

Fe_2O_3 nanostructures have also been employed as effective vaccine delivery agents, exhibiting enhanced therapeutic efficacy. The integration of small interfering RNA (siRNA) onto nanostructured carriers was one strategy utilized to successfully suppress virus reproduction [101]. The utilization of a *trans*-activator or transcriptional protein as an auxiliary to a targeted cellular unit for intercellular distribution of MO nanomaterials and the utilization of such nanomaterials to prevent virus-based glycoproteins that are crucial for attaching to the host cell and thereby preventing infection are some other important techniques worth mentioning [102]. Owing to such disparate features, iron oxide nanoparticles have been employed as covert transporters to administer antiretroviral medications to a latent variant of the AIDS virus while documenting the localization of pharmaceuticals [103]. Enfuvirtide, a relatively complicated antihuman immunodeficiency virus (HIV) medication that cannot pass the blood–brain barrier (BBB), has been successful in doing so after it was mounted on PMA amphiphilic polymer-coated Fe_2O_3 NPs [104]. Nanomaterials have now been found to trigger antigen-presenting cells and stimulate the production of cytokines and chemokines, among other aspects of the immunological response [105]. Iron oxide nanoparticles with superparamagnetic properties may encourage the proinflammatory profile in macrophages whenever they are absorbed by them [106]. In order to promote a cancer-specific response by T cells in glioma-bearing mice, Shevtsov et al. [107] developed a vaccine using superparamagnetic iron oxide nanoparticles encapsulated within a recombinant heat shock protein 70 (Hsp 70) antigenic peptide. Antigen transmission to antigen-presenting cells (APC) was observed to be easily achieved, which resulted in slower tumor growth and higher survival rates. Internalized Fe_2O_3 nanomaterials encapsulated within dimercaptosuccinic acid, aminopropyl silane, or aminodextran demonstrated no cytotoxicity effect in M2-polarized macrophages, which have been known to be primarily engaged in tissue regeneration, addressing inflammatory response, angiogenesis, and tissue remodeling [108].

MO nanostructures have been employed to improve the overall, longer-lasting immune reaction from the low immunostimulatory, however safer, constituent vaccination compared to live attenuated vaccines. Researchers have proven that $MnFe_2O_4$ nanoparticles coated with citrate possess a protein corona from fusion protein (CMX) consisting of antigens for *Mycobacterium tuberculosis* that have the ability to facilitate the overall development of the particular immune cell response (T-helper 1, T-helper 17, and TCD8) [109]. Iron oxide nanoparticles have also been employed to create virus-like particulates (VLPs), which have demonstrated potentially extraordinary value as vaccinations, gene-carrying nanocarriers, contrast agents for magnetic resonance imaging, as well as delivery platforms for different drugs. These VLPs were created in addition to bacterial antigens [110]. Throughout this method, bacterial viruses have been employed as a platform in order to carry an increased amount of Fe_2O_3 nanostructures that could strongly attach onto the tumor cell interface in comparison with Fe_2O_3 nanomaterials themselves, in addition to serving as bacteriophages linked with functionalized MO NPs functionalized with carboxyl and amino groups, respectively, to ensure enhanced loading capacity for antimicro applications [111,112]. Furthermore, bi-specific aggregates could be generated by antibody and antibody like conjugates of nanoparticles, employing several methods of action. These methods include the transportation of poisons or chemicals targeting cancerous tissues, the suppression of twin ligands or receptors, and the crossover of two receptors [113]. Nonetheless, there are still a lot of unknowns regarding the immunological reaction caused by MO NPs.

Investigations have revealed that ZnO NPs cause the generation of cytokines and chemokines, which are employed as indicators for immunotoxicity. These substances have been found to possess axillary benefits via a variety of methods, such as intrinsic immunological enhancement, increased antigen acquisition via antigen-presenting lymphocytes, and cytokine network modulation [73]. For example, exposing dendritic cells to ZnO nanostructures with negative zeta potential increased the effect of CD80 and CD86, inducing the secretion of IL-6 and TNF-α without having any negative effects [114]. Moreover, it has been shown that ZnO nanostructures stimulate NF-B activation by causing I_kB_α, an NF_kB blocker, to degrade [115]. Once administered intravaginally, specially created ZnO tetrapod nanostructures showed that viruses are able to effectively inhibit genital infections in female BALB/c rats [116]. HSV-2 culture using ZnO nanostructures previously induced a localized immunological response similar to infection but with reduced inflammatory and clinical symptoms. The researchers attribute this decreased infectiousness to the fact that it is easy to transfer antigen for viruses into APCs intracellularly and thereby activate antigen-specific host defenses. It has also been discovered that greenly synthesized nanomaterials are less harmful to living tissue as compared to nanoparticles produced in a lab via any chemical method, apparently due to phytochemical encapsulation that could reduce the generation of ROS [117] and hence the inflammatory responses. Thatoi et al. [117] investigated the ability of different nanoparticles to prevent protein denaturation by preventing albumin denaturation caused by high temperatures in order to assess the antiinflammatory effect of bioreduced zinc oxide nanoparticles synthesized using Heritiera fomes and Sonneratia apetala. ZnO nanomaterials produced by *S. apetala* were observed to possess the highest level of inhibitory potential ($63.26\,\mu g\,mL^{-1}$; IC_{50}), followed by ZnO nanostructures produced by H. fomes ($72.35\,\mu g\,mL^{-1}$ IC_{50}), whose results were comparable to those of the common drug diclofenac ($61.37\,\mu g\,mL^{-1}$ IC_{50}). ZnO nanowire-coated poly-L-lactic acid (PLLA) microfibers have been shown to greatly increase inflammatory cytokines such as IL-10, IL-6, IL-1β, and TNF-α after exposing a tumor antigen inside a dendritic cell [118]. Comparatively to animals inoculated with PLLA fibers directly coupled to tumor antigen, the nanocarriers incorporating ZnO boosted the overall penetration of T cells into the malignant cells. Research conducted by Hu et al. [119] revealed improved anticancer treatment efficacy in in vivo animal subjects receiving injections of 4T1 cancerous cells, particularly when ZnO nanoparticles and DOX were administered together. The fact that ZnO nanostructures enhanced Atg5-regulated autophagy flux indicates that ZnO's ability to induce autophagy strengthens the death of tumor cells by increasing Zn ion production and ROS formation if paired with DOX. It could be argued that, as a result of the unique properties of these nanocarrier structures, MO nanomaterials are used to boost the efficacy of immunotherapeutic applications. Various adjustments, such as variations of therapeutic agents, biologically active compounds, or medicines, can alter the targeting capacity of nanoparticles. Such nanoparticle frameworks are the subject of extensive study that is currently going on, even today. Immunotherapy that is used in conjunction with certain other therapeutic modalities, including chemotherapeutic treatments and photothermal therapy, has also attracted significant interest from the research community. Problems persist, nevertheless, because little is known about how nanoparticles that could function as both immune system-stimulating adjuvants and/or delivery agents to stimulate antigenic processing behave in vivo.

3.3 Metal oxide-based diagnosis

MO nanomaterials have been extensively used for diagnostic applications because of their biophotonic characteristics attributed to their fluorescent properties as well as their magnetic behavior. One of the very important applications of the use of nanostructured metal oxides is that disease survival and advancement can be accurately examined and quantitatively measured by their use. Strongly luminescent nanomaterials could be used to mark a variety of biomedical targets, including tumors, stem cells, microbes, or specific biomolecules. Quantum dots (QDs) are colloidal, semiconducting nanostructures with a diameter ranging from 2 to 10 nm and the property of being extremely luminescent and highly stable [120], thus making them an effective tool for in vivo bioimaging of cellular components and processes. It is possible to track cellular proliferation in this way, keep the cells at their designated sites, or assess the vitality of the cells. When activated, QDs emit photons with improved quantum yield and bright, uniform electromagnetic radiation emission. Such primary qualities are controlled electron or ion exchange action, chemical and photostability, as well as catalytic abilities. Furthermore, whenever such QDs are injected into a host, safety is of the utmost importance. Because of this, Wierzbinski et al. [121] examined Fe_2O_3 nanoparticles and employed them as labeling materials for in vitro studies of human skeletal myoblast cells and in experimentally controlled rats following subcutaneous and intracardial injection. As compared to unlabeled cell lines, research has revealed that labeling myoblasts utilizing dimercaptosuccinate (DMSA) encased within iron oxide nanoparticles had no influence on their differentiation ability. It was observed that treated and untreated cells exhibited comparable levels of FTL gene expression, which is an indicator of aging and codes for the transferrin receptor's light chain, which transports exogenous iron ions into cytosolic suspension. The homeostasis of stem cells was observed to remain unaffected by DMSA-coated nanomaterials, despite certain gene expressions being affected. Other researchers enhanced ZnO nanoparticle cell labeling functionalization by covering the particles with silica that was attached to a biotin (vitamin) amino group [122]. A green fluorescence emitted by nerve fibers was seen while they were subjected to biotin-conjugated ZnO nanoparticles and quantum dots having a diameter of 125 nm; however, no fluorescence was seen when the nanoparticles were encapsulated within silica without the presence of biotin. Luminescent dyes were also employed as nanoparticle indicators for in vivo ophthalmic monitoring [123]. It is essential to conduct more studies to investigate the movement, interconnections, cytotoxicity, and destiny of each QD in specific tissues, organs, and species. According to a 2019 study on $Fe_5C_2@Fe_2O_4$ nanoparticles with magnetic properties for cancer therapy conducted by Yu et al. [124], such nanomaterials were able to produce an imaging-traceable impact that allowed for both diagnosis and curing. This was found to be in contrast to other frequently used trackable therapeutic agents, where the image-generating elements and therapeutic components are typically distinctive. The outcomes demonstrated the ability of $Fe_5C_2@Fe_3O_4$ nanoparticles for image-guided therapies that are spontaneous, tumor specific, and at the same time based on ROS production with swappable MRI without exogenous variables (Fig. 2 [124]).

Because of its exceptional soft-tissue contrasting property, superior spatial resolution, 3D anatomic data, and nonionizing irradiation, MRI is now a very frequently utilized noninvasive imaging procedure. With the aid of MRI techniques, diverse molecular alterations associated with the onset, as well as progression of pathologic conditions may be quantified, allowing for the rapid diagnosis and treatment of carcinoma and other diseases [125]. In most cases, Fe_2O_3 nanoparticles or ferrites constitute the basis for the contrast agents employed during MRI. One of the most common stabilizing compounds for Fe_2O_3 nanoparticles is polymeric materials, but collectively, these two ingredients create negatively acting imaging components that reduce signal strength, along with the lingering effects of circulating nanoparticles that cause MRI to be delayed by 24–72 h. In contrast to other tissues, solid cancerous cells show that constrained lymphatic outflow facilitates nanoparticle entrapment. Alloy-based nanostructures that produce substances are classified as ferrites, such as Zn-ferrite, which improve the overall magnetization of nanoparticles and have also been employed to improve MRI diagnoses and at the same time boost signal responsiveness [126,127]. Moreover, because of the monodispersity of the particles,

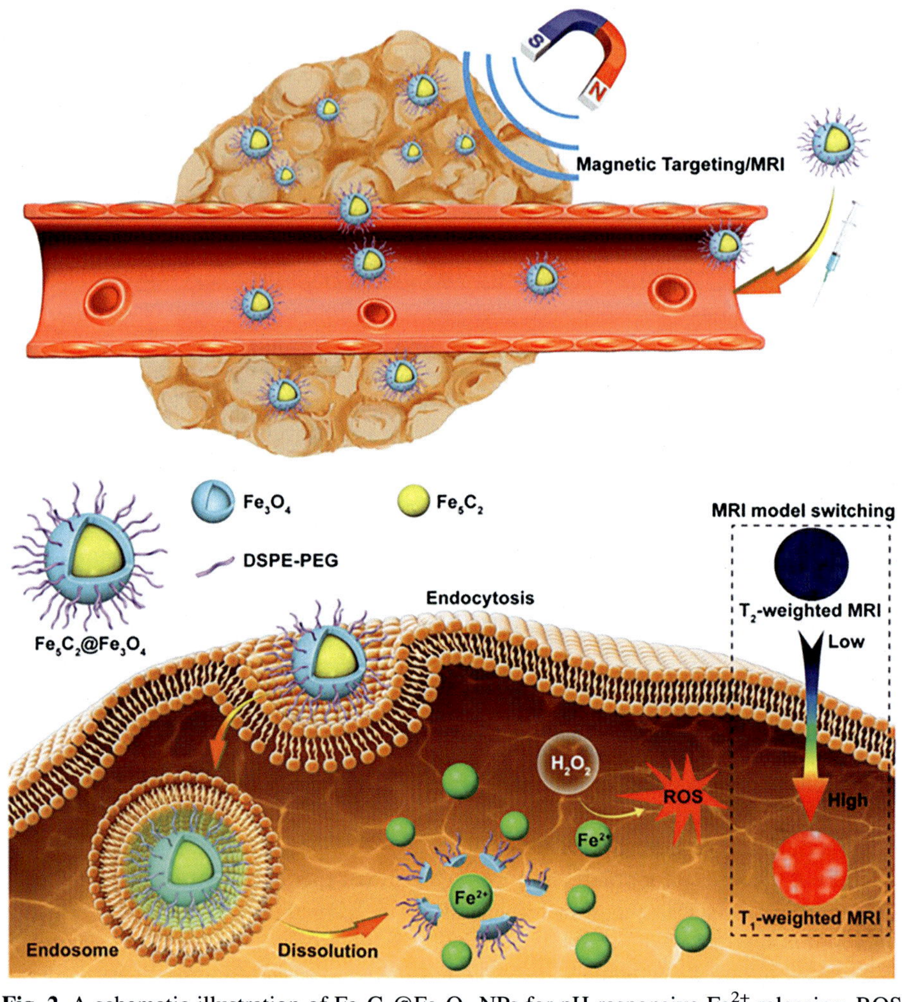

Fig. 2 A schematic illustration of $Fe_5C_2@Fe_3O_4$ NPs for pH-responsive Fe^{2+} releasing, ROS generation, and T2/T1 signal conversion.
Reprinted with permission from J. Yu, F. Zhao, W. Gao, X. Yang, Y. Ju, L. Zhao, W. Guo, J. Xie, X.J. Liang, X. Tao, J. Li, Y. Ying, W. Li, J. Zheng, L. Qiao, S. Xiong, X. Mou, S. Che, Y. Hou, Magnetic reactive oxygen species Nanoreactor for switchable magnetic resonance imaging guided Cancer therapy based on pH-sensitive Fe5C2@Fe3O4 nanoparticles, ACS Nano 13 (2019) 10002–10014. https://doi.org/10.1021/ACSNANO.9B01740/SUPPL_FILE/NN9B01740_SI_001.PDF.

liposomes possessing magnetic properties and a solitary Fe_2O_3 nanoparticle, vesicles with water-dispersible iron oxides, or iron oxides encased in PEG [128] might enhance imaging contrast. Such diagnostic data can be improved by combining Fe_2O_3 nanomaterials with certain other metals or by developing core-shell-structured iron oxide nanoparticle structures.

3.4 Use of metal oxides for dentistry

One of the main objectives of endodontics is to eradicate the infectious diseases caused by bacteria inside the root canal area for the purpose of stopping microbial infection from obstructing periapical regeneration. In order to assess the permanence, as well as the sealing abilities of ZnO nanoparticles as an antimicrobial root canal sealing agent, Javidi et al. [129] performed microleakage assessments at the interval durations of 3, 45, and 90 days. The microleakage increased as the annealing temperature increased from 500°C to 700°C, owing to the increase in nanoparticle diameter and subsequent reduction in functional surface area. They reported that minimal microleakage was visible throughout all developed groups of ZnO nanoparticles as compared to routinely utilized ZOE sealing agents. Regarding polycarboxylate-based dental cementing agents employed in dental applications as a luting agent, orthodontics, cavity liners along with bases, and tooth replacement [130], a different team of investigators generated significantly porous and spongier ZnO/MgO nanoparticles. The overall mechanical stability of zinc oxide and magnesium oxide nanostructures, as well as ZnO/MgO composite type material, was found to be greater than that of standard zinc polycarboxylate type material, and the drying time for the fabrication of such dental concrete was comparable to that of commercially available samples. The histopathology, genotoxicity, and cytotoxicity of antimicrobial nanostructured films generated electrochemically on TiO_2 nanoparticles on the exterior of NiTi arches have been assessed in vivo in Long-Evans rats [131]. After being deposited, this coated surface was found to be vulnerable to breakdown in the presence of a phosphate buffer environment once ingests were digested and titanium dioxide nanoparticles reached the blood circulatory system, causing acute toxicity on the organs, with the liver parenchyma becoming the most affected among all organs, resulting in severe ulcer-type lesions. After being internalized, titanium dioxide nanoparticles aggregated inside lysosomes, where they ruptured and released their contents, activating caspase pathways that cause apoptosis. Cierech et al. [132] combined polymethyl methacrylate (PMMA) with ZnO nanoparticles to generate a biological material intended to be used for dentures that could prevent the binding of microbes to its interface and hence inhibit the proliferation of denture stomatitis. This decreased microbial development upon the denture base was attributed to its improved hydrophilic nature, toughness, and overall absorbability. The antifungal capabilities increased as the amount of ZnO nanoparticles inside the polymeric material increased. Likewise, during UV illumination, a thin coating of nanomaterial-based TiO_2 film on the surface of a Co-Cr alloy demonstrated considerable antifungal properties that may be used in the near future for the treatment of denture stomatitis. Considering the fact that nanotechnology possesses the ability to save both money and time while simultaneously shielding the patient from undesirable side effects, it provides increased efficiency over currently existing methods.

3.5 Metal oxide-based biosensors

Nanomaterial-based biosensors, whose general scheme is shown in Fig. 3 [133], function by attaching a ligand to a binding site called a receptor, which allows a signaling transducer to respond. Biosensors are classified on the basis of their signal

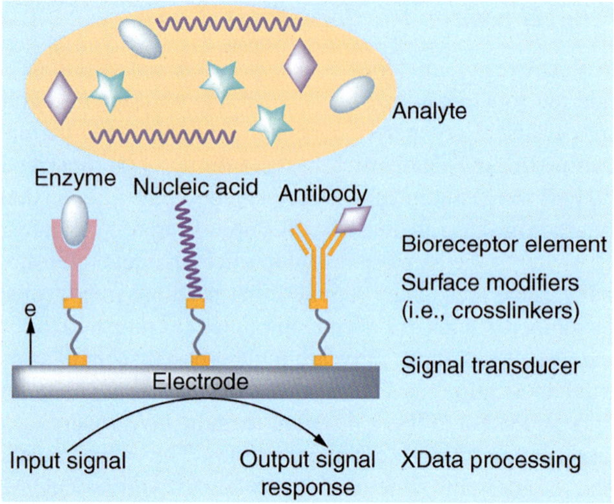

Fig. 3 General schematic of a biosensors [133].

measurement and detection mechanisms into piezoelectric, electrochemical, semiconductor, optical, and calorimetric categories [134]. Each one of these nanosized biosensors actually converts information into an electrical signal that can be detected externally. As a result, the material of the electrode used in the detection process is critical for the purpose of developing high-performance electrochemical sensors capable of identifying specific substances using various cutting-edge analytical techniques [135]. Such biosensors, which are used for identifying small molecules, can sometimes be enzyme-based, gene-based immunosensors, cytosensors, or biosensors. Hydrogen peroxide, glucose, or dopamine could all be detected using electrochemical analysis of small molecules. α-Fe_2O_3 nanoparticles with cubical morphology, for example, have been used as glucose sensing agents for nonenzyme catalytic oxidation, exhibiting excellent detection capability as well as rapid reaction. They were reported to be generated from water-repelling, iron-containing fluids in the presence of a hydrothermal environment [136]. Due to their great adsorption capacity, improved kinetics of electron transfer, and enhanced biosensing properties, MO nanomaterials are being extensively employed to immobilize enzymes [137]. A very thin layer of immobilized enzyme covers the surface of the working electrode in the case of an enzyme-based sensor. As an example, we can consider how the shape of the nanomaterials employed affects how the enzyme lactase is immobilized, thus enabling the detection of dopamine [138]. When three different types of phytic acid-enhanced silica nanostructures with various shapes, such as spherical, rod-shaped, and helical type morphology, were compared for effectiveness in the identification of dopamine, the maximum electrochemical activity was determined for nanoparticles with helical morphology. Lactate dehydrogenase-immobilized TiO_2 nanoparticles on silica-based sol–gel enhanced gold-based electrodes were used as lactic acid biosensors, with a limit of detection of $0.4\,\mu mol\,L^{-1}$ and a low Michaelis–Menten constant ($K_m^{app} = 2.2\,\mu mol\,L^{-1}$) [139]. Zinc oxide nanorod structures functionalized using the glucose oxidase enzyme have been employed as miniature nanobiosensors for intracellular detection of glucose levels due to their substantial surface area and biocompatibility [140]. Such nanosized electrodes provide a surface that is responsive to electrical stimuli for immobilizing enzymes that are biocompatible and have improved biological functions [141]. Their active surface, which enables protein binding, grows if the nanomaterials are of porous structure, and this creates a safe environment for such enzymes to remain stable and also preserve their activity [142].

One of the basic differences between immune sensors and genosensors is the former work on the basis of unique antigen–antibody identification, whereas in the case of the latter, singularly strained fragments of DNA are fixed on the exterior of the electrode. Owing to their excellent specificity, mobility, and cost-effectiveness, electrochemical biosensors are frequently used in contemporary DNA identification methods [143]. Utilizing nanosized materials enables the synthesis of microfluidic-based genosensing systems possessing higher responsiveness while also lowering the overall size, reagent, and specimen utilization. For instance, the combination of carbon nanotubes with CeO_2 nanoshuttles was observed to demonstrate a significant synergic activity that improved DNA immobilization and, consequently, increased the overall efficiency of the process of DNA detection through the use of hybridization [144]. Regarding electrochemical impedance detection of DNA immobilization along with hybridization, another study coupled the substantial DNA adsorption properties of Fe_2O_3 microspheres with the outstanding conductivity of self-doped polyaniline nanofibers (SPAN) over a carbon ionic liquid electrode. This suggested method positively distinguished overall target DNA from mismatched genomes and provided a broad range of detection along with a low detection limit ($2.1 \times 10^{-14}\,\mathrm{mol\,L^{-1}}$) [145]. Immunosensors typically pair electrochemical transducers with antibodies or antigens acting as biorecognition components [137]. Because the number of immobilized antibodies and/or antigens determines the overall sensitivity and selectivity of a biosensor, MO nanoparticles are a good choice for enhancing the sensor's effectiveness. ZrO_2 nanoparticles over reduced graphene oxide were complexed with 3-aminopropyl triethoxy saline (APTES), electrophoretically coated on ITO-coated glass, and afterward biofunctionalized with the proteinaceous biomarker CYFRA-21-1, which is generated in enhanced amounts in the case of patients suffering from oral cancer [146]. Such an immunosensor system demonstrated a larger uniform range of detection ($2\text{–}22\,\mathrm{ng\,mL^{-1}}$), outstanding selectivity ($0.756\,\mathrm{\mu A\,mL\,ng^{-1}}$), and a fairly low limit of detection ($0.122\,\mathrm{ng\,mL^{-1}}$). According to research, such immune-electrodes have an 8-week life span. The fungus-based toxin ochratoxin-A was detectable after it was immobilized on nanosized CeO_2 films over ITO-coated glass using rabbit immunoglobulin antibodies and bovine serum [147]. Such an immune-electrode displays improved properties within the specified range ($0.5\text{–}6\,\mathrm{ng\,dL^{-1}}$), minimal detection limit ($0.25\,\mathrm{ng\,dL^{-1}}$), rapid response (30 s), better responsivity ($1.27\,\mathrm{\mu A\,ng^{-1}dL^{-1}sm^{-2}}$), and an increased value of the association constant ($K_a = 0.9 \times 10^{11}\,\mathrm{Lmol^{-1}}$). During the research, it was observed that surface plasmon resonance (SPR) improved the detection of cancer markers by using gold/ZnO nanocomposite-based films [148]. The detection limit was lowered to $0.025\,\mathrm{U\,mL^{-1}}$, along with the uniform response range for carbohydrate antigen, which was observed to increase from 1 to 40 U mL^{-1}. Compared to Au/Cr layers, the overall charged intensity of the Au/ZnO sensor was roughly two times higher; however, the limit of detection was four times lower. Antibodies, receptors, glycans, and some other substances that are overexpressed over the membrane of the target cell are identified with the help of cytosensors. For instance, it was shown that TiO_2 nanorods with their surfaces modified by the use of monoclonal antibodies and immobilized on gold electrodes via mask-welded joints have been observed to be sensitive, precise, and quick in determining the presence of the bacteria *Listeria monocytogenes* at concentration levels as low as 102 cfu mL^{-1} for a time period of 1 h in the absence of interference from a number of other pathogenic organisms present in food [149]. The strengths of light and electrochemical-based monitoring have been found to be present in photoelectrochemical experiments [150]. Porphyrin-conjugated TiO_2 nanoparticles were used to identify biomarkers such as glutathione at comparatively low potentials. Similarly, porphyrin-conjugated ZnO nanoparticles have been developed to identify cysteine in physiological fluids, exhibiting broad linearity inside the range of 0.6–157 mmol L^{-1}. Nanobiosensors have become increasingly sensitive as a result of extensive research, allowing them to detect even the breakdown of a single hydrogen bond as they are rapidly moved and distorted as a result of stresses as low as 10 pN [151]. Thus, it can be understood that the primary characteristics of MO nanomaterials, including their high specificity and selectivity, quick reaction and recuperation periods, reversibility, and unification at diverse levels, make them appropriate for tracking infectious diseases, drug pharmacokinetic properties, detecting biological markers (cancer and disease), and small molecules.

4 Conclusions

MO nanomaterials have enormous capabilities to replace antibiotic drugs in the battle against multidrug-resistant bacteria. The capacity of the redox-active MO nanoparticles to modify an individual's innate and adaptive immune system responses could be employed to good use in boosting antibody responses following vaccination or regulating immune tolerance against autoimmune disorders, allergies, or malignancies. Although MO nanostructures have great potential as biomedical imaging agents, drug delivery systems, and tumor treatments, there remain a number of obstacles and unresolved issues that must be overcome well before clinical or commercial utilization. One main problem with ROS-based therapies is the failure to control how ROSs are delivered to tissues or cells, in spite of the enormous capability of MO nanomaterials for biological applications. However, there are some undiscovered aspects of MO nanostructure-based cellular and extracellular activity that have not yet been successfully tested and have been observed to be unconnected to ROS production. MO nanomaterials have been extensively discussed in relation to their potential use in biomedicine as biosensing materials, tissue and immunotherapeutics, quantum dots, tissue regeneration, dental care, and tissue repair. Their in-depth discussion also covered their potential antiviral, antifungal, and antibacterial activities. Most of the numerous varieties of MO nanostructures have been discussed here while highlighting the main benefits of such materials. Their biological functions and toxicity should indeed be addressed in relation to the type, size, concentration, fabrication technique, and other aspects in light of specific usage. Additionally, in vivo analyses of the long-term potential health hazards should also be included. Both benefits and drawbacks present challenges for researchers working in material science as well as biologists who must create and investigate the relationship between a specific therapeutic MO nanomaterial architecture and biological efficiency. A thorough understanding of the relationship between regulation and activation variables, as well as cytotoxicity processes, can serve as the foundation for the design of effective functional, biocompatible, and nontoxic MO nanomaterials used in the treatment of human disorders using nanotechnology.

References

[1] C. Tagusagawa, A. Takagaki, A. Iguchi, K. Takanabe, J.N. Kondo, K. Ebitani, S. Hayashi, T. Tatsumi, K. Domen, Highly active mesoporous Nb-W oxide solid-acid catalyst, Angew, Angew. Chem. Int. Ed. Engl. 122 (2010) 1146–1150. https://doi.org/10.1002/ANGE.200904791.

[2] J.N. Kondo, T. Yamashita, K. Nakajima, D. Lu, M. Hara, K. Domen, Preparation and crystallization characteristics of mesoporous TiO_2 and mixed oxides, J. Mater. Chem. 15 (2005) 2035–2040. https://doi.org/10.1039/b418331b.

[3] L.V. Wang, H.-I. Wu, Biomedical Optics, John Wiley & Sons, Inc., Hoboken, NJ, USA, 2009. https://doi.org/10.1002/9780470177013.

[4] J. Popp, V.V. Tuchin, A. Chiou, S.H. Heinemann, Handbook of Biophotonics, Volume 2: Photonics for Health Care, Wiley-VCH, 2012, p. 108.

[5] S.L. Jacques, Optical properties of biological tissues: a review, Phys. Med. Biol. 58 (2013). https://doi.org/10.1088/0031-9155/58/11/R37.

[6] K. Kulikov, Laser interaction with biological material: mathematical modeling, in: Biological and Medical Physics, Biomedical Engineering, 2013, p. 215. https://link.springer.com/content/pdf/10.1007/978-3-319-01739-6.pdf.

[7] K. Kong, I. Notingher, Frontiers in Biophotonics for Translational Medicine, Springer, 2016. https://link.springer.com/content/pdf/10.1007/978-981-287-627-0.pdf.

[8] L.D. Carlos, Theranostic of orthotopic gliomas by core-shell structured nanoplatforms, Light Sci. Appl. 11 (2022). https://doi.org/10.1038/s41377-022-00852-2.

[9] E.J. Chung, L. Leon, C. Rinaldi, Nanoparticles for Biomedical Applications: Fundamental Concepts, Biological Interactions and Clinical Applications, Elsevier, 2019. https://doi.org/10.1016/C2017-0-04750-X.

[10] A.K. Biswas, M.R. Islam, Z.S. Choudhury, A. Mostafa, M.F. Kadir, Nanotechnology based approaches in cancer therapeutics, Adv. Nat. Sci. Nanosci. Nanotechnol. 5 (2014) 043001. https://doi.org/10.1088/2043-6262/5/4/043001.

[11] S.A.A. Rizvi, A.M. Saleh, Applications of nanoparticle systems in drug delivery technology, Saudi Pharm. J. 26 (2018) 64–70. https://doi.org/10.1016/j.jsps.2017.10.012.

[12] P. Sanchez-Moreno, J.L. Ortega-Vinuesa, J.M. Peula-Garcia, J.A. Marchal, H. Boulaiz, Smart drug-delivery systems for cancer nanotherapy, Curr. Drug Targets 19 (2016) 339–359. https://doi.org/10.2174/1389450117666160527142544.

[13] Z. Rafiei-Sarmazdeh, S. Morteza Zahedi-Dizaji, A. Kafi Kang, Two-Dimensional Nanomaterials, in: Nanostructures, IntechOpen, 2020. https://doi.org/10.5772/intechopen.85263.

[14] M. Nikzamir, Y. Hanifehpour, A. Akbarzadeh, Y. Panahi, Applications of dendrimers in nanomedicine and drug delivery: a review, J. Inorg. Organomet. Polym. Mater. 31 (2021) 2246–2261. https://doi.org/10.1007/S10904-021-01925-2/TABLES/3.
[15] Y.K. Mishra, R. Adelung, ZnO tetrapod materials for functional applications, Mater. Today 21 (2018) 631–651. https://doi.org/10.1016/J.MATTOD.2017.11.003.
[16] M. Sharma, M. Joshi, S. Nigam, S. Shree, D.K. Avasthi, R. Adelung, S.K. Srivastava, Y. Kumar Mishra, ZnO tetrapods and activated carbon based hybrid composite: adsorbents for enhanced decontamination of hexavalent chromium from aqueous solution, Chem. Eng. J. 358 (2019) 540–551. https://doi.org/10.1016/j.cej.2018.10.031.
[17] M.S. Chavali, M.P. Nikolova, Metal oxide nanoparticles and their applications in nanotechnology, SN Appl. Sci. 1 (2019) 1–30. https://doi.org/10.1007/s42452-019-0592-3.
[18] H. Sung, M. Choi, Assembly of nanoparticles: towards multiscale three-dimensional architecturing, KONA Powder Part. J. 30 (2012) 31–46. https://doi.org/10.14356/kona.2013008.
[19] A. Sukhanova, S. Bozrova, P. Sokolov, M. Berestovoy, A. Karaulov, I. Nabiev, Dependence of nanoparticle toxicity on their physical and chemical properties, Nanoscale Res. Lett. 13 (2018) (2018) 1–21. https://doi.org/10.1186/S11671-018-2457-X.
[20] D. Chen, Q. Xu, W. Wang, J. Shao, W. Huang, X. Dong, Type I photosensitizers revitalizing photodynamic oncotherapy, Small 17 (2021) 2006742. https://doi.org/10.1002/SMLL.202006742.
[21] Z. Meng, H. Xue, T. Wang, B. Chen, X. Dong, L. Yang, J. Dai, X. Lou, F. Xia, Aggregation-induced emission photosensitizer-based photodynamic therapy in cancer: from chemical to clinical, J. Nanobiotechnol. 20 (2022) 1–35. https://doi.org/10.1186/S12951-022-01553-Z.
[22] K. Gulia, A. James, S. Pandey, K. Dev, D. Kumar, A. Sourirajan, Bio-inspired smart nanoparticles in enhanced cancer theranostics and targeted drug delivery, J. Funct. Biomater. 13 (2022) 207. https://doi.org/10.3390/JFB13040207.
[23] A. Sirelkhatim, S. Mahmud, A. Seeni, N.H.M. Kaus, L.C. Ann, S.K.M. Bakhori, H. Hasan, D. Mohamad, Review on zinc oxide nanoparticles: antibacterial activity and toxicity mechanism, Nano-Micro Lett. 7 (2015) 219–242. https://doi.org/10.1007/S40820-015-0040-X/TABLES/2.
[24] M. Elsabahy, K.L. Wooley, Design of polymeric nanoparticles for biomedical delivery applications, Chem. Soc. Rev. 41 (2012) 2545–2561. https://doi.org/10.1039/C2CS15327K.
[25] Q. Feng, Y. Liu, J. Huang, K. Chen, J. Huang, K. Xiao, Uptake, distribution, clearance, and toxicity of iron oxide nanoparticles with different sizes and coatings, Sci. Rep. 8 (2018) 1–13. https://doi.org/10.1038/s41598-018-19628-z.
[26] Z.G. Yue, W. Wei, Z.X. You, Q.Z. Yang, H. Yue, Z.G. Su, G.H. Ma, Iron oxide nanotubes for magnetically guided delivery and pH-activated release of insoluble anticancer drugs, Adv. Funct. Mater. 21 (2011) 3446–3453. https://doi.org/10.1002/ADFM.201100510.
[27] R.G.D. Andrade, S.R.S. Veloso, E.M.S. Castanheira, Shape anisotropic iron oxide-based magnetic nanoparticles: synthesis and biomedical applications, Int. J. Mol. Sci. 21 (2020) 2455. https://doi.org/10.3390/IJMS21072455.
[28] P.P. Wyss, S. Lamichhane, M. Rauber, R. Thomann, K.W. Krämer, V.P. Shastri, Tripod USPIONs with high aspect ratio show enhanced T2 relaxation and cytocompatibility, Nanomedicine (Lond.) 11 (2016) 1017–1030. https://doi.org/10.2217/NNM.16.8.
[29] G. Hemery, C. Genevois, F. Couillaud, S. Lacomme, E. Gontier, E. Ibarboure, S. Lecommandoux, E. Garanger, O. Sandre, Monocore vs. multicore magnetic iron oxide nanoparticles: uptake by glioblastoma cells and efficiency for magnetic hyperthermia, Mol. Syst. Des. Eng. 2 (2017) 629–639. https://doi.org/10.1039/C7ME00061H.

[30] R.A. Agarwal, N.K. Gupta, R. Singh, S. Nigam, B. Ateeq, Ag/AgO nanoparticles grown via time dependent double mechanism in a 2D layered Ni-PCP and their antibacterial efficacy, Sci. Rep. 7 (2017) 1–9. https://doi.org/10.1038/srep44852.
[31] A. Baun, P. Sayre, K.G. Steinhäuser, J. Rose, Regulatory relevant and reliable methods and data for determining the environmental fate of manufactured nanomaterials, NanoImpact. 8 (2017) 1–10. https://doi.org/10.1016/J.IMPACT.2017.06.004.
[32] J.R. Morones, J.L. Elechiguerra, A. Camacho, K. Holt, J.B. Kouri, J.T. Ramírez, M.J. Yacaman, The bactericidal effect of silver nanoparticles, Nanotechnology 16 (2005) 2346. https://doi.org/10.1088/0957-4484/16/10/059.
[33] W. Utembe, K. Potgieter, A.B. Stefaniak, M. Gulumian, Dissolution and biodurability: important parameters needed for risk assessment of nanomaterials, Part. Fibre Toxicol. 12 (2015) 1–12. https://doi.org/10.1186/S12989-015-0088-2/TABLES/2.
[34] M. Michaelis, C. Fischer, L. Colombi Ciacchi, A. Luttge, Variability of zinc oxide dissolution rates, Environ. Sci. Technol. 51 (2017) 4297–4305. https://doi.org/10.1021/ACS.EST.6B05732/SUPPL_FILE/ES6B05732_SI_001.PDF.
[35] H. He, J. Cao, X. Fei, N. Duan, High-temperature annealing of ZnO nanoparticles increases the dissolution magnitude and rate in water by altering O vacancy distribution, Environ. Int. 130 (2019) 104930. https://doi.org/10.1016/J.ENVINT.2019.104930.
[36] L. Zhang, Y. Jiang, Y. Ding, M. Povey, D. York, Investigation into the antibacterial behaviour of suspensions of ZnO nanoparticles (ZnO nanofluids), J. Nanopart. Res. 9 (2006) 479–489. https://doi.org/10.1007/S11051-006-9150-1.
[37] S. Lousinian, D. Missopolinou, C. Panayiotou, Fibrinogen adsorption on zinc oxide nanoparticles: a micro-differential scanning calorimetry analysis, J. Colloid Interface Sci. 395 (2013) 294–299. https://doi.org/10.1016/J.JCIS.2013.01.007.
[38] Y. He, S. Ingudam, S. Reed, A. Gehring, T.P. Strobaugh, P. Irwin, Study on the mechanism of antibacterial action of magnesium oxide nanoparticles against foodborne pathogens, J. Nanobiotechnol. 14 (2016) 1–9. https://doi.org/10.1186/S12951-016-0202-0/FIGURES/6.
[39] B.T.T. Pham, E.K. Colvin, N.T.H. Pham, B.J. Kim, E.S. Fuller, E.A. Moon, R. Barbey, S. Yuen, B.H. Rickman, N.S. Bryce, S. Bickley, M. Tanudji, S.K. Jones, V.M. Howell, B.S. Hawkett, Biodistribution and clearance of stable superparamagnetic maghemite iron oxide nanoparticles in mice following intraperitoneal administration, Int. J. Mol. Sci. 19 (2018) 205. https://doi.org/10.3390/IJMS19010205.
[40] F. Alexis, E. Pridgen, L.K. Molnar, O.C. Farokhzad, Factors affecting the clearance and biodistribution of polymeric nanoparticles, Mol. Pharm. 5 (2008) 505–515. https://doi.org/10.1021/MP800051M/ASSET/IMAGES/LARGE/MP-2008-00051M_0001.JPEG.
[41] T. Osaka, T. Nakanishi, S. Shanmugam, S. Takahama, H. Zhang, Effect of surface charge of magnetite nanoparticles on their internalization into breast cancer and umbilical vein endothelial cells, Colloids Surf. B Biointerfaces 71 (2009) 325–330. https://doi.org/10.1016/J.COLSURFB.2009.03.004.
[42] K. Xiao, Y. Li, J. Luo, J.S. Lee, W. Xiao, A.M. Gonik, R.G. Agarwal, K.S. Lam, The effect of surface charge on in vivo biodistribution of PEG-oligocholic acid based micellar nanoparticles, Biomaterials 32 (2011) 3435–3446. https://doi.org/10.1016/J.BIOMATERIALS.2011.01.021.
[43] M.L. Avramescu, P.E. Rasmussen, M. Chénier, H.D. Gardner, Influence of pH, particle size and crystal form on dissolution behaviour of engineered nanomaterials, Environ. Sci. Pollut. Res. 24 (2017) 1553–1564. https://doi.org/10.1007/S11356-016-7932-2/TABLES/4.

[44] M. Zimbone, M.A. Buccheri, G. Cacciato, R. Sanz, G. Rappazzo, S. Boninelli, R. Reitano, L. Romano, V. Privitera, M.G. Grimaldi, Photocatalytical and antibacterial activity of TiO2 nanoparticles obtained by laser ablation in water, Appl. Catal. Environ. 165 (2015) 487–494. https://doi.org/10.1016/j.apcatb.2014.10.031.
[45] J. Podporska-Carroll, E. Panaitescu, B. Quilty, L. Wang, L. Menon, S.C. Pillai, Antimicrobial properties of highly efficient photocatalytic TiO2 nanotubes, Appl. Catal. Environ. 176–177 (2015) 70–75. https://doi.org/10.1016/j.apcatb.2015.03.029.
[46] R. Augustine, A.P. Mathew, A. Sosnik, Metal oxide nanoparticles as versatile therapeutic agents modulating cell signaling pathways: linking nanotechnology with molecular medicine, Appl. Mater. Today 7 (2017) 91–103. https://doi.org/10.1016/J.APMT.2017.01.010.
[47] B.C. Schanen, S. Das, C.M. Reilly, W.L. Warren, W.T. Self, S. Seal, D.R. Drake, Immunomodulation and T Helper TH1/TH2 response polarization by CeO2 and TiO2 nanoparticles, PloS One 8 (2013) e62816. https://doi.org/10.1371/JOURNAL.PONE.0062816.
[48] X.Q. Zhang, L.H. Yin, M. Tang, Y.P. Pu, ZnO, TiO2, SiO2, and Al2O3 nanoparticles-induced toxic effects on human fetal lung fibroblasts, Biomed. Environ. Sci. 24 (2011) 661–669. https://doi.org/10.3967/0895-3988.2011.06.011.
[49] S. Lanone, F. Rogerieux, J. Geys, A. Dupont, E. Maillot-Marechal, J. Boczkowski, G. Lacroix, P. Hoet, Comparative toxicity of 24 manufactured nanoparticles in human alveolar epithelial and macrophage cell lines, Part. Fibre Toxicol. 6 (2009) 1–12. https://doi.org/10.1186/1743-8977-6-14/TABLES/3.
[50] D. Sahu, G.M. Kannan, M. Tailang, R. Vijayaraghavan, In vitro cytotoxicity of nanoparticles: a comparison between particle size and cell type, J. Nanosci. 2016 (2016) 1–9. https://doi.org/10.1155/2016/4023852.
[51] H. Chhabra, R. Deshpande, M. Kanitkar, A. Jaiswal, V.P. Kale, J.R. Bellare, A nano zinc oxide doped electrospun scaffold improves wound healing in a rodent model, RSC Adv. 6 (2015) 1428–1439. https://doi.org/10.1039/C5RA21821G.
[52] A.V. Abega, H.M. Ngomo, I. Nongwe, H.E. Mukaya, P.M.A. Kouoh Sone, X. Yangkou Mbianda, Easy and convenient synthesis of CNT/TiO2 nanohybrid by in-surface oxidation of Ti3+ ions and application in the photocatalytic degradation of organic contaminants in water, Synth. Met. 251 (2019) 1–14. https://doi.org/10.1016/J.SYNTHMET.2019.03.012.
[53] O.S. Dahham, N.N. Zulkepli, Robust interface on ENR-50/TiO2 nanohybrid material based sol-gel technique: insights into synthesis, characterization and applications in optical, Arab. J. Chem. 13 (2020) 6568–6579. https://doi.org/10.1016/J.ARABJC.2020.06.013.
[54] M. Safavipour, M. Kharaziha, E. Amjadi, F. Karimzadeh, A. Allafchian, TiO2 nanotubes/reduced GO nanoparticles for sensitive detection of breast cancer cells and photothermal performance, Talanta 208 (2020) 120369. https://doi.org/10.1016/J.TALANTA.2019.120369.
[55] P. Biswas, A. Adhikari, S. Mondal, M. Das, S. Sankar Bhattacharya, D. Pal, S. Shyam Choudhury, S. Kumar Pal, Synthesis and spectroscopic characterization of a zinc oxide-polyphenol nanohybrid from natural resources for enhanced antioxidant activity with less cytotoxicity, Mater. Today Proc. 43 (2021) 3481–3486. https://doi.org/10.1016/J.MATPR.2020.09.567.
[56] Z. Rafiee, S. Mallakpour, S. Khalili, Preparation and characterization of polyimide/titania nanohybrid films, Polym. Compos. 35 (2014) 1486–1493. https://doi.org/10.1002/PC.22802.
[57] S.A. Arifin, S. Jalaludin, R. Saleh, Preparation, characterization of Fe3O4/TiO2/CuO Nanohybrid and its potential performance for chromium removal from aqueous solution,

Mater. Sci. Forum 827 (2015) 49–55. https://doi.org/10.4028/WWW.SCIENTIFIC.NET/MSF.827.49.

[58] S. Ghosh, V.S. Goudar, K.G. Padmalekha, S.V. Bhat, S.S. Indi, H.N. Vasan, ZnO/Ag nanohybrid: synthesis, characterization, synergistic antibacterial activity and its mechanism, RSC Adv. 2 (2012) 930–940. https://doi.org/10.1039/C1RA00815C.

[59] Q.Y. Siew, S.Y. Tham, H.S. Loh, P.S. Khiew, W.S. Chiu, M.T.T. Tan, One-step green hydrothermal synthesis of biocompatible graphene/TiO2 nanocomposites for non-enzymatic H2O2 detection and their cytotoxicity effects on human keratinocyte and lung fibroblast cells, J. Mater. Chem. B 6 (2018) 1195–1206. https://doi.org/10.1039/C7TB02891A.

[60] R. Nandi, S. Mishra, T.K. Maji, K. Manna, P. Kar, S. Banerjee, S. Dutta, S.K. Sharma, P. Lemmens, K. Das Saha, S.K. Pal, A novel nanohybrid for cancer theranostics: folate sensitized Fe2O3 nanoparticles for colorectal cancer diagnosis and photodynamic therapy, J. Mater. Chem. B 5 (2017) 3927–3939. https://doi.org/10.1039/C6TB03292C.

[61] Z. Hu, J. Li, C. Li, S. Zhao, N. Li, Y. Wang, F. Wei, L. Chen, Y. Huang, Folic acid-conjugated graphene–ZnO nanohybrid for targeting photodynamic therapy under visible light irradiation, J. Mater. Chem. B 1 (2013) 5003–5013. https://doi.org/10.1039/C3TB20849D.

[62] B. Janani, A. Syed, B. Hari Kumar, A.M. Elgorban, A.H. Bahkali, B. Ahmed, A. Das, S. Sudheer Khan, High performance MnO2–Al2O3 nanocomposite as white light photocatalyst and bactericidal agent: insights on photoluminescence and intrinsic mechanism, Opt. Mater. (Amst). 120 (2021) 111438. https://doi.org/10.1016/j.optmat.2021.111438.

[63] P. Kundu, N. Singhania, G. Madras, N. Ravishankar, ZnO–Au nanohybrids by rapid microwave-assisted synthesis for CO oxidation, Dalton Trans. 41 (2012) 8762–8766. https://doi.org/10.1039/C2DT30882G.

[64] F. Luo, M. Wang, L. Huang, Z. Wu, W. Wang, A. Zafar, Y. Tian, M. Hasan, X. Shu, Synthesis of zinc oxide Eudragit FS30D Nanohybrids: structure, characterization, and their application as an intestinal drug delivery system, ACS, Omega 5 (2020) 11799–11808. https://doi.org/10.1021/ACSOMEGA.0C01216/ASSET/IMAGES/MEDIUM/AO0C01216_M003.GIF.

[65] Y. Haldorai, J.J. Shim, Novel chitosan-TiO2 nanohybrid: preparation, characterization, antibacterial, and photocatalytic properties, Polym. Compos. 35 (2014) 327–333. https://doi.org/10.1002/PC.22665.

[66] R.K. Jain, T. Stylianopoulos, Delivering nanomedicine to solid tumors, Nat. Rev. Clin. Oncol. 7 (2010) 653–664. https://doi.org/10.1038/nrclinonc.2010.139.

[67] M.P. Nikolova, M.S. Chavali, Metal oxide nanoparticles as biomedical materials, Biomimetics. 5 (2020). https://doi.org/10.3390/BIOMIMETICS5020027.

[68] Z. Zhu, S. Sheng-Nan, W. Chao, Z. Zan-Zan, H. Yang-Long, S.S. Venkatraman, X. Zhi-Chuan, Magnetic iron oxide nanoparticles: synthesis and surface coating techniques for biomedical applications, Chin. Phys. B 23 (2014) 37503. https://doi.org/10.1088/1674-1056/23/3/037503.

[69] L.S. Arias, J.P. Pessan, A.P.M. Vieira, T.M.T. De Lima, A.C.B. Delbem, D.R. Monteiro, Iron oxide nanoparticles for biomedical applications: a perspective on synthesis, drugs, antimicrobial activity, and toxicity, Antibiotica 7 (2018) 46. https://doi.org/10.3390/ANTIBIOTICS7020046.

[70] L. Peng, M. He, B. Chen, Q. Wu, Z. Zhang, D. Pang, Y. Zhu, B. Hu, Cellular uptake, elimination and toxicity of CdSe/ZnS quantum dots in HepG2 cells, Biomaterials 34 (2013) 9545–9558. https://doi.org/10.1016/J.BIOMATERIALS.2013.08.038.

[71] J. Estelrich, E. Escribano, J. Queralt, M.A. Busquets, Iron oxide nanoparticles for magnetically-guided and magnetically-responsive drug delivery, Int. J. Mol. Sci. 16 (2015) 8070–8101. https://doi.org/10.3390/IJMS16048070.

[72] T.K. Indira, P.K. Lalshmi, Magnetic nanoparticles—a review, Int. J. Pharm. Sci. Nanotechnol. 3 (2010) 1035–1042. https://doi.org/10.37285/IJPSN.2010.3.3.1.

[73] R. Roy, D. Kumar, A. Sharma, P. Gupta, B.P. Chaudhari, A. Tripathi, M. Das, P.D. Dwivedi, ZnO nanoparticles induced adjuvant effect via toll-like receptors and Src signaling in Balb/c mice, Toxicol. Lett. 230 (2014) 421–433. https://doi.org/10.1016/J.TOXLET.2014.08.008.

[74] A. Hanini, A. Schmitt, K. Kacem, F. Chau, S. Ammar, J. Gavard, Evaluation of iron oxide nanoparticle biocompatibility, Int. J. Nanomedicine 6 (2011) 787. https://doi.org/10.2147/IJN.S17574.

[75] A.J. Giustini, A.A. Petryk, S.M. Cassim, J.A. Tate, I. Baker, P.J. Hoopes, Magnetic nanoparticle hyperthermia in cancer treatment, Nano Life 01 (2012) 17–32. https://doi.org/10.1142/S1793984410000067.

[76] I. Hilger, R. Hiergeist, R. Hergt, K. Winnefeld, H. Schubert, W.A. Kaiser, Thermal ablation of tumors using magnetic nanoparticles: an in vivo feasibility study, Invest. Radiol. 37 (2002) 580–586. https://doi.org/10.1097/00004424-200210000-00008.

[77] M. Hurbankova, K. Volkovova, D. Hraskova, S. Wimmerova, S. Moricova, Respiratory toxicity of Fe3O4 nanoparticles: experimental study, Rev. Environ. Health 32 (2017) 207–210. https://doi.org/10.1515/REVEH-2016-0022/MACHINEREADABLECITATION/RIS.

[78] S.F. Shi, J.F. Jia, X.K. Guo, Y.P. Zhao, D.S. Chen, Y.Y. Guo, T. Cheng, X.L. Zhang, Biocompatibility of chitosan-coated iron oxide nanoparticles with osteoblast cells, Int. J. Nanomedicine 7 (2012) 5593. https://doi.org/10.2147/IJN.S34348.

[79] V. Kalidasan, X.L. Liu, T.S. Herng, Y. Yang, J. Ding, Bovine serum albumin-conjugated ferrimagnetic iron oxide nanoparticles to enhance the biocompatibility and magnetic hyperthermia performance, Nano-Micro Lett. 8 (2016) 80–93. https://doi.org/10.1007/S40820-015-0065-1/FIGURES/14.

[80] E. Vismara, C. Bongio, A. Coletti, R. Edelman, A. Serafini, M. Mauri, R. Simonutti, S. Bertini, E. Urso, Y.G. Assaraf, Y.D. Livney, Albumin and hyaluronic acid-coated superparamagnetic iron oxide nanoparticles loaded with paclitaxel for biomedical applications, Molecules 22 (2017) 1030. https://doi.org/10.3390/MOLECULES22071030.

[81] P. Yu, X.M. Xia, M. Wu, C. Cui, Y. Zhang, L. Liu, B. Wu, C.X. Wang, L.J. Zhang, X. Zhou, R.X. Zhuo, S.W. Huang, Folic acid-conjugated iron oxide porous nanorods loaded with doxorubicin for targeted drug delivery, Colloids Surf. B Biointerfaces 120 (2014) 142–151. https://doi.org/10.1016/J.COLSURFB.2014.05.018.

[82] G. Bisht, S. Rayamajhi, ZnO nanoparticles: a promising anticancer agent, Nanomedicine 3 (2016). https://doi.org/10.5772/63437/ASSET/IMAGES/LARGE/10.5772_63437-FIG2.JPEG.

[83] J. Jiang, J. Pi, J. Cai, The advancing of zinc oxide nanoparticles for biomedical applications, Bioinorg. Chem. Appl. 2018 (2018). https://doi.org/10.1155/2018/1062562.

[84] A. Boroumand Moghaddam, M. Moniri, S. Azizi, R. Abdul Rahim, A. Bin Ariff, M. Navaderi, R. Mohamad, Eco-friendly formulated zinc oxide nanoparticles: induction of cell cycle arrest and apoptosis in the MCF-7 cancer cell line, Genes (Basel) 8 (2017) 281. https://doi.org/10.3390/genes8100281.

[85] L. García-Hevia, R. Valiente, R. Martín-Rodríguez, C. Renero-Lecuna, J. González, L. Rodríguez-Fernández, F. Aguado, J.C. Villegas, M.L. Fanarraga, Nano-ZnO leads to tubulin macrotube assembly and actin bundling, triggering cytoskeletal catastrophe and cell necrosis, Nanoscale 8 (2016) 10963–10973. https://doi.org/10.1039/C6NR00391E.

[86] Y. Deng, H. Zhang, The synergistic effect and mechanism of doxorubicin-ZnO nanocomplexes as a multimodal agent integrating diverse anticancer therapeutics, Int. J. Nanomedicine 8 (2013) 1835. https://doi.org/10.2147/IJN.S43657.
[87] N. Puvvada, S. Rajput, B. Prashanth Kumar, S. Sarkar, S. Konar, K.R. Brunt, R.R. Rao, A. Mazumdar, S.K. Das, R. Basu, P.B. Fisher, M. Mandal, A. Pathak, Novel ZnO hollow-nanocarriers containing paclitaxel targeting folate-receptors in a malignant pH-microenvironment for effective monitoring and promoting breast tumor regression, Sci. Rep. 5 (2015) 1–15. https://doi.org/10.1038/srep11760.
[88] R. Dhivya, J. Ranjani, J. Rajendhran, J. Mayandi, J. Annaraj, Enhancing the anti-gastric cancer activity of curcumin with biocompatible and pH sensitive PMMA-AA/ZnO nanoparticles, Mater. Sci. Eng. C 82 (2018) 182–189. https://doi.org/10.1016/J.MSEC.2017.08.058.
[89] E. Sarugeri, N. Dozio, F. Meschi, M.R. Pastore, E. Bonifacio, Cellular and humoral immunity against cow's milk proteins in type 1 diabetes, J. Autoimmun. 13 (1999) 365–373. https://doi.org/10.1006/jaut.1999.0327.
[90] J. Hussein, M. El-Banna, T.A. Razik, M.E. El-Naggar, Biocompatible zinc oxide nanocrystals stabilized via hydroxyethyl cellulose for mitigation of diabetic complications, Int. J. Biol. Macromol. 107 (2018) 748–754. https://doi.org/10.1016/J.IJBIOMAC.2017.09.056.
[91] H.J. Haugen, M. Monjo, M. Rubert, A. Verket, S.P. Lyngstadaas, J.E. Ellingsen, H.J. Rønold, J.C. Wohlfahrt, Porous ceramic titanium dioxide scaffolds promote bone formation in rabbit peri-implant cortical defect model, Acta Biomater. 9 (2013) 5390–5399. https://doi.org/10.1016/J.ACTBIO.2012.09.009.
[92] P. Martin, S.J. Leibovich, Inflammatory cells during wound repair: the good, the bad and the ugly, Trends Cell Biol. 15 (2005) 599–607. https://doi.org/10.1016/J.TCB.2005.09.002.
[93] C. Wang, S. Cao, X. Tie, B. Qiu, A. Wu, Z. Zheng, Induction of cytotoxicity by photoexcitation of TiO2 can prolong survival in glioma-bearing mice, Mol. Biol. Rep. 38 (2011) 523–530. https://doi.org/10.1007/S11033-010-0136-9/FIGURES/5.
[94] P.F. Zeni, D.P. Dos Santos, R.R. Canevarolo, J.A. Yunes, F.F. Padilha, R.L.C. de Albuquerque Júnior, S.M. Egues, M.L. Hernández-Macedo, Photocatalytic and cytotoxic effects of nitrogen-doped TiO2 nanoparticles on melanoma cells, J. Nanosci. Nanotechnol. 18 (2017) 3722–3728. https://doi.org/10.1166/JNN.2018.14621.
[95] W. Ni, M. Li, J. Cui, Z. Xing, Z. Li, X. Wu, E. Song, M. Gong, W. Zhou, 808 nm light triggered black TiO2 nanoparticles for killing of bladder cancer cells, Mater. Sci. Eng. C 81 (2017) 252–260. https://doi.org/10.1016/J.MSEC.2017.08.020.
[96] Z. Wang, C. Xie, F. Luo, P. Li, X. Xiao, P25 nanoparticles decorated on titania nanotubes arrays as effective drug delivery system for ibuprofen, Appl. Surf. Sci. 324 (2015) 621–626. https://doi.org/10.1016/J.APSUSC.2014.10.147.
[97] T. Kumeria, H. Mon, M.S. Aw, K. Gulati, A. Santos, H.J. Griesser, D. Losic, Advanced biopolymer-coated drug-releasing titania nanotubes (TNTs) implants with simultaneously enhanced osteoblast adhesion and antibacterial properties, Colloids Surf. B Biointerfaces 130 (2015) 255–263. https://doi.org/10.1016/J.COLSURFB.2015.04.021.
[98] M. Masoudi, M. Mashreghi, E. Goharshadi, A. Meshkini, Multifunctional fluorescent titania nanoparticles: green preparation and applications as antibacterial and cancer theranostic agents, Artif. Cells Nanomed. Biotechnol. 46 (2018) 248–259. https://doi.org/10.1080/21691401.2018.1454932.
[99] S. Hussain, J.A.J. Vanoirbeek, P.H.M. Hoet, Interactions of nanomaterials with the immune system, Wiley Interdiscip. Rev. Nanomed. Nanobiotechnol. 4 (2012) 169–183. https://doi.org/10.1002/WNAN.166.

[100] M. Elsabahy, K.L. Wooley, Cytokines as biomarkers of nanoparticle immunotoxicity, Chem. Soc. Rev. 42 (2013) 5552–5576. https://doi.org/10.1039/C3CS60064E.
[101] L. Singh, H.G. Kruger, G.E.M. Maguire, T. Govender, R. Parboosing, The role of nanotechnology in the treatment of viral infections, Ther. Adv. Infect. Dis. 4 (2017) 105–131. https://doi.org/10.1177/2049936117713593/ASSET/IMAGES/LARGE/10.1177_2049936117713593-FIG2.JPEG.
[102] T. Yadavalli, D. Shukla, Role of metal and metal oxide nanoparticles as diagnostic and therapeutic tools for highly prevalent viral infections, nanomedicine nanotechnology, Biol. Med. 13 (2017) 219–230. https://doi.org/10.1016/J.NANO.2016.08.016.
[103] S. Gorantla, H. Dou, M. Boska, C.J. Destache, J. Nelson, L. Poluektova, B.E. Rabinow, H.E. Gendelman, R.L. Mosley, Quantitative magnetic resonance and SPECT imaging for macrophage tissue migration and nanoformulated drug delivery, J. Leukoc. Biol. 80 (2006) 1165–1174. https://doi.org/10.1189/JLB.0206110.
[104] L. Fiandra, M. Colombo, S. Mazzucchelli, M. Truffi, B. Santini, R. Allevi, M. Nebuloni, A. Capetti, G. Rizzardini, D. Prosperi, F. Corsi, Nanoformulation of antiretroviral drugs enhances their penetration across the blood brain barrier in mice, nanomedicine nanotechnology, Biol. Med. 11 (2015) 1387–1397. https://doi.org/10.1016/J.NANO.2015.03.009.
[105] L.M. Marques Neto, A. Kipnis, A.P. Junqueira-Kipnis, Role of metallic nanoparticles in vaccinology: implications for infectious disease vaccine development, Front. Immunol. 8 (2017) 239. https://doi.org/10.3389/FIMMU.2017.00239/BIBTEX.
[106] A. Laskar, J. Eilertsen, W. Li, X.M. Yuan, SPION primes THP1 derived M2 macrophages towards M1-like macrophages, Biochem. Biophys. Res. Commun. 441 (2013) 737–742. https://doi.org/10.1016/J.BBRC.2013.10.115.
[107] M.A. Shevtsov, B.P. Nikolaev, L.Y. Yakovleva, M.A. Parr, Y.Y. Marchenko, I. Eliseev, A.V. Dobrodumov, O. Zlobina, A. Zhakhov, A.M. Ischenko, E. Pitkin, G. Multhoff, 70-kDa heat shock protein coated magnetic nanocarriers as a nanovaccine for induction of anti-tumor immune response in experimental glioma, J. Control. Release 220 (2015) 329–340. https://doi.org/10.1016/J.JCONREL.2015.10.051.
[108] J.M. Rojas, L. Sanz-Ortega, V. Mulens-Arias, L. Gutiérrez, S. Pérez-Yagüe, D.F. Barber, Superparamagnetic iron oxide nanoparticle uptake alters M2 macrophage phenotype, iron metabolism, migration and invasion, nanomedicine nanotechnology, Biol. Med. 12 (2016) 1127–1138. https://doi.org/10.1016/J.NANO.2015.11.020.
[109] L.M.M. Neto, N. Zufelato, A.A. de Sousa-Júnior, M.M. Trentini, A.C. da Costa, A.F. Bakuzis, A. Kipnis, A.P. Junqueira-Kipnis, Specific T cell induction using iron oxide based nanoparticles as subunit vaccine adjuvant, Hum. Vaccin. Immunother. 14 (2018) 2786–2801. https://doi.org/10.1080/21645515.2018.1489192/SUPPL_FILE/KHVI_A_1489192_SM5284.ZIP.
[110] A.A. Mieloch, M. Kręcisz, J.D. Rybka, A. Strugała, M. Krupiński, A. Urbanowicz, M. Kozak, B. Skalski, M. Figlerowicz, M. Giersig, The influence of ligand charge and length on the assembly of brome mosaic virus derived virus-like particles with magnetic core, AIP Adv. 8 (2018) 035005. https://doi.org/10.1063/1.5011138.
[111] L.L. Li, P. Yu, X. Wang, S.S. Yu, J. Mathieu, H.Q. Yu, P.J.J. Alvarez, Enhanced biofilm penetration for microbial control by polyvalent phages conjugated with magnetic colloidal nanoparticle clusters (CNCs), Environ. Sci. Nano 4 (2017) 1817–1826. https://doi.org/10.1039/C7EN00414A.
[112] D. Ghosh, Y. Lee, S. Thomas, A.G. Kohli, D.S. Yun, A.M. Belcher, K.A. Kelly, M13-templated magnetic nanoparticles for targeted in vivo imaging of prostate cancer, Nat. Nanotechnol. 7 (2012) 677–682. https://pubmed.ncbi.nlm.nih.gov/22983492/.

[113] C. Schütz, J.C. Varela, K. Perica, C. Haupt, M. Oelke, J.P. Schneck, C. Schütz, J.C. Varela, K. Perica, C. Haupt, M. Oelke, J.P. Schneck, Antigen-specific T cell redirectors: a nanoparticle based approach for redirecting T cells, Oncotarget 7 (2016) 68503–68512. https://doi.org/10.18632/ONCOTARGET.11785.

[114] R. Simón-Vázquez, T. Lozano-Fernández, A. Dávila-Grana, A. González-Fernández, Metal oxide nanoparticles interact with immune cells and activate different cellular responses, Int. J. Nanomedicine 11 (2016) 4657–4668. https://doi.org/10.2147/IJN.S110465.

[115] A. Agelidis, L. Koujah, R. Suryawanshi, T. Yadavalli, Y.K. Mishra, R. Adelung, D. Shukla, An intra-vaginal zinc oxide tetrapod nanoparticles (zoten) and genital herpesvirus cocktail can provide a novel platform for live virus vaccine, Front. Immunol. 10 (2019) 500. https://doi.org/10.3389/FIMMU.2019.00500/BIBTEX.

[116] C. Wang, X. Hu, Y. Gao, Y. Ji, ZnO nanoparticles treatment induces apoptosis by increasing intracellular ROS levels in LTEP-a-2 cells, Biomed. Res. Int. 2015 (2015). https://doi.org/10.1155/2015/423287.

[117] P. Thatoi, R.G. Kerry, S. Gouda, G. Das, K. Pramanik, H. Thatoi, J.K. Patra, Photo-mediated green synthesis of silver and zinc oxide nanoparticles using aqueous extracts of two mangrove plant species, Heritiera fomes and Sonneratia apetala and investigation of their biomedical applications, J. Photochem. Photobiol. B Biol. 163 (2016) 311–318. https://doi.org/10.1016/J.JPHOTOBIOL.2016.07.029.

[118] P. Sharma, J.B. Shin, B.C. Park, J.W. Lee, S.W. Byun, N.Y. Jang, Y.J. Kim, Y. Kim, Y.K. Kim, N.H. Cho, Application of radially grown ZnO nanowires on poly-L-lactide microfibers complexed with a tumor antigen for cancer immunotherapy, Nanoscale 11 (2019) 4591–4600. https://doi.org/10.1039/C8NR08704K.

[119] Y. Hu, H.R. Zhang, L. Dong, M.R. Xu, L. Zhang, W.P. Ding, J.Q. Zhang, J. Lin, Y.J. Zhang, B.S. Qiu, P.F. Wei, L.P. Wen, Enhancing tumor chemotherapy and overcoming drug resistance through autophagy-mediated intracellular dissolution of zinc oxide nanoparticles, Nanoscale 11 (2019) 11789–11807. https://doi.org/10.1039/C8NR08442D.

[120] C.A. Charitidis, P. Georgiou, M.A. Koklioti, A.F. Trompeta, V. Markakis, Manufacturing nanomaterials: from research to industry, Manuf. Rev. 1 (2014) 11. https://doi.org/10.1051/MFREVIEW/2014009.

[121] K.R. Wierzbinski, T. Szymanski, N. Rozwadowska, J.D. Rybka, A. Zimna, T. Zalewski, K. Nowicka-Bauer, A. Malcher, M. Nowaczyk, M. Krupinski, M. Fiedorowicz, P. Bogorodzki, P. Grieb, M. Giersig, M.K. Kurpisz, Potential use of superparamagnetic iron oxide nanoparticles for in vitro and in vivo bioimaging of human myoblasts, Sci. Rep. 8 (2018) 1–17. https://doi.org/10.1038/s41598-018-22018-0.

[122] K. Matsuyama, N. Ihsan, K. Irie, K. Mishima, T. Okuyama, H. Muto, Bioimaging application of highly luminescent silica-coated ZnO-nanoparticle quantum dots with biotin, J. Colloid Interface Sci. 399 (2013) 19–25. https://doi.org/10.1016/J.JCIS.2013.02.047.

[123] H.C. Huang, S. Barua, G. Sharma, S.K. Dey, K. Rege, Inorganic nanoparticles for cancer imaging and therapy, J. Control. Release 155 (2011) 344–357. Elsevier https://doi.org/10.1016/j.jconrel.2011.06.004.

[124] J. Yu, F. Zhao, W. Gao, X. Yang, Y. Ju, L. Zhao, W. Guo, J. Xie, X.J. Liang, X. Tao, J. Li, Y. Ying, W. Li, J. Zheng, L. Qiao, S. Xiong, X. Mou, S. Che, Y. Hou, Magnetic reactive oxygen species nanoreactor for switchable magnetic resonance imaging guided Cancer therapy based on pH-sensitive Fe5C2@Fe3O4 nanoparticles, ACS Nano 13 (2019) 10002–10014. https://doi.org/10.1021/ACSNANO.9B01740/SUPPL_FILE/NN9B01740_SI_001.PDF.

[125] J. Estelrich, M.J. Sánchez-Martín, M.A. Busquets, Nanoparticles in magnetic resonance imaging: from simple to dual contrast agents, Int. J. Nanomedicine 10 (2015) 1727. https://doi.org/10.2147/IJN.S76501.

[126] D. Yoo, J.H. Lee, T.H. Shin, J. Cheon, Theranostic magnetic nanoparticles, Acc. Chem. Res. 44 (2011) 863–874. https://doi.org/10.1021/AR200085C/ASSET/IMAGES/MEDIUM/AR-2011-00085C_0012.GIF.

[127] J.T. Jang, H. Nah, J.-H.H. Lee, S.H. Moon, M.G. Kim, J. Cheon, Critical enhancements of MRI contrast and Hyperthermic effects by dopant-controlled magnetic nanoparticles, Angew. Chem. Int. Ed. 48 (2009) 1234–1238. https://onlinelibrary.wiley.com/doi/full/10.1002/anie.200805149.

[128] S.J.H. Soenen, U. Himmelreich, N. Nuytten, T.R. Pisanic, A. Ferrari, M. De Cuyper, Intracellular nanoparticle coating stability determines nanoparticle diagnostics efficacy and cell functionality, Small 6 (2010) 2136–2145. https://doi.org/10.1002/SMLL.201000763.

[129] M. Javidi, M. Zarei, N. Naghavi, M. Mortazavi, A. Nejat, Zinc oxide nano-particles as sealer in endodontics and its sealing ability, Contemp. Clin. Dent. 5 (2014) 20. https://doi.org/10.4103/0976-237X.128656.

[130] M.A. Karimi, S.H. Roozbahani, R. Asadiniya, A. Hatefi-Mehrjardi, M. Hossein Mashhadizadeh, R. Behjatmanesh-Ardakani, M. Mazloum-Ardakani, H. Kargar, M. Zebarjad, Synthesis and characterization of nanoparticles and nanocomposite of ZnO and MgO by Sonochemical method and their application for zinc Polycarboxylate dental cement preparation, Nano Lett. 1 (2011) 43–51.

[131] J. Morán-Martínez, R. Beltrán Del Río-Parra, N.D. Betancourt-Martínez, R. García-Garza, J. Jiménez-Villarreal, M.S. Niño-Castañeda, L.E. Nava-Rivera, J.A. Facio Umaña, P. Carranza-Rosales, R.D. Arellano Pérez-Vertti, Evaluation of the coating with TiO2 nanoparticles as an option for the improvement of the characteristics of NiTi Archwires: histopathological, cytotoxic, and genotoxic evidence, J. Nanomater. 2018 (2018). https://doi.org/10.1155/2018/2585918.

[132] M. Cierech, I. Osica, A. Kolenda, J. Wojnarowicz, D. Szmigiel, W. Łojkowski, K. Kurzydłowski, K. Ariga, E. Mierzwińska-Nastalska, Mechanical and physicochemical properties of newly formed ZnO-PMMA nanocomposites for denture bases, Nanomater. 8 (2018) 305. https://doi.org/10.3390/NANO8050305.

[133] N.R. Shanmugam, S. Muthukumar, S. Prasad, A review on ZnO-based electrical biosensors for cardiac biomarker detection, Futur. Sci. OA. 3 (2017). https://doi.org/10.4155/FSOA-2017-0006/ASSET/IMAGES/LARGE/FIGURE5.JPEG.

[134] R.N. AlKahtani, The implications and applications of nanotechnology in dentistry: a review, Saudi, Dent. J. 30 (2018) 107–116. https://doi.org/10.1016/J.SDENTJ.2018.01.002.

[135] C. Zhu, G. Yang, H. Li, D. Du, Y. Lin, Electrochemical sensors and biosensors based on nanomaterials and nanostructures, Anal. Chem. 87 (2015) 230–249. https://doi.org/10.1021/AC5039863/ASSET/IMAGES/LARGE/AC-2014-039863_0006.JPEG.

[136] L. Xu, J. Xia, L. Wang, J. Qian, H. Li, K. Wang, K. Sun, M., He, α-Fe2O3 cubes with high visible-light-activated Photoelectrochemical activity towards glucose: hydrothermal synthesis assisted by a hydrophobic ionic liquid, Chem. – A Eur. J. 20 (2014) 2244–2253. https://doi.org/10.1002/CHEM.201304312.

[137] P.R. Solanki, A. Kaushik, V.V. Agrawal, B.D. Malhotra, Nanostructured metal oxide-based biosensors, NPG Asia Mater. 3 (2011) 17–24. https://doi.org/10.1038/asiamat.2010.137.

[138] W. Zhao, K. Wang, Y. Wei, Y. Ma, L. Liu, X. Huang, Laccase biosensor based on phytic acid modification of nanostructured SiO2 surface for sensitive detection of dopamine, Langmuir 30 (2014) 11131–11137. https://doi.org/10.1021/LA503104X/ASSET/IMAGES/MEDIUM/LA-2014-03104X_0010.GIF.
[139] J. Cheng, J. Di, J. Hong, K. Yao, Y. Sun, J. Zhuang, Q. Xu, H. Zheng, S. Bi, The promotion effect of titania nanoparticles on the direct electrochemistry of lactate dehydrogenase sol–gel modified gold electrode, Talanta 76 (2008) 1065–1069. https://doi.org/10.1016/J.TALANTA.2008.05.006.
[140] C. Xia, N. Wang, L. Lidong, G. Lin, Synthesis and characterization of waxberry-like microstructures ZnO for biosensors, Sens. Actuators B 129 (2008) 268–273. https://doi.org/10.1016/J.SNB.2007.08.003.
[141] M.M. Rahman, A.J.S. Ahammad, J.H. Jin, S.J. Ahn, J.J. Lee, A comprehensive review of glucose biosensors based on nanostructured metal-oxides, Sensors 10 (2010) 4855–4886. https://doi.org/10.3390/S100504855.
[142] X. Lu, H. Zhang, Y. Ni, Q. Zhang, J. Chen, Porous nanosheet-based ZnO microspheres for the construction of direct electrochemical biosensors, Biosens. Bioelectron. 24 (2008) 93–98. https://doi.org/10.1016/J.BIOS.2008.03.025.
[143] K.M. Abu-Salah, S.A. Alrokyan, M.N. Khan, A.A. Ansari, Nanomaterials as analytical tools for genosensors, Sensors 10 (2010) 963–993. https://doi.org/10.3390/S100100963.
[144] W. Zhang, T. Yang, X. Zhuang, Z. Guo, K. Jiao, An ionic liquid supported CeO2 nanoshuttles–carbon nanotubes composite as a platform for impedance DNA hybridization sensing, Biosens. Bioelectron. 24 (2009) 2417–2422. https://doi.org/10.1016/J.BIOS.2008.12.024.
[145] W. Zhang, T. Yang, X. Li, D. Wang, K. Jiao, Conductive architecture of Fe2O3 microspheres/self-doped polyaniline nanofibers on carbon ionic liquid electrode for impedance sensing of DNA hybridization, Biosens. Bioelectron. 25 (2009) 428–434. https://doi.org/10.1016/J.BIOS.2009.07.032.
[146] S. Kumar, J.G. Sharma, S. Maji, B.D. Malhotra, Nanostructured zirconia decorated reduced graphene oxide based efficient biosensing platform for non-invasive oral cancer detection, Biosens. Bioelectron. 78 (2016) 497–504. https://doi.org/10.1016/j.bios.2015.11.084.
[147] A. Kaushik, P.R. Solanki, A.A. Ansari, S. Ahmad, B.D. Malhotra, A nanostructured cerium oxide film-based immunosensor for mycotoxin detection, Nanotechnology 20 (2009) 055105. https://doi.org/10.1088/0957-4484/20/5/055105.
[148] C.-C. Chang, N.-F. Chiu, D.S. Lin, Y. Chu-Su, Y.-H. Liang, C.-W. Lin, High-sensitivity detection of carbohydrate antigen 15–3 using a gold/zinc oxide thin film surface plasmon resonance-based biosensor, Anal. Chem. 82 (2010) 1207–1212. https://doi.org/10.1021/AC901797J.
[149] R. Wang, C. Ruan, D. Kanayeva, K. Lassiter, Y. Li, TiO2 nanowire bundle microelectrode based impedance immunosensor for rapid and sensitive detection of listeria monocytogenes, Nano Lett. 8 (2008) 2625–2631. https://doi.org/10.1021/NL080366Q/SUPPL_FILE/NL080366Q_SI_001.PDF.
[150] J. Lei, H. Ju, Signal amplification using nanomaterials for biosensing, in: Appl. Nanomater, Sensors Diagnostics, Springer, Berlin, Heidelberg, 2013, pp. 17–41. https://doi.org/10.1007/5346_2012_46.
[151] J.L. Arlett, E.B. Myers, M.L. Roukes, Comparative advantages of mechanical biosensors, Nat. Nanotechnol. 6 (2011) 203–215. https://doi.org/10.1038/nnano.2011.44.

Metal oxides for plasmonic applications

16

Vishnu Chauhan[a,b], *Garima Vashisht*[c], *Deepika Gupta*[d], *Sonica Upadhyay*[e], *and Rajesh Kumar*[d]
[a]Materials Science Group, Inter-University Accelerator Centre, New Delhi, India, [b]Materials Research Department, GSI Helmholtz Centre for Heavy Ion Research, Darmstadt, Germany, [c]Department of Physics and Astrophysics, University of Delhi, Delhi, India, [d]University School of Basic and Applied Sciences, Guru Gobind Singh Indraprastha University, New Delhi, India, [e]Department of Computer Science & Engineering, Maharaja Surajmal Institute of Technology, New Delhi, India

Chapter outline

Metal Oxides for Next-Generation Optoelectronic, Photonic, and Photovoltaic Applications
https://doi.org/10.1016/B978-0-323-99143-8.00013-4

1 Introduction

With the rapid growth in nanotechnology, distinct materials with significant chemical and physical characteristics have been developed and explored in many applications. Among the various promising phenomena described in nanomaterials, localized surface plasmon resonance (LSPR) and surface plasmon resonance (SPR) are particularly significant optical characteristics that play an important role in nanoelectronics, nanophotonics, and theranostics. The research on metal oxides, particularly for potential plasmonic applications, has increased over the past few decades, exhibiting a wide range of technical and industrial implications [1]. In metal oxides, the study in the field of plasmonic applications can be explained as the study of collective charge and plasmons oscillations. In plasmonics phenomena, there is an interaction of collective charge oscillations with electromagnetic (EM) radiation at the conducting material's surface. Typically, at a given frequency, the hybrid excitations are known as surface plasmons, with the choice of conducting material and surrounding dielectric that permits the subwavelength confinement of optical modes. The structures and metal oxides that support surface plasmon excitation have a significant role in next generation applications in optical interconnects, optical and gas sensing, photovoltaic and optoelectronic technologies, water splitting and purification process, etc. [2,3]. At a fundamental level, the association of strong electric field with the subwavelength confinement of EM radiation can offer a testbed exploration of the light-matter interaction in nanoscale systems.

LSPR is generally described in nanomaterials based on noble metals. It is instigated when the free electrons interact with electromagnetic light [4]. In the 1990s, surface plasmons attracted significant attention from the scientific community. Plasmonic nanomaterials behave as antenna for the collection of surrounding light energy, which gives rise to increments in coupled electric field intensity and enhances the fields present between the plasmonic nanoparticles. The significant characteristic of these plasmonic metals is that they interact with the concentrated electromagnetic field and scatter light, then absorb the light present in the visible region, transforming the light energy into heat and behaving as sinks for electrons, making them ideal substitutes for photovoltaics [5,6], photocatalysis, surface enhanced Raman spectroscopy [7], biosensors [8,9], and other domains. These applications utilize the plasmonic metals' interaction with light, premised on components of multiple plasmonic metals or hybrids of metals and semiconductors.

Surface plasmons are considered to be of two kinds, i.e., propagating surface plasmon polaritons and local surface plasmon polaritons. Materials that exhibit small and real positive dielectric constants are efficient in displaying surface plasmon resonance qualities. Surface plasmon resonance involves surface conduction electrons' coherent oscillations instigated by electromagnetic radiation. Thus, plasmonics involves the understanding of these specific light-matter interactions, which provide a wide range of applications in lithographic fabrication, chemical and biological sensing, and surface enhanced spectroscopies [10]. In the case of localized surface plasmon resonance, interaction of light occurs with specific particles that are smaller than the accompanying wavelength. This brings about local oscillations of plasmons around the nanoparticle with a specific frequency, known as localized surface plasmon resonance (LSPR). Equivalent to surface plasmon resonance (SPR), LSPR displays sensitivity towards the alterations that occur in the local dielectric environment. Mainly, researchers detected alterations occurring in the local environment by studying the wavelength shift of LSPR [11–13].

In surface plasmon polaritons, propagation of plasmas occurs in the x and y directions in the interface of dielectric and metal, exhibiting distances in the order of microns and, eventually, decay along the z direction with a decay length of 1/e. The reciprocity between the layer of surface molecules and metal surface-bound EM wave gives rise to shifting in the condition of plasmon resonance, generally detected through three modes imaging, shifting in wavelength, and angle resolved. In the modes associated with the wavelength shift and angle resolved, measurement is taken with the light reflectivity from the surface of metal as a function of incidence angle and wavelength, keeping wavelength and incidence angle constant. The method based on the imaging utilizes light of both constant incident angle and wavelength to illustrate the 2D region of the respective sample; also surface reflectivity study is done with the function of position [14,15]. The appreciable small ohmic losses and free carrier density of the Ag and Au noble metals are highly efficient in the domain of field-enhanced spectroscopy [16,17].

With the evolution of research in the plasmonics domain and their enhanced significance in current technology, novel combinations based on plasmonic materials with significant characteristics are being pursued, which could enhance efficiency or give rise to new application domains. Distinct materials for plasmonics include a broad range of doped semiconductors, metals, nitrides, metal oxides, and 2D graphene materials. In addition to the typical applications in the near infrared and visible spectral range some of the illustrated materials focus on the tetra hertz and mid infrared range in the fields of security and chemical sensing [18]. The investigation of metal oxide nanoparticles and heavily doped semiconductors that demonstrate the LSPR phenomenon is a comparatively fast growing research topic. This group of materials represents a suitable alternative to conventional metal nanostructures in many applications and displays suitability in the field where mid-IR and IR wavelength is required. Many research papers has been published in recent years involving synthesis of good quality colloidal nanocrystals and primarily illustrating the optical peculiarities of these metal oxide nanomaterials. The exploration of these materials has also been studied in the fields of sensing, optoelectronics, and theranostics. Metal oxides exhibit significant LSPR phenomena and are considered to be important for different applications, including bioimaging, smart windows, and photothermal therapy.

Doping of ZnO nanocrystals with group 3 elements achieved significant free charge carrier density. ZnO is considered to be an Earth abundant and low cost element. This could be significant for applications in the fields of displays and photovoltaics, etc. ZnO colloidal nanocrystals have been synthesized by the chemical solution technique [19]. Tungsten oxide ($WO_{3-\delta}$) is an electrochromic material that is suitable for applications in the fabrication of smart windows. The free charge carriers are efficacious for the electrochemically regulated optical absorbance emerges due to oxygen vacancies in $WO_{3-\delta}$ [20]. As compared to tungsten, rhenium (Re) exhibits one more valence outer d electron. The optical peculiarities of rhenium oxide (ReO_3) were illustrated by Biswa et al. ReO_3 colloidal nanocrystals contain LSPR in the visible wavelength range (488 to 534 nm) [21]. Nanocrystals based on colloidal molybdenum trioxide (MO_{3-x}) are considered to be plasmonic transition metal oxide nanomaterials. The previous research reports reveal that $MO_{3\text{-}x}$ acquires LSPR ranges from 1200 to 600 nm [22].

Titanium oxide (TiO_2) is a semiconductor material with a wide optical bandgap that is frequently explored in relation to cosmetics, coatings, and paper, and also for applications based on dye sensitized solar cells and photocatalysis, etc. Gordon et al. noted the LSPR phenomenon in the mid IR wavelength range in nanocrystals based on TiO_2 formed by TiF_4. The LSPR was observed on account of oxygen vacancies instigated by fluorine, illustrated by X-ray photoelectron spectroscopy (XPS) and electron paramagnetic resonance (EPR). Metal nanocrystals that are extrinsically doped, for example, doping of indium with cadmium oxide (CdO), display significant LSPR [23]. Synthesis of different morphologies of indium cadmium oxide was done by organic solution phase synthesis. On account of free conduction electrons, indium cadmium oxide nanocrystals display the LSPR phenomenon [24]. For the exploration of metal oxides materials in plasmonics applications, various synthesis techniques have been followed, such as atomic layer deposition, hydrothermal technique [25], solvothermal processing technique [26], etc. Some of the synthesis routes have been described in the following section.

2 Synthesis of plasmonic materials

Plasmonic materials containing hybrids of metal oxides and noble metals can be synthesized either in the form of thin films or nanocomposites. Thin films of metal oxides can be deposited as under-layer or over-layer to the layer of noble metals grown on different substrates. Various techniques, such as magnetron sputtering, atomic layer deposition, molecular beam epitaxy, electrochemical deposition, etc., can be used for the fabrication of metal oxide/noble metal heterojunctions in the form of thin films. The nanocomposite hybrids of metal oxides and noble metals can be synthesized using numerous chemical routes, such as sol-gel, hydrothermal, polyol, and chemical colloidal methods, etc. Brief descriptions of these techniques are provided below.

2.1 Magnetron sputtering

Magnetron sputtering belongs to the category of physical vapor deposition techniques used to grow layers of a sample with varying thicknesses over a substrate. The technique involves ejection of atoms from the target material to be deposited by bombardment with energetic ionic species constituting plasma [27,28]. A basic schematic of the sputtering process is shown in Fig. 1. The magnetron sputtering unit mainly comprises a sputtering chamber equipped with vacuum system, power supply, and inert

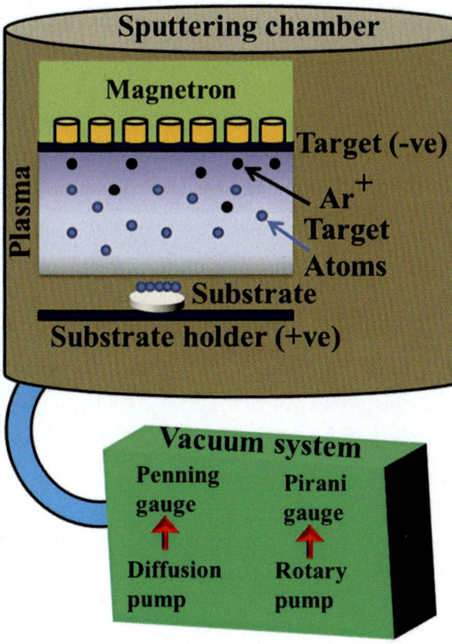

Fig. 1 Schematics of the sputtering process.

gas supply. A diffusion pump, rotary pump, Pirani gauge, and Penning gauge, along with the accessories for their functioning, make up the essential parts of the vacuum system.

The sputtering chamber is maintained under high vacuum conditions with pressure of not less than 10^{-6} mbar, which is monitored by the corresponding gauges. A high vacuum or a low pressure inside the sputtering chamber is mandatory for high quality deposition since it helps to get rid of the ambient air molecules interfering with the deposition. After attaining the desired vacuum conditions, the inert gas (specifically, argon gas) is injected into the sputtering chamber. A negative bias is then applied to the target, which eventually ionizes the inert gas atoms into positively charged ionic species with the generation of secondary electrons. This ionized inert gas constitutes the glowing discharge known as plasma. The positively charged ions of inert gas are accelerated towards the negatively biased target material. After gaining sufficient energy, the inert gas ions transfer their momentum to the target atoms, which subsequently results in sputtering out of target atoms from the surface of the material.

The ejected atoms deposit on the substrate material after drifting between the electrodes while undergoing successive collisions with the ionic and atomic species present in the path. The magnetron generally placed behind the target material helps in plasma confinement, so as to enhance the sputtering yield by constraining the electrons to travel in a spiral path. The power supply used to bias the electrodes in the sputtering chamber could be a direct current (DC) supply or a radio frequency (RF) supply, and the sputtering modes are commonly referred to as DC magnetron sputtering and RF magnetron sputtering, respectively. The drawback of DC sputtering is that it can only be used for the deposition of conducting materials. This is because of the charge accumulation at the surface of insulating target materials due to electric charge neutralization, which consequently ceases the sputtering process in insulating materials. The problem of charge accumulation is resolved by periodically varying the polarity of electrodes with a fixed frequency (mostly 13.56 MHz). Hence, RF magnetron sputtering is commonly employed for the deposition of metal oxide layers, while DC magnetron sputtering is utilized for depositing conducting noble metals.

Torrell et al. deposited TiO_2/Au films by reactive magnetron co-sputtering by powering a titanium target with gold pellets using DC current density of 100 A/m^2 and maintaining the substrate temperature at 150°C [29]. The homogeneous composites of TiO_2 and Au were made by postannealing the samples at distinct temperatures from 200°C to 800°C. The as-deposited sample shows clustering of Au nanoparticles to form nonhomogeneous composites (Fig. 2A). Nanoparticle coalescence after annealing leads to the formation of homogeneously dispersed Au nanoparticles in the TiO_2 matrix (Fig. 2B and C). The group were able to effectively modify the characteristics of plasmonic peak, i.e., shift the position of the peak from 610 to 710 nm, and also alter the peak width by varying the annealing temperature. Veziroglu et al. also fabricated Au decorated TiO_2 thin film structures where reactive DC magnetron sputtering was utilized to fabricate TiO_2 thin films and a photocatalytic deposition method was used for Au synthesis [30]. The group demonstrated the enhanced photocatalytic performance of Au/TiO_2 plasmonic structures in comparison to pure TiO_2 nanostructures.

2.2 Molecular beam epitaxy

Molecular Beam Epitaxy (MBE) is a significant thin film deposition technique utilized to grow epitaxial layers of a desired material on a suitable substrate, generally at higher temperatures. In this technique, the atomic or molecular beams produced by evaporating a solid source at high temperatures diffuse towards a thermally heated substrate to grow an epitaxial thin film. MBE is performed in ultra-high vacuum (UHV) conditions so as to increase the mean free path of the atomic or molecular beam drifting between the source and substrate. The major concern for the MBE process is

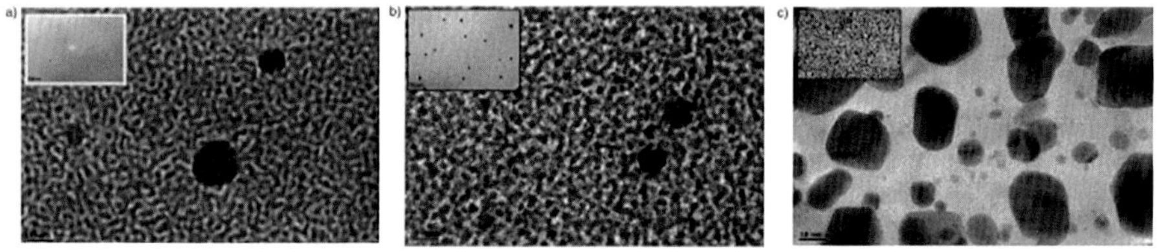

Fig. 2 Planar TEM view of the (A) as-deposited sample, (B) sample annealed at 500°C showing growing gold NPs, and (C) sample annealed at 700°C, showing the formation of larger and elongated NPs.
Reused with permission from M. Torrell et al., Tuning of the surface plasmon resonance in TiO_2/Au thin films grown by magnetron sputtering: the effect of thermal annealing. J. Appl. Phys. 109(7) (2011) 074310.

low deposition rate, which restricts its usage for the mass production of thin films to be used for commercial purposes. Nevertheless, a high quality deposition with precise control over the lattice parameters, interface, textured growth, thickness, and composition require the MBE technique for thin film deposition.

A typical MBE system comprises a deposition chamber made of stainless steel, a load-lock chamber to transfer the substrate from air to main deposition chamber, gate valves to connect different chambers, a substrate holder with heater, effusion cells to evaporate the solid sources at high temperature, liquid N_2 cryopanels, and a vacuum system [31,32]. The substrate holder is rotated at a fixed rpm so as to attain uniform and homogeneous deposition. The effusion cells form the primary constituent of the MBE system, since they are responsible for providing outstanding flux stability resulting in a uniform and high purity film deposition. The material and geometry for the crucibles used in effusion cells must be carefully chosen (commonly used crucible material is pyrolitic boron nitride) because these materials need to withstand a high temperature (~1400°C) for a long duration. Ta filament is used to provide heating in the effusion cells. These effusion cells are generally 6 to 10 in number for modern MBE systems. These cells, in proper orientation, are simultaneously focused over the substrate holder while retaining optimum flux uniformity. Furthermore, the orientation of effusion cells must be carefully designed in such a way that the material flux does not show significant fluctuations as the source becomes partially exhausted. The UHV environment in an MBE system also provides an additional advantage for using electron diffraction probes, specifically, reflection high energy electron diffraction (RHEED), to study in situ growth mechanisms and also measure the film thickness precisely.

Cheng et al. fabricated heavily Ga-doped ZnO (GaZnO) plasmonic metamaterials using MBE and varied the Ga doping concentration as 6.8, 6.3, and 3.2 at% [33]. Fig. 3 shows the variation in morphology for decreasing concentration of Ga in GaZnO as observed by Atomic force microscopy. The plasmonic peak absorbed by this system showed a wide shift in the wavelength from 1333 to 1515 nm with increasing Ga doping concentration. Sadofev et al. prepared ZnO:Ga and ZnMgO:Ga layers using a

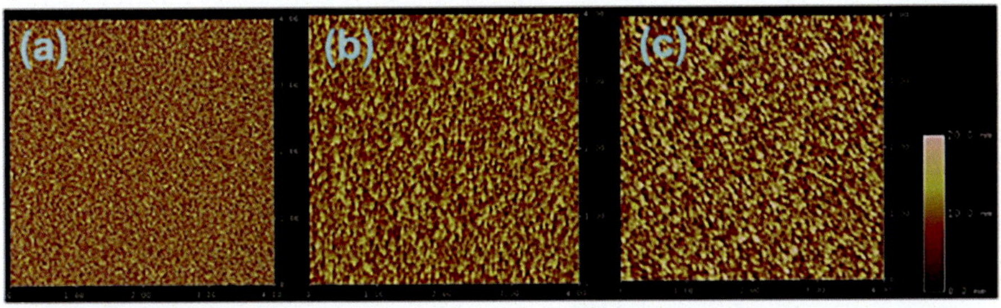

Fig. 3 AFM images of GaZnO with varied the Ga doping concentration as (A) 6.8 at%, (B) 6.3 at%, and (C) 3.2 at%.
Reused with permission from Y.C. Cheng et al., Wide range variation of resonance wavelength of GaZno plasmonic metamaterials grown by molecular beam epitaxy with slight modification of Zn effusion cell temperatures, J. Alloys Compd. 870 (2021) 159434.

DCA 450 MBE system with standard solid-source effusion cells (Zn, Mg, Ga) and an Addonrf plasma source providing active O [34]. The group demonstrated the fabricated system as a low damping plasmonic material at 1.55 μm wavelength, which is commonly used for telecommunications.

2.3 Atomic layer deposition

Atomic layer deposition (ALD) is an advanced form of traditional chemical vapor deposition (CVD) technique. ALD differs from CVD in terms of sequential insertion of precursors in ALD, as opposed to the simultaneous reaction between different precursors used in CVD. This results in a more controlled deposition using ALD. ALD proves to be superior compared to CVD due to the ease of deposition at (or close to) room temperature, precise control over thickness, stoichiometry, and conformal degree of freedom in the former technique; however, it is limited by its low deposition rate.

In the ALD process, the precursor in gaseous phase is introduced to the substrate, resulting in the adsorption of precursor molecules on the substrate surface [35,36]. The gaseous precursor is purged out of the chamber, followed by a flow of inert gas. The second precursor, in gas phase, is then introduced into the chamber and similar reactions occur on the substrate surface. This procedure is repeated to attain a layer of desired thickness and stoichiometry. The reaction by-products are removed by a flow of inert gas after the surface reaction between each precursor and substrate and also after the completion of the growth process. The sequential insertion of gaseous precursors separated by inert gas avoids the inter-mixing and self-reaction of two precursors.

The exclusive characteristic of the ALD process is that the chemical reactions occur only on the surface to form a thin film, i.e., the reactions are self-limiting in nature. Therefore, an appropriate choice of precursors is essential for a controlled ALD process, such that the precursors exhibit high reactivity towards the surface sites while having low reactivity within themselves. Thus, highly crystalline stable structures can be achieved even for the ultra-thin films deposited using ALD.

Various ALD processes are possible, based on the type of surface reactions taking place between substrate and precursors. Some of the ALD processes are thermal ALD, plasma-enhanced ALD, photo-assisted ALD, spatial ALD, etc., depending on whether the surface reactions occur by providing temperature, RF power supply, UV light, or localized exposure of substrate to precursor, etc., respectively. The ALD reactor chamber is maintained under high vacuum conditions to eradicate the possibility of reaction by interfering gases. An ALD reactor chamber primarily constitutes a vacuum pump, pressure gauge heads, inlet for gaseous precursors and inert gas, and a substrate holder equipped with a heater and water circulation system. Each ALD process requires a unique design of ALD reactor. Major ALD reactors include flow-type reactors, showerhead ALD reactors, batch ALD reactors, energy enhanced ALD reactors, spatial ALD reactors, etc.

ALD has been used as a technique for the deposition of numerous metal oxide-based plasmonic materials. Several groups have made an attempt to fabricate

Al-doped ZnO plasmonic hetrostructures using ALD at temperatures less than 250°C using diethylzinc, trimethylaluminum, and water as the precursors for Zn, Al, and O_2, respectively [37,38]. It is well demonstrated that the formation of OH^- radicals on the substrate surface due to water decomposition provides as active sites for the growth of Zn^+ and Al^+ ions. Santangelo et al. prepared α-Fe_2O_3/NiO core shell nanofibers, where NiO layers were grown using ALD and the α-Fe_2O_3 layers were deposited using electrospinning followed by calcination [39]. The fabrication of NiO was done in a commercial ALD reactor by Arradiance [40], where nickelocene was used as a metal precursor and ozone acted as precursor for oxygen. The deposition could be achieved at a low temperature of 200°C by monitoring the pulse and exposure time of the two precursors. The desired thickness of NiO could be achieved by controlling the ALD cycles' number (100–1150). They established that polycrystalline NiO shell deposited by ALD onto the electrospun crystalline α-Fe_2O_3 serves as a plasmonic layer, which enables two-magnon mode detection of α-Fe_2O_3 at 1585 cm^{-1} in the micro-Raman spectrum at excitation wavelength of 532 nm.

2.4 Electrochemical deposition

Electrochemical deposition is a thin film fabrication technique in physical chemistry involving electrical and chemical methods of synthesis. The process involves solid coating of metallic ions immersed in an electrolyte solution being deposited over the surface of a conducting substrate acting as cathode [41]. The electrochemical setup typically consists of an electrochemical cell containing the electrodes and electrolyte solution. The electrolyte solution is usually a homogeneous aqueous solution of dissolved salts containing ions of the material to be deposited. Two or three electrodes can be used for the electrochemical deposition, namely counter electrode, working electrode, and reference electrode (optional). The positively charged ionic species dispersed in the electrolyte solution gets deposited over the negatively biased working electrode. The working electrode must be a substrate made up of a conducting material so as to allow the flow of charge carriers. The counter electrode is essential to make a closed circuit mediated by ions in the electrolyte solution. Pt electrode is most commonly utilized as counter electrode due to its inert nature, such that it does not alter the desired electrochemical reaction taking place. The working electrode is considered to be the cathode, while the counter electrode is considered as the anode in the electrochemical reactions. An optional reference electrode is also equipped in modern day electrochemical apparatus, which is considered as a reference to quantify the potential of working and counter electrodes. Therefore, a reference electrode should be made of a material with well-defined and stable equilibrium potential. The most common examples of reference electrodes are saturated calomel electrode (SCE), standard hydrogen electrode (SHE), and AgCl/Ag electrode, etc., immersed in an aqueous media.

A material can be deposited over a particular substrate by applying either a suitable potential difference between the two electrodes or by supplying an appropriate current to the two electrodes viz. working and counter electrode. The two modes of electrochemical deposition are known as potentiostatic mode and galvanostatic mode, respectively [42]. The reduction of ionic species present in the aqueous electrolyte solution occurs beyond a certain threshold potential between the cathode and the anode. This threshold potential, also known as redox potential, is distinct for each chemical reaction taking place, since each element becomes reduced and oxidized at a specific redox potential. Cyclic voltammetry is an extensively used technique to identify the redox potential of elements to be deposited with respect to the reference electrode. In this technique, a linear potential sweep is applied to the working electrode, which allows the flow of electronic and ionic charges across the circuit established between the two electrodes and intermediate electrolyte solution. The current is measured at the working electrode simultaneously. The current flowing through the cathode (or working electrode) shows maxima at redox potential, which can be used to identify the potential at which the analyte is reducing or oxidizing. The pH of electrolyte and temperature of deposition are the two crucial factors determining the current flowing through the electrodes. Fig. 4 shows a typical electrochemical deposition setup and the typical cyclic voltammogram obtained using this setup.

The electrochemical deposition technique has been widely utilized for the fabrication of plasmonic oxides. For instance, Khan et al. illustrated the electrochemical synthesis of Ag/α-Fe_2O_3/TiO_2 nanoarrays implementing a sonochemically assisted electrochemical anodization technique [43]. A two-electrode electrochemical cell was used, having Ti foil and Pt as anode and cathode, respectively, with a distance of ~1 cm between them. The electrolyte consisted of 0.5 wt% NaF solution in ethylene glycol and 5% DI water. They claimed that DI water facilitated the dissolution of NaF in ethylene glycol while decreasing the viscosity of the solution and ethylene glycol augmented the formation of homogeneous nanotubes due to its mild etching environment and polar protic nature. $FeCl_3$ and $AgNO_3$ precursors were subsequently added to the electrolyte solution after appropriate anodization time and specific voltage at

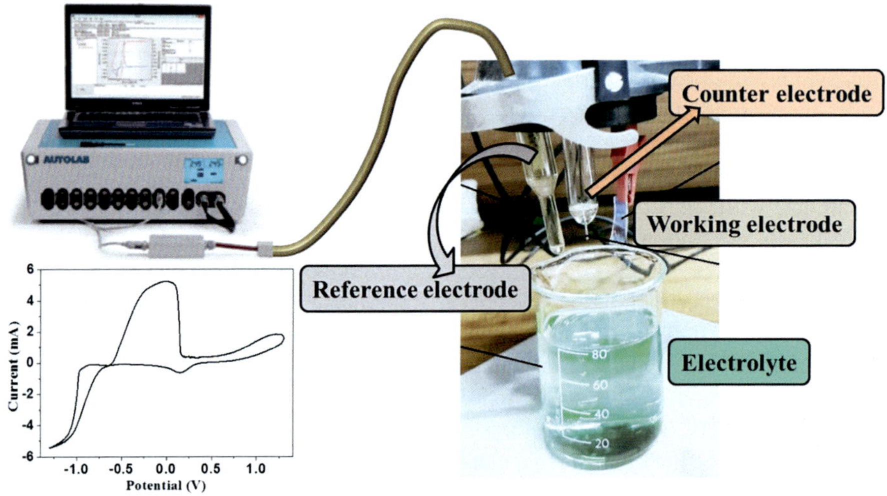

Fig. 4 Electrochemical deposition setup and a representative cyclic voltammogram.

each step. The crystallinity of prepared nanostructures was increased by washing, drying, and calcination procedures at 400°C. The group demonstrated reduction in band gap along with enhanced absorption towards higher wavelength (470 nm) for Ag/α-Fe_2O_3/TiO_2 nanoarrays in comparison to α-Fe_2O_3/TiO_2 (424 nm) and TiO_2 structures (386 nm). Owing to the optical characterizations, the group suggested Ag/α-Fe_2O_3/TiO_2 nanoarrays for efficient solar-driven water splitting applications.

Cabezas et al. directed the electrochemical synthesis of Nb_2O_5 and Ta_2O_5 using a three-electrode electrochemical cell with Ti doped indium oxide (ITO) coated glass substrates as working electrode, Pt foil as counter electrode, and Ag/AgCl as the reference electrode [44]. The electrolyte comprised an aqueous solution of hexaniobate or hexatantalate clusters (10 mg/mL) dissolved in 0.1 M tetramethylammonium chloride solution. The deposition was carried out by fixing the potential ranging between 2.4 and 4.0 V (with respect to Ag/AgCl reference electrode) for a specific time. The hybrids of Nb_2O_5 or Ta_2O_5 with WO_{3-x} or ITO nanocrystals are beneficial for tuning near infrared transmittance without causing an impact on visible transparency due to the plasmonic response of nanocrystals.

2.5 Polyol method

The polyol technique is a widely used liquid-phase synthesis technique for obtaining various inorganic compounds, such as metal, oxide, and semiconductor nanoparticles in polyalcohols. In this process, liquid organic compounds, such as 1, 2-diols or ether glycols, are used as a solvent and reducing agent [45]. Over the years, the polyol synthesis route has attracted much attention due to the controlled synthesis of nanoparticles of different shapes, sizes, and compositions. Several important parameters, such as the choice of solvent, reducing agent, and capping agent, can stimulate both nucleation and growth of nanoparticles. This is because, at certain operating temperatures, liquid organic compounds are oxidized to ketone and aldehyde species that can allow reduction of solid precursors. Furthermore, the capping agent facilitates the reduction at specific crystal faces through surface adsorption, which further allows the anisotropic growth of nanoparticles along with highly-index facets. Therefore, a careful selection of precursors, solvents. and capping agents can lead to a nanoparticle of explicit shape and size with a specific physicochemical property.

The different steps involved in the polyol process are as follows: (i) dissolving the precursor at the optimal reflux temperature, (ii) the nucleation step, and (iii) the growth step leading to the formation of nanoparticles. Qi et al. synthesized TiO_2@Ag core shell structures where the polyol method was used to produce Ag nanostructures [46]. 0.1 mmol $AgNO_3$ solution was dissolved in 25 mL ethylene glycol solution containing polyvinylpyrrolidone (PVP) with constant stirring at ambient temperature. The solution was then stirred at 120°C for 1 h followed by centrifugation to separate out the nanoparticles and coat over TiO_2 nanoparticles. Fig. 5 shows TEM images of synthesized TiO_2@Ag nanoparticles and the corresponding absorption spectra. Higher dielectric constant of TiO_2 than that of PVP results in the shift of plasmonic peak towards higher wavelengths from 403 nm in Ag nanoparticles to 421 nm in TiO_2@Ag nanoparticles.

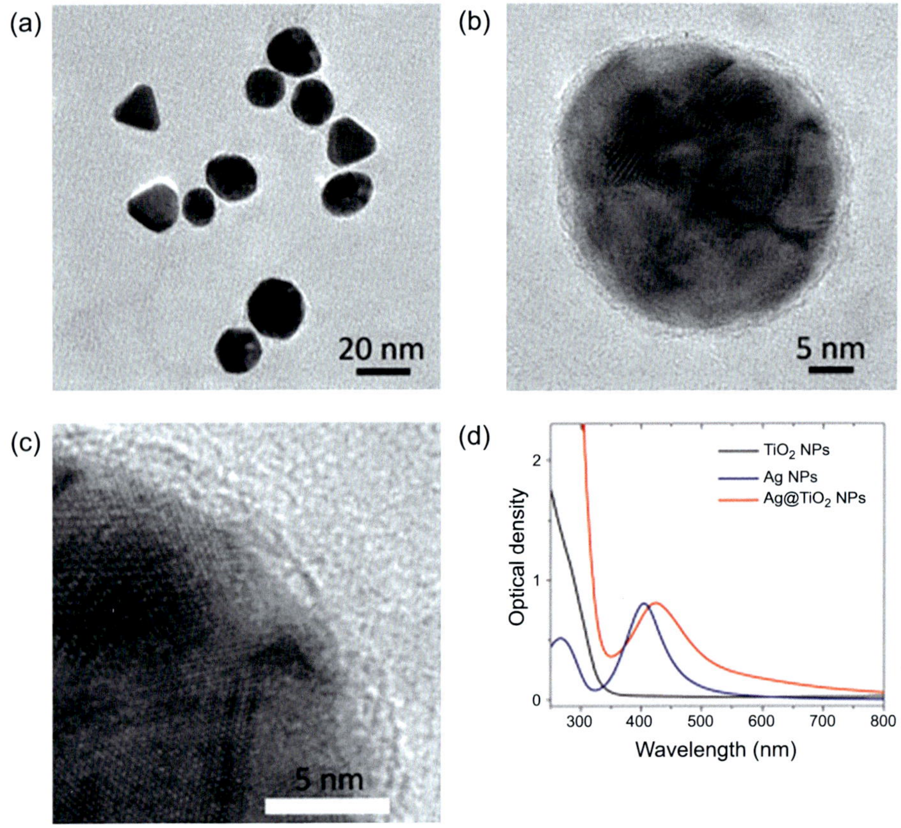

Fig. 5 Ag@TiO_2 NPs. (A–C) TEM and HRTEM images (D) Optical absorption spectra of solutions of Ag NPs stabilized by PVP, TiO_2, and Ag@TiO_2 NPs.
Reused with permission from J. Qi, X. Dang, P.T. Hammond, A.M. Belcher, Highly efficient plasmon-enhanced dye-sensitized solar cells through Metal@Oxide core-shell nanostructure. ACS Nano, 5(9) (2011) 7108–7116. https://doi.org/10.1021/nn201808g.

2.6 Sol-gel method

The sol-gel method, which is a considered to be a wet chemical method, is used for the synthesis of different metal and metal oxide nanostructures at low reaction temperature. It is a cost-effective approach to obtain a high surface area material with a uniform and stable surface. In this method, the precursors are dissolved in a solvent and turned into a gel by continuous magnetic stirring and heating [47]. Following this, the resulting wet gel is dried by an appropriate method, i.e., thermal or supercritical drying, depending on the desired properties and application. Subsequently, the produced gel is pulverized and calcined at a particular temperature. Basically, the sol-gel method involves three main steps: (i) hydrolysis/alcoholysis of the precursors, (ii) poly-condensation, (iii) drying of the obtained products. Finally, thermal treatment is carried out to expel the residues and the water molecules from the sample.

Giancaterini et al. synthesized ZnO thin films with 5% Au and Pt additives using the sol-gel method to evaluate their sensing performance for detecting H_2, CO, and NO_2 gases [48]. ZnO was prepared by sol-gel method using zinc acetate (Zn $(CH_3COO)_2$), ethanol (C_2H_5OH), and monoethanolamine (MEA, (CH_2CH_2OH) NH_2), while Au and Pt were prepared by the polyol method using chloroplatinic acid hydrate (H_2PtCl_6), ethylene glycol, polyvinyl pyrrolidone (PVP), and sodium nitrate ($NaNO_3$). The group demonstrated the role of Pt additions in enhancing the gas sensing characteristics. Kambhampati et al. also investigated the basic properties and sensing characteristics of sol-gel prepared SiO_2 films on noble metals [49]. They presented the sol-gel method as being superior to the other deposition techniques on the basis of careful monitoring of the nanostructures and thickness of films offered by the sol-gel technique. This is because the group used highly diluted Si precursors to allow complete hydrolysis of the initial alkoxide and to constrain the condensation reactions of hydrolyzed monomers and oligomers. Thus, good control could be attained over the sizes of resulting sol particles resulting in a large packing density.

2.7 Chemical colloidal method

The chemical colloidal method is another cost-effective technique for synthesizing various crystalline nanomaterials due to its fast nucleation and controlled growth. Size control is achieved through the use of coating agents and controlling processing temperature and processing time. Fig. 6 shows the basic schematic of the experimental apparatus used for synthesis using the chemical colloidal method [50]. In this synthesis approach, deionized water acts as a solvent that limits the growth and agglomeration of nanoparticles using reducing and stabilizing agents. The configuration used for this technique is also known as "reflux" because the chemical reaction involved in this method takes place at the boiling point of the solvent. In this configuration, cold water is also circulated in a condenser tube in order to condense the vapor. After heating for fixed time intervals, the prepared solution is cooled to ambient temperature and the resulting precipitate is removed after centrifugation. Subsequently, the resulting precipitate is dried in a vacuum oven and ground to a fine powder with an electric agate mortar.

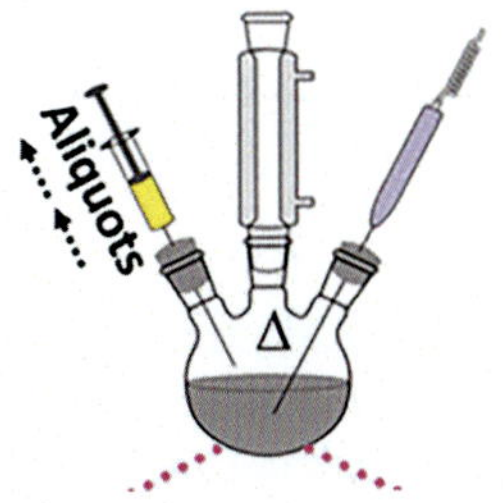

Fig. 6 Schematic illustration of a colloidal synthesis setup.
Reused with permission from F. Horani, E. Lifshitz, Unraveling the growth mechanism forming stable γ-In2S3 and β-In2S3 colloidal nanoplatelets, Chem. Mater., 31(5) (2019) 1784-93.

The chemical colloidal method has been extensively utilized for the deposition of metal oxide-based plasmonic materials. Choudhary et al. synthesized ZnO@Ag plasmonic nanostructures using the chemical colloidal method and demonstrated the enhanced photocatalytic response of the system as compared to pure ZnO nanorods [51]. The group used zinc acetate dehydrate ($Zn(CH_3COO)_2.2H_2O$, Alfa Aesar) as Zn precursor along with PVP($(C_6H_9NO)_n$, alfa aesar) of 8000 molecular weight to avoid agglomeration of particles and 0.15 M hexamethylenetetramine [HMTA, ($C_6H_{12}N_4$)], all uniformly mixed in an aqueous solution. $AgNO_3$ salt was used as the precursor for Ag, which was added in different amounts to prepare composites with different concentrations of Ag. They reported the formation of dumbbell-shaped ZnO nanostructures for ZnO@Ag composites and explained systematically that Ag stops the growth of ZnO along the central diameter of the dumbbells (Fig. 6). The optical absorption at 440nm confirmed the SPR effect of Ag nanoparticles. The photoluminescence spectra also showed enhancement in the emission peak with increasing Ag concentration.

3 Methods to observe the plasmonic effect

After a brief introduction to various synthesis techniques that can be employed for the fabrication of plasmonic materials, it is important to learn about the characterization techniques utilized to demonstrate the plasmonic response of metal oxide/noble metal hybrids in thin film and nanocomposite forms. Some of the characterization techniques, including UV-visible spectroscopy, surface enhanced Raman spectroscopy, photoluminescence spectroscopy, electron energy loss spectroscopy, etc., have been discussed here. A brief outline for these techniques in observing the plasmonic effect is presented in this section.

3.1 UV-visible spectroscopy

Plasmonic materials are known to absorb radiations with energy ranging in the UV-visible region. The amount energy of radiation absorbed depends on the noble metals used along with the morphology and size of plasmonic materials. This is because of the resonance occurring between the incident radiation frequency and frequency of collective oscillations of electrons present on the surface of nanoparticles, commonly known as the surface plasmon resonance (SPR) effect. The absorption of radiation at particular energy gives rise to a peak in the absorption spectra in the UV-visible region. Hence, UV-visible spectroscopy is a prevalent technique used to characterize plasmonic nanoparticles.

Choudhary et al. used UV-visible spectroscopy to demonstrate the plasmonic effect of shuttlecock-shaped ZnO@Ag hybrid nanocomposites [52]. The group scanned the absorption spectra of pure Ag, pure ZnO, and ZnO@Ag hybrid nanocomposites and observed that pure ZnO only exhibited the absorption edge at 374nm, corresponding to its band gap, and did not show any SPR peak; whereas, pure Au spherical nanoparticles of 43nm diameter showed a narrow SPR peak at 414nm. Interestingly,

ZnO@Ag nanocomposites were reported to exhibit a broad SPR peak, centered at 423 nm, along with the intrinsic absorption edge of ZnO. A perceived redshift in the wavelength of SPR peak in ZnO@Ag hybrid nanocomposites was reported due to the decrease in charge density of Ag in ZnOA@Ag nanocomposites compared to pure Ag nanoparticles. This reduced charge density of Ag was explained by the charge transfer from Ag to ZnO, while realignment of Fermi levels occurred in composites. The absorption wavelength of plasmonic nanoparticles (λ_p) depends on the effective mass of electrons (m_{eff}) and the electron density of the material (N), as given in Eq. (1).

$$\lambda_p = \sqrt{\frac{4\pi^2 c^2 m_{eff}}{Ne^2}} \tag{1}$$

where c is the speed of light.

Similarly, in another work, same group also reported enhanced absorption in UV-visible spectroscopy with the increase in concentration of Ag in ZnO@Ag nanocomposites [52,53]. Gorelikov et al. characterized a polymer loaded with Au nanorods using UV-visible spectroscopy and observed two SPR peaks at 400 and 810 nm (Fig. 7), ascribed to the transverse and longitudinal plasmonic behavior [54]. The group concluded that the appearance of a single SPR peak centered at ~400 nm is due to the spherical particles and the double SPR peaks signifies the

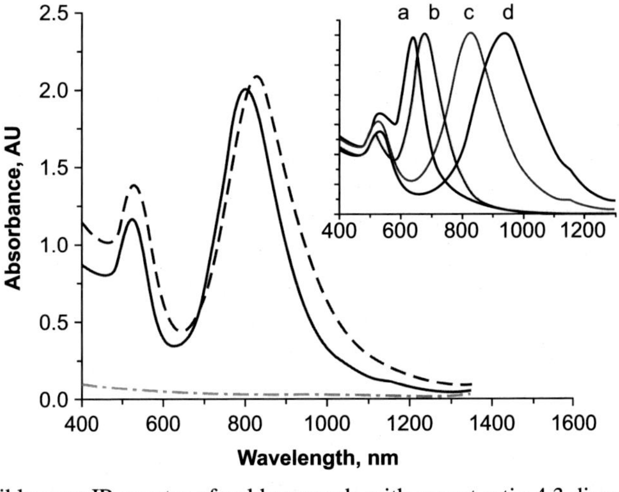

Fig. 7 UV-visible-near-IR spectra of gold nanorods with aspect ratio 4.3 dispersed in an aqueous solution (s) and localized in the interior of microgels (- - -). (Inset) Variation in absorption spectra of gold nanorods dispersed in an aqueous solution as a function of their aspect ratio: (A) 2, (B) 2.5, (C) 4.3, and (D) 6.
Reused with permission from I. Gorelikov, L. M. Field, E. Kumacheva, Hybrid microgels photoresponsive in the near-infrared spectral range, J. Am. Chem. Soc., 126 (49) (2004) 15938–15939. https://doi.org/10.1021/ja0448869.

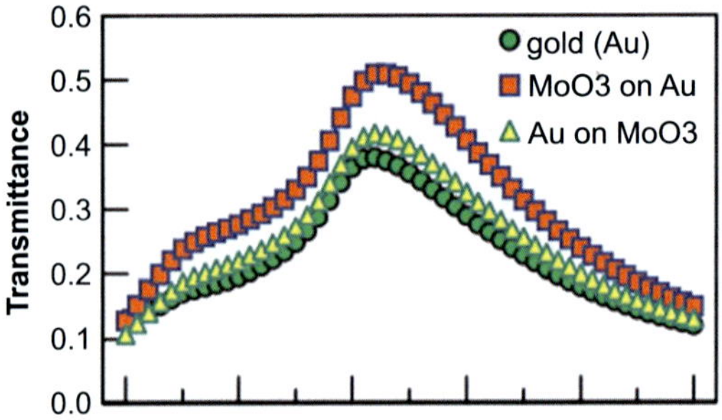

Fig. 8 Comparison of the transmittance of a pure single gold layer with that of a bi-layer consisting of gold on MoO_3 or MoO_3 on gold. The thickness of each layer was 25 nm. Reused with permission from M.F. Al-Kuhaili, Enhancement of plasmonic transmittance of porous gold thin films via gold/metal oxide bi-layers for solar energy-saving applications, Sol. Energy 181 (2019) 456–463.

nonspherical, elongated or rod-like morphology of metallic nanoparticles. Therefore, it can be justified that the position and intensity of SPR peaks can provide information regarding the morphology of plasmonic nanoparticles. Fig. 8 shows the plasmonic transmittance spectra in the range of 300–800 nm with the plasmonic peak of pure gold and gold on MoO_3 or MoO_3 on gold. Here, the dielectric is chosen to be MoO_3 thin films to enhance the gold transmittance, and it is preferred to apply this on the top of the gold layer as an antireflection coating [55].

3.2 Photoluminescence spectroscopy

After irradiating a metal oxide nanoparticles-based system with electromagnetic radiation of appropriate frequency, the electrons are excited to high energy levels, which then relax to the ground state through radiative or nonradiative transitions. These transitions result in either the enhancement or quenching of photoluminescence (PL) signal. In a metal oxide/noble metal plasmonic system, the resonance may occur between the frequency of collective oscillations and exciting frequency, resulting in the enhancement of the local electric field of the plasmonic material [56]. This results in the enhancement of PL intensity in the plasmonic system.

Nakaji et al. revealed a remarkable increase in PL intensity for Si nanodisks hybridized with Au nanoplates (Fig. 9), and reported that the intensity enhancement was greater for an increasing area of Au plates [57]. They reported PL enhancement factor up to 5, subject to the morphology and dimension of Au nanostructures along with the excitation wavelength. The group suggested enhanced PL intensity associated with strong plasmonic effects owing to the plasmon-enhanced electric field localized in the vicinity of surface of metallic nanostructures. The plasmonic effects can be strengthened further when the optically active materials are placed at a certain distance, ranging from 5 to 15 nm, based on the morphology of metallic nanoparticles.

Rashed et al. demonstrated enhanced PL intensity of LDS750 dye in the presence of Au plasmonic nanoparticles of different sizes [58]. The group ascribed higher PL

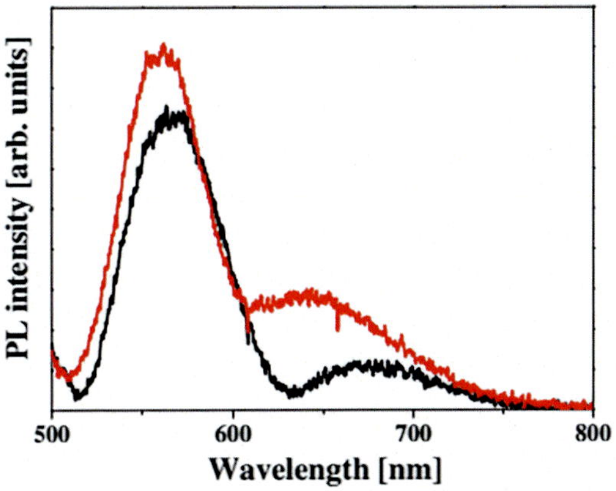

Fig. 9 Micro-PL spectra from Si ND arrays.
Reused with permission from K. Nakaji, H. Li, T. Kiba, M. Igarashi, S. Samukawa, A. Murayama, Plasmonic enhancements of photoluminescence in hybrid Si nanostructures with Au fabricated by fully top-down lithography, Nanoscale Res. Lett. 7 (2012) 629.

intensity in the plasmonic hybrids to the enhanced Purcell factor owing to the near-field coupling between plasmonic oscillations and incident excitation frequency. They confirmed that the LSPR band of the plasmonic nanoparticles (referred to as nanoresonators) can be specifically matched with the emission spectrum of LDS 750 dye by tailoring the diameter of Au nanodisks. After illuminating the sample with a 532-nm laser, the PL intensity was observed for different diameters of Au nanodisks, ranging from 102 to 180 nm, and the PL intensity was found to be maximum for Au nanodisks with diameter 135 nm. Therefore, the PL emission from plasmonic materials can be used as an efficient technique to determine the plasmonic response of a specimen.

3.3 Surface enhanced Raman spectroscopy

Surface enhanced Raman spectroscopy (SERS) is a significant technique devised from the local surface plasmon resonance (LSPR) effect. SERS offers outstanding capacity to examine photo-induced plasmonic reactions with a high resolution of energy and surface sensitivity [59]. Typically, Raman spectroscopy involves inelastic scattering of photons by molecules inside a sample due to the transitions between vibrational modes. This results in Stokes or anti-Stokes Raman lines, depending on the decrease or increase in frequency of emitted radiation with respect to the incident frequency. The intensity of Raman lines increases distinctly for isolated metallic nanostructures owing to the LSPR effect. The technique used to measure such high resolution and highly sensitive Raman signals is commonly termed SERS. Raman intensity is observed to increase from 10^3 to 10^5 times for single spheres or 10^6 to 10^{11} times for single polygons in coupled particle-ensembles [60]. This increase in Raman intensity is due to the increase in the local electric field due to plasmonic oscillations by metallic nanoparticles. The enhancement in Raman intensity acts to the fourth power of the enhancement in the local electric field induced by plasmonic nanocomposites. Thus, if enhancement in the localized electric field is 10^3 times, then the SERS signal is enhanced by 10^{12} times.

Besides the contribution of electric field enhancement in SERS signal, a significant impact is also provided by chemical enhancement, owing to the increase in Raman scattering cross-section and charge transfer mechanisms mediated by photons illumination. Zhou et al. showed highly intense Raman signal at $1583\,cm^{-1}$ in the SERS spectra of ZnO-Au composites and this Raman peak intensity in ZnO-Au was double that of pure Au [61]. The group reasonably attributed the enhanced SERS signal to the chemical enhancement mechanism, since the resonant excitation of plasmons could not be seen in ZnO nanorods. They also suggested that the charge transfer from Au to conduction band of ZnO could further enhance the SERS signal. Zhao et al. demonstrated the SERS activity applications in self-cleaning TiO_2/Au nanowire arrays to be used as highly sensitive and recyclable SERS sensors [62]. Fig. 10A shows the SERS spectra of (10^{-4} M) R6 G molecules over TiO_2/Au nanowire arrays before and after self-cleaning. The spectra shows Raman signals disappearing after UV light irradiation, which are absorbed again on the cleaned TiO_2/Au system, exhibiting a strong SERS signal. The recyclable SERS activity for detecting R6 G molecules is shown in Fig. 10B. The group successfully concluded that the prepared TiO_2/Au nanowire arrays could be utilized as efficient SERS sensors.

4 Remarkable plasmonic applications of metal oxides

4.1 Gas sensing

The chemical sensing of the gases dependent on metal oxide for escalated temperature plasmonics-based sensors is important for the optimization of different combustion systems. The detection of emission gases such as H_2, CO, and No_x is technologically significant for the reduction of the effects of these gases in the environment and for maintaining the increased emission regulatory requirements [63–65]. There are some potential metal oxide candidates, such as ZnO, YSZ, WO_3, SnO_2, TiO_2, ITO, etc., used for plasmonic gas sensors. Degenerately doped metal oxides (MOs) nanocrystals (NCs) are used for plasmonic and chemoresistive gas sensing. The optical properties shown by near-IR LSPR degenerately doped MOs have never been sensed as real-time gas sensors.

Sturaro et al. reported the application of GZO nanocrystals for novel plasmonic gas sensors for the detection of H_2 and NO_2 gases. The change in the dopant concentration resulted in optical detection of oxidizing and reducing gases from GZO films at temperatures lower than 100°C. The gas sensitivity of NO_2 is improved by activating the response of the sensor through combined purple-blue (λ=430 nm) light radiation and heating with a temperature of 75°C [66]. Accountably, the aliovalent dopants are used to give rise to localized surface plasmon resonance (LSPR) in the near IR region, which is aroused from increasing the free electron concentration in the material.

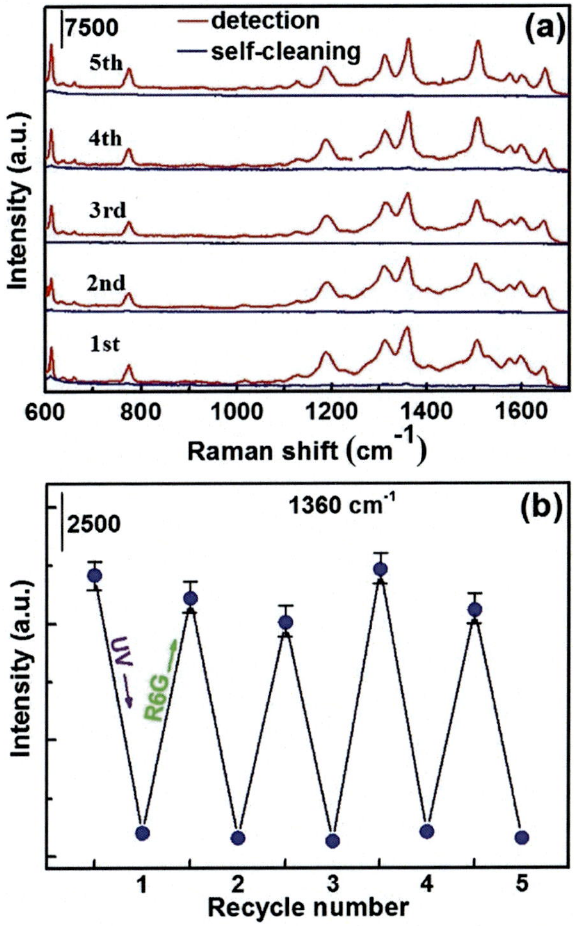

Fig. 10 (A) SERS spectra of the R6G molecules with 10^{-4} M on the TiO_2/Au NWAs before and after self-cleaning over five cycles; (B) Raman intensities of R6G molecules with 10^{-4} concentration during repeated measurement and self-cleaning with UV light irradiation. Reused with permission from X. Zhao, W. Wang, Y. Liang, J. Fu, M. Zhu, H. Shi, S Lei, C. Tao, Visible-light-driven charge transfer to significantly improve surface-enhanced Raman scattering (SERS) activity of self-cleaning TiO_2/Au nanowire arrays as highly sensitive and recyclable SERS sensor. Sensors Actuators B Chem. 279 (2019) 313-9.

Recently, transparent conductive oxide (TCO) has been demonstrated as an interesting material for gas sensing applications. For instance, indium tin oxide has been tested for sensing of hydrogen, and zinc oxide doped with gallium (GZO) and aluminum has been used for detection of hydrogen, nitrogen oxide, and methanol [67–71]. The LSPR of metal nanostructures is intensely responsible for interesting application of sensors that depends on many factors such as role of chemical reaction on the surface of metal oxide and dielectric features of surrounding medium. Generally, the plasmonic metal nanostructures are coupled to metal oxide, which accounts for the reaction with the target materials analyte for gas sensing applications. In this case, the metal acts as an optical probe that provides the optical detectable signal.

Dharmalingam et al. reported the sensing response of H_2, CO, and NO_2 dependent on different metal oxide thickness for high temperature plasmonics-based sensors on Au nanorods encapsulated with yttria stabilized zirconia (YSZ) using electron beam lithography. The samples greater than 60 nm showed the expected good sensing response while the samples greater than 100 nm were found to have reduced response because of the compounding and competing reaction channel [72]. Yang et al. reported the surface plasmon resonance of SnO_2/Au thin films with different thicknesses of SnO_2 and Au deposited by pulsed laser deposition technique for NO gas sensing application. The sensing response was carried out at 150°C. It is observed that the reflectance intensity is increased with the introduction of 50 or 100 ppm NO/N_2 gas mixing and reflectance intensity is decreased with switching back of NO/N_2 gas mixture to pure N_2 gas. Reasonably good sensing response is obtained with the second and third runs of NO introduction, and the intensity is not significantly changed with the change in gas composition, as compared to the first run. It is significant that NO molecules were absorbed irreversibly on the surface of SnO_2 at 150°C. Fig. 11 shows the SPR response of SnO_2/Au films with different thickness of SnO_2 and Au at the detection angles of 41.5, 40.3, and 43.5 degree with alternate exposition of pure N_2 and 100 ppm NO with N_2 gas mixture [73].

Ohodnicki Jr. et al. reported plasmonic transparent Al-doped ZnO nanoparticles-based thin films synthesized by drop-casting and sol-gel method for optical sensing applications. The sensing treatments were performed at 500°C with good, rapid, and reproducible near-IR optical sensing response to hydrogen gas, confirming that Al-doped ZnO demonstrate good optical gas sensor properties [74]. In conducting metal oxides, the relevance of this class of materials for high temperature gas sensing application is due to a large near-IR optical absorptance response to high temperature gas treatments. Moreover, the choice of adequate doping in a metal oxide system may produce a large near IR optical absorption in thin films based on nanoparticles, which can be very useful for good sensing response for high temperature gas treatments. Experimental and theoretical study signifies that metal oxides are potential candidates for high temperature gas sensing applications.

4.2 Solar energy

In semiconductor devices, a significant alternative to the conventional electron-hole separation process is plasmonic energy conversion. This evolving technique relies on generation of hot electrons in plasmonic nanostructures using metal-metal oxides through EM decay of surface plasmons. Recent studies have shown that plasmonic nanostructures can also convert the collected light into electrical energy by the generation of hot electrons. The new pathway for solar conversion is proving to be a new way for the realization of photovoltaic devices, the performance of which may be equivalent to or better than conventional devices. Moreover, the use of plasmonic metal oxides, such as conducting and semiconductor materials, is proposed to

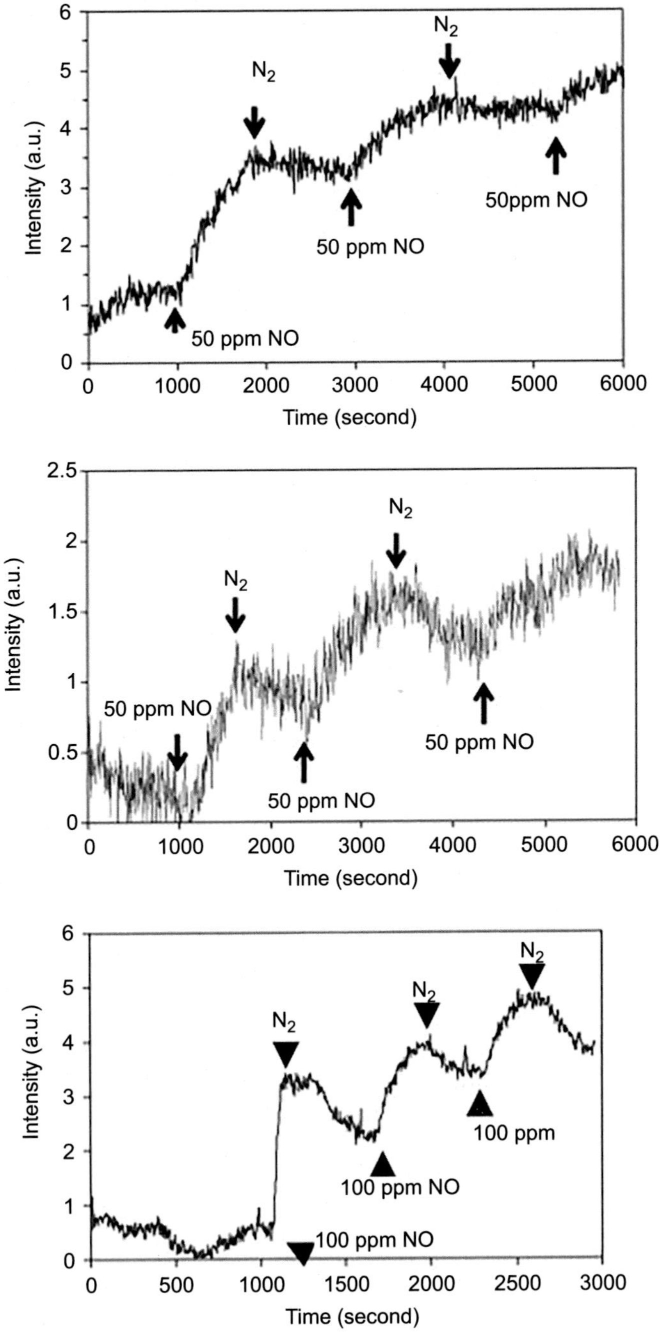

Fig. 11 SPR response of SnO_2/Au films with different thicknesses of SnO_2 and Au at the detection angles of 41.5, 40.3, and 43.5 degree, with alternate exposition of pure N_2 and 100 ppm NO with N_2 gas mixture.
Reused with permission from D. Yang, H. Lub, B. Chena, C. Lin, Surface plasmon resonance of SnO_2/Au Bi-layer films for gas sensing applications, Sensors Actuators B 145 (2010) 832–838.

encompass the spectral range of hot-electron generation and light absorption that enables the ideal use of the solar spectrum [75,76].

Al-Kuhaili reported increment of plasmonic transmittance of old/metal oxide for solar energy-saving applications [55]. Plasmonic nanoparticles (typically with Ag or Au NPs used as plasmonic core) are encapsulated with metal oxides like TiO_2 and SnO_2 (used as the shell) for use in dye-sensitized solar cells (DSSCs). These DSSCs are third generation solar cells that are less expensive than other, second generation solar cells. In DSSCs, there are four major components: a semiconductor coated photoanode, an electrolyte, a counter electrode, and a sensitizer. Significantly, the photoanode is made up of metal oxide (preferably SiO_2 and TiO_2) coated on fluorine-doped tin oxide. These nanoparticles with these metal oxides are preferred to form the core shell structure and to improve the chemical and thermal stability of DSSCs.

There are some factors that affect the performance of DSSCs, such as resonant energy transfer, increase in local EM field, and hot electron injection. Here, some important characteristic are reported, such as photocurrent-voltage, transporting process, extinction coefficient, quantum efficiency and absorbance for the DSSCs applications using plasmonic noble metal (Ag, Au) and metal oxides (SiO_2 and TiO_2) [77–80]. Clavero reported the generation of plasmon-induced hot-electron at NPs/MOs interface for photovoltaic and photocatalytic devices. An overall system consisting of Au or Ag NPs absorbed on TiO_2 or ZrO_2 was investigated. The action spectra presented the incident photon-to-electron conversion efficiency (IPCE) as a function of wavelength, obtaining that the extinction spectra that signify the LSPR (laser pulse with the duration of 150 fs and 550 nm wavelength used for the LSPR excitation in Au NPS on TiO_2) plays an important role in encouraging the generation of hot-electron. The action curves showed the reproduction of extinction spectra with maximum IPCE ~8.4%. Strikingly, the photoelectric conversion efficiency is increased with increasing temperature with this plasmonic system. Hence, the phenomenon of energy conversion efficiency could resolve the problem of overheating that occurs in conventional photovoltaic cells [81].

4.3 Photoelectrochemical biosensing

Photoelectrochemical (PEC) sensing is recognized as a particularly promising technique for biomolecules sensing. This sensing has many advantages over other techniques it includes low cost method, electron transfer photoactive material, environmental monitoring and analyte, etc. [82]. These sensors show high sensitivity and photocurrent and low background. The efficiency and performance of PEC sensors mainly rely on the PC ability of active materials. Recent studies have reported PEC sensing of biomolecules such as nicotinamide adenine dinucleotide hydrogen (NADH), DNA, and glucose, etc. [83–87].

Devadoss et al. reported synergistic metal-metal oxide nanoparticles for PEC glucose sensing applications. The hybrid membrane system was fabricated using grapheme-WO_3-Au hybrid membranes for glucose sensing. The reported results suggest sustainable improvement in WO_3-based electrode application and it is considered the best choice to

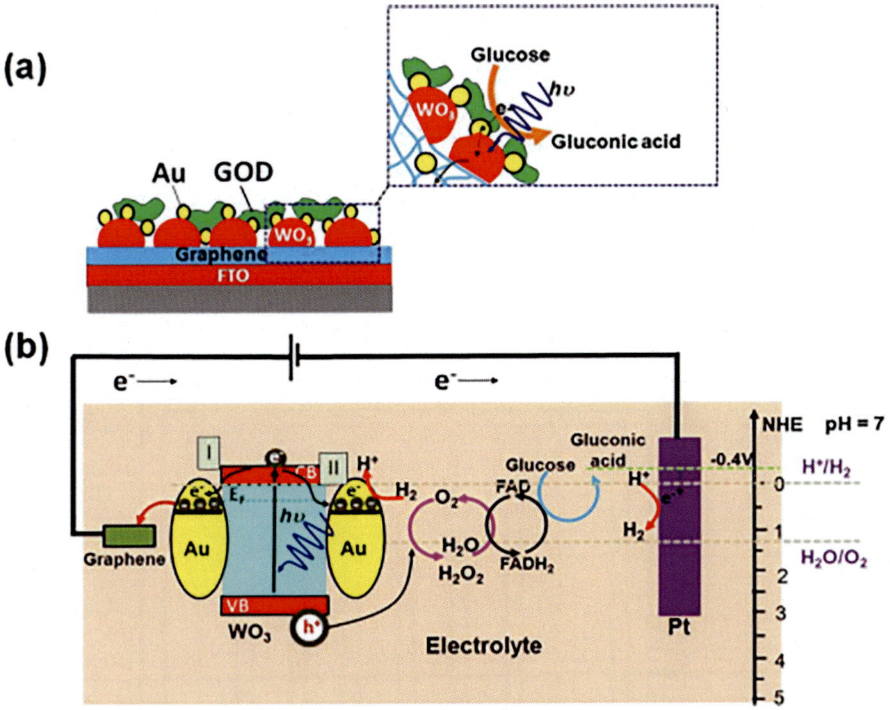

Fig. 12 Schematic diagram of (A) graphene-WO_3-Au triplet junction for glucose sensing; (B) glucose oxidation mechanism and energy levels at graphene-WO_3-Au photoelectrode under light illumination.
Reused with permission from A. Devadoss, P. Sudhagar, S. Das, S. Y. Lee, C. Terashima, K. Nakata, A. Fujishima, W. Choi, Y. S. Kang, U. Paik, Synergistic metal–metal oxide nanoparticles supported electrocatalytic graphene for improved photoelectrochemical glucose oxidation, ACS Appl. Mater. Interfaces 6 (2014) 4864–4871.

replace TiO_2 nanoparticles in PEC applications. Fig. 12 shows a schematic diagram of graphene-WO_3-Au triplet junction for glucose sensing application [88]. Chen et al. reported the formation of a photoelectrochemical biosensor using mesoporous zinc oxide for glucose sensing. The mesoporous ZnO with oxygen vacancies (Vo) was synthesized in a controlled manner with variation of low temperature (250°C to 450°C) using an annealing method. Glucose oxidase was used as a model enzyme for PEC biosensor and the relationship between PEC biosensor and Vo concentration determined the performance of the sensor. The high concentration (ZnO_{350}) of oxygen vacancies resulted in excellent performance of the sensor due to high carrier intensity and active sites [89].

A PEC immunosensor has been reported for the detection of amyloid β-protein based on SnO_2 nanocomposites. An ultrasensitive immunosensor containing high visible-light activity was fabricated on Ag_2S sensitized SnO_2/SnS_2 nanocomposites. Specifically, a SnO_2 flower-like porous nanostructure was used as basal material in the PEC sensor. The hetrostructure with SnS_2 increased the separation of photogenerated electrons and holes and claimed the sensitization of SnO_2. Significantly, the growth of Ag_2S on SnO_2/SnS_2 surface increased the photocurrent response. The content of Aβ was detected under optimum conditions of sensor and showed a linear concentration ranging from 0.5 pg mL^{-1} to 100 ng mL^{-1} with low limit of sensing (0.17 pg mL^{-1}). In addition, the fabricated PEC sensor demonstrated stability, good reproducibility and specificity, which may be advantageous for significant application in biosensing, clinical diagnosis, food safety, and environmental monitoring systems, detection of tumor markers, biomedicine, and photocatalysis [90,91]. PEC responses of the sensor with different concentrations of Aβ and time-based photocurrent responses under several on/off irradiation cycles are shown in Fig. 13 [90].

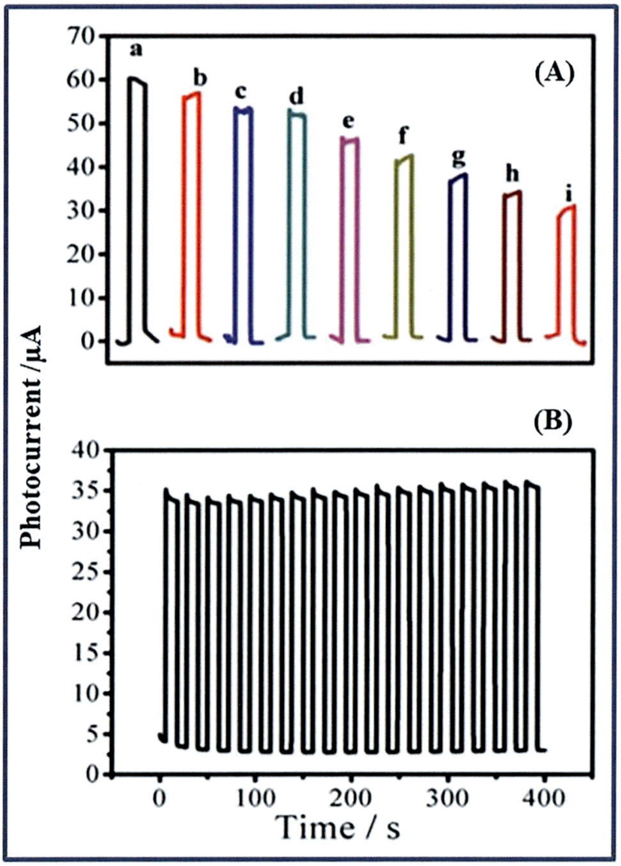

Fig. 13 (A) PEC responses with different concentrations of Aβ: (A–I) 0.0005 to 100 ng mL^{-1}; (B) time-based photocurrent responses of the PEC immunosensor under several on/off irradiation cycles for 400 s, $c_{A\beta}$=1.0 ng mL^{-1}.
Reused with permission from Y. Wanga, D. Fana, G. Zhaoa, J. Fenga, D. Weib, N. Zhanga, W. Caoa, B. Dua, Q. Wei, Ultrasensitive photoelectrochemical immunosensor for the detection of amyloid β-protein based on SnO_2/SnS2/Ag2S nanocomposites, Biosens. Bioelectron. 120 (2018) 1–7.

4.4 Photocatalysis

Transition metal oxides, including ZnO, TiO_2, MnO_2, WO_3, and MoO_3, are potential candidates to serve as surface plasmon resonance hosts because of their fascinating properties, which arise due to exhibiting outer-d valence electrons characteristics. Coupling of metal nanoparticles into the host matrix of transition metal oxides has been explored extensively on account of significant optical, catalytic, and electrical properties. Particularly, incorporation of Au nanoparticles into WO_3 and MoO_3 thin films are an interesting subject for investigation of tunable plasmon resonances in NPs because these metal oxides exist with stability under a wide range of conditions and exhibit band gaps of 2.6–3.1 eV, which is considered ideal for surface plasmon resonance absorption [92,93].

Kumar et al. demonstrated the synthesis of plasmonic Au-NPs-transition metal oxides (WO_3 and MoO_3) thin films for optical and electronic applications [94]. The plasmonic metal oxides induce tphotocatalytic behavior and that impacts the interaction of QDs with metal nanoparticles. The Au nanostructures were coated with a thin layer of Al_2O_3, CuO, CrO, and TiO_2 and further covered with CdSe/ZnS quantum dots. It was observed that the photocatalytic properties of these metal oxides were modified by plasmon near fields. The effect of Al_2O_3, CrO, and TiO_2 signified the direct influence of plasmon field with increment in optical excitations of the QDs while CuO signified unique outcomes. It was observed that the CuO not only provided plasmonic photocatalytic behavior with the direct impact of plasmon near fields but also the increment in interband transitions in CuO nanoparticles. Fig. 14 shows the emission spectra of QDs on SiO_2, Al_2O_3, CrO, TiO_2, and CuO in the absence and presence of metallic nanoislands (NISs) [95].

An investigation was carried out for a comparative ultrafast spectroscopy study with Au NPs in direct contact with metal oxides, such as TiO_2, ZnO, SnO_2, and Al:ZnO. This study was carried out for investigation of the role of MOx electron acceptor on Au plasmon hot carrier dynamics for applications in photocatalysis [96]. The coupling of zinc oxide nanostructures with Ag NPs resulted in enhanced UV and visible light photocatalytic behavior. The increased photocatalytic activity of ZnO with Ag NPs can be described by the formation of localized electric field

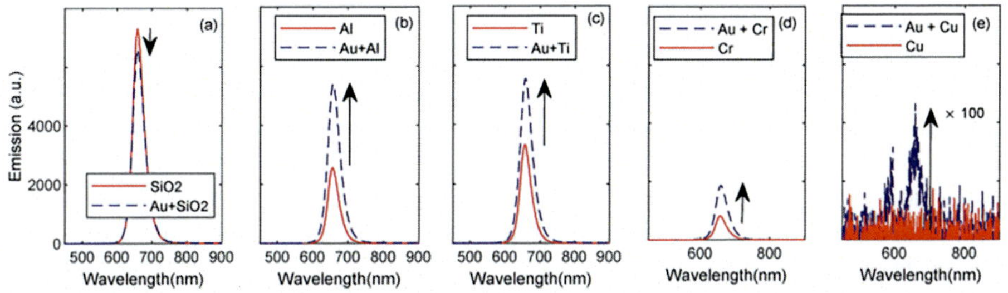

Fig. 14 Emission spectra of CdSe/ZnS quantum dots on (A) SiO2, (B) Al oxide, (C) Ti oxide, (D) Cr oxide, and (E) Cu oxide in the absence (solid lines) and presence of the NISs.
Reused with permission from S. M. Sadeghi, R. R. Gutha, W.J. Wing, Impact of the plasmonic metal oxide-induced photocatalytic processes on the interaction of quantum dots with metallic nanoparticles, J. Phys. Chem. C 124 (2020) 4261–4269.

and optical vibration by plasmonic particle surface plasmon resonance at the metal-SC interface [97,98].

4.5 Combined diverse applications

The combination of plasmonic nanoparticles with narrow band gap oxide materials has been reported comparatively less, although these systems enable metal and metal oxides supports to be sensitized by visible light [99]. The transfer of the electrons to metal oxides drive reduction processes, and holes left in metal nanoparticles can induce oxidative alterations, making them suitable for applications in water splitting, CO_2 reduction, degradation of environment pollutants, aerobic oxidation, preventing fogging of glass, and hydrogenation of cinnamaldehyde, etc. [25,100–103].

Metal oxides with unique phenomena of SPR and LSPR play important roles due to optical properties that are significant for theranostics, nanophotonics, and nanoelectronics [104]. A catalyst has been synthesized using visible-light-responsive vanadium oxide (V_2O_5), which displayed over an order of magnitude enhancement in rate of reaction with excitation wavelength (λ=532 nm, green light), as compared to no illumination. The good performance of Au NPs-V_2O_5-functionalized inverse opal (IO) is ascribed to spectral overlap of electronic bandgap and the LSPR phenomenon [105]. From the class of oxide materials, the properties of electrochromic materials can be reversibly modified by applying voltage. Electrochromic metal oxides are known as very promising materials because of their applications in smart windows or mirrors, smart sunglasses, optical information, and storage devices [106]. Llordés et al. reported the tunability in nanocrystalline glass composites near-IR and visible-light transmittance. They reported the advancement in nanocrystals-based electrochromism, demonstrating that optical switching behavior can be modified over a wide spectral range, from the visible to NIR spectral region, by coupling plasmonic In_2O_3 NCs with niobium oxide glass (NbOx) [107].

Dope metal oxides such as indium tin oxide (ITO) and Al doped zinc oxide are widely used to synthesize transparent electrodes in industries using vapor phase method. It is possible for the transparent conducting films (TCO) or electrodes to be fabricated by the colloidal dispersion of nanocrystals using low solution-based techniques [22,108]. An emerging and exciting new domain of potential applications of thin films and plasmonic semiconductors is in optical metamaterials. In these applications, researchers have reported that heavily doped metal oxides are advantageous over noble metals [109]. The first report has been published on surface plasmon polaritons on unpatterened ITO/dielectric surface using reflectance measurements. Whereas, the plasmonic nanostructures based on transparent conducting oxide have been studied, including NPs, discs, pillars, split-ring resonators, and MIM stacks, with their applications in metamaterials, surface-enhanced IR absorption and telecommunication [110].

Cheng et al. reported epitaxial Al on Al_2O_3 films, demonstrated as a plasmonic material system for UV and full visible spectral regions. This plasmonic structure was fabricated by focused ion beam milling for plasmonic white light applications. A clear surface wave interference pattern signified the low-loss nature of this plasmonic system. Moreover, low-threshold zinc oxide (ZnO-UV) and indium gallium nitride (InGaN-blue and green) plasmonic nanolasers on Al/Al2O3 structure using drop casting method was reported. The lasing measurements signified that the plasmonic material system has a significant potential to be applied in UV and full visible range [111]. Surface depletion layers play an important role in plasmonic metal oxide NCs, making them a compelling choice for IR applications due to the combination of strong IR light-matter interaction and spectral tunability. Particularly, this combination has driven the interest in electrochromic smart windows, these nanocrystals being suitable for surface-enhanced IR sensing, enhanced photocatalysis, and photothermal therapy due to their plasmonic-driven process. The phenomenon of surface depletion in plasmonic MOx nanocrystals is leading to a new area of materials design. These insights have resulted in significant improvements in efficiency that can be applied to new opportunities in resonant infrared energy transduction and hot carrier extraction [112].

5 Conclusion and future outlook

Metal oxides nanoparticles as promising nanomaterials have been effectively applied in technological as well as industrial applications, including biomedical, energy, sensing, and information technology, due to their significant plasmonic and optical properties, including absorption, transmittance, and local surface plasmon resonance (LSPR) effect. This chapter focuses on the fundamental aspects of plasmonic metal oxides, explaining their method of synthesis, characteristic plasmonic behavior, and their detailed application in particular application-oriented disciplines. Moreover, metal oxides for plasmonic applications with tunable optical features have been rapidly advanced, although their utilization at large scale is still limited due to various aspects. The deep understanding of their properties and plasmonic behavior can provide a major contribution to effective optical and other platforms.

For application prospects, the synthesis of plasmonic metal oxides has to be controlled because of their dependency on plasmonic properties of size, morphology, and composition, etc. Metal oxide nanoparticles are considered to enhance and improve devices performance based on different synthesis methods, including magnetron sputtering, molecular beam epitaxy, atomic layer deposition, electrochemical deposition, polyol method, sol-gel method, and chemical colloidal method. Metal oxides exhibit plasmonic application in different technologies, including solar energy, enhanced absorption properties for detectors in sensing windows, gas sensing, optical sensing, photocatalytics, photovoltaics, optoelectronics, water splitting, photothermal therapy, and RI sensors in optical communication. Tuning optical and scattering properties of plasmonic metal oxides and their derived assemblies provides a broad range of colors and saturation displays. Moreover, the easy synthesis, optical and chemical stability, and functionalization strategies of plasmonic metal oxides with their significant properties can address critical problems and challenges to obtain good performance in information, energy, optical, and biomedical technologies.

The lattice mismatch between the noble metal core and the metal oxide shell presents one challenge in the fabrication of metal oxide nanocomposites. Additionally, the plasmonic interaction that occurs when particles are in close proximity to one another and have overlapping plasmon resonances presents challenges in the fabrication of metal nanoparticle films, since it changes the characteristics in the solid state in comparison to dispersions [113]. Typically, low dimensional noble metals like silver and gold are used to achieve LSPR. Even so, it turns out that tuning the LSPR wavelengths is very challenging after these materials have been synthesized [114]. The fundamental research going on into plasmonic metal oxides will boost opportunities for designing significant applications-oriented plasmonic metal oxides.

Acknowledgments

One of the authors, Dr. Rajesh Kumar, acknowledges the Faculty Research Grant Scheme (FRGS) Project-2021 provided by Guru Gobind Singh Indraprastha University, New Delhi, India, and Inter University Accelerator Center (IUAC), New Delhi, India, for project (Ref: IUAC/XIII.3A/68308/2020). One of the authors, Vishnu Chauhan, is indebted to Inter-University Accelerator Centre, New Delhi, India, for providing a Senior Research Associate Fellowship.

References

[1] S.A. Maier, Plasmonics: Fundamentals and Applications, Springer, 2007.

[2] F. Caruso, T. Hyeon, V. Rotello, The golden age: gold nanoparticles for biomedicine, Chem. Soc. Rev. 41 (2012) 2740–2779.

[3] E. Ozbay, Plasmonics: merging photonics and electronics at nanoscale dimensions, Science 311 (2006) 189–193.

[4] S.D. Standridge, G.C. Schatz, J.T. Hupp, Distance dependence of plasmon-enhanced photocurrent in dye-sensitized solar cells, J. Am. Chem. Soc. 131 (2009) 8407.

[5] M.D. Brown, T. Suteewong, R.S.S. Kumar, V. D'Innocenzo, A. Petrozza, M.M. Lee, U. Wiesner, H.J. Snaith, Plasmonic dye-sensitized solar cells using core−shell metal−insulator nanoparticles, Nano Lett. 11 (2011) 438.

[6] H.A. Atwater, A. Polman, Plasmonics for improved photovoltaic devices, Nat. Mater. 9 (2010) 205.

[7] Y. Guo, L. Kang, S. Chen, X. Li, High performance surface-enhanced Raman scattering from molecular imprinting polymer capsulated silver spheres, Phys. Chem. Chem. Phys. 17 (2015) 21343–21347.

[8] P.K. Jain, X. Huang, I.H. El-sayed, M.A. El-sayed, Noble metals on the nanoscale: optical and photothermal properties and some applications in imaging, sensing, biology, and medicine, Acc. Chem. Res. 41 (2008) 7–9.

[9] W. Law, K. Yong, A. Baev, P.N. Prasad, L.A.W.E.T. Al, Sensitivity improved surface plasmon resonance biosensor for cancer biomarker detection based on plasmonic enhancement, ACS Nano 6 (2011) 4858–4864.

[10] M.A. Butt, S.N. Khonina, N.L. Kazanskiy, Plasmonics: a necessity in the field of sensing—a review, Fiber Integr. Optics 40 (2021) 14–47.

[11] J.M. Brockman, B.P. Nelson, R.M. Corn, Surface plasmon resonance imaging measurements of ultrathin organic films, Annu. Rev. Phys. Chem. 51 (2000) 41–63.

[12] W. Knoll, Interfaces and thin films as seen by bound electromagnetic waves, Annu. Rev. Phys. Chem. 49 (1998) 569–638.
[13] H. Knobloch, H. Brunner, A. Leitner, F. Aussenegg, W. Knoll, H. Knobloch, H. Brunner, A. Leitner, F. Aussenegg, W. Knolla, Probing the evanescent field of propagating plasmon surface polaritons by fluorescence and Raman spectroscopies, J. Chem. Phys. 10093 (1993) 10–13.
[14] K.L. Kelly, E. Coronado, L.L. Zhao, G.C. Schatz, The optical properties of metal nanoparticles: the influence of size, shape, and dielectric environment, J. Phys. Chem. B 107 (2003) 668–677.
[15] T.R. Jensen, M.L. Duval, K.L. Kelly, A.A. Lazarides, G.C. Schatz, R.P. Van Duyne, Nanosphere lithography: effect of the external dielectric medium on the surface plasmon resonance spectrum of a periodic array of silver nanoparticles, J. Phys. Chem. B 103 (1999) 9846–9853.
[16] S. Maier, Surface plasmon polaritons at metal/insulator interfaces, in: Plasmonics: Fundamentals and Applications, Springer, New York, NY, 2007.
[17] M. Faraday, The Bakerian Lecture: Experimental Relations of Gold (and Other Metals) to Light, 1857, pp. 145–181.
[18] H.J. Chen, L. Shao, Q. Li, J.F. Wang, Gold nanorods and their plasmonic properties, Chem. Soc. Rev. 42 (2013) 2679–2724.
[19] R. Buonsanti, A. Llordes, S. Aloni, B.A. Helms, D.J. Milliron, Tunable infrared absorption and visible transparency of colloidal aluminum-doped zinc oxide nanocrystals tunable infrared absorption and visible transparency of colloidal, Nano Lett. 11 (2011) 4706–4710.
[20] K. Manthiram, A.P. Alivisatos, Tunable localized surface plasmon resonances in tungsten oxide nanocrystals, J. Am. Chem. Soc. 8 (2012) 8–11.
[21] K. Biswas, C.N.R. Rao, Metallic ReO_3 nanoparticles, J. Phys. Chem. B 110 (2006) 842–845.
[22] Q. Huang, S. Hu, J. Zhuang, X. Wang, MoO_{3-x}-based hybrids with tunable localized surface plasmon resonances: chemical oxidation driving transformation from ultrathin nanosheets to nanotubes, Chem. Eur. J. 18 (2012) 15283–15287.
[23] T.R. Gordon, T. Paik, D.R. Klein, G.V. Naik, H. Caglayan, A. Boltasseva, C.B. Murray, Shape-dependent plasmonic response and directed self-assembly in a new semiconductor building block, indium-doped cadmium oxide (ICO), Nano Lett. 13 (2013) 2857–2863.
[24] M. KarbalaeiAkbari, Z. Hai, Z. Wei, C. Detavernier, E. Solano, F. Verpoort, S. Zhuiykov, ALD-developed plasmonic two-dimensional Au-WO_3-TiO_2 heterojunction architectonics for design of photovoltaic devices, ACS Appl. Mater. Interfaces 10 (2018) 10304–10314.
[25] Q.J. Xiang, J.G. Yu, B. Cheng, H.C. Ong, Microwave-hydrothermal preparation and visible-light photoactivity of plasmonic photocatalyst Ag-TiO_2 nanocomposite hollow spheres, Chem. Asian J. 5 (2010) 1466–1474.
[26] H. Rigneault, S. Monneret, R. Quidant, J. Polleux, G. Ba, Light-assisted solvothermal chemistry using plasmonic nanoparticles, ACS Omega 1 (2016) 2–8.
[27] D.M. Mattox, Handbook of Physical Vapor Deposition (PVD) Processing, Elsevier, 2010, ISBN: 978-0-8155-2037-5, https://doi.org/10.1016/C2009-0-18800-1.
[28] D. Lundin, T. Minea, J.T. Gudmundsson, High Power Impulse Magnetron Sputtering, Elsevier, 2020, ISBN: 9780128124550, https://doi.org/10.1016/C2016-0-02463-4.
[29] M. Torrell, et al., Tuning of the surface plasmon resonance in TiO_2/Au thin films grown by magnetron sputtering: the effect of thermal annealing, J. Appl. Phys. 109 (7) (2011) 074310.

[30] S. Veziroglu, Plasmonic and non-plasmonic contributions on photocatalytic activity of Au-TiO_2 thin film under mixed UV-visible light, Surf. Coat. Technol. 389 (2020) 125613.
[31] L. Morresi, Molecular Beam Epitaxy (MBE) Silicon Based Thin Film Solar Cells, Bentham Books, 2013, pp. 81–107, ISBN: 978-1-60805-518-0, https://doi.org/10.2174/97816080551801130101.
[32] S. Franchi, Molecular beam epitaxy: fundamentals, historical background and future prospects, in: Molecular Beam Epitaxy, Elsevier, 2013, pp. 1–46. ISBN 9780123878397 https://doi.org/10.1016/B978-0-12-387839-7.00001-4.
[33] Y.C. Cheng, et al., Wide range variation of resonance wavelength of GaZno plasmonic metamaterials grown by molecular beam epitaxy with slight modification of Zn effusion cell temperatures, J. Alloys Compd. 870 (2021) 159434.
[34] S. Sadofev, et al., Molecular beam epitaxy of n-Zn (Mg) O as a low-damping plasmonic material at telecommunication wavelengths, Appl. Phys. Lett. 102 (18) (2013) 181905.
[35] R.W. Johnson, A. Hultqvist, S.F. Bent, A brief review of atomic layer deposition: from fundamentals to applications, Mater. Today 17 (5) (2014) 236–246, https://doi.org/10.1016/j.mattod.2014.04.026.
[36] S.M. George, J.D. Ferguson, J.W. Klaus, Atomic layer deposition of thin films using sequential surface reactions, Mater. Res. Soc. Symp. Proc. 616 (2000) 93–101.
[37] A. Anopchenko, S. Gurung, L. Tao, C. Arndt, H.W.H. Lee, Gate-tunable Epsilon-Near-Zero Nanophotonics, Mater. Res. Express 5 (2018) 014012.
[38] E. Shkondin, O. Takayama, M.E.A. Panah, P. Liu, P.V. Larsen, M.D. Mar, F. Jensen, A.V. Lavrinenko, High aspect ratio titanium nitride trench structures as plasmonic biosensor, Opt. Mater. Express 7 (2017) 1606.
[39] S. Santangelo, M.H. Raza, N. Pinna, S. Patanè, On the plasmon-assisted detection of a 1585 cm−1 mode in the 532 nm Raman spectra of crystalline α-Fe_2O_3/polycrystalline NiO core/shell nanofibers, Appl. Phys. Lett. 118 (25) (2021) 251105.
[40] A. Ponti, M.H. Raza, F. Pantò, A.M. Ferretti, C. Triolo, S. Patanè, N. Pinna, S. Santangelo, Structure, defects, and magnetism of electrospun hematite nanofibers silica-coated by atomic layer deposition, Langmuir 36 (2020) 1305, https://doi.org/10.1021/acs.langmuir.9b03587.
[41] B. Bhattacharyya, Electrochemical Micromachining for Nanofabrication, MEMS and Nanotechnology, Elsevier, 2015, ISBN: 978-0-323-32737-4, https://doi.org/10.1016/C2014-0-00027-5.
[42] A.I. Pruna, N.M. Rosas-Laverde, D.B. Mataix, Effect of deposition parameters on electrochemical properties of polypyrrole-graphene oxide films, Materials 13 (2020) 624, https://doi.org/10.3390/ma13030624.
[43] I. Khan, A. Qurashi, Sonochemical-assisted in situ electrochemical synthesis of Ag/α-Fe_2O_3/TiO_2 nanoarrays to harness energy from photoelectrochemical water splitting, ACS Sustain. Chem. Eng. (2018), https://doi.org/10.1021/acssuschemeng.7b02848.
[44] C.A. SaezCabezas, K. Miller, S. Heo, A. Dolocan, G. LeBlanc, D.J. Milliron, Direct electrochemical deposition of transparent metal oxide thin films from polyoxometalates, Chem. Mater. 32 (2020) 4600–4608, https://doi.org/10.1021/acs.chemmater.0c00849.
[45] T. Arun, R. Justin Joseyphus, Prussian blue modified FePt nanoparticles for the electrochemical reduction of H_2O_2, Ionics 22 (2016) 877–883, https://doi.org/10.1007/s11581-015-1617-6.
[46] J. Qi, X. Dang, P.T. Hammond, A.M. Belcher, Highly efficient plasmon-enhanced dye-sensitized solar cells through Metal@Oxide core-shell nanostructure, ACS Nano 5 (9) (2011) 7108–7116, https://doi.org/10.1021/nn201808g.

[47] M. Parashar, V.K. Shukla, R. Singh, Metal oxides nanoparticles via sol-gel method: a review on synthesis, characterization and applications, J. Mater. Sci. Mater. Electron. 31 (2020) 3729–3749, https://doi.org/10.1007/s10854-020-02994-8.

[48] L. Giancaterini, C. Cantalini, M. Cittadini, M. Sturaro, M. Guglielmi, A. Martucci, U.A. Tamburini, Au and Pt nanoparticles effects on the optical and electrical gas sensing properties of sol-gel-based ZnO thin-film sensors, IEEE Sensors J. 15 (2) (2015) 1068–1076, https://doi.org/10.1109/jsen.2014.2356252.

[49] D.K. Kambhampati, T.A.M. Jakob, J.W. Robertson, M. Cai, J.E. Pemberton, W. Knoll, Novel silicon dioxide sol-gel films for potential sensor applications: a surface plasmon resonance study, Langmuir 17 (4) (2001) 1169–1175, https://doi.org/10.1021/la001250w.

[50] F. Horani, E. Lifshitz, Unraveling the growth mechanism forming stable γ-In_2S_3 and β-In_2S_3 colloidal nanoplatelets, Chem. Mater. 31 (5) (2019) 1784–1793.

[51] S. Choudhary, V. Kumar, V. Malik, R. Nagarajan, S. Annapoorni, R. Malik, Synthesis of ZnO@ Ag dumbbells for highly efficient visible-light photocatalysts, J. Phys. Condens. Matter 32 (40) (2020) 405202.

[52] S. Choudhary, G. Vashisht, R. Malik, C.L. Dong, C.L. Chen, A. Kandasami, S. Annapoorni, Photo generated charge transport studies of defects-induced shuttlecock-shaped ZnO/Ag hybrid nanostructures, Nanotechnology 32 (30) (2021) 305708.

[53] T. Hirakawa, P.V. Kamat, Charge separation and catalytic activity of Ag@TiO_2 core-shell composite clusters under UV-irradiation, J. Am. Chem. Soc. 127 (2005) 3928–3934.

[54] I. Gorelikov, L.M. Field, E. Kumacheva, Hybrid microgels photoresponsive in the near-infrared spectral range, J. Am. Chem. Soc. 126 (49) (2004) 15938–15939, https://doi.org/10.1021/ja0448869.

[55] M.F. Al-Kuhaili, Enhancement of plasmonic transmittance of porous gold thin films via gold/metal oxide bi-layers for solar energy-saving applications, Sol. Energy 181 (2019) 456–463.

[56] R. Begum, Z.H. Farooqi, K. Naseem, F. Ali, M. Batool, J. Xiao, A. Irfan, Applications of UV/Vis spectroscopy in characterization and catalytic activity of noble metal nanoparticles fabricated in responsive polymer microgels: a review, Crit. Rev. Anal. Chem. 48 (6) (2018) 503–516.

[57] K. Nakaji, H. Li, T. Kiba, M. Igarashi, S. Samukawa, A. Murayama, Plasmonic enhancements of photoluminescence in hybrid Si nanostructures with Au fabricated by fully top-down lithography, Nanoscale Res. Lett. 7 (2012) 629.

[58] A.R. Rashed, M. Habib, N. Das, E. Ozbay, H. Caglayan, Plasmon-modulated photoluminescence enhancement in hybrid plasmonic nano-antennas, New J. Phys. 22 (2020) 093033.

[59] X.J. Chen, G. Cabello, D.Y. Wu, Z.Q. Tian, Surface-enhanced Raman spectroscopy towards application in plasmonic photocatalysis on metal nanostructures, J Photochem Photobiol C: Photochem Rev 21 (2014) 54–80.

[60] W. Kiefer, Surface Enhanced Raman Spectroscopy: Analytical, Biophysical and Life Science Applications, John Wiley & Sons, 2011, ISBN: 978-3-527-32567-2.

[61] J. Zhou, J. Zhang, H. Yang, Z. Wang, J.A. Shi, W. Zhou, N. Jiang, G. Xian, Q. Qi, Y. Weng, C. Shen, Plasmon-induced hot electron transfer in Au-ZnO heterogeneous nanorods for enhanced SERS, Nanoscale 11 (24) (2019) 11782–11788.

[62] X. Zhao, W. Wang, Y. Liang, J. Fu, M. Zhu, H. Shi, S. Lei, C. Tao, Visible-light-driven charge transfer to significantly improve surface-enhanced Raman scattering (SERS)

activity of self-cleaning TiO_2/Au nanowire arrays as highly sensitive and recyclable SERS sensor, Sensors Actuators B Chem. 279 (2019) 313–319.

[63] United States Environmental Protection Agency, National Air Pollution Trends, 1900–1993, 1994.

[64] Federal Register, Environmental Protection Agency, Rules and Regulations, vol. 77, 2012, pp. 36342–36386.

[65] United States Environmental Protection Agency, Direct Final Rule for Control of Air Pollution From Aircraft and Aircraft Engines: Emission Standards and Test Procedures, 2012, Retrieved from: *https://www/epa.gov*.

[66] M. Sturaro, E.D. Gaspera, N. Michieli, C. Cantalini, S.M. Emamjomeh, M. Guglielmi, A. Martucci, Degenerately doped metal oxide nanocrystals as plasmonic and chemoresistive gas sensors, ACS Appl. Mater. Interfaces 8 (2016) 30440–30448.

[67] K. Yoo, S. Park, J. Kang, Nano-grained thin film indium tin oxide gas sensors for H_2 detection, Sensors Actuators B Chem. 108 (2005) 159–164.

[68] P.P. Sahay, R.K. Nath, Al-doped ZnO thin films as methanol sensors, Sensors Actuators B Chem. 134 (2008) 654–659.

[69] P.P. Sahay, R.K. Nath, Al-doped zinc oxide thin films for liquid petroleum gas (LPG) sensors, Sensors Actuators B Chem. 133 (2008) 222–227.

[70] N. Vorobyeva, M. Rumyantseva, D. Filatova, E. Konstantinova, D. Grishina, A. Abakumov, S. Turner, A. Gaskov, Nanocrystalline ZnO (Ga): paramagnetic centers, surface acidity and gas sensor properties, Sensors Actuators B Chem. 182 (2013) 555–564.

[71] Y. Hou, A.M. Soleimanpour, A.H. Jayatissa, Low resistive aluminum doped nanocrystalline zinc oxide for reducing gas sensor application via sol–gel process, Sensors Actuators B Chem. 177 (2013) 761–769.

[72] G. Dharmalingam, M.A. Carpenter, Chemical sensing dependence on metal oxide thickness for high temperature plasmonics-based sensors, Sensors Actuators B 251 (2017) 1104–1111.

[73] D. Yang, H. Lub, B. Chena, C. Lin, Surface plasmon resonance of SnO_2/Au Bi-layer films for gas sensing applications, Sensors Actuators B 145 (2010) 832–838.

[74] P.R. Ohodnicki Jr., C. Wang, M. Andio, Plasmonic transparent conducting metal oxide nanoparticles and nanoparticle films for optical sensing applications, Thin Solid Films 539 (2013) 327–336.

[75] M.J. Mendes, A. Luque, I. Tobías, A. Martí, Plasmonic light enhancement in the near-field of metallic nanospheroids for application in intermediate band solar cells, Appl. Phys. Lett. 95 (2009) 071105.

[76] G. Zhao, H. Kozuka, T. Yoko, Sol-gel preparation and photoelectrochemical properties of TiO2 films containing Au and Ag metal particles, Thin Solid Films 277 (1996) 147–154.

[77] P. Rai, Plasmonic noble metal@metal oxide core-shell nanoparticles for dye-sensitized solar cell applications, Sustainable, Energy Fuel 3 (2019) 63–91.

[78] F. Bella, C. Gerbaldi, C. Barolob, M. Graetzel, Aqueous dye-sensitized solar cells, Chem. Soc. Rev. 44 (2015) 3431–3473.

[79] A. Yella, H.-W. Lee, H.N. Tsao, C. Yi, A.K. Chandiran, M. Nazeeruddin, E.W.-G. Diau, C.-Y. Yeh, S.M. Zakeeruddin, M. Graetzel, Porphyrin-sensitized solar cells with Cobalt (II/III)—based redox electrolyte exceed 12 percent efficiency, Science 334 (2012) 629–634.

[80] W.R. Erwin, H.F. Zarick, E.M. Talbert, R. Bardhan, Light trapping in mesoporous solar cells with plasmonic nanostructures Energy Environ, Science 9 (2016) 1577–1601.

[81] C. Clavero, Plasmon-induced hot-electron generation at nanoparticle/metal-oxide interfaces for photovoltaic and photocatalytic devices, nature photonics, Nat. Photonics 8 (2014) 95–103.
[82] Y. Zang, J. Lei, H. Ju, Principles and applications of photoelectrochemical sensing strategies based on biofunctionalized nanostructures, Biosens. Bioelectron. 96 (2017) 8–16.
[83] Y.V. Stebunov, D.I. Yakubovsky, D.Y. Fedyanin, A.V. Arsenin, V.S. Volkov, Superior sensitivity of copper-based plasmonic biosensors, Langmuir 34 (2018) 4681–4687.
[84] G.-L. Wang, J.-J. Xu, H.-Y. Chen, Dopamine sensitized nanoporous TiO_2 film on electrodes: photoelectrochemical sensing of NADH under visible irradiation, Biosens. Bioelectron. 24 (2009) 2494–2498.
[85] K. Wang, J. Wu, Q. Liu, Y. Jin, J. Yan, J. Cai, Ultrasensitive photoelectrochemical sensing of nicotinamide adenine dinucleotide based on graphene-TiO_2 nanohybrids under visible irradiation, Anal. Chim. Acta 745 (2012) 131–136.
[86] Y. Hu, Z. Xue, H. He, R. Ai, X. Liu, X. Lu, Photoelectrochemical sensing for hydroquinone based on porphyrin functionalized Au nanoparticles on graphene, Biosens. Bioelectron. 47 (2013) 45–49.
[87] M. Zheng, Y. Cui, X. Li, S. Liu, Z. Tang, Photoelectrochemical sensing of glucose based on quantum dot and enzyme nanocomposites, J. Electroanal. Chem. 656 (2011) 167–173.
[88] A. Devadoss, P. Sudhagar, S. Das, S.Y. Lee, C. Terashima, K. Nakata, A. Fujishima, W. Choi, Y.S. Kang, U. Paik, Synergistic metal−metal oxide nanoparticles supported electrocatalytic graphene for improved photoelectrochemical glucose oxidation, ACS Appl. Mater. Interfaces 6 (2014) 4864–4871.
[89] D. Chena, L. Lva, L. Penga, J. Pengd, Y. Caoa, X. Wanga, X. Wanga, Q. Wub, J. Tua, Controlled synthesis of mesoporous zinc oxide containing oxygen vacancies in low annealing temperature for photoelectrochemical biosensor, Ceram. Int. 45 (2019) 18044–18051.
[90] Y. Wanga, D. Fana, G. Zhaoa, J. Fenga, D. Weib, N. Zhanga, W. Caoa, B. Dua, Q. Wei, Ultrasensitive photoelectrochemical immunosensor for the detection of amyloid β-protein based on $SnO_2/SnS_2/Ag_2S$ nanocomposites, Biosens. Bioelectron. 120 (2018) 1–7.
[91] H. Wang, Y. Wang, Y. Zhang, Q. Wang, X. Ren, D. Wu, Q. Wei, Photoelectrochemical immunosensor for detection of carcinoembryonic antigen based on 2D TiO_2 nanosheets and carboxylated graphitic carbon nitride, Sci. Rep. 6 (2016) 27385, https://doi.org/10.1038/srep27385.
[92] J.B. Goodenough, Metallic oxides, NBS Special Publication, Prog. Solid State Chem. 5 (1971) 145–399.
[93] H. Liao, W. Wen, G.K.L. Wong, Photoluminescence from Au nanoparticles embedded in Au:oxide composite films, J. Opt. Soc. Am. B 23 (2006) 2518–2521.
[94] N. Kumar, H.B. Lee, S. Hwang, T.-W. Kim, J.-W. Kang, Fabrication of plasmonic gold-nanoparticle-transition metal oxides thin films for optoelectronic applications, J. Alloys Compd. 775 (2019) 39–50.
[95] S.M. Sadeghi, R.R. Gutha, W.J. Wing, Impact of the plasmonic metal oxide-induced photocatalytic processes on the interaction of quantum dots with metallic nanoparticles, J. Phys. Chem. C 124 (2020) 4261–4269.
[96] Y. Hattori, S.G. Alvarez, J. Meng, K. Zheng, J. Sa, Role of the metal oxide electron acceptor on Gold−Plasmon hot carrier dynamics and its implication to photocatalysis and photovoltaics, ACS Appl. Nano Mater. 4 (2021) 2052–2060.
[97] M.J. Sampaio, M.J. Lima, D.L. Baptista, A.M.T. Silva, C.G. Silva, J.L. Faria, Ag-loaded ZnO materials for photocatalytic water treatment, Chem. Eng. J. 318 (2017) 95–102.

[98] Q. Deng, X. Duan, D.H.L. Ng, H. Tang, Y. Yang, M. Kong, Z. Wu, W. Cai, G. Wang, Ag nanoparticle decorated nanoporous ZnO microrods and their enhanced photocatalytic activities, ACS Appl. Mater. Interfaces 4 (2012) 6030–6037.

[99] D.B. Ingram, P. Christopher, J.L. Bauer, S. Linic, Predictive model for the design of plasmonic metal/semiconductor composite photocatalysts, ACS Catal. 1 (2011) 1441–1447.

[100] Z.K. Zheng, W. Xie, B.B. Huang, Y. Dai, Plasmon-enhanced solar water splitting on metal-semiconductor photocatalysts, Chem. Eur. J. 24 (2018) 18322.

[101] J.L. White, M.F. Baruch, J.E. Pander, Y. Hu, I.C. Fortmeyer, J.E. Park, T. Zhang, K. Liao, J. Gu, Y. Yan, T.W. Shaw, E. Abelev, A.B. Bocarsly, Light-driven heterogeneous reduction of carbon dioxide: photocatalysts and photoelectrodes, Chem. Rev. 115 (2015) 12888.

[102] D. Tsukamoto, Y. Shiraishi, Y. Sugano, S. Ichikawa, S. Tanaka, T. Hirai, Gold nanoparticles located at the interface of anatase/rutile TiO_2 particles as active plasmonic photocatalysts for aerobic oxidation, J. Am. Chem. Soc. 134 (2012) 6309.

[103] C.H. Hao, X.N. Guo, Y.T. Pan, S. Chen, Z.F. Jiao, H. Yang, X.Y. Guo, Visible-light-driven selective photocatalytic hydrogenation of cinnamaldehyde over Au/SiC catalysts, J. Am. Chem. Soc. 138 (2016) 9361.

[104] X. Liu, M.T. Swihart, Heavily-doped colloidal semiconductor and metal oxide nanocrystals: an emerging new class of plasmonic nanomaterials, Chem. Soc. Rev. 43 (2014) 3908.

[105] G. Collins, A. Lonergan, D. McNulty, C. Glynn, D. Buckley, C. Hu, C. O'Dwyer, Semiconducting metal oxide photonic crystal plasmonic photocatalysts, Adv. Mater. Interfaces 7 (2020) 1901805.

[106] J.K. Fink, High Performance Polymer, A Volume in Plastic Desing Library, second ed., Elsevier, 2014, ISBN: 978-0-323-31222-6, https://doi.org/10.1016/C2013-0-18803-4.

[107] A. Llordes, G. Garcia, J. Gazquez, D.J. Milliron, Tunable near-infrared and visible-light transmittance in nanocrystal-in-glass composites, Nature 500 (2013) 323.

[108] M. Kanehara, H. Koike, T. Yoshinaga, T. Teranishi, Indium tin oxide nanoparticles with compositionally tunable surface plasmon resonance frequencies in the near-IR region, J. Am. Chem. Soc. 131 (2009) 17736–17737.

[109] G.V. Naik, V.M. Shalaev, A. Boltasseva, Alternative plasmonic materials: beyond Gold and Silver, Adv. Mater. 25 (2013) 3264–3294.

[110] F. D'apuzzo, M. Esposito, M. Cuscuna, A. Cannavale, S. Gambino, G.E. Lio, A. De Luca, G. Gigli, S. Lupi, Mid-infrared plasmonic excitation in indium tin oxide microhole arrays, ACS Photon. 5 (2018) 2431–2436.

[111] C. Cheng, Y.J. Liao, C.-Y. Liu, B.H. Wu, S.S. Raja, C.Y. Wang, X. Li, C.K. Shih, L.J. Chen, S. Gwo, Epitaxial aluminum-on-sapphire films as a plasmonic material platform for ultraviolet and full visible spectral regions, ACS Photon. 5 (2018) 2624–2630.

[112] S.L. Gibbs, C.M. Staller, D.J. Milliron, Surface depletion layers in plasmonic metal oxide nanocrystals, Acc. Chem. Res. 52 (2019) 2516–2524, https://doi.org/10.1021/acs.accounts.9b00287.

[113] N. Cathcart, et al., Selective plasmonic sensing and highly ordered metallodielectrics via encapsulation of plasmonic metal nanoparticles with metal oxides, ACS Appl. Nano Mater. 2 (11) (2018) 6514–6524.

[114] J. Luther, et al., Localized surface plasmon resonances arising from free carriers in doped quantum dots, Nat. Mater. 10 (2011) 361–366.

Metal oxide nanomaterials-dispersed liquid crystals for advanced electro-optical devices

17

S. Anas[a], T.K. Abhilash[a,b], Harris Varghese[a,b], and Achu Chandran[a,b]
[a]Materials Science and Technology Division, CSIR-National Institute for Interdisciplinary Science and Technology (NIIST), Thiruvananthapuram, India, [b]Academy of Scientific and Innovative Research (AcSIR), Ghaziabad, India

Chapter outline

1 Introduction to liquid crystals

Liquid crystals (LCs), all began in 1888 when Friedrich Reinitzer, a botanist from Austria, observed two distinct melting points on a material, cholesteryl benzoate [1]. That is, this material when in the solid phase showed a unique stable intermediate semi-liquid phase before attaining the liquid phase. Specifically, LC is a thermodynamically stable intermediate state between liquids and solids exhibited by some materials; In other words, it shows anisotropic nature without the presence of a

Metal Oxides for Next-Generation Optoelectronic, Photonic, and Photovoltaic Applications
https://doi.org/10.1016/B978-0-323-99143-8.00008-0

three-dimensional crystal lattice [2]. The idea of a state of matter being liquid and crystalline simultaneously might feel ludicrous at best. However, LC mesophases have, in one or more spatial directions, periodical arrangement of molecules, optical anisotropy, and electric and magnetic properties which are observed for crystal structures [3]. In the meantime, they exhibit liquid-like behaviors such as fluidity, formation as well as the agglomeration of droplets, and the inability to support shear. These point to the fact that the LCs are not structurally ordered as that of a solid; however, there is some degree of alignment exists in them, and they are more like liquids than solids. The molecules in an LC, mesogens, are usually rod- or disc-shaped molecules that are easily polarizable, and the LC state is also termed as mesogenic phase. One thing that differentiates LC from an isotropic liquid is the tendency of the mesogens to point along a common axis, known as the director, and this leads to the anisotropy in LCs. Two parameters mainly decide the structure or various phase behaviors of LCs, orientational order and positional order [4]. The tendency of the molecule to align with the director is the orientational order and it is a long-range order, whereas the positional order is the translational symmetry between the group of molecules.

The nematic LC phase [5] is where the molecules exhibit no positional order while having orientational order. That is, there is hardly any long-range order in the positions of the centers of mass of the LC molecules. As in a liquid, the aligned molecules are free to randomly drift around without any restrictions. The smectic phase is another distinct mesophase where additional translational symmetry is present, and the molecules align as planes or layers. Here, the molecules are restricted within these layers, and each layer seems to glide over the other as soap and hence the name smectic [5]. Here the two important types of smectic phases are the smectic A and C phases. In the former, molecules orient along the line perpendicular to the plane, while in the smectic C phase, the director is slightly tilted with respect to the layer normal [6]. A special case of the nematic and smectic phase is called the chiral (or cholesteric) phase which exists in either right-handed or left-handed form. It differs from the nematic phase in that, even in an unstrained state, the director changes throughout the LC in an orderly manner. That is, the chiral phase is what would result if nematic LC is twisted along the Y-axis about the X-axis [7]. In this phase, along the long axis, successive molecules or layers of molecules are rotated along the axis in a methodical means. It gives a helical periodicity that repeats itself after a certain distance, known as pitch. These different phases are represented in Fig. 1 for a better understanding. With the increasing degree of molecular arrangement, from nematic to smectic to cholesteric, the liquid becomes opaque. Howbeit, a direct comparison is not probable since most of the time only one of these phases is formed by most LC compounds [8]. Additionally, according to the chemical properties, LC materials can be categorized into three: thermotropic [9], lyotropic [10], and metallotropic LCs [11]. It is based on the methods in the LC, which drive the self-organization in them. In thermotropic liquid crystals, the transition into the mesophase is instigated thermally, that is, by heating the solids or cooling the liquids. Here mesophase formation occurs owing to the packing interactions and dispersion forces between the molecules. Further, thermotropic LC is further classified into enantiotropic, where LC state can be achieved by both heating or cooling, and monotropic phases, where only either heating

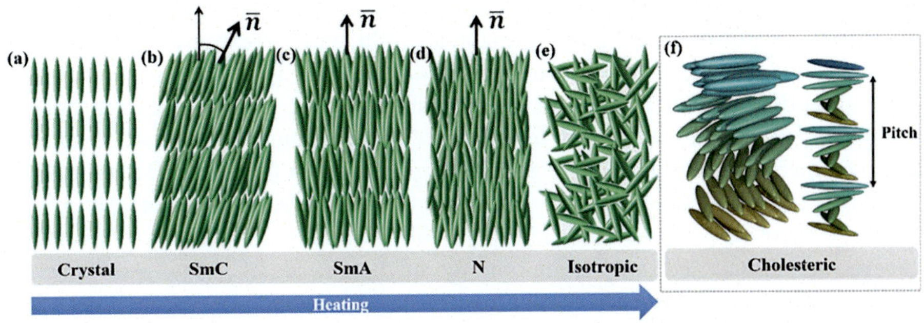

Fig. 1 Schematic illustration of different states of matter and liquid crystal (LC) phases. (A) Crystal, (B) smectic C phase, (C) smectic A phase, (D) nematic phase, (E) isotropic liquid, and (F) cholesteric phase.

or cooling is possible and not both. On the contrary, micellar formation in lyotropic LC results from the aggregation of mesogens because of the lyophilic and lyophobic interactions with the solvent. Coalesce of these micelle increases as the temperature decreases and as the solution concentration is increased. In metallotropic LCs, the composition is of both organic and inorganic molecules; here, the organic-inorganic composition ratio also plays a vital role in the LC transition in addition to temperature and concentration.

Most liquid crystals are organic molecules which are composed of π-conjugated systems with flexible alkyl chain attachments. Owing to the anisotropic structures, many unique electro-optical properties are exhibited by LC. One of them is birefringence [12], that is, the light entering the LC splits into an oppositely polarized pair of rays that travel with different velocities. A combination of this ordinary ray and extraordinary ray is the reason for the patterns and colors of LC when observed via a crossed polarizer. That is, in polarized white light, it can result in multicolored images of liquid crystals. The capability of material in the solid phase to exist in multiple forms of the crystal structure is called polymorphism—LC materials exhibit polymorphism too. Additionally, LC materials are mainly organic in nature, and they are called soft materials since relatively modest forces such as electric, magnetic, pressure, or temperature can change their physical properties. Therefore, owing to the weak intermolecular forces in LC, they can be easily polarized by applying an external electric field. Furthermore, LC materials that tend to exhibit spontaneous polarization are classified as ferroelectric liquid crystals (FLC) owing to the reduction in the symmetry of mesogens [13]. In normal liquid crystalline phases, such as N, SmA, and SmC, ferroelectricity is absent due to the high rotational symmetry around the long molecular axis. The bistability shown by the FLC material opens up a new world of opportunities in nonvolatile memory applications. Moreover, these materials tend to demonstrate much better properties for display applications as compared to the present LC technologies [14].

Liquid crystal materials are used in many applications in the fields of science and engineering, as well as device technology. The most famous and common application is that of liquid crystal displays (LCD) [15]. The invention of the modern transistor in

1947 set off the miniaturization of electronics. This facilitates the requirement for displays, which are power efficient, lightweight, and thin, which reenergized the research and developments in the field of electro-optical properties of LC materials. Now, it is a multi-billion-dollar industry with many ongoing scientific and engineering breakthroughs as well as developments in various display technologies. Another one is LC thermometers, which utilize the change in pitch and in turn the reflected light's wavelength change, corresponding to the change in temperature to accurately gauge any temperature range with color cheaply [16]. This helps to identify connection problems in a circuit and detect tumors in the medical field. LC materials dispersed in the polymer matrix, as sheets or rolls, are used as smart windows to switch between opaque and transparent states electrically in households and other buildings. Spatial light modulators are a different arena where LCs are used to control the wavelength, polarization, intensity, phase, wavefront, and coherence of a light [17]. Similarly, in chromatographic separation, LCs are used to study molecular arrangements and kinetics by employing them as a solvent to direct the course of chemical reactions [18]. Some other places where LC is used for optical imaging [19] such as, electronic slides for computer-aided drawing [20], visualization of radiofrequency waves in waveguides [21], nondestructive mechanical testing of materials [22], and optical antennas [23].

2 Metal oxide nanomaterials and their applications

As we know, the world is made up of materials whose properties are determined by their complex and heterogeneous structures. Materials in a size range of 1–100 nm with internal or surface structures in one or more dimensions are called nanomaterials. These materials can be nanoscale in one-, two-, and three dimensions. The nanomaterials can exist in single, aggregated, and agglomerated forms in different shapes. When these nanomaterials are assembled in a certain pattern in the nanoscale range, they are called nanostructures which include nanospheres, nanotubes, nanorods, nanowires, nanobelts, etc. Nanostructured materials' physical and chemical properties are extremely different from their bulk counterparts. The properties of the bulk materials can be improved in the nanoscale domain because of the small size of the building units, high density of the surfaces, grain and phase boundaries, defects, and presence of pores in the nanomaterials. Nanomaterials are classified based on their dimensionality, composition, morphology, uniformity, and agglomeration. There are four categories based on dimensionality, they are 0D materials, 1D materials, 2D materials, and 3D materials. Based on the chemical composition, nanomaterials can be classified as organic, inorganic, and biological. Inorganic nanomaterials consist of metals, alloys, metal oxides, semimetal oxides, semiconductor quantum dots, carbon structures, etc. Organic nanostructures include polymer particles and dendrimers. The stabilization and functionalization of inorganic nanoparticles require organic nanostructures; thus, the above classifications cannot be regarded as complete ones, and the nanomaterials are generally hybrid in nature [24,25].

Metal oxides have a significant role in the world of material science [26–28]. The metal elements are blessed in such a way that they can form a vast number of oxide compounds. Oxide particles can display distinct properties owing to the limited size and high density of corner sites. They demonstrate different properties than bulk materials or isolated molecules. Understanding their morphologies and features will open a new avenue for the synthesis and application point of view.

2.1 Salient features of metal oxide nanoparticles

The physical and chemical properties of metal oxide nanoparticles(MONPs) depend on their size to a large extent. The size effect has two interrelated faces in oxide chemistry. They are structural/electronic quantum size and size defects or nonstoichiometry effects. The main physical and chemical properties of metal oxides are described here.

2.1.1 Morphological and structural properties

Metal oxide nanomaterials gained particular interest because their properties are determined by different sizes, shapes, porosities, surface areas, crystals, and aggregation. The characterization techniques such as scanning electron microscopy (SEM), transmission electron microscopy (TEM), dynamic light scattering (DLS), static light scattering (SLS), and atomic force microscopy (AFM) are feasible for the primary size, size distribution, agglomeration state, porosity, and shape characterization of the MONPs [29–31]. The surface X-ray diffraction technique (XRD) can be used to analyze the arrangement of atoms in bulk. The crystallographic properties can be analyzed using small-angle X-ray scattering (SAXS) [32]. The crystalline structure is vital in determining the morphology of the NPs. The spinel iron oxides are characterized by a face-centered cubic stacking of oxide ions where the metal cations take up the tetrahedral and octahedral sites. The cubic symmetry of the structure corresponds to the pseudo-spherical nanoparticles and the truncated cubes leading to octahedra for large crystals. Hematite forms the rhombohedral nanocrystals while elongated nanorods are formed by goethite (Pnma), corresponding to the growth of the crystals along the direction of double chains of octahedra. The qualitative description of the particle size is described by considering the minimization of the total surface energy [33]. The pH of the medium and proton adsorption also influences their particle morphology [34,35].

2.1.2 Optical properties

Optical conductivity is one of the elemental properties of metal oxides, and it is related to the measurable parameter reflectivity which is size-dependent. The absorption of light is discrete-like and size-dependent due to quantum-size confinement. In the case of direct bandgap and indirect bandgap metal oxide semiconductors, optical bandgap energy shows an inverse squared dependence against the primary particle size [36,37]. But the deviations from the inverse square behavior are reported for ZnO and SnO_2 [38,39]. The nonstoichiometry size-dependent defects had a vital role in influencing the optical absorption features of nanosized oxides.

2.1.3 Transport properties

The nanostructure size determines the oxide materials' conductivity. They can deliver ionic or mixed ionic/electronic conductivity, and both are significantly influenced by the size of the NPs. According to Boltzmann statistics, the bandgap energy is related to the number of electronic charge carriers in a metal oxide. Depending on the nature of principal charge carriers, the conduction is referred to as n-type or p-type. By introducing nonstoichiometry, the number of free electrons/holes can be enhanced. The charge carrier (defect) distribution is modified due to the presence of the shielded electrostatic potential depletion at the surface layers of nanosized materials. As a consequence of these nanoscale effects, CeO_2 exhibits an improved n-type conductivity which may be four orders of magnitude higher than the bulk contribution to conductivity [40].

2.1.4 Mechanical properties

Mechanical properties include yield stress (σ), hardness (H), and super plasticity. In bulk materials, the yield stress (σ) and harness (H) follow Hall-Petch (H-P) equation.

$$\frac{\sigma}{H} = \frac{\sigma_o}{H_o} + kd^{\frac{-1}{2}} \tag{1}$$

where the constants σ_o and H_o describe friction stress and hardness, respectively, d is the primary particle/grain size, and k is the corresponding slope. The H-P slope decreases the particle/grain size to a few ten nanometres, which is positive, gets smaller values. However, below a critical point, the traditional dislocation mechanism stops, and a reversal H-P mechanism becomes effective. The mechanical properties are also strain-rate dependent. By decreasing particle/grain size, the strain rate sensitivity at room temperature is enhanced for TiO_2 and ZrO_2. The oxide materials like Al_2O_3, ZrO_2, CeO_2, and TiO_2 sintered under vacuum or using the spark plasma technique exhibit enhanced strength and hardness compared to the bulk materials [41,42]. Super plasticity is the capacity of oxide materials to undergo extensive tensile deformation without necking or fracture. The super plasticity strain rate can be increased by reducing the particle at a constant temperature.

2.1.5 Chemical properties

Metal oxides show acid and base properties. The acid or base characteristics depend on the nature of the element present in the oxide. The Lewis acidity is a function of solid-state variables and it is a characteristic of ionic oxides and not present in covalent oxides. The surface OH groups of most ionic oxides have a basic character to acidic ones. However, strong Bronsted acidity appears in oxides of elements with formal valence five or higher such as WO_3, V_2O_5, etc.

2.1.6 Electronic properties

The electronic properties of MONPs can be used to address their valance gap energy and bandgap energy. The unusual properties like high-yield strength, hardness, flexibility, rigidity, and ductility are related to their high surface-to-volume ratio of the nanoparticles. Their nanosize is responsible for properties like the quantum tunneling effect and quantum size effect [43].

2.2 Synthesis of metal oxide nanoparticles

The synthesis of metal oxide nanomaterials is a very challenging one. Their preparation methods are mainly classified into two major groups based on the nature of the transformations. They are liquid-solid transformations and gas-solid transformations.

2.2.1 Liquid-solid transformation

This method is widely used to control morphological characteristics with some chemical diversity which follows the bottom-top approach. Commonly used methods for the synthesis are discussed here.

(a) *Coprecipitation methods*: In this method, with the help of a base, a salt precursor is dissolved in water to precipitate the oxohydroxide form. Generally, it is very difficult to regulate the size and chemical homogeneity of metal oxides; however, with the introduction of surfactants and sonochemical methods, it is possible to optimize the solid morphological characteristics [44,45].

(b) *Sol-gel processing*: Here, the metal oxides are prepared through the hydrolysis of precursors. Usually, when alkoxides are dissolved in an alcoholic solution, they result in the corresponding oxohydroxide. The molecules are condensed by giving off water and a network of metal hydroxide is formed. A dense porous gel is formed by the polymerizing hydroxyl species by condensation mechanism. The drying and calcination are given appropriately to form porous oxides [46].

(c) *Microemulsion technique*: It is an approach based on forming micro/nanoreaction vessels under a ternary mixture containing water, a surfactant, and oil. In the water, the metal precursors will start precipitation as oxohydroxides within the aqueous droplets, which head to the monodispersed materials with size limited by the surfactant hydroxide contact [47].

(d) *Solvothermal methods*: Here, with the help of pressure, the metal complexes are thermally decomposed either by boiling in an inert atmosphere or using an autoclave. A suitable surfactant agent is usually added to the reaction media in order to control the particle size growth and agglomeration limit.

(e) *Template or Surface derivatized methods*: It associates the involvement of two types of tools: soft templates (surfactants) and hard templates (porous solids such as carbon or silica). The self-assembly systems can be synthesized with the help of template and surface-mediated nanoparticle precursors.

2.2.2 Gas-solid transformation methods

These methods are widely used to produce ultrafine-oxide powder synthesis. Mainly it involves chemical vapor deposition (CVD) and pulses laser deposition (PLD). For the formation of NPs, there are plenty of CVD processes used. The most common methods are metal organic, plasma-assisted, and photo-CVD methods [48]. These methods can provide uniform NPs, and films. However, it requires a careful setting up of experimental parameters. In multiple-pulse laser deposition method, the target sample is heated at a high temperature (4000 K) which leads to instantaneous evaporation, ionization, and decomposition, followed by subsequent mixing of desired atoms. From subsequent pulses, the gaseous entities absorb radiation energy and acquire kinetic energy perpendicularly to the target to be deposited in a substrate generally heated to allow crystalline growth [49].

2.3 Applications of metal oxide nanoparticles

The unique physical and chemical properties of MONPs are arising due to the high density and limited size of corners and edges on their surface. These particles are potential candidates for applications in sensing, agriculture, information technology, biomedical, optical, electronics, environment, and energy. The size of the NPs plays a vital role in its applications in different areas. The reduction in the size of NPs leads to an enhancement in the number of surface and interface atoms which could create strain or stress and adjoining structural perturbations [50]. The size of the NPs can modulate magnetic, chemical, conducting, and electronic properties [51]. The magnetic MONPs are of considerable interest among researchers because their properties can be tuned based on their shape and size in a considerable manner. The magnetic, electronic, and chemical properties of the NPs depend on their size. The size dependency was established in γ-Fe_2O_3 nanoparticles where the ferromagnetic behavior of the NPs with a size of 55 nm changes to superparamagnetic behavior without hysteresis when its size is reduced to 12 nm. The change to superparamagnetic behavior is induced by a decrease in the total magnetic anisotropy due to a reduction in the particle size [52,53]. The electrical properties are also significantly influenced by the size of the NPs, which enhances their applicability in gas sensors. TiO_2 is a pioneering candidate in gas sensing, optoelectronics, and photovoltaic applications because of its tunable conductivity with its size. Its size can be reduced easily at high-temperature treatment and electrical conductivity can be enhanced in a considerable way. It is clear that the properties of MONPs are significantly influenced by their size and shape. Plenty of factors influence the structural properties and functional characteristics of the nanoparticles for advanced applications.

2.3.1 Gas sensing

The presence of more active sites and large surface-to-volume ratios, and high specific surface areas with high surface reactivity makes the MONPs an ideal candidate for gas sensing applications [54,55]. Metal oxide nanomaterials based on sensing of gases such as carbon dioxide, carbon monoxide, oxygen, ozone, methane, nitric oxide, nitrogen oxide, hydrogen sulfide, ammonia, and hydrogen are well studied [56].

2.3.2 Batteries

Rechargeable batteries have a vital role in the field of automobiles for the past few years because of their efficiency and environmental concerns. Lithium-ion batteries (LIBs) have a great impact in this area due to their high energy density and long lifetime. MONPs are extensively used in LIBs as electrode materials. The size of the NPs has a fascinating role in the electrochemical properties. The rate of capability and reversibility can be increased by using smaller-sized oxides. The nanosized metal oxides and along with graphene provide a more beneficial integrated structure for shortening lithium diffusion pathways and plummeting polarization within the electrode and it leads to enhancing the performance of LIBs [57].

2.3.3 Solar cells

Metal oxide semiconductors are low cost, eco-friendly, and highly reliable. The outstanding flexible properties, low cost, and scalable methods make metal oxides an extraordinary material in new-generation photovoltaics. The presence of large bandgap energy and its tunability makes them an ideal candidate for use in photoharvesters. They are widely used in photovoltaics as photoelectrodes in dye solar cells and are also used to develop metal oxide p-n junctions [58]. Solar cells based on various MONPs like Cu_2O, ZnO, ITO, and amorphous ITO are reported [56].

2.3.4 Antennas

Antennas convert electromagnetic radiation in space into electric current. Traditional antennas involve rigid materials and are conductors subjected to oxidation in several ways. Metal oxides are stable materials and are good enough to be used as applications in antennas. Nowadays, rectifier antennas and optically transparent antennas are widely used rather than conventional ones due to their improved efficiency [56].

2.3.5 Optoelectronic and electronics

Transparent conducting oxides (TCOs) are widely used in optoelectronic applications because of their optical transparency and electrical conductivity. They have significant applications in producing LEDs, touch screens, flat panel displays, low emissivity windows, etc. Ga_2O_3 nanoparticles have been widely used in electronic and optical applications, specifically at low temperatures. The material provides a broad range of light emissions due to its wider band gap [59]. Recently, they have been widely used in LCs to enhance the electro-optical properties for display applications and the development of memory devices [60].

2.3.6 Catalyst

The larger surface-to-unit mass ratio of the NPs can improve the chemical reactivity. Many MONPs are used as a catalyst in various fields due to their size, shape, and structural properties. The CuO nanoparticles are used as a redox catalyst and catalyst in various oxidation processes in photoconductive and photothermal applications in organic synthesis [61].

The Al_2O_3 is used as a support for active phases in the area of catalysis coated with other material because of its novel structure [62]. The ZrO_2 nanoparticles are used as a catalyst as well as solid electrolytes because of their unique structural properties. CeO_2 nanoparticles are also used as a catalyst in several reactions. The TiO_2 nanoparticles are used as a photocatalyst for pollutant elimination in various industries [63].

3 Metal oxide nanomaterials-doped liquid crystal composites

As discussed, exploring the LC nature of materials is not a new area for researchers. Still, in the last few years, the advancement of nanotechnology created a boom for introducing nanoparticles into different types of LCs for widespread applications. Considerable efforts have been reported by doping various kinds of nanomaterials, especially ferroelectric NPs, magnetic NPs, carbon nanomaterials, semiconductor NPs, metal NPs, and MONPs into the LCs (Fig. 2) for improving their electro-optical characteristics [64–68]. The dielectric and optical anisotropic nature of LCs makes them the most versatile material for making switchable electro-optical devices, and their properties can be easily tuned by controlling electric field, temperature, pressure, etc. The importance of doping nanomaterials into LC matrix for sophisticated applications is well established in recent research activities around the globe, including cheap and reliable displays for electronic equipment, flexible optical devices for the healthcare industry, chemical and gas sensors, beam steering devices for wireless communications, etc. [68–71]. As we have seen, the FLCs have favorable characteristics for future displays showing faster

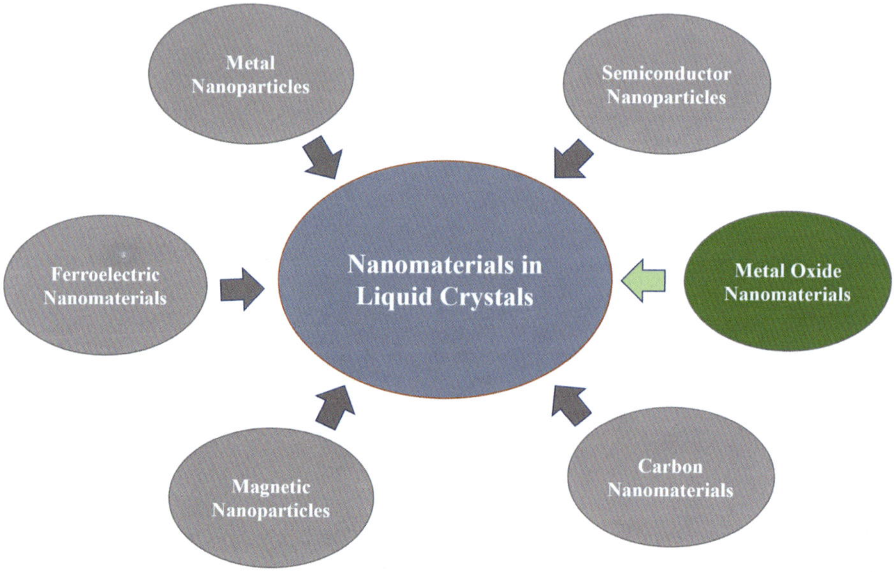

Fig. 2 Various nano dopants utilized in LC systems.

response time, high contrast, lower operating voltage, and improved bistability. But still, the difficulty to orient FLC mesogens over different substrate surfaces and constraints for tuning their material properties for individuate device application results in unscalable hurdles in industrial-level usages. Furthermore, the quality of LC devices is affected by the presence of undesired ionic impurities, which may come from various sources such as the alignment layer, through atmospheric interactions, or during the synthesis part of LCs. When applying an electric field, these ions will align, and the resulting electric field will act opposite to the applied field, thus adversely affecting the performance of LC devices. This depolarization field causes image sticking, grey-level shifting, slow response, and degradation of bistability.

Even though these challenges exist, MONPs-dispersed LCs show excellent optical and dielectric properties due to their unique physical and chemical structures. MONPs can tailor the optical and dielectric properties and be used for tuning the anisotropic nature of LC-based electronic and optical devices with a wide range of applications. Here, we analyze the importance of doping various MONP-doped LC composites for advanced electro-optical applications, mainly based on MgO, ZrO_2, NiO, CuO, TiO_2, Al_2O_3, Fe_3O_4, SiO_2, ZnO, CeO_2, and SnO_2 nanomaterials.

3.1 Magnesium-oxide nanoparticles (MgO)

Magnesium-oxide nanoparticles (MgO NPs) are odorless and nontoxic substances used in various areas, including electronics, catalysis, and dielectric applications [72]. MgO NPs enhance the electro-optical and dielectric properties of LCs on doping with it, which have potential applications in the field of display and optical memory devices. Chandran and his coworkers have reported significant work by dispersing different concentrations of MgO NPs in W301 host FLC resulting in substantial improvements in its electro-optical properties [73]. An optimum concentration of 0.5 wt% MgO NPs/FLC composites showed ~50% improvement in electro-optical response time compared to its pristine counterpart (Fig. 3). Faster response time has resulted from decreased rotational viscosity and increased surface anchoring energy, which agrees with Eq. (2).

$$\tau = \frac{\gamma}{P_s E} \tag{2}$$

where τ is the switching time, γ is the rotational viscosity, P_s is the spontaneous polarization, and E is the applied electric field. The same concentration of 0.5 wt% MgO-doped FLC showed a considerable increase of ~2.5 degree in optical tilt angle for all applied electric fields above 1 V and a faster switching time which is a great advantage in making fast and high-contrast displays. Photoluminescence (PL) characterization study on MgO NPs showed surface defects due to the oxygen vacancies, thereby more chemical activity and absorption of ionic impurities on their surface. Once an electric field is applied to the matrix, the interaction of dipolar mesogens with MgO NPs having inherent dipole moment produces strong anchoring strength, which in turn improves the optical tilt angle. On top of that, by comparing the bright and dark states,

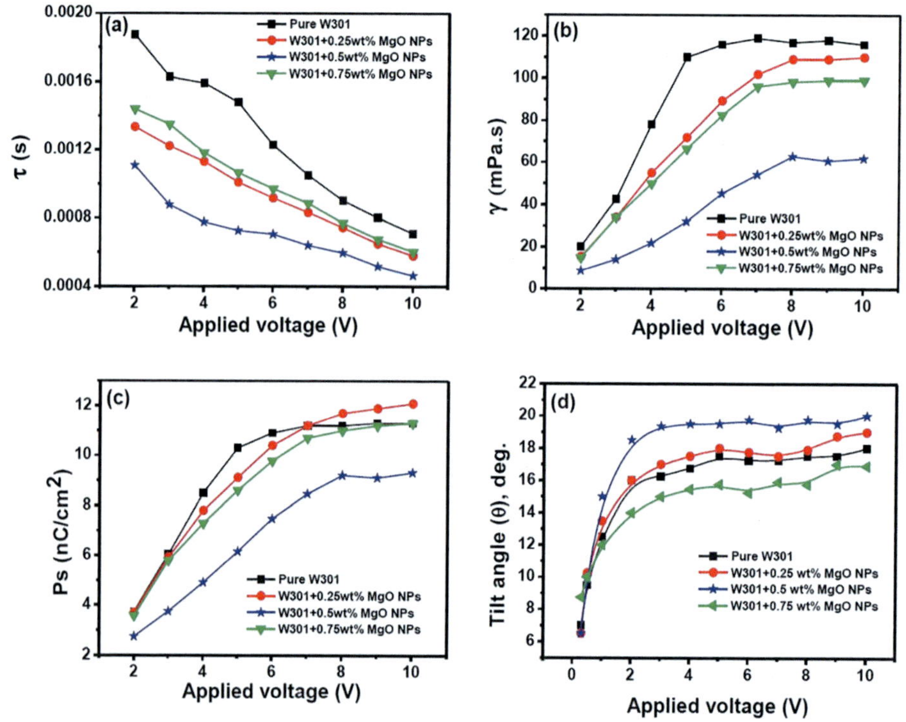

Fig. 3 Enhanced electro-optical behavior of pure FLC material W301 that was doped with different concentrations of MgO NPs at room temperature(27°C). (A) Response time (τ), (B) rotational viscosity (γ), (C) spontaneous polarization (P_s), and (D) optical tilt angle (θ) of pure FLC material W301 and that doped with different concentrations of MgO NPs at room temperature(27°C).
Reused with permission from Chandran A, Prakash J, Naik KK, Srivastava AK, Dąbrowski R, Czerwiński M, et al. Preparation and characterization of MgO nanoparticles/ferroelectric liquid crystal composites for faster display devices with improved contrast. J. Mater. Chem. C 2014;2:1844-53. https://doi.org/10.1039/c3tc32017k.

light leakage centers had been reduced substantially; the bright state became brighter, and the dark state became darker by the dispersion of MgO NPs.

Similarly, doping of MgO nanoparticles in 6CHBT nematic LC showed improvement of nonlinear effect, particularly Kerr constant and birefringence above the nematic-isotropic phase transition temperature [74]. Also, introduction of the MgO NPs of ~10 nm size results in the reduction of threshold voltages owing to the increase in order parameters, suitable for the fabrication of tunable low-voltage optoelectronic devices.

3.2 Zirconium-dioxide nanomaterials (ZrO_2)

Nowadays, soft materials exhibiting memory effects are of great interest owing to their potential applications in various data storage devices. FLCs are one such material that exhibits optical memory effects, which can be used in advanced display and data storage devices. Retention of its stretched state even after the removal of the electric field is referred as the optical memory effect in FLCs, which can be used for data storage. FLCs' ability to provide flicker-free high contrast displays coupled with this memory effect can pronounce the next-generation display technology. The FLC material, KCFLC 10S doped with zirconia nanoparticles, exhibited an electro-optical memory effect as a result of strong adsorption/trapping of ionic impurities [75]. Zirconium nanoparticles (ZNPs) doped KCFLC 10S mixture retains its stretched state up to some extent even after the removal of external stimulus, which was confirmed by dielectric spectroscopic and optical characteristics (Fig. 4). A low-frequency relaxation peak that appeared in pure material was entirely diminished by the presence of ZNPs, indicating the adsorption/trapping of ionic impurities by ZNPs. Excess ions present in the pure FLC material were accountable for the depolarization field, which causes nonlinearity in output responses, but this depolarization field was minimized by the dispersion of ZNPs which resulted in memory effect.

The study on ZrO_2 NPs with nematic LCs exhibited a high optical transmittance ratio and better alignment, playing a pivotal role in trapping undesired ionic impurities from LCs, its capturing ability was enhanced by increasing doping concentration under an applied voltage [76]. Experimental results confirmed that the optimum concentration of 1.5 wt% ZNP doped with LC cells showed the lowest threshold voltage and slowest response time due to its strong dipole interaction between NPs and LC molecules. The NPs' larger size than LC molecules provide a higher dipole moment, generate strong mesogen anchoring, and show higher electrical torque, making it practicable to design LC devices with a desired electro-optical performance by varying the concentration of ZNPs.

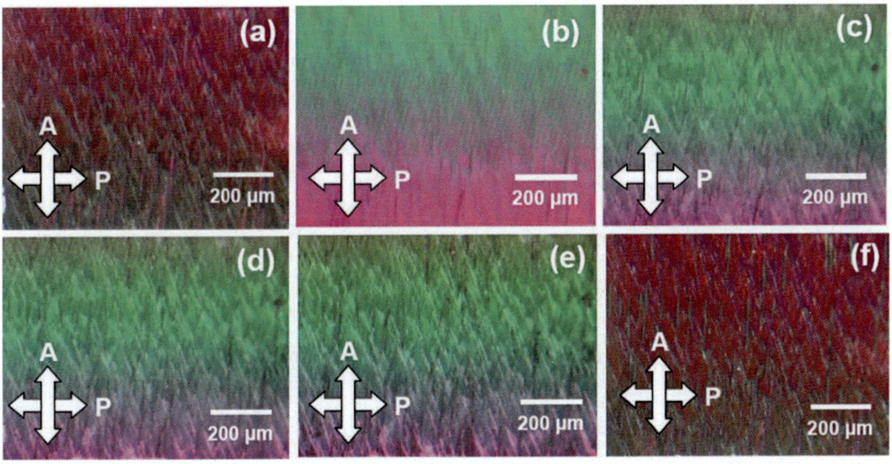

Fig. 4 Memory effect of ZNPs-doped FLCs in a 6-μm cell were captured using polarizing optical microscopy at room temperature (A) 0 V, (B) 15 V bias, (C) 10 s after removal of bias, (D) 1 min after removal of bias, (E) 10 min after removal of bias, and (F) 0 V again. scale bar: 200 mm. Crossed arrows represent crossed polarizer (P) and analyzer (A).
Reused with permission from Chandran A, Prakash J, Ganguly P, Biradar AM. Zirconia nanoparticles/ferroelectric liquid crystal composites for ionic impurity-free memory applications. RSC Adv. 2013;3:17166-173. https://doi.org/10.1039/c3ra41964a.

3.3 Nickel-oxide nanomaterials (NiO)

Nickel-oxide nanoparticles (NiO NPs) are used in various technological and industrial applications, including catalysis, high-performance supercapacitors, and gas sensors [77,78]. NiO NPs have also proven important for fabricating ionic impurity-free memory devices for advanced electro-optical devices. A recent published work by Chandran et al. on short-length NiO nanorods in ZLI3654 resulted in prolonged optical memory over 10 min and an almost 50% reduction in operating voltage [79]. Dispersion of different concentrations of NiO NPs produced better alignment, which effectively produced better dark and bright states. Along with a 50% fall in the operating voltage, there is a reduction in saturation values of spontaneous polarization, rotational viscosity, and switching time. The presence of dipolar NiO NPs allows mesogens to align with the applied electric field and retains their switched state even after removing bias. By applying an external stimulus, the value of dielectric permittivity set foot into a minimum value due to the phase fluctuation of molecules in phason mode. After removing the bias, the retention of stretched state was for more than 10 min and then gradually reverted to a relaxed state in around 20 min, where it was 1 min for the pristine material. Then this memory effect was also evident in polarizing optical microscopic images by retaining the switched state for the same time interval (Fig. 5).

The delayed optical response that conferred explicit confirmation about the optical memory behavior of NPs/FLC composite is shown in Fig. 6, and confinement of the ionic impurities by the presence of NiO nanorods is depicted in the schematic representation in Fig. 7. Upon application of an electric field, the addition of nanomaterials causes more ordering to the molecules by providing anchoring strength and

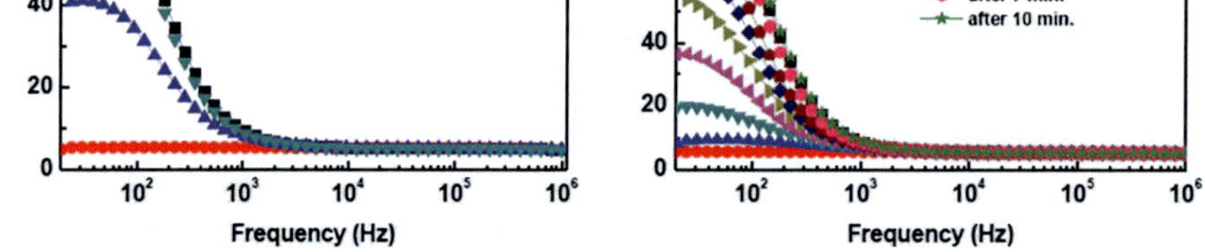

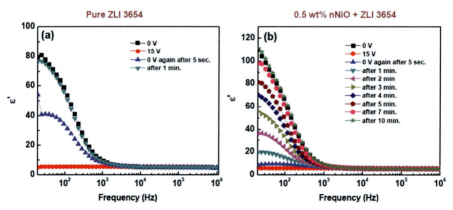

Fig. 5 Dielectric dispersion spectra (dielectric permittivity, as a function of frequency) of (A) pure FLC material, ZLI 3654 and (B) NiO nanorods/FLC composite-filled sample cell. Reused with permission from Chandran A, Prakash J, Gangwar J, Joshi T, Srivastava AK, Haranath D, Birder AM. Low-voltage electro-optical memory device based on NiO nanorods dispersed in a ferroelectric liquid crystal. RSC Adv. 2016;6:53873-881. https://doi.org/10.1039/C6RA04037C.

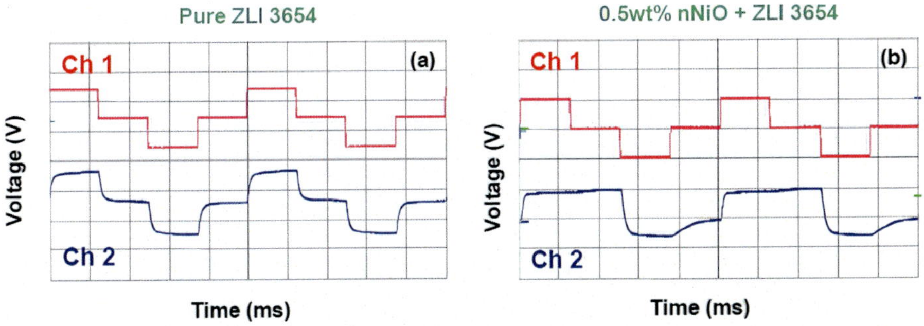

Fig. 6 Confirmation of memory behavior via the optical response of (A) pure FLC material, ZLI 3654 and (B) nNiO/FLC composite material filled sample cells to a driving time delayed positive and negative square pulse of 20 V_{pp} at 100 Hz. Ch_1 (Ch_1: 10 V/div, Ch_2: 5 V/div, time: 5 ms) shows the input square pulse and Ch_2 shows its optical response.
Reused with permission from Chandran A, Prakash J, Gangwar J, Joshi T, Srivastava AK, Haranath D, Birder AM. Low-voltage electro-optical memory device based on NiO nanorods dispersed in a ferroelectric liquid crystal. RSC Adv. 2016;6:53873-881. https://doi.org/10.1039/C6RA04037C.

subsequently diminishing the depolarization field. Dielectric absorption spectra confirmed that NiO NPs/FLC composite could trap impurity ions present in the FLC mixture compared to pristine material and showed a suppressed dielectric absorption peak at the low-frequency region corresponding to ionic impurities.

NiO NPs-doped nematic LCs provided lower threshold voltages, faster response time, and increased dielectric anisotropy with the increased amount of NPs concentration [80]. Thus, improvement in dielectric anisotropy can significantly reduce the threshold voltages in nematic LCs without the disturbance of ionic impurities, and the increase in the anchoring strength between LC molecules and the alignment layer causes molecules to switch easier than before. Subsequently, faster response times have been observed, which is an essential factor for diminishing the barrier of motion blur in conventional LCDs. The trapping of ionic impurities also addresses the problem of image sticking caused by the internal charges accumulating on localized defect regions. Thus, cost-effective and more reliable ionic impurity-free memory devices can be manufactured from the LC-NiO nanomaterial dispersion and proved to be a suitable material for future displays and memory devices.

3.4 Cupric-oxide nanoparticles (CuO)

CuO nanoparticles (CuO NPs) possess a wide range of potential applications in various areas, including catalysis, superconducting materials, thermoelectric materials, sensing materials, glass, antimicrobial, and ceramics [81–83]. CuO NPs in ferroelectric and nematic LCs enhanced the dielectric and electro-optical parameters, which help design advanced electro-optical devices. Doping of 0.25 and 0.5 wt% CuO with W302 FLC improved dielectric properties due to the effective coupling between CuO NPs and FLC molecules; the low-frequency relaxation mode observed in the pristine

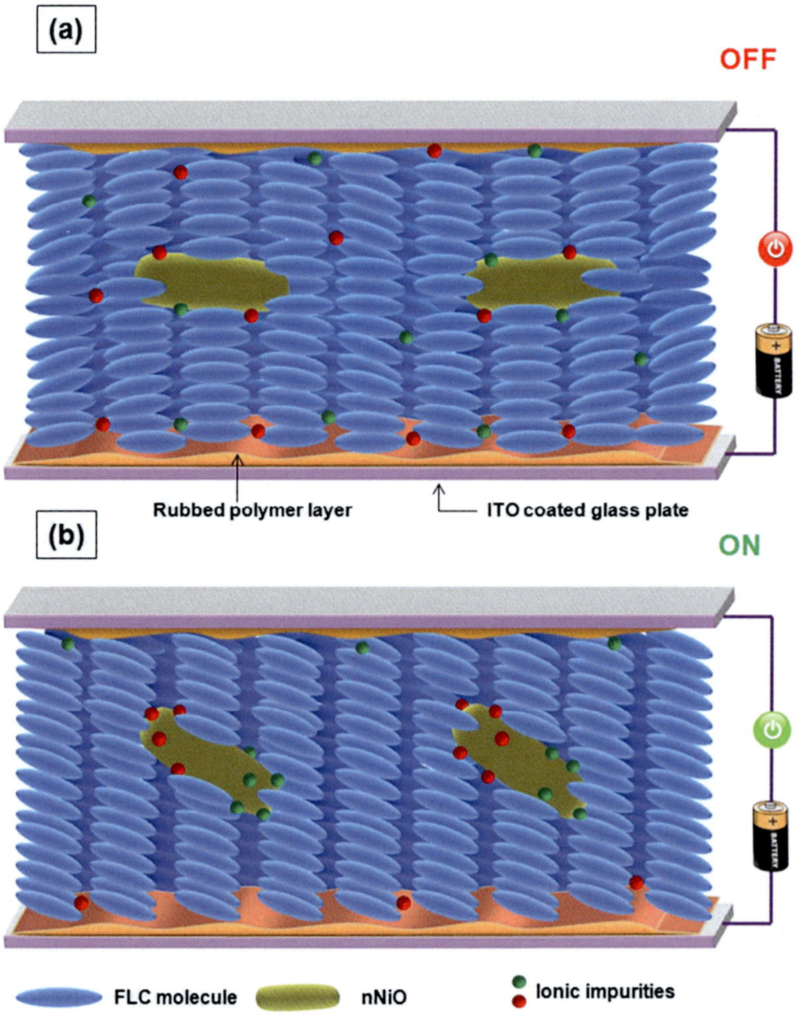

Fig. 7 Schematic depiction of the mechanism of memory behavior in nNiO/FLC composite material-filled sample cell. The homogeneously aligned composite sample cell at (A) OFF and (B) ON state of external DC bias.
Reused with permission from Chandran A, Prakash J, Gangwar J, Joshi T, Srivastava AK, Haranath D, Birder AM. Low-voltage electro-optical memory device based on NiO nanorods dispersed in a ferroelectric liquid crystal. RSC Adv. 2016;6:53873-881. https://doi.org/10.1039/C6RA04037C.

FLC material also completely disappeared [84]. The partial unwinding of helical structure (pUHM) caused peaks at the low-frequency region that were obliterated either by the presence of nanomaterials in the host FLC or by lowering the value of surface anchoring of the substrate. The material parameters such as rotational viscosity, spontaneous polarization, and response time had shown almost the same saturation value but comparatively higher values at lower voltages. In much the same way CuO nanorod in PDLCs out-turn 20 times increase in the conductivity, 40 times increase in the dielectric constant, and an enhancement in the material parameters, which make it a preferable material for designing advanced optical devices like holographic grating and switchable windows [85–87].

3.5 Aluminum-oxide nanomaterials (Al_2O_3)

Nanosized aluminum-oxide particles (AO NPs) are unique metal oxide NPs commonly used to remove pollutants due to their active sites and greater surface area. These properties of AO NPs make it capable of absorbing undesired ionic molecules, showing enhancement in the quality of electro-optical devices based on LCs. The presence of ionic impurities inside LC can be in many ways as discussed earlier, which degrades the performance of LC devices like image sticking, grey level shifting, and slow response time. Reports suggested that these ionic impurity effects could be suppressed by incorporating AO NPs with LCs. Joshi et al. reported the impact of AO NPs in KCFLC 7S mixtures by analyzing dielectric spectroscopy and electrical resistivity/conductivity measurements for both doped and pristine FLC mixtures [88]. Doping one wt% of AO NPs, size less than 30 nm, didn't disturb the alignment of FLC, but a size larger than 30 nm caused degradation of the material property by forming large aggregates. A low-frequency mode was observed for pure KCFLC 7S LC material owing to the ionization-recombination assisted diffusion of ionic impurities, which completely disappeared in the case of AO NPs-doped FLC mixtures owing to its trapping abilities and the size range between 20 and 30 nm is most effective for suppressing the impact of ionic impurities. The presence of alumina nanowires (AO NW) increased the transition temperature by 2°C in 5CB nematic LCs for both in-planar and in-plane switching (IPS) cell configurations [89]. AO NW-dispersed LCs minimized the light leakage centers, improving the contrast ratio, which is beneficial for making sharp-quality displays. Planar configuration is suitable for designing low-power consumption and high transmission-intensity electronic devices, but a better contrast ratio in IPS produces wide viewing angles. Energy bands of doped samples for both configurations increased due to the charge carrier mobility in the composite mixture, independent of its cell's configuration. In a similar manner, study on AO NP-doped polymer-dispersed LC films showed a decrease in drive voltages, which indicates that the optimized concentration of AO NPs is a promising way of achieving superior-quality electro-optical devices [90].

3.6 Zinc-oxide nanomaterials (ZnO)

Zinc-oxide (ZnO) NPs are popular owing to their applications in various fields, such as biomedical, gas sensors, electronics, photovoltaic, and optoelectronics [91–93]. Joshi et al. examined the impact of different concentrations of ZnO NPs in KCFLC7S material [94]. The dipolar interaction between ZnO NPs and KCFLC7S resulted in decrease in operating voltage and enhanced optical contrast. A larger dipole moment of ZnO provides strong anchoring strength and improved order parameters, consequently providing more defect-free alignment and decreasing the number of light leakage centers (Fig. 8).

The synergic interaction between spherical ZnO NPs and polymer-dispersed FLC matrixes could significantly contribute to the dielectric and electro-optical characteristics [95]. A change of morphology in the polymer dispersed FLC by doping ZnO was observed, because ZnO alters the refractive index of materials, thereby modifying scattering behavior. A lower frequency shift in the Goldstone mode relaxation peak and a greater dielectric strength value were monitored by varying the concentration of dispersed ZnO NPs from 0 to 1 wt%. Photoluminescence (PL) spectra suggested

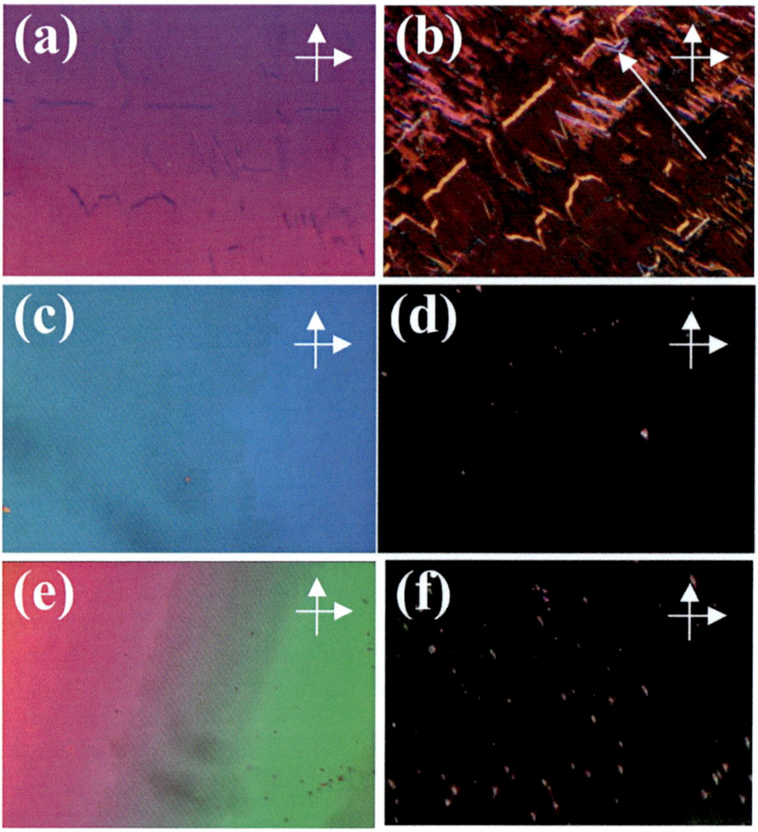

Fig. 8 Polarizing optical micrographs showing better alignment in ZnO-doped KCFLC 7S material. (A) Bright, (B) dark state of pure cell; (C) bright, (D) dark state of 1 wt % ZnO-NPs doped and (E) bright, (F) dark state of 2 wt% ZnO-NPs-doped KCFLC 7S material are represented. Crossed arrows depict the crossed polarizers, while the single arrow portrays the rubbing direction.
Reused with permission from Joshi T, Kumar A, Prakash J, Biradar AM. Low power operation of ferroelectric liquid crystal system dispersed with zinc oxide nanoparticles. Appl. Phys. Lett. 2010;96:235109. https://doi.org/10.1063/1.3455325.

a more significant number of FLC molecules absorb light by the presence of ZnO NPs due to the improvement in the ordering inside the domain; hence a two-fold increase in the PL intensity is observed. ZnO NPs have the ability to capture impurity ions present in the LC mixture and a larger dipole moment provides strong electrical torque, which is effective for better anchoring. The combination of ZnO NPs improved the ordering in 8CB LC molecules and out-turn in three orders of the current generation, which is convenient for fabricating photovoltaic cells based on LC materials [96].

3.7 Titanium-oxide nanomaterials (TiO_2)

Titanium dioxide (TiO_2) NPs are renowned for their applications in photovoltaic and visible-light photocatalysis [97,98]. These unique features of TiO_2 motivated researchers to conduct a series of studies through doping TiO_2 NPs in ferroelectric and nematic LCs. Conspicuous reduction in free ion concentration and thereby decreased conductivity was observed by adding TiO_2 NPs with W206 FLC [99]. The decrease in ionic concentrations was due to NPs' trapping of ionic impurities; interaction between charged particles and higher viscosity resulted in a reduction in diffusion constant. FLC W206E with 1.0 wt% TiO_2 NPs showed higher viscosity compared to the FLC W206E with 0.5 wt% TiO_2 NPs sample, but FLC W206E with 1.0 wt% TiO_2 NPs appeared higher conductivity values than the rest of the sample cells. The presence of TiO_2 nanoparticles with nematic LCs effectively interacted with ionic impurities and resulted in a reduced screening effect [100]. This generates larger effective voltage and reduced threshold voltage while maintaining a voltage holding ratio above 95%. The size of the NPs is also an essential factor in determining the screening effect of ionic impurities, LC cells had doped with 5, 10, and 30–40 nm-sized TiO_2 NPs and measured their threshold voltage drop from the voltage-transmittance characteristic curves as 2.31, 2.21, and 1.73 V, respectively. The measured and calculated values of threshold voltage drop closely confirmed the relationship between ionic concentration and threshold voltage. The smaller NPs could trap more ionic impurities than larger ones due to the difference in the total surface area; also result indicates that smaller NPs are the better choice for tailoring the optical properties of LCs. These studies confirmed that the optimum concentration of TiO_2 NPs can control undesired screening effects due to mobility-ions and thus results in depletion in threshold voltage.

3.8 Iron-oxide nanomaterials (Fe_3O_4)

To improve the characteristics of LC devices, researchers have examined the importance of doping magnetic nanoparticles into the LC mixture. In particular, magnetite (Fe_3O_4) nanoparticles already proved their importance in innumerable areas exclusively in biomedical applications, including magnetic resonance imaging, thermal therapy, and drug delivery [101,102]. Spherical Fe_3O_4 NPs doped into the nematic LC exhibited enhancement in dielectric parameters due to the strong orientational coupling between magnetic NPs and LC molecules and an average size range of 14–18 nm, thereby showing superparamagnetic behavior [103]. The dielectric constant temperature dependence for different concentrations (0, 1, 5, 10 wt%) of NPs had been studied, and a notable increase in the dielectric constant compared to the pristine nematic LC was observed. The mean dielectric constant can be measured by using the equation:

$$\bar{\varepsilon} = \frac{1}{3}\left(\varepsilon_{\parallel} + 2\varepsilon_{\perp}\right) \tag{3}$$

Where ($\varepsilon_{\parallel}$) and ($\varepsilon_{\perp}$) are parallel and perpendicular dielectric constants. The addition of NPs resulted in the increment of the mean dielectric constant, which increases with the increase in doping concentration, and a concentration of 1 wt% exhibited considerable increase in the dielectric anisotropy and a positive shift in nematic-isotropic transition temperature of 2.5 K. The strong dipole-dipole interaction between superparamagnetic Fe_3O_4 NPs and the nematic LC molecules gives rise to better particle directional orientation and improvement in dielectric parameters.

Zakerhamidi and his coworkers observed a linear increase in the dielectric permittivity value of Fe_3O_4 NPs-doped E7 NLCs in the nematic phase and this variation is due to modifications in molecular orientations and high permittivity values of dispersed NPs [104]. Sample with 1 wt% concentration exhibited higher permittivity values; however, a rise in the concentration of Fe_3O_4 resulted in a reduction in the permittivity values due to a decrease in molecular orientation and induced dipole orientation (Fig. 9). When doping magnetic nanoparticles into the FLC mixture by forming a multiferroic system [105], the magnetic field generated inside the LC mixture can minimize the screening effect caused by the ionic impurities and thus, enhancing electro-optical characteristics.

3.9 Silicon-dioxide nanomaterials (SiO_2)

Silicon-dioxide nanoparticles or silica NPs were extensively used in biomedical applications due to their unique physicochemical properties such as larger surface area, porosity, and the ability for functionalization [106]. There are many possibilities for doping different sizes of silica NPs in various LC systems for forthcoming electro-optical applications. Chaudhary et al. suggested doping silica NPs can improve the electro-optical properties of LC materials by increasing anchoring strength between LC molecules and silica NPs [107]. A faster switching time, decreased spontaneous polarization, and rotational viscosity were observed, which were explained based on anchoring phenomena. The equation connecting anchoring energy coefficient with switching time can be expressed as:

$$\tau_s = \frac{\eta}{K\pi^2}\left(d^2 + \frac{4dK}{W}\right) \text{ (for strong anchoring)} \tag{4}$$

$$\tau_s = \frac{4\eta d}{W\pi^2} \text{ (for weak anchoring)} \tag{5}$$

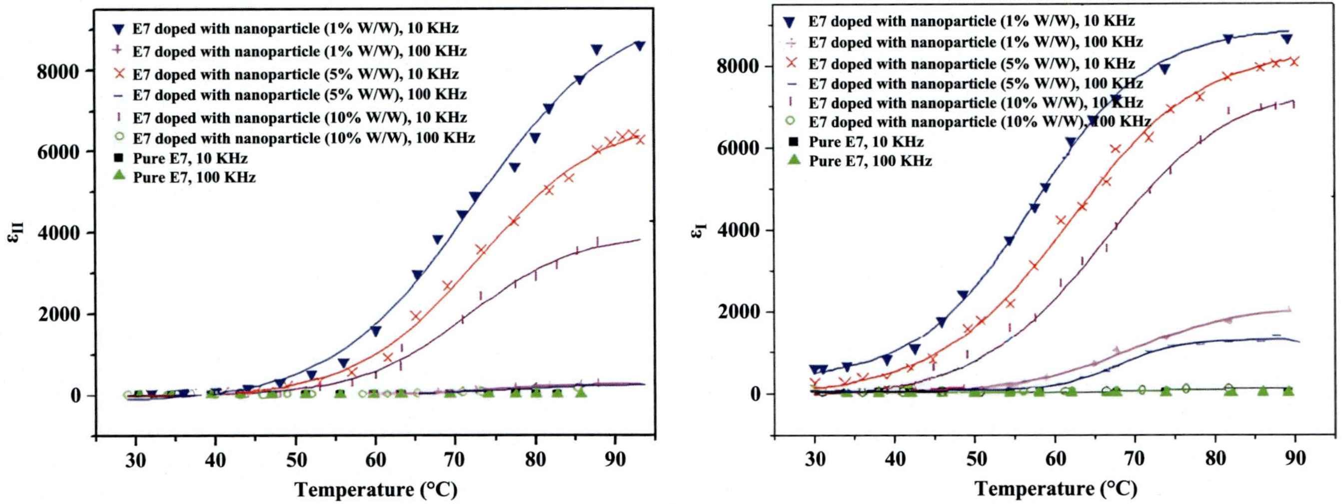

Fig. 9 Dependences of permittivity of the homeotropic and homogeneous oriented ferro-nematics at different temperatures.
Reused with permission from Zakerhamidi MS, Shoarinejad S, Mohammadpour S. Fe3O4 nanoparticle effect on dielectric and ordering behavior of nematic liquid crystal host. J. Mol. Liq. 2014;191:16-19. https://doi.org/10.1016/j.molliq.2013.11.020.

Where η is the rotational viscosity, d is cell thickness, K is bend elastic constant, and W is the anchoring energy coefficient. The equations imply strong anchoring can out-turn faster switching time observed with silica NPs-doped FLC mixture and considerably minimizes image ghosting in display applications. The presence of silica NPs can enhance disorders of LC molecules making it strenuous for LC molecules to reorient in a specified direction. The argument agrees with the experimental result by showing ~52% reduction in dielectric strength in 0.03 wt% silica NPs-doped samples than the pristine material. The size of the silica NPs is also an essential factor when considering it for specified applications, electro-optical properties are greatly influenced by smaller-sized particles (~12 nm) but dielectric properties by larger ones (~40 nm) [108]. The smaller silica NPs can positively affect the molecular alignment, resulting in strong anchoring energy coefficients and electro-optical properties. The ~23% increase in the value conductivity for smaller and ~38% increase for larger silica NPs-doped samples reveals the generation of excess mobile ions and mutual coupling between optical anisotropies of silica NPs (5–15 nm). Silica NPs in KCFLC10R material exhibited 75% improvement in transmittance and 77% improvement in optical contrast and also a remarkable increase in the optical birefringence with a decrease in operating voltage [109]. The significant reduction in dielectric relaxation time was observed due to the interfacial polarization effect between silica NPs and LC molecules which can be utilized for designing conventional LCDs with fast response time [110]. Silica NPs-dispersed nematic and ferroelectric LCs have substantial potential applications, especially with the growing demand for advanced display and electro-optical devices.

3.10 Other metal oxides

Yttrium oxide (Y_2O_3) NPs are air-stable materials with a wide bandgap of 5.6 eV as well as large thermal conductivity and can be used for fabricating superior-quality LC devices. The sample with the most effective concentration of 2 wt% showed homogeneous alignment without agglomeration or crack, a significant reduction in threshold voltages from 4.2 to 2.6 V, and a 53% improvement in response time [111]. Y_2O_3 NPs have the ability to trap ionic impurities, thereby reducing the field screening effect, which results in a slight improvement in the pretilt angle, decreased operating voltage, and enhanced electro-optical characteristics, which has future applications in LC-based electro-optical devices. Similarly, CeO_2 is highly recommended for its applications in numerous areas owing to its material properties like high electrical conductivity, diffusivity, and thermal conductivity. Recent research indicates that dispersing CeO_2 NPs remarkably improved the electro-optical properties and thermal stability of host LC medium. The optimum concentration of 0.8 wt% CeO_2 in PDLC films considerably reduced its threshold voltage from 26.6 to 16.8 V and appreciably increased the contrast ratio [112]. Noticeable improvement in the thermal stability was reported by doping CeO_2 NPs to NLC without aggregation of molecules. The charged CeO_2 NPs effectively suppressed the ionic impurities while applying voltage and out-turn faster response time with lower threshold voltage [113]. Dispersion of tin-oxide nanomaterials (SnO_2) in FLC material W343 resulted in a decrease in ac conductivity, relative permittivity, and dielectric loss due to the absorption of ionic impurities [114]. SnO_2 NPs were forming self-assembled 2D arrays, which significantly changed the alignment properties in FLC, higher doping concentrations showed the lowest photoluminescent intensity due to the close aggregation of NPs in LC molecules. Indium tin oxide dispersed in polymer-dispersed liquid crystal (PDLC) exhibited the lowest transmittance, low-driven voltage, and relatively high contrast ratio, indicating potential PDLC applications in the energy-saving area [115].

4 Conclusions

The doping of MONPs scale-up the applications of LCs in electronic devices by delivering superior electro-optical and dielectric characteristics. MONP could become or is a paradigmatic material for tuning LC's optical and anisotropic characteristics by utilizing its distinctive physical and chemical properties. Based on the molecular arrangement, metal oxides have metallic, semiconducting, and insulating properties that determine LC molecules' interaction with nanomaterials. The smaller-sized dopant materials (<30 nm) could effectively improve the electro-optical characteristics without disturbing molecular alignment, but larger ones are suitable for achieving superior dielectric properties. MONPs with inherent dipole moments could interact electrostatically with LC molecules, creating better anchoring energy and alignment to the LC matrix. As far as the electro-optical properties are concerned, an enhancement in the same is seen by doping with MONPs through trapping of ionic impurities, which brings flicker-free good contrast displays and high-speed optical shutters. Importantly, the depolarization field caused by ionic impurities in FLCs can be minimized by adding MONPs, which results in optical memory effect suitable for fabricating next-generation display and data storage devices. Thus, dispersion of MONPs in LCs enhances the order parameter eventuating better electro-optical and dielectric performance, and is an ideal choice for designing tunable electronic devices based on LCs. Despite the practical applications of LC being prevalent in many areas, the tuning of material properties and achieving monodomain alignment over various substrate surfaces for device applications face limitations. However, the strategic utilization of optimized metal oxide nanoparticle dispersion with liquid crystal emerges as a compelling solution to surmount the existing constraints. That is, in the present research pursuits for stable, highly efficient, and fast-switching LC devices for advanced electro-optical applications and reliable optical memory devices, the MONPs incorporated in LC matrix might prove to be a solution.

References

[1] M. Mitov, Liquid-crystal science from 1888 to 1922: building a revolution, ChemPhysChem 15 (2014) 1245–1250, https://doi.org/10.1002/cphc.201301064.
[2] D. Andrienko, Introduction to liquid crystals, J. Mol. Liq. 267 (2018) 520–541, https://doi.org/10.1016/j.molliq.2018.01.175.

[3] A. de Vries, On the difference between solid crystals and liquid crystals with three-dimensional order, Mol. Cryst. Liq. Cryst. 49 (1978) 19–25, https://doi.org/10.1080/00268947808070321.

[4] G. Friedel, Les états mésomorphes de la matière, Ann. Phys. 9 (1922) 273–474, https://doi.org/10.1051/anphys/192209180273.

[5] H.R. Brand, P.E. Cladis, H. Pleiner, Macroscopic properties of smectic liquid crystals, Eur. Phys. J. B 6 (1998) 347–353, https://doi.org/10.1007/s100510050560.

[6] A. De Vries, Experimental evidence concerning two different kinds of smectic C to smectic A transitions, Mol. Cryst. Liq. Cryst. 41 (1977) 27–31, https://doi.org/10.1080/01406567708071949.

[7] D. Coates, Development and applications of cholesteric liquid crystals, Liq. Cryst. 42 (2015) 653–665, https://doi.org/10.1080/02678292.2015.1020454.

[8] M.J. Stephen, J.P. Straley, Physics of liquid crystals, Rev. Mod. Phys. 46 (1974) 617–704, https://doi.org/10.1103/RevModPhys.46.617.

[9] N. Popov, L.W. Honaker, M. Popova, N. Usoltseva, E.K. Mann, A. Jákli, et al., Thermotropic liquid crystal-assisted chemical and biological sensors, Materials 11 (2017) 20, https://doi.org/10.3390/ma11010020.

[10] I. Dierking, A. Martins Figueiredo Neto, Novel trends in lyotropic liquid crystals, Crystals 10 (2020) 604, https://doi.org/10.3390/cryst10070604.

[11] J.D. Martin, C.L. Keary, T.A. Thornton, M.P. Novotnak, J.W. Knutson, J.C.W. Folmer, Metallotropic liquid crystals formed by surfactant templating of molten metal halides, Nat. Mater. 5 (2006) 271–275, https://doi.org/10.1038/nmat1610.

[12] R. Oldenbourg, Analysis of edge birefringence, Biophys. J. 60 (1991) 629–641, https://doi.org/10.1016/S0006-3495(91)82092-6.

[13] C. Tschierske, Mirror symmetry breaking in liquids and liquid crystals, Liq. Cryst. 45 (2018) 2221–2252, https://doi.org/10.1080/02678292.2018.1501822.

[14] W.J.A.M. Hartmann, Ferroelectric liquid crystal displays for television application, Ferroelectrics 122 (1991) 1–26, https://doi.org/10.1080/00150199108226025.

[15] J.A. Castellano, Liquid crystal display applications: past, present & future, Liquid Crystals Today 1 (1991) 4–6, https://doi.org/10.1080/13583149108628568.

[16] L. Mauta, J. Vince, P. Ripa, Comparison of the use of liquid crystal thermometers with glass mercury thermometers in febrile children in a Children's Ward at Port Moresby General Hospital, Papua New Guinea, J. Trop. Pediatr. 55 (2009) 368–373, https://doi.org/10.1093/tropej/fmp027.

[17] G.B. Cohen, R. Pogreb, K. Vinokur, D. Davidov, Spatial light modulator based on a deformed-helix ferroelectric liquid crystal and a thin a-Si:H amorphous photoconductor, Appl. Opt. 36 (1997) 455, https://doi.org/10.1364/AO.36.000455.

[18] G.M. Janini, J. Kevin, W.L. Zielinski, Use of a nematic liquid crystal for gas-liquid chromatographic separation of polyaromatic hydrocarbons, Anal. Chem. 47 (1975) 670–674,-https://doi.org/10.1021/ac60354a016.

[19] S.L. Lai, S. Huang, X. Bi, K.-L. Yang, Optical imaging of surface-immobilized oligonucleotide probes on DNA microarrays using liquid crystals, Langmuir 25 (2009) 311–316,-https://doi.org/10.1021/la802672b.

[20] R. Devi, Liquid crystals—the 'fourth' phase of matter (review), Rec. Res. Asp. 2 (2015) 18–20.

[21] V.M. Dolgov, L.G. Likholetova, Use of thermooptical effects in liquid crystals to visualize electromagnetic fields, Radiophys. Quant. Electron. 22 (1979) 330–335, https://doi.org/10.1007/BF01035359.

[22] L.J. Broutman, T. Kobayashi, D. Carrillo, Determination of fracture sites in composite materials with liquid crystals, J. Compos. Mater. 3 (1969) 702–704, https://doi.org/10.1177/002199836900300411.
[23] C.M. Swenson, C.A. Steed, I.A. De La Rue, R.Q. Fugate, Low-power FLC-based retromodulator communications system, Free-Space Laser Commun. Technol. (1997) 296–310, https://doi.org/10.1117/12.273706.
[24] I. Khan, K. Saeed, I. Khan, Nanoparticles: properties, applications and toxicities, Arab. J. Chem. 12 (2019) 908–931, https://doi.org/10.1016/j.arabjc.2017.05.011.
[25] S.K. Narendra Kumar, Essentials in Nanoscience & Nanotechnology, Wiley, 2016.
[26] W.F. Maier, Transition metal oxides: surface chemistry and catalysis, Angew. Chem. 102 (1990) 965–966, https://doi.org/10.1002/ange.19901020838.
[27] M. Fernandez-Garcia, J.A. Rodriguez, Metal oxide nanoparticles, in: Encyclopedia of Inorganic Chemistry, 2009, https://doi.org/10.1002/adma.19950070122.
[28] J.A. Rodriguez, F. Garcia, Metal oxide nanoparticles, in: Encyclopedia of Inorganic Chemistry, 2007, https://doi.org/10.1002/0470862106.ia377.
[29] B.J. Shaw, R.D. Handy, Physiological effects of nanoparticles on fish: a comparison of nanometals versus metal ions, Environ. Int. 37 (2011) 1083–1097, https://doi.org/10.1016/j.envint.2011.03.009.
[30] A. Bootz, V. Vogel, D. Schubert, J. Kreuter, Comparison of scanning electron microscopy, dynamic light scattering and analytical ultracentrifugation for the sizing of poly(butyl cyanoacrylate) nanoparticles, Eur. J. Pharm. Biopharm. 57 (2004) 369–375,-https://doi.org/10.1016/S0939-6411(03)00193-0.
[31] J. Ying, T. Zhang, M. Tang, Metal oxide nanomaterial QNAR models: available structural descriptors and understanding of toxicity mechanisms, Nanomaterials 5 (2015) 1620–1637, https://doi.org/10.3390/nano5041620.
[32] A. Harano, K. Shimada, T. Okubo, M. Sadakata, Crystal phases of TiO_2 ultrafine particles prepared by laser ablation of solid rods, J. Nanopart. Res. 4 (2002) 215–219, https://doi.org/10.1023/A:1019935427050.
[33] J.-P. Jolivet, C. Froidefond, A. Pottier, C. Chanéac, S. Cassaignon, E. Tronc, P. Euzen, Size tailoring of oxide nanoparticles by precipitation in aqueous medium. a semi quantitative modelling, J. Mater. Chem. 14 (2004) 3281–3288, https://doi.org/10.1039/B407086K.
[34] D. Chiche, M. Digne, R. Revel, C. Chanéac, J.-P. Jolivet, Accurate determination of oxide nanoparticle size and shape based on X-ray powder pattern simulation: application to boehmite AlOOH, J. Phys. Chem. C 112 (2008) 8524–8533, https://doi.org/10.1021/jp710664h.
[35] J.P. Jolivet, S. Cassaignon, C. Chanéac, D. Chiche, O. Durupthy, D. Portehault, Design of metal oxide nanoparticles: control of size, shape, crystalline structure and functionalization by aqueous chemistry, C. R. Chim. 13 (2010) 40–51, https://doi.org/10.1016/j.crci.2009.09.012.
[36] M. Iwamoto, T. Abe, Y. Tachibana, Control of bandgap of iron oxide through its encapsulation into SiO_2-based mesoporous materials, J. Mol. Catal. A Chem. 155 (2000) 143–153, https://doi.org/10.1016/S1381-1169(99)00330-1.
[37] O. Vigil, F. Cruz, A. Morales-Acevedo, G. Contreras-Puente, L. Vaillant, G. Santana, Structural and optical properties of annealed CdO thin films prepared by spray pyrolysis, Mater. Chem. Phys. 68 (2001) 249–252, https://doi.org/10.1016/S0254-0584(00)00358-8.
[38] R. Viswanatha, S. Sapra, B. Satpati, P.V. Satyam, B.N. Dev, D.D. Sarma, Understanding the quantum size effects in ZnO nanocrystals, J. Mater. Chem. 14 (2004) 661–668,-https://doi.org/10.1039/b310404d.

[39] H. Deng, J.M. Hossenlopp, Combined X-ray diffraction and diffuse reflectance analysis of nanocrystalline mixed Sn(II) and Sn(IV) oxide powders, J. Phys. Chem. B 109 (2005) 66–73, https://doi.org/10.1021/jp047812s.
[40] E.B. Lavik, I. Kosacki, H.L. Tuller, et al., Nonstoichiometry and electrical conductivity of nanocrystalline CeO_{2-x}, J. Electroceram. 1 (1997) 7–14, https://doi.org/10.1023/A:1009934829870.
[41] S.C. Tjong, H. Chen, Nanocrystalline materials and coatings, Mater. Sci. Eng. 45 (2004) 1–88, https://doi.org/10.1016/j.mser.2004.07.001.
[42] Z. Wang, S. Seal, S. Patil, C. Zha, Q. Xue, Anomalous quasihydrostaticity and enhanced structural stability of 3 nm nanoceria, J. Phys. Chem. C 111 (2007) 11756–11759, https://doi.org/10.1021/jp074909g.
[43] M.T.D. Cronin, Quantitative structure-activity relationships (QSARs)—applications and methodology, in: Recent Advances in QSAR Studies, 2010, pp. 3–11, https://doi.org/10.1007/978-1-4020-9783-6_1.
[44] K.S. Susiiik, S. Choet, A.A. Cichowlass, M.W. Grinstaff, Sonochemical synthesis of amorphous iron, Lett. Nat. 353 (1991) 414–416, https://doi.org/10.1038/353414A0.
[45] J.-F. Chen, Y.-H. Wang, F. Guo, X.-M. Wang, C. Zheng, Synthesis of nanoparticles with novel technology: high-gravity reactive precipitation, Ind. Eng. Chem. Res. 39 (2000) 948–954, https://doi.org/10.1021/ie990549a.
[46] D.J. Cole-Hamilton, Chemistry of advanced materials: an overview, Appl. Organomet. Chem. 14 (2) (2000) 127–132, https://doi.org/10.1002/(SICI)1099-0739(200002)14:2<127::AID-AOC898>3.0.CO;2-R.
[47] V. Uskokovic, M. Drofenik, Synthesis of materials within reverse micells, Surf. Rev. Lett. 12 (2005) 239–277, https://doi.org/10.1142/S0218625X05007001.
[48] M. Ohring, Film structure, in: Materials Science of Thin Films, 2002, https://doi.org/10.1016/b978-0-12-524975-1.x5000-9.
[49] M.S. Hegde, Epitaxial oxide thin films by pulsed laser deposition: retrospect and prospect, J. Chem. Sci. 113 (2001) 445–458, https://doi.org/10.1007/BF02708783.
[50] V. Bansal, P. Poddar, A. Ahmad, M. Sastry, Room-temperature biosynthesis of ferroelectric barium titanate nanoparticles, J. Am. Chem. Soc. 128 (2006) 11958–11963, https://doi.org/10.1021/ja063011m.
[51] W.-T. Liu, Nanoparticles and their biological and environmental applications, J. Biosci. Bioeng. 102 (2006) 1–7, https://doi.org/10.1263/jbb.102.1.
[52] Y. Jun, J. Seo, J. Cheon, Nanoscaling laws of magnetic nanoparticles and their applicabilities in biomedical sciences, Acc. Chem. Res. 41 (2008) 179–189, https://doi.org/10.1021/ar700121f.
[53] M.T. Klem, D.A. Resnick, K. Gilmore, M. Young, Y.U. Idzerda, T. Douglas, Synthetic control over magnetic moment and exchange bias in all-oxide materials encapsulated within a spherical protein cage, J. Am. Chem. Soc. 129 (2007) 197–201, https://doi.org/10.1021/ja0667561.
[54] X. Liu, J. Zhang, S. Wu, D. Yang, P. Liu, H. Zhang, et al., Single crystal α-Fe2O3 with exposed {104} facets for high performance gas sensor applications, RSC Adv. 2 (2012) 6178, https://doi.org/10.1039/c2ra20797d.
[55] X. Li, W. Wei, S. Wang, L. Kuai, B. Geng, Single-crystalline α-Fe2O3 oblique nanoparallelepipeds: high-yield synthesis, growth mechanism and structure enhanced gas-sensing properties, Nanoscale 3 (2011) 718–724, https://doi.org/10.1039/C0NR00617C.
[56] M.S. Chavali, M.P. Nikolova, Metal oxide nanoparticles and their applications in nanotechnology, SN Appl. Sci. (2019) 1, https://doi.org/10.1007/s42452-019-0592-3.

[57] Z.-S. Wu, G. Zhou, L.-C. Yin, W. Ren, F. Li, H.-M. Cheng, Graphene/metal oxide composite electrode materials for energy storage, Nano Energy 1 (2012) 107–131, https://doi.org/10.1016/j.nanoen.2011.11.001.
[58] R. Jose, V. Thavasi, S. Ramakrishna, Metal oxides for dye-sensitized solar cells, J. Am. Ceram. Soc. 92 (2009) 289–301, https://doi.org/10.1111/j.1551-2916.2008.02870.x.
[59] S.-Y. Lee, X. Gao, H. Matsui, Biomimetic and aggregation-driven crystallization route for room-temperature material synthesis: growth of β-Ga_2O_3 nanoparticles on peptide assemblies as nanoreactors, J. Am. Chem. Soc. 129 (2007) 2954–2958, https://doi.org/10.1021/ja0677057.
[60] J. Prakash, S. Khan, S. Chauhan, A.M. Biradar, Metal oxide-nanoparticles and liquid crystal composites: a review of recent progress, J. Mol. Liq. 297 (2020) 112052, https://doi.org/10.1016/j.molliq.2019.112052.
[61] W. Zhang, D. Zhang, T. Fan, J. Ding, Q. Guo, H. Ogawa, Fabrication of ZnO microtubes with adjustable nanopores on the walls by the templating of butterfly wing scales, Nanotechnology 17 (2006) 840–844, https://doi.org/10.1088/0957-4484/17/3/038.
[62] J. Aizenberg, J. Hanson, T.F. Koetzle, S. Weiner, L. Addadi, Control of macromolecule distribution within synthetic and biogenic single calcite crystals, J. Am. Chem. Soc. 119 (1997) 881–886, https://doi.org/10.1021/ja9628821.
[63] J.S. Wilkes, M.J. Zaworotko, Air and water stable 1-ethyl-3-methylimidazolium based ionic liquids, J. Chem. Soc. Chem. Commun. 13 (1992) 965–967, https://doi.org/10.1039/c39920000965.
[64] M. Rao Darla, S. Hegde, S. Varghese, Effect of $BaTiO_3$ nanoparticle on electro-optical properties of polymer dispersed liquid crystal displays, J. Crystall. Process Technol. 04 (2014) 60–63, https://doi.org/10.4236/jcpt.2014.41008.
[65] P. Ganguly, A. Kumar, S. Tripathi, D. Haranath, A.M. Biradar, Faster and highly luminescent ferroelectric liquid crystal doped with ferroelectric $BaTiO_3$ nanoparticles, Appl. Phys. Lett. 102 (2013) 222902, https://doi.org/10.1063/1.4809515.
[66] F.V. Podgornov, M. Gavrilyak, A. Karaawi, V. Boronin, W. Haase, Mechanism of electrooptic switching time enhancement in ferroelectric liquid crystal/gold nanoparticles dispersion, Liq. Cryst. 45 (2018) 1594–1602, https://doi.org/10.1080/02678292.2018.1458256.
[67] J. Prakash, A. Choudhary, A. Kumar, D.S. Mehta, A.M. Biradar, Nonvolatile memory effect based on gold nanoparticles doped ferroelectric liquid crystal, Appl. Phys. Lett. 93 (2008) 112904, https://doi.org/10.1063/1.2980037.
[68] A. Chaudhary, P. Malik, R. Mehra, K.K. Raina, Observation of memory behaviour in cadmium sulphide nanorods doped ferroelectric liquid crystal mixture, Phase Transit. 86 (2013) 1256–1266, https://doi.org/10.1080/01411594.2012.748908.
[69] C. Esteves, E. Ramou, A.R.P. Porteira, A.J. Moura Barbosa, A.C.A. Roque, Seeing the unseen: the role of liquid crystals in gas-sensing technologies, Adv. Opt. Mater. 8 (2020) 1902117, https://doi.org/10.1002/adom.201902117.
[70] R. Morris, C. Jones, M. Nagaraj, Liquid crystal devices for beam steering applications, Micromachines 12 (2021) 247, https://doi.org/10.3390/mi12030247.
[71] I. Abdulhalim, R. Moses, R. Sharon, Biomedical optical applications of liquid crystal devices, Acta Phys. Pol. A 112 (2007) 715–722, https://doi.org/10.12693/APhysPolA.112.715.
[72] J. Hornak, Synthesis, properties, and selected technical applications of magnesium oxide nanoparticles: a review, Int. J. Mol. Sci. 22 (2021) 12752, https://doi.org/10.3390/ijms222312752.
[73] A. Chandran, J. Prakash, K.K. Naik, A.K. Srivastava, R. Dąbrowski, M. Czerwiński, et al., Preparation and characterization of MgO nanoparticles/ferroelectric liquid crystal

composites for faster display devices with improved contrast, J. Mater. Chem. C 2 (2014) 1844–1853, https://doi.org/10.1039/c3tc32017k.
[74] D. Pourmostafa, H. Tajalli, A. Vahedi, K. Milanchian, Electro-optical Kerr effect of 6CHBT liquid crystal doped with MgO nanoparticles in different concentration, Opt. Mater. (Amst.) 107 (2020) 110061, https://doi.org/10.1016/j.optmat.2020.110061.
[75] A. Chandran, J. Prakash, P. Ganguly, A.M. Biradar, Zirconia nanoparticles/ferroelectric liquid crystal composites for ionic impurity-free memory applications, RSC Adv. 3 (2013) 17166–17173, https://doi.org/10.1039/c3ra41964a.
[76] H.-J. Kim, Y.-G. Kang, H.-G. Park, K.-M. Lee, S. Yang, H.-Y. Jung, et al., Effects of the dispersion of zirconium dioxide nanoparticles on high performance electro-optic properties in liquid crystal devices, Liq. Cryst. 38 (2011) 871–875, https://doi.org/10.1080/02678292.2011.584637.
[77] J. Cheng, B. Zhao, W. Zhang, F. Shi, G. Zheng, D. Zhang, et al., High-performance supercapacitor applications of nio-nanoparticle-decorated millimeter-long vertically aligned carbon nanotube arrays via an effective supercritical CO_2-assisted method, Adv. Funct. Mater. 25 (2015) 7381–7391, https://doi.org/10.1002/adfm.201502711.
[78] S.R. Gawali, V.L. Patil, V.G. Deonikar, S.S. Patil, D.R. Patil, P.S. Patil, et al., Ce doped NiO nanoparticles as selective NO_2 gas sensor, J. Phys. Chem. Solids 114 (2018) 28–35,-https://doi.org/10.1016/j.jpcs.2017.11.005.
[79] A. Chandran, J. Prakash, J. Gangwar, T. Joshi, A.K. Srivastava, D. Haranath, A.M. Birder, Low-voltage electro-optical memory device based on NiO nanorods dispersed in a ferroelectric liquid crystal, RSC Adv. 6 (2016) 53873–53881, https://doi.org/10.1039/C6RA04037C.
[80] H.M. Lee, H.-K. Chung, H.-G. Park, H.-C. Jeong, J.-H. Kim, T.-K. Park, et al., Nickel oxide nanoparticles doped liquid crystal system for superior electro-optical properties, J. Nanosci. Nanotechnol. 15 (2015) 8139–8143, https://doi.org/10.1166/jnn.2015.11263.
[81] A. Rydosz, The use of copper oxide thin films in gas-sensing applications, Coatings 8 (2018) 425, https://doi.org/10.3390/coatings8120425.
[82] G. Ren, D. Hu, E.W.C. Cheng, M.A. Vargas-Reus, P. Reip, R.P. Allaker, Characterisation of copper oxide nanoparticles for antimicrobial applications, Int. J. Antimicrob. Agents 33 (2009) 587–590, https://doi.org/10.1016/j.ijantimicag.2008.12.004.
[83] M. Mumtaz, A.I. Bhatti, K. Nadeem, N.A. Khan, A. Saleem, S.T. Hussain, Study of CuO nano-particles/CuTl-1223 superconductor composite, J. Low Temp. Phys. 170 (2013) 185–204, https://doi.org/10.1007/s10909-012-0741-1.
[84] S. Khan, S. Chauhan, A. Chandran, M. Czerwiński, J. Herman, A.M. Biradar, et al., Enhancement of dielectric and electro-optical parameters of a newly prepared ferroelectric liquid crystal mixture by dispersing nano-sized copper oxide, Liq. Cryst. 47 (2020) 263–272, https://doi.org/10.1080/02678292.2019.1643506.
[85] Z. Jiang, J. Zheng, Y. Liu, Q. Zhu, Investigation of dielectric properties in polymer dispersed liquid crystal films doped with CuO nanorods, J. Mol. Liq. 295 (2019) 11167, https://doi.org/10.1016/j.molliq.2019.111667.
[86] S.S. Parab, M.K. Malik, P.G. Bhatia, R.R. Deshmukh, Investigation of liquid crystal dispersion and dielectric relaxation behavior in polymer dispersed liquid crystal composite films, J. Mol. Liq. 199 (2014) 287–293, https://doi.org/10.1016/j.molliq.2014.09.013.
[87] Y.-T. Lai, J.-C. Kuo, Y.-J. Yang, Polymer-dispersed liquid crystal doped with carbon nanotubes for dimethyl methylphosphonate vapor-sensing application, Appl. Phys. Lett. 102 (2013) 191912, https://doi.org/10.1063/1.4804297.
[88] T. Joshi, J. Prakash, A. Kumar, J. Gangwar, A.K. Srivastava, S. Singh, et al., Alumina nanoparticles find an application to reduce the ionic effects of ferroelectric liquid crystal,

J. Phys. D. Appl. Phys. 44 (2011) 315404, https://doi.org/10.1088/0022-3727/44/31/315404.

[89] A. Sharma, P. Kumar, P. Malik, Textural and electro-optical study of a room temperature nematic liquid crystal 4′-pentyl-4-biphenylcarbonitrile doped with metal oxide nanowires in planar and in-plane switching cell configurations, Liq. Cryst. 47 (2020) 1–15,-https://doi.org/10.1080/02678292.2020.1755467.

[90] M. Jia, Y. Zhao, H. Gao, D. Wang, Z. Miao, H. Cao, et al., The electro-optical study of Al_2O_3 nanoparticles doped polymer dispersed liquid crystal films, Liq. Cryst. 49 (2021) 39–49, https://doi.org/10.1080/02678292.2021.1943024.

[91] J. Jiang, J. Pi, J. Cai, The advancing of zinc oxide nanoparticles for biomedical applications, Bioinorg. Chem. Appl. (2018) 1–18, https://doi.org/10.1155/2018/1062562.

[92] Q. Zhang, C. Xie, S. Zhang, A. Wang, B. Zhu, L. Wang, Z. Yang, Identification and pattern recognition analysis of Chinese liquors by doped nano ZnO gas sensor array, Sensors Actuators B Chem. 110 (2005) 370–376, https://doi.org/10.1016/j.snb.2005.02.017.

[93] M. Godlewski, E. Guziewicz, K. Kopalko, G. Łuka, M.I. Łukasiewicz, T. Krajewski, B.S. Witkowski, S. Gieraltowska, Zinc oxide for electronic, photovoltaic and optoelectronic applications, Low Temp. Phys. 37 (2011) 235–240, https://doi.org/10.1063/1.3570930.

[94] T. Joshi, A. Kumar, J. Prakash, A.M. Biradar, Low power operation of ferroelectric liquid crystal system dispersed with zinc oxide nanoparticles, Appl. Phys. Lett. 96 (2010) 235109, https://doi.org/10.1063/1.3455325.

[95] D. Jayoti, P. Malik, S.K. Prasad, Effect of ZnO nanoparticles on the morphology, dielectric, electro-optic and photo luminescence properties of a confined ferroelectric liquid crystal material, J. Mol. Liq. 250 (2018) 381–387, https://doi.org/10.1016/j.molliq.2017.12.035.

[96] L.J. Martínez-Miranda, K.M. Traister, I. Meléndez-Rodríguez, L. Salamanca-Riba, Liquid crystal-ZnO nanoparticle photovoltaics: role of nanoparticles in ordering the liquid crystal, Appl. Phys. Lett. 97 (2010) 223301, https://doi.org/10.1063/1.3511736.

[97] Y. Bai, I. Mora-Seró, F. de Angelis, J. Bisquert, P. Wang, Titanium dioxide nanomaterials for photovoltaic applications, Chem. Rev. 114 (2014) 10095–10130,-https://doi.org/10.1021/cr400606n.

[98] Y. Nam, J.H. Lim, K.C. Ko, J.Y. Lee, Photocatalytic activity of TiO_2 nanoparticles: a theoretical aspect, J. Mater. Chem. A 7 (2019) 13833, https://doi.org/10.1039/C9TA03385H.

[99] P. Kumar, A. Sinha, Effect of TiO2 nanoparticle doping on the electrical properties of ferroelectric liquid crystal, Adv. Opt. Sci. Eng. 194 (2017) 499–505, https://doi.org/10.1007/978-981-10-3908-9_61.

[100] T.-R. Chou, J. Hsieh, W.-T. Chen, C.-Y. Chao, Influence of particle size on the ion effect of TiO_2 nanoparticle doped nematic liquid crystal cell, Jpn. J. Appl. Phys. 53 (2014) 071701, https://doi.org/10.7567/JJAP.53.071701.

[101] L.S. Ganapathe, M.A. Mohamed, R. Mohamad Yunus, D.D. Berhanuddin, Magnetite (Fe_3O_4) nanoparticles in biomedical application: from synthesis to surface functionalisation, Magnetochemistry 6 (2020) 68, https://doi.org/10.3390/magnetochemistry6040068.

[102] J. Li, S. Wang, C. Wu, Y. Dai, P. Hou, C. Han, et al., Activatable molecular MRI nanoprobe for tumor cell imaging based on gadolinium oxide and iron oxide nanoparticle, Biosens. Bioelectron. 86 (2016) 1047–1053, https://doi.org/10.1016/j.bios.2016.07.044.

[103] A. Maleki, M.H.M. Ara, F. Saboohi, Dielectric properties of nematic liquid crystal doped with Fe_3O_4 nanoparticles, Phase Transit. 90 (2017) 371–379, https://doi.org/10.1080/01411594.2016.1201821.

[104] M.S. Zakerhamidi, S. Shoarinejad, S. Mohammadpour, Fe_3O_4 nanoparticle effect on dielectric and ordering behavior of nematic liquid crystal host, J. Mol. Liq. 191 (2014) 16–19, https://doi.org/10.1016/j.molliq.2013.11.020.

[105] P.N. Romero-Hasler, L.K. Kurihara, L.O. Mair, I.N. Weinberg, E.A. Soto-Bustamante, L.J. Martínez-Miranda, Nanocomposites of ferroelectric liquid crystals and FeCo nanoparticles: towards a magnetic response via the application of a small electric field, Liq. Cryst. 47 (2020) 169–178, https://doi.org/10.1080/02678292.2019.1633429.

[106] A. Bitar, N.M. Ahmad, H. Fessi, A. Elaissari, Silica-based nanoparticles for biomedical applications, Drug Discov. Today 17 (2012) 1147–1154, https://doi.org/10.1016/j.drudis.2012.06.014.

[107] A. Chaudhary, P. Malik, R. Mehra, K.K. Raina, Electro-optic and dielectric studies of silica nanoparticle doped ferroelectric liquid crystal in SmC* phase, Phase Transit. 85 (2012) 244–254, https://doi.org/10.1080/01411594.2011.624274.

[108] G. Kaur, P. Kumar, A.K. Singh, D. Jayoti, P. Malik, Dielectric and electro-optic studies of a ferroelectric liquid crystal dispersed with different sizes of silica nanoparticles, Liq. Cryst. 47 (2020) 2194–2208, https://doi.org/10.1080/02678292.2020.1759154.

[109] A. Chaudhary, P. Malik, R.K. Shukla, R. Mehra, K.K. Raina, Role of SiO_2 optically active mediators to tailor optical and electro-optical properties of ferroelectric liquid crystalline nanocomposites, J. Mol. Liq. 314 (2020) 133580, https://doi.org/10.1016/j.molliq.2020.113580.

[110] C.-Y. Huang, C.-C. Lai, Y.-H. Tseng, Y.-T. Yang, C.-J. Tien, K.-Y. Lo, Silica-nanoparticle-doped nematic display with multistable and dynamic modes, Appl. Phys. Lett. 92 (2008) 221908, https://doi.org/10.1063/1.2938880.

[111] H.-Y. Jung, H.-J. Kim, S. Yang, Y.-G. Kang, B.-Y. Oh, H.-G. Park, D.S. Seo, Enhanced electro-optical properties of Y_2O_3 (yttrium trioxide) nanoparticle-doped twisted nematic liquid crystal devices, Liq. Cryst. 39 (2012) 789–793, https://doi.org/10.1080/02678292.2012.681073.

[112] L. Jinqian, Y. Zhao, H. Gao, D. Wang, Z. Miao, H. Cao, Z. Yang, W. He, Polymer dispersed liquid crystals doped with CeO_2 nanoparticles for the smart window, Liq. Cryst. 49 (2022) 29–38, https://doi.org/10.1080/02678292.2021.1942573.

[113] H.-Y. Mun, H.-G. Park, H.-C. Jeong, J.H. Lee, B.-Y. Oh, D.-S. Seo, Thermal and electro-optical properties of cerium-oxide-doped liquid-crystal devices, Liq. Cryst. 44 (2017) 538–543, https://doi.org/10.1080/02678292.2016.1225838.

[114] K. Agrahari, T. Vimal, A. Rastogi, K.K. Pandey, S. Gupta, K. Kurp, et al., Ferroelectric liquid crystal mixture dispersed with tin oxide nanoparticles: study of morphology, thermal, dielectric and optical properties, Mater. Chem. Phys. 237 (2019) 121851, https://doi.org/10.1016/j.matchemphys.2019.121851.

[115] Y. Zhang, J. Yang, L. Zhou, Y. Gao, M. Hai, L. Zhang, et al., Preparation of polymer-dispersed liquid crystal doped with indium tin oxide nanoparticles, Liq. Cryst. 45 (2018) 1068–1077, https://doi.org/10.1080/02678292.2017.1408864.

Section D

Metal oxides for solar-cell applications

Metal oxides for dye-sensitized solar cells

18

N.J. Shivaramu, J. Divya, E. Coetsee, and Hendrik C. Swart
Department of Physics, University of the Free State, Bloemfontein, Free State, South Africa

Chapter outline

1 Introduction

Currently, the global plan is to effectively resolve environmental pollution and global warming using clean, renewable energy consumption [1,2]. The most abundant renewable energy source, solar energy, may be directly converted into electrical energy through photovoltaic (PV) technology [3]. A green and renewable energy source that can substitute conventional fossil fuels is solar energy [4]. The majority of the solar energy market is made up of silicon-based solar cells, which are described to be the first generation of PV devices. It exhibits a conversion efficiency of greater than 20%; however, the comparatively high expense of these cells confines their various applications.

Metal Oxides for Next-Generation Optoelectronic, Photonic, and Photovoltaic Applications
https://doi.org/10.1016/B978-0-323-99143-8.00017-1

Cadmium telluride (CdTe) and copper indium gallium selenide (CIGS) thin-film technologies are second generation solar cells. However, the problems of expensive and ecologically harmful PV devices led to endless growth in the progress of novel PV devices. Third generation solar cells, which often have double or triple junctions, and voltaic system nanotechnology in solar cell production methods have started to emerge in the photovoltaic market to avoid these issues. Perovskite solar cells are characteristically based on organic-inorganic halide perovskites and have been extensively used due to their high conversion efficiency, low making cost, and reduction of fabrication time and complicity [5]. The real-world application of these cells is stuck due to the presence of toxic elements, humidity instability, and poor reproducibility. Dye solar cells (DSCs) are known as dye-sensitized solar cells (DSSCs) or Gratzel cells, after Professor Michael Gratzel who was significantly engaged in the fabrication of new devices. DSSCs, the third generation PV cells, are considered inexpensive and have simple fabrication procedures, limited environmental impact, excellent stability, and comparatively high conversion efficiency [6,7]. Since the prototype of a DSSC was described in 1991 by Gratzel [8], it has one of the best hopeful applicants to replace silicon analogues in the solar energy field because of its environmentally friendly production processes and inexpensive fabrication [9–12]. The transformation of photons into carriers and the passage of photogenerated carriers through the device are two of the primary tasks of DSSCs. Even with the 100 mW cm^{-2} air mass global illumination, they have shown power conversion efficiencies (PCEs) between 10% and 14%. The liquid electrolyte's temperature sensitivity is, however, constrained by this. As a result, much research is being done to improve the electrolyte's functionality and stability of PV performance.

According to O'Regan and Gratzel's 1991 paper, nanostructured titanium dioxide (TiO_2) was coated with ruthenium-derived dyes, and iodide/tri-iodide was utilized as the electrolyte in the DSC [8]. The benefit of nanostructured TiO_2 was a noticeably larger surface area supplied for the adsorption of dye molecules, which led to an order of magnitude improvement in conversion yield up to 7.1%–7.9%. Generally speaking, most widely used Ru dyes, for example, N719 (Di-tetrabutylammonium cis-bis (isothiocyanato) bis(2,2′-bipyridyl-4,4′-dicarboxylato) ruthenium (II)) and N749 (Tris(*N*,*N*,*N*-tributyl-1butanaminium)[[2,2″,6′,2″-terpyridine]-4,4′4″ tricarboxylato (3-)-*N*1,*N*1′,*N*1″] tris(thiocyanato-N) hydrogen ruthenate) only absorbs in wavelengths from 300 to 800 nm due to its 1.8 eV bandgap [10,13]. Therefore, the ultraviolet and infrared wavelengths, which make up 50% of solar light, are not used. Cosensitization of the photoanode has been employed to boost the light absorption capacity of DSSCs to get over this drawback [14,15].

The Shockley-Queisser limit [16] states that solar cells can convert energy at a maximum determined efficiency of 30%. One of the main causes of energy losses is spectral mismatch, which prevents photons with energies below the optical bandgap (E_g) from being absorbed and causes them to lose a considerable portion of their energy as heat. There are two methods for reducing energy losses brought on by spectral mismatch. Either the solar cell can be modified to utilize the solar spectrum more effectively or it can be modified to be modified prior to it is absorbed by the solar cell. Heterogeneous bandgap semiconductor materials, where each semiconductor efficiently converts a different portion of the solar spectrum, can be used to construct solar cells that make better use of the spectrum. Solar cells have been successfully tandemized using this technique, and energy efficiency of over 40% have been reported [17]. Upconversion (UC), the first choice, is particularly beneficial for solar cells with a wide bandgap where transmission losses predominate. In order to create one higher-energy photon that can be absorbed by the solar cell, upconversion involves combining two photons with a lower-energy that is transmitted to obtain one photon with higher energy that can be absorbed by the solar cell. The application of UC materials has resulted in appreciable advancements in device performance and broadband light absorption [18,19]. The second choice is to divide a single, highly energetic photon into two, less energetic photons. Each of these photons can then be absorbed by the solar cell, producing an exciton. When thermalization losses are the main cause of loss, this is known as down-conversion (DC), and it is advantageous for solar cells with a smaller bandgap.

Due to their absurd energy level structure, which enables effective spectral conversion, and the fact that the transitions between the 5d and 4f orbitals are protected from the host lattice, rare earth (RE) ions are well-suited for usage in both DC and UC. Lanthanides are used in a number of effective UC and DC processes, either with single- or double-doped lanthanides [20,21]. Due to their 4f-4f transitions, RE-doped UC and DC materials are excellent luminous conversion materials and have been broadly used in PV devices [22,23]. The performance of DSSCs is primarily impacted by the use of DC or UC materials and the capacity to gather light. The largest amount of light is absorbed by DSSCs with dye-sensitized photo-anodes, nevertheless, between the wavelengths of 400 and 750 nm [21]. When using UC and DC nanophosphors to reduce energy losses in DSSC, infrared and ultraviolet radiation energy was transformed to visible light, which enhanced the PCEs of the solar cell. Another strategy that uses optical enrichment effects to modify the photoelectrode's capacity to capture light is light scattering. The internal surface area of the photoelectrode; however, may be reduced and the light's distance is increased, if submicron-sized particles in the photo anode sheets are present. Within photo anodes, redoped semiconductors and metal oxides can scatter light and convert to luminescence.

1.1 Metal oxide nanomaterials

Metal oxides are a crucial group of useful materials due to their diverse physical, magnetic, electrical, and optical properties [24]. These materials display metallic, semiconductor, or insulator behavior at ambient temperature depending on their bandgap. Metal oxides utilized in the production of photoelectrochemical cells, fuel cells, sensors, piezoelectric devices, corrosion-resistant coatings, field emission, and magnetic memory. The physical and chemical properties of a material are influenced by the particle size, particularly the nanodomain, of metal oxides. As particle size decreases from bulk to nanometers, the metal oxide's surface area increases. The material's surface now has more active sites and a different density of electronic states as a result of the findings. The use of metal oxides as photoelectrodes in DSSCs is the specific emphasis of this chapter.

2 Construction and working of DSSCs

The working electrode, sensitizer (dye), electrolyte (redox mediator), and counter electrode are the four main parts of DSSCs. Working electrode in close contact with counter electrode through cramp with sensitizer or dye soaking. The DSSC was then created by adding electrolyte through a dropper to the space between a pair of electrodes.

2.1 Transparent and conductive substrate

Glass electrodes with two sheets of transparent conducting oxide (TCO) coating, commonly fluorine- and indium-doped tin oxide (FTO, F-SnO_2 and ITO, Sn-In_2O_3), are used and are the typical building blocks for DSSCs (Fig. 2). One of these glass substrates has a photoelectrode (PE) in the form of a thick layer (10–15 μm) of semiconducting metal oxide [25,26]. In order to permit the passage of perfect sunlight to the active component of the cell, the FTO films revealed a lower transmittance of 75% in the visible wavelengths with a sheet resistance of 8.5/cm^2 [27], but the ITO films displayed a transmittance of >80% with a sheet resistance of 18/cm^2. As a counter electrode (CE), platinum (Pt), or carbon (C) are put on a second glass substrate.

2.2 Working electrode

On an FTO or ITO transparent conducting glass substrate, working electrodes (WEs) are prepared by depositing a thin layer of metal oxide semiconductors (MOSs), such as TiO_2, Nb_2O_5, ZnO, SnO_2 (n-type), and p-type NiO [28–30]. These MOSs have an optical bandgap with a wide energy range of 3–3.2 eV. [28–30]. TiO_2 is frequently used in the WE due to its low environmental effects, inexpensive production costs, and wide availability. However, this WE only absorbs a tiny amount of UV-visible light. Therefore, the dye is covalently linked to the WE surface when this WE is immersed in the dye solution. Numerous dye molecules stick to the WE surface of the electrode because of its large surface area and very porous structure, which increases light absorption at the MOS surface.

2.3 Photosensitizer or dye

Dyes play a key role in absorbing sunlight's UV to NIR wavelengths and transforming them into electricity [31,32]. As a result, the type of dye molecules used as a sensitizer affects the PCEs of DSSCs [33]. The dye is responsible for the incident light's maximal absorption in DSSCs. For a material to be a dye it exhibits the given electrochemical and photophysical responses: (1) the dye exhibited a luminescent material, (2) it should absorbed from ultraviolet-visible (UV-Vis) to NIR wavelengths, (3) the highest occupied molecular orbital (HOMO) should be located lower than the redox electrolyte, (4) the dye's periphery should be hydrophobic naturally, and (5) the HOMO should be located far from the surface of the conduction band of the electrode/MOS and the lowest unoccupied molecular orbital (LUMO) should be positioned as nearby surface of the electrode/MOS [31–33]. Ruthenium (Ru) complex dyes are generally used in the industry [34]. However, Ru has some limits due to expensive [35], rarity [36] and complexity for synthesis [37]. Therefore, instead of Ru dyes, scientists are currently focusing on natural dyes derived from plants [38]. Natural dyes provide advantages to Ru dyes, such as being cheap, nontoxic, environmentally safe, easily accessible, and abundant [39,40]. The flowers and leaves of the plants are typically used to make natural dyes. The varied colors in dyes are provided by pigments like betalains, anthocyanins, and chlorophyll. To be suitable for energy conversion, plant pigments must absorb in the visible to NIR light. Various pigments in natural dyes have been emphasized in other reports, such as anthocyanin [41,42], chlorophyll [43], and betalain [44,45].

The presence of functional groups including hydroxyl (—OH), ester, and carbonyl (C═O) groups [46–48] present in the dye molecules causes the natural dyes' chemical adhesion to or adsorption on the electrode's film surface. The C═O and —OH groups on the anthocyanin molecule, for instance, allow it to bond to the electrode. These anchors transfer the photoexcited dye's electrons from dye to the electrode layer. They used various techniques where colors are either blended to make a cocktail or applied side by side [49,50]. To stop the dye from adhering to the electrode surface, coabsorbents like chenodeoxycholic acid, alkoxy-silyl [32], phosphoric[33], and carboxylic acid groups [34,35] were utilized between the dye and the electrode. Thus, the recombination between the electrode's electrons and those in the redox electrolyte was constrained. Additionally, a stable connection was formed as a result [36].

N3, cis-di(isothiocyanato) bis-(2,2′-bipyridyl-4,4′ ruthenium (II), and N719 are the most frequently employed dyes in DSSCs. Additionally, Ru505 and Ru470 demonstrated electrode sensitization potential [51,52]. Natural dyes like mercurochrome (20, 70-dibromo-50-(hydroxymercurio)) and Rose Bengal dye (4,5,6,7-tetrachloro-20, 40, 50, 70-tetraiodofluorescein) have also been investigated as DSSC sensitizers, which were low-cost but produced low photocurrents and less efficiencies [53]. The chemical structure of organic and natural dyes is given in Fig. 1.

2.4 Electrolyte

The electrolyte, which acts as a conduit for the flow of charge carriers—in this example, ions—between the photo electrode and counter electrode, is one of the most important components of DSSCs. They are used in DSSCs to constantly replenish both the oxidized dye molecules and themselves during operation. The electrolyte has a good effect on both the PCEs and the long-period stability of the devices. The electrolyte in an electrochemical cell offers pure ionic conductivity between the positive and negative electrodes. The same electrolyte is used in capacitors, supercapacitors, electrolytic cells, fuel cells, and batteries. The electrolyte should possess the following properties: (1) The redox couples must be able to regenerate the oxidized dye as well as themselves, (2) they must have good stabilities, including interfacial, chemical, thermal, optical, and electrochemical, (3) they must be corrosion-resistant with DSSC apparatuses, (4) they must ensure rapid charge carrier diffusion, improve conductivity,

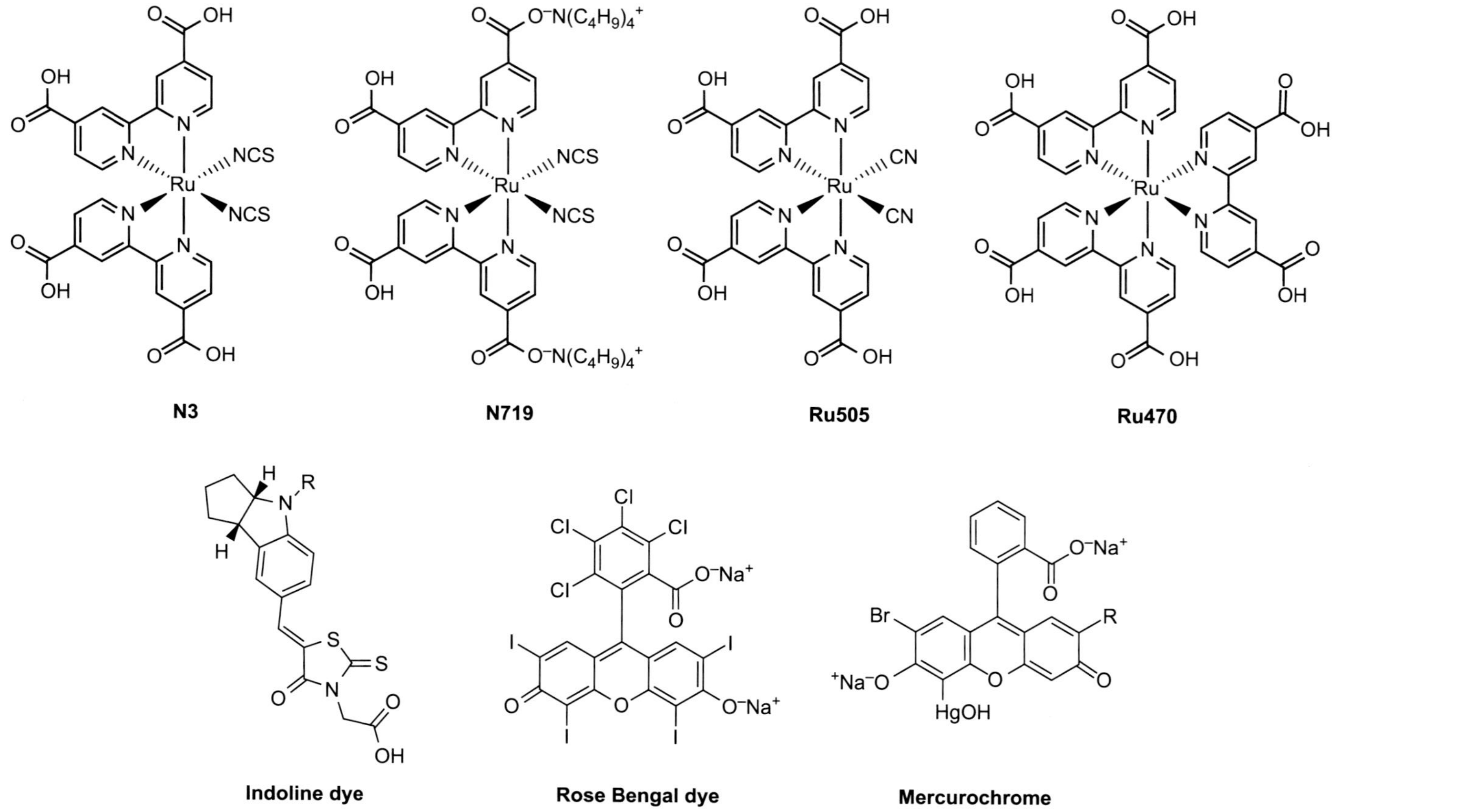

Fig. 1 Chemical structure of inorganic and organic dyes utilized in DSSCs.
Reused with permission from B.N. Nunes, L.A. Faustino, A.V. Muller, A.S. Polo, A.O.T. Patrocinio, Chapter 8. Nb2O5 dye-sensitized solar cells, Editor(s): Sabu Thomas, El Hadji Mamour Sakho, Nandakumar Kalarikkal, Samuel Oluwatobi Oluwafemi, Jihuai Wu, Nanomaterials for Solar Cell Applications (2019) 287-322.

and create effective interfacial contact between the pair electrodes, (5) there should not be a considerable correlation between the range of an electrolyte's absorption spectra and that of a dye [54,55].

In the DSSCs, electrolytes such as iodide/tri-iodide (I^-/I_3^-), Br^-/Br_2^- [56], SCN^-/SCN_2 [54] and Co(II)/Co(III) [55] were utilized. Among these electrolytes, I^-/I_3^- has been used as a highly effective electrolyte [57], although it has some limitations when used in DSSCs. The I^-/I_3^- electrolyte corrodes, is exceedingly volatile, promotes photodegradation and dye desorption, and causes device instability [58]. A liquid electrolyte typically consists of three basic modules: solvent, additives, and an ionic conductor. Acetonitrile (ACN), N-methylpyrrolidine (NMP), and solvent mixtures with high dielectric constants, such as ACN/valeronitrile, have been utilized as solvents. TBP (tert-butylpyridine) is a popular addition in electrolytes that increases cell performance [59]. This raises the open-circuit voltage (V_{oc}), lowering the photocurrent (J_{sc}) of the cell and resulting in a low injection driving force. TBP on a working electrode (TiO_2) surface is thought to minimize charge carrier recombination via back transfer to an electrolyte [60]. To get electrolytes, sodium iodide (NaI), lithium iodide (LiI), and tetra alkylammonium iodide (R_4NI) are well-known compounds of a combination of iodide commonly dissolved in nonprotonic solvents such as propylene carbonate, acetonitrile, and propionitrile. The long-term stability of DSSCs is greatly determined by ion conductivity in the electrolyte, which is influenced by solvent viscosity. To minimize charge carrier transport resistance, a lower viscosity solvent is greatly advised. Furthermore, Li^+, Na^+, and R_4N^+ have an effect on the operation of DSSCs, owing to their adsorption on the electrodes or the electrolyte's ion conductivity. Despite this, one of the main disadvantages of ionic liquids is their leakage factor, which reduces the lifetime of the cell. Another disadvantage is technical challenges with the device's sealing and, as a result, long-term stability. Many researchers are studying the use of ionic liquids, polymers, and hole conductor electrolytes in an effort to reduce the requirement for organic solvents in liquid electrolytes. Solid-state electrolytes have been designed to prevent ionic liquid electrolytes from leaking [61].

2.5 Counter electrode (CE)

The platinum (Pt) or carbon (C) commonly used as a CE in the DSSCs. The WE and CE are wrapped together and then introduced an electrolyte between pair of electrodes. The cathode is comprised of an ITO or FTO conductive glass sheet covered with Pt or C components. The Pt and C operate as a catalyst to reduce the I^-/I_3^- liquid electrolytes and accumulate electrons from the external circuit. Pt is commonly employed as a counter electrode due to its greater efficiencies [62]. However, it has some disadvantages due to its expansive and low availability. Thus, Pt is replaced by carbon [63] or carbonylsulfide (CoS) [11] or Au/GNP [64] or FeSe [65].

2.6 The working mechanism of DSSC

The four essential functioning principles of DSSCs are light absorption, electron injection, carrier conveyance, and current collection. Fig. 2 is a schematic illustration of a DSSC. Sunlight enters the cell via the TCO top contact and strikes the dye on the

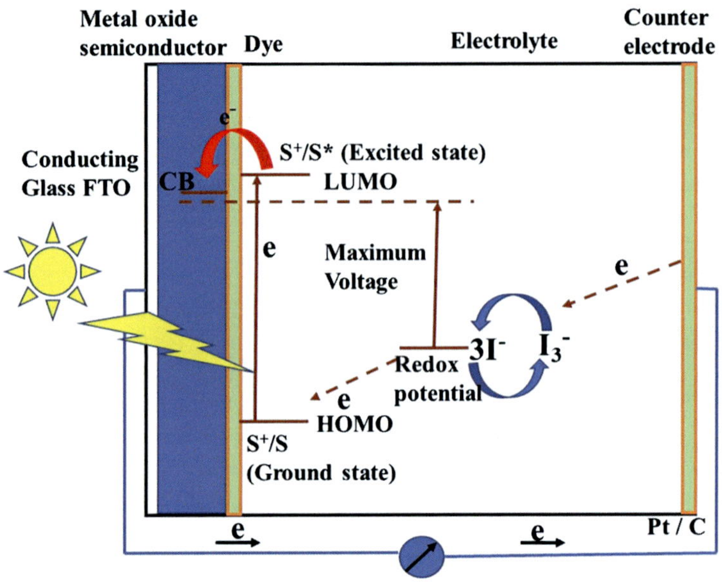

Fig. 2 Schematic design of construction and working principle of the DSSC.

MOS. MOS provides the necessary surface area to adsorb dye molecules and also functions as a sensitizer. TCO is typically made up of ITO or FTO. Photons with sufficient energy reach the dye and cause the dye electrons to be stimulated from the HOMO (S) to the LUMO (S^*) states (see Eq. 1), as seen in Fig. 2. The dye molecules' excited electrons are oxidized, and one electron is injected into the MOS conduction band (CB) (Eq. 2). The energized electrons then go from the MOS to the TCO through the external circuit from anode to cathode, completing the cycle and producing photocurrent (Eq. 3). Electrolytes (I^-/I_3^- redox ions) are used as an electron mediator between the working electrode and the CE. The electrons at the CE reduce from I_3^- to I^-; consequently, regeneration or the regeneration of the dye takes place because of the electron acceptance from I^- ion that is oxidized to I_3^- (Eq. 4). At the CE, oxidized I_3^- is reduced to I^- ion once more (Eq. 5). As a result, no lasting material alteration or transformation happened during DSSC power generation [11].

$$\text{Excitation process} : S + h\nu \rightarrow S^* \tag{1}$$

$$\text{Injection process} : S^* + \text{MOS} \rightarrow S^+ + \text{e}^- \ (\text{MOS}) \tag{2}$$

$$\text{Energy generation} : \text{e}^- \ (\text{MOS}) + \text{C.E.} \rightarrow \text{MOS} + \text{e}^- \ (\text{C.E.}) \tag{3}$$

$$\text{Regeneration of dye} : S^+ + \frac{3}{2}\,\text{I}^- \rightarrow S + \frac{1}{2}\,\text{I}_3^- \tag{4}$$

$$\text{e}^- \ \text{Recapture reaction} : \frac{1}{2}\,\text{I}_3^- + \text{e}^- \ (\text{C.E.}) \rightarrow \frac{3}{2}\,\text{I}^- + \text{C.E.} \tag{5}$$

3 Evaluation of dye-sensitized solar cell performance

A solar cell generates current by accepting photons from the sun. The current density is highly dependent on the cell's potential as well as the intensity of the incident light. A DSSC can be identified by current density voltage curves (also known as *J-V* curves), which plot the resultant current density (*J*) at increasing voltage (*V*) (Fig. 3). The short circuit current (J_{sc}) is measured at a bias of 0 V, and the open-circuit voltage (V_{oc}) is described when the current reaches 0 A.

The maximum output power (P_{max}) produced by a DSSC is achieved when the product of the current and the voltage is a maximum corresponding to the expression bellow [66]

$$P_{max} = I_{mp} \times V_{mp} \tag{6}$$

where, I_{mp} and V_{mp} are the current and the voltage at maximum power, respectively. The PCE of DSSC was determined from the *J-V* plot of the cell under the irradiation of visible light. The factor (FF) was determined from the *J-V* plot (Fig. 3), as follows:

$$\text{FF} = \frac{J_m \times V_m}{J_{sc} \times V_{oc}} \tag{7}$$

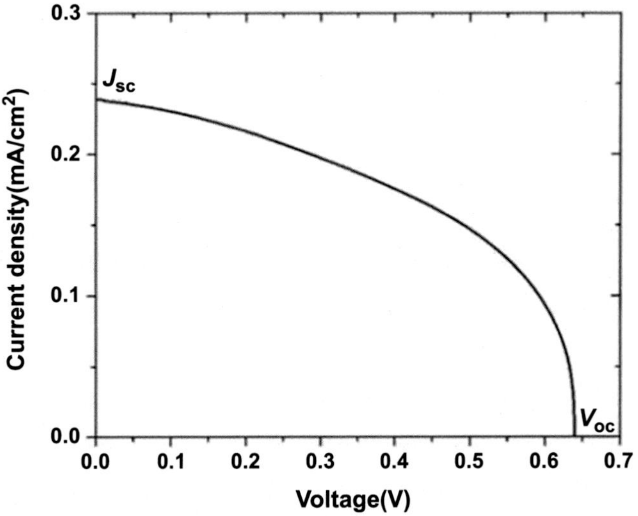

Fig. 3 The *J-V* curve of a cladode DSSC, recorded under the irradiation by the sun simulator, indicating the V_{oc} and short-circuit current density (J_{sc} along the *x*- and *y*-axes, respectively). Reused with permission from D. Ganta, J. Jara, R. Villanueva, Dye-sensitized solar cells using Aloe Vera and Cladode of Cactus extracts as natural sensitizers, Chem. Phys. Lett. 679 (2017) 97-101.

Where, J_m and V_m are the maximum current density and voltage, respectively. J_{sc} is the short circuit current (at zero cell voltage) expressed in $mA\,cm^{-2}$, V_{oc} (V) is the open circuit voltage.

Solar light converting into electricity and its efficiency (η) is determined with the following equation [12]:

$$\eta\% = \frac{J_{sc} \times V_{oc} \times FF}{I_0} \times 100 \tag{8}$$

where, I_0 is the incident light intensity stated in W/cm^2. The incident photon-to-current conversion efficiency (IPCE) and it was determined at the wavelength (λ) of each excitation using the given expression:

$$\text{IPCE}\% = \frac{1240 \times J_{Photo}(\text{Acm}^{-2})}{\lambda\ (\text{nm}) \times I_0(\text{Wcm}^{-2})} \times 100 \tag{9}$$

the photocurrent (J_{photo}) occurred with incident monochromatic light of intensity I_0 and wavelength λ.

3.1 TiO_2

TiO_2 is a favored material for solar energy applications because to its remarkable physical, chemical, and spectroscopic properties, the nonpoisonous and low-cost wide bandgap metal semiconductor, the optimum band edge levels for charge injection and extraction, the extended lifetime of excited electrons, and the better photo corrosion resistance. Three crystal types of TiO_2 occur naturally are tetragonal (anatase), tetragonal (rutile), and orthorhombic (brookite). Due to its exceptional charge transport properties, anatase phase is the type of crystalline structure that is most frequently utilized in DSSCs [66–69]. A single-crystal anatase TiO_2 is exhibited n-type semiconductor with optical bandgap ~3.2 eV and the resistivity is $10^{15}\,\Omega$cm [70]. Mesoporous TiO_2 nanoparticle (NP) films are used as photoanodes in DSSCs for high efficiency [71]. TiO_2 has a high recombination rate due to its poor electron mobility, fast electron transit rate, and low electron mobility. [72,73].

TiO_2's rutile phase is not widely employed in DSSC electrodes since it typically has a lower CB than anatase and obtain a smaller value of V_{oc} and J_{sc} [74–77]. In TiO_2 strong covalency and the CB comprising of vacant Ti^{4+} 3d bands and the VB comprising mostly oxygen 2p states is filled [78]. Titanium interstitials, bulk oxygen vacancies, and decreased TiO_2 crystal structure all create shallow-level electron traps that can increase TiO_2's conductivity. [70,79]. The first report on Ru-bipyridyl was used as a sensitization on TiO_2 performed in 1980 [80]. In 1991, O'Regan and Gratzel [8] utilized this principle to construct a complete DSSC, in which the photon-absorber layer was made of TiO_2 NPs that were dye-coated.

3.1.1 Dopant

Dopant addition alters the optical and the electrical characteristics of TiO_2. The rise in charge carriers and subsequent increase in conductivity, which is brought by the donation of electrons for dopants with a greater valence than the host matrix (n-type) or holes for dopants with a smaller valence (p-type), are the key explanations for the outcome. The mechanism is highly complex since TiO_2 has a higher percentage of defects, and doping mostly impacts the trap levels and electronic structure of TiO_2 [81]. Doping can be accomplished by adding Ti^{4+} or O_2 in place of TiO_2. Compared to cationic dopants, which are often metals, anionic dopants are non-metals. Since a dopant cation is replacing the Ti^{4+} in the CB, this has a significant impact on how the CB is configured. The VB configuration is impacted by the O^{2-} in the VB being replaced by a dopant anion. To prevent lattice deformation, the dopant's ionic radius should be close to that of the host replacement ion. In TiO_2, the pigment molecules are joined to the Ti atoms [82]. Because the dopant and the dye have different binding energies or the dopant creates oxygen vacancies, substituting Ti with another cation will also impact the dye adsorption [83]. Dopants frequently slow down the formation of TiO_2 crystals, which reduces the size of the particles [84] because smaller particles have more TiO_2 surface area than larger particles. The more dye molecules that can fit on the surface, the more light may be absorbed, and the higher current densities that result. The major benefit of light absorption is that it allows for the use of thinner films in DSSCs, which reduces charge recombination and benefits both J_{sc} and V_{oc}.

Another element that affects the characteristics of TiO_2 is the shape and distribution of the particle shapes. Although they have smaller surface surfaces and may adsorb fewer dyes than nanoparticle assemblies, one-dimensional crystal structures of nanotubes demonstrated higher charge transfer [85]. Additionally, doping affects the phase transition from anatase to rutile [86]. TiO_2 is incorporated with dopants such as Al, Cr, Ta, Zn, W, Cu, Nb, Sr, Eu and Sm to improve their photoelectrochemical performance [85–92].

Subramanian et al. [93] demonstrated that the higher particle sizes of Li^+-doped TiO_2 exhibited a reduced device efficiency due to the formation of Ti^{3+} from Ti^{4+} after trapping electrons at Ti^{4+} species in the bulk and acting as electron traps. This is then resulted in decreased photocurrents. While the Li^+ ions at the surface functioning as transient electron traps, which enhanced the lifetime of electron and reduced recombination, was thought to be the cause of efficiency amplification for the lowest particle size [94]. When Mg^{2+} ions were substitutionally introduced into a TiO_2 system the particle sizes decreased [95]. UV/Vis spectroscopy then confirmed a broadening of the bandgap that suggested an upward shift in the CB. This lead to rise in V_{oc} of up to 1.2 V [95]. Zhang et al. [96] studied the influence of Mg^{2+} ions switched into the anatase phase of TiO_2 on the rate of recombination and band-edge shift in DSSCs. Their results decided that the reduced yield of electron injection for DSCs with Mg^{2+}-doped TiO_2 was ascribed to the negative move of the band edge in the TiO_2:Mg^{2+} electrode to obtain a reduced J_{sc} [96]. Q. Liu et al. [97] presented an improved photovoltaic response of DSSCs by co-doping with Zn and Mg into a TiO_2 electrode.

Using a hydrothermal route, the co-doped and pure TiO_2 NPs were created. It was noted that when the dopant content to TiO_2 ratio was 2.0 mol%, the PCE of DSSCs reached 9.07%. Comparing this to pure TiO_2 (7.16%), there was an improvement of 26.7%. The increase in PCE of the Zn and Mg co-doped TiO_2 was attributed to the flat band potentials' positive motion and the quick electron transport in TiO_2 films.

TiO_2:Ca^{2+}-electrodes were synthesized via the hydrothermal route showed a rise in J_{sc} that also resulted in a PCE of 8.35%, which was greater than that of the pure TiO_2 thin film [98]. This is because of an enhanced injection efficiency of electrons from the LUMO of the dye to the CB of TiO_2 and the rapid electron transport rate [98]. Other studies revealed comparable increases in J_{sc} at low dopant concentrations due to the faster electron transport and the shorter lifetimes that existed in the Ca-doped TiO_2 DSSC [99]. In addition, the introduction of Ca^{2+} in the TiO_2 lattice lead to the creation of oxygen vacancy and electrons, which can be stated in the equation described below [100]

$$Ca^{2+} + TiO_2 \rightarrow Ca''_{Ti} + 2V_O \tag{10}$$

Thus, the electron concentration increased for the Ca-doped TiO_2 and improved the conductivity of the films. Furthermore, the doping of Ca^{2+} in TiO_2 lead to the positive movement of the flat band potential that increased the energy gap between the LUMO of the dye and the CB of TiO_2. The injection driving force was enhanced and this lead to an enhanced photocurrent. Ca^{2+}-doping; therefore, resulted in the reduction of electron transfer resistance because of a decreased electron-hole recombination rate [100]. Motlak et al. [101] gave a similar explanation on a Ca-doped TiO_2 photoanode that attained a high PCE of 2.7% which was markedly improved by 40% as compared to the pure TiO_2 nanofibers-based DSSCS (1.54%). The biggest reduction in the rate of electron-hole recombination at the electrode and electrolyte contact was attributed to the improvement of the PCE.

The positive shift of the CB of the Ge-doped TiO_2 electrode enhanced the PCE of the Ge-doped TiO_2 DSSCs by 20% compared to the undoped TiO_2. When porphyrin was introduced onto the surface of a Ge-doped TiO_2 electrode, it demonstrated a maximum PCE of 3.5% [102]. The PCE of DSSC for the TiO_2:1%Sb favorably reaches 8.13%, offering efficiency enhanced of 9.5% over the undoped TiO_2. Because of the quicker electron transport in the Sb-doped TiO_2 sheets, this DSSC had a higher J_{sc}. The Sb-doped TiO_2 electrode caused a positive shift in the flat band potential (E_{fb}), which was also responsible for the significant improvement in J_{sc} [103]. The increased PCE caused by the usage of Sb_2S_3 grown on TiO_2 composite photoanode film has been attributed to the combined effect of greater surface area, improved light harvesting, effective transport, and photocurrent production [104]. At 0.2% at Sc in anatase, the highest PCEs of 9.6% were discovered that is 6.7% higher than the DSSCs, with pure anatase. The swapping of Sc on the Ti site caused the establishment of the atomic and electrical disorder. Changes in pore diameter and nanoparticle size were implied by the atomic disorder. Because of the oxygen vacancies in anatase, holes in the valence band are often formed, which counteract the donor levels, resulting in electrical instability [105].

Lanthanide (Ln) ions typically exhibit a +3 oxidation state with an electronic configuration $[Xe]^4f_n$, where n varies between 0 and 14 as the series progresses from left to right (from La^{3+} to Lu^{3+}). When introduced into diverse host materials, this group of elements exhibits distinctive luminescent capabilities, producing radiation with a wide range of energies between the UV and IR regions of the electromagnetic spectrum. The transitions are mostly intra-configurational f-f transitions, however a few lanthanide ions also exhibit f-d transitions from $4f^n$ ground state to $4f^{n-1}5d^1$ excited state [106]. Unlike f-f transitions, the allowed f-d transition produces broad spectral bands and strong absorption cross-sections as well as crystal field and the nephelauxetic effects [107–110]. Due to the intriguing optic and electrical features that lanthanides have due to their 4f bands, such as up-conversion or down-conversion luminescence [20,111], it is feasible to generate photons that are not absorbed by practically all dyes (400–800nm). Despite the fact that several studies [111–113] have indicated active up- or down-conversion, these processes are not yet well-organized in DSCCs, and it is unclear how they contribute to greater device efficiency. It is difficult to determine the precise role of conversion since several optical conversion experiments do not account for the impact of doping on the lifetime and electron transport properties. When TiO_2 is doped with Ln^{3+}, the oxygen will be scavenged off the surface, creating vacancies.

The microwave prepared TiO_2 nanoparticles for the DSSCs, displayed a maximum PCE of 7.44%. The PCEs of the DSSCs constructed with commercially purchased TiO_2 and TiO_2 prepared by a solvothermal method were 4.12% and 5.98%, respectively [114], which could also be due to the occurrence of extreme oxygen vacancies [114]. The dye-impregnated mesoscopic TiO_2 films with a Cu(II)/Cu(I) electrolyte achieved a notable PCE of 13.1% [115]. It also achieved a new record for reprocessing the electricity ambient interior light with a PCE of 32% under 1000 lux intensity, which succeeded the performance of all recognized PV converters. A summary of some of the developments attained by modification in the photoanode of DSSCs is described in Table 1.

3.2 Niobium(V) oxide (Nb_2O_5)

In comparison to conventional TiO_2 nanocrystals, Nb_2O_5 is an n-type transition MOS with a bandgap varies from 3.2 to 4 eV and can produce a higher conduction band energy level as a photoanode [123,143]. In theory, it can also guarantee that the V_{oc} and PCE in DSSCs are greater. Additionally, it has improved chemical stability and minimal interfacial electron recombination [123], which are crucial for liquid and solid-state DSSCs. It also has higher electron properties. Nb_2O_5 can arise in the amorphous or in crystalline polymorphs. Commonly, all the Nb_2O_5 polymorphs are in white powder form or it is transparent (in the form of single crystals). The physical characteristics of Nb_2O_5 depend on its polymorphism and the preparation technique and parameters [144–147]. The most common phases of Nb_2O_5 are H-Nb_2O_5 (pseudo-hexagonal), M-Nb_2O_5 (monoclinic), T-Nb_2O_5 (tetragonal) and O-Nb_2O_5 (orthorhombic). The M phase is thermodynamically more stable compared to all the other polymorphs of

Table 1 A summary of published representative results of PV parameters deduced from the *J-V* curves.

WE	CE	Electrolyte	Dye	V_{oc} (mV)	J_{sc} (mA cm^{-2})	*FF* (%)	η (%)	Ref.
TiO_2	Pt	I^-/I_3^-	N719	730	14.52	70	7.44	[114]
TiO_2	Pt	0.6 M dimethylpropylimidazolium iodide, 0.1 M LiI, 0.05 M I_2, and 0.5 M TBP in ACN	N719	640	22.4	72	10.3	[116]
TiO_2	Pt	0.25 M Co(bpy)3(TFSI)2, 0.06 M Co(bpy)3(TFSI)3, 0.1 M LiTFSI	SM315	910	18.1		13	[117]
TiO_2:W^{6+}	Pt	1 M 1-propyl-3-methylimidazolium iodide (PMII), 0.03 MI_2, 0.05 M LiI, 0.1 M GNCS, and 0.5 M TBP in mixed solvent of propylene carbonate and ACN	N719	730	15.10	67	7.42	[118]
TiO_2:Sc	pt	0.6 M 1,2- dimethyl-3-propylimidazolium iodide (DMPII), 0.1 M LiI, 0.03 MI_2, and 0.5 M TBP in ACN	N719	750	19.10	67.5	9.6	[105]
TiO_2:In	Pt	0.6 M butylmethylimidazolium iodide, 0.05 M I2, 0.1 M LiI, and 0.5 M TBP in 1:1 ACN/ valeronitrile	N719	720	16.97	61.4	7.48	[119]
TiO_2:B	Pt	I^-/I_3^-	N719	660	7.85	66	3.44	[120]
YbF_3-Ho@TiO_2		I^-/I_3^-	N719	750	17.06	60	7.73	[121]
TiO_2/YbF_3@TiO_2		I^-/I_3^-	N719	750	16.69	60	7.54	[121]
1% WO_3 added TiO_2	Pt	0.5 M KI, 0.05 M I2, and 0.5 M TBP	N719	670	12.94	57.9	5.04	[117]

Nb_2O_5	Pt	1.0 M DMPII, 0.1 M LiI, 0.12 M I2, and 0.5 M 4-TBP in methoxypropionitrile (MPN)	N719	738	6.81	58.9	2.97	[122]
H (pseudohexagonal)-Nb_2O_5	Pt	ACN containing 0.1 M LiI, 0.03 M iodine, 0.5 M TBP, and 0.6 M PMII	N3	770	6.68	59.06	3.05	[123]
CM-Nb_2O_5	Pt	I^-/I_3^-	N719	104	23.2	0.72		[124]
C-Nb_2O_5	Pt	I^-/I_3^-	N719	104	23.9	0.73		[124]
Nb_2O_5	Pt	I^-/I_3^-	N3	710	6.65	50	2.3	[125]
Nb_2O_5	Pt	I^-/I_3^-	N719	701	10	58.5	4.10	[126]
$Nb_3O_7(OH)$ Nanorod	Pt	I^-/I_3^-	N719	740	15	61	6.77	[51]
Nb_2O_5 Nanorod	Pt	I^-/I_3^-	N719	749	12.2	66	6.03	[51]
Nb_2O_5 NP	Pt	I^-/I_3^-	N719	662	11.57	61.1	4.68	[51]
SnO_2 nanoparticles (NP)	Pt	I^-/I_3^-	N719	356	11	36	2	[127]
SnO_2 nanowire coated with TiO_2 NP	Pt	I^-/I_3^-	N719	686	8.56	49	4.1	[127]
SnO_2	Pt	I^-/I_3^-	Ruthenium 535	350	7.63	43	1.14	[128]
Cd-SnO_2	Pt	I^-/I_3^-	Ruthenium 535	450	8.76	46	1.82	[128]
Zn-SnO_2	Pt	I^-/I_3^-	Ruthenium 535	420	8.46	48	1.71	[128]
Ni-SnO_2	Pt	I^-/I_3^-	Ruthenium 535	460	7.03	58	1.85	[128]
Cu-SnO_2	Pt	I^-/I_3^-	Ruthenium 535	470	6.73	57	1.80	[128]
Pb-SnO_2	Pt	I^-/I_3^-	Ruthenium 535	380	5.78	50	1.10	[128]

Continued

Table 1 Continued

WE	CE	Electrolyte	Dye	V_{oc} (mV)	J_{sc} (mA cm^{-2})	*FF* (%)	η (%)	Ref.
SnO_2	C	I^-/I_3^-	Rose Bengal	690	16.91	51	5.73	[129]
ZnO- SnO_2	C	I^-/I_3^-	Rose Bengal	650	15.93	47	4.72	[129]
SnO_2 nanofibers	Pt	0.6 M PMII, 0.03 M I_2, 0.05 M LiI, 0.1 M GuNCS, and 0.5 M TBP in ACN and valeronitrile (85:15 *v*/v)	N719	702	14.9	50	5.44	[130]
SnO_2 nanorod-TiO_2	Pt	0.6 M 1-hexyl-2,3-dimethyl imidazolium iodide (C6DMI), 0.2 M LiI, 0.04 M I2 and 0.5 M TBP in MPN /ACN (1:1 *v*/*v*)	N719	644	19.61	68	8.61	[131]
P25-5S-SnO_2-HMSs	Pt	0.6 M PMII, 0.03 M I_2, 0.05 M LiI, 0.1 M GuSCN, , and 0.5 M TBP in ACN and valeronitrile (85:15 *v*/v)	N719	750	20.10	63.2	9.53	[132]
SnO_2:Ga	Pt	I^-/I_3^-	N719 and D149	740	7.41	73.7	4.05	[133]
SnO_2:Zn	Pt	I^-/I_3^-	N719	780	6.06	62	3.0	[134]
SnO_2-decorated graphene oxide (SnO_2/GO)	Pt	0.5 M LiI, 0.05 M I_2, and TBP, 0.2 M $C_9H_{13}N$ in ACN	N719	670	14.04	49	4.57	[135]
SnO_2/MgO	Pt	I^-/I_3^-	N719	654	15.4	65	6.5	[136]
SnO_2/TiO_2	Pt	0.5 M LiI/0.05 M I_2 in 1:1 ACN-NMO (3-methyl-2-oxazolidinone)	Z907	452	6.7	34	1.12	[137]
ZnO	Pt	I^-/I_3^-	N719	420	1.10	43	0.76	[138]
ZnO	Pt	I^-/I_3^-	N719	620	8.9	46	2.50	[139]
ZnO nanoporous spheres	Pt	I^-/I_3^-	N719	580	7.47	67	2.91	[140]

ZnO	C	I^-/I_3^-	Rose Bengal	715	17.52	56	6.74	[129]
ZnO aggregates		0.5 M tetrabutylammonium iodide, 0.1 M LiI, 0.1 M iodine, and 0.5 M TBP in ACN	N3	635	18.7	45.1	5.4	[141]
ZnO nanosheet	Pt	I^-/I_3^-	N3 and N719	580	19.53	63	7.07	[142]
ZnO:Ga	Pt	I^-/I_3^-	N719	770	11.12	49	4.30	[127]
ZnO	Pt	0.5 M LiI, 0.05 M I_2 and 0.1 M TBP in 1:1 acetonitrile-propylene carbonate (APC)	N719	494	5.20	54	1.37	[142]
ZnO:La	Pt	0.5 M LiI, 0.05 M I_2 and 0.1 M TBP in 1:1 APC	N719	499	4.27	57	1.22	[142]
ZnO:Ce	Pt	0.5 M LiI, 0.05 M I_2 and 0.1 M TBP in 1:1 APC	N719	450	0.66	28	0.09	[142]
ZnO:Nd	Pt	0.5 M LiI, 0.05 M I_2 and 0.1 M TBP in 1:1APC	N719	527	4.39	58	1.31	[142]
ZnO:Sm	Pt	0.5 M LiI, 0.05 M I_2 and 0.1 M TBP in 1:1 APC	N719	530	5.02	57	1.52	[142]
ZnO:Gd	Pt	0.5 M LiI, 0.05 M I_2 and 0.1 M TBP in 1:1 APC	N719	536	5.64	65	1.98	[142]

Nb_2O_5 [123]. The H-Nb_2O_5 phase exhibited the maximum PCE because it has greater surface area [123].

Since high substrate temperatures (above 600°C) in DSSCs have a substantial impact on the optical and the electrical properties, the FTO regulates the annealing temperature [148,149]. By electrochemically anodizing metallic niobium, Liu et al. [150] created Nb_2O_5 powders that were then doctor-bladed onto FTO glass substrates. The film underwent a 2-h air annealing process at 550°C. The Nb_2O_5 solar cell had a high V_{oc} and noticeably long electron lifetimes, according to the results [150]. A similar structural arrangement of Nb_2O_5 was seen by Ok et al. [151]. The Nb_2O_5 mesoporous spheres made using a solvothermal technique had a PCE of 2.97% [122]. Nb_2O_5 thin films used as a blocking layer between TiO_2 and FTO nanocrystalline films have been working in ionic liquid electrolytes. This enhanced V_{oc} and resulted in an improved conversion efficiency of the DSSCs of TiO_2 [152].

Nb_2O_5 nanocrystals were prepared by Ghosh et al. [125] using the pulsed laser deposition method for use as a photoanode material in DSSC. During deposition, the photoanode produced a high dye load and improved photoelectrochemical performance at an optimal background oxygen pressure of 13.3 Pa. With the N_3 dye and I_3^-/I^- pair in acetonitrile with a V_{oc} of 0.71 V and a PCE of 2.41%, CE values were determined to be 40% for optimum structures [125]. In order to improve the PCE of DSSCs under various sun intensities, Kai Wen Chen et al. [153] examined a Nb_2O_5 layer that was deposited on the TiO_2-coated photoanode by a facial dip-coating process. This demonstrated that the Nb_2O_5 thin layer effectively inhibited electron-hole recombination on the photoanode and served as a blocking layer. Zhang et al. [51] investigate the effects of various morphologies on Nb_3O_5's DSSC performance. When compared to Nb_2O_5 nanorods (6.03%) and Nb_2O_5 NPs (4.68%), the simple hydrothermal approach used to create a single crystal $Nb_3O_7(OH)$ nanorod photoanode showed a higher light conversion efficiency of 6.77%. The higher dye filling capacity of the photoanode, which was brought on by the photoanode's huge surface area, was attributed to the high DSSCs' performance. This study showed that alternative metal oxide photoanodes, such as single crystal nanorods of $Nb_3O_7(OH)$, could be used to assemble a higher-performance DSSC [51,126].

Nb_2O_5 has been serving as a blocking layer to stop the back electron transfer processes from happening in these interfaces. According to Zaban et al. [154], a mesoporous electrode with an inner nanoporous TiO_2 layer encased in a Nb_2O_5 layer performed better in DSSCs than electrodes made of plain TiO_2. Other researchers noticed comparable outcomes. [155–157].

3.3 Tin oxide (SnO_2)

Due to its wide bandgap (E_g=3.6 eV at 300 K), great electron mobility (250 $cm^2\ V^{-1}S^{-1}$), and chemical stability, tin oxide (SnO_2) is a suitable n-type MOS for use in a variety of electronic devices, including DSSCs [158,159], photocatalysis [129], gas sensors [160], and photoconductive elements. Additionally, because of its high bandgap, it is long-period stable and has a low sensitivity to UV degradation [127]. The rutile-type (P42/mnm), $CaCl_2$-type (Pnnm), pyrite-type (Pa3), α-PbO_2-type (Pbcn), and ZrO_2-type orthorhombic phase I (Pbca) are the five primary polymorphs found in SnO_2 crystals [160,161]. The most significant naturally occurring polymorph is cassiterite SnO_2, which crystallizes in the rutile tetragonal phase with space group P42/mnm at normal temperature. Six atoms make up the SnO_2 unit cell: two Sn and four O atoms, with each Sn atom forming the center of an octahedron with six oxygen atoms surrounding it.

The undoped-SnO_2 nanowire showed a lower efficiency (2.1%), but the SnO_2 nanowire coated with TiO_2 nanoparticles displayed a V_{oc} of 720 mV and an high efficiency of 4.1% [127]. In comparison to a single-layer SnO_2 NPs film, the PV performance of the DSSCs derived on a bilayered photoanode of SnO_2 NPs/SnO_2 nanofibers (NFs) was 5.44% better. This might be primarily a result of the SnO_2 NFs layer's better light scattering capability, decreased charge carriers recombination, and increased electron transit rate. With an additional SnO_2 blocking layer, the SnO_2 NPs/NFs bilayered photoanode also managed to obtain an overall PCE of 6.31%, reducing the amount of electron recombination between the substrate and the electrolyte [130]. In a different study, manufactured SnO_2 NRs were protected by a TiO_2 particle coating and employed as a photoanode in DSSCs [131]. The photoanode's PCE of 8.61% was increased by 14.2% when compared to the photoanode with a pure TiO_2 particle layer. [131]. Photoanodes coated with 5S-SnO_2-HMSs-CDS scattering layers on P25 films produced a PCE of up to 9.53%, according to Zhenghong Dong et al. research [132]. They continued to research the PCE of the DSSC throughout time. SnO_2-based DSSC were continually monitored for 60 days, and after 20 days in a dark environment, the PCE of the DSSC reached its peak performance. After 20 days, the average PCE of 10 devices grew to 3.3%, while the average V_{oc} and fill factor increased to 0.57 V and 49%, respectively [162]. Surface oxidation has the ability to drastically reduce chemical capacitance (C_μ), which facilitates charge extraction [162]. The improved transfer resistance is likely to lower the photogenerated electron recombination on the SnO_2/electrolyte interface[162].

3.3.1 Doped SnO_2

By doping transition elements with equal ionic size to increase the Fermi level and reducing electron recombination with holes in the electrolyte by employing insulating MOS (Al_2O_3 and MgO) layers, the V_{oc} in SnO_2 photoanodes can be improved. These insulating MOS layers prevent photoexcited electron recombination. A few of them have SnO_2 that has been doped with transition metals (Zn, Pb, Cu, Ni, and Mg), which causes the Fermi level to rise [163]. SnO_2's band edge can be raised by gallium doping, which results in improved open circuit potential (0.74 V), fill factor (73.7%), and PCE (4.05%) [133]. The DSSC efficiency of a nanocrystalline SnO_2 electrode with MgO coating was enhanced from 1.3% to 6.5% [136]. According to Chappel et al. [137], the synthetic TiO_2 that was coated on SnO_2 nanoporous was used for DSSC. A distinctive energy barrier between the electrolyte and electrode interface made up these core-shell electrodes. When compared to an electrode made of pure SnO_2, this barrier reduced the recombination process and raised the device's PCE. Zn-doped SnO_2 nanoflowers were studied by Xincun Dou et al. [134] as a photoanode for efficient DSSCs. With a V_{oc} of 0.78 V, the PCE for the Zn-doped SnO_2 nanoflower

DSSC achieved 3.0%, and it rose to 6.78% when $TiCl_4$ was added. Due to the film's high inherent electron mobility, rapid charge transport is encouraged [134].

NiO, ZnO, CdO, PbO, and CuO semiconductors were employed by Min-Hye Kim and Young-Zuk Kwon as the blocking layer for SnO_2-derived DSSCs [128]. Additionally, the coated divalent metal oxides prevent recombination process. As a result, when CuO, NiO, ZnO, and CdO were used as the blocking layer, the total PCEs of DSSCs increased [128]. While the J_{sc} and V_{oc} increased significantly when employing PbO, the conversion efficiency did not differ much from that of the pure SnO_2. Materials must have very large band gaps, a high dielectric constant, and large isoelectric points in order to be effective blocking layers [128].

3.4 Zinc oxide (ZnO)

ZnO materials have been the subject of in-depth research for DSSC applications. Due to its large bandgap, ZnO has band structures and physical characteristics that are similar to those of TiO_2 [164]. Due to its greater electron diffusivity and high electron mobility of 115–155 $cm^2\ V^{-1}\ s^{-1}$, it has effective electron transport and low recombination rate when employed in DSSCs [164]. Zinc oxide is inexpensive, readily accessible, and it displays improved stability against photocorrosion. The wide bandgap semiconductor ZnO [165] provided the first experimental demonstration of irreversible electron injection from organic components into the conduction band. Comparing ZnO to other metal oxides, a wider range of structural morphologies is probably present. TiO_2 and ZnO crystals promote anisotropic nature [166]. The advantages of ZnO materials for use in DSSCs have been described in numerous studies in the past. Because of its anisotropic evolution and effortless crystallization, ZnO is still regarded as a remarkable TiO_2 alternative even though it only obtained PCEs of 0.4%–5.8% as opposed to TiO_2's (11%).

The nearest alternative to TiO_2 for a DSSC's semiconductor material is ZnO. It is reported that using liquid electrolyte, a ZnO-based DSSC may convert electricity at a maximum efficiency of 7.5%. The effectiveness of ZnO-derived DSSCs was improved through the creation of ZnO photoelectrodes with novel designs, inhibition of Zn^{2+}/dye aggregation, and suppression of the recombination rate [167]. The ZnO nanostructures for DSSCs, gained more attention due to their special morphological parameters, for example, the electron diffusion coefficient and the dye adsorption on the semiconductor surface are strongly influenced by the surface area, porosity, pore diameter, particle size, and particle shape. 1D shapes like nanotubes and nanowires are particularly advantageous for the quick electron transport and decrease in recombination rate in a DSSC (when vertically oriented on the substrate).

ZnO materials have been doped with a variety of dopants, such as I, C, B, Mg, Al, F, Ga, Li, N, Sn, Sb, Y, and polyoxometalates to improve the electron transport in the photoanodes or to control the recombination rate [168–175]. The electrode surfaces' lanthanides (La, Ce, Nd, Gd, and Sm) can produce energy barriers and lower the rate of recombination. A ZnO electrode's surface states can be passivated by certain lanthanide ion modifications [176]. The typical *J-V* characterizations of cells crammed between ZnO films altered by various lanthanide ions under 100 mW/cm^2 light

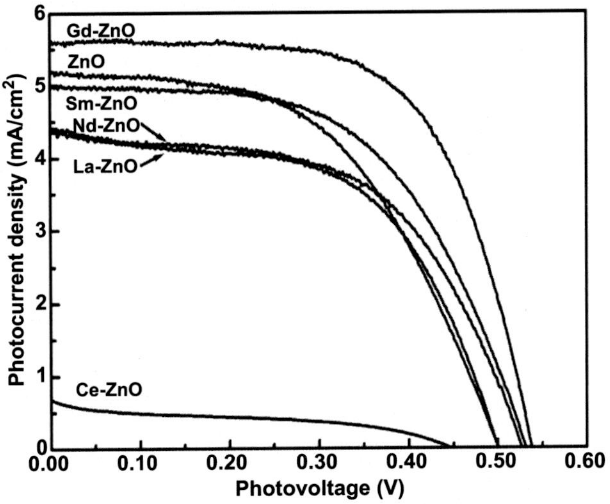

Fig. 4 *J-V* curve of the ZnO-based DSSCs by films altered with various RE elements under 100 mW/cm^2 simulated AM1.5 light irradiation.
Reused with permission from L. Lu, R. Li, T. Peng, K. Fan, K. Dai, Effects of rare earth ion modifications on the photoelectrochemical properties of ZnO-based dye-sensitized solar cells, Renew. Energy, 36(12) (2011) 3386–3393.

irradiation are shown in Fig. 4. The J_{sc} decreased with RE-ion doping, with the exception of Gd-ion [176]. However, the short-circuit photocurrent is reduced as a result of the changes made to La, Nd, Sm, and Ce. The adjustments with Nd, Gd, and Sm among these RE elements have improved the V_{oc} and FF of the cells. An effective light scattering effect from a dual-layer electrode led to higher light absorption and the production of additional photoexcited electrons.

4 Advantages of DSSCs

(1) Capable of construction in an easy way: DSSCs do not require a vacuum system for production and are economical compared to silicon solar cells' manufacturing costs [62,177–179].

(2) Colorable, transparent: A wide variety of colored and translucent cells are possible thanks to the application of different dyes. The transparency and variety of colors of DSSCs could be employed as decorative elements in windows and sunroofs [180].

(3) Thin and elastic structure: The elastic thin films of DSSCs can be constructed using mixtures of tiny photoelectric material particles.

(4) Generation attributes that are unaffected by the strength and angle of the sunlight's incident beam: Even under low light situations, generation traits can still be preserved.

(5) Lighter weight: The weight of the DSSCs panels was reduced by using plastic substrates due to their light weight. The DSSCs can be installed all over the place whereas installing solar cells in the exterior and interior walls of a home, the exterior panels of a car, the sunroof, and the enclosure of mobiles is challenging. This permits an establishment of modern markets with high requirements [178,179].

(6) Environmentally friendly: DSSCs are environmentally friendly and since they are simple to separate and retrieve, the materials are helpful in a reprocessing and reuse framework for solar cell panels [27,177].

5 Applications of DSSCs

In order to make a thin-film solar cell, sometimes referred to as a flexible solar cell, a photoelectrode material is deposited in extremely thin layers over a substrate made of paper, tissue, plastic, metal, glass, etc. The accessibility of lightweight flexible DSSC or modules are voltaic desirable for applications in indoor or outdoor, and electronic gadgets (see Figs. 5 and 6) [181,182]. This is due to the product's adaptability, extreme thinness, light weight, low installation costs, and ease of installation everywhere. Due to their flexibility, DSC panels can now be integrated with a variety of portable objects, including clothing, gear, and bags. In 2010, Sony announced that it was possible to produce modules with an efficiency of up to 10%, opening up the possibility of DSSC module commercialization.

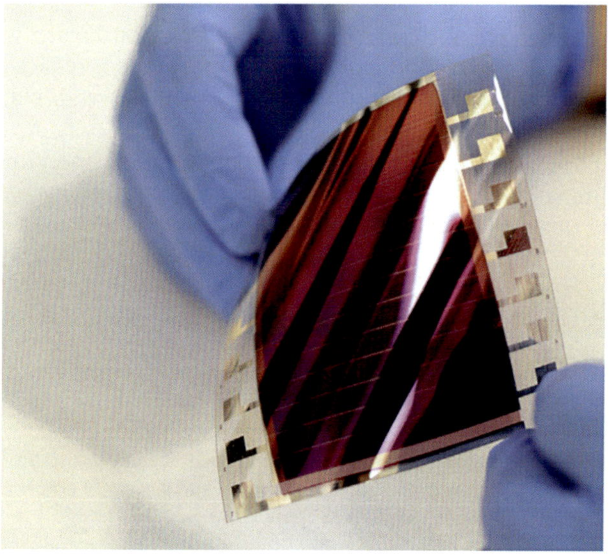

Fig. 5 Image of flexible component using IMI sputtered stack as a transparent electrode manufactured in i-MEET laboratories.
Reused with permission from P. Kubis, L. Lucera, F. Machui, G. Spyropoulos, J. Cordero, A. Frey, J. Kaschta, M.M. Voigt, G.J. Matt, E. Zeira, C.J. Brabec, High precision processing of flexible P3HT/PCBM modules with geometric fill factor over 95%. Org. Electron., 15(10) (2014) 2256-2263.

Fig. 6 Full size 2×32-cell prototype modules in roof station.
Reused with permission from J. Goldstein, I. Yakupov, B. Breen, Development of large-area photovoltaic dye cells at 3GSolar. Sol. Energy Mater. Sol. Cells 94 (4) (2010) 638-641.

6 Research and development challenges in DSSCs improvement

The development of DSSCs by researchers revealed that their current conversion efficiencies are not particularly high. By expanding the photoelectrodes' surface area for usage in DSSCs, numerous research efforts have attempted to alleviate this issue [62]. The primary tasks for the industrial adoption of DSSCs are low stability and efficiency [21]. Below is a list of the primary causes of poor PCEs in DSSCs:

(1) Minimum absorption at UV and NIR regions.
(2) A greater surface area requires lowering the extinction coefficient.
(3) DSSCs has a large over potential, but I^-/I_3^- redox has low recombination rate.
(4) Weak contact between the working and the contour electrodes.
(5) The electrolyte properties can be degraded by the absorption of UV light.

Stable cells are a highly significant concern when examining these cells [60]. A long lifespan is the foundation for good stability [59]. The aspects that are most important to DSSCs' stability and robustness are:

(1) Although solid-state hole conductors have lower efficiency, electrolyte solution is harmful [66].
(2) I^-/I_3^- is highly corrosive.

Additionally, there is substantial interest in the plant-derived dyes' volatility in sunlight [183]. Finding techniques to incorporate new natural dyes without affecting the dye's adherence to photoelectrodes would be necessary to overcome this challenge.

Modifying the dye-surface chemistry is another way to stop photodegradation. Natural colors made from plants are typically consumed within 48 hours. To estimate the lifespan of the cell, rigorous research is needed to look at how exposure to sunshine affects the way that chemical reactions alter over time.

7 Conclusions

In the present chapter, the latest applications of metal oxide electrodes utilized in DSSCs and their working mechanisms were summarized. The basic investigations on the various Nb_2O_5, SnO_2, and ZnO nanostructures evidenced this substance's usefulness for DSSCs applications. It becomes obvious that ongoing research is required to understand and control how different structural morphologies and crystalline forms of metal oxides form. These are key characteristics for modifying the electronic and the optical properties. Recent works on the films made by using anodization and pulsed-laser deposition methods have produced nanostructures, such as wires, rods, and spherical particles on the substrates. And found to be beneficial for enhancing PCEs and providing superior electron transport properties. The resistance in the charge transport; however, visibly rises with an increasing Nb_2O_5 film thickness, which then limits the photocurrent. In contrast to the TiO_2 anatase phase, Nb_2O_5 acts as an electron blocking layer in DSSCs, improving performance and reducing electron recombination. In order to get over the photocurrent's physical limitations, doping, and creation of hybrid materials using Nb_2O_5 and other metal oxides are both viable options.

Acknowledgments

The authors express their sincere thanks to the South African Research Chairs Initiative of the Department of Science and Technology (84415). The financial support from the University of the Free State, South Africa is highly recognized.

References

[1] N. Lior, Energy resources and use: the present situation and possible paths to the future, Energy 33 (2008) 842–857.
[2] A. Kumar, T.C. Kandpal, Renewable energy technologies for irrigation water pumping in India: a preliminary attempt towards potential estimation, Energy 32 (2007) 861–870.
[3] B. Parida, S. Iniyan, R. Goic, A review of solar photovoltaic technologies, Renew. Sust. Energ. Rev. 15 (2011) 1625–1636.
[4] T.R. Cook, D.K. Dogutan, S.Y. Reece, Y. Surendranath, T.S. Teets, D.G. Nocera, Solar energy supply and storage for the legacy and nonlegacy worlds, Chem. Rev. 110 (2010) 6474–6502.
[5] S. Shi, Y. Li, X. Li, H. Wang, Advancements in all-solid-state hybrid solar cells based on organometal halide perovskites, Mater. Horizons 2 (2015) 378–405.
[6] C.Y. Chen, M. Wang, J.Y. Li, N. Pootrakulchote, L. Alibabaei, C.H. Ngoc-Le, J.D. Decoppet, J.H. Tsai, C. Grätzel, C.G. Wu, S.M. Zakeeruddin, M. Grätzel, Highly efficient light-harvesting ruthenium sensitizer for thin-film dye-sensitized solar cells, ACS Nano 3 (2009) 3103–3109.

[7] T.A. Heimer, E.J. Heilweil, C.A. Bignozzi, G.J. Meyer, Electron injection, recombination and halide oxidation dynamics at dye-Sensitized metal oxide interfaces, J. Phys. Chem. A 104 (2000) 4256–4262.
[8] B.O. Regan, M. Grätzel, A low-cost, high-efficiency solar cell based on dye-sensitized colloidal TiO_2 films, Nature 353 (1991) 737–740.
[9] H.P. Anders Hagfeldt, G. Boschloo, L. Sun, L. Kloo, Dye-sensitized solar cells, Chem. Rev. 110 (2010) 6595–6663.
[10] M. Grätzel, Solar cells to dye for, Nature 421 (2003) 586–587.
[11] M. Grätzel, Solar energy conversion by dye-sensitized photovoltaic cells, Inorg. Chem. 44 (2005) 6841–6851.
[12] D. Ganta, J. Jara, R. Villanueva, Dye-sensitized solar cells using Aloe Vera and Cladode of Cactus extracts as natural sensitizers, Chem. Phys. Lett. 679 (2017) 97–101.
[13] H. Yang, F. Peng, Q. Zhang, W. Liu, D. Sun, Y. Zhao, X. Wei, Strong upconversion luminescence in $LiYMo_2O_8$:Er, Yb towards efficiency enhancement of dye-sensitized solar cells, Opt. Mater. (Amst.) 35 (2013) 2338–2342.
[14] C.M. Lan, H.P. Wu, T.Y. Pan, C.W. Chang, W.S. Chao, C.T. Chen, C.L. Wang, C.Y. Lin, E.W.G. Diau, Enhanced photovoltaic performance with co-sensitization of porphyrin and an organic dye in dye-sensitized solar cells, Energy Environ. Sci. 5 (2012) 6460–6464.
[15] A. Yella, H.W. Lee, H.N. Tsao, C. Yi, A.K. Chandiran, M.K. Nazeeruddin, E.W.G. Diau, C.Y. Yeh, S.M. Zakeeruddin, M. Grätzel, Porphyrin-sensitized solar cells with cobalt (II/III)-based redox electrolyte exceed 12 percent efficiency, Science 334 (2011) 629–634.
[16] T. Trupke, M.A. Green, P. Würfel, Improving solar cell efficiencies by down-conversion of high-energy photons, J. Appl. Phys. 92 (2002) 1668–1674.
[17] M. Green, K. Emery, Y. Hishikawa, W. Warta, E. Dunlop, D. Barkhouse, O. Gunawan, T. Gokmen, T. Todorov, D. Mitzi, The path to 25% silicon solar cell efficiency: history of silicon cell evolution martin, Prog. Photovolt. Res. Appl. 17 (2009) 183–189.
[18] J. Zhang, H. Shen, W. Guo, S. Wang, C. Zhu, F. Xue, J. Hou, H. Su, Z. Yuan, An upconversion $NaYF_4:Yb^{3+}$, Er^{3+}/TiO_2 core-shell nanoparticle photoelectrode for improved efficiencies of dye-sensitized solar cells, J. Power Sources 226 (2013) 47–53.
[19] X. Huang, S. Han, W. Huang, X. Liu, Enhancing solar cell efficiency: the search for luminescent materials as spectral converters, Chem. Soc. Rev. 42 (2013) 173–201.
[20] F. Auzel, Upconversion and anti-stokes processes with f and d ions in solids, Chem. Rev. 104 (2004) 139–173.
[21] R.T. Wegh, H. Donker, K.D. Oskam, A. Meijerink, Visible quantum cutting in $LiGdF_4$: Eu^{3+} through downconversion, Science 283 (1999) 663–666.
[22] H. Lin, G. Meredith, S. Jiang, X. Peng, T. Luo, N. Peyghambarian, E. Yue-Bun Pun, Optical transitions and visible upconversion in Er^{3+} doped niobic tellurite glass, J. Appl. Phys. 93 (2003) 186–191.
[23] L. Liang, Y. Yulin, Z. Mi, F. Ruiqing, Q. Lele, W. Xin, Z. Lingyun, Z. Xuesong, H. Jianglong, Enhanced performance of dye-sensitized solar cells based on TiO_2 with NIR-absorption and visible upconversion luminescence, J. Solid State Chem. 198 (2013) 459–465.
[24] M. Fernández-Garcia, J. Rodriguez, Metal oxide nanoparticles, in: Encyclopedia of Inorganic and Bioinorganic Chemistry, 2011.
[25] R. Baby, P.D. Nixon, N.M. Kumar, M.S.P. Subathra, N. Ananthi, A Comprehensive Review of Dye-Sensitized Solar Cell Optimal Fabrication Conditions, Natural Dye Selection, and Application-Based Future Perspectives, Springer, Berlin Heidelberg, 2021.
[26] A. Andualem, S. Demiss, Review on dye-sensitized solar cells (DSSCs), Edelweiss Appl. Sci. Technol. (2018) 145–150.

[27] U. Mehmood, S.U. Rahman, K. Harrabi, I.A. Hussein, B.V.S. Reddy, Recent advances in dye sensitized solar cells, Adv. Mater. Sci. Eng. 2014 (2014) 1–12.
[28] F.A. Grant, Properties of rutile (titanium dioxide), Rev. Mod. Phys. 31 (1959) 646–674.
[29] K. Tennakone, G.R.R.A. Kumara, I.R.M. Kottegoda, V.P.S. Perera, An efficient dye-sensitized photoelectrochemical solar cell made from oxides of tin and zinc, Chem. Commun. 1 (1999) 15–16.
[30] K. Sayama, H. Sugihara, H. Arakawa, Photoelectrochemical properties of a porous Nb_2O_5 electrode sensitized by a ruthenium dye, Chem. Mater. 10 (1998) 3825–3832.
[31] E. Jonathan, M.Y. Onimisi, D. Eli, Natural pigments as sensitizers for dye sensitized solar cells, Adv. Mater. 5 (2016) 31–34.
[32] O. Adedokun, K. Titilope, A.O. Awodugba, Review on natural dye-sensitized solar cells (DSSCs), Int. J. Eng. Technol. 2 (2016) 34–41.
[33] I.S. Mohamad, S.S. Ismail, M.N. Norizan, S.A.Z. Murad, M.M.A. Abdullah, ZnO photoanode effect on the efficiency performance of organic based dye sensitized solar cell, IOP Conf. Ser. Mater. Sci. Eng. 209 (2017) 012028.
[34] H.M. Upadhyaya, S. Senthilarasu, M.H. Hsu, D.K. Kumar, Recent progress and the status of dye-sensitised solar cell (DSSC) technology with state-of-the-art conversion efficiencies, Sol. Energy Mater. Sol. Cells 119 (2013) 291–295.
[35] M. Alhamed, A. Issa, A. Doubal, Studying of natural dyes properties as photo-sensitizer for dye sensitized solar cells (DSSC), J. Electron Devices 16 (2012) 1370–1383.
[36] M. Hamadanian, J. Safaei-Ghomi, M. Hosseinpour, R. Masoomi, V. Jabbari, Uses of new natural dye photosensitizers in fabrication of high potential dye-sensitized solar cells (DSSCs), Mater. Sci. Semicond. Process. 27 (2014) 733–739.
[37] R. Kushwaha, P. Srivastava, L. Bahadur, Natural pigments from plants used as sensitizers for TiO_2 based dye-sensitized solar cells, J. Energy 2013 (2013) 1–8.
[38] N.A.M. Ahmad Hambali, N. Roshidah, M. Norhafiz Hashim, I.S. Mohamad, N. Hidayah Saad, M.N. Norizan, Dye-sensitized solar cells using natural dyes as sensitizers from Malaysia local fruit "Buah Mertajam", AIP Conf. Proc. 1660 (2015) 070050.
[39] R. Hemmatzadeh, A. Mohammadi, Improving optical absorptivity of natural dyes for fabrication of efficient dye-sensitized solar cells, J. Theor. Appl. Phys. 7 (2013) 57.
[40] C.Y. Chien, B.D. Hsu, Optimization of the dye-sensitized solar cell with anthocyanin as photosensitizer, Sol. Energy 98 (2013) 203–211.
[41] S. Rajkumar, K. Suguna, Analysis of natural sensitizers to enhance the efficiency in dye sensitized solar cell, J. Eng. Res. Appl. 6 (2016) 41–46.
[42] I.C. Maurya, Neetu, A.K. Gupta, P. Srivastava, L. Bahadur, Natural dye extracted from saraca asoca flowers as sensitizer for TiO_2-based dye- sensitized solar cell, ASME. J. Sol. Energy Eng. Trans. 138 (2016) 051006.
[43] M.R. Faiz, D. Widhiyanuriyawan, E. Siswanto, I.N.G. Wardana, Theoretical study on the application of natural green pigment for sensitizer in dye-sensitized solar cell (DSSC), in: Proceeding—2017 5th Int. Conf. Electr. Electron. Inf. Eng. Smart Innov. Bridg. Futur. Technol. ICEEIE 2017, 2018, pp. 32–37.
[44] N. Prabavathy, S. Shalini, R. Balasundaraprabhu, D. Velauthapillai, S. Prasanna, N. Muthukumarasamy, Enhancement in the photostability of natural dyes for dye-sensitized solar cell (DSSC) applications: a review, Int. J. Energy Res. 41 (2017) 1372–1396.
[45] M.R. Narayan, Review: dye sensitized solar cells based on natural photosensitizers, Renew. Sust. Energ. Rev. 16 (2012) 208–215.

[46] S. Meng, J. Ren, E. Kaxiras, Natural dyes adsorbed on TiO_2 nanowire for photovoltaic applications: enhanced light absorption and ultrafast electron injection, Nano Lett. 8 (2008) 3266–3272.

[47] I.C. Maurya, P. Srivastava, L. Bahadur, Dye-sensitized solar cell using extract from petals of male flowers Luffa cylindrica L. as a natural sensitizer, Opt. Mater. 52 (2016) 150–156.

[48] W. Maiaugree, S. Lowpa, M. Towannang, P. Rutphonsan, A. Tangtrakarn, S. Pimanpang, P. Maiaugree, N. Ratchapolthavisin, W. Sang-Aroon, W. Jarernboon, A dye sensitized solar cell using natural counter electrode and natural dye derived from mangosteen peel waste, Sci. Rep. 5 (2015) 15230.

[49] H. Hug, M. Bader, P. Mair, T. Glatzel, Biophotovoltaics: natural pigments in dye-sensitized solar cells, Appl. Energy 115 (2014) 216–225.

[50] N.T.R.N. Kumara, A. Lim, C.M. Lim, M.I. Petra, P. Ekanayake, Recent progress and utilization of natural pigments in dye sensitized solar cells: a review, Renew. Sust. Energ. Rev. 78 (2017) 301–317.

[51] H. Zhang, Y. Wang, D. Yang, Y. Li, H. Liu, P. Liu (Eds.), Directly hydrothermal growth of single crystal $Nb_3O_7(OH)$ nanorod film for high performance dye-sensitized solar cells, Adv. Mater. 24 (12) (2012) 1598–1603.

[52] X. Ai, J. Guo, N.A. Anderson, T. Lian, Ultrafast electron transfer from Ru polypyridyl complexes to Nb_2O_5 nanoporous thin films, J. Phys. Chem. B 108 (34) (2004) 12795–12803.

[53] N.I. Beedri, S.A.A.R. Sayyed, S.R. Jadkar, H.M. Pathan, Rose Bengal sensitized niobium pentaoxide photoanode for dye sensitized solar cell application, AIP Conf. Proc. 1832 (2017) 040022.

[54] G. Oskam, B.V. Bergeron, G.J. Meyer, P.C. Searson, Pseudohalogens for dye-sensitized TiO_2 photoelectrochemical cells, J. Phys. Chem. B 105 (2001) 6867–6873.

[55] H. Nusbaumer, J.E. Moser, S.M. Zakeeruddin, M.K. Nazeeruddin, M. Grätzel, CoII(dbbip)22+ complex rivals tri-iodide/iodide redox mediator in dye-sensitized photovoltaic cells, J. Phys. Chem. B 105 (2001) 10461–10464.

[56] B.N. Nunes, L.A. Faustino, A.V. Muller, A.S. Polo, A.O.T. Patrocinio, Chapter 8. Nb_2O_5 dye-sensitized solar cells, in: S. Thomas, E.H.M. Sakho, N. Kalarikkal, S.O. Oluwafemi, J. Wu (Eds.), Nanomaterials for Solar Cell Applications, 2019, pp. 287–322.

[57] F. Gao, Y. Wang, D. Shi, J. Zhang, M. Wang, X. Jing, R. Humphry-Baker, P. Wang, S.M. Zakeeruddin, M. Grätzel, Enhance the optical absorptivity of nanocrystalline TiO_2 film with high molar extinction coefficient ruthenium sensitizers for high performance dye-sensitized solar cells, J. Am. Chem. Soc. 130 (2008) 10720–10728.

[58] M. Toivola, F. Ahlskog, P. Lund, Industrial sheet metals for nanocrystalline dye-sensitized solar cell structures, Sol. Energy Mater. Sol. Cells 90 (2006) 2881–2893.

[59] M.K. Nazeeruddin, A. Kay, I. Rodicio, R. Humpbry-Baker, E. Miiller, P. Liska, N. Vlachopoulos, M. Gratzel, Conversion of light to electricity by cis-xzbis(2,2'-bipyridyl-4,4'-dicarboxylate)ruthenium(11) charge-transfer sensitizers (X = C1-, Br-, I-, CN-, and SCN-) on nanocrystalline TiO_2 electrodes, J. Am. Chem. Soc. 115 (1993) 6382–6390.

[60] A. Kay, M. Grätzel, Low cost photovoltaic modules based on dye sensitized nanocrystalline titanium dioxide and carbon powder, Sol. Energy Mater. Sol. Cells 44 (1996) 99–117.

[61] M. Wang, A.M. Anghel, B. Marsan, N.L.C. Ha, N. Pootrakulchote, S.M. Zakeeruddin, M. Grätzel, CoS supersedes Pt as efficient electrocatalyst for triiodide reduction in dye-sensitized solar cells, J. Am. Chem. Soc. 131 (2009) 15976–15977.

[62] M. Grätzel, J.E. Moser, Solar energy conversion, in: I. Gould, V. Balzani (Eds.), Electron Transfer in Chemistry, Energy and the Environment, vol. 5, 2001, pp. 589–644.

[63] H.J. Wang, C.P. Chen, R.J. Jeng, Polythiophenes comprising conjugated pendants for polymer solar cells: a review, Materials (Basel) 7 (2014) 2411–2439.
[64] K. Kakiage, Y. Aoyama, T. Yano, K. Oya, J.I. Fujisawa, M. Hanaya, Highly-efficient dye-sensitized solar cells with collaborative sensitization by silyl-anchor and carboxy-anchor dyes, Chem. Commun. 51 (2015) 15894–15897.
[65] J. Liu, Q. Tang, B. He, L. Yu, Cost-effective, transparent iron selenide nanoporous alloy counter electrode for bifacial dye-sensitized solar cell, J. Power Sources 282 (2015) 79–86.
[66] K. Sharma, V. Sharma, S.S. Sharma, Dye-sensitized solar cells: fundamentals and current status, Nanoscale Res. Lett. 13 (2018) 1–46.
[67] D.Y.C. Leung, X. Fu, C. Wang, M. Ni, M.K.H. Leung, X. Wang, X. Fu, Hydrogen production over titania-based photocatalysts, ChemSusChem 3 (2010) 681–694.
[68] M.A. Rauf, M.A. Meetani, S. Hisaindee, An overview on the photocatalytic degradation of azo dyes in the presence of TiO_2 doped with selective transition metals, Desalination 276 (2011) 13–27.
[69] H. Park, Y. Park, W. Kim, W. Choi, Surface modification of TiO_2 photocatalyst for environmental applications, J Photochem Photobiol C: Photochem Rev 15 (2013) 1–20.
[70] H.K. Ardakani, Electrical and optical properties of in situ "hydrogen-reduced" titanium dioxide thin films deposited by pulsed excimer laser ablation, Thin Solid Films 248 (1994) 234–239.
[71] J. Augustynski, N. Vlachopoulos, P. Liska, M. GräTzel, Very efficient visible light energy harvesting and conversion by spectral sensitization of high surface area polycrystalline titanium dioxide films, J. Am. Chem. Soc. 110 (1988) 1216–1220.
[72] A.K. Chandiran, M. Abdi-Jalebi, M.K. Nazeeruddin, M. Grätzel, Analysis of electron transfer properties of ZnO and TiO_2 photoanodes for dye-sensitized solar cells, ACS Nano 8 (2014) 2261–2268.
[73] J. Fan, Y. Hao, A. Cabot, E.M.J. Johansson, G. Boschloo, A. Hagfeldt, Cobalt(II/III) redox electrolyte in ZnO nanowire-based dye-sensitized solar cells, ACS Appl. Mater. Interfaces 5 (2013) 1902–1906.
[74] G. Li, C.P. Richter, R.L. Milot, L. Cai, C.A. Schmuttenmaer, R.H. Crabtree, Synergistic effect between anatase and rutile TiO_2 nanoparticles in dye-sensitized solar cells, Dalton Trans. (2009) 10078–10085.
[75] D.O. Scanlon, C.W. Dunnill, J. Buckeridge, S.A. Shevlin, A.J. Logsdail, S.M. Woodley, C.R.A. Catlow, M.J. Powell, R.G. Palgrave, I.P. Parkin, G.W. Watson, T.W. Keal, P. Sherwood, A. Walsh, A.A. Sokol, Band alignment of rutile and anatase TiO_2, Nat. Mater. 12 (2013) 798–801.
[76] N.G. Park, J. Van De Lagemaat, A.J. Frank, Comparison of dye-sensitized rutile- and anatase-based TiO_2 solar cells, J. Phys. Chem. B 104 (2000) 8989–8994.
[77] J. Li, X. Yang, X. Yu, L. Xu, W. Kang, W. Yan, H. Gao, Z. Liu, Y. Guo, Rare earth oxide-doped titania nanocomposites with enhanced photocatalytic activity towards the degradation of partially hydrolysis polyacrylamide, Appl. Surf. Sci. 255 (2009) 3731–3738.
[78] D.M. de los Santos, J. Navas, T. Aguilar, A. Sánchez-Coronilla, C. Fernández-Lorenzo, R. Alcántara, J.C. Piñero, G. Blanco, J. Martín-Calleja, Tm-doped TiO_2 and $Tm_2Ti_2O_7$ pyrochlore nanoparticles: enhancing the photocatalytic activity of rutile with a pyrochlore phase, Beilstein J. Nanotechnol. 6 (2015) 605–616.
[79] A. Paxton, L. Thiên-Nga, Electronic structure of reduced titanium dioxide, Phys. Rev. B Condens. Matter Mater. Phys. 57 (1998) 1579–1584.
[80] M.P. Dare-Edwards, J.B. Goodenough, A. Hamnett, K.R. Seddon, R.D. Wright, Sensitisation of semiconducting electrodes with ruthenium-based dyes, Faraday Discuss. Chem. Soc. 70 (1980) 285–298.

[81] Y.L. Yandong Duan, F. Nianqing, Q. Liu, Y. Fang, X. Zhou, J. Zhang, Sn-doped TiO_2 photoanode for dye-sensitized solar cells, Aust. J. Chem. 116 (2012) 8888–8893.
[82] F. De Angelis, S. Fantacci, A. Selloni, M.K. Nazeeruddin, First-principles modeling of the adsorption geometry and electronic structure of Ru(II) dyes on extended TiO_2 substrates for dye-sensitized solar cell applications, J. Phys. Chem. C 114 (2010) 6054–6061.
[83] S. Meng, E. Kaxiras, Electron and hole dynamics in dye-sensitized solar cells: influencing factors and systematic trends, Nano Lett. 10 (2010) 1238–1247.
[84] C.J. Brinker, G.W. Scherer, Sol–Gel Science, The Physics and Chemistry of Sol–Gel Processing, Academic Press, Boston, 1990.
[85] S. Iwamoto, Y. Sazanami, M. Inoue, T. Inoue, T. Hoshi, K. Shigaki, M. Kaneko, A. Maenosono, Fabrication of dye-sensitized solar cells with an open-circuit photovoltage of 1 V, ChemSusChem 1 (2008) 401–403.
[86] K.P. Wang, H. Teng, Zinc-doping in TiO_2 films to enhance electron transport in dye-sensitized solar cells under low-intensity illumination, Phys. Chem. Chem. Phys. 11 (2009) 9489–9496.
[87] K.H. Ko, Y.C. Lee, Y.J. Jung, Enhanced efficiency of dye-sensitized TiO_2 solar cells (DSSC) by doping of metal ions, J. Colloid Interface Sci. 283 (2005) 482–487.
[88] C. Kim, K.S. Kim, H.Y. Kim, Y.S. Han, Modification of a TiO_2 photoanode by using Cr-doped TiO_2 with an influence on the photovoltaic efficiency of a dye-sensitized solar cell, J. Mater. Chem. 18 (2008) 5809–5814.
[89] X. Feng, K. Shankar, M. Paulose, C.A. Grimes, Tantalum-doped titanium dioxide nanowire arrays for dye-sensitized solar cells with high open-circuit voltage, Angew. Chem. Int. Ed. 48 (2009) 8095–8098.
[90] L. Wei, Y. Yang, X. Xia, R. Fan, T. Su, Y. Shi, J. Yu, L. Li, Y. Jiang, Band edge movement in dye sensitized Sm-doped TiO_2 solar cells: a study by variable temperature spectroelectrochemistry, RSC Adv. 5 (2015) 70512–70521.
[91] H.F. Mehnane, C. Wang, K.K. Kondamareddy, W. Yu, W. Sun, H. Liu, S. Bai, W. Liu, S. Guo, X.Z. Zhao, Hydrothermal synthesis of TiO_2 nanoparticles doped with trace amounts of strontium, and their application as working electrodes for dye sensitized solar cells: tunable electrical properties & enhanced photo-conversion performance, RSC Adv. 7 (2017) 2358–2364.
[92] B. Roose, S. Pathak, U. Steiner, Doping of TiO_2 for sensitized solar cells, Chem. Soc. Rev. 44 (2015) 8326–8349.
[93] A. Subramanian, J. Bow, H. Wang, The effect of Li^+ intercalation on different sized TiO_2 nanoparticles and the performance of dye-sensitized solar cells, Thin Solid Films 520 (2012) 7011–7017.
[94] D.F. Watson, G.J. Meyer, Cation effects in nanocrystalline solar cells, Coord. Chem. Rev. 248 (2004) 1391–1406.
[95] K. Kakiage, T. Tokutome, S. Iwamoto, T. Kyomen, M. Hanaya, Fabrication of a dye-sensitized solar cell containing a Mg-doped TiO_2 electrode and a Br^{3-}/Br^- redox mediator with a high open-circuit photovoltage of 1.21 V, Chem. Commun. 49 (2013) 179–180.
[96] C. Zhang, S. Chen, L.E. Mo, Y. Huang, H. Tian, L. Hu, Z. Huo, S. Dai, F. Kong, X. Pan, Charge recombination and band-edge shift in the dye-sensitized Mg^{2+}-doped TiO_2 solar cells, J. Phys. Chem. C 115 (2011) 16418–16424.
[97] Q. Liu, Y. Zhou, Y. Duan, M. Wang, Y. Lin, Improved photovoltaic performance of dye-sensitized solar cells (DSSCs) by Zn + Mg co-doped TiO_2 electrode, Electrochim. Acta 95 (2013) 48–53.

[98] Q. Liu, Y. Zhou, Y. Duan, M. Wang, X. Zhao, Y. Lin, Enhanced conversion efficiency of dye-sensitized titanium dioxide solar cells by Ca-doping, J. Alloys Compd. 548 (2013) 161–165.

[99] M. Pan, H. Liu, Z. Yao, X. Zhong, Enhanced efficiency of dye-sensitized solar cells by trace amount Ca-doping in TiO_2 photoelectrodes, J. Nanomater. 2015 (2015) 3.

[100] W. Li, J. Yang, J. Zhang, S. Gao, Y. Luo, M. Liu, Improve photovoltaic performance of titanium dioxide nanorods based dye-sensitized solar cells by Ca-doping, Mater. Res. Bull. 57 (2014) 177–183.

[101] M. Motlak, N.M. Barakat, M.S. Akhtar, A.M. Hamza, A. Taha, O.-B. Yang, H.Y. Kim, Enhancement the conversion efficiency of the dye-sensitized solar cells using novel Ca-doped TiO_2 nanofibers, Energy Environ. Focus 2 (2013) 217–221.

[102] H. Imahori, S. Hayashi, T. Umeyama, S. Eu, A. Oguro, S. Kang, Y. Matano, T. Shishido, S. Ngamsinlapasathian, S. Yoshikawa, Comparison of electrode structures and photovoltaic properties of porphyrin-sensitized solar cells with TiO_2 and Nb, Ge, Zr-added TiO_2 composite electrodes, Langmuir 22 (2006) 11405–11411.

[103] M. Wang, S. Bai, A. Chen, Y. Duan, Q. Liu, D. Li, Y. Lin, Improved photovoltaic performance of dye-sensitized solar cells by Sb-doped TiO_2 photoanode, Electrochim. Acta 77 (2012) 54–59.

[104] V. Sharma, T.K. Das, P. Ilaiyaraja, A.C. Dakshinamurthy, Growth of Sb_2S_3 semiconductor thin film on different morphologies of TiO_2 nanostructures, Mater. Res. Bull. 131 (2020) 110980.

[105] A. Latini, C. Cavallo, F.K. Aldibaja, D. Gozzi, D. Carta, A. Corrias, L. Lazzarini, G. Salviati, Efficiency improvement of DSSC photoanode by scandium doping of mesoporous titania beads, J. Phys. Chem. C 117 (2013) 25276–25289.

[106] G. Blasse, B.C. Grabmaier, Luminescent Materials, Springer Verlag, Berlin Heidelberg, 1994.

[107] P. Dorenbos, 5D Level positions of the trivalent lanthanides in inorganic compounds, J. Lumin. 91 (2000) 155–176.

[108] P. Dorenbos, J. Andriessen, C.W.E. Van Eijk, $4F^n$-5D centroid shift in lanthanides and relation with anion polarizability, covalency, and cation electronegativity, J. Solid State Chem. 171 (2003) 133–136.

[109] P. Dorenbos, Lanthanide 4f-electron binding energies and the nephelauxetic effect in wide band gap compounds, J. Lumin. 136 (2013) 122–129.

[110] Y. Wen, Y. Wang, B. Liu, F. Zhang, Luminescence properties of $Ca_4Y_6(SiO_4)_6O:RE^{3+}$ (RE = Eu, Tb, Dy, Sm and Tm) under vacuum ultraviolet excitation, Opt. Mater. (Amst.) 34 (2012) 889–892.

[111] C. Strümpel, M. McCann, G. Beaucarne, V. Arkhipov, A. Slaoui, V. Švrček, C. del Cañizo, I. Tobias, Modifying the solar spectrum to enhance silicon solar cell efficiency—an overview of available materials, Sol. Energy Mater. Sol. Cells 91 (2007) 238–249.

[112] J. Wang, T. Ming, Z. Jin, J. Wang, L.D. Sun, C.H. Yan, Photon energy upconversion through thermal radiation with the power efficiency reaching 16%, Nat. Commun. 5 (2014) 1–9.

[113] Y. Masuda, M. Yamagishi, K. Koumoto, Site-selective deposition and micropatterning of visible-light-emitting europium-doped yttrium oxide thin film on self-assembled monolayers, Chem. Mater. 19 (2007) 1002–1008.

[114] V. Madurai Ramakrishnan, S. Pitchaiya, N. Muthukumarasamy, K. Kvamme, G. Rajesh, S. Agilan, A. Pugazhendhi, D. Velauthapillai, Performance of TiO_2 nanoparticles synthesized by microwave and solvothermal methods as photoanode in dye-sensitized solar cells (DSSC), Int. J. Hydrog. Energy 45 (2020) 27036–27046.

[115] Y. Cao, Y. Liu, S.M. Zakeeruddin, A. Hagfeldt, M. Grätzel, Direct contact of selective charge extraction layers enables high-efficiency molecular photovoltaics, Joule 2 (2018) 1108–1117.
[116] S. Chuangchote, T. Sagawa, S. Yoshikawa, Efficient dye-sensitized solar cells using electrospun TiO_2 nanofibers as a light harvesting layer, Appl. Phys. Lett. 93 (2008) 2012–2015.
[117] N. Prabhu, S. Agilan, N. Muthukumarasamy, T.S. Senthil, Enhanced photovoltaic performance of WO_3 nanoparticles added dye sensitized solar cells, J. Mater. Sci. Mater. Electron. 25 (2014) 5288–5295.
[118] Z. Tong, T. Peng, W. Sun, W. Liu, S. Guo, X.Z. Zhao, Introducing an intermediate band into dye-sensitized solar cells by W^{6+} doping into TiO_2 nanocrystalline photoanodes, J. Phys. Chem. C 118 (2014) 16892–16895.
[119] A.M. Bakhshayesh, N. Farajisafiloo, Efficient dye-sensitised solar cell based on uniform In-doped TiO_2 spherical particles, Appl. Phys. A Mater. Sci. Process. 120 (2015) 199–206.
[120] D. Chen, D. Yang, Q. Wang, Z. Jiang, Effects of boron doping on photocatalytic activity and microstructure of titanium dioxide nanoparticles, Ind. Eng. Chem. Res. 45 (2006) 4110–4116.
[121] J. Yu, Y. Yang, C. Zhang, R. Fan, T. Su, Preparation of YbF_3-Ho@TiO_2 core-shell submicrocrystal spheres and their application to the electrode of dye-sensitized solar cells, New J. Chem. 44 (2020) 10545–10553.
[122] X. Jin, C. Liu, J. Xu, Q. Wang, D. Chen, Size-controlled synthesis of mesoporous Nb_2O_5 microspheres for dye sensitized solar cells, RSC Adv. 4 (2014) 35546–35553.
[123] A. Le Viet, R. Jose, M.V. Reddy, B.V.R. Chowdari, S. Ramakrishna, Nb_2O_5 photoelectrodes for dye-sensitized solar cells: choice of the polymorph, J. Phys. Chem. C 114 (2010) 21795–21800.
[124] D. Shen, W. Zhang, Y. Li, A. Abate, M. Wei, Facile deposition of Nb_2O_5 thin film as an electron-transporting layer for highly efficient perovskite solar cells, ACS Appl. Nano Mater. 1 (2018) 4101–4109.
[125] R. Ghosh, M.K. Brennaman, T. Uher, M.-R. Ok, E.T. Samulski, L.E. McNeil, T.J. Meyer, R. Lopez, Nanoforest Nb_2O_5 photoanodes for dye-sensitized solar cells by pulsed laser deposition, ACS Appl. Mater. Interfaces 3 (2011) 3929–3935.
[126] J.Z. Ou, R.A. Rani, M.-H. Ham, M.R. Field, Y. Zhang, H. Zheng, P. Reece, S. Zhuiykov, S. Sriram, M. Bhaskaran, R.B. Kaner, K. Kalantar-zadeh, Elevated temperature anodized Nb_2O_5: a photoanode material with exceptionally large photoconversion efficiencies, ACS Nano 6 (2012) 4045–4053.
[127] S. Gubbala, V. Chakrapani, V. Kumar, M.K. Sunkara, Band-edge engineered hybrid structures for dye-sensitized solar cells based on SnO_2 nanowires, Adv. Funct. Mater. 18 (2008) 2411–2418.
[128] M.-H. Kim, Y.-U. Kwon, Semiconducting divalent metal oxides as blocking layer material for SnO_2-based dye-sensitized solar cells, J. Phys. Chem. C 115 (2011) 23120–23125.
[129] S.A. Mahmoud, O.A. Fouad, Synthesis and application of zinc/tin oxide nanostructures in photocatalysis and dye sensitized solar cells, Sol. Energy Mater. Sol. Cells 136 (2015) 38–43.
[130] Y.F. Wang, K.N. Li, W.Q. Wu, Y.F. Xu, H.Y. Chen, C.Y. Su, D. Bin Kuang, Fabrication of a double layered photoanode consisting of SnO_2 nanofibers and nanoparticles for efficient dye-sensitized solar cells, RSC Adv. 3 (2013) 13804–13810.
[131] H. Song, K.H. Lee, H. Jeong, S.H. Um, G.S. Han, H.S. Jung, G.Y. Jung, A simple self-assembly route to single crystalline SnO_2 nanorod growth by oriented attachment for dye sensitized solar cells, Nanoscale 5 (2013) 1188–1194.

[132] Z. Dong, H. Ren, C.M. Hessel, J. Wang, R. Yu, Q. Jin, M. Yang, Z. Hu, Y. Chen, Z. Tang, H. Zhao, D. Wang, Quintuple-shelled SnO_2 hollow microspheres with superior light scattering for high-performance dye-sensitized solar cells, Adv. Mater. 26 (2014) 905–909.
[133] J.J. Teh, S.L. Ting, K.C. Leong, J. Li, P. Chen, Gallium-doped tin oxide nano-cuboids for improved dye sensitized solar cell, ACS Appl. Mater. Interfaces 5 (2013) 11377–11382.
[134] X. Dou, D. Sabba, N. Mathews, L.H. Wong, Y.M. Lam, S. Mhaisalkar, Hydrothermal synthesis of high electron mobility Zn-doped SnO_2 nanoflowers as photoanode material for efficient dye-sensitized solar cells, Chem. Mater. 23 (2011) 3938–3945.
[135] M.S. Mahmoud, M. Motlak, A.M. Barakat, Facile synthesis and characterization of two dimensional SnO_2-decorated graphene oxide as an effective counter electrode in the DSSC, Catalysts 9 (2019) 139.
[136] K. Tennakone, J. Bandara, P.K.M. Bandaranayake, G.R.A. Kumara, A. Konno, Enhanced efficiency of a dye-sensitized solar cell made from MgO-coated nanocrystalline SnO_2, Jap. J. Appl. Phys. 40 (2001) L732.
[137] S. Chappel, S.G. Chen, A. Zaban, TiO_2-coated nanoporous SnO_2 electrodes for dye-sensitized solar cells, Langmuir 18 (2002) 3336–3342.
[138] S. Khadtare, A.S. Bansode, V.L. Mathe, N.K. Shrestha, C. Bathula, S.H. Han, H.M. Pathan, Effect of oxygen plasma treatment on performance of ZnO based dye sensitized solar cells, J. Alloys Compd. 724 (2017) 348–352.
[139] A. Amala Rani, S. Ernest, Structural, morphological, optical and compositional characterization of spray deposited Ga doped ZnO thin film for dye-sensitized solar cell application, Superlattice. Microst. 75 (2014) 398–408.
[140] C. Kim, H. Choi, J.I. Kim, S. Lee, J. Kim, W. Lee, T. Hwang, S. Kang, T. Moon, B. Park, Improving scattering layer through mixture of nanoporous spheres and nanoparticles in ZnO-based dye-sensitized solar cells, Nanoscale Res. Lett. 9 (2014) 1–6.
[141] Q. Zhang, T.P. Chou, B. Russo, S.A. Jenekhe, G. Cao, Aggregation of ZnO nanocrystallites for high conversion efficiency in dye-sensitized solar cells, Angew. Chem. 120 (2008) 2436–2440.
[142] C.-Y. Lin, Y.-H. Lai, H.-W. Chen, J.-G. Chen, R. Vittal, K.-C. Ho, Highly efficient dye sensitized solar cell with a ZnO nanosheet-based photoanode, Energy Environ. Sci. 4 (2011) 3448–3455.
[143] F. Lenzmann, J. Krueger, S. Burnside, K. Brooks, M. Grätzel, D. Gal, S. Rühle, D. Cahen, Surface photovoltage spectroscopy of dye-sensitized solar cells with TiO_2, Nb_2O_5, and $SrTiO_3$ nanocrystalline photoanodes: indication for electron injection from higher excited dye states, J. Phys. Chem. B 105 (2001) 6347–6352.
[144] C. Nico, M.R.N. Soares, J. Rodrigues, M. Matos, R. Monteiro, M.P.F. Graça, et al., Sintered NbO powders for electronic device applications, J. Phys. Chem. C 115 (2011) 4879–4886.
[145] L.A. Reznichenko, V.V. Akhnazarova, L.A. Shilkina, O.N. Razumovskaya, S.I. Dudkina, Invar effect in n-Nb_2O_5, aht-Nb_2O_5, and LNb_2O_5, Crystallogr. Rep. 54 (2009) 483–491.
[146] C. Valencia-Balvín, S. Pérez-Walton, G.M. Dalpian, J.M. Osorio-Guillén, First-principles equation of state and phase stability of niobium pentoxide, Comput. Mater. Sci. 81 (2014) 133–140.
[147] C. Nico, T. Monteiro, M.P.F. Graça, Niobium oxides and niobates physical properties: review and prospects, Prog. Mater. Sci. 80 (2016) 1–37.
[148] S. Suresh, T.G. Deepak, C.S. Ni, C.N.O. Sreekala, M. Satyanarayana, A.S. Nair, V.P.P. M. Pillai, The role of crystallinity of the Nb_2O_5 blocking layer on the performance of dye-sensitized solar cells, New J. Chem. 40 (7) (2016) 6228–6237.

[149] S. Suresh, G.E. Unni, C. Ni, R.S. Sreedharan, R.R. Krishnan, M. Satyanarayana, M. Shanmugam, V.P.M. Pillai, Phase modification and morphological evolution in Nb_2O_5 thin films and its influence in dye-sensitized solar cells, Appl. Surf. Sci. 419 (2017) 720–732.
[150] X. Liu, R. Yuan, Y. Liu, S. Zhu, J. Lin, X. Chen, Niobium pentoxide nanotube powder for efficient dye-sensitized solar cells, New J. Chem. 40 (7) (2016) 6276–6280.
[151] M.-R. Ok, R. Ghosh, M.K. Brennaman, R. Lopez, T.J. Meyer, E.T. Samulski, Surface patterning of mesoporous niobium oxide films for solar energy conversion, ACS Appl. Mater. Interfaces 5 (8) (2013) 3469–3474.
[152] J. Xia, Naruhiko Masaki, Kejian Jiang, Shozo Yanagida, Fabrication and characterization of thin Nb_2O_5 blocking layers for ionic liquid-based dye-sensitized solar cells, J. Photochem. Photobiol. A Chem. 188 (2007) 120–127.
[153] K.W. Chen, L.S. Chen, C.M. Chen, Post-treatment of Nb_2O_5 compact layer in dye-sensitized solar cells for low-level lighting applications, J. Mater. Sci. Mater. Electron. 30 (2019) 15105–15115.
[154] A. Zaban, S. Chen, S. Chappel, B.A. Gregg, Bilayer nanoporous electrodes for dye sensitized solar cells, Chem. Commun. 22 (2000) 2231–2232.
[155] J. Xia, S. Yanagida, Strategy to improve the performance of dye-sensitized solar cells: interface engineering principle, Sol. Energy 85 (12) (2011) 3143–3159.
[156] K.-S. Ahn, M.-S. Kang, J.-K. Lee, B.-C. Shin, J.-W. Lee, Enhanced electron diffusion length of mesoporous TiO_2 film by using Nb_2O_5 energy barrier for dyesensitized solar cells, Appl. Phys. Lett. 89 (1) (2006) 013103.
[157] E. Barea, X. Xu, V. González-Pedro, T. Ripollés-Sanchis, F. Fabregat-Santiago, J. Bisquert, Origin of efficiency enhancement in Nb_2O_5 coated titanium dioxide nanorod based dye sensitized solar cells, Energy Environ. Sci. 4 (9) (2011) 3414–3419.
[158] Q. Wali, A. Fakharuddin, R. Jose, Tin oxide as a photoanode for dye-sensitised solar cells: current progress and future challenges, J. Power Sources 293 (2015) 1039–1052.
[159] P.V.K. Idriss Bedja, S. Hotchandani, Preparation and photoelectrochemical characterization of thin SnO_2 nanocrystalline semiconductor films and their sensitization with bis(2,2'-bipyridine) (2,2'-bipyridine-4,4'-dicarboxylic acid) ruthenium (11) complex idriss, J. Phys. Chem. B 98 (1994) 4133–4140.
[160] S. Das, V. Jayaraman, SnO_2: a comprehensive review on structures and gas sensors, Prog. Mater. Sci. 66 (2014) 112–255.
[161] L. Gracia, A. Beltrán, J. Andrés, Characterization of the high-pressure structures and phase transformations in SnO_2. A density functional theory study, J. Phys. Chem. B 111 (2007) 6479–6485.
[162] S. Zou, L. Song, J. Duan, L. Huang, W. Liu, H. Wu, W. Xiao, Spontaneous enhancement of power conversion efficiency in SnO_2-based dye-sensitized solar Cells, IEEE J. Photovoltaics 11 (2021) 674–678.
[163] M.H. Kim, Y.U. Kwon, Semiconducting divalent metal oxides as blocking layer material for SnO_2-based dye-sensitized solar cells, J. Phys. Chem. C 115 (2011) 23120–23125.
[164] Q. Zhang, C.S. Dandeneau, X. Zhou, G. Cao, ZnO nanostructures for dye-sensitized solar cells, Adv. Mater. 21 (2009) 4087–4108.
[165] J.A. Anta, E. Guillén, R. Tena-Zaera, ZnO-based dye-sensitized solar cells, J. Phys. Chem. C 116 (2012) 11413–11425.
[166] J.B. Baxter, A.M. Walker, K. van Ommering, E.S. Aydil, Synthesis and characterization of ZnO nanowires and their integration into dye-sensitized solar cells, Nanotechnology 17 (11) (2006) S304–S312.

[167] R. Vittala, K.-C. Ho, Zinc oxide based dye-sensitized solar cells: a review, Renew. Sust. Energ. Rev. 70 (2017) 920–935.
[168] R. Tao, T. Tomita, R.A. Wong, K. Waki, Electrochemical and structural analysis of Al doped ZnO nanorod arrays in dye-sensitized solar cells, J. Power Sources 214 (2012) 159–165.
[169] L. Luo, W. Tao, X. Hu, T. Xiao, B. Heng, W. Huang, et al., Mesoporous F-doped ZnO prism arrays with significantly enhanced photovoltaic performance for dyesensitized solar cells, J. Power Sources 196 (23) (2011) 10518–10525.
[170] A.S. Gonalves, M.S. Goes, F. Fabregat-Santiago, T. Moehl, M.R. Davolos, J. Bisquert, et al., Doping saturation in dye-sensitized solar cells based on ZnO:Ga nanostructured photoanodes, Electrochim. Acta 56 (18) (2011) 6503–6509.
[171] Q. Zhang, C.S. Dandeneau, S. Candelaria, D. Liu, B.B. Garcia, X. Zhou, et al., Effects of lithium ions on dye-sensitized ZnO aggregate solar cells, Chem. Mater. 22 (8) (2010) 2427–2443.
[172] L. Zhang, Y. Yang, R. Fan, H. Chen, R. Jia, Y. Wang, et al., The charge-transfer property and the performance of dye-sensitized solar cells of nitrogen doped zinc oxide, Mater. Sci. Eng. B-Solid 177 (12) (2012) 956–961.
[173] C.J. Raj, K. Prabakar, S.N. Karthick, K.V. Hemalatha, M.K. Son, H.J. Kim, et al., Banyan root structured Mg-doped ZnO photoanode dye-sensitized solar cells, J. Phys. Chem. C 117 (6) (2013) 2600–2607.
[174] R.K. Chava, M. Kang, Improving the photovoltaic conversion efficiency of ZnO based dye sensitized solar cells by indium doping, J. Alloys Compd. 692 (2017) 67–76.
[175] L. Li, Y. Shen, Q. Wu, M. Cao, F. Gu, J. Zhang, Influence of ion-doping on the photoelectric properties of mesoporous ZnO thin films, Proc. SPIE Int. Soc. Opt. Eng. 7995 (2011) 79950M.
[176] L. Lu, R. Li, T. Peng, K. Fan, K. Dai, Effects of rare earth ion modifications on the photoelectrochemical properties of ZnO-based dye-sensitized solar cells, Renew. Energy 36 (12) (2011) 3386–3393.
[177] H. Arakawa, Recent Advances in Research and Development for Dye-Sensitized Solar Cells II, CMC Publishing, 2007.
[178] S. Suhaimi, M. Mohamad Shahimin, Z.A. Alahmed, J. Chyský, A.H. Reshak, Materials for enhanced dye-sensitized solar cell performance: electrochemical application, Int. J. Electrochem. Sci. 10 (2015) 2859–2871.
[179] S. Yoon, S. Tak, Y. Jinsoo Kim, K.K. Jun, et al., Application of transparent dye-sensitized solar cells to building integrated photovoltaic systems, Build. Environ. 46 (2011) 1899–1904.
[180] J. Gong, J. Liang, K. Sumathy, Review on dye-sensitized solar cells (DSSCs): fundamental concepts and novel materials, Renew. Sust. Energ. Rev. 16 (2012) 5848–5860.
[181] P. Kubis, L. Lucera, F. Machui, G. Spyropoulos, J. Cordero, A. Frey, J. Kaschta, M.M. Voigt, G.J. Matt, E. Zeira, C.J. Brabec, High precision processing of flexible P3HT/PCBM modules with geometric fill factor over 95%, Org. Electron. 15 (10) (2014) 2256–2263.
[182] J. Goldstein, I. Yakupov, B. Breen, Development of large-area photovoltaic dye cells at 3GSolar, Sol. Energy Mater. Sol. Cells 94 (4) (2010) 638–641.
[183] H. Hug, M. Bader, P. Mair, T. Glatzel, Biophotovoltaics: natural pigments in dye-sensitized solar cells, Appl. Energy 115 (2014) 216–225.

Metal oxides in organic solar cells **19**

Swadesh Kumar Gupta[a], *Asmita Shah*[b], *and Dharmendra Pratap Singh*[b,c]
[a]Department of Physics, DBS PG College, Kanpur, Uttar Pradesh, India, [b]Unité de Dynamique et Structure des Matériaux Moléculaires (UDSMM), Université du Littoral Côte d'Opale, Calais, France, [c]Department of Industrial Engineering, EIL Côte d'Opale, Longuenesse, France

Chapter outline

1 Organic solar cell: Introduction and architecture

The endowed achievement and zest of life for us emanate from being open to our earth's intricacy, beauty, power, and splendid richness of resources, which have been deposited for millions of years beneath the earth's crust. Energy is vital to a country's socioeconomic development and energy security. According to the 2021 annual review of the U.S. Energy Information Administration (EIA), the primary sectors for energy usage are industrial, transportation, residential, and commercial, which account for approximately 32.4%, 26.1%, 22.4%, and 18.0% of total energy consumption, respectively [1]. In 2020, the European Energy Commission issued an almost identical report in which it was revealed that transportation, households, industry, services, agriculture and forestry, and others consumed 30.9%, 26.3%, 25.6%, 13.7%, 3.0%, and 0.5% of the total energy consumption, respectively. Renewable and finite resources (whose depletion is faster than generation) have been utilized to meet the energy demand. As per the consequence of the deviation in the ecological functioning of the earth due to hackneyed mining and the consumption of nonrenewable resources,

Metal Oxides for Next-Generation Optoelectronic, Photonic, and Photovoltaic Applications
https://doi.org/10.1016/B978-0-323-99143-8.00018-3

climate change, and global warming have become major challenges for sustainable human growth. Global warming is caused by CO_2 and greenhouse gases (GHGs) released by the combustion of fossil fuels, which results in climate change. Reducing our net carbon emissions to net zero to mitigate the effects of global warming is one of the 21st century's most critical problems. Effective measures and a rapid shift away from fossil fuels in the electricity, transportation, and thermal fields, along with the adaptation of sustainable technologies, will be vital to reduce 1.5°C global temperature target in the remaining time. Due to the disruptive potential of conventional fossil fuels, there is a critical demand for renewable energy and its storage devices. Nevertheless, simply replacing fossil fuels with renewables is insufficient; even renewable energy sources generate CO_2 and GHGs throughout their development and operations. The green transformation will require the widespread adoption of renewable energy technologies, most notably wind and solar. Solar energy has long been considered as a highly promising source of energy. Solar energy received by the earth's surface on an annual basis is approximately 120,000 TW (1 TW $= 10^{12}$ W) or approximately 6–7 times the world's current energy consumption. All solar energy converters must be constructed to efficiently harvest the sun's energy. This energy is released in the sun's core by a nuclear fusion reaction, and after being emitted as thermal radiation from the surface of the sun, it can reach the earth's surface in about 8 min. The solar spectrum is defined as the spectral energy flux density radiated by the sun and measured at the sun-earth distance. Spectra are frequently studied as the light intensity per unit wavelength, as shown in Fig. 1.

The spectrum outside of the earth's atmosphere is known as air mass (AM) 0, and its overall radiant energy flux density is around $1360.8 \pm 0.5\,W/m^2$ (i.e., solar constant) [2]. The air mass determines the distance the sun's rays travel via the atmosphere in multiples of the atmosphere's radial extension. Consequently, on the earth's surface, the sun's ray AM is one or greater, depending on the angle of

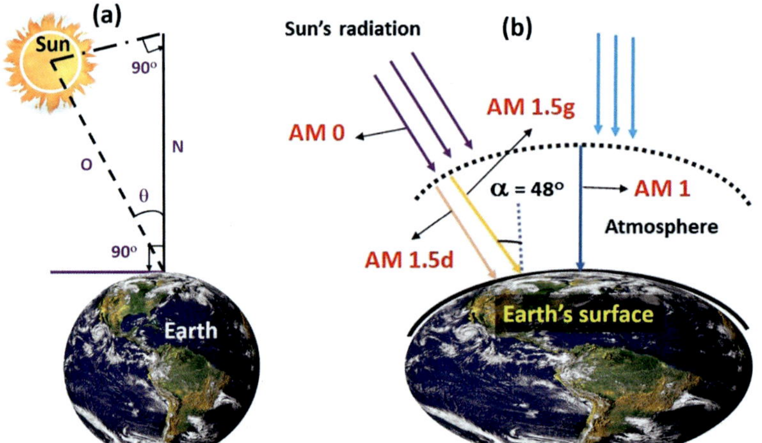

Fig. 1 (A) Position of the sun at solar zenith angle θ and (B) representation of air-mass (AM) spectra.

incidence. AM 1 represents the case of perpendicular incidence. From the illustration of the air mass denotation (Fig. 1), the AM can be approximated as 1/cos(α) with the incident angle determined vertically. The atmospheric absorption of different materials is the main factor for separating various AM spectra. AM 1.5 has been adopted as the reference spectrum for solar cell testing, with an incidence angle of 48 degrees. At noon on the equinox (i.e., March 21st and September 23rd), AM 1.5 is probably attained in southern Germany (for example, Münsingen). Because of the absorption in the earth's atmosphere, the total energy density is lower than the solar constant. Additionally, the AM spectrum can be represented in two ways for terrestrial use. AM 1.5g (global spectrum) is designed for flat modules with an integrated power density of 100 mW/cm^2. On the other hand, AM 1.5d (standard direct solar spectrum) constitutes the direct beam from the sun accompanied by the circumsolar component (in a disk 2.5 degrees around the sun). AM 1.5d shows an integrated power density of 90 mW/cm^2, which can be used for solar concentrator work. The value is proportional to the maximum photocurrent which could be harvested at a given wavelength. Therefore, solar energy can be regarded as one of the most promising alternatives for addressing human society's energy requirements [3,4].

In general, the photovoltaic effect refers to the transformation of light into electricity, which was first time realized in 1839 by Edmond Becquerel once he shed light onto an AgCl electrode present in an electrolytic solution. The word "*photovoltaic*" was coined in 1849. This word is a combination of a Greek word "*phōs*" which means light and "*volt*" which means the unit of electromotive force, which in turn is taken from the last name of the Italian physicist Alessandro Volta, the inventor of the electrochemical battery. In 1876, William Grylls Adams and his student Richard Evans Day created the first solid-state photovoltaic device using sintered selenium and platinum electrode [5]. Afterward, Charles Fritts demonstrated the first solid-state solar cell in 1883, wherein he deposited a thin coating of gold onto a selenium (Se) semiconductor to build a junction and obtained a power conversion efficiency (PCE) of roughly 1% [6]. The first US patent on thermopile solar cells was credited to Edward Weston in 1888. After countless efforts, the demonstration of modern solar cells was presented by Russell Ohl in 1946 [7].

Despite an excellent room temperature power conversion efficiency of roughly 33% for a single-junction c-Si inorganic solar cell, researchers are also paying attention to organic material-based solar cells. Organic solar cells (OSCs) are the simplest and least expensive source of energy, emphasizing their potential to be the least carbon-intensive energy generation technology. Additionally, OSCs demonstrate captivating properties, including lightweight, outstanding mechanical flexibility with ease of printing, and roll-to-roll (R2R) processing mechanisms without high-temperature processing. Due to vibrant colors and transparency, numerous trendy designs could be printed on the dwellings' walls, windows, and roofs. Furthermore, OSCs are well suited for constructing solar power plants in space, as their low weight enables them to be easily launched into orbit. Pochettino discovered photoconductivity in anthracene in 1906 [8], and Tang et al. created the first heterojunction photovoltaic device in 1986 [9]. Yu et al. [10] reported bulk polymer/C60 systems in 1994, while Schmidt et al. demonstrated perylene-based PV systems in 2001 [11]. Various attempts were made from 2007 to 2018 for improving the power conversion

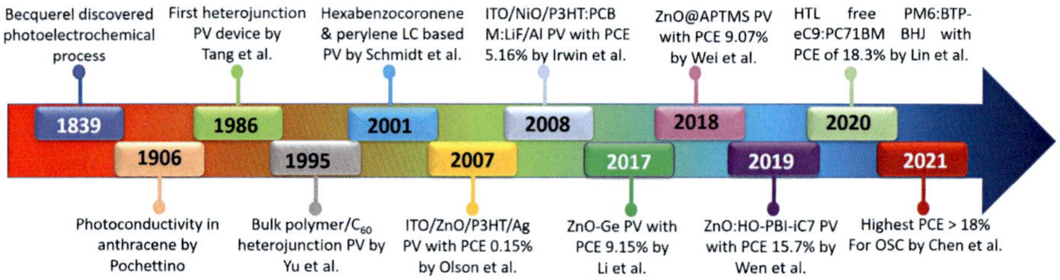

Fig. 2 Timeline evolution of organic solar cells (OSCs).

efficiency (PCE) from 0.15% to 9.29% [12–16]. For the past decade, OSCs have been breaking efficiency records, achieving up to 16%–18% efficiency because of the non-fullerene acceptors (NFAs) [17]. Approximately, 1% per year of consistent efficiency gain has been realized in the last two decades without any saturation. The steady improvement in efficiency is encouraging for researchers, as it reached up to 18.3% in 2020–21 [18–21], with a 20% increase expected by 2025. The timeline progress for OSCs is represented in Fig. 2.

In comparison to Si-based PV systems, OSCs have emerged as the least expensive ones in which carbon-based semiconducting material is used as an active layer. Organic semiconductors are easily tunable and could be prepared using abundant and nontoxic raw materials. Large-scale solution processing at low temperatures, vacuum coating, compatibility with flexible substrates, and low material consumption (1 g of organic material/m^2) made them excellent for developing a cost-effective socioeconomic energy source.

The most commonly used OSCs architectures along with their charge transport are illustrated in Fig. 3. A conventional OSC device is composed of indium tin oxide (ITO)/poly(3,4-ethylenedioxythiophene):(styrene sulfonate) (PEDOT:PSS)/active organic layer/metallic electrode. Many researchers have suggested opting the solution-processed p-i-n architectures consisting of transparent electron and hole injection layers which enable for different layer sequencing to be introduced. Another option is an inverted layer architecture, in which the anode and cathode positions are reversed. At present, the inverted structures have been extensively used in OSCs because of their technical superiority in terms of better air stability. The use of PEDOT:PSS and easily oxidized low work-function metal are repudiated in the inverted structure and electron transport layer (ETL) plays a vital role in determining the stability and device performance because it enables the formation of ohmic contact between the active organic layer and the cathode causing an improvement in electron collecting efficiency. Usually, two kinds of charge transport materials are employed in OSCs, viz. cathode interfacial layer (CIL, also abbreviated as ETL) materials, and anode interfacial layer (AIL, also abbreviated as HTL) materials for hole collection. In the current state of OSCs, AIL materials are trailing behind ETL materials in terms of advancement. The inorganic and organic materials could be used as CILs. The inorganic materials like low-work-function metals (Ca, Ba, etc.) and transition metal oxides (e.g., ZnO, TiO_x, Al_2O_3) are often used as CILs. Organic materials such as

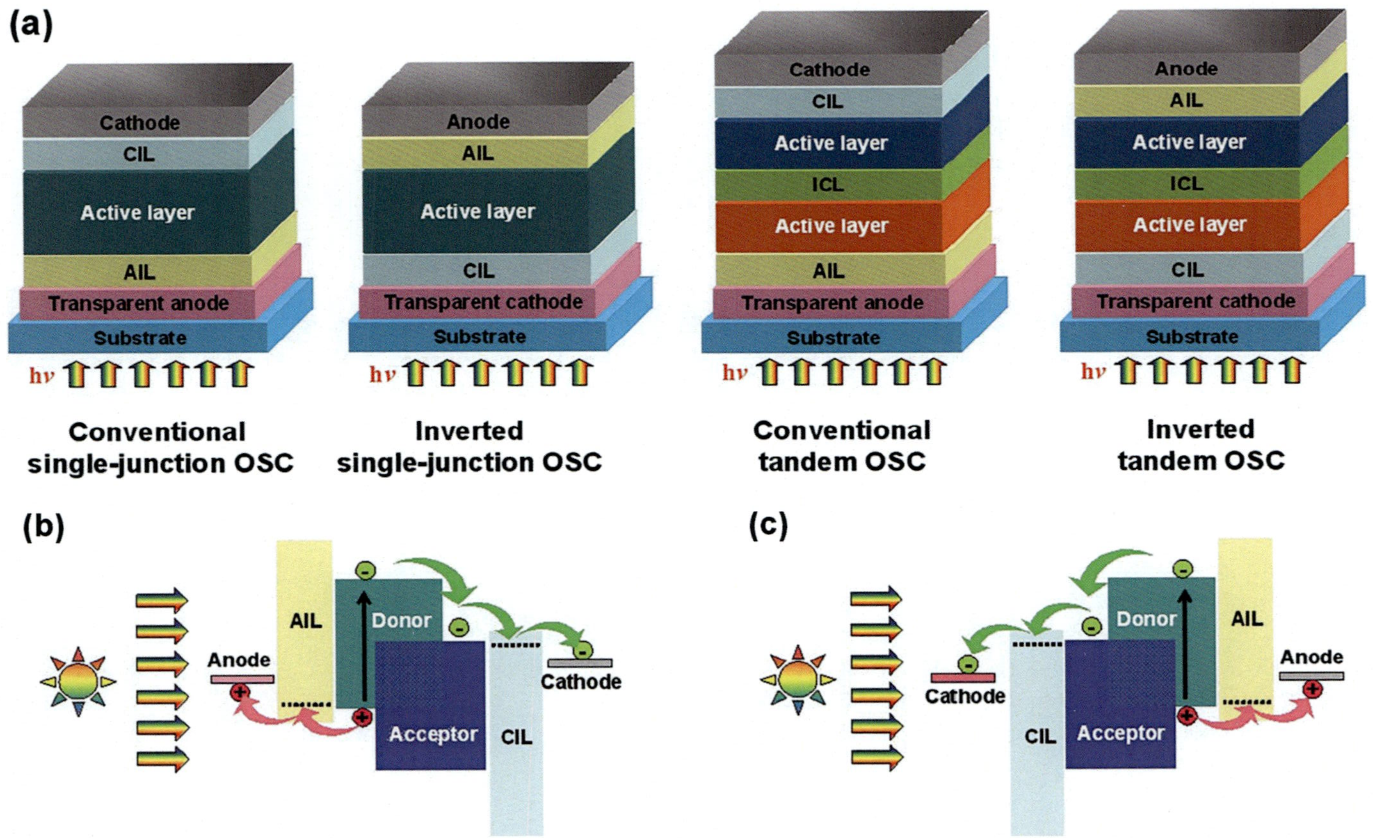

Fig. 3 (A) Representation of architectures of conventional and inverted single-junction/tandem OSCs. *AIL*, anode interface layer; *CIL*, cathode interface layer; *ICL*, interconnecting layer. Illustration of the energy level diagrams and the main charge-transporting processes in (B) conventional and (C) inverted OSCs.

Reproduced with permission from Z. Yin, J. Wei, Q. Zheng, Interfacial materials for organic solar cells: recent advances and perspectives, Adv. Sci. 3 (8) (2016) 1500362, https://doi.org/10.1002/advs.201500362.

small molecule-based electrolytes or water/alcohol conjugated and nonconjugated polymers have exhibited their efficacy as CILs for increasing electron collection efficiency [22]. The use of inorganic materials in the preparation of CIL needs high-temperature treatment that causes the formation of surface defects and poor contact between the active layer and the cathode. These two limitations produce an increase in series resistance (R_s) and charge carrier recombination that reduces the PCE of OSCs. The AILs are used to serve two purposes: (1) to improve the hole transport efficiency and hole collection, and (2) to improve the contact between the active layers and the anode. The hole transport efficiency and collection can be improved by three ways: first, the work function (WF) of anode can be efficiently increased by utilizing AIL, which supports in the construction of a built-in electric field in OSC when combined with a low-WF cathode. The built-in electric field acts as a primary driving force for transferring holes to the anode, where they are collected; second, the WF of the anode can be increased by modifying the AILs to be greater than the donor's E_{ICT+} (positive integer charge-transfer state), where, the Fermi-level pinning to the E_{ICT+} level occurs, lowering the hole collection barrier at the anode interface [23]; and third, AILs must effectively block electrons at the interface to decrease the charge recombination. As far as the improvement in physical contact is concerned, the modification of anodes' hydrophilic surfaces may increase the wettability and film quality of the active layers that are deposited on the anode, Besides, AIL promotes the smoothness of the anode surface leading to passivate surface defects and pinholes that minimize the leakage current in OSCs. AIL should exhibit certain characteristics such as outstanding hole transportation and great electrical conductivity, a suitable WF value, higher optical transparency, neutral pH to avoid corrosion at ITO/active layer interface, orthogonal solvent processability, etc. Only a few materials like poly(3,4-ethylenedioxythiophene):(styrenesulfonate) (PEDOT:PSS) and MoO_3 have been extensively used as AILs in OSCs. In general, conjugated polymers are better to serve as AILs in OSCs due to easy and smooth fabrication characteristics [24]. More detail about metal oxides-based ET/HT layers is provided in the next section.

The creation of useful charge transport layers between the organic active layer and the electrodes is one of the most important steps in the fabrication of highly effective and stable OSC devices. This problem led to the evolution of tandem solar cells (TSCs) which are the bulk-heterojunction OSCs that pave the way toward high performance by controlling the two major losses viz. the thermalization of hot charge carriers and subbandgap transmission [25]. In tandem OSCs, an interconnecting layer (ICL) is created in between two active layers of organic material and the highest PCE of 19.6% has been reported for TSCs, which is the maximum value achieved so for [18].

Due to the large bandgap, metal oxides such as ZnO_x, SnO_x, TiO_x, V_2O_x, NiO_x, MoO_x, WO_x, etc. act as barrier for holes or electrons for providing extremely selective contacts for extracting other types of carriers. The usage of these metal oxides in OSCs improves the PCE [i.e., $\eta = P_{max}/P_{in} = (V_{oc} \times J_{sc} \times FF)/P_{in}$, where V_{oc} is the open-circuit voltage, J_{sc} is short-circuit current density, and FF is fill factor] by controlling the charge recombination and surface imperfections. Furthermore, the state-of-the-art of OSCs in combination with metal oxides has been thoroughly covered in the following sections of this chapter.

2 Types of active organic layers

Conventionally, OSC device comprises layer of single organic material between a transparent electrode and metallic reflecting electrode (Fig. 4). Due to the poor exciton dissociation, these devices typically have very low efficiency because light absorption, exciton separation, and charge transport all take place in a single organic layer. To liberate free charges from excitons, an additional exciton dissociation energy (binding energy) is needed in OSCs. For different organic semiconductors, the dissociation energy ranges between 0.2 and 1.0eV. Therefore, a second layer of organic material is needed to add with the original organic layer for fast dissociation of excitons at the boundary of two organic semiconductors. This forms a heterojunction consisting of an electron donor material and an electron acceptor material, with the offset matching of their HOMO and LUMO energy levels (Fig. 4B). The appearance of holes in a material with a low ionization potential (donor) and electrons in a material with a high electron affinity results from the charge transfer process that takes place at the donor-acceptor interface (acceptor). These carriers (electrons and holes) are still associated by the Coulomb force but can be separated by an internal electric field of the OSC, which is created due the difference in the WFs of the two different electrodes, i.e., anode and cathode. Holes travel through the donor material to the electrode with a high WF (cathode) and electrons through the acceptor layer to the electrode with a low WF (anode). In organic semiconductors, the characteristic distance, i.e., the diffusion length (l_d), traveled by the exciton during its lifetime is about 10nm due to their low mobility and short lifetime [26]. Thus, it is evident that only excitons appearing at distances $\approx l_d$ can move toward the interface and ensure the generation of charge carriers. Practically, in the binary structure of OSCs, about 0.01% absorbed photons can participate to the photocurrent. On the basis of structure of these two organic materials; OSCs can be classified in two main groups. The first group involves binary structure of photoactive layer where the acceptor and donor type are deposited in separate layer. This type of structure was first demonstrated by C. Tang in 1986 [27]. He introduced a second layer of organic material as acceptor which efficiently enhances the dissociation of exciton. It was demonstrated that in a two-layer donor-acceptor system, metal phthalocyanine/perylene compound enhances the dissociation of exciton and hence photovoltaic effect occurs with a rather high efficiency. However, the

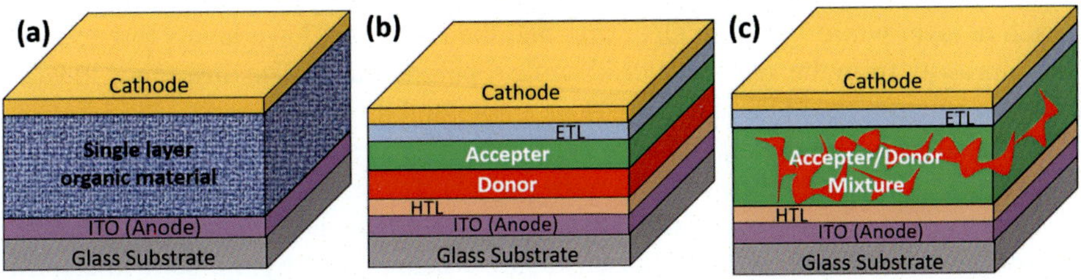

Fig. 4 Illustration of active organic layers in (A) single layer and binary structure of (B) planar and (C) bulk heterojunction photoactive layers in typical OSC. ETL and HTL represent the electron and hole transport layers, respectively.

improvement in the device efficiency was not sufficient due to the small diffusion length of exciton of about 10nm. Thus, the exciton generated beyond this distance from the heterojunction recombines and decays without dissociation of charges.

In another group of OSCs, i.e., bulk heterojunction, the donor and acceptor are blended to form a single photoactive layer (Fig. 4C). The OSCs based on bulk heterojunction work on the principle of the phase separation at the nanoscale in polymer materials. In such OSCs, the donor-acceptor interface is dispersed over entire volume of active layer, offering higher number of dissociation center for excitons around them. The first volumetric heterojunction was obtained from solutions in 1995 [28,29]. Afterward, numerous materials have been reported for preparing bulk heterojunction [30–33]. In recent years, poly(3-hexyl-thiophene) (P3HT) has extensively been used as donor organic material for OSCs. P3HT has a bandgap energy (E_g) of $\approx$1.85 eV and absorbs light with a wavelength of less than 675 nm. On the other hand, for the acceptor material, fullerene composites are most popular choice. [6,6]-phenyl-$C_{61/71}$-butyric acid methyl ester ($PC_{60}BM$ or $PC_{70}BM$) are common fullerene derivatives to use as acceptor material to mix with P3HT. In 2002, the first promising results for P3HT:$PC_{60}BM$ (in 1:3 weight ratio) solar cells were published in which the largest short-circuit current density of 8.7 mA/cm^2 has been reported [34]. However, further work on donor polymer material was needed to investigate to bring the energy gap between HOMO and LUMO below the 2 eV to cover a longer wavelength range of the sunlight, i.e., higher than 675 nm, to increase the number of exciton generation and efficiency of OSCs.

3 Metal oxide in OSCs: Role as the hole transport layer (HTL) and the electron transport layer (ETL)

The inclusion of transport layers in OSCs is crucial because the active donor-acceptor blend, which is termed as essential layer, has a low charge carrier concentration. In this situation, OSC device is constructed like a PIN diode in which the donor-acceptor material is sandwiched between the HTL and the ETL. The HTL and ETL must be p- and n-type materials, respectively, with a high conductivity value. The energy levels of the HTL and ETL must be chosen precisely according to the active organic layer in order to control the carrier separation (or to avoid charge carrier recombination) and to realize optimal PCE (Fig. 5). The HTL must have energy levels that pull holes from the active layer while repelling electrons, and the ETL must have levels that transfer electrons away from the active material while repelling holes. The difference in work functions between the HTL and the ETL is a potential difference across the junction. The ETL has a higher potential than the HTL, which generates an internal electric field across the active layer that is directed from the ETL to the HTL. When a photon with enough energy is incident on OSC, it creates an exciton (a quantum bound state of the electron-hole pair which is thermally unstable), which is split by the built-in electric field, forcing holes, and electrons to travel toward the ETL and HTL, respectively. An OSC device with no HTL and ETL could suffer from the charge carrier recombination, energy loss at layer interfaces, and the pinning effect, resulting in poor performance.

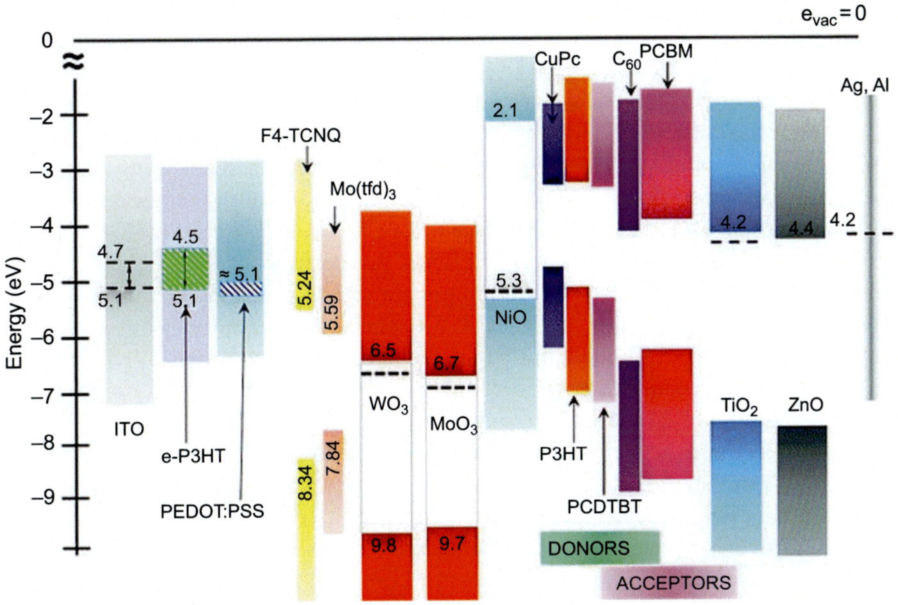

Fig. 5 Illustration of the energy gaps and frontier orbital energies for some of common OSCs. Reproduced with permission from E.L. Ratcliff, B. Zacher, N.R. Armstrong, Selective interlayers and contacts in organic photovoltaic cells, J. Phys. Chem. Lett. 2 (11) (2011) 1337–1350, https://doi.org/10.1021/jz2002259.

Most often, metal oxides like ZnO, TiO_2, etc., are used as ETL [35]. In OSCs, ZnO is most frequently used as electron-transporting layer (ETL). It's Fermi level aligns well with the LUMO of the cathode layer, i.e., electron acceptors [36]. Furthermore, it works as an effective hole blocking material due to its high ionization potential and high energy bandgap, thereby increasing the shunt resistance of OSC [37,38]. ZnO serves as a low band pass filter for the photoactive layer because it is transparent to visible light and gets absorbed in the UV spectrum. Besides, ZnO is also used to create an interconnecting unit in tandem solar cells. In a high efficient tandem cell, ZnO NPs are utilized as the bottom cathode interlayer, P3HT:indene-C60 bisadduct (ICBA) [39] as the front cell, ZnO NPs/PEDOT:PSS as an interconnecting unit, and a low bandgap active layer as the back cell. Such devices have shown PCEs varying from 8% to 11% in which the V_{oc} is almost the sum of the two single junction cells, demonstrating a negligible photovoltage loss through the interconnecting unit [40]. Use of ZnO nanowires across the ZnO nanoparticles has shown an increased electron lifetime, reduced recombination, and enhanced charge collection at the suitable electrodes. It is observed that the insertion of electrospun ZnO nanowires could efficiently increase the carrier lifetime by twofold [41].

On the other side, TiO_2 is an n-type semiconductor with wide bandgap ($E_g = 3.2\,eV$) with Ti 3d band possessed by its conduction band minima and O 2p states by its valence band maxima [42]. TiO_2 is less transparent to visible light in comparison to ZnO layer but acts as better low pass filter for active layer due to higher UV absorption. In OSCs, TiO_2 films are in amorphous phase which contribute to a higher density of defect states. TiO_2 exhibits a strong dependence on UV light. [43]. Under constant illumination and typical operating conditions, the UV-activated TiO_2 films deteriorate much more quickly. Charge extraction is hampered by the UV light-generated photodoping of TiO_2, which results in critical band bending at the sites where oxygen is chemisorbed [44]. UV photodoping is one of the major issues to be resolved to utilize TiO_2 efficiently in OSCs as an ETL layer. The method for synthesis and coating on the cathode layer also affects the performance of the ETL layer. Annealing of these layers also plays an important role to passivate the surface defect states for better performance of OSCs. UV-ozone (UVO) treatment has also been found suitable to passivate these surface defects [36]. However, the UVO treatment only works with photoactive polymers with a deep HOMO energy with oxidization threshold energy of 5.27 eV and above [45].

HTLs are commonly prepared by Molybdenum oxide (MoO_3), Tungsten trioxide (WO_3), and Vanadium oxide (V_2O_5) [35]. MoO_3 has attracted noteworthy consideration for enhancing the device performance and stability of OSCs. By inhibiting exciton quenching and resistance at the photoactive layer/anode interface, MoO_3 HTL reduces charge recombination [46]. MoO_3 is generally used as hole injection material and was initially considered as a p-type semiconductor. Besides, the ultraviolet photoelectron spectroscopy (UPS) studies revealed it as n-type semiconductor and charge transport takes place via the Fermi level being pinned with the valence band of the polymers. MoO_3 is very sensitive to oxygen and moisture; even the residues of oxygen in the nitrogen-filled glove box during device fabrication showed degrading effects on its electronic levels, imposing severe inadequacies in the device stability [47]. Typically, the narrow defect states in the bandgap of a nanostructured MoO_3 layer that result from oxygen vacancies facilitates hole transport [48,49]. These oxygen vacancies act as n-type dopants and lead to Fermi-level pinning at the photoactive layer-MoO_3 interface [50]. The use of MoO_3 in place of PEDOT:PSS as the HTL has shown comparable performance with an enhanced stability of OSC devices. In both conventional and inverted devices, PEDOT:PSS can be replaced with high-work-function n-type metal oxides (MOs). According to reports, P3HT:PCBM cells treated with solution-processed MoO_3 showed a PCE of 3.1%. Despite low-temperature processing and acceptable PCE values, aggregation of s-MoO_x is one of the main obstacles preventing its use in large-scale processing. However, the s-MoO_x films show lower work function values as compared to e-MoO_3 and thus affects the quality of ohmic contacts in the device.

Tungsten oxide (WO_3) is also n-type MO with high work function value alike MoO_3. The electronic functionality of WO_3 is determined by its stoichiometry, crystalline structure, and processing/deposition conditions. Thermally evaporated WO_3 amorphous films are usually oxygen-deficient, thereby possessing a large amount of gap states. Oxygen deficiency also promotes the n-type semiconducting behavior in WO_3. The optical bandgap of thermally evaporated WO_3 films is $\approx$3.2–3.4 eV. Oxygen exposure increases this gap up to 4.7 eV. As HTL, its performance is significantly acceptable as PEDOT: PSS, with the devices exhibiting open circuit voltage (V_{oc}) of 0.6 V and fill factor (FF) of 60% [51,52]. Studies showed that the solution-processed WO_3 with a larger work function has improved the efficiency up to 3.4% with $J_{sc} \approx 8.6\,mA/cm^2$; $V_{oc} \approx 0.6\,V$; FF ≈ 0.6 for a P3HT:PCBM cell. Devices showed improved lifetime/stability by maintaining 90% of the initial value after being exposed to ambient conditions for close to 200 h without any encapsulation [53,54]. These features make s-WO_3 a viable contender than other solution-processed high-work-function oxides for large-scale coating.

V_2O_5 is another n-type semiconducting oxide that has been used as the HTL. It shows a bandgap of ~2.8 eV as assessed by UPS and inverse photoemission spectroscopy (IPES) studies [55], revealing that its absorption band partially overlaps with that of the $PC_{71}BM$. The band structure of thermally evaporated V_2O_5 is extremely sensitive to the environment, just like MoO_3 and WO_3. The P3HT:PCBM devices fabricated using e-V_2O_5 (~10^{-6} Torr) as a HTL have shown an optimum PCE of ~3%. Under air exposure, the work function of e-V_2O_5 further decreases to 5.3 eV along with a substantial decrease of electron affinity and growth of defects. In contrast to MoO_3, research on V_2O_5 is immature. Thus, more investigations are required for better understanding of V_2O_5 as HTL in OSCs.

4 Atomic layer deposition (ALD) of metal oxides for OSCs

The thin-film deposition process for OSCs is one of the vital factors that determines device performance and stability. ALD stands for atomic layer deposition, a chemical vapor phase deposition process that enables for the precise fabrication of thin films of various materials. It is a well-known adopted technique for semiconducting material and energy conversion devices [56]. The ALD method is more suitable and superior to its chemical and vapor counterparts because it allows for perfect control of the material's thickness, composition, and stability. Typically, the ALD process involves successively alternating pulses of gaseous chemical precursors reacting with the substrate. An ALD system permits a certain number of precursor lines and is useful to deposit high-quality pinhole-free films of uniform thickness on various substrates with an extremely high aspect ratio [56]. As previously explained in the architecture of OSCs, an organic absorber layer (active layer) is sandwiched between the CIL and AIL. The role of active layer is to facilitate the generation of the excitons (solely by active layer or by donor and acceptor materials). Subsequently, electrons or holes are transported to electron and hole-selective contacts, respectively. P3HT:PCBM is usually spin-coated on the substrate which acts as an absorber layer. ZnO and TiO_2 are used as selective electron contact; whereas, PEDOT:PSS, WO_x, NiO, MoO_x, V_2O_5 are preferred to use as selective hole contact. Using ALD technique, one can obtain the thickness of 1–1000 Å with the rates of <0.1–1 Å/s. Fig. 6A shows the vacuum thermal evaporation (VTE) approach, which includes heating an organic material in vacuum. The substrate is set at several centimeters away from the source, allowing evaporated material to fall straight onto it. This approach can be used to deposit multiple layers of various materials without causing chemical reactions between them. However, issues with film thickness homogeneity and consistent doping over wide areas of substrates,

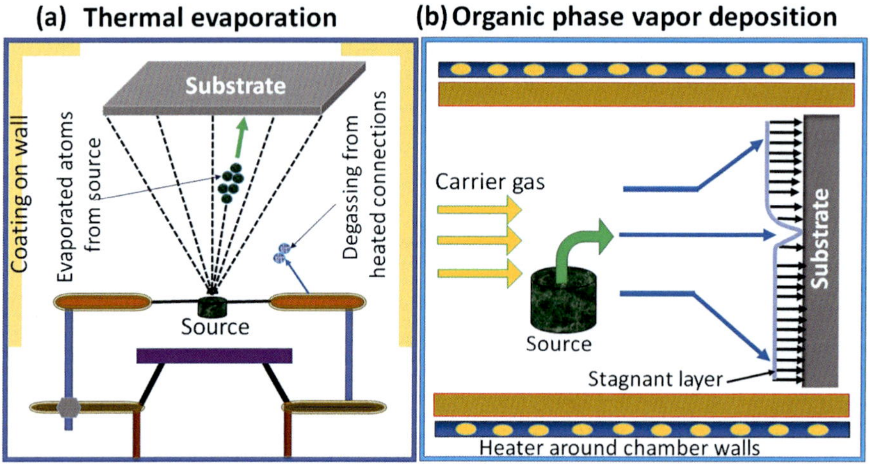

Fig. 6 Schematics of (A) thermal evaporation and (B) organic phase vapor deposition systems.

as well as contamination for later depositing materials, can occur. Fig. 6B shows an organic vapor phase deposition (OVPD) system that allows contamination-free control of the film's structure and shape than vacuum thermal evaporation. The organic material is evaporated over a substrate in the presence of inert carrier gas, such as N_2 or Ar. The shape of the resultant film can be altered by adjusting the gas flow rate and source temperature. One can grow uniform films by reducing the pressure of the carrier gas, which results in an increase in velocity and mean free path of the gas, leading to a decrease in layer thickness.

Transition metal oxides (TMOs)-based thin films are nowadays employed in OSCs which are listed here:

(a) ZnO and TiO_2 as an electron transporting layer:

ZnO and TiO_2 are employed as electron transport layers in OSCs. The Ag-capped ZnO thin film of thickness 36 nm deposited at 80°C for the P3HT:PCBM device configuration had a PCE of 4.18% [57]. During the ZnO deposition process, lowering the growth temperature or increase in the purging time may result in a lower fill factor (FF) due to a rise in series resistance, i.e., $R_s = (V_{oc}/J_{sc})(1 - FF/pFF)$, where, pFF is pseudo fill factor. In another study, Stakhira et al. [58] deposited an ultrathin ZnO layer onto nickel phthalocyanine (NiPc) without using an anode layer. This process has increased the quality of the NiPc/Al interface resulting in an improvement in J_{sc} and subsequently an increase of 2.5 fold in PCE. Frankenstein et al. [59] recently deposited ZnO at 60°C using ALD and surface treatment, allowing precursor diffusion into the substrate for the device structure of MoO_x (anode layer)/PCDTBT:PCBM (active layer). The PCE obtained using this method was 3.5%. The use of an ultrathin ALD ZnO layer with a thickness of 2 nm improves the stability and performance of OSC. The P3HT:PCBM layer can be replaced by TiO_2 due to the similar photoactive characteristics. The use of TiO_2 for an inverted OSC with P3HT:PCBM and PEDOT:PSS has been demonstrated by Waldauf et al. [60] and Kuwabara et al. [61] The hydrolysis and sol-gel synthesis processes also played a role in determining the PCE, which were found to be 3.1% and 2.47% for the abovementioned synthesis processes, respectively. The sol-gel process lowers the PCE to 2.47% that could be due to film quality and surface defects. The air exposure to OSCs can affect the FF significantly with increasing oxygen content.

(b) NiO as a hole conduction layer

Park et al. [62] deposited a 5-nm thick NiO film on P3HT:PCBM, and the OSC device performed similarly to a standard PEDOT:PSS anode layer with improved stability. Hsu et al. [63] used a thermal ALD process with nickel bis(*N*,*N*′-di-tert-butylacetamidinate) and water to deposit NiO onto P3HT:PCBM at 200°C with Ca and Al as cathode and electrode, respectively. This NiO film with a thickness of 4 nm had a PCE of 3.38%, which is similar to the PCE of PEDOT:PSS.

(c) V_2O_5 and WO_3 as a buffer layer

The work functions of V_2O_5 and WO_3 are around 5.6 and 6.7 eV, respectively. V_2O_5 has already demonstrated its utility as a corrosion-resistant alternative to PEDOT:PSS [64]. Shrotriya et al. [65] introduced a 3-nm thick layer of V_2O_5 into OSCs and they noticed a decrease in FF due to the increase of R_s, but the PCE remained constant at 3.1%, as compared to the PCE of 3.18% for PEDOT:PSS. Furthermore, using V_2O_5/P3HT:PCBM in combination with ZnO nanorods as a cathode layer results an increase in PCE to 3.9% [66]. It is observed that in the presence of WO_x, with an optimized thickness of 4 nm, J_{sc} and V_{oc} were found to be increased from 7.80 to 10.1 mA/cm^2 and 0.34 to 0.42 V, respectively [67]. In this situation, authors have recorded an increased PCE from 2.64% to 3.33%. On increasing the thickness of WO_x, it leads to a decrease in PCE due to an increase in R_s. In comparison to PEDOT:PSS, some remarkable results have been obtained for the spin-coated WO_x films which render superior hole conductivity attributing via the incorporation of oxygen vacancies. This situation disclosed a superior PCE of 7.5% when compared with annealed WO_x films with PCE of 6.19% and PEDOT:PSS having PCE of 5.91% [68]. The annealed WO_x films in OSCs face another challenge of sensitivity to heat, however Kawano et al. stated that the electrical properties could be enhanced through thermal annealing up to 140°C [69].

(d) MoO_x as a hole-selective contact

Oxygen vacancies are sometimes critical for OSCs. Under the thermal evaporation growth, MoO_x is usually oxygen-deficient with a work function of 6.8 eV which makes it a better hole-selective contact. MoO_x and V_2O_5 could be used as an alternative for PEDOT:PSS. It has been observed that the use of MoO_x enhances the PCE from 3.18% to 3.33% [65]. Henceforth, it has been well accepted that the incorporation of these TMOs leads to a decrement in J_{sc} due to an increase in R_s. The role and performance of different metal oxides in OSCs have been tabulated in Table 1.

Table 1 Performance of OSCs with ALD-assisted metal-oxide layers.

Material	Heterojunction	J_{sc} (mA/cm^2)	V_{oc} (V)	FF (%)	η (%)	Ref.
Cathode						
ZnO	P3HT:PCBM	11.90	0.59	60	4.18	[57]
TiO_2	P3HT:PCBM	10.80	0.56	59	3.60	[60]
Anode						
WO_x	CuPc:C60	10.10	0.42	36	3.33	[67]
	P3HT:PCBM	17.08	0.70	63.01	7.50	[68]
MoO_x	P3HT:PCBM	16.75	0.77	61.57	7.94	[70]
V_2O_5	P3HT:PCBM	10.10	0.57	67	3.90	[64]
NiO	pDTG-TPD: PCBM	13.90	0.82	68.40	7.82	[71]

5 Characteristics of metal oxide in OSCs

Metal oxides in OSCs are used as electrode layer, cathode buffer layer, and are also used in hybrid OSCs as an acceptor for electron transport. On the basis of their application in different parts of OSCs, the requirement of MOs properties also differs. Some of the major characteristics to identify the required MOs in OSCs are as follows:

(a) Crystallinity

Crystallinities are important to determine the phase and structure of synthesized MOs. To estimate the crystal structure, XRD study of crystalline material is performed using a XRD profilometer. The distinct diffraction peaks in XRD pattern provide the knowledge of MO crystalline phase. The phase of MO provides the preliminary idea about the material properties like conductivity, absorbance, luminescence, etc., which are prerequisite before performing further characterization. The observed phase and structure of MOs also confirm the correct synthesis process.

(b) Conductivity and mobility

In a solar cell, electrical conductivity (σ) and mobility (μ) are two main parameters to determine the charge transportation ability. Electrical conductivity can be determined by the Van der Pauw method, while the mobility can be calculated from the product of electrical conductivity and charge density (n_c). Charge density can be obtained from the Hall-effect measurement. The relation between these parameters can be represented as: $\sigma = qn_c\mu$, where q is the elementary charge. In numerous works, it has been observed that the mobility of MOs has less effect on the charge carrier transfer in organic solar cells. It mostly depends on the mobility of organic material. However, the conductivity of MOs can influence the charge transportation ability due to different recombination rates. The PCE in donor-acceptor blends-based OSCs is usually controlled by the polaron pair dissociation and bimolecular pair recombination, which are considerably dependent on charge carrier mobility. Normally, the dissociation increases with faster charge transport. By enhancing the hole mobility from 10^{-4} cm^2/V s to 10^{-2} cm^2/V s, the PCE of polymer:fullerene-based OSCs can be improved by 20%.

(c) Surface morphology

Surface morphology of MOs layer in OSCs provides the information of surface roughness, grain size of nanocrystallites, and size distribution of these crystallites with increasing thickness of MOs. Atomic force microscopy (AFM) is the best tool for characterizing the surface morphology of these MOs layer. AFM is a type of scanning probe microscopy (SPM) which has resolutions that are more than a thousand times better than the optical diffraction limit. MO layer with small grain size of crystallites shows smooth surface with small interface area between electron conducting layer and organic layer. However, large grain size with increasing roughness indicates wide distribution of grain size with larger surface area available to provide a large interface between MOs and organic material. It enhances the charge separation that boosts the overall charge transportation ability of the device.

(d) Absorption profile

Absorption spectra of MO layer is important to study to determine the transparency range of MO layer. An ideal MO layer should be completely transparent in visible and IR region but should be entirely opaque in the UV region. To ensure that the organic photoactive layer in the device receives the proper amount of light for a high photogenerated excitation, a high transparency of the MO layer in the visible-IR region is essential. The blockage of UV light inhibits from the degradation of OSCs and improves the illumination stability (discussed later in the chapter) of device. However, in real practice no MO can provide complete transparency with full blockage to UV light. The two most popular MOs, ZnO and TiO_2, show limited blockage and limited transmittance, respectively. ZnO has high transparency in visible range, however, provides poor absorption of UV light. On the other hand, the TiO_2 provides good absorption of UV light but the transparency in visible range reduces with growing thickness.

6 Metal oxides (e.g., ZnO, TiO_2, MoO_x, NiO, and SnO_x)-based OSCs

The high conductivity, photo-corrosion resistance, electron mobility, and low cost of ZnO have established it as a preferred material for solar cells [72]. In 2017, Lin et al. have fabricated inverted OSCs with an aqueous solution-processed zinc oxide (ZnO) layer that acts as an electron transport layer [73]. For this OSC device, the V_{oc} and J_{sc} were recorded to be 0.54 V and 8.82 mA/cm^2, respectively. Besides, they introduced a 1.2-nm thin layer of Al and an enhancement in the V_{oc} and J_{sc} were noticed with their values of 0.60 V and 9.73 mA/cm^2, respectively. In this way, the authors have effectively increased the OSC performance from 2.7% to 3.9% for the poly (3-hexylthiophene):phenyl-C_{61}-butyric acid methyl ester (P3HT:PC_{61}BM) blend.

A high PCE of 8% was obtained for the device by incorporating the Poly({4,8-bis[(2-ethylhexyl)oxy]benzo[1,2-*b*:4,5-*b*′]dithiophene-2,6-diyl}{3-fluoro-2-[(2-ethylhexy) carbonyl]thieno[3,4-*b*]thiophenediyl}):[6,6]-phenyl-C_{71}-butyric acid methyl ester(PTB-7:$PC_{71}BM$) active layer. The increased build-in potential and the decreased cathode surface work-function were responsible for the rise in PCE. The FF was increased from 56% to 66% due to an improvement in charge carrier mobility and reduced surface defects (Fig. 7A). In a recent study, TiO_2 nanostructures are used in PTB7:$PC_{71}BM$-based OSCs. The incorporation of 10 wt% TiO_2 nanoparticles with annealing treatment increases the optical absorption leading to a PCE of around 5.6% [74] (Fig. 7B).

NiO is a potential HTL material in inverted OSCs because of its intrinsic p-type behavior and attractive thermal and chemical stabilities; however, the wettability problem persists with material that requires high-temperature annealing. To overcome the wettability problem, Tran et al. [75] utilized a mixture of DI water and isopropyl alcohol with a small amount of 2-butanol additive in place of only DI water. This mixture of solvents allows an easy deposition of NiO NPs suspension (s-NiO) on a hydrophobic active layer surface. Nonfullerene-based inverted solar cells using a mixture of p-type polymer poly[4,8-bis(5-(2-ethylhexyl)thiophen-2-yl)benzo[1,2-*b*:4,5-*b*′]dithiophene-alt-3-fluorothieno[3,4-*b*]thiophene-carboxylate] (also known as PCE10) and the non-fullerene acceptor 2,2′-[[4,4,9,9-tetrakis(4-hexylphenyl)-4,9-dihydro-s-indaceno[1,2-*b*:5,6-*b*′] dithiophene-2,7-diyl] bis[[4-[(2-ethylhexyl)oxy]-5,2thiophenediyl] methylidyne (5,6-difluoro-3-oxo-1*H*-indene-2,1(3*H*)-diylidene]]bis[propanedinitrile] (IEICO-4F) had a high efficiency of 11.23% with a s-NiO HTL, equivalent to an inverted solar cell with a MoO_x HTL deposited by thermal evaporation. Furthermore, traditionally fabricated OSC devices with s-NiO layer have demonstrated similar efficiency to that of a PEDOT:PSS HTL device (Fig. 7C).

For the first time, Tran et al. demonstrated the use of alkali carbonates (Li_2CO_3, K_2CO_3, and Rb_2CO_3) solution processed at a low temperature, as an interfacial layer for SnO_2 as an ETL for inverted OSCs [76]. The interfacial properties of the alkali carbonates/SnO_2 as ETLs are superior as compared to pristine SnO_2 and offered efficient charge transport with a reduced charge recombination rates for OSCs. The 5.49% of PCE with pristine SnO_2 was found to be increased to 6.70%, 6.85%, and 7.35%, respectively, with Li_2CO_3, K_2CO_3, and Rb_2CO_3-modified SnO_2 as ETLs (Fig. 7D).

7 Stability of metal oxides-based OSCs

Various factors are responsible for the performance and stability of OSC devices [77]. Some of the factors are described below:

(a) Illumination stability

OSCs are susceptible to the oxygen, which causes performance degradation. Since the solar cells works under high illumination power of sunlight; the UV light present in sun light easily activates the oxygen near interface between metal oxides and organics. If the MO layer is not sufficiently thick to block UV light, the super oxide or hydrogen peroxide formed within will then aggressively attack the organic compounds resulting

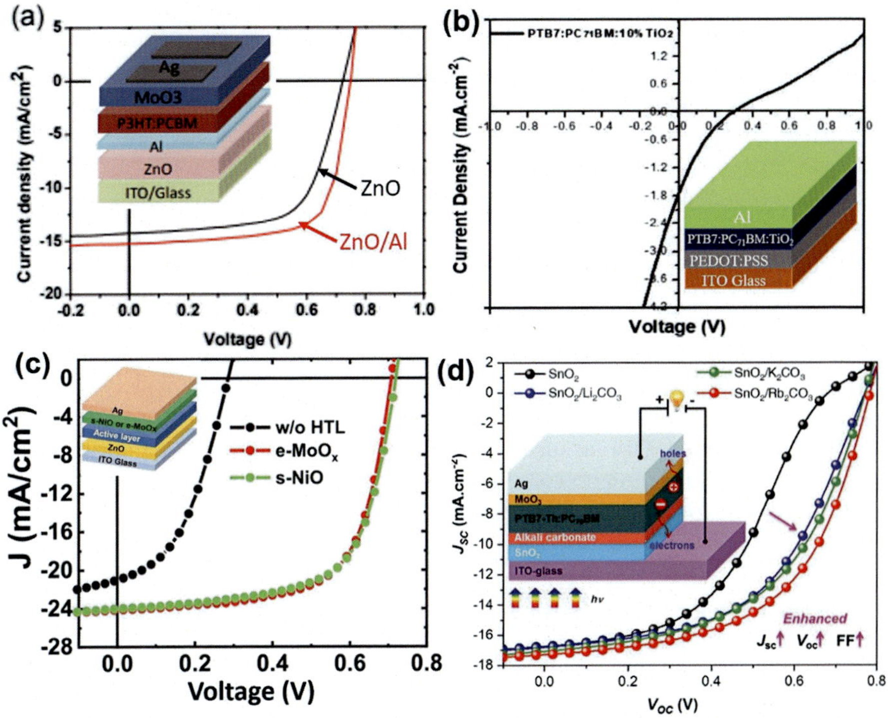

Fig. 7 Schematic views and *J-V* characteristics of inverted solar cells based on (A) PTB-7: $PC_{71}BM$ material system with ZnO or ZnO/Al cathode interlayer, and (B) PTB7:$PC_{71}BM$:TiO_2, (C) PTB7-Th:IEICO-4F blend films with s-NiO and e-MoO_x as HTLs and (D) device performance using bare ITO, alkali carbonates only, SnO_2, and SnO_2/alkali carbonates as ETLs. Reproduced with permission from Z. Lin, J. Chang, C. Zhang, D. Chen, J. Wu, Y. Hao, Enhanced performance and stability of polymer solar cells by in situ formed AlO_x passivation and doping, J. Phys. Chem. C 121 (19) (2017) 10275–10281, https://doi.org/10.1021/acs.jpcc.6b12459, N. Al-Shekaili, et al., The impact of TiO_2 nanostructures on the physical properties and electrical performance of organic solar cells based on PTB7:$PC_{71}BM$ bulk heterojunctions, Mater. Today Proc. 42 (2021) 1921–1927, https://doi.org/10.1016/j.matpr.2020.12.234, H.N. Tran, D.Q. Dao, Y.J. Yoon, Y.S. Shin, J.S. Choi, J.Y. Kim, S. Cho, Inverted polymer solar cells with annealing-free solution-processable NiO, Small 17 (31) (2021) 2101729, https://doi.org/10.1002/smll.202101729, V.-H. Tran, H. Park, S.H. Eom, S.C. Yoon, S.-H. Lee, Modified SnO_2 with alkali carbonates as robust electron-transport layers for inverted organic solar cells, ACS Omega 3 (12) (2018) 18398–18410, https://doi.org/10.1021/acsomega.8b02773.

in a degradation of solar cell [78]. Therefore, the illumination stability is an important parameter for solar cells to determine the degradation rate under working environments. TiO_2 and ZnO both are used as hole blocking layer which also prevents from UV illumination. However, TiO_2 shows higher absorption in UV region for the comparable thickness of ZnO layer. Thus, the solar cell with TiO_2 is more stable under solar light illumination. Though, TiO_2 layers also absorb some amount of visible spectrum which results in reduced saturation current and PCE of solar cell with increasing thickness of MO layer. Therefore, a tradeoff between stability and PCE is required to estimate the optimize thickness of TiO_2 layer. Normally, the thickness of 100 nm of TiO_2 layer provides good stability with small effect on PCE and saturation current of solar cell. Although, the ZnO layer is nearly transparent to visible light and the PCE of ZnO solar cells is not much influenced by the thickness of ZnO layer; however, the dependence on the saturation current, PCE, and illumination stability is still evident.

(b) Air Stability

As discussed, OSCs are vulnerable under the presence of oxygen in ambient. The OSC's performance is also harmed by moisture in the environment. Due to moisture, insulated islands are formed at organic/electrode interface [79]. It is widely acknowledged that in oxidizing environments, ZnO is reactive and unstable [80,81] because of O_2 adsorption on the ZnO polar surfaces, which causes an electrical instability [82,83]. The polymer layer's sealing process lowers the possibility of ZnO layer's exposure to air, which significantly slows down the oxidation process.

(c) Performance, efficiency, and environmental issues

Due to their excellent optical translucency across the entire visible spectral range and their wide bandgap with easy doping ability to active ions, metal oxides like TiO_2 and ZnO are used in many applications [84,85]. According to research by Hadipour et al., using temperature-dependent solution-processed TiO_x as a buffer layer in an inverted solar cell offers better performance and processable benefits than using an evaporated buffer layer made of Ca [86]. Gopinath et al. synthesized cost-effective nanoparticles (NPs) of TiO_2, Au, and Pt from plant extraction of Terminalia arjuna bark [87] and used in OSC. The efficiency of a pristine TiO_2 layer was 2.79%, but it has increased to 3.44% in the presence of Au NPs and 3.14% in the presence of Pt. The decreased recombination rate caused by the metallic NPs was the reason of the increased V_{oc} in Au- and Pt-doped TiO_2. Örnek et al. demonstrated that the use of an interfacial layer of Eu-doped TiO_2 in P3HT:PCBM inverted solar cell allows to record an efficiency of ~2.47%, in comparison to the pure TiO_2 which rendered an efficiency of 1.16% [88]. This improvement was ascribed to the accumulation of electron transport of Eu-doped TiO_2 as compared to pristine TiO_2 (Fig. 8).

Yang et al. described that the aggregation of TiO_2 nanotubes could decrease the absorbance of device; however, the collection and transportation of charges increased with slowed recombination rate [89]. The presence of TiO_2 nanotubes aggregates produced an efficiency of about 3.2% due to an enhancement of electron transport. Tai et al. used electrospinning technique to produce a nanofibrous film of TiO_2 blending with P3HT-based hybrid solar cells [90]. Due to their continuous carrier pathway and substantial interfacial area, the designed electrospun nanofibers with controllable alignment and size demonstrated an increase in electrical contact between the cathode electrode and the TiO_2 NPs. The comodification of 3-phenylphronic acid (PPA) and ruthenium dye (N719) on the TiO_2 nanofibers' surface influenced the ordered backbone of P3HT layer and further enhanced the device performance. This process

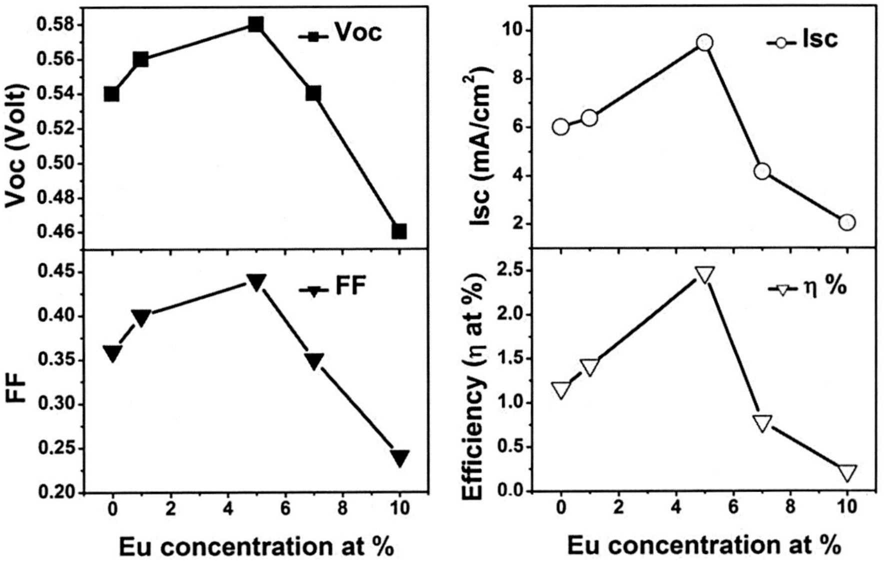

Fig. 8 Solar cell performance of Eu-doped and undoped TiO_2 electron transport layer (A) V_{oc}; (B) I_{sc}; (C) FF; and (D) η (%).
Reproduced with permission from O. Örnek, Z.A. Kösemen, S. Öztürk, B. Canımkurbey, Ş. Fındık, M. Erkovan, A. Kösemen, Performance enhancement of inverted type organic solar cells by using Eu doped TiO_2 thin film, Surf. Interfaces 9 (2017) 64–69, https://doi.org/10.1016/j.surfin.2017.08.003.

increased the PCE of hybrid OSC from ~0.59% to 1.09%, attributed to the enhanced coupling of electrons within the TiO_2 and absorbed dye and to the impeded recombination rate of charge carriers that diminish the number of traps at the interface. Tu et al. established that the lower efficiency of stable P3HT:TiO_2-based solar cell can be improved by doping of boron or bismuth [91]. The W4-dye modified TiO_2 for B-doped and Bi-doped TiO_2 showed the PCE≈0.87% and 0.89%, respectively. The improved performance in case of B-doped TiO_2 is credited to the higher electron mobility and the improved crystalline order due to smaller ionic radius of B^{3+} than that of Ti^{4+}. However, in the case of Bi-doped TiO_2, the highest V_{oc} and J_{sc} was attributed to an enhanced electron density in the presence of Bi electron lone pairs.

ZnO is known as n-type semiconductor with a wide bandgap of 3.37 eV with conduction band position similar to TiO_2; however, the charge carrier mobility in ZnO is higher as compared to anatase TiO_2 [92]. Additionally, the doping ZnO with aluminum at low temperatures can increase its carrier density. There are a variety of ways to prepare ZnO films, making it an appealing option for improved efficiency of OSCs [93–95]. Through the use of the sol-gel technique, Tan et al. established the performance of the P3HT:PCBM layer combined with crystalline ZnO NP in the buffer layer of heterojunction solar cells [96]. They achieved a greater efficiency of ≈3.72% in the inverted solar cell consisting of ZnO NP buffer layer as compared to MoO_3 and PEDOT:PSS buffer layers with efficiencies around 2.07% and 3.04%, respectively.

This enhanced efficiency was noticed due to the better matching of energy levels of the ZnO NP and the PCBM, whereas in the absence of ZnO buffer layer resulted in the S curve shape accompanied with a decreased V_{oc}. Mohan et al. compared the effect of the sol-gel method and mixed solvent on ZnO buffer layer in P3HT- and PTB7-based OSC devices [97]. The mixed solvent method on both active layer of P3HT:$PC_{71}BM$ and PTB7:$PC_{71}BM$ has shown better efficiency of 5.6% and 9.1%, respectively. However, using sol-gel method at high temperature, P3HT:$PC_{71}BM$ and PTB7:$PC_{71}BM$-based OSC devices resulted in PCE $\approx$ 4.1% and 6.5%, respectively. Sharma et al. analyzed the PffBT4T-2OD:$PC_{70}BM$-based OSCs with a buffer layer of Ga-doped ZnO (GZO) using sol-gel method [98]. Contrary to prior studies, they noticed an enhanced V_{oc}, by using a low process temperature before the spin coating of active layer. They also revealed that the highly ordered crystalline PffBT4T-2OD increased the FF, with an efficiency around 9.74% due to GZO's improved transport properties over pristine ZnO. Compared to TiO_2's high crystallization temperature, ZnO's capacity to crystallize at low temperatures has garnered a lot of interest. However, ZnO with its wide bandgap absorbs only in the UV light [99,100] and therefore the modification of ZnO is required to absorb in the visible region. Aqoma et al. prepared a high-performance ZnO-based electron transport layers (ETLs) for OSCs with a novel γ-ray-assisted solution process [101]. The γ-ray-assisted ZnO (ZnO-G) films show a significantly low defect density in contrast to the traditionally prepared ZnO films. The ZnO-G films have the ability to improve charge extraction efficiency of ETL with no further treatment. Using ZnO-G ETLs, the PCE of device is 11.09% with $V_{oc} \sim 0.80\,V$, $J_{sc} \sim 19.54\,mA/cm^2$, and FF $\sim$ 0.71, which is preeminent among widely studied PTB7-Th:$PC_{71}BM$-based devices (Fig. 9).

In a P3HT-based OSC, Spoerke et al. used the CdS-modified electrospun ZnO nanofibers on ITO substrate [102]. They noted an efficiency of 0.65% in the presence of a three-layered modified CdS, in contrast to 0.3% PCE for the device without using CdS. This was attributed to the prominent CdS absorption in the visible region and suppressed recombinations. Ajuria et al. spin coated ZnO NPs on the ZnO nanowire arrays in the active layer of P3HT:PCBM solar devices [103]. From the SEM, they noticed a poor mixing on the infiltration of different ZnO nanostructures on P3HT:PCBM surface, causing a low efficiency of 3.63%. This effect was ascribed to the fact that ZnO nanostructure was covered by P3HT:PCBM only, on the top surface which resulted in potential loss of optoelectronic properties of OSC. But, good infiltration of P3HT:PCBM resulted an efficiency of $\approx$4.1%, because of the smooth coverage of nanowire array by P3HT:PCBM that has enlarged the dissociation area, giving an increased exciton splitting as switched to planar structured.

All OSCs that have been studied so far contain a small amount of inorganic entity in the form of oxides. This amount is negligible when compared to inorganic solar cells, making it environmentally friendly with little to no adverse impact. OSCs are well packaged in a thin plastic barrier that is designed to be impermeable to liquid water. During the working operation, no leaching or emission to the environment should thus occur, even when the solar cells are installed outdoors. However, over time, the encapsulation's damage, such as crack formation or edge delamination, creates openings for air and water to enter. In this case, OSC compounds will be exposed to the environment, either to surface waters or soils.

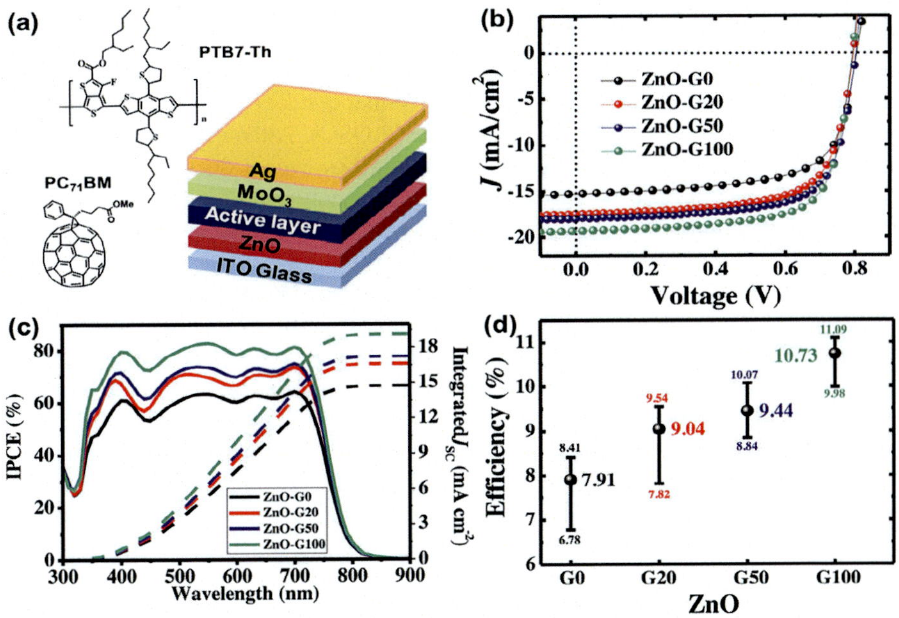

Fig. 9 Device performance of OSC devices using ZnO-G ETLs; (A) Schematic representation of an inverted OPV using different active materials, (B) *J-V* characteristics of the devices, (C) IPCE spectra of the devices, and (D) statistical results of the performances of the devices. Reproduced with permission from H. Aqoma, S. Park, H.-Y. Park, W.T. Hadmojo, S.-H. Oh, S. Nho, D.H. Kim, J. Seo, S. Park, D.Y. Ryu, S. Cho, S.-Y. Jang, 11% organic photovoltaic devices based on PTB7-Th: $PC_{71}BM$ photoactive layers and irradiation-assisted ZnO electron transport layers, Adv. Sci. 5 (7) (2018) 1700858, https://doi.org/10.1002/advs.201700858.

In studies, the only elements found in elevated quantities in leachates were Zn and Ag. In heavy rain scenario, the leaching of metals from the OSCs were seen in modules with a visibly damaged barrier and in OSCs that delaminated. Nevertheless, with a few exceptions, the metal concentrations did not go above the WHO limit for water consumption [104]. However, for outdoor use of OSCs, the encapsulation and edge sealing quality are crucial, both for lifetime performance and environmental concerns. It would be interesting to examine the environmental influences on the organic molecules found in OSCs in addition to metals. These organic molecules, such as P3HT, PEDOT:PSS, and PCBM, are highly susceptible to degradation by light, oxygen, and water and are not yet listed in databases [105].

8 Current problems and future perspective

Use of organic solar cells has shown promising future for the solar energy harvesting with more sustainable approach. However, the low PCE and small operating life of these devices limit their application and thus have a huge scope for improvement. For practical applications, in addition to increased PCE, environmental stability is crucial for the long-term reliable operation of OSCs in challenging circumstances. However, higher susceptibility toward oxygen and moisture of OSCs requires encapsulation of these devices which increases the cost of OSCs. So far, reports of inorganic/organic hybrid solar cells (HSCs) performances are still inferior to OSCs, particularly in case of the PCBM/polymer based OSCs. Although theoretically, the inorganic phase renders higher electron mobility, however it does not always turn to higher PCE due to various factors. Due to the poor polymer infiltration in the nanostructure, low polymer/inorganic wettability, and the narrow interfacial area between donor and acceptor, lower current densities are often observed across HSCs when compared to OSCs. Low electronic mobility is also detected in HSCs utilizing P3HT as the organic constituent. P3HT was utilized because of its excellent performance in PCBM/P3HT OSCs. Although the P3HT may be constrained by the MO morphology in the active layer, this results in substantially decreased hole mobility and poor PCE of the HSCs. Similar findings were also reported with poly[2-methoxy-5-(3′,7′-dimethyloctyloxy)-1,4-phenylenevinylene] MDMO-PPV as an organic p-type semiconductor [106]. Therefore, coupling ZnO nanostructures with additional, less-studied p-type organic semiconductors may be a further area of interest. The random dispersion of inorganic nanocrystals in the specific situation of randomly dispersed nanocrystal solar cell architecture may be attributed to the reduced current density caused by the lack of a direct electron transport pathway. When using a presynthesized inorganic nanocrystals, other issues may arise, as solvent evaporation during synthesis can cause the aggregation of nanocrystals, which could lower the current density. The fundamental issue with the bilayer structure is the tiny interfacial region between the two phases. Although in theory, vertically aligned structures are the best configuration for OPV application because they can provide a straight conduit for charge carrier collection, however there haven't been many reports on this specific architecture up to this point. This is mainly due to the poor polymer infiltration and crystallinity, which can be relatively low even after thermal annealing, as well as unoptimized properties, such as unfavorable nanostructure morphologies. Nevertheless, it is possible that cells made with this architecture can still be optimized and improved significantly in the future with proper morphology control (e.g., the spacing between nanorods should match the exciton diffusion length of the polymer, optimized nanorod/nanowire length), as well as the optimized blocking layer thickness. Despite several signs of progress so far in terms of ZnO growth, there is still plenty of room for progress. A larger surface area and better light-harvesting capacity are two essential improvements. The interface should be more effective than the existing innovation in solar cell technologies [107]. Overall, such advances would lead to a higher PCE and prolonged lifetime under various environments with a diverse thermal condition, mechanical loading, and humidity.

9 Conclusions

Organic solar cell (OSC) technology is an emerging field of energy harvesting leading to low-cost processing over wide regions, semitransparency, mechanical flexibility, and lightweight, which renders a low-cost technique as compared to Si-based photovoltaics. In modern OSCs, various transition metal oxides (TMOs) have been used as electron and hole contact layers in order to achieve maximum PCE of around 19%. Usually, ZnO and TiO_2 have been employed as electron transport layers, whereas TMOs with high work functions such as WO_x, NiO_x, MoO_x, and V_2O_5 are used as hole contact layers. It has been demonstrated that MoO_x and V_2O_5 could be used as an alternative for PEDOT:PSS. The presence of TMOs can influence the V_{oc} and J_{sc} attributing to change in R_s and subsequent modification in PCE. The current OSC devices also suffer from some specific problems that restrict their large-scale production. The first constraint is the lifetime of OSCs; however, it is observed that the inverted structure of OSCs has an extended lifetime. The other challenge is associated with the use of chlorinated solvents (like chloroform) in the active layer. Usually, the active layer of OSCs, composed of a conjugated polymer and fullerene mixture, is deposited using any chlorinated solvent. These solvents are volatile in nature and could affect the environment and human health. The abovementioned problem can be solved by using eco-friendly water-based polymer nanoparticles in the inverted OSCs device. It is also evinced that the performance of OSCs is limited due to their poor red absorption, low stability, poor charge transport, and electronic conductivity. To tackle these problems, the synthesis of new materials, the use of composite materials, development of self-assembling pi-conjugated systems, optimization of molecular architecture, and controlled morphology with precise conditions of fabrications and processing are essentially required. The primary objective of researchers and engineers in the following decade will be to improve the PCE of OSCs by applying various new and hybrid techniques.

Acknowledgments

DPS would like to thank all funding sources, specifically Commission Permanente de Coopération Franco-Québécoise (CPCFQ) for the program "Samuel de Champlain" with Concordia University and INRS, Canada.

References

[1] Annual Energy Review, U.S. Energy Information Administration (EIA), 2021.
[2] G. Kopp, J.L. Lean, A new, lower value of total solar irradiance: evidence and climate significance, Geophys. Res. Lett. 38 (1) (2011) https://doi.org/10.1029/2010GL045777.
[3] M. Grätzel, Powering the planet, Nature 403 (2000) 363, https://doi.org/10.1038/35000273.
[4] V. Esen, Ş. Sağlam, B. Oral, Light sources of solar simulators for photovoltaic devices: a review, Renew. Sustain. Energy Rev. 77 (2017) 1240–1250, https://doi.org/10.1016/j.rser.2017.03.062.
[5] W.G. Adams, R.E. Day, V. The action of light on selenium, Proc. R. Soc. Lond. 25 (171–178) (1877) 113–117, https://doi.org/10.1098/rspl.1876.0024.
[6] C.E. Fritts, On a new form of selenium cell, and some electrical discoveries made by its use, Am. J. Sci. 26 (156) (1883) 465, https://doi.org/10.2475/ajs.s3-26.156.465.
[7] R.S. Ohl, Light-Sensitive Electric Device, USA, 1946.
[8] A. Pochettino, A. Sella, Photoelectric behavior of anthracene, Acad. Lincei Rend. 15 (1906) 355–363.

[9] C.W. Tang, Two-layer organic photovoltaic cell, Appl. Phys. Lett. 48 (1986) 183, https://doi.org/10.1063/1.96937.

[10] G. Yu, J. Gao, J.C. Hummelen, F. Wudl, A.J. Heeger, Polymer photovoltaic cells: enhanced efficiencies via a network of internal donor-acceptor heterojunctions, Science 270 (5243) (1995) 1789–1791, https://doi.org/10.1126/science.270.5243.1789.

[11] M. Thelakkat, P. Pösch, H.-W. Schmidt, Synthesis and characterization of highly fluorescent main-chain copolyimides containing perylene and quinoxaline units, Macromolecules 34 (21) (2001) 7441–7447, https://doi.org/10.1021/ma010615w.

[12] D.C. Olson, S.E. Shaheen, R.T. Collins, D.S. Ginley, The effect of atmosphere and ZnO morphology on the performance of hybrid poly(3-hexylthiophene)/ZnO nanofiber photovoltaic devices, J. Phys. Chem. C 111 (44) (2007) 16670–16678, https://doi.org/10.1021/jp0734225.

[13] D. Irwin Michael, D.B. Buchholz, W. Hains Alexander, P.H. Chang Robert, J. Marks Tobin, p-Type semiconducting nickel oxide as an efficiency-enhancing anode interfacial layer in polymer bulk-heterojunction solar cells, Proc. Natl. Acad. Sci. 105 (8) (2008) 2783–2787, https://doi.org/10.1073/pnas.0711990105.

[14] X. Wen, A. Nowak-Król, O. Nagler, F. Kraus, N. Zhu, N. Zheng, M. Müller, D. Schmidt, Z. Xie, F. Würthner, Tetrahydroxy-perylene bisimide embedded in a zinc oxide thin film as an electron-transporting layer for high-performance non-fullerene organic solar cells, Angew. Chem. Int. Ed. 58 (37) (2019) 13051–13055, https://doi.org/10.1002/anie.201907467.

[15] J. Wei, G. Ji, C. Zhang, L. Yan, Q. Luo, C. Wang, Q. Chen, J. Yang, L. Chen, C.-Q. Ma, Silane-capped ZnO nanoparticles for use as the electron transport layer in inverted organic solar cells, ACS Nano 12 (2018) 5518, https://doi.org/10.1021/acsnano.8b01178.

[16] C. Li, X. Sun, J. Ni, L. Huang, R. Xu, Z. Li, H. Cai, J. Li, Y. Zhang, J. Zhang, Ternary organic solar cells based on ZnO-Ge double electron transport layer with enhanced power conversion efficiency, Sol. Energy 155 (2017) 1052–1058, https://doi.org/10.1016/j.solener.2017.07.053.

[17] Y. Lin, B. Adilbekova, Y. Firdaus, E. Yengel, H. Faber, M. Sajjad, X. Zheng, E. Yarali, A. Seitkhan, O.M. Bakr, A. El-Labban, U. Schwingenschlögl, V. Tung, I. McCulloch, F. Laquai, T.D. Anthopoulos, 17% efficient organic solar cells based on liquid exfoliated WS_2 as a replacement for PEDOT:PSS, Adv. Mater. 31 (46) (2019) 1902965, https://doi.org/10.1002/adma.201902965.

[18] J. Wang, Z. Zheng, Y. Zu, Y. Wang, X. Liu, S. Zhang, M. Zhang, J. Hou, A tandem organic photovoltaic cell with 19.6% efficiency enabled by light distribution control, Adv. Mater. 33 (39) (2021) 2102787, https://doi.org/10.1002/adma.202102787.

[19] Y. Cui, Y. Xu, H. Yao, P. Bi, L. Hong, J. Zhang, Y. Zu, T. Zhang, J. Qin, J. Ren, Z. Chen, C. He, X. Hao, Z. Wei, J. Hou, Single-junction organic photovoltaic cell with 19% efficiency, Adv. Mater. 33 (41) (2021) 2102420, https://doi.org/10.1002/adma.202102420.

[20] H. Chen, R. Zhang, X. Chen, G. Zeng, L. Kobera, S. Abbrent, B. Zhang, W. Chen, G. Xu, J. Oh, S.-H. Kang, S. Chen, C. Yang, J. Brus, J. Hou, F. Gao, Y. Li, Y. Li, A guest-assisted molecular-organization approach for >17% efficiency organic solar cells using environmentally friendly solvents, Nat. Energy 6 (11) (2021) 1045–1053, https://doi.org/10.1038/s41560-021-00923-5.

[21] Y. Lin, Y. Firdaus, F.H. Isikgor, M.I. Nugraha, E. Yengel, G.T. Harrison, R. Hallani, A. El-Labban, H. Faber, C. Ma, X. Zheng, A. Subbiah, C.T. Howells, O.M. Bakr, I. McCulloch, S.D. Wolf, L. Tsetseris, T.D. Anthopoulos, Self-assembled monolayer enables hole transport layer-free organic solar cells with 18% efficiency and improved operational stability, ACS Energy Lett. 5 (9) (2020) 2935–2944, https://doi.org/10.1021/acsenergylett.0c01421.

[22] N. Ahmad, H. Zhou, P. Fan, G. Liang, Recent progress in cathode interlayer materials for non-fullerene organic solar cells, EcoMat 4 (1) (2022) e12156, https://doi.org/10.1002/eom2.12156.

[23] Q. Bao, O. Sandberg, D. Dagnelund, S. Sandén, S. Braun, H. Aarnio, X. Liu, W.M. Chen, R. Österbacka, M. Fahlman, Trap-assisted recombination via integer charge transfer states in organic bulk heterojunction photovoltaics, Adv. Funct. Mater. 24 (40) (2014) 6309–6316, https://doi.org/10.1002/adfm.201401513.

[24] B. Xu, J. Hou, Solution-processable conjugated polymers as anode interfacial layer materials for organic solar cells, Adv. Energy Mater. 8 (20) (2018) 1800022, https://doi.org/10.1002/aenm.201800022.

[25] T. Ameri, G. Dennler, C. Lungenschmied, C.J. Brabec, Organic tandem solar cells: a review, Energy Environ. Sci. 2 (4) (2009) 347–363, https://doi.org/10.1039/B817952B.

[26] G. Dennler, C. Lungenschmied, H. Neugebauer, N.S. Sariciftci, A. Labouret, Flexible, conjugated polymer-fullerene-based bulk-heterojunction solar cells: basics, encapsulation, and integration, J. Mater. Res. 20 (2005) 3224–3233, https://doi.org/10.1557/jmr.2005.0399.

[27] C.W. Tang, Two-layer organic photovoltaic cell, Appl. Phys. Lett. 48 (2) (1986) 183–185, https://doi.org/10.1063/1.96937.

[28] J.J.M. Halls, C.A. Walsh, N.C. Greenham, E.A. Marseglia, R.H. Friend, S.C. Moratti, A. B. Holmes, Efficient photodiodes from interpenetrating polymer networks, Nature 376 (6540) (1995) 498–500, https://doi.org/10.1038/376498a0.

[29] G. Yu, A.J. Heeger, Charge separation and photovoltaic conversion in polymer composites with internal donor/acceptor heterojunctions, J. Appl. Phys. 78 (7) (1995) 4510–4515, https://doi.org/10.1063/1.359792.

[30] J.Y. Kim, S.H. Kim, H.-H. Lee, K. Lee, W. Ma, X. Gong, A.J. Heeger, New architecture for high-efficiency polymer photovoltaic cells using solution-based titanium oxide as an optical spacer, Adv. Mater. 18 (5) (2006) 572–576, https://doi.org/10.1002/adma.200501825.

[31] J. Peet, J.Y. Kim, N.E. Coates, W.L. Ma, D. Moses, A.J. Heeger, G.C. Bazan, Efficiency enhancement in low-bandgap polymer solar cells by processing with alkane dithiols, Nat. Mater. 6 (7) (2007) 497–500, https://doi.org/10.1038/nmat1928.

[32] J.Y. Kim, K. Lee, N.E. Coates, D. Moses, T.-Q. Nguyen, M. Dante, A.J. Heeger, Efficient tandem polymer solar cells fabricated by all-solution processing, Science 317 (5835) (2007) 222–225, https://doi.org/10.1126/science.1141711.

[33] R. Gaudiana, C. Brabec, Fantastic plastic, Nat. Photonics 2 (5) (2008) 287–289, https://doi.org/10.1038/nphoton.2008.69.

[34] P. Schilinsky, C. Waldauf, C.J. Brabec, Recombination and loss analysis in polythiophene based bulk heterojunction photodetectors, Appl. Phys. Lett. 81 (20) (2002) 3885–3887, https://doi.org/10.1063/1.1521244.

[35] D. Zhang, W.C.H. Choy, F.-x. Xie, X. Li, Large-area, high-quality self-assembly electron transport layer for organic optoelectronic devices, Org. Electron. 13 (10) (2012) 2042–2046, https://doi.org/10.1016/j.orgel.2012.06.012.

[36] S. Chen, C.E. Small, C.M. Amb, J. Subbiah, T.-h. Lai, S.-W. Tsang, J.R. Manders, J.R. Reynolds, F. So, Inverted polymer solar cells with reduced interface recombination, Adv. Energy Mater. 2 (11) (2012) 1333–1337, https://doi.org/10.1002/aenm.201200184.

[37] Z. Liang, Q. Zhang, O. Wiranwetchayan, J. Xi, Z. Yang, K. Park, C. Li, G. Cao, Effects of the morphology of a ZnO buffer layer on the photovoltaic performance of inverted polymer solar cells, Adv. Funct. Mater. 22 (10) (2012) 2194–2201, https://doi.org/10.1002/adfm.201101915.

[38] I. Gonzalez-Valls, M. Lira-Cantu, Vertically-aligned nanostructures of ZnO for excitonic solar cells: a review, Energy Environ. Sci. 2 (1) (2009) 19–34, https://doi.org/10.1039/B811536B.
[39] G. Li, R. Zhu, Y. Yang, Polymer solar cells, Nat. Photonics 6 (3) (2012) 153–161, https://doi.org/10.1038/nphoton.2012.11.
[40] L. Dou, J. You, J. Yang, C.-C. Chen, Y. He, S. Murase, T. Moriarty, K. Emery, G. Li, Y. Yang, Tandem polymer solar cells featuring a spectrally matched low-bandgap polymer, Nat. Photonics 6 (3) (2012) 180–185, https://doi.org/10.1038/nphoton.2011.356.
[41] N.K. Elumalai, T.M. Jin, V. Chellappan, R. Jose, S.K. Palaniswamy, S. Jayaraman, H.K. Raut, S. Ramakrishna, Electrospun ZnO nanowire plantations in the electron transport layer for high-efficiency inverted organic solar cells, ACS Appl. Mater. Interfaces 5 (19) (2013) 9396–9404, https://doi.org/10.1021/am4013853.
[42] S. Ito, S.M. Zakeeruddin, R. Humphry-Baker, P. Liska, R. Charvet, P. Comte, M.K. Nazeeruddin, P. Péchy, M. Takata, H. Miura, S. Uchida, M. Grätzel, High-efficiency organic-dye-sensitized solar cells controlled by nanocrystalline-TiO_2 electrode thickness, Adv. Mater. 18 (9) (2006) 1202–1205, https://doi.org/10.1002/adma.200502540.
[43] R. Steim, S.A. Choulis, P. Schilinsky, C.J. Brabec, Interface modification for highly efficient organic photovoltaics, Appl. Phys. Lett. 92 (9) (2008) 093303, https://doi.org/10.1063/1.2885724.
[44] H. Xue, X. Kong, Z. Liu, C. Liu, J. Zhou, W. Chen, TiO_2 based metal-semiconductor-metal ultraviolet photodetectors, Appl. Phys. Lett. 90 (20) (2007) 201118, https://doi.org/10.1063/1.2741128.
[45] D.M. de Leeuw, M.M.J. Simenon, A.R. Brown, R.E.F. Einerhand, Stability of n-type doped conducting polymers and consequences for polymeric microelectronic devices, Synth. Met. 87 (1) (1997) 53–59, https://doi.org/10.1016/S0379-6779(97)80097-5.
[46] H.D. Irfan, Y. Gao, D.Y. Kim, J. Subbiah, F. So, Energy level evolution of molybdenum trioxide interlayer between indium tin oxide and organic semiconductor, Appl. Phys. Lett. 96 (7) (2010) 073304, https://doi.org/10.1063/1.3309600.
[47] M.C. Gwinner, et al., Doping of organic semiconductors using molybdenum trioxide: a quantitative time-dependent electrical and spectroscopic study, Adv. Funct. Mater. 21 (8) (2011) 1432–1441, https://doi.org/10.1002/adfm.201002696.
[48] M.A. Khilla, Z.M. Hanafi, B.S. Farag, A.A.-e. Saud, Transport properties of molybdenum trioxide and its suboxides, Thermochim. Acta 54 (1) (1982) 35–45, https://doi.org/10.1016/0040-6031(82)85062-4.
[49] M. Vasilopoulou, L.C. Palilis, D.G. Georgiadou, P. Argitis, S. Kennou, L. Sygellou, I. Kostis, G. Papadimitropoulos, N. Konofaos, A.A. Iliadis, D. Davazoglou, Reduced molybdenum oxide as an efficient electron injection layer in polymer light-emitting diodes, Appl. Phys. Lett. 98 (12) (2011) 123301, https://doi.org/10.1063/1.3557502.
[50] D.Y. Kim, J. Subbiah, G. Sarasqueta, F. So, H.D. Irfan, Y. Gao, The effect of molybdenum oxide interlayer on organic photovoltaic cells, Appl. Phys. Lett. 95 (9) (2009) 093304, https://doi.org/10.1063/1.3220064.
[51] J. Meyer, S. Hamwi, M. Kröger, W. Kowalsky, T. Riedl, A. Kahn, Transition metal oxides for organic electronics: energetics, device physics and applications, Adv. Mater. 24 (40) (2012) 5408–5427, https://doi.org/10.1002/adma.201201630.
[52] H. Höchst, R.D. Bringans, Electronic structure of evaporated and annealed tungsten oxide films studied with UPS, Appl. Surf. Sci. 11–12 (1982) 768–773, https://doi.org/10.1016/0378-5963(82)90119-2.
[53] H. Choi, et al., Solution processed WO_3 layer for the replacement of PEDOT:PSS layer in organic photovoltaic cells, Org. Electron. 13 (6) (2012) 959–968, https://doi.org/10.1016/j.orgel.2012.01.033.

[54] S. Han, W.S. Shin, M. Seo, D. Gupta, S.-J. Moon, S. Yoo, Improving performance of organic solar cells using amorphous tungsten oxides as an interfacial buffer layer on transparent anodes, Org. Electron. 10 (5) (2009) 791–797, https://doi.org/10.1016/j.orgel.2009.03.016.
[55] H.-L. Yip, A.K.Y. Jen, Recent advances in solution-processed interfacial materials for efficient and stable polymer solar cells, Energy Environ. Sci. 5 (3) (2012) 5994–6011, https://doi.org/10.1039/C2EE02806A.
[56] A.F. Palmstrom, P.K. Santra, S.F. Bent, Atomic layer deposition in nanostructured photovoltaics: tuning optical, electronic and surface properties, Nanoscale 7 (29) (2015) 12266–12283, https://doi.org/10.1039/C5NR02080H.
[57] J.-C. Wang, et al., Highly efficient flexible inverted organic solar cells using atomic layer deposited ZnO as electron selective layer, J. Mater. Chem. 20 (5) (2010) 862–866, https://doi.org/10.1039/B921396A.
[58] P.I. Stakhira, G.L. Pakhomov, V.V. Cherpak, D. Volynyuk, G. Luka, M. Godlewski, E. Guziewicz, Z.Y. Hotra, Photovoltaic cells based on nickel phthalocyanine and zinc oxide formed by atomic layer deposition, Cent. Eur. J. Phys. 8 (5) (2010) 798–803, https://doi.org/10.2478/s11534-009-0159-9.
[59] H. Frankenstein, M.D. Losego, G.L. Frey, Atomic layer deposition of ZnO electron transporting layers directly onto the active layer of organic solar cells, Org. Electron. 64 (2019) 37–46, https://doi.org/10.1016/j.orgel.2018.10.002.
[60] C. Waldauf, M. Morana, P. Denk, P. Schilinsky, K. Coakley, S.A. Choulis, C.J. Brabec, Highly efficient inverted organic photovoltaics using solution based titanium oxide as electron selective contact, Appl. Phys. Lett. 89 (23) (2006) 233517, https://doi.org/10.1063/1.2402890.
[61] T.N. Takayuki Kuwabara, K. Uozumi, T. Yamaguchi, K. Takahashi, Highly durable inverted-type organic solar cell using amorphous titanium oxide as electron collection electrode inserted between ITO and organic layer, Sol. Energy Mater. Sol. Cells 92 (11) (2008) 1476–1482, https://doi.org/10.1016/j.solmat.2008.06.012.
[62] S.-Y. Park, Y.-J. Kang, D.-H. Kim, J.-W. Kang, Organic solar cells employing magnetron sputtered p-type nickel oxide thin film as the anode buffer layer, Sol. Energy Mater. Sol. Cells 94 (12) (2010) 2332–2336, https://doi.org/10.1016/j.solmat.2010.08.004.
[63] C.-C. Hsu, H.-W. Su, C.-H. Hou, J.-J. Shyue, F.-Y. Tsai, Atomic layer deposition of NiO hole-transporting layers for polymer solar cells, Nanotechnology 26 (38) (2015) 385201, https://doi.org/10.1088/0957-4484/26/38/385201.
[64] C.-P. Chen, Y.-D. Chen, S.-C. Chuang, High-performance and highly durable inverted organic photovoltaics embedding solution-processable vanadium oxides as an interfacial hole-transporting layer, Adv. Mater. 23 (33) (2011) 3859–3863, https://doi.org/10.1002/adma.201102142.
[65] V. Shrotriya, G. Li, Y. Yao, C.-W. Chu, Y. Yang, Transition metal oxides as the buffer layer for polymer photovoltaic cells, Appl. Phys. Lett. 88 (7) (2006) 073508, https://doi.org/10.1063/1.2174093.
[66] K. Takanezawa, K. Tajima, K. Hashimoto, Efficiency enhancement of polymer photovoltaic devices hybridized with ZnO nanorod arrays by the introduction of a vanadium oxide buffer layer, Appl. Phys. Lett. 93 (6) (2008) 063308, https://doi.org/10.1063/1.2972113.
[67] M.Y. Chan, C.S. Lee, S.L. Lai, M.K. Fung, F.L. Wong, H.Y. Sun, K.M. Lau, S.T. Lee, Efficient organic photovoltaic devices using a combination of exciton blocking layer and anodic buffer layer, J. Appl. Phys. 100 (9) (2006) 094506, https://doi.org/10.1063/1.2363649.

[68] M. Qiu, D. Zhu, X. Bao, J. Wang, X. Wang, R. Yang, WO_3 with surface oxygen vacancies as an anode buffer layer for high performance polymer solar cells, J. Mater. Chem. A 4 (3) (2016) 894–900, https://doi.org/10.1039/C5TA08898D.
[69] K. Kawano, C. Adachi, Evaluating carrier accumulation in degraded bulk heterojunction organic solar cells by a thermally stimulated current technique, Adv. Funct. Mater. 19 (24) (2009) 3934–3940, https://doi.org/10.1002/adfm.200901573.
[70] X. Li, W.C.H. Choy, F. Xie, S. Zhang, J. Hou, Room-temperature solution-processed molybdenum oxide as a hole transport layer with Ag nanoparticles for highly efficient inverted organic solar cells, J. Mater. Chem. A 1 (22) (2013) 6614–6621, https://doi.org/10.1039/C3TA10531H.
[71] J.R. Manders, S.-W. Tsang, M.J. Hartel, T.-H. Lai, S. Chen, C.M. Amb, J.R. Reynolds, F. So, Solution-processed nickel oxide hole transport layers in high efficiency polymer photovoltaic cells, Adv. Funct. Mater. 23 (23) (2013) 2993–3001, https://doi.org/10.1002/adfm.201202269.
[72] A. Wibowo, M.A. Marsudi, M.I. Amal, M.B. Ananda, R. Stephanie, H. Ardy, L.J. Diguna, ZnO nanostructured materials for emerging solar cell applications, RSC Adv. 10 (70) (2020) 42838–42859, https://doi.org/10.1039/D0RA07689A.
[73] Z. Lin, J. Chang, C. Zhang, D. Chen, J. Wu, Y. Hao, Enhanced performance and stability of polymer solar cells by in situ formed AlO_x passivation and doping, J. Phys. Chem. C 121 (19) (2017) 10275–10281, https://doi.org/10.1021/acs.jpcc.6b12459.
[74] N. Al-Shekaili, S. Hashim, F.F. Muhammadsharif, M.Z. Al-Abri, K. Sulaiman, M.Y. Yahya, M.R. Ahmed, The impact of TiO_2 nanostructures on the physical properties and electrical performance of organic solar cells based on PTB7:$PC_{71}BM$ bulk heterojunctions, Mater. Today Proc. 42 (2021) 1921–1927, https://doi.org/10.1016/j.matpr.2020.12.234.
[75] H.N. Tran, D.Q. Dao, Y.J. Yoon, Y.S. Shin, J.S. Choi, J.Y. Kim, S. Cho, Inverted polymer solar cells with annealing-free solution-processable NiO, Small 17 (31) (2021) 2101729, https://doi.org/10.1002/smll.202101729.
[76] V.-H. Tran, H. Park, S.H. Eom, S.C. Yoon, S.-H. Lee, Modified SnO_2 with alkali carbonates as robust electron-transport layers for inverted organic solar cells, ACS Omega 3 (12) (2018) 18398–18410, https://doi.org/10.1021/acsomega.8b02773.
[77] L. Duan, A. Uddin, Progress in stability of organic solar cells, Adv. Sci. 7 (11) (2020) 1903259, https://doi.org/10.1002/advs.201903259.
[78] M. Jørgensen, K. Norrman, F.C. Krebs, Stability/degradation of polymer solar cells, Sol. Energy Mater. Sol. Cells 92 (7) (2008) 686–714, https://doi.org/10.1016/j.solmat.2008.01.005.
[79] H.-T. Chien, P.W. Zach, B. Friedel, Short-term environmental effects and their influence on spatial homogeneity of organic solar cell functionality, ACS Appl. Mater. Interfaces 9 (33) (2017) 27754–27764, https://doi.org/10.1021/acsami.7b08365.
[80] D.F. Wang, T.J. Zhang, Study on the defects of ZnO nanowire, Solid State Commun. 149 (43) (2009) 1947–1949, https://doi.org/10.1016/j.ssc.2009.07.038.
[81] J.Y. Park, J.-J. Kim, S.S. Kim, Ambient air effects on electrical transport properties of ZnO nanorod transistors, J. Nanosci. Nanotechnol. 8 (11) (2008) 5929–5933, https://doi.org/10.1166/jnn.2008.257.
[82] V. Khranovskyy, J. Eriksson, A. Lloyd-Spetz, R. Yakimova, L. Hultman, Effect of oxygen exposure on the electrical conductivity and gas sensitivity of nanostructured ZnO films, Thin Solid Films 517 (6) (2009) 2073–2078, https://doi.org/10.1016/j.tsf.2008.10.037.

[83] M. Liu, H.K. Kim, Ultraviolet detection with ultrathin ZnO epitaxial films treated with oxygen plasma, Appl. Phys. Lett. 84 (2) (2004) 173–175, https://doi.org/10.1063/1.1640468.
[84] S. Anandan, Y. Ikuma, K. Niwa, An overview of semi-conductor photocatalysis: modification of TiO_2 nanomaterials, Solid State Phenom. 162 (2010) 239–260, https://doi.org/10.4028/www.scientific.net/SSP.162.239.
[85] X. Kong, Y. Hu, W. Pan, Enhancement of photocatalytic activity for fold-like ZnO via hybridisation with graphene, Micro Nano Lett. 13 (2) (2018) 232–236, https://doi.org/10.1049/mnl.2017.0395.
[86] A. Hadipour, R. Müller, P. Heremans, Room temperature solution-processed electron transport layer for organic solar cells, Org. Electron. 14 (10) (2013) 2379–2386,-https://doi.org/10.1016/j.orgel.2013.05.028.
[87] K. Gopinath, S. Kumaraguru, K. Bhakyaraj, S. Thirumal, A. Arumugam, Eco-friendly synthesis of TiO_2, Au and Pt doped TiO_2 nanoparticles for dye sensitized solar cell applications and evaluation of toxicity, Superlattice. Microst. 92 (2016) 100–110, https://doi.org/10.1016/j.spmi.2016.02.012.
[88] O. Örnek, Z.A. Kösemen, S. Öztürk, B. Canımkurbey, Ş. Fındık, M. Erkovan, A. Kösemen, Performance enhancement of inverted type organic solar cells by using Eu doped TiO_2 thin film, Surf. Interfaces 9 (2017) 64–69, https://doi.org/10.1016/j.surfin.2017.08.003.
[89] T. Yang, W. Cai, D. Qin, E. Wang, L. Lan, X. Gong, J. Peng, Y. Cao, Solution-processed zinc oxide thin film as a buffer layer for polymer solar cells with an inverted device structure, J. Phys. Chem. C 114 (14) (2010) 6849–6853, https://doi.org/10.1021/jp1003984.
[90] Q. Tai, X. Zhao, F. Yan, Hybrid solar cells based on poly(3-hexylthiophene) and electrospun TiO_2 nanofibers with effective interface modification, J. Mater. Chem. 20 (35) (2010) 7366–7371, https://doi.org/10.1039/C0JM01455A.
[91] Y.-C. Tu, H. Lim, C.-Y. Chang, J.-J. Shyue, W.-F. Su, Enhancing performance of P3HT:TiO_2 solar cells using doped and surface modified TiO_2 nanorods, J. Colloid Interface Sci. 448 (2015) 315–319, https://doi.org/10.1016/j.jcis.2015.02.015.
[92] Q. Zhang, C.S. Dandeneau, X. Zhou, G. Cao, ZnO nanostructures for dye-sensitized solar cells, Adv. Mater. 21 (41) (2009) 4087–4108, https://doi.org/10.1002/adma.200803827.
[93] T. Pauporté, Synthesis of ZnO nanostructures for solar cells—a focus on dye-sensitized and perovskite solar cells, in: M. Lira-Cantu (Ed.), The Future of Semiconductor Oxides in Next-Generation Solar Cells, Elsevier, 2018, pp. 3–43 (Chapter 1).
[94] T. Stubhan, I. Litzov, N. Li, M. Salinas, M. Steidl, G. Sauer, K. Forberich, G.J. Matt, M. Halik, C.J. Brabec, Overcoming interface losses in organic solar cells by applying low temperature, solution processed aluminum-doped zinc oxide electron extraction layers, J. Mater. Chem. A 1 (19) (2013) 6004–6009, https://doi.org/10.1039/C3TA10987A.
[95] R.M. Pasquarelli, D.S. Ginley, R. O'Hayre, Solution processing of transparent conductors: from flask to film, Chem. Soc. Rev. 40 (11) (2011) 5406–5441, https://doi.org/10.1039/C1CS15065K.
[96] M.J. Tan, S. Zhong, J. Li, Z. Chen, W. Chen, Air-stable efficient inverted polymer solar cells using solution-processed nanocrystalline ZnO interfacial layer, ACS Appl. Mater. Interfaces 5 (11) (2013) 4696–4701, https://doi.org/10.1021/am303004r.
[97] M. Mohan, V. Nandal, S. Paramadam, K.P. Reddy, S. Ramkumar, S. Agarwal, C.S. Gopinath, P.R. Nair, M.A.G. Namboothiry, Efficient organic photovoltaics with

improved charge extraction and high short-circuit current, J. Phys. Chem. C 121 (10) (2017) 5523–5530, https://doi.org/10.1021/acs.jpcc.7b01314.

[98] R. Sharma, H. Lee, K. Borse, V. Gupta, A.G. Joshi, S. Yoo, D. Gupta, Ga-doped ZnO as an electron transport layer for PffBT4T-2OD: $PC_{70}BM$ organic solar cells, Org. Electron. 43 (2017) 207–213, https://doi.org/10.1016/j.orgel.2017.01.028.

[99] S. Wu, J. Li, S.-C. Lo, Q. Tai, F. Yan, Enhanced performance of hybrid solar cells based on ordered electrospun ZnO nanofibers modified with CdS on the surface, Org. Electron. 13 (9) (2012) 1569–1575, https://doi.org/10.1016/j.orgel.2012.04.018.

[100] Z. Liu, F. Yan, Photovoltaic effect of $BiFeO_3$/poly(3-hexylthiophene) heterojunction, Phys. Status Solidi Rapid Res. Lett. 5 (10–11) (2011) 367–369, https://doi.org/10.1002/pssr.201105411.

[101] H. Aqoma, S. Park, H.-Y. Park, W.T. Hadmojo, S.-H. Oh, S. Nho, D.H. Kim, J. Seo, S. Park, D.Y. Ryu, S. Cho, S.-Y. Jang, 11% organic photovoltaic devices based on PTB7-Th: $PC_{71}BM$ photoactive layers and irradiation-assisted ZnO electron transport layers, Adv. Sci. 5 (7) (2018) 1700858, https://doi.org/10.1002/advs.201700858.

[102] E.D. Spoerke, M.T. Lloyd, Y.-J. Lee, T.N. Lambert, B.B. McKenzie, Y.-B. Jiang, D.C. Olson, T.L. Sounart, J.W.P. Hsu, J.A. Voigt, Nanocrystal layer deposition: surface-mediated templating of cadmium sulfide nanocrystals on zinc oxide architectures, J. Phys. Chem. C 113 (37) (2009) 16329–16336, https://doi.org/10.1021/jp900564r.

[103] J. Ajuria, I. Etxebarria, E. Azaceta, R. Tena-Zaera, N.F. Montcada, E. Palomares, R. Pacios, Novel ZnO nanostructured electrodes for higher power conversion efficiencies in polymeric solar cells, Phys. Chem. Chem. Phys. 13 (46) (2011) 20871–20876, https://doi.org/10.1039/C1CP22830G.

[104] N. Espinosa, Y.-S. Zimmermann, G.A. dos Reis Benatto, M. Lenz, F.C. Krebs, Outdoor fate and environmental impact of polymer solar cells through leaching and emission to rainwater and soil, Energy Environ. Sci. 9 (5) (2016) 1674–1680, https://doi.org/10.1039/C6EE00578K.

[105] R.K. Rosenbaum, T.M. Bachmann, L.S. Gold, M.A.J. Huijbregts, O. Jolliet, R. Juraske, A. Koehler, H.F. Larsen, M. MacLeod, M. Margni, T.E. McKone, J. Payet, M. Schuhmacher, D. van de Meent, M.Z. Hauschild, USEtox—the UNEP-SETAC toxicity model: recommended characterisation factors for human toxicity and freshwater ecotoxicity in life cycle impact assessment, Int. J. Life Cycle Assess. 13 (7) (2008) 532, https://doi.org/10.1007/s11367-008-0038-4.

[106] W.J.E. Beek, M.M. Wienk, R.A.J. Janssen, Efficient hybrid solar cells from zinc oxide nanoparticles and a conjugated polymer, Adv. Mater. 16 (12) (2004) 1009–1013, https://doi.org/10.1002/adma.200306659.

[107] Z. Yin, J. Wei, Q. Zheng, Interfacial materials for organic solar cells: recent advances and perspectives, Adv. Sci. 3 (8) (2016) 1500362, https://doi.org/10.1002/advs.201500362.

Metal oxides for hybrid photoassisted electrochemical energy systems

20

Noé Arjona[a], Jesús Adrián Díaz-Real[a], Catalina González-Nava[b], Lorena Alvarez-Contreras[c], and Minerva Guerra-Balcázar[d]
[a]Centro de Investigación y Desarrollo Tecnológico en Electroquímica, Pedro Escobedo, Querétaro, Mexico, [b]Universidad Politécnica de Guanajuato, Cortázar, Guanajuato, Mexico, [c]Centro de Investigación en Materiales Avanzados, Chihuahua, Chihuahua, Mexico, [d]Facultad de Ingeniería, División de Investigación y Posgrado, Universidad Autónoma de Querétaro, Santiago de Querétaro, Querétaro, Mexico

Chapter outline

1 Introduction

Electrochemical devices are those involving the conversion of chemical energy to electrical energy and vice versa. In this manner, chemical energy can be used to produce electricity, which is the case with electrochemical devices like fuel cells and batteries. On the other hand, electricity can be used to produce chemical reactions, which have been used for the development of new molecules and galvanostatic applications like the deposition of metals like zinc, copper, and others. An important matter in

Metal Oxides for Next-Generation Optoelectronic, Photonic, and Photovoltaic Applications
https://doi.org/10.1016/B978-0-323-99143-8.00016-X

electrochemistry is the energy saving, which is greatly influenced by the electrolytic cells for generation of molecules of interests of electrodeposits. In this way, the use of solar energy is of high interest because it is a free and highly available energy source. Moreover, solar energy can be used to assist typical electrochemical process increasing their performance, durability, and even selectivity. The great success of electrochemical systems assisted by photocatalysis is that energy can be produced, while organic compounds from wastewater can be treated. There are different photoassisted electrochemical energy systems like direct alcohol fuel cell, microbial fuel cells, photocatalytic fuel cells, Z-scheme photocatalytic fuel cells, and different hybrid designs of batteries as energy storage systems. The functionality of these systems greatly depends on the cell/battery design and, mainly in the electrode configuration involving the nanomaterial with unique properties, which make it to behave as a photoelectrocatalyst, which can be acting as photoanode or photocathode. In this chapter, a revision of photoelectrocatalysis principles, types of electrochemical energy conversion/storage systems, which are photoassisted, and the most recently reported nanomaterials based on metal oxide semiconductors (MOS) are revised and discussed. In terms of nanomaterials, we will provide a novel viewpoint of the most recently reported materials involving interesting ways to improve the activity: through surface engineering, morphology engineering, facet engineering or defect engineering. In addition, because of the vast literature for metal oxide semiconductors for photoassisted applications, selected electrochemical devices such as photoassisted fuel cells for those operating with methanol, ethanol, and glycerol were discussed in detail. Also, the introduction of photoassisted microfluidic fuel cells, as well as microbial fuel cells, Z-scheme cells, and metal-air batteries are also discussed.

2 Principles of photoelectrocatalysis

To discuss photoelectrocatalysis, we will build the concept from its roots. We can define catalysis as the increase in the reaction rate that leads to obtain specific products using a substance (catalyst) that does not undergo changes. These catalysts can be found either in a homogeneous or heterogeneous phase. The word electrocatalysis is an extension of catalysis, where heterogeneous catalysis with a solid/liquid interface is involved. For this reason, the catalysis of electrochemical reactions is often called electrocatalysis. Derived from the above, when an electrochemical reaction is catalyzed by an agent that interacts with light (photons, specifically), this receives the name of photoelectrocatalysis.

Thus, it is notable to mention that photoactive materials are also referred to as photoelectrocatalysts. These photonically excitable materials are, mostly but not limited to, semiconductors. Hence, when we discuss about photoelectrocatalysis, semiconductors are naturally involved. Semiconductors that interact with visible light (400–700 nm) are of particular interest, since this is the spectral range that contains the highest energy density in the solar spectrum. Apart from MOS, there is a number of material subfamilies in the semiconductors, such as chalcogenides and perovskites [1–10], that will not be addressed in the present text. The processes associated with photoelectrocatalysis are detailed with far more depth in dedicated literature [11,12], but some of the relevant aspects that must be considered for optimal selection and use in energy conversion systems are discussed as follows.

The generalities of the photoelectrocatalysis can be classified in two main groups: those related with internal processes (photon interactions, electronic configuration, and crystallinity) [11], and those associated with the specific electrolyte/reaction. In regard to the first group, MOS interact with light due to their electronic configuration, followed after photonic excitation (process 1, Fig. 1A); they can promote electrons residing in their highest occupied energy level (valence band) toward a higher, electronically-unoccupied energy level (conduction band). The difference in energy that separates these energy states is known as the "forbidden band" or bandgap. Consequently, photons require a minimum energy equal to or greater than the bandgap to promote electrons from the valence band to the conduction band of MOS. Thus, the photoactive materials of interest have bandgaps ranging from 3.3 eV (such as TiO_2) to 1.1 eV (such as Si). As a result of the electronic promotion, an electronic vacancy ("hole") will have been left in the valence band, thus maintaining the electroneutrality of the system, and these pair of entities are called "charge carriers". If there is no impeding force, such as an internal or external electric field, the electron-hole pair will recombine and the excess energy is dissipated by thermal or momentum mechanisms, as shown in Fig. 1A (process 2 and 3). Therefore, the density of charge carriers will be directly related to other properties, such as the intrinsic conductivity of the material, in order to internally move these charge carriers.

The second group of considerations strongly involves the chemical environment immediately to the MOS: the electrolyte and the reaction of interest. For charge carriers to be exploitable in specific reactions, the band position of MOS must be aligned with the energy level of the redox couple in the electrolyte for the reaction of interest. Fig. 1A depicts these reactions in step 4, for the evolution of oxygen (H_2O/O_2) and hydrogen (H^+/H_2) cases. In electrochemistry, the reactions of interest in systems for the conversion and use of solar energy are associated with photoelectrolysis of water (water splitting), the oxidation of organic fuels (e.g., methanol, ethanol, glycerol, etc.), or waste in aqueous media, or with the reduction of some component dissolved in the electrolyte (e.g., reduction of CO_2).

Finally, one extra but important aspect has to be accounted for in photoelectrocatalysis: the electrochemistry of semiconductors. Given the fact that semiconductors typically have a charge carrier density lower than that of the typical electrolytes, the electrode will deplete its charge carrier density after developing the equilibrium with the interface. This depletion modifies the energy value of both the conduction and the valence bands from the interface to the inner part of the MOS. This region, where the charge carrier density is affected, is known as the space-charge region and corresponds to the volume that is juxtaposed to the electrolyte up to the inner part of the electrode where no changes are observed [11]. Since the charge is a parameter that will be constantly changing due to the external polarization, the reactions occurring across the interface will not only depend on the applied potential but also on the intrinsic value of the charge carrier density of the semiconductor. The latter translates into a not energetic (like in metals or metalloids) but probabilistic charge

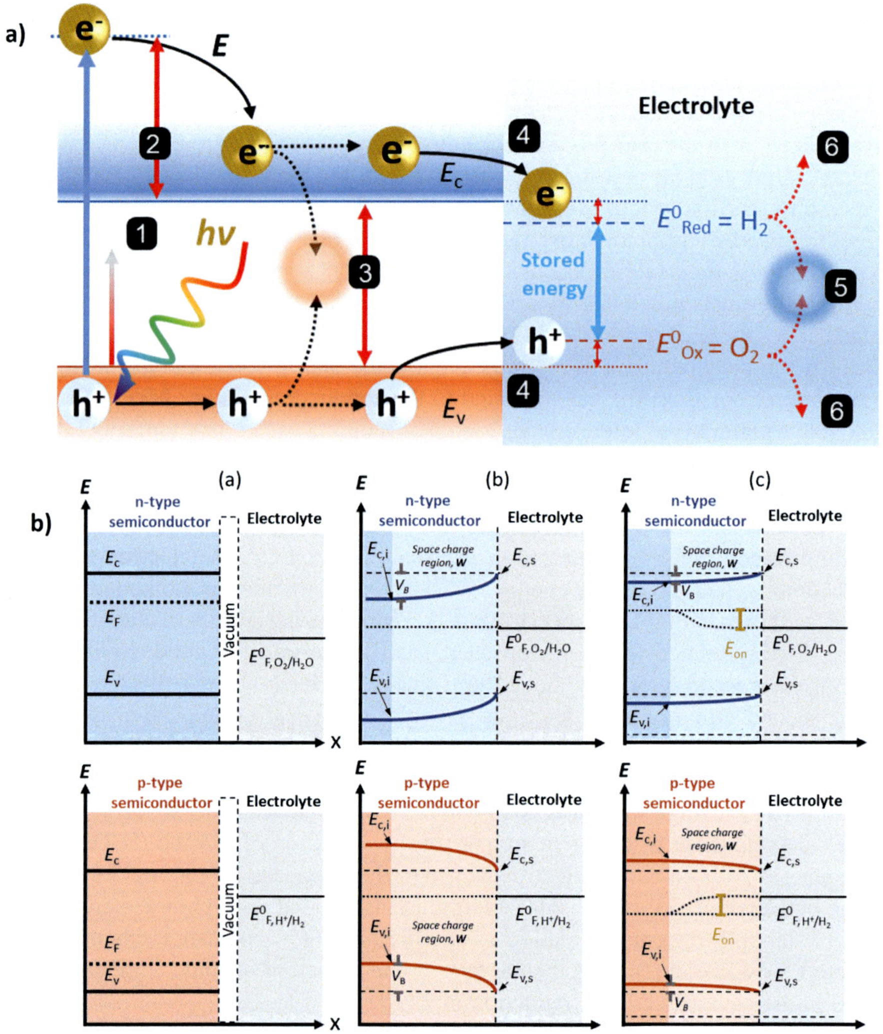

Fig. 1 (A) Energy barriers decay mechanisms to be overcome in a photosynthetic system at the photoanode/electrolyte interface: (1) insufficient photon energy, (2) thermalized excess photon energy, (3) recombination of charges, (4) overpotential for charge transfer at the interface, (5) spatial separation of products, and (6) separation of products. (B) The band position and band bending process for n-, p-type semiconductors. (a) before contact, (b) after reaching the equilibrium, and (c) at the equilibrium and under illumination. E_F = Fermi level for the semiconductor, E_F^0 = Fermi level for the redox couple in the electrolyte, E_{on} = photovoltage generated from the irradiation of the interface, V_B = magnitude of the band bending. E_c, $E_{c,i}$, $E_{c,s}$ denote the energy level of the conduction band before the equilibrium, the level at the interior, and surface after the equilibrium, respectively. The latter applies to the valence band levels (E_v, $E_{v,i}$, $E_{v,s}$).

Reprinted with permission from J.A. Díaz-Real, T. Holm, N. Alonso-Vante, Photoelectrochemical hydrogen production (PEC H2), in: Current Trends and Future Developments on (Bio-)membranes, 2020, Elsevier, pp. 255–289.

transfer where the compensation of the space-charge capacitance has to be accounted for. Fortunately, photonic excitation of the semiconductor reduces the energy required to drive these charge transfers. All these phenomena are summarized in Fig. 1B. Thus, the space-charge region is a critical aspect to consider in photoelectrocatalysis as it will define the applicability of the MOS.

3 Photoassisted fuel cells

Fuel cells are electrochemical devices, which convert the chemical energy into electrical energy. Fuel cells are typically classified according to the operating temperature in low-, medium-, and high-temperature fuel cells [13]. Among the low-temperature fuel cells (<120°C), direct alcohol fuel cells (DAFCs) are promising energy conversion technologies because the facility of handling, storage, and transport of liquid fuels [14]. As with any fuel cell, DAFCs are constituted by the anode, cathode, membrane, and the external electrical conductor (Fig. 2A). The anode is typically a metallic electrocatalyst like Pd and Pt or bimetallic materials composed of these precious metals and a metal oxide [15,16]. The cathode is typically based on Pt and Pt-based materials [17]. However, the latest advances in alkaline membranes have promoted the use of Pt-free electrocatalysts like metal oxides, layered double hydroxides, spinels, perovskites, metal-organic frameworks, and nonmetallic materials like carbon allotropes

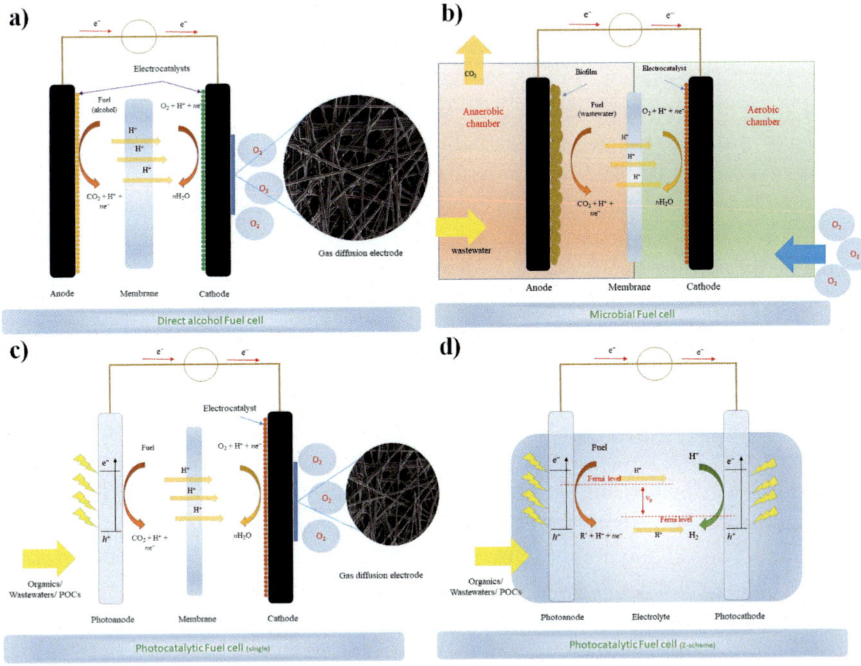

Fig. 2 Schematic representation of the components of (A) a direct alcohol fuel cell, (B) a microbial fuel cell, (C) a photocatalytic fuel cell and (D) a Z-scheme photocatalytic fuel cell.

doped with heteroatoms [18–20]. In acidic conditions, the fuel (alcohols like methanol, glycerol, ethanol, etc.) is oxidized in the anode to CO_2-releasing protons and electrons [21]. Complex alcohols like ethanol, glycerol, and ethylene glycol tend to produce different intermediates instead of CO_2 [22]. On the other hand, the typical oxidant is the oxygen from air because it is abundant and free. Likewise, the use of gas diffusion electrodes in air-breathing fuel cells allows the direct use of oxygen from air, resulting in a decrease of the cell volume and weight.

Microbial fuel cells (MFCs) as well as DAFCs are electrochemical energy conversion systems, and their greatest advantages are related to the combined effect of wastewater treatment and energy conversion from using wastewater as fuel (Fig. 2B). In the MFC, bacteria are used as catalysts in the anode (named as bioanode) to produce electricity, while organic and inorganic compounds from wastewater are mineralized (oxidized to CO_2) [23]. Bacteria are grown in the bioanode forming a biofilm in an anaerobic chamber, and the released protons and electrons are then employed for the oxygen reduction (ORR) as the cathodic reaction. In the absence of oxygen, the cathodic reaction is the hydrogen evolution reaction (HER), providing an extra renewable character to MFCs due to the production of biohydrogen [24–26].

The multiple issues presented in MFCs like limitations in the electron transfer process, biofouling in the membrane, long start-up due to the formation of the biofilm and low-cell performance [27], has motivated the search for alternatives to the bioanode, with photoanodes being one of these alternatives. A photoanode is capable of to produce electricity while degrading organic compounds from the generation of electron-hole pairs (e^-, h^+) through the light irradiation. In this manner, the photoassisted (light-driven) fuel cells named as photocatalytic fuel cells (PFCs) emerged as an alternative to MFCs. A PFC consists of n-type semiconductor as photoanode and a conventional GDE as cathode (Fig. 2C). The photoanode produces electrons, which can be used to supply the energetic requirements of different electronic apparatus, while the pollutants can be degraded directly by the holes, or indirectly by reactive oxygenated species formed from the light irradiation. As mentioned, the MFCs can produce hydrogen. In PFCs, in order to achieve higher efficiencies, the cathode is replaced by a photocathode consisted of a p-type semiconductor (Fig. 2D). The PFCs with both photoelectrodes are denominated as dual PFC or as Z-scheme PFCs [28].

In this manner, PFCs can be used for energy conversion (using oxygen in the cathode) or for hydrogen production in anaerobic chambers (in these cells the open-circuit potential is almost zero, making impossible to have a galvanic cell) while treating wastewater. In the single PFC, there are challenges related to the photoanode, the membrane and the cathode. In the case of membranes, the use of benchmarked Nafion membranes increases costs making the PFCs commercially inviable. The main issues to solve are related to the ionic conductivity and mechanical properties of membranes [29,30]. The oxygen reduction reaction (ORR) is usually the limiting step in fuel cells due to its sluggish reaction kinetics, and as mentioned, Pt is the typical electrocatalyst reported for the ORR in PFCs [27].

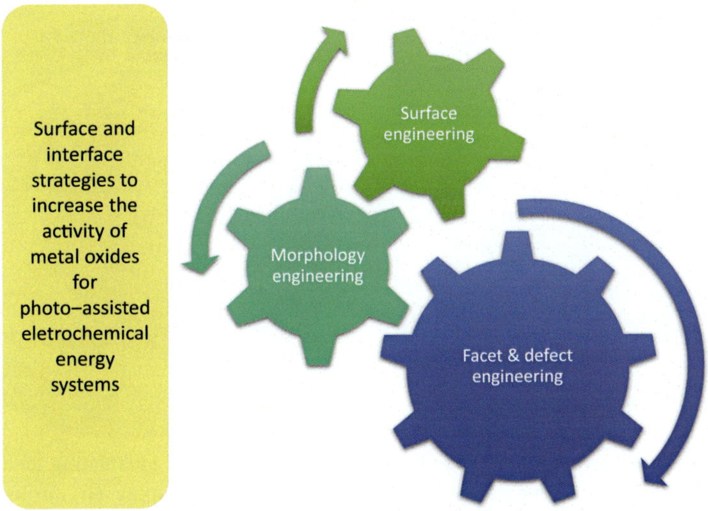

Fig. 3 Overall strategies to improve the activity of materials for photocatalytic fuel cells.

However, since a materials viewpoint, there are different strategies, which can be used to improve the sluggish reaction kinetics, while decreasing the use of Pt or eliminating it (Fig. 3) [31]. These general strategies are also employed during the development of new photoanodes [27,32].

Surface engineering consists of modifying or doping one or both, cathodic material and/or n-type semiconductor as photoanode, with noble, non-noble metals, or compounds to shift their electronic properties [33–36]. The morphology engineering consists of changing the shape and size of materials because the intrinsic activity depends on the micro/nanostructure. Shapes can be categorized in 0D, 1D, 2D, and 3D materials comprising nanoplates, nanowires, nanoflakes, nanobelts, nanopores, nanotubes, nanorods, nanorings, nanoflowers, hollow particles, hemispheres, hierarchical structures, among others. Faceting engineering is related to the shift in the crystallite orientation toward particular facets to boost the reactivity and selectivity due to changes in the surface energy [37,38].

Defect engineering is perhaps the most fashionable way to improve the activity. It consists of adding surface defects to materials like metal oxides and carbonaceous materials to break the periodicity of the crystalline structure, causing a redistribution of the chemical and electrical properties of nanomaterials [39]. Defect engineering pursues to modify the species, contents, and location of the surface defects. There are different types of surface defects categorized in point, line, pattern, and interlayer defects [40]. The vacancies, dislocations, grain boundaries, chemical functionalization, and metallic/nonmetallic doping are the most studied surface defects [41]. The defect engineering also promotes the enhancement of different reactions like the oxygen reduction reaction (Fig. 4A), and there are different methods to promote defects like plasma, ball milling, annealing, etching, through electrochemical

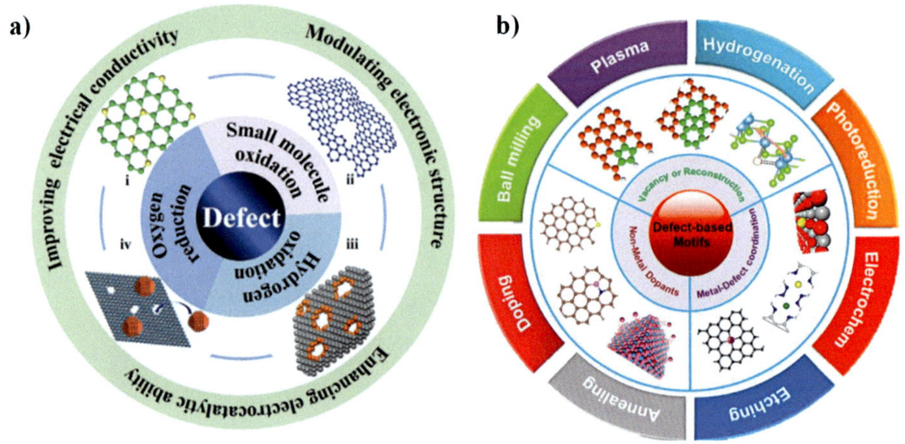

Fig. 4 (A) Surfaces defects to improve the electrical conductivity, modulating the electronic structure to improve the activity of nanomaterials (i), heteroatom doping, (ii) intrinsic defects, (iii) defective metallic catalysts, and (iv) defective supports. (B) Techniques to enable surface defects on nanomaterials.
(A) Reprinted with permission from W. Li, D. Wang, Y. Zhang, L. Tao, T. Wang, Y. Zou, Y. Wang, R. Chen, S. Wang, Adv. Mater. 32 (2020) 1907879. (B) Reprinted with permission from Y. Jia, K. Jiang, H. Wang, X. Yao, Chem 5 (2019) 1371–1397.

techniques, and others (Fig. 4B) [42]. Some of these strategies have been used to improve the performance of PFCs [43–45].

As above mentioned, the purpose of these modifications is to achieve a higher PFC performance. In this manner, the cell efficiency of PFCs is calculated by the incident photon-to-current efficiency (IPCE, Eq. 1):

$$\text{IPCE} = \frac{1240 \times J_{ph}}{\lambda \times P} \times 100\% \tag{1}$$

where J_{ph} is the highest current density obtained with the minimum polarization possible, reported in mA cm^{-2}; P is the incident irradiance in mW cm^{-2} at a given wavelength (λ). The number 1240 represents a conversion factor that includes the Planck constant, the speed of light, and the fundamental charge to match the units. This equation allows obtaining the IPCE in percentages from 0 to 100%.

Different photoanodic materials have been synthesized and tested in a single PFC operated with different fuels. The most recent works are described in Table 1, being observed that the mixture of RuO_2 and TiO_2 offers the highest photocurrent density.

Mikrut et al. [37], tested PFCs using different fuels like methanol, ethanol, 2-propanol, and others (Fig. 5), finding that methanol and ethanol are promising fuels for the photoassisted fuel cells, and it is related to the complexity to break the chain of larger fuels like glycerol. In the following two sections are devoted to analyzing photoanodes for methanol and ethanol oxidations.

Table 1 Photocatalytic fuel cell performances of recently reported photoanodes in single-chamber PFCs.

Photoanode	Photoanode loading (mg cm^{-2})	Cathode	Cathode loading (mg cm^{-2})	Electrolyte	Fuel	Photocurrent density (mA cm^{-2})[a]	OCV	IPCE (%)	Ref.
RuO_2-TiO_2	3.5	Pt black	1	1 M Na_2SO_4	10% v/v ethanol	32.0		–	[46]
Nanocrystalline TiO_2	NR	Co/N-RGO	5	0.5 M NaOH	10% v/v ethanol	~5.5	1.12	–	[35]
Belt-shaped TiO_2 crystals with exposed {100} facets	NR	NR	NR	0.1 M KOH	5% w/w methanol	0.496	1.00	0.19	[37]
Burr-like Ag-TiO_2	NR	Pt/C on carbon cloth	NR	1 M H_2SO_4	1 M methanol	6.2	0.81[b]	–	[47]
Burr-like Ag-TiO_2	NR	Pt/C on carbon cloth	NR	None	Brewery effluent (alcohols, sugars)	1.3	0.88	9.4[c]	[47]
$SrTiO_3$/TiO_2	NR	Pt foil	Solid electrode	0.2 M KCl	Triethanolamine	0.32	1.3	0.24	[48]
W-$BiVO_4$/V_2O_5	NR	Pt plate	Solid electrode	0.5 M Na_2SO_4/0.5 M Na_2SO_3 (anode)	Sulfate species	8.79	0.85	1.35	[49]
Coral-like WO_3/W	NR	Pt foil	Solid electrode	0.5 M Na_2SO_4	Perfluoro octanoic acid	0.31	0.51	5.72%	[50]

[a] The photocurrent density was reported at the short-circuit condition. Nomenclature: N is Nitrogen doped, RGO = reduced-graphene oxide, NR = not reported.
[b] Data obtained under UV irradiation.
[c] Coulombic efficiency.

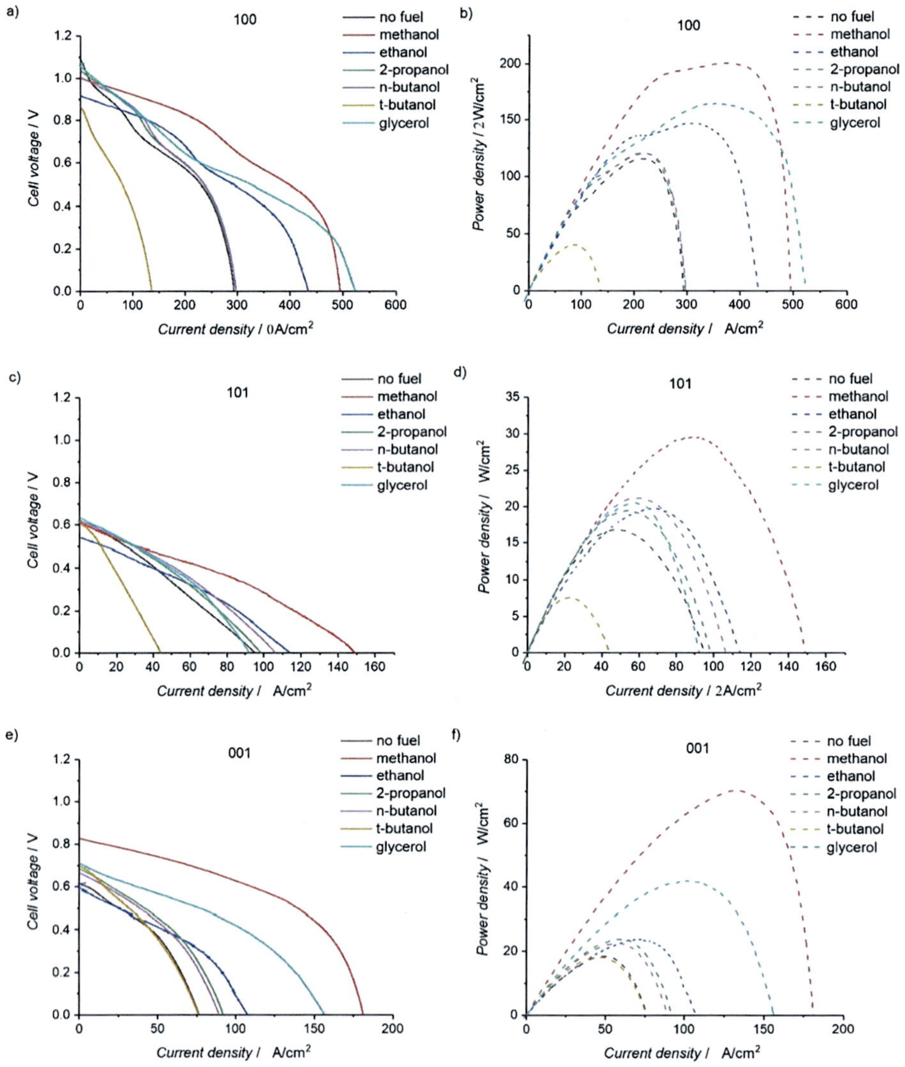

Fig. 5 Polarization curves and power densities for different fuels using anatase TiO_2 with different preferentially exposed facet: belts {100}, truncated bipyramids {101} and sheets {001}.

Reprinted with permission from P. Mikrut, D. Mitoraj, R. Beranek, W. Macyk, Appl. Surf. Sci. 566 (2021) 150662.

4 Nanomaterials for the photoassisted methanol oxidation

Some oxides like TiO_2, SnO_2, and WO_3 have been proposed as excellent photoanodes for the methanol photooxidation due to their optical properties and high resistance to corrosion. Among different inorganic metal oxides, titania is widely used as a support/promoter in fuel cell anode materials due to its photocatalytic properties and excellent corrosion resistance at different electrolytic media. However, the disadvantage of using TiO_2 as a catalyst support is its poor electrical conductivity and small surface area, resulting in low dispersion of cometallic nanoparticles. To overcome this obstacle, titania is typically mixed with highly conductive materials like carbonaceous-based materials in order to increase conductivity and superficial area [51].

Another strategy for the development of nanomaterials for methanol photooxidation was the use of multiple dimensions of TiO_2, including zero-dimensional (0D), one-dimensional (1D), two-dimensional (2D), and three-dimensional (3D). In particular, well-designed 1D TiO_2 nanorods (NRs) displayed superior photoelectron transfer an enhanced methanol photooxidation [52]. One-dimensional TiO_2 nanorods not only prevents aggregation of Pt active sites and improve the adsorption and desorption of target molecules during the methanol oxidation reaction, but this special structure with high slenderness ratio also retains the advantages of excellent photogenerated carrier transport and low recombination rate, which is beneficial to the activity and stability of Pt catalysts.

The methanol photooxidation mechanism in TiO_2-based materials occurs in the following way [51]:

When, the photocatalyst is irradiated with UV an electron/hole pair (e^-/h^+) is generated by promotion of an electron from the valence band (V_B) to the conduction band (C_B) (Eq. 2):

$$TiO_2 + h\nu \rightarrow h^+ + e^- \tag{2}$$

Then, hydroxyl radicals are formed by the oxidation of H_2O molecules or OH^- ions adsorbed on TiO_2 surface, according to (Eq. 3):

$$h^+ + H_2O \rightarrow OH^\cdot + H^+ \tag{3}$$

$$h^+ + OH^- \rightarrow OH^\cdot \tag{4}$$

Finally, methanol can be oxidized by two paths, the direct oxidation that involves oxidation by hydroxyl radicals [53] (Eqs. 5 and 6):

$$CH_3OH + OH^\cdot \rightarrow CH_2OH + H_2O \tag{5}$$

$$CH_2OH^\cdot \rightarrow HCHO + H^+ + e^- \tag{6}$$

Or, the indirect oxidation in which the holes oxidize methoxy species from methanol as the following equations [54] (Eqs. 6 and 7):

$$CH_3O^- + h^+ \rightarrow CH_3O^\circ \tag{7}$$

$$CH_3O^\cdot \rightarrow HCHO + H^+ + e^- \tag{8}$$

In the literature, it is reported that direct oxidation has an order of magnitude more reactive than molecularly adsorbed methanol for hole-mediated photooxidation, this oxidation occurs by the cleavage of methoxy photo decomposition through of C-H bond forming adsorbed formaldehyde.

In some works, 1D was employed the evidence shows that the morphology of 1D TiO_2 NRs expands the dispersion of Pt while decreasing the poisoning effect. The photocatalyst acts forming photoelectrons and holes and then, the hydroxyl radicals formed can oxidize methanol to CO_2. Additionally, they can oxidize adsorbed molecules on the photocatalyst surface, being these processes capable to improve the methanol oxidation [52].

Additional materials, which have been tested for the methanol photoelectrooxidation includes two-dimensional Fe_2O_3 nanosheets, which facilitates the photoassisted MOR activity by providing abundant catalytically active sites as well as thin flat structures, where the photogenerated carriers are directly transferred from inside to the surface. In addition, when the thickness of the semiconductor is reduced to a certain extent, the density of states and surface charge density near the Fermi level increase, which is favorable for the transfer of photogenerated electrons to Pt and decreases the overpotential for the formation of Pt-OH_{ads}. It was found that under exposure to visible light, and under simulated sunlight illumination, the MOR activity of 2D Fe_2O_3 nanoplates is one-fold higher than in the dark conditions. This may be attributed to the unique 2D structure of Fe_2O_3, shortening the time required for carrier migration, while decreasing the rate of carrier recombination. Therefore, holes with stronger oxidizing properties are available to form OH radicals in 2D Fe_2O_3 nanosheets to enhance methanol oxidation. On the other hand, the large specific surface area, and more active sites of the 2D Fe_2O_3 nanosheets also play an important role in the photocatalytic MOR [55].

Other strategy to improve the photoactivity is through the combination of some metals like PtNi alloy nanoparticles with different compositions supported on carbon-doped TiO_2 nanotube array (PtNiCTiO_2NTs). It was found that the atomic composition of Pt and Ni have a strong influence on methanol photooxidation. In this material, it is possible to observe that in presence of light the methanol was oxidized more efficiently with less CO_{ads} on the catalyst [56]. $La_2Ti_2O_7$ has also been used for the photocatalytic MOR in combination with Pt [57].

5 Nanomaterials for the photoassisted ethanol oxidation

For ethanol photooxidation, it was found than conventional photocatalyst like TiO_2 are not efficient. In this manner, some 2D inorganic semiconductors like bismuth oxyhalides (BiOX) are employed [58]. BiOX has great chemical stability, and

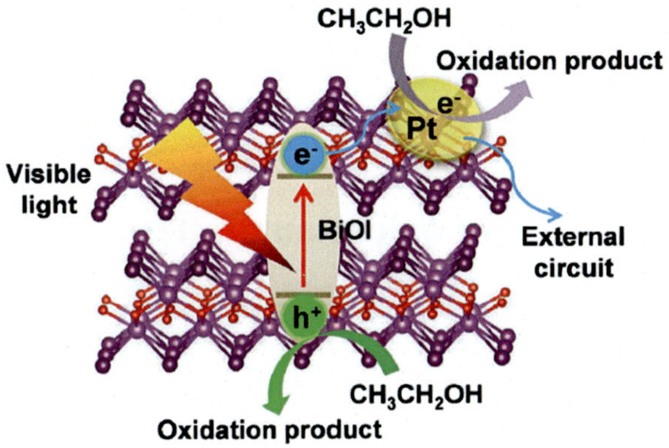

Fig. 6 Graphic illustration for the ethanol electro- and photocatalytic oxidation with Pt-BiOI electrode below visible light.
Reprinted with permission from C. Zhai, J. Hu, M. Sun, M. Zhu, Appl. Surf. Sci. 430 (2018) 578–584.

corrosion resistance, being BiOI the most efficient bismuth oxyhalide because its band gap (1.7 eV) [59]. According to literature, 2D BiOI nanoplates synthesized by hydrothermal method and decorated with Pt nanoparticles have shown good activity for the ethanol photooxidation in alkaline medium under visible light; where, the combined effect of photocatalytic properties and enhanced area from its 2D structure has led to increased activity and stability. In Fig. 6, the synergistic effect of this photocatalyst is schematized.

In order to increase the electrical conductivity, it has been reported the combination of BiOI with carbon-based materials like carbon nitride (g-C_3N_4) nanosheets [60]. This 2D/2D heterostructure was decorated with platinum. This novel material shows an increased activity as shows in Fig. 7, where Pt/g-C_3N_4/BiOI photocatalyst is excellent for the photoassisted ethanol oxidation, the synergistic effect between 2D

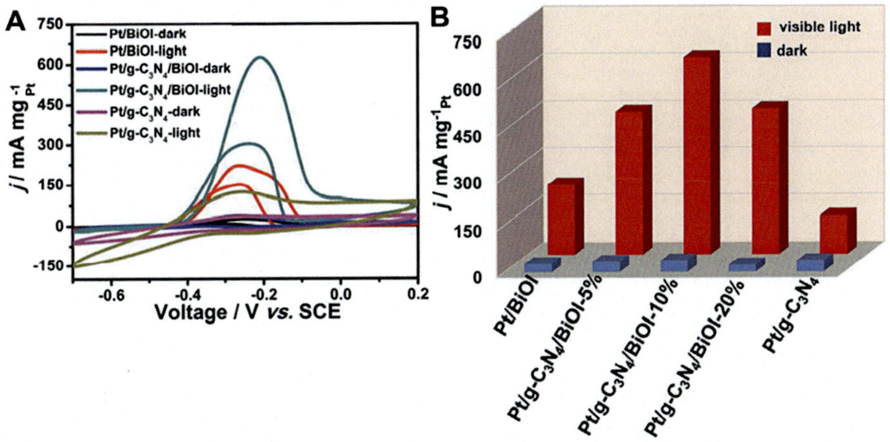

Fig. 7 Photocatalytic activity comparison between several materials for ethanol photooxidation evaluated with cyclic voltammetry (A), and peak current maxima (B).
Reprinted with permission from J. Hu, C. Zhai, H. Gao, L. Zeng, Y. Du, M. Zhu, Sustain. Energy Fuel 3 (2019) 439–449.

g-C_3N_4 and 2D BiOI leads to an efficient charge separation on both materials, resulting in greater photoactivity performance. On the other hand, the wide-ranging absorption of visible light of BiOI, has consequence a great photoactivity for ethanol photooxidation.

6 Photoassisted electrochemical oxidation of glycerol

Glycerol is an important raw material because its oxidation produces several high-value products. Glycerol valorization can be done in several ways, including the photocatalytic oxidation. The photoassisted glycerol oxidation can be achieved in absence and presence of oxygen according to the following reactions (Eqs. 9 and 10):

$$C_3H_8O_3 + 3H_2O \rightarrow 3\ CO_2 + 7\ H_2 \tag{9}$$

$$C_3H_8O_3 + 7/2\ O_2 \rightarrow 3\ CO_2 + 4\ H_2O \tag{10}$$

It has been reported that the glycerol photooxidation in presence of oxygen can occur directly or indirectly by the photogenerated holes (Fig. 8). In absence of oxygen, the photogenerated electrons can oxidize the water to produce hydrogen (Fig. 8d). It is evident from the mechanism that regardless of whether the reaction takes place in the presence of oxygen or not, the generated holes contribute to the oxidation of glycerol and electrons react with either oxygen or water [61].

The photocatalytic value-added of glycerol to hydrogen and value-added liquid products can be carried through its direct reaction with photogenerated holes (hole-mediated mechanism) or indirect reaction with photogenerated hydroxyl groups (radical-mediated mechanism). Direct photocatalysis occurs on the catalyst surface and thus, the substrate should first adsorb to the surface. However, indirect

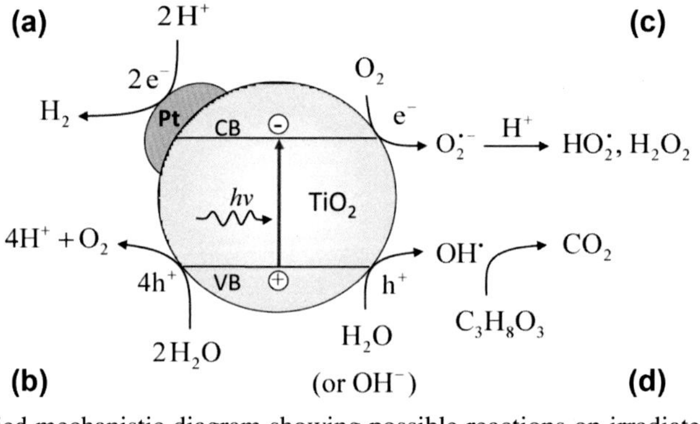

Fig. 8 Simplified mechanistic diagram showing possible reactions on irradiated Pt/ TiO_2 suspensions under conditions of water splitting (a and b), glycerol oxidation (c and d), and glycerol photoreforming (a and d).
Reprinted with the permission of P. Panagiotopoulou, E.E. Karamerou, D.I. Kondarides, Catal. Today 209 (2013) 91–98.

photocatalysis can take place on the catalyst surface or in the bulk of the liquid. In addition to the reaction mechanism, the deep oxidation of glycerol eventually leads to the production of hydrogen and CO_2 or water and CO_2, representing photo reformation and photooxidation of glycerol, respectively.

From a sustainability perspective, photocatalytic value-adding of glycerol to hydrogen and value-added liquid products is a promising approach. It has been proven that a stoichiometric amount of hydrogen can be produced from glycerol, showing that the maximum possible chemical energy of glycerol can be converted into this valuable energy carrier. It must be emphasized that, in order to study the ability of photocatalysts for glycerol valorization, their performance must be investigated under optimized operating parameter conditions. Therefore, in addition to synthesizing new photocatalysts and their enhancement, various studies have been carried out on the optimization of synthesis and operating parameters [62].

A wide variety of photoelectrocatalytic materials have been used for the photoreaction of glycerol, among which are titanium dioxide or oxide-metal mixtures where the metal can be Pt, Cu, Pd, Au, or Bi, among others. It was found that the percentage of cocatalyst in the photocatalyst was very important because it can create a barrier that prevents the incidence of photons, increasing the rate of recombination, or avoiding the access of glycerol to the surface of the catalyst [63].

7 Photoassisted microfluidic fuel cells

Microfluidic PFCs (μPFCs) consist of a miniaturized version of a PFC. The particularity of microfluidic fuel cells (μFFCs) consists in the range of volumes that these devices use, typically in microliters, hence their name. Initially, microfluidics was mainly used for actuators and sensing applications and dedicated literature has been published dealing with such application [64–66]. Recently, the use of microfluidic devices has enabled researchers with enormous possibilities for the study of particular conditions that were not possible to assess in larger systems, such as the elimination of ohmic drops due to the separators, the fine control of mass flow in the species avoiding turbulence, and enhanced reaction rates in small volumes overflowing the electrodes. These platforms are excellent idealized reactors that allow for the exploration of substantially different concepts in PFCs while maintaining control over the parameters that define the system. In addition to these advantages, μPFCs include the light-stimuli to the parameters that can be controlled, triggering the reactions mentioned in previous sections.

As a direct heritage from PEC cells, it is often found in μPFCs, the use of transparent conductive glasses, such as fluorine or indium-doped tin oxide (FTO and ITO, respectively), to deposit MOS onto their surface. Several considerations for the optimal assembly and design in microfluidic sensor can be rationalized for μPFCs [64,67]. Li et al., reported the assembly of a cell where a photoanode comprising CdS–ZnS/TiO_2 deposited onto FTO glass, and a Pt/C air-breathing cathode was used for the degradation of methylene blue (MB) of up to 83% and maximum power density of $0.58\,mW\,cm^{-2}$ (Fig. 9A) [68]. A similar system oriented toward CO_2 reduction was presented by Xie et al., with an FTO/TiO_2 photoanode and Pt/C cathode

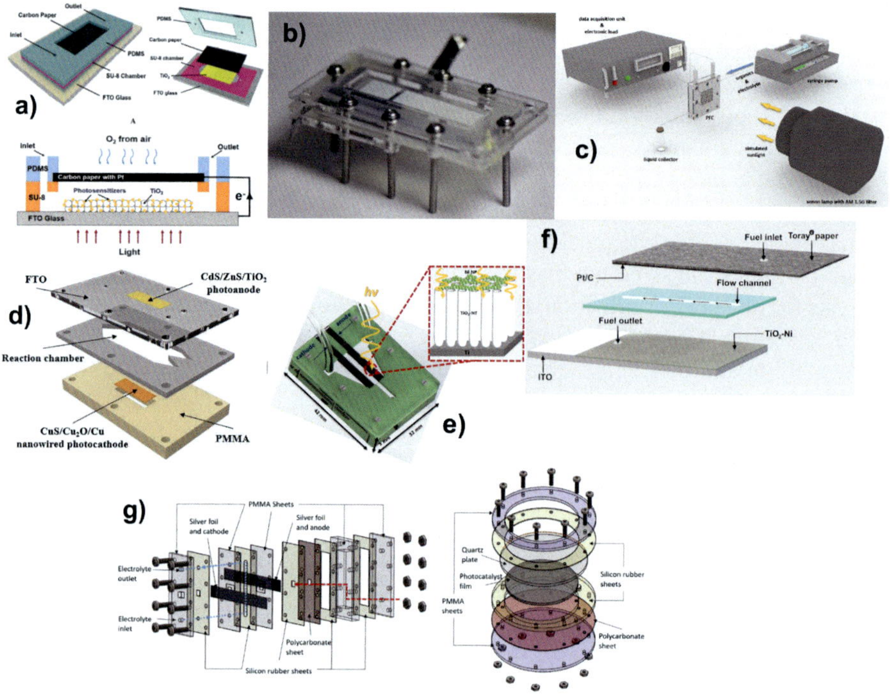

Fig. 9 Examples of μPFCs using (A) a CdS–ZnS/TiO_2/FTO photoanode, (B) a TiO_2/FTO photoanode, (C) a CdS–ZnS/TiO_2/C photoanode. (D) A dual μPFCs profiting from a photoanode and a photocathode. Single anodic photoelectrode μPFCs based on (E) Ti/TiO_2/Ni, and (F) FTO/TiO_2/Ni. (G) A dual photoelectrocatalytic system for hydrogen production/ consumption from methanol/water mixtures.
Reproduced with permission from L. Li, G. Wang, R. Chen, X. Zhu, H. Wang, Q. Liao, Y. Yu, Lab Chip 14 (2014) 3368–3375; F. Xie, R. Chen, X. Zhu, Q. Liao, D. Ye, B. Zhang, Y. Yu, J. Li, J. CO2 Util. 32 (2019) 31–36; X. He, M. Chen, R. Chen, X. Zhu, Q. Liao, D. Ye, B. Zhang, W. Zhang, Y. Yu, J. Hazard. Mater. 358 (2018) 346–354; H.-N. Wang, X. Chen, R. Chen, X. Zhu, Q. Liao, D.-D. Ye, B. Zhang, Y.-X. Yu, W. Zhang, J.-W. Li, J. Power Sources 435 (2019) 226766; J. A. Díaz-Real, E. Ortiz-Ortega, M. P. Gurrola, J. Ledesma-García, L. G. Arriaga, Electrochim. Acta 206 (2016) 388–399; D. Dector, D. Ortega-Díaz, J.M. Olivares-Ramírez, A. Dector, J.J. Pérez-Bueno, D. Fernández, D.M. Amaya-Cruz, A. Reyes-Rojas, Int. J. Hydrogen Energy 46 (2021) 26163–26173, Y.H. Kwok, Y. Wang, Y. Zhang, H. Zhang, F. Li, W. Pan, D.Y.C. Leung, Int. J. Hydrogen Energy 45 (2020) 21796–21807.

(Fig. 9B), obtaining up to circuit current density of $0.175\,mA\,cm^{-2}$ and a maximum power output of $0.110\,mW\,cm^{-2}$ [69].

However, carbonaceous materials can also be used for the deposition of MOS, as reported by He et al. [70], and as is shown in Fig. 9C, who prepared the same CdS–ZnS/TiO_2 via spray deposition but over carbon paper for ethanol oxidation as fuel. The system used in this work was very similar to a typical fuel cell. The integration of photocathodes has also been explored for the cases of CuO [71] and $CuS/Cu_2O/Cu$ for O_2 reduction (Fig. 9D) [72]. Another approach for MOS support was the one proposed by Diaz-Real et al. and depicted in Fig. 9E, where a TiO_2 nanotube array was grown over Ti foil by electrochemical anodization [73]. Subsequently, the photoanode was decorated with Ni nanoparticles and paired with a Pt/C Toray carbon paper as cathode for methanol oxidation. Similarly, Dector et al., used an $FTO/TiO_2/Ni$ photoanode but innovating by using as fuel real-human urine (Fig. 9F), achieving a current density of up to $1.7\,mA\,cm^{-2}$ [74]. Lastly, a novel coupling of two devices: one involving a TiO_2/Pt photoanodic disk chamber evolving and feeding directly H_2 (from methanol/water electrolytes) to a secondary microfluidic cell (Fig. 9G) [75]. This type of array was capable of generating a maximum of $213\,mW\,cm^{-2}$ and achieving up to 78% of methanol utilization. These examples demonstrate the flexibility of the μPFCs to design particular experimental setups and serve as advanced proof-of-concept platforms. Possible future technologies can profit from fast screening by using initial μPFCs prototypes.

8 Photoassisted microbial fuel cells: Principles and fundamentals

Microbial fuel cells (MFCs) are bioelectrochemical devices for pollutants degradation and power generation by employing microorganisms as catalysts (Fig. 2C). MFCs have relatively low overall performance because of the problems found in the electrodic materials, membrane employed in MFCs, cell configuration, and electron transport mechanisms. A MFCs is essentially a photobioelectrochemical system (photoassisted microbial fuel cells), which is a revolutionary green technology for waste treatment, hydrogen generation, and electricity production. The literature refers different configurations of photobioelectrochemical system: photoelectrode MFC, photoelectrocatalytic MFC, photoelectrocatalytic MFC hybrids, and photosynthetic MFC [71,76]. In these MFCs, the microorganism is grafted to the semiconductor electrode, where photogenerated h^+/e^- pairs are generated under irradiation. The microorganisms are embedded as catalysts to perform the oxidation/reduction reactions, while the photocatalyst is capable of performing photolysis of water molecules to generate active radicals of oxygen species, promoting the oxidation of organic pollutants. Concomitantly, exoelectrogens produce electrons from pollutants oxidation on the surface of the semiconductor, quenching the photogenerated holes. The current is conducted through the external circuit to the cathode for power production [71]. The microorganisms can strengthen the excitation, separation, and transportation of electrons in the semiconductor and the biodegradation process for contaminants in the anolyte is promoted [76].

Photocatalytic materials used in MFCs offer a higher generation of electrons and holes, strong absorption of photon energy, and presence of $OH^•$ and -€OO$^•$ (super-oxide-radicals) to boost the activity for pollutants, degradation, and energy conversion [77]. The chemical composition, specific surface area, porosity, direct charge transportation, light adsorption capacity, morphological alignment, band gap, binding energy, and crystallography of photoactive semiconductors are very important to improve the activity. Different semiconductor nanomaterials have been developed including single, binary, ternary, and quaternary heterojunction of zero-, one-, two-, and three-dimensional materials. Geometrical structures as nanorod, nanowire, nanosheet, dumb-bell shape, nanoflower, and nanosphere has been also reported as found elsewhere [71], a summary of materials for MFCs are enlisted in Table 2.

Table 2 Photoactive semiconductor materials to photoelectrocatalytic MFC.

Geometry/ morphology	Photoactive Semiconductor materials	Ref.
Zero-dimensional photocatalysts	Quantum dots (QDs) structure of metal semiconductors ($CuInS_2$, $CuInSe_2$, PbS (QDs)/TiO_2), Nitrogen-doped graphene oxide QDS (NGOQDs), NGQDs-BiOI/$MnNb_2O_6$, CQDs/CdS, ZnSQDs, Graphene oxide/Mg-doped, ZnO/tungsten oxide QDs, graphene oxide/WO_3, QDs/ TiO_2@SiO_2 (GOWTS), TiO_2/Au QDs, CdS, WS_2QDs, carbon QDs (CQDs).	[28,77]
2D-photocatalyst	SiO_2, nanorods, Phosphorus-doped Co_3O_4 nanowire array, nanosheets of Ni_3S_2, NiCoP, $NiCo_2O_4$/Ni $(OH)_2$ hybrid nanosheet, $MnMoO_4H_2O$@MnO_2, KCu_7S_4@NiMn, Ni_2P, g-C_3N_4 Nanosheets and SnS_2 hexagonal nanosheets	[28]
3D-photocatalyst	Titanate nanotube, carbon nanotube/reduced graphene oxide (AC/CNT/RGO), cellulose nanofibril (CNF)/ reduced graphene oxide (RGO)/carbon nanotube (CNT), Fullerene nanosphere, nanosphere of ZnO, CuO, Fe_2O_3, $BiVO_4$, and $ZnFe_2O_4$, NiO hollow nanosphere, $CoFe_2O_4$ hollow sphere, $CoFe_2O_4$/ZnO core shell, $CoSe_2$ nano-vesicle and $CoS_2NiFe_2O_4$-nanoflower, Bi_2O_3, $CuCo_2O_4$, SnO_2, MVO_4, WO_3, $CuWO_4$, $CuCo_2$-S_4/ $CuCo_2O_4$ nanoflowers, $NiCo_2O_4$ hollow micro-cuboids, Codoped Fe_3C nano-onions	[28,71,77]
Metal oxides	WO_3, $ZnFe_2O_4$, NiO, $CoFe_2O_4$, CeO_2, $NiFe_2O_4$, Bi_2O_3, $CuCo_2O_4$, Fe_2O_3, $BiVO_4$	[28,71,77]
Sulfides	ZnS, FeS_2, WS_2, CdS, SnS_2, PbS	[28]

8.1 Degradation of dyes in photoassisted microbial fuel cells

Several metal oxides have been used for dye degradation in MFCs and other biochemical systems; TiO_2-coated materials have been proposed as photoactive cathodes since these photocatalysts mainly respond to the ultraviolet component of the solar radiation spectrum [78]. Materials such as Pd-modified graphite carbon nitride/BiOBr heterojunction have been further developed to visible component, but the conduction band potential in the materials is lower than theoretical electrode potential for hydrogen evolution, and the consumption of electrons for hydrogen generation decrease the rate of azo dye decolorization [79]. Materials like W and Mo oxides exhibit a conduction band potential that suppresses the hydrogen evolution reaction enhancing the rate of azo dye degradation [80]. W and Mo oxides also have been used with visible light and electro-Fenton conditions to obtain efficient systems for azo dye decoloration and mineralization [81].

8.2 Removal of heavy metals in photoassisted microbial fuel cells

Heavy metals in the environment are pollutants due to their toxicity, bioaccumulation, and nonbiodegradability. Several methods have been developed for treatment, one of them are MFCs where metal recovery by bioelectrocatalysis is possible in combination with power generation. The metal ions in soil or wastewater can be reduced and deposited [82,83]; these metals with high redox potentials can be utilized as electron acceptors in the cathode compartment and then, they are reduced and precipitated [82]. Some heavy metals recovered using MFCs includes Cr, V, As, Au, Ag, Cu, Co, U, Hg, Se, Ba, Sr, Zn, and Mo. The typically reported metal-reducing efficiency is >90%. Different electrode materials have been reported, for example, plates, felt carbon cathodes materials, and spherical nanoparticles of oxides [82], and material with high absorption in the visible range such as polyaniline nanofiber (PANI)-CdS Quantum Dots (QDs) [84].

Hexavalent chromium has been recovered via photocatalyst, $NaVO_3$ as the cathodic electron acceptor was precipitated and recovered. For some metallic pollutants, such as arsenic, strong oxidants derived from H_2O_2 produced in MFC can oxidize arsenite into arsenate [85]. Metallic Ag can be recovered using Ag(I) ions or Ag(I) thiosulfate complexes [86], while metallic Cu and/or cuprous oxide are recovered using $CuSO_4$ solutions [87]. Microbial fuel cells can remove and recover metals even in very low concentration and eliminate the energy needed for the treatment process, but extensive research is required to optimize and enhance the process efficiency, the several metal oxides as photocatalyst material should be tested [82].

8.3 Degradation of antibiotics in photoassisted microbial fuel cells

Removal of antibiotics from wastewater, is another application of photoassisted microbial fuel cells, the biodegradation of this pollutant can be enhanced on the anode and photocatalytic degradation on the cathode. The electrons generated by biological anode can be transferred to the photocathode through an external circuit and combined with the photogenerated holes generated at the photocathode. For this purpose the metal oxides reported are AgBr/CuO, TiO_2, $LiNbO_3$/CF, and other materials as g-C_3N_4/CdS, ternary nanocomposite system of RGO/TiO_2/Ag nanoparticle-decorated nickel mesh [88], several antibiotics such as nitrofurazone, ofloxacin, several estrogens, have been degraded while generating electricity.

In the mechanism of degradation O_2^- was the dominant reaction active species followed by ·OH [88,89]. Photoassisted microbial fuel cells represent an optional technology for refractory pollutants treatment with many advantages compared with other technologies like AOP, photocatalysis and Fenton, integrate metal oxides in this photoassisted microbial fuel cells is important because it increases Coulombic efficiency (E_c), energy conversion recover [89].

9 Photoassisted rechargeable Zn-air batteries

The combination of technology in solar cells and rechargeable metallic batteries unveils a new alternative for the development of solar-assisted rechargeable batteries, in which solar energy can be used to generate a partial photocharge with or without external electrical polarization that allows energy saving in photoenergy-assisted charging mode. Currently, the development of metal-air batteries focus on the use of light metals such as lithium, sodium, and zinc as anode; however, this chapter focuses mainly on the latter mentioned due to the additional advantages that this metal presents such as low toxicity, environmentally friendly, low cost because of Zn abundance and its high theoretical energy density ($1086\,Wh\,kg^{-1}$) [90].

The rechargeable metal-air battery can store and deliver energy by employing the oxygen reduction/evolution reactions (ORR/OER) in air electrodes, but these reactions present large overpotentials caused by slow reaction kinetics. Therefore, integrated systems that synergize the harvest and storage of solar energy in a well-regulated way are needed, which theoretically guarantees a clean, functional, and energetically efficient system. In 1976, Hodes and coworkers built the first-generation photovoltaic rechargeable batteries [91], and in recent years, with the rise of network-scale systems, artificial intelligence devices, portable electronic devices, and wearable self-powered devices, integrated units of photoassisted rechargeable batteries that reconcile harvesting, storage, and the use of solar energy have renewed academic and industrial interests to solve practical needs [92]. Furthermore, photoassisted battery charging through solar radiation represents a project with a great impact on sustainable development [93–101]. There are reports involving different battery types such as Li-air [93–96], Zn-air [97], vanadium flow redox batteries [98,99], Li photointercalation [100], and bifunctional electrodes [101].

Typically, photoassisted rechargeable battery system is mainly composed of two parts: an energy harvesting module (photovoltaic cells and semiconductor photoelectrode), and an energy storage module (supercapacitors, metal-ion batteries, metal-air batteries, redox flow batteries, lithium metal batteries, etc.) [102–105]. Generally, two forms of integration are reported: (i) discrete connection, and (ii) photoelectrode incorporation. Each arrangement obtains different general energy conversion, storage efficiency ($\eta_{general}$), and size of energy harvesting-storage system. On the one hand, the discrete mode requires external cables and/or shared appliances that are usually connected between separate modules. This integration is an option that allows high $\eta_{general}$; however, these systems exhibit bulky connection, higher ohmic resistance, and lack of a highly integrated design [106,107]. While the photoelectrode concept is based on the use of photosensitive materials as $BiVO_4$ or α-Fe_2O_3, which have dual functions of photovoltaic conversion and electrical energy storage [108]. This allows miniaturization of modern functional devices [109,110], as well as simpler integration, but nevertheless has a low $\eta_{general}$ due to the intrinsically limited photocarrier concentration [111].

A simplified way of describing photoelectrode integration is considering that it is formed by a photoelectrochemical cell with battery in this particular case is Zn-air battery (Fig. 10). The photoelectrochemical cell is constructed using a photoanode

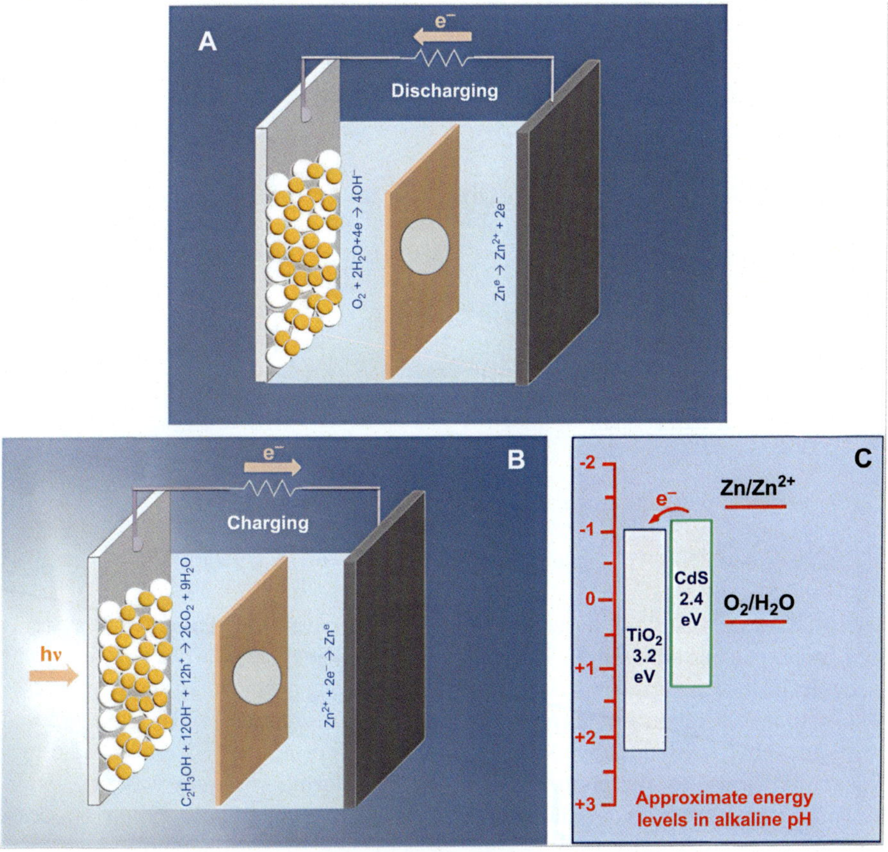

Fig. 10 Schematic illustration of the PEB reactor and of the discharging (A) and charging (B) operations; (C) shows the approximate energy levels in alkaline pH.
Reproduced with permission from T.S. Andrade, M.C. Pereira, P. Lianos, J. Electroanal. Chem. 878 (2020) 114559.

loaded with a semiconductor photocatalyst, which usually absorbs visible radiation, and the counter electrode is Zn. Both parts are in contact with an electrolyte, which interact with the metal and oxidizes it, generating a rechargeable photoelectrochemical battery (PEB). The discharge is the first-operation stage of the device and begins as soon as the electrodes come into contact with the electrolyte. During the discharge, which takes place in the dark conditions, the metal is oxidized by the electrolyte, producing, such as metal hydroxide. Oxidation leads to the release of electrons that are conducted through an external circuit to the photocatalyst loaded electrode, which acts as an air electrode and functions as an oxygen reduction cathode (Fig. 10A). During charging (associated with OER), which is done under lighting conditions, the OER kinetics increases significantly due to the formation of photogenerated holes that have strong oxidative capacity and are favorable for proton removal. In this stage, the electrodes roles are inverted. The photons are absorbed by the photocatalyst generating electron-hole pairs. The holes will be replenished during the oxidation reactions. The water can be oxidized producing oxygen, or a sacrificial agent can be added to increase efficiency. The photogenerated electrons are conducted through the external circuit to the oxide metal electrode generated during discharged, which now acts as the cathode, where the metal oxidation products are reduced to metallic original state (Fig. 10B). This completes the discharge-charge cycle (Eqs. 11–14) [112].

Discharge step:

$$\text{Zn electrode: Zn} + 4\text{OH}^- \rightarrow \text{Zn(OH)}_4{}^{2-} + 2\text{e}^- \ (\text{E}^\circ = -1:25\ \text{V}) \tag{11}$$

$$\text{Air-electrode: O}_2 + 2\text{H}_2\text{O} + 4\text{e}^- \rightarrow 4\text{OH}^- \ (\text{E}^\circ = +0.40\ \text{V}) \tag{12a}$$

Or

$$\text{O}_2 + 2\text{H}_2\text{O} + 2\text{e}^- \rightarrow \text{H}_2\text{O}_2 + 2\text{OH}^- \ (\text{E}^\circ = +0.40\ \text{V}) \tag{12b}$$

Photocharging step:

$$\text{Photoanode: photocatalysts} + h\nu \rightarrow \text{e}^- + h^+ \tag{13}$$

$$\text{Zn electrode: Zn(OH)}_4 + 2\text{e}^- \rightarrow \text{Zn} + 4\text{OH}^- \ (\text{E} = -1.25\text{V}) \tag{14}$$

The search for the coupling of the best materials to increase the efficiency of systems based on photoelectrodes/Zn-air batteries, which is however, poorly explored so far has, evolved since the primary systems proposed by Tong group using semiconductive polyterthiophene/carbon paper (pTTh/CP) as the photocathode [113], which Li and coworkers [114], subsequently used to demonstrate its stability (95 cycles) and rechargeability; however, the photoresponse of pTTh during charging was very poor. Inorganic semiconductors as air photoelectrodes for rechargeable Zn-air batteries have recently been explored [90]. Despite the achievements obtained nowadays, their performance is still seriously limited by low bifunctional electrocatalytic activity for oxygen reactions (ORR and OER) and/or photoresponsiveness. An alternative to overcome this problem was the use of a photocatalyst generated by a dual system based on

Table 3 Characteristics of photoassisted rechargeable Zn-air batteries.

Photoelectode	Charge voltage with light	Charge voltage in dark	Cycling numbers	η (%)	Ref.
α-Fe_2O_3	1.64	2.11	–	70.3	[90]
Co_3O_4	2.0	2.1	17	–	[117]
$NiCo_2S_4$	1.92	–	230	68.8	[118]
$Ni_{12}P_5$@NCNT	1.90	1.94	320	64.2	[119]
Polytrithiophene	1.8–2.0	–	60	–	[114]
PEDOT/PEO/ CNTs/polyurethane foam	1.88	1.98	–	53.2	[116]

a p-type ($Ni_{12}P_5$) and n-type semiconductor (N-doped carbon nanotubes), developing activity for the OER and ORR, respectively. Additionally, this system's fabricated all-solid-state battery also exhibited excellent photoresponsiveness, providing the way to smart modern electronics. Similarly, a solid-state intelligent Zn-air rechargeable battery (SRZAB) based on a dual photosensitive air cathode fabricated by multiscale block-copolymer-CNTs-polyurethane foam conjugate was reported [115,116]. This compound not only possesses good bifunctional activities, but moreover, generates a large photocurrent under ultraviolet light (UV) ($\lambda = 365$ nm), widening the range of harnessed sunlight. Further, the battery not only showed good discharge/charge capacity for 110 cycles, but also stability and reproducibility. Table 3 shows some photoassisted rechargeable Zn-air batteries reported and some of their main characteristics. As it is a very recent research area, there are no standardized techniques for the evaluation of photoassisted rechargeable Zn-air batteries, such as source wavelength and power density and the effect of environmental stimuli (e.g., heat generated by light, pressure, and moisture). Therefore, it is not possible to make a direct comparison based on the efficiencies reported in Table 3.

10 Conclusions

Metal oxides represent an alternative to increase the power density of photoassisted electrochemical energy conversion/storage systems. Many strategies to increase the photoactivity have been proposed like changing the particle size, shape, and by doping the metal oxide with metallic nanoparticles. However, for real applications, the improvement of activity and durability of photocatalytic fuel cells still requires the enhancement of the intrinsic properties of nanomaterials. In the case of photosensitive batteries, they have shown a promising future in realizing large-scale solar energy applications. However, more efforts are required in this field to systematize test conditions, such as those mentioned above to accurately compare the performance of different systems and discuss the development trend in the future. Furthermore, the integration of photosensitive batteries through the photovoltaic effect with other environmental stimuli will be a promising research direction and could further expand the potential applications of such energy conversion and storage systems.

11 Challenges and perspectives

As discussed above, these emerging photoassisted technologies represent an advanced method to simultaneously utilize different energy inputs to reduce energy barriers in complicated reactions that are, otherwise, prohibitive. Thus, several processes must be optimized not only in terms of thermodynamics, which is the most common approach in these systems, but also in their kinetics. Thermodynamic properties are usually projected via in silico calculations, but kinetics is often disregarded. Thus, a great improvement in research would be the study of these the kinetics using computational modeling. Moreover, most of the current studies usually do not consider mechanistic approaches to the understanding of the interplay of kinetics at different interfaces. Fundamental research in kinetics is still required to understand and distinguish contributions from the complexity at these interfaces, where morphology, crystallinity, composition, and other properties individually play a role.

From the side of material science, a paramount challenge is the feasibility of synthesizing large quantities of the optimized materials with the required reproducibility while being economically favorable. Currents methods are oriented to the specific tailoring of precise interfaces using multistep synthetic methods. Unfortunately, these methods also are cost-prohibitive scalable processes. Therefore, novel single-step synthetic methods are important to be developed in obtaining such highly efficient materials. Another aspect related to the previous point is the development of newer technologies for the large-scale production of nanosized materials. Newer methods that are only possible to reproduce at laboratory scale should be explored and further developed. To this end, techno-economic analyses could assess the impact and possibly give a stronger rationale for the deployment of these emerging technologies.

The development of light-assisted rechargeable metal batteries is only an intermediate stage to achieve completely photorechargeable batteries as a sustainable energy storage and conversion system. Undoubtedly, the success stories obtained on the operation mechanism, active materials, and design structure of the photoassisted rechargeable Zn-air batteries will be the core for the manufacturing of completely photorechargeable Zn-air batteries. In this way, the use of new approaches to improve the activity and durability like defect engineering can promote the photodegradation of pollutants while producing electrical energy.

References

[1] J. Wen, J. Xie, X. Chen, X. Li, Appl. Surf. Sci. 391 (2017) 72–123.

[2] M. Pelaez, N.T. Nolan, S.C. Pillai, M.K. Seery, P. Falaras, A.G. Kontos, P.S.M. Dunlop, J.W.J. Hamilton, J.A. Byrne, K. O'Shea, M.H. Entezari, D.D. Dionysiou, Appl Catal B 125 (2012) 331–349.

[3] P. Kanhere, Z. Chen, Molecules 19 (2014) 19995–20022.

[4] E. Grabowska, Appl Catal B 186 (2016) 97–126.
[5] K. Sivula, R. Van de Krol, Nat. Rev. Mater. 1 (2016) 15010.
[6] G. Zhang, G. Liu, L. Wang, J.T.S. Irvine, Chem. Soc. Rev. 45 (2016) 5951–5984.
[7] C.C.L. McCrory, S. Jung, I.M. Ferrer, S.M. Chatman, J.C. Peters, T.F. Jaramillo, J. Am. Chem. Soc. 137 (2015) 4347–4357.
[8] F.E. Osterloh, Chem. Soc. Rev. 42 (2013) 2294–2320.
[9] N. Zheng, X. Bu, P. Feng, Angew. Chem. Int. Ed. 44 (2005) 5299–5303.
[10] A.I. Korin, Physico-chemical properties of novel nanocrystalline ruthenium based chalcogenide materials, in: A.I. Kokorin, D.W. Bahnemann (Eds.), Chemical Physics of Nanostructured Semiconductors, first ed., CRC Press, 2003. 18 pages.
[11] J.A. Díaz-Real, T. Holm, N. Alonso-Vante, Photoelectrochemical hydrogen production (PEC H_2), in: Current Trends and Future Developments on (Bio-)membranes, Elsevier, 2020, pp. 255–289.
[12] H.O. Finklea, Semiconductor Electrodes, Elsevier, 1988. pages 520.
[13] U. Lucia, Renew. Sustain. Energy Rev. 30 (2014) 164–169.
[14] B.C. Ong, S.K. Kamarudin, S. Basri, Int. J. Hydrogen Energy 42 (2017) 10142–10157.
[15] Y. Zheng, X. Wan, X. Cheng, K. Cheng, Z. Dai, Z. Liu, Catalysts 10 (2020) 166.
[16] J. Bai, D. Liu, J. Yang, Y. Chen, ChemSusChem 12 (2019) 2117–2132.
[17] M. Liu, Z. Zhao, X. Duan, Y. Huang, Adv. Mater. 31 (2019) 1802234.
[18] M. Kuang, G. Zheng, Small 12 (2016) 5656–5675.
[19] B. Zhu, D. Xia, R. Zou, Coord. Chem. Rev. 376 (2018) 430–448.
[20] W. Xia, A. Mahmood, Z. Liang, R. Zou, S. Guo, Angew. Chem. Int. Ed. 55 (2016) 2650–2676.
[21] G.L. Soloveichik, Beilstein J. Nanotechnol. 5 (2014) 1399–1418.
[22] N.A. Karim, S.K. Kamarudin, Introduction to direct alcohol fuel cells (DAFCs), in: R.G. Akay, A.B. Yurtcan (Eds.), Direct Liquid Fuel Cells, Academic Press, 2020, pp. 49–70.
[23] A.J. Slate, K.A. Whitehead, D.A.C. Brownson, C.E. Banks, Renew. Sustain. Energy Rev. 101 (2019) 60–81.
[24] S.S. Kumar, V. Kumar, R. Kumar, S.K. Malyan, A. Pugazhendhi, Fuel 55 (2019) 115682.
[25] S. Khandaker, S. Das, M.T. Hossain, A. Islam, M.R. Miah, M.R. Awual, J. Mol. Liq. 344 (2021) 117795.
[26] A. Almatouq, A.O. Babatunde, M. Khajah, G. Webster, M. Alfodari, J. Water Process Eng. 34 (2020) 101140.
[27] M. Li, Y. Liu, L. Dong, C. Shen, F. Li, M. Huang, C. Ma, B. Yang, X. An, W. Sand, Sci. Total Environ. 668 (2019) 966–978.
[28] P. Mishra, P. Saravanan, G. Packirisamy, M. Jang, C. Wang, Int. J. Hydrogen Energy 46 (2021) 22877–22906.
[29] P.H. Cyril, G. Saravanan, New J. Chem. 44 (2020) 19977–19995.
[30] Y. Zhang, Q. Chen, L. Liu, Y. Wang, M.K.H. Leung, Chem. Eng. J. 399 (2020) 125731.
[31] R. Ma, G. Lin, Y. Zhou, Q. Liu, T. Zhang, G. Shan, M. Yang, J. Wang, npj Comput. Mater. 5 (2019) 78.
[32] Y. Vasseghian, A. Khataee, E.-N. Dragoi, M. Moradi, S. Nabavifard, G.O. Conti, A.M. Khaneghah, Arab. J. Chem. 13 (2020) 8458–8480.
[33] D. Raptis, A. Ploumistos, E. Zagoraiou, E. Thomou, M. Daletou, L. Sygellou, D. Tasis, P. Lianos, Catal. Today 315 (2018) 31–35.
[34] Q. Zeng, S. Chang, A. Beyhaqi, S. Lian, H. Xu, J. Xie, F. Guo, M. Wang, C. Hu, J. Hazard. Mater. 394 (2020) 121425.
[35] Y.-P. Ong, L.-N. Ho, S.-A. Ong, J. Banjuraizah, A.H. Ibrahim, S.-L. Lee, N. Nordin, Chemosphere 219 (2019) 277–285.

[36] H. Yao, Y. Xu, D. Zhong, Y. Zeng, N. Zhong, Ionics 27 (2021) 4875–4884.
[37] P. Mikrut, D. Mitoraj, R. Beranek, W. Macyk, Appl. Surf. Sci. 566 (2021) 150662.
[38] W.-J. Ong, L.L. Tan, S.-P. Chai, S.-T. Yong, A.R. Mohamed, ChemSusChem 7 (2014) 690–719.
[39] W. Li, D. Wang, Y. Zhang, L. Tao, T. Wang, Y. Zou, Y. Wang, R. Chen, S. Wang, Adv. Mater. 32 (2020) 1907879.
[40] H. Qin, V. Sorkin, Q.-X. Pei, Y. Liu, Y.-W. Zhang, J. Appl. Mech. 87 (2020) 030802.
[41] N. Khossossi, D. Singh, A. Ainane, R. Ahuja, Chem. Asian J. 15 (2020) 3390–3404.
[42] Y. Jia, K. Jiang, H. Wang, X. Yao, Chem 5 (2019) 1371–1397.
[43] L. Hu, Y. Liao, D. Xia, F. Peng, L. Tan, S. Hu, C. Zheng, X. Lu, C. He, D. Shu, Chem. Eng. J. 385 (2020) 123824.
[44] Y. Wang, Y.M. Wang, X.–M. Song, Y. Zhang, T. Ma, Appl. Surf. Sci. 506 (2020) 144949.
[45] H. Hirakawa, M. Hashimoto, Y. Shiraishi, T. Hirai, J. Am. Chem. Soc. 139 (2017) 10929–10936.
[46] B. Jiang, J. Bai, L. Li, N. He, Q. Zhang, B. Wang, D. Tang, Chemosphere 286 (2022) 131657.
[47] G. Lui, G. Jiang, M. Fowler, A. Yu, Z. Chen, J. Power Sources 425 (2019) 69–75.
[48] X. Lv, F.L.-Y. Lam, X. Hu, Chem. Eng. J. 427 (2022) 131602.
[49] I. Campos Sena, D. Oliveira Sales, T. Santos Andrade, M. Rodriguez, A. Cândido Silva, F.G. Esteves Nogueira, J. Lisboa Rodrigues, J.P. Mesquita, M.C. Pereira, J. Electroanal. Chem. 881 (2021) 114940.
[50] D. Pan, S. Xiao, X. Chen, R. Li, Y. Cao, D. Zhang, S. Pu, Z. Li, G. Li, H. Li, Environ. Sci. Technol. 53 (2019) 3697–3706.
[51] E. Antolini, Appl Catal B 237 (2018) 491–503.
[52] S. Hua, B. Wanga, H. Jua, L. Jianga, Y. Maa, Y. Fan, J. Taiwan Inst. Chem. Eng. 80 (2017) 533–539.
[53] T.L. Villareal, R. Gomez, M. Neumann-Spallart, N. Alonso-Vante, P. Salvador, J. Phys. Chem. B 108 (2004) 15172–15181.
[54] K.-W. Park, S.-B. Han, J.-M. Lee, Electrochem. Commun. 9 (2007) 1578–1581.
[55] S. Hu, L. Jiang, B. Wang, Y. Ma, Int. J. Hydrogen Energy 44 (2019) 13214–13220.
[56] H. He, P. Xiao, M. Zhou, F. Liu, S. Yu, L. Qiao, Y. Zhang, Electrochim. Acta 88 (2013) 782–789.
[57] J. Hu, M. Sun, X. Cai, C. Zhai, J. Zhang, M. Zhu, J. Taiwan Inst. Chem. Eng. 80 (2017) 231–238.
[58] H.F. Cheng, B.B. Huang, Y. Dai, Nanoscale 6 (2014) 2009–2026.
[59] C. Zhai, J. Hu, M. Sun, M. Zhu, Appl. Surf. Sci. 430 (2018) 578–584.
[60] J. Hu, C. Zhai, H. Gao, L. Zeng, Y. Du, M. Zhu, Sustain, Energy Fuel 3 (2019) 439–449.
[61] P. Panagiotopoulou, E.E. Karamerou, D.I. Kondarides, Catal. Today 209 (2013) 91–98.
[62] M.R.K. Estahbanati, M. Feilizadeh, F. Attar, M.C. Iliuta, React. Chem. Eng. 6 (2021) 197–219.
[63] M.R. Karimi Estahbanati, M. Feilizadeh, M.C. Iliuta, Appl Catal B 209 (2017) 483–492.
[64] T. Li, J.A. Díaz-Real, T. Holm, Adv. Mater. Technol. 6 (2021) 2100569.
[65] J.A. Díaz-Real, M. Guerra-Balcázar, N. Arjona, F. Cuevas-Muñiz, L.G. Arriaga, J. Ledesma-García, Microfluidics in membraneless fuel cells, in: X.-Y. Yu (Ed.), Advances in Microfluidics - New Applications in Biology, Energy, and Materials Sciences, Intech Open, 2016. pages 273.
[66] N.T. Nguyen, S.T. Wereley, Fundamentals and applications of microfluidics, third ed., Artech House, 2019. pages 548.

[67] T. Li, J.A. Díaz-Real, T. Holm, ACS Sensors 7 (2022) 2934–2939.
[68] L. Li, G. Wang, R. Chen, X. Zhu, H. Wang, Q. Liao, Y. Yu, Lab Chip 14 (2014) 3368–3375.
[69] F. Xie, R. Chen, X. Zhu, Q. Liao, D. Ye, B. Zhang, Y. Yu, J. Li, J. CO2 Util. 32 (2019) 31–36.
[70] X. He, M. Chen, R. Chen, X. Zhu, Q. Liao, D. Ye, B. Zhang, W. Zhang, Y. Yu, J. Hazard. Mater. 358 (2018) 346–354.
[71] J. Liu, R. Chen, X. Zhu, Q. Liao, D. Ye, B. Zhang, W. Zhang, Y. Yu, Sep. Purif. Technol. 229 (2019) 115821.
[72] H.-N. Wang, X. Chen, R. Chen, X. Zhu, Q. Liao, D.-D. Ye, B. Zhang, Y.-X. Yu, W. Zhang, J.-W. Li, J. Power Sources 435 (2019) 226766.
[73] J.A. Díaz-Real, E. Ortiz-Ortega, M.P. Gurrola, J. Ledesma-García, L.G. Arriaga, Electrochim. Acta 206 (2016) 388–399.
[74] D. Dector, D. Ortega-Díaz, J.M. Olivares-Ramírez, A. Dector, J.J. Pérez-Bueno, D. Fernández, D.M. Amaya-Cruz, A. Reyes-Rojas, Int. J. Hydrogen Energy 46 (2021) 26163–26173.
[75] Y.H. Kwok, Y. Wang, Y. Zhang, H. Zhang, F. Li, W. Pan, D.Y.C. Leung, Int. J. Hydrogen Energy 45 (2020) 21796–21807.
[76] Z. Xu, S. Chen, S. Guo, D. Wan, H. Xu, W. Yan, X. Jin, J. Feng, J. Power Sources 501 (2021) 230000.
[77] P. Xu, H. Xu, ACS Omega 4 (2019) 5848–5851.
[78] Y. Hou, R. Zhang, Z. Yu, L. Huang, Y. Liu, Z. Zhou, Bioresour. Technol. 224 (2017) 63–68.
[79] Y. Hou, Y. Gan, Z. Yu, X. Chen, L. Qian, B. Zhang, L. Huang, J. Huang, J. Power Sources 371 (2017) 26–34.
[80] D. Robert, Catal. Today 122 (2007) 20–26.
[81] Q. Wang, L. Huang, X. Quan, P.G. Li, Appl Catal B 245 (2019) 672–680.
[82] A.S. Mathuriya, J.V. Yakhmi, Environ. Chem. Lett. 12 (2014) 483–494.
[83] X. Song, W. Yang, Z. Lin, L. Huang, X. Quan, Sci. Total Environ. 651 (2019) 1698–1708.
[84] E.F. Talooki, M. Ghorbani, M. Rahimnejad, M.S. Lashkenari, J. Electroanal. Chem. 873 (2020) 114469.
[85] A. Xue, Z.Z. Shen, B. Zhao, H.Z. Zhao, J. Hazard. Mater. 261 (2013) 621–627.
[86] H.C. Tao, Z.Y. Gao, H. Ding, N. Xu, W.M. Wu, Bioresour. Technol. 111 (2012) 92–97.
[87] M. Liang, H.C. Tao, S.F. Li, W. Li, L.I. Zhang, J.R. Ni, Environ. Sci. Technol. 32 (2011) 179–185.
[88] Y. Dai, S. Li, Y. Guo, J. Ren, L. Zhang, X. Wang, P. Zhang, X. Liu, J. Power Sources 497 (2021) 229876.
[89] P. Xu, D. Zheng, Q. He, J. Yu, Sep. Purif. Technol. 250 (2020) 117106.
[90] X. Liu, Y. Yuan, J. Liu, B. Liu, X. Chen, J. Ding, X. Han, Y. Deng, C. Zhong, W. Hu, Nat. Commun. 10 (2019) 4767–4777.
[91] G. Hodes, J. Manassen, D. Cahen, Nature 261 (1976) 403–404.
[92] Y. Wua, C. Lib, Z. Tianb, J. Suna, J. Power Sources 478 (2020) 228762.
[93] J. Rugolo, M.J. Aziz, Energ. Environ. Sci. 5 (2012) 7151–7160.
[94] Y. Hou, R. Vidu, P. Stroeve, Ind. Eng. Chem. Res. 50 (2011) 8954–8964.
[95] B. Dunn, H. Kamath, J.M. Tarascon, Science 334 (2011) 928–935.
[96] K.A. Bush, A.F. Palmstrom, Z.S.J. Yu, M. Boccard, R. Cheacharoen, J.P. Mailoa, D.P. McMeekin, R.L.Z. Hoye, C.D. Bailie, T. Leijtens, I.M. Peters, M.C. Minichetti,

N. Rolston, R. Prasanna, S. Sofia, D. Harwood, W. Ma, F. Moghadam, H.J. Snaith, T. Buonassisi, Z.C. Holman, S.F. Bent, M.D. McGehee, Nat. Energy 2 (2017) 17009.
[97] P. Caprioglio, C.M. Wolff, O.J. Sandberg, A. Armin, B. Rech, S. Albrecht, D. Neher, M. Stolterfoht, Adv. Energy Mater. 10 (2020) 2000502.
[98] Q. Liu, Y. Jiang, K. Jin, J. Qin, J. Xu, W. Li, J. Xiong, J. Liu, Z. Xiao, K. Sun, S. Yang, X. Zhang, L. Ding, Sci. Bull. 65 (2020) 272–275.
[99] L.X. Meng, Y.M. Zhang, X.J. Wan, C.X. Li, X. Zhang, Y.B. Wang, X. Ke, Z. Xiao, L.M. Ding, R.X. Xia, H.L. Yip, Y. Cao, Y.S. Chen, Science 361 (2018) 1094–1098.
[100] H. Yao, J. Wang, Y. Xu, S. Zhang, J. Hou, Acc. Chem. Res. 53 (2020) 822–832.
[101] B. Lei, G.-R. Li, P. Chen, X.-P. Gao, ACS Appl. Energy Mater. 2 (2019) 1000–1005.
[102] P. Liu, H.X. Yang, X.P. Ai, G.R. Li, X.P. Gao, Electrochem. Commun. 16 (2012) 69–72.
[103] N.F. Yan, W.H. Zhang, H.M. Cui, X.J. Feng, Y.W. Liu, J.S. Shi, Sustain, Energy Fuel 2 (2018) 353–356.
[104] L. Cao, M. Skyllas-Kazacos, D.-W. Wang, Adv. Sustain. Syst. 2 (2018) 1800031.
[105] P. Liu, Y.-L. Cao, G.-R. Li, X.-P. Gao, X.-P. Ai, H.-X. Yang, ChemSusChem 6 (2013) 802–806.
[106] N.F. Yan, G.R. Li, X.P. Gao, J. Mater. Chem. A 1 (2013) 7012–7015.
[107] A. Das, S. Deshagani, R. Kumar, M. Deepa, A.C.S. Appl, Mater. Interfaces. 10 (2018) 35932–35945.
[108] H. Meng, S. Pang, G. Cui, ChemSusChem 12 (2019) 3431–3447.
[109] C.H. Ng, H.N. Lim, S. Hayase, I. Harrison, A. Pandikumar, N.M. Huang, J. Power Sources 296 (2015) 169–185.
[110] K. Poonam, A. Sharma, A. Arora, S.K. Tripathi, J. Energy Storage 21 (2019) 801–825.
[111] N.-F. Yan, X.-P. Gao, Energy Environ. Mater. 5 (2022) 439–451.
[112] T.S. Andrade, M.C. Pereira, P. Lianos, J. Electroanal. Chem. 878 (2020) 114559.
[113] K. Wang, Z. Mo, S. Tang, M. Li, H. Yang, B. Long, Y. Wang, S. Song, Y. Tong, J. Mater. Chem. A 7 (2019) 14129–14135.
[114] D. Zhu, Q. Zhao, G. Fan, S. Zhao, L. Wang, F. Li, J. Chen, Angew. Chem. Int. Ed. 58 (2019) 12460–12464.
[115] Z. Fang, X. Hu, D. Yu, ChemPlusChem 85 (2020) 600–612.
[116] Z. Fang, Y. Zhang, X. Hu, X. Fu, L. Dai, D. Yu, Angew. Chem. Int. Ed. 58 (2019) 9248–9253.
[117] C. Tomon, S. Sarawutanukul, S. Duangdangchote, Chem. Commun. 55 (2019) 5855–5858.
[118] S. Sarawutanukul, C. Tomon, S. Duangdangchote, N. Phattharasu-pakun, M. Sawangphruk, Batteries Supercaps. 3 (2020) 541–547.
[119] J. Lv, S.C. Abbas, Y. Huang, Q. Liu, M. Wu, Y. Wang, L. Dai, Nano Energy 43 (2018) 130–137.

Index

Note: Page numbers followed by *f* indicate figures and *t* indicate tables.

C

J

K

L

M

N

Q

R

W

X

Y

Z

Printed in the United States
by Baker & Taylor Publisher Services